F.J. Pettijohn P.E. Potter R. Siever

Sand and Sandstone

Second Edition

With 355 Figures

Springer-Verlag
New York Berlin Heidelberg
London Paris Tokyo

F.J. Pettijohn
Professor Emeritus of Geology, The Johns Hopkins University,
Baltimore, MD 21218/USA

Paul Edwin Potter
Professor of Geology, University of Cincinnati,
Cincinnati, OH 45221/USA

Raymond Siever
Professor of Geology, Harvard University,
Cambridge, MA 02138/USA

Library of Congress Cataloging in Publication Data
Pettijohn, F.J. (Francis John), 1904–
Sand and sandstone.
Includes bibliographical references and index.
1. Sand. 2. Sandstone. I. Potter, Paul Edwin.
II. Siever, Raymond. III. Title.
QE471.2.P47 1987 552′.5 86-17925

Typeset by Bi-Comp, Inc., York, Pennsylvania.
Printed and bound by Arcata Graphics/Halliday Lithograph, West Hanover, Massachusetts.
Printed in the United States of America.

9 8 7 6 5 4 3 2 1

ISBN 0-387-96355-3 Springer-Verlag New York Berlin Heidelberg (hard cover)
ISBN 3-540-96355-3 Springer-Verlag Berlin Heidelberg New York (hard cover)
ISBN 0-387-96350-2 Springer-Verlag New York Berlin Heidelberg (soft cover)
ISBN 3-540-96350-2 Springer-Verlag Berlin Heidelberg New York (soft cover)

Sand and Sandstone—Illustrious Forefathers

Henry C. Sorby, 1826–1908

The first to study rocks in thin section and a pioneer in the study of sand: sedimentary structures, paleocurrents, provenance, and diagenesis. He also founded carbonate petrology and metallography. Truly an outstanding innovator and observer and the ultimate model of what a creative individual working alone can accomplish.

Johan A. Udden, 1859–1932

Pioneer Swedish–American frontier scientist at Augustana College in the American Midwest. Systematically studied the size distribution of sand and was one of the first to relate it to process and environment in *The Mechanical Composition of Wind Deposits,* published in 1898. A devoted teacher and a great contributor to a broad vista of sedimentology.

Lucien Cayeux, 1864–1944

Emphasized the "natural history" of a sediment from its earliest deposition to its most recent diagenetic event, placing great reliance on thin section petrography. A prolific publisher with a long career as a teacher. His monographs, such as *Les Roches Sedimentaires de France, Roches Siliceuses,* set a high standard.

Marcus Goldman, 1881–1965

An American who studied with Thoulet at Nancy, Cayeux at Paris, and Walther at Halle, he was the pioneer sedimentary petrologist in America. His Ph.D. thesis on the Cretaceous of Maryland and his study of the Catahoula (Tertiary) sandstone of Texas were milestone papers.

Paul D. Krynine, 1901–1964

An imaginative thinker, he emphasized careful thin section petrology and deductive thinking. A leader in linking tectonics and sandstone compositions. Because he was never afraid to speculate, many of today's controversies about sandstones would have been instantly familiar to Krynine.

Philip H. Kuenen, 1902–1976

The origin and properties of sand and the origin and characteristics of turbidity currents in both the modern and ancient environment were only a few of his favorite topics. Both were studied in the field, on the oceans, and experimentally in the laboratory. He had imagination, ingenuity in experimentation, and a farsightedness in sedimentary dynamics possessed by few.

William C. Krumbein, 1902–1979

The size distribution of sand, shape, roundness, porosity, and permeability, sand mineralogy, the petrographic classification of sandstones, diagenesis via Eh and pH, beach processes, facies maps and their interpretation, statistical and computer analysis all show his sustained innovation and application of quantitative methods and statistical approaches.

Edwin D. McKee, 1906–1984

Studies of modern and ancient sandstones on five continents plus field and experimental investigations of stratification, trace fossils, and ancient paleo-environments were the contributions of this long-term member of the U.S. Geological Survey. An authority on the classical sections of the Grand Canyon region, he applied his special knowledge of that area to many general problems of sandstones, especially to eolian sedimentation.

Preface to the Second Edition

The first edition appeared fourteen years ago. Since then there have been significant advances in our science that warrant an updating and revision of *Sand and Sandstone*.

The main framework of the first edition has been retained so that the reader can begin with the mineralogy and textural properties of sands and sandstones, progress through their organization and classification and their study as a body of rock, to consideration of their origin—provenance, transportation, deposition, and lithification—and finally to their place in the stratigraphic column and the basin.

The last decade has seen the rise of facies analysis based on a closer look at the stratigraphic record and the recognition of characteristic bedding sequences that are the signatures of some geologic process—such as a prograding shallow-water delta or the migration of a point bar on an alluvial floodplain. The environment of sand deposition is more closely determined by its place in such depositional systems than by criteria based on textural characteristics—the "fingerprint" approach. Our revision reflects this change in thinking.

As in the geological sciences as a whole, the concept of plate tectonics has required a rethinking of our older ideas about the origin and accumulation of sediments—especially the nature of the sedimentary basins. These had been categorized as geosynclines of one kind or another. We now have to redefine these in terms of plate tectonics. Plate motions generate depositional basins. How do we classify these and recognize them in the ancient record? What does the study of sandstones contribute to this problem? We are still feeling our way and the criteria for recognition of the several types of basins and the characteristics of their fill are only partially understood. A number of papers have appeared that focus on the sands in particular.

We now know a good deal more about the relations between bedforms and the internal current structures of sandstone and modern and ancient flow regimes; hence we have incorporated this new knowledge in the second edition. We have also added a chapter on paleocurrents. It is not enough to study the process of sand transport and the bedforms but we need also to reconstruct the transport pattern to better understand the paleogeography at the time of deposition. Sandstones are the prime record of these paleocurrents.

Also a great deal of progress has been made in sedimentary geochemistry, especially of diagenesis. These advances have led to extended revision of the subject as it relates to sandstones. We are now learning to use new tools for investigating sandstone composition—the scanning electron microscope, the electron microprobe, and others. These tools provide data we did not have and the means to refine our interpretations. We take cognizance of these new data in our treatment of provenance and diagenesis.

We note also that just as thin-section studies of ancient sandstones are routine, so also this technique is now being applied to the study of modern sands. Such sections provide a much better means for identification of rock and mineral grains and for point-counting. They have greatly enhanced our ability to compare ancient and modern sands.

We also became aware that sedimentology has become truly cosmopolitan. Whereas formerly most of the relevant literature came from the English-speaking world, especially the United States, Canada, and Great Britain and from western Europe, it is now truly international. We have taken account of this expanded literature in our revision.

During this revision we were made acutely aware of the great quantity of excellent work on sand and sandstone that is represented by the flood of literature on the subject that has appeared since the first edition. It became obvious that we could neither completely survey all the world's work nor even refer to all of the new developments in the geology of sands, the application of other disciplines to sand study or the application of sand studies to practical matters such as the search for oil or mineral deposits. What we hope is that we have covered most of the major advances that have become part of the body of knowledge we call the geology of sand and sandstone.

Acknowledgments

We thank all those who helped with the second edition of *Sand and Sandstone*.

Those who read text include: H.E. Clifton, U.S. Geological Survey, Menlo Park, California; Jorge della Farvara, PETROBRAS, Rio de Janeiro, Brazil; J.A. Gilreath, Schlumberger Offshore Services, New Orleans, Louisiana; R.V. Ingersoll, University of California at Los Angeles, California; J. Barry Maynard, University of Cincinnati, Cincinnati, Ohio; J.H. McGowen, ARCO Research, Plano, Texas; Rafael Unrug, Wright State University, Dayton, Ohio; and W. Zimmerle, Deutsche Texaco, Celle, West Germany. Richard Spohn, Geology Librarian of the University of Cincinnati, was most helpful.

We also thank our typists for their care and patience: Wanda Osborne and Joan Harman of Cincinnati, Christine Levitt of Harvard, and Kate Francis of Johns Hopkins.

November 1, 1986

F.J. Pettijohn
P.E. Potter
R. Siever

Preface to the First Edition

This book is the outgrowth of a week-long conference on sandstone organized by the authors, first held at Banff, Alberta, in 1964 under the auspices of the Alberta Association of Petroleum Geologists and the University of Alberta, and again, in 1965, at Bloomington, Indiana, under the sponsorship of the Indiana Geological Survey and the Department of Geology, Indiana University. A 200-page syllabus was prepared for the second conference and published by the Indiana Geological Survey. Continuing interest in and demand for the syllabus prompted us to update and expand its contents. The result is this book.

We hope this work will be useful as a text or supplementary text for advanced undergraduate and graduate courses in sedimentation, sedimentary petrology, or general petrology and perhaps will be helpful to the teachers of such courses. Though we have focussed on sandstones we have necessarily included much of interest to students of all sediments. We hope also that it will be a useful reference work for the professional geologist, especially those concerned with petroleum, ground-water, and economic geology either in industry or government. Because the subject is so closely tied to surface processes it may also be of interest to geomorphologists and engineers who deal with beaches and rivers where sand is in transit.

This work presupposes a general knowledge of the elements of mineralogy, chemistry and statistics on the part of the reader. As no investigation of sediments—especially sandstones—can be considered adequate or complete without careful microscopical analysis, we also presume, therefore, that the user of this book has the knowledge and skills needed to study sands and sandstones under the microscope.

On the other hand, some cognate fields of knowledge are less familiar to geologists and while we did not include a section on statistics or thermodynamics, we did include a section on the principles underlying fluid flow and the propulsion of granular materials. We feel that some knowledge of this subject will become increasingly important in understanding physical sedimentation and the resulting textures and structures of sands.

The book is organized in such a manner as to lead the reader from consideration of the component grains in a sandstone to the analysis of sandstones in the sedimentary basin as a whole. The first half is largely descriptive, a summary of what is known about sandstones beginning with the components, their composition (Chapter 2) and geometrical properties (Chapter 3), progressing to the larger organization and structure (Chapter 4) to the whole rock itself (Chapters 5, 6 and 7). The second half of the book is more largely interpretative and process-oriented. It includes the processes of sand formation (Chapter 8), transportation and deposition (Chapter 9), and post-depositional alteration (Chapter 10). The book concludes with a résumé of the relation of sands to their environment of deposition and to other sediments (Chapter 11) and a summary of their distribution in space and time (Chapter 12). We have included a synoptic review of several better-known sedimentary basins in which an integrated approach—involving stratigraphy, sedimentary petrology, and paleocurrents—was used to unravel geologic history.

For the most part, analytical techniques are omitted. They are adequately covered in several modern texts and manuals (see references, p. 19). Exceptionally, however, we have included a short appendix on the art of petrographical description and analysis which, like field work, is best learned perhaps from experience under the guidance of a skilled master of the subject. We felt it worthwhile, however, to set down some guiding principles as these are seldom made explicit in most published works.

We did not include many "case histories" because, unlike in law or psychiatry, we feel that the student can turn to no better source of instruction than the rocks themselves. No course on this subject can be considered adequate or complete without a well-integrated program of field and laboratory studies. The student, under the supervision of his teacher, should work out his own problems. The clinic is a better guide to practice than the case book.

References to the literature are of two kinds—actual citations in the text to specific papers and a collection of annotated references. The latter for the most part supplement rather than repeat the text citations. Both are placed at the end of the appropriate chapters. In general, our references are selective, that is, although they include some older classic papers, emphasis is on the more recent ones. In many cases, such as the chapter on sedimentary structures, we did not feel the need of an in-depth review of the literature inasmuch as several specialized modern works which contain an extensive bibliography are readily available.

As is inevitable in a work of this kind, much of what is contained therein is a compilation from many sources which transcend and go beyond the immediate and direct experience of the authors. We have tried to acknowledge our debt to these sources at the appropriate places. We also wish to acknowledge the helpful criticism of those who read sections of this work when it was in manuscript form. In particular, we are indebted to Earle McBride, University of Texas, for checking our glossary of rock names applied to sandstones, to Robert L. Smith, U.S. Geological Survey, William F. Jenks, University of Cincinnati, and Richard V. Fisher, University of California at Santa Barbara, for reading the chapter on volcaniclastic sands, to Lee Suttner for criticism of Chapters 4 and 6, to Gerald V. Middleton of McMaster University, Yaron M. Sternberg of the University of Maryland and John B. Southard of the Massachusetts Institute of Technology for their help with the chapter on transport and deposition, to S.V. Hrabar of the Humble Oil Company for reading all of Chapter 11 and Donald A. Holm of Williams, Arizona, and Richard Mast of the Illinois Geological Survey for reading parts of it, to D.A. Pretorius of the University of Witwatersrand and R.W. Ojakangas of the University of Minnesota at Duluth for their comments on portions of Chapter 12, and to Miriam Kastner for help in the X-ray and electron probe analysis of the Trivoli Sandstone. Alan S. Horowitz of Indiana University read and helpfully edited many of the chapters. We wish to thank Mrs. Susan Berson, Miss Kathleen Feinour, Miss Jean Dell'Uomo, Mrs. Debby Powell, and Miss Cynthia Worswick for the final typing of the manuscript and our publishers for their help in the preparation of the illustrations and seeing the work through the press.

To emphasize our spirit of teamwork we have listed our names in alphabetical order.

January 1, 1972

F. J. Pettijohn
P. E. Potter
R. Siever

Contents

Part III. Processes that Form Sand and Sandstone

Sand and Sandstone

CHAPTER 1
Introduction and Source Materials

Sand and Sandstone Defined

Sands and clastic sediments, in general, differ from the igneous and other crystalline rocks in possessing a framework of grains—a framework stable in the earth's gravitational field. Unlike the grains of the igneous and related rocks, which are in continuous contact with their neighbors, the grains in a sand are generally in tangential contact only and thus form an open, three-dimensional network. As a consequence, sands have a high porosity—have a fluid-filled pore system. The unequal distribution of stress along grain boundaries may lead to solution at points of pressure and deposition elsewhere increasing the surfaces of contact and decreasing the pore space. Such action, coupled in some cases with the introduction of cementing materials, leads to the ultimate end-product—a rock with grains in continuous contact and without porosity. In this manner, a sand with tangential contacts and a porosity of 35 to 40 percent is converted to an interlocking crystalline mosaic with zero porosity.

Sand is loose, noncohesive granular material, the grains or framework elements of which must by definition be sand-sized. Various attempts have been made to define sand more precisely. These attempts are largely directed toward expressing grain size in terms of grain "diameter" of some specified magnitude. Inasmuch as sand grains are non-regular solids, it is first necessary to define the term "diameter" as applied to such solids (see Chap. 3). Attempts to codify the meaning of "sand" as a size term are many. The effort to do so is usually part of a larger effort to codify all size terms and to construct a "grade scale" (see Chap. 3). The various choices made for the size class "sand" in some of these grade scales are shown in Fig. 1-1. We shall here adopt the diameter limits 0.0625 (1/16) and 2.0 mm for "sand"—limits which have become generally accepted among sedimentologists.

Sand, although restricted by definition to the 0.0625 to 2.0 mm range of diameters, actually encompasses a vast range in grain size. A grain 2 mm in diameter, as a sphere, has a volume of about 4.2 mm^3. A grain 1/16 (0.0625) mm in diameter has a volume of about 0.00012 mm^3. The larger volume is 34,688 times greater than the smaller. In short, while sand has a 32-fold range in diameter, it has nearly a 35,000-fold range in volume—the truest measure of size.

Definitions of "sand" as a deposit—as distinct from a size term—are diverse. No generally accepted usage is apparent from a review of the literature. The questions are: should the average, median, or modal size of the material designated "sand" fall in the sand range? Or must 50 percent or some other specified proportion of the material be within these limits? Or to put the question another way, what are the permissible proportions of oversized or undersized material in "sand"? Two of the various alternatives are shown in Fig. 1-2, together with suggested nomenclatural solutions to the problem of sands with various admixtures of other grades.

It is clear from the above discussion that "sand," both as a size term and as a deposit, is defined without reference to composition or to genesis. It could be a quartz sand or a carbonate sand. It could arise (Fig. 1-3) as the indestructible residue from decomposition of a granite (quartz sand) or the product of chemical precipitation (oolitic sand). In practice, we tend to call all these materials of diverse compositions and origins "sand." However, many tend to restrict the term "sandstone" to those indurated sands of siliceous character. The lithified carbonate sands would be termed limestones—not sandstones. Even the term "sand" without adjectival modifiers tends to imply a siliceous compo-

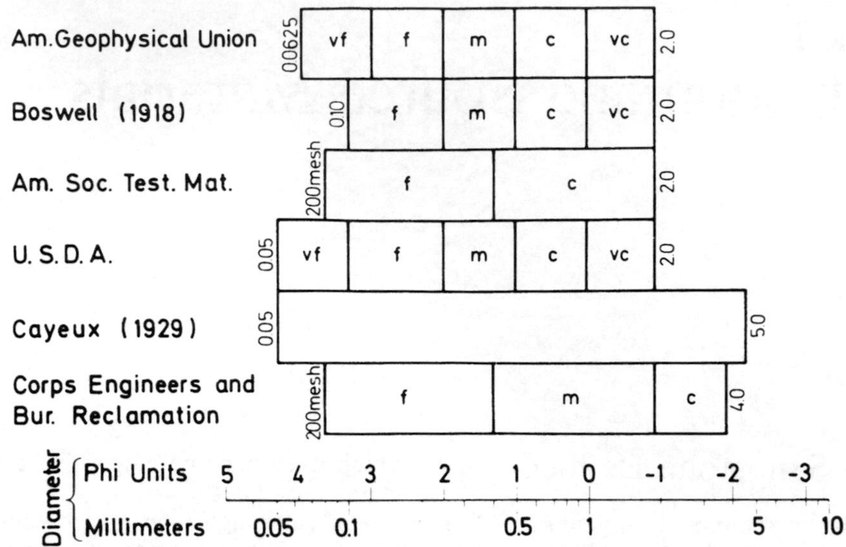

FIGURE 1-1. Grade scales and size limits of sand. Note the diverse meanings of the size terms and the variations in the limits of sand. *vf*—very fine; *f*—fine; *m*—medium; *c*—coarse; *vc*—very coarse.

sition. The terms "arenite" and "psammite" have been proposed as size terms devoid of any compositional connotations to avoid this ambiguity. In this book we largely exclude the carbonate and non-siliceous sands from our consideration.

Sand may be defined in terms of certain arbitrarily agreed upon size limits as indicated above. Some investigators, however, have supposed that there are some "natural" limits which set sand apart from other materials. Wentworth (1933), for example, presumed that the size limits of the several principal classes of clastic sediments—sand, silt, clay—were genetically circumscribed because of the mode of derivation from the parent rock and because of certain fundamental modes of transport by running water. Sand is a product of breakdown of coarse-grained source rocks and the range of sizes (primarily of the quartz) is limited by the original texture of the source rock. There is a presumed dearth of material in those size grades transitional to gravel and to silt (for a review of this problem see Russell, 1968). Sand, unlike the finer materials, is largely transported by rolling and sliding along the bottom or by saltation and only to a smaller extent by turbulent suspension.

Bagnold (1941, p. 6) places the lower limit of sand as that at which the terminal fall velocity is

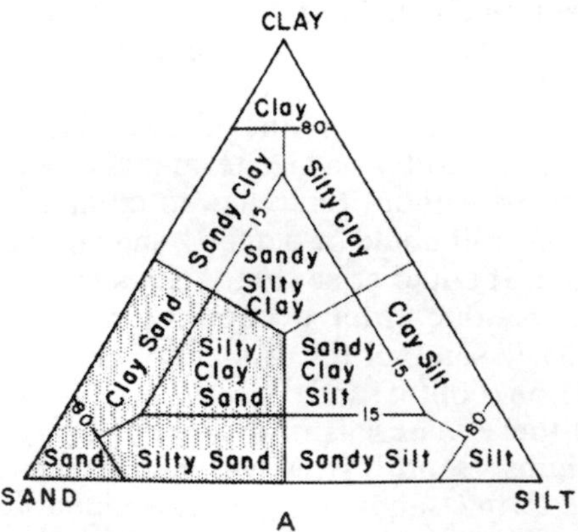

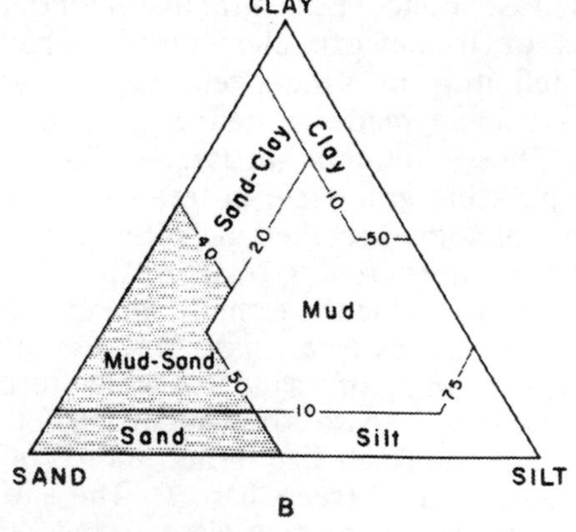

FIGURE 1-2. Two examples of nomenclature of mixed clastic sediments: A) symmetrical conceptual scheme and B) asymmetrical scheme, the latter based on actual usage of marine geologists. (Modified from Shepard, 1954, Figs. 4 and 5).

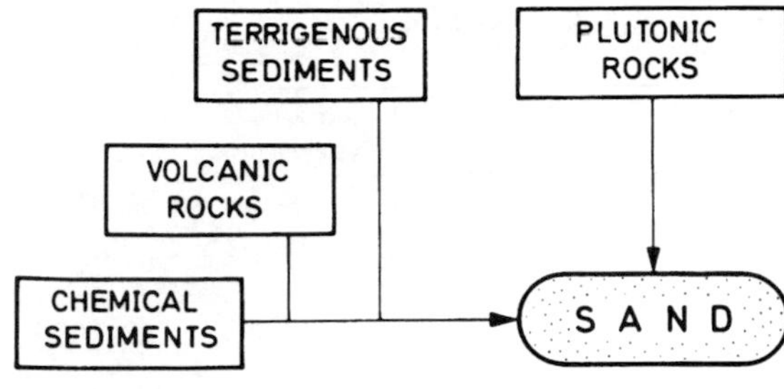

FIGURE 1-3. Genesis of sand. Plutonic rocks supply mostly quartz and feldspar, terrigenous sediments mostly quartz and rock fragments, volcanic rocks mostly rock fragments and glass, and chemical sediments mostly carbonate debris.

less than the upward eddy currents and the upper limit as that size such that a grain resting on the surface ceases to be movable either by direct pressure of the fluid or by the impact of other moving grains. This kind of behavioral definition of sand depends on the nature of the flowing medium (air or water) and must be valid for "average" conditions of flow. The size limits thus defined approximate those set by tradition. Bagnold notes further that materials designated "sand" have one peculiar characteristic which is not shared by coarser or finer materials—namely, the power of self-accumulation—of utilizing the energy of the transporting medium to collect their scattered components together in definite heaps, leaving the intervening surface free of grains. The common mode of transport of sand is by the migration of such heaps or "dunes," be they subaerial or subaqueous.

Relative and Absolute Abundance

Several methods have been employed to estimate the abundance of sandstone relative to the other common sedimentary rocks. These consist either of actual measurement of many stratigraphic sections (Table 1-1) or of calculation of the relative proportion of sandstone based on some geochemical considerations (Table 1-2). The results obtained by these two approaches are somewhat different. In general, the proportions of sandstone and limestone determined by actual measurement are greater than those derived by calculations. This may in part be due to loss of the finer clayey fraction to the deep sea so that shales are underrepresented in the stratigraphic column with a compensatory increase in sandstone and limestone. In summary, sandstones constitute 14 percent of all sediments according to Kuenen (1941), about 20 percent according to Ronov *et al.* (1963, p. 212), and about 32–37 percent according to Leith and Mead (1915) and to Schuchert (1931). Kuenen's estimate is based solely on Indonesian data; Ronov's most recent estimate, 26 percent, apparently applies only to the Russian platform (Ronov *et al.*, 1969, p. 192).

Poldervaart (1955) considered the estimates of Leith and Mead, Schuchert, and Krynine to be applicable to the continental-shield areas, whereas Kuenen's Indonesian estimate was thought to be more characteristic of the younger folded belts of the world. Because the sediments of the former have a volume of about 52.5×10^6 km^3 and the latter a volume of about 126×10^6 km^3, the weighted proportion of sand-

TABLE 1-1. Percentage of common sedimentary rocks based on stratigraphic measurements.

					Horn and Adams (1966)[5]		Ronov (1968)[6]	
	Leith and Mead (1915)[1]	Schuchert (1931)[2]	Kuenen (1941)[3]	Krynine (1948)[4]	Continent-shield	Mobile belt-shelf	Platforms	Geo-synclines
Shale	46	44	57	42	53	59	49	39
Sandstone	32	37	14	40	28	36	24	19
Limestone	22	19	29	18	19[a]	5[a]	21	16

[a] Includes 3 percent evaporite.
1. Leith and Mead (1915, p. 60) based these figures on the average of North American sections aggregating 520,000 ft; an average of sections totalling 188,000 ft in Eurasia gives the proportions 49, 32, and 19, respectively.
2. Schuchert (1931, p. 12) based his figures on measurement of North American Paleozoic maximum of 259,000 ft.
3. Kuenen (1941, p. 168) derives his figures from measurements in the East Indies.
4. Krynine (1948, p. 156) did not indicate how these estimates were obtained.
5. Horn and Adams (1966, p. 282) utilize the data from several sources, both published and unpublished. They do not explain how their estimates were made.
6. Summarized in Ronov, 1968, p. 30. Volcanic rocks are estimated to form an additional 25 percent of the geosynclinal fill. Of the total volume of sediments, 75 percent are geosynclinal and 25 percent platform. The weighted proportion of sandstone is 20 or, if volcanics are excluded, 25 percent.

TABLE 1-2. Percentage of common sedimentary rocks based on geochemical and other calculations.

	Mead[1] (1907)	Clarke[2] (1924)	Holmes[3] (1913)	Wickman[a4] (1954)	Horn[5] (1966)	Garrels and Mackenzie (1971)
Shale	82	80	70	83	73	74
Sandstone	12	15	16	8	20	11
Limestone	6	5	14	9[b]	7	15

[a] Percentage values calculated from Wickman's data.
[b] "Carbonate rock."
1. Mead's figures were derived from bulk chemical analyses by calculating the proportions of average shale, sandstone, and limestone which combined would be as nearly like the average igneous rock as possible.
2. Clarke (1924, p. 34) obtained his figures by assigning all of the free quartz of the average crystalline rock to the production of sandstone and half of the calcium to the formation of limestone. The figure for quartz was obtained from a statistical examination of 700 igneous rocks.
3. Holmes' (1913, p. 60) estimates are the proportions of sediment now deposited annually.
4. Wickman's calculations are "a modernized version of Mead's ideas."
5. Cited by Garrels and Mackenzie (1971, p. 207).

stone is about 26 percent (based on 43 percent sand in continental-shield regions and 18 percent in the younger folded belt). Horn and Adams (1966) likewise noted that the proportion of sand in the continental-shield area was different from that in the mobile belts. They presumed that the sand content in the latter was higher than in the former—a conclusion at variance with other estimates (Table 1-1). The volume of the continental-shield sediments was estimated to be 127 × 10^6 km^3 and that of the mobile belts 395 × 10^6 km^3—estimates considerably higher than those of Poldervaart. The weighted proportion of sandstone (based on 28 percent in the continental-shield areas and 36 percent in the sediments of the mobile belts and shelf) is, therefore, 34 percent. Ronov (1968, p. 30), on the other hand, estimated 25 percent of the total volume of sediment to be sand. Because Horn and Adams did not indicate how their estimates of the percentage of sandstone were obtained nor how the respective volumes were determined, it is difficult to evaluate their results. Hence we will adopt a rounded figure of 25 percent—a result more in accord with both the estimates of Poldervaarat and Ronov. One-fourth the total volume of sediment (the deep oceans and suboceanic materials excluded) is, therefore, sandstone.

What is the total volume of all sedimentary material in the earth's crust? Various estimates have been made (Table 1-3). These estimates have been arrived at in different ways.

Clarke (1924, p. 32) estimated the total volume of average igneous rock which must be weathered to provide the sodium in the sea plus that retained in the sediments. He calculated this to be 84 × 10^6 mi^3 or 350× 10^6 km^3. To this is added 10 percent for porosity and additions from the atmosphere to give the 3.7 × 10^8 km^3. Goldschmidt (1933) likewise calculated the amount of weathered igneous rocks and that of the sediments formed therefrom based on the sodium content of the oceans. Kuenen (1941) applied various corrections to Clarke's data and obtained a figure of 8 × 10^8 km^3 to which he added an estimated 5 × 10^8 km^3 of disintegrated but undecomposed material (tuffs, graywackes, etc.) for a total of 13.0 × 10^8 km^3.

Wickman's (1954) calculations are similar to those of Mead, but based on better data.

Poldervaart's (1955, p. 124, etc.) estimates are based on the thickness of the sediments in the continental-shield areas, the younger folded belts, the ocean basins, and the suboceanic areas (shelves, etc.). Poldervaart used Kay's estimates (Kay, 1951, p. 92) for the first two and estimates based on geophysics and rates of sedimentation for the last two. Horn and Adams (1966) approached the problem in somewhat the same manner as Poldervaart but, using somewhat different data, arrived at a figure of 10.8 × 10^8 km^3—an estimate surpassed only by those of Kuenen and of Blatt (1970). The estimates of

TABLE 1-3. Total volume of sedimentary deposits.

Authority	Kilometers^3
1. Clarke, F.W. (1924, p. 32)	3.7 × 10^8
2. Kuenen (1941, p. 188)	13.0 × 10^8
3. Goldschmidt (1933)	3.0 × 10^8
4. Wickman (1954)	4.1 ± 0.6 × 10^8
5. Poldervaart (1955, pp. 126–130)	6.3 × 10^8
6. Horn and Adams (1966, p. 282)	10.8 × 10^8
7. Ronov (1968, p. 29)	9.0 × 10^8
8. Blatt (1970, p. 259)	42 × 10^8

Poldervaart and of Horn and Adams include deep oceanic and suboceanic sediments of which sand forms a negligible part. Sand is concentrated on the continental blocks rather than in the oceanic basins. Sand forms a fourth to a third of the sediments on the continental platform (including the mobile belts and contiguous shelves). A recent estimate of total sediment volume is that of Blatt (1970, p. 259) who estimated 10×10^8 mi^3 or about 42×10^8 km^3. This estimate is by far the largest on record.

Using Poldervaart's estimate of volume for the continental sediments (176×10^6 km^3) and assuming one-fourth of it to be sand, the total volume of sand in the world is 44.0×10^6 km^3. If we use the larger estimate of Horn and Adams (522×10^6 km^3) and assume one third of it is sand, the total volume of sand is about 174×10^6 km^3. The total *mass* of sediments on the continental block is estimated by Poldervaart to be 480×10^{15} tons(4800 Gg) of which, therefore, 120×10^{15} metric tons (1200 Gg) is sand. If an individual sand grain (1 mm in diameter and density of 2.7) has a mass of about 0.0014 grams, there would be some 85.7×10^{24} grains of sand in the crust of the earth. As most sand grains are of smaller size, the total number would be much larger (eight times as many if the grains were 0.5 mm in diameter and sixteen times as many for grains 0.25 mm in diameter).

Is this quantity a fixed sum or is it constantly being added to? Weathering breaks rocks down. One product of this breakdown is sand, thereby increasing the total quantity of sand on earth. But sand grains are subject to abrasion and other size-reduction processes. Conceivably such action would so reduce the size of the material acted on as to eliminate it from the sand grade. Moreover, sandstones are, in turn, owing to deep burial, being transferred to those zones within the earth where they undergo metasomatic transformations into granites, gneisses, and other rocks no longer recognizable as sediments. Do the processes of sand formation and sand destruction balance? Is there a steady state or is there a net increase in the total quantity of sand? Certainly these are fundamental questions that deserve our attention.

Kuenen (1959) has estimated that the yearly production of quartz sand is of the order of 0.05 cubic kilometers. He has shown also that the loss of sand by abrasion is incredibly slow. Such rounding of sand as has taken place is thought to be largely due to eolian action in desert areas. Balancing new sand production against losses led Kuenen to the conclusion that the total quantity of sand is increasing. He estimated that during each and every second in the incredibly long past, the number of quartz grains on earth has increased by 1,000 million grains!

As Garrels and Mackenzie (1971, p. 260) note, if the present rate of sediment production and deposition prevailed during and since the Cambrian, the accumulated sediment would be eightfold greater than what is present in the geologic record. This observation suggests destruction of the accumulated sediment, in part by erosion, in part by deep burial and conversion to rocks no longer recognizable as sediments (i.e., gneisses and granites, and in part by subduction into the upper mantle). Possibly, therefore, the quantity of sand is constant or increasing only very slowly.

In both the foregoing tables and discussion we have considered only sands of nonvolcanic origin.

Distribution, Past and Present

Just where is sand in the world today and where did the sand accumulations of the past take place?

Where is sand found in the world today? The most obvious places are the rivers and beaches and, to a lesser extent, the dunes and shallow shelf seas. The fluvial sands include those found on alluvial fans, in river channels, and on floodplains, and those of the deltas of both lakes and the oceans. A little sand also escapes the river channel and finds its way into the backwater swamps and bayous. Shoreline sand includes not only that of the beaches but also that found on offshore bars, in lagoons, and on tidal flats. Many windblown dunes are closely associated with beaches and also with major rivers but the most impressive eolian sands are those of the dune fields of some desert basins. Marine sands are largely shelf sands though some sand occurs on the continental slopes, on most of the continental rises as well as on the oceanic abyssal plains and isolated sediment ponds that are found in hilly or mountainous subsea topography.

In short, there seems to be no large geomorphic region of the earth where sand is not found. The deep oceanic basins, the most extensive geomorphic elements, have the least, being almost devoid of sand—and containing only the scattered grains of eolian origin, the turbidite sands on the continental rise and the abyssal plains, and, in some cases, volcaniclastic sands generated by subaqueous eruptions. Clearly the principal environments of sand accumulation are on the continent. The absence of sand in any particular environment is probably due more to an absence of supply rather than to conditions unfavorable for its accumulation.

Not all of the environments of sand accumulation are of equal importance. Not only are some of lesser importance than others but many are places of temporary lodgement of sand which ultimately will be eroded and the sand retrieved to be redeposited elsewhere. Much of the present-day detritus from the western cordillera of North America, for example, is trapped in intermontaine basins and will, in time, be recycled and redeposited. One should not presume, however, that the ultimate destination of all sand is the sea. The ultimate destination of most sand is a geosyncline where it may be deposited in an alluvial rather than a marine environment. It has been estimated that three-quarters of all the sediments of the geologic past are in geosynclines and that only one-fourth is found on the cratonic platform. Sands

TABLE 1-4. Famous sandstones.

Athabasca (Cretaceous): central Alberta

Very famous tar sand with reserves estimated at 625 × 10^9 barrels. Mostly fluvial and deltaic paleoenvironments. Also called Lower Manville.

Botucatú (Jurassic–Cretaceous): southwestern Brazil and adjacent Paraguay and Argentina.

Widespread, mature eolianite of Paraná Basin that is now a prolific aquifer and has been much studied in Brazil.

Buntsandstein (Triassic): Germany, France, Netherlands, Poland, England, and North Sea.

One of the most thoroughly studied sandstones with complete data on paleocurrents and much petrology. First systematic regional paleocurrent map was made on the Buntsandstein from which significant environmental, petrographic, and provenance conclusions were drawn. Vast literature mostly, but not totally, in German. Minor petroleum production.

Grès Amoricain (Ordovician): France, Portugal, Morocco, and Mauretania.

Transgressive, mature shallow-water sandstone up to 1000 m. Castle of William the Conqueror in Falise, Normandy, is built on an outcrop of the Gres Amoricain. Also called Arenig Formation.

Navajo (Jurassic): western United States

Abundant crossbedding, much of it spectacularly thick (over 30 m); "Crossbedded on a scale and perfection . . . difficult to exaggerate" (Gregory, 1917, p. 58). An often photographed and much cited example of an ancient eolianite—an interpretation which is challenged from time to time.

New Red (Triassic): Occurs around North Atlantic basin in isolated, "half graben" rifted basins in Great Britain, Norway, Greenland, and along much of the eastern coastline of Canada and the United States. Many different local names including Newark Series in eastern United States.

Locally very thick and formed during the initial rifting and breakup of the North Atlantic Ocean basin. Much studied classic molasse. See Buntsandstein.

Nubian (Jurassic to late Cretaceous): Egypt Sudan, Libya, and Chad, and parts of adjacent Asia Minor.

First recognized in 1847 as the "Nubia," this thick widespread (1000 to 3000 m) sandstone sequence contains but minor shales. Alluvial, coastal plain, and marine shelf environments.

Old Red (Devonian): North Atlantic Basin including Great Britain, Germany, Belgium, the Baltic, Norway, Spitzbergen, Greenland, and eastern Canada and U.S. Many different local names of which the most famous in North America is the Catskill. Early description by Hugh Miller in 1841.

Post-orogenic Caledonian molasse derived from the post-orogenic uplift of the "Old Red Sandstone Continent." Thick alluvial sections of sandstone, commonly but not always red, and locally imbedded with volcanics, shales, and paleosoils. Compare to New Red Sandstone, a much studied molasse.

Peninsula (Lower Paleozoic): Republic of South Africa

Up to 400 m of mostly quartz arenites commonly with prodigious facies stacking chiefly of littoral (beach, washover fans, and tidal channel) and shelf sandstones. The Peninsula Formation is part of the vast, largely sand-filled Cape Basin (Ordovician to Carboniferous) of southern Africa which extends over 1000 miles along the coast. Spectacular coastal scenery such as Table Mountain is formed by the sandstones of this basin.

Roraima (Middle Precambrian): eastern Venezuela, Guyana, Surinam, and Roraima (Brazil).

Very thick, widespread (200,000 km^2) but still inadequately studied unit forms spectacular table mountains and plateaus covered by tropical rain forest. A South American equivalent of the Nubian—because it is widespread, thick, and spans a wide time range? Diamonds occur near base.

Rotliegendes (Permian): greater North Sea region of England, Low Countries, Germany, Poland, and Norway.

Thin to moderately thick redbeds of eolian, fluvial–wadi, and sabkha origin plus some desert lakes. Much studied and now an important petroleum reservoir. Spectacular crossbedding in eolian facies was early studied.

St. Peter (Ordovician): Central United States.

In the Upper Mississippi valley and mid-continent region of the United States, extending into North Texas is a thin, transgressive, widespread, supermature crossbedded and much studied unit; often cited as the archtype of a cratonic sandstone.

form a significant part of the accumulation in both these tectonic elements.

It is noteworthy that most common modern sites of sand accumulation—the beaches and rivers—are linear features and the sand associated with them is confined to a narrow zone. Yet the sands of the past commonly occur in areally extensive stratiform sheets. This discrepancy between the essentially linear loci of sand deposition in the present day and the extensive sand sheets of the past suggests that the latter are the product of lateral shifting of sand deposition through time by the lateral migration of streams or by transgressive or regressive shift of the shoreline.

Exceptions to this rule are the more extensive dune fields of some deserts and the broad expanse of sand on some shelf areas. Much of the sand on modern continental shelves, however, turns out to be a relict sand not in equilibrium with the present regime and is probably a fluvial deposit inherited from the low stand of the sea in glacial times (Emery, 1966, p. 12).

In summary, sand is the most continental of all sediments. It is produced in grains that are too large to be blown or washed far off the continents. Therefore, it remains as an ever-increasing cover on the continental blocks. See Table 1-4 for some famous sandstones all of which are on the continents. Sand is produced on the continent, is shifted from the higher places to the lower sites of accumulation. The only "leakage" from the continents is due to a trifling amount carried to the deep sea in dust storms or by turbidity currents which transport sand down the continental slope to the abyssal plains.

Where did sand accumulation take place in the geologic past? Presumably it could and did accumulate in the same environments then as now. But the relative importance of each of these environments was vastly different. In the central Appalachians, for example, sandstone constitutes about 23 percent of the whole section (Colton, 1970, p. 11). Of this about 55 or 60 percent is believed to be alluvial, about 25 percent marine turbidite; the balance, no more than 20 percent, is probably littoral or shallow marine. None is identifiable as eolian. These figures emphasize the importance of the alluvial sediments in miogeosynclines, of which the Appalachians is perhaps a fairly typical example.

Some sands display an extraordinary roundness. If Kuenen is right in believing that such roundness is most likely acquired by eolian action, many of the sand grains in the world's accumulation have had a desert eolian stage at some time in their history. Kuenen (1959, p. 23) has estimated that 2×10^6 km^2 of desert is needed to keep the world average roundness constant (to offset the new, sharp-cornered sand added each year).

Nothing has been said here about the distribution of the several kinds of sand (arkosic, lithic, etc.) in the present-day world or about their distribution in the geologic past. It may be that the bulk composition of sand has changed with time. These problems are treated in the last chapter in this book.

According to Ronov (1964), sands have been a significant and nearly constant part of the sedimentary record since earliest recorded time (3.5 b.y. ago). He suggests, however, that the types of sand may have changed with time. On the other hand, volcaniclastic sands may have had a greater role in earlier times.

It is of interest that we now have evidence of sand on Mars. Photographs show both dune and ripple forms characteristic of sand. Mars seems to be the only other place, Venus possibly excepted, where sand occurs in our solar system.

History of Investigation

The scientific investigation of sands and sandstones goes back nearly two centuries. The earliest work on sands—mainly river and beach sands—was directed toward determining their mineral composition—work inspired in part by the fact that sands may contain useful materials such as gold. This activity was greatly stimulated and advanced by the use of the polarizing microscope. These early researches were largely descriptive and attempted only to record what was there. The indurated sands—the sandstones—were looked on mainly as stratigraphic entities—formations in a geological column. They were described by the field geologist, named, and then placed in the proper position in the geologic section. Names such as the Millstone Grit, the Old Red Sandstone, and the Buntsandstein appeared in the early literature and bear witness to the earliest field studies involving sandstones. Interpretations of their origin were based largely on field observations of sedimentary structures, such as ripple marks, and on the contained fossils, if any.

Not until the thin-section technique was available was there serious study of the fabric and composition of these sandstones and utilization of their microscopical characteristics to elucidate the natural history of the rock. Henry Clifton Sorby, whose Presidential address in

1879 before the Geological Society of London, "On the Structure and Origin of Non-calcareous Stratified Rocks," was a milestone, initiated the modern approach to the study of sandstones. One of the earliest papers on the petrography of sandstones was his paper on the Millstone Grit (Sorby, 1859b). In addition to his microscopical investigations, Sorby made many significant field observations on the structure of sandstones, particularly on crossbedding—researches which anticipated the paleocurrent analyses a century later.

Despite Sorby's brilliant demonstration of the usefulness of the polarizing microscope in the study of the sedimentary rocks, it was to the igneous rocks that the thin-section technique was largely applied—especially by the German school of petrographers (Rosenbusch, Zirkel, and others). A major exception was the work of Lucien Cayeux, whose monographs on the sedimentary rocks of France remain unsurpassed even to this day. Among these was his work on the Tertiary Sandstones of the Paris Basin (1906) and his volume on the siliceous rocks of France (1929).

The work on the mineralogy of sands, much of it on the minor accessory minerals—the "heavy minerals"—was greatly expedited by the microscopic techniques using polarized light. The utility of heavy minerals in stratigraphic correlation, especially subsurface correlation in oil field exploration and development, led to widespread interest in these constituents of sand. This interest culminated in the appearance of H.B. Milner's "Introduction to Sedimentary Petrography" in 1922—a revised fourth edition appeared forty years later—which was oriented largely toward the use of heavy minerals in subsurface correlation. For the most part, heavy minerals of sands now play only a minor role in correlation—having been superseded by microfossils and to an even greater extent by geophysical logging techniques.

The mineralogy of sands—ancient as well as Recent—continues to be of interest, however, in provenance studies. Studies of the mineralogy led further to paleogeographic analyses represented, for example, by the modern work of Füchtbauer (1964) on the Molasse of southern Germany. Such studies owe much to the pioneer work of Mackie on the sands and sandstones of Scotland. Mackie attempted to work out the principles of interpretation of sand mineralogy—such as the use of quartz varieties in provenance (1896), the climatic significance of feldspar (1899), and the use of mineral analyses of ancient sandstones to unravel their natural history.

Excepting, perhaps, Hadding's work on the sandstones of Sweden (1929), there have been few monographic studies of sandstones in the Cayeux tradition. There have been, however, noteworthy classical studies of particular sandstone formations. Well-known examples are Krynine's study of the Devonian Third Bradford Sand of Pennsylvania (1940), Dake's earlier study of the Ordovician St. Peter Sandstone of the Upper Mississippi Valley (1921), and Gilligan's study of the Millstone Grit (1920) in Great Britain. Mention should also be made of Marcus Goldman's study of the Miocene Catahoula Sandstone of Texas (1915), which is an unsurpassed example of how much information and understanding can be extracted from a small sample of sandstone.

Interest in the mineral composition of sands, particularly modern beach, dune, and river sands, was superseded, or, perhaps more properly, supplemented, by an interest in their "mechanical" composition. Grain-size analyses, and later the measurement of grain shape and roundness, led to an era of "quantitative sedimentation" and to the study of sands as gross particulate systems. A major contribution to this approach was J.A. Udden's paper on "Mechanical Constitution of Clastic Sediments" published in 1914. Although including silts and gravels, this work was largely a compilation of grain-size analyses of sands. The concept of sand as a population of grains led to the application of statistical methods of population analysis. The pioneer work of Wentworth (1929) and Trask (1932) was followed by a flood of papers. Most influential in this approach, perhaps, are the many papers of W.C. Krumbein (beginning in 1936). Efforts continue to find ways of utilizing the grain size distribution to discriminate between differing environments and/or agents of deposition.

It is presumed that the surfaces of sand grains bearing microscopic to near submicroscopic features—etching, scratches, percussion scars, and the like—are indicative of the agent of transport and deposition. There is a considerable earlier literature on the subject (Williams, 1937). In more recent years the scanning electron microscope (SEM) has rekindled interest in the subject (see, for example, Krinsley and Donahue, 1968). The SEM has also proved to be a useful tool for studying the fabric of sandstone cements.

The last twenty years have seen the introduction and widespread adoption of new tools other than the SEM. The electron microprobe gives precise chemical data on individual mineral grains, especially the compositions of feldspars. Cathodo-luminescence has been increasingly applied to the study of detrital provenance and diagenetic cement. New materials and methods of impregnation with colored dyes are now used to study the nature of porosity in sandstones. Sophisticated organic analytical instruments determine the nature of organic fractions of sandstones, especially liquid hydrocarbons. All of these analytical devices serve to augment and refine the basic mineralogic and textural information coming from microscopic petrography, one of the most durable basics of the study of sandstones.

Most recent studies of sand deposits have emphasized the anisotropic fabric of such accumulations. Sand accumulations show a response to the earth's magnetic and gravitational fields and to fluid flow systems. Such response extends from the orientation of individual nonspherical grains through primary sedimentary structures of the deposit to the shape and orientation of the sand body itself. Such "paleocurrent analyses" early made by Sorby (1859a) have been much in vogue during the past decade (see Potter and Pettijohn, 1977).

With the growth and interest in geochemistry since the Second World War, there has been renewed interest in sediments as the products of a complex process of chemical differentiation and fractionation. The sands constitute such a differentiate—the "resistates." As a result of the pioneer efforts of geochemists such as F.W. Clarke, data on the chemical composition of sands and other sediments has accumulated (Clarke, 1924). Sandstones have a bulk composition reflecting their degree of maturity, that is, the degree to which they have evolved toward the stable end-product of the sand-forming processes. Chemical compositions also reflect the mineralogic differences among the main groups of sandstones, the arenites and wackes, as well as specific provenances such as volcaniclastics. Isotopic geochemistry has been applied to sandstones, mainly for diagenetic studies using oxygen isotopes.

In many basins sandstones are a component of a well-defined "package" or cycle characterized by its own sequence of lithologies and structures—the "vertical profile." There are a limited number of such sequences. Studies of modern deposits have shown that they are the product of a particular process and environment such as a prograding delta, lateral migration of an offshore bar, etc. Our interpretation of the environment of deposition of sandstones is greatly enhanced by recognition of these autocyclic sequences (Selley, 1970).

The study of vertical sequences has been joined by analysis of cross sections of seismic stratigraphy (Vail *et al.*, 1977). Continuous seismic profiles reveal horizontal and vertical patterns of depositional sequences such as those of a prograding deltaic sequence with associated shelf and submarine fan deposits. Worldwide correlations have led to the differentiation of sequences affected by eustatic sea level changes from those that are the result of local tectonic movements. These techniques and ideas have revitalized subsurface studies of deeply buried sandstones on the continents and continental margins.

We also need to take a larger view of the environment of sand deposition. Krynine (1945) and others have emphasized the role of tectonics and sedimentation and its relation to sandstone composition. Interest in the subject has been much accelerated since the concept of plate tectonics has come to dominate geologic thought. There is now a considerable literature on the composition of sandstones in relation to the various kinds of basins created by plate motions (see, for example, Dickinson and Suczek, 1979). The structural style and evolution of diverse basins, from fore-arc to rift valley and cratonic to exotic terrane, are subjects that have become close partners to sandstone sedimentology.

The geophysical evolution of basins has become linked to sedimentology as the origin of continental margin basins such as the North Sea is cast in terms of lithospheric stretching, subsidence, and consequent continental and marine sedimentation. The origin of many intracratonic basins is now being sought in deeply buried former rifts that initiated subsidence in basins whose earliest sediments were rift valley sandstones.

Though the focus of sandstone sedimentology continues to be on provenance and environment, the diagenesis of sandstones has received renewed and intensive interest, especially from those interested in oil and gas or diagenetic mineral deposits. The diagenetic histories of sandstones are now cast in terms of the primary depositional input and the subsequent burial regime. Elements of these histories are thermal regime and tectonic environment of the basin, on the one hand, and detailed mechanisms of pore fluid migration and geochemical evolution that mediate cement and authigenic mineral precipitation or alterations on the other.

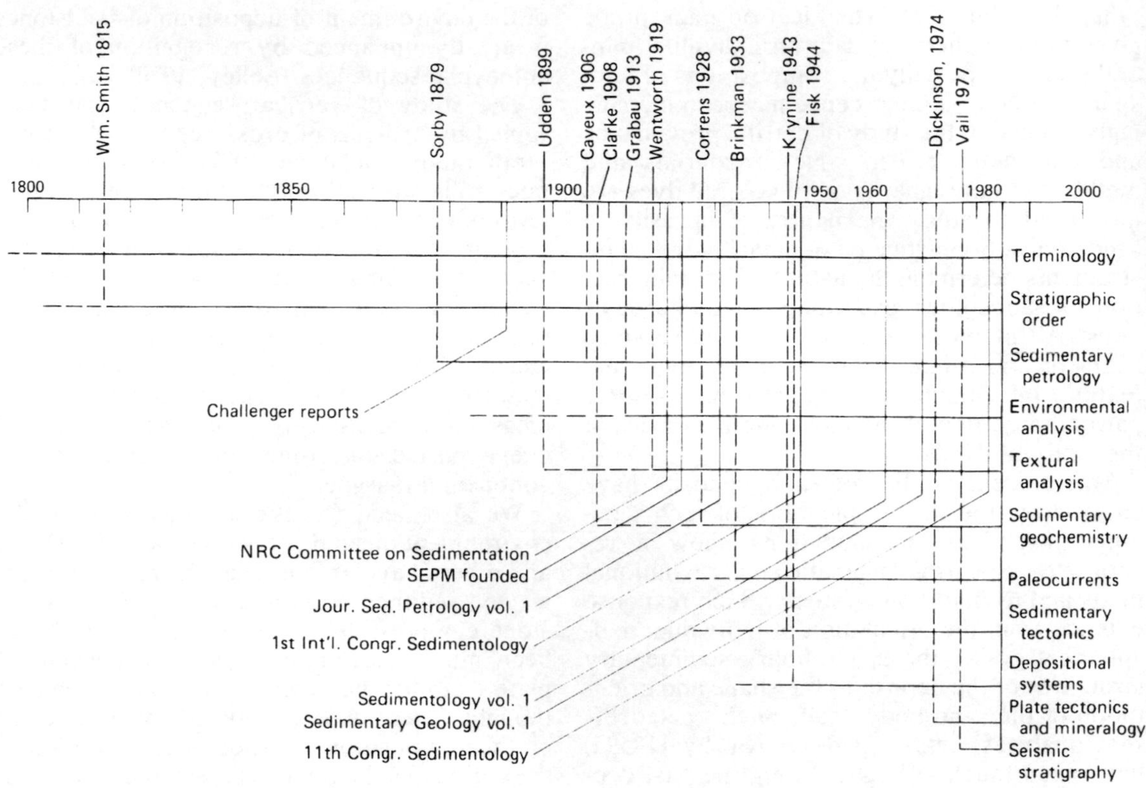

FIGURE 1-4. Development of "arenology".

In summary, the history of the investigations of sands and sandstones reveals the shifting emphasis in ways of looking at sandstones. It is our aim in this book to look at sands and sandstones from all of these points of view and to utilize the several approaches to an overall understanding and interpretation of the natural history of these deposits. The several points of view and the evolution of the science of "arenology" are summarized in Fig. 1-4.

Economic Value of Sand

Sand is an important economic resource. The uses of sand are many. For some purposes, such as an abrasive to clean a skillet or mess kit, most any sand will do; other uses require a particular kind and quality of sand (Boswell, 1919). Sands are a source of silica for making sodium silicate, for manufacture of carborundum (silicon carbide), for silica brick, and for manufacture of both common and optical glass. Sand is an ingredient in plaster, in concrete, is added to clays to reduce shrinkage and cracking in brick manufacture, and is mixed with asphalt to make "road dressing." It is used in foundries as molding and parting sand, is used as an abrasive (sandpaper, sandblast), and is employed in the grinding of marble, plate glass, and metal. Some sands are used as soil conditioners (lime sand) or fertilizers (glauconitic greensand). Sand is used in filtration and as friction sand (on locomotives). And perhaps most important of all, it is what every child loves to play in.

Sands are exploited for rare minerals and rare elements which they contain. Some are gold-bearing; others contains gems, platinum, uranium, tin (cassiterite), and tungsten (wolframite). Others are exploited for monazite (containing thorium and rare earths), zircon (for zirconium), rutile (for titanium), and others. The search for and exploitation of these sands is a special art—that of alluvial prospecting (Raeburn and Milner, 1927; Smirnov, 1965). Sandstones may, in some cases, be the host rock during mineralization. Mineral deposits thus formed may be unrelated to the depositional environment but a part of the diagenetic history of the sandstone. The detailed treatment of these deposits are beyond the scope of this book; the reader is referred to texts such as Maynard (1983).

In addition to the uses to which sands are put and in addition to the rarer constituents which are recovered from sands, these materials constitute most important reservoirs for the storage of valuable fluids. Large volumes of fluid are contained in the pore systems of sands and sandstones. They are important reservoirs of fresh waters, of brines, and of petroleum and natural gas. Sand strata are also the conduits for artesian flow. Knowledge of the shape and attitude of these sand reservoirs and their porosity and permeability is necessary for extraction of the fluids therein. Fluids may be injected into sands also. Sands are thus utilized for natural gas storage and for recharge with fresh waters for future use or are injected with water to drive out the contained oil.

Sandstones are used for building stone construction, as flagstone, and, if crushed, as road fill, road metal, and railroad ballast.

The economic aspects of sands are not confined to their value as raw material and their various uses. Sand production, movement, and deposition is of concern to the engineering geologist and to the geomorphologist, especially those concerned with shore erosion and harbor development. Some understanding of the sand budget—sand supply and accretion and sand deficit or removal—is necessary if the problems of shore engineering are to be solved. Similarly, dune stabilization and prevention of sand encroachment on cultivated lands and forests or on roads and other structures requires an understanding of the geology of sand. Many of the problems of river management likewise involve sand.

To the three of us sands are perhaps the most rewarding of all sediments to study. They are more amenable to study than the shales and have greater value in determination of provenance, paleocurrents, and paleogeographic reconstruction than either shales or limestones. Surely if one had to pick a *single* lithology from which he had to reconstruct the earth's geologic history, it would be sandstone: it goes back further in time than limestone, is less susceptible to diagenesis than shales and carbonate rocks, and is widespread all over the continents.

References

Bagnold, R.A.: The physics of blown sand and desert dunes, 265 pp. London: Methuen 1941.

Blatt, Harvey: Determination of mean sediment thickness in the crust: A sedimentologic model. Geol. Soc. America. Bull. 81, 255–262 (1970).

Boswell, P.G.H.: Sands, considered geologically and industrially, under war conditions. Inaugural Lecture, 38 pp. Liverpool: Univ. Liverpool Press 1919.

Cayeux, Lucien: Structure et origine des grès du Tertiare parisien. Etudes des gites minéraux de la France, 160 pp. Paris: Impr. Nationale 1906.

Cayeux, Lucien: Les roches sédimentaires de France. Roches siliceuses, 774 pp. Paris: Impr. Nationale 1929.

Clarke, F.W.: The data of geochemistry, 5th Ed., 841 pp. U.S. Geol. Survey Bull. 770 (1924).

Colton, G.W.: The Appalachian basin—Its depositional sequences and their geologic relationships. In: Fisher, G.W.; Pettijohn, F.J.; Reed, J.C.; and Weaver, K.N. (Eds.): Studies of Appalachian geology, central and southern, pp. 5–47. New York: Interscience Publ. 1970.

Dake, C.L.: The problem of the St. Peter sandstone. Univ. Missouri School Mines and Metall. Bull., Tech. Ser., 6, 158 pp. (1921).

Dickinson, W.R., and Suczek, C.A.: Plate tectonics and sandstone composition. Am. Assoc. Petroleum Geologists Bull. 63, 2164–2182 (1979).

Emery, K.O.: Geologic background. In: The Atlantic continental shelf and slope of the United States. U.S. Geol. Survey Prof. Paper 529-A, A1–A23 (1966).

Füchtbauer, H.: Sedimentpetrographische Untersuchungen in der älteren Molasse nördlich der Alpen. Eclogae géol. Helvetiae 57, 157–298 (1964).

Garrels, R.M., and Mackenzie, F.T.: Evolution of sedimentary rocks, 397 pp. New York: Norton and Company 1971.

Gilligan, A.: The petrography of the Millstone Grit of Yorkshire. Geol. Soc. London Quart. Jour. 75, 251–294 (1920).

Goldman, M.I.: Petrographic evidence on the origin of the Catahoula Sandstone of Texas. Am. Jour. Sci., ser. 4, 39, 261–287 (1915).

Goldschmidt, V.M.: Grundlagen der quantitativen Geochemie. Fortschr. Mineral. Krist. Petrog. 17, 112–156 (1933).

Gregory, H.E.: Geology of the Navajo country. U.S. Geol. Survey Prof. Paper 93, 161 p (1917).

Hadding, A.: The pre-Quaternary sedimentary rocks of Sweden: Part III, The Paleozoic and Mesozoic sandstones of Sweden. Lunds Univ. Arsskr., N. F., Avd. 2, 25, nr. 3, 287 pp. (1929).

Holmes, A.: The age of the earth, 196 pp. London: Nelson 1913.

Horn, M.K., and Adams, J.A.S.: Computer-derived geochemical balances and element abundances. Geochim. et Cosmochim. Acta 30, 279–290 (1966).

Kay, Marshall: North American geosynclines. Geol. Soc. America, Mem. 48, 143 pp., (1951).

Krinsley, D., and Donahue, J.: Environmental interpretation of sand grain surface texture by electron microscopy. Geol. Soc. America Bull. 79, 743–748 (1968).

Krynine, P.D.: Petrology and genesis of the Third Bradford Sand. Pennsylvania State College Bull. 29, 134 pp. (1940).

Krynine, P.D.: Sediments and the search for oil: Producers Monthly 9, 17–22 (1945).

Krynine, P.D.: The megascopic study and field classification of sedimentary rocks. Jour. Geology 56, 130–165 (1948).

Krumbein, W.C.: Application of logarithmic moments to size frequency distribution of sediments. Jour. Sed. Petrology 6, 35–47 (1936).

Kuenen, Ph.H.: Geochemical calculations concerning total mass of sediments in the earth. Am. Jour. Sci. 239, 161–190 (1941).

Kuenen, Ph.H.: Sand—its origin, transportation, abrasion and accumulation. Geol. Soc. South Africa, 62, Annexure, 33 pp. (1959).

Leith, C.K., and Mead, W.J.: Metamorphic geology, 337 pp. New York: Henry Holt 1915.

Mackie, W.: The sands and sandstone of eastern Moray. Trans. Edinburgh Geol. Soc. 7, 148–172 (1896).

Mackie, W.: The feldspars present in sedimentary rocks as indicators of contemporaneous climates. Trans. Edinburgh Geol. Soc. 7, 443–468 (1899).

Maynard, J. Barry: Geochemistry of sedimentary ore deposits, 305 pp. New York: Springer Verlag, 1983.

Mead, W.J.: Redistribution of elements in the formation of sedimentary rocks. Jour. Geology 15, 238–256 (1907).

Milner, H.B.: An introduction to sedimentary petrography, 125 pp. London: Murby 1922.

Milner, H.B.: Sedimentary petrography, Vol. 1, 643 pp.; Methods in sedimentary petrography, Vol. 2, 715 pp. Principles and applications. New York: Macmillan 1962.

Poldervaart, A.: Chemistry of earth's crust. In: Poldervaart, A. (Ed.): Crust of the earth—a symposium. Geol. Soc. America Spec. Paper 62, 119–144 (1955).

Potter, P.E., and Pettijohn, F.J.: Paleocurrents and basin analysis, 2nd Ed., 425 pp. Berlin–Göttingen–Heidelberg: Springer 1977.

Raeburn, C., and Milner, H.B.: Alluvial prospecting, 478 pp. London: Murby 1927.

Ronov, A.B.: Common tendencies in the chemical evolution of the earth's crust, ocean, and atmosphere. Geochem. Int. 4, 713–737 (1964).

Ronov, A.B.: Probable changes in the composition of sea water during the course of geologic time. Sedimentology 10, 25–43 (1968).

Ronov, A.B.; Migdisov, A.A.; and Barskaya, N.V.: Tectonic cycles and regularities in the development of sedimentary rocks and paleogeographic environments of sedimentation of the Russian platform (an approach to a quantitative study). Sedimentology 13, 179–212 (1969).

Ronov, A.B.: Mikhailovskaya, M.S.: and Solodkova, I.I.: Evolution of the chemical and mineralogical composition of arenaceous rocks. In: Vinogradov, A.P. (Ed.): Chemistry of the earth's crust, Vol. 1. U.S.S.R. Acad. Sci., Israel Progr. Sci., Translations, 1966, 212–262 (1963).

Russell, R.J.: Where most grains of very coarse sand and gravel are deposited. Sedimentology 11, 31–38 (1968).

Schuchert, Chas.: Geochronology or the age of the earth on the basis of sediments and life. In: The age of the earth. Bull. Natl. Research Council 80, 10–64 (1931).

Selley, R.C.: Ancient sedimentary environments, 237 pp. Ithaca, New York: Cornell Univ. Press 1970.

Shepard, F.P.: Nomenclature based on sand–silt–clay ratios. Jour. Sed. Petrology 24, 151–158 (1954).

Smirnov, V.I.: Geologiia rossypei (Geology of placer deposits), 440 pp. Div. Earth Sci., Sci. Council Oreformation, Acad. Sci. U.S.S.R., Moscow (1965).

Sorby, H.C.: On the structure and origin of the Millstone-Grit of South Yorkshire. Proc. West Yorkshire Geol. Soc. 3, 669–675 (1859a).

Sorby, H.C.: On the structures produced by the current present during the deposition of stratified rocks. The Geologist 2, 137–147 (1859b).

Sorby, H.C.: On the structure and origin of noncalcareous stratified rocks (Presidential address). Geol. Soc. London Proc. 35, 56–77 (1880).

Trask, P.D.: Origin and environment of source sediments of petroleum. 321 pp. Houston: Gulf Publ. Co. 1932.

Trefethen, J.M.: Classification of sediments. Am. Jour. Sci. 248, 55–62 (1950).

Udden, J.A.: Mechanical composition of clastic sediments. Geol. Soc. America Bull. 25, 655–744 (1914).

Vail, P.R.; Mitchum, R.M. Jr.; Todd, R.G.; Widmier, J.M.; Thompson, S., III; Sangree, J.B.; Bubb, J.N.; and Hatlelid, W.G.: Seismic stratigraphy and global changes in sea level. In: Payton, C.E. (Ed.): Seismic stratigraphy-applications to hydrocarbon exploration: Amer. Assoc. Petroleum Geologists Mem. 26, 49–212 (1977).

Wentworth, C.K.: Method of computing mechanical composition types of sediments. Geol. Soc. America Bull. 40, 771–790 (1929).

Wentworth, C.K.: Fundamental limits to the sizes of clastic grains, Science 77, 633–634 (1933).

Wickman, F.F.: The "total" amount of sediment and the composition of the "average igneous rock". Geochim. et Cosmochim. Acta 5, 97–110 (1954).

Williams, Lou: Classification and selected bibliography of the surface texture of sedimentary fragments. In: Rept. Committee on Sedimentation, 1936–1937, pp. 114–128. Washington, D.C.: Natl. Res. Council 1937.

General Sources for the Study of Sand and Sandstone

Included below are references that reflect the many different ways sand and sandstone are studied. Most of the pre-1972 texts and major references, some from as early as the 1930s, are reported in the first edition of *Sand and Sandstone*, and hence only post-1972 references are listed.

General

The references below include all the general texts, with much on sand and sandstone.

Blatt, Harvey; Middleton, Gerard; and Murray, Raymond C.: Origin of sedimentary rocks, 2nd Ed., 782 pp. Englewood Cliffs, New Jersey: Prentice-Hall, Inc. 1979.	Much on sandstones, chiefly in Chaps. 3, 4, 5, 7, 8, 9, 10, and 19. Basic reference also useful for its insights into other sediments.
Briggs, David: Sediments, 192 pp. London: Butterworths 1977.	Well-written overview with five chapters and a glossary. Chiefly about sand and gravel. Elementary.
Conybeare, C.E.B.: Lithostratigraphic analysis of sedimentary basins, 555 pp. New York: Academic Press 1979.	Basin analysis in six chapters, but includes much sedimentology, source rock analysis, and structure.
Corrales, I.; Rosell, J.; Sanchez de la Torre, L.M.; Vera, J.A.; and Vitas, L.: Estratigrafia, 718 pp. Madrid: Editorial Rueda 1977.	Comprehensive, elementary text of 24 chapters with emphasis on processes and products of sedimentation. Much on sand and sandstone. Five appendices and a glossary.
Engelhardt, Wolf von: The origin of sediments and sedimentary rocks, Part III, 2nd Ed., 359 pp. New York: John Wiley and Sons 1977.	Five chapters—parent materials, weathering, transport and deposition, chemical sediments, and diagenesis. Chemistry and equations play significant role. About 600 references. Translated from the German by W.D. Johns.
Folk, R.L.: Sandstones. Encyclopedia Britannica, Macropedia 16, 212–216 (1974).	Short summary by experienced petrologist provides excellent overview.
Friedman, G.M., and Sanders, J.E.: Principles of sedimentology, 792 pp. New York: John Wiley and Sons 1978.	Fourteen chapters in four parts, plus five appendices and a glossary. Comprehensive.
Füchtbauer, Hans: Sediments and sedimentary rocks 1, Part II, 464 pp. Stuttgart: E. Schweizerbart'sche Verlagsbuchhandlung 1974.	Conglomerates, breccias, sandstones, pyroclastics, carbonate rocks and diagenesis and pore space. Over 1700 references. Section on pyroclastics by H.W. Schmincke notable.
Füchtbauer, Hans, and Müller, German: Sedimente und Sedimentgesteine (Sediment-Petrologie, Teil 2), 3rd Ed., 782 pp. Stuttgart: Schweizerbart'sche Verlagsbuchhandlung 1977.	Seven long chapters cover all major rock types and are followed by a supplement of 57 pages. Sparingly illustrated but well referenced with complete index.
Gradzinski, R.; Kostecka, A.; Radomski, A.; and Unrug, R.: Sedymentologia, 613 pp. Warsaw: Wydawnictwa Geologicne 1976.	Eleven chapters, many illustrations, and much on sandstones and turbidites. Good source for introduction to the Polish literature of sedimentology.
Leeder, M.R.: Sedimentology, 378 pp. Winchester, Massachusetts: Allen and Unwin 1982.	Eight parts that trace the history of clastic grains from their origin, through transport and deposition, into final burial. Author stresses chemical and physical processes. Almost 400 illustrations.
Lewis, Douglas W.: Practical sedimentology, 288 pp. New York: Van Nostrand Reinhold 1983.	This paperback of 13 chapters has much on sands and sandstone including suggestions for field work, sampling, introduction to well logging, the development of sedimentary associations, final reports, evenings, and impossible weather, etc. Practical field companion.

Lombard, Augustin: Séries sédimentaires, genèse–évolution, 425 pp. Paris: Masson 1972.

Three major parts: sedimentary rocks, stratigraphy and vertical sequences, and facies and their environments. Major emphasis on lithostratigraphy. Sparingly illustrated, but has good sampling of relevant pre-1970 European literature.

Pettijohn, F.J.: Sedimentary rocks, 3rd Ed., 628 pp. New York: Harper and Row 1975.

Completely revised with many newly drafted figures, an old classic continues to provide useful insights, especially to sandstones. Sixteen chapters.

Ricci Lucchi, Franco: Sedimentologia, Partes I, II, and III, 217, 210, and 504 pp. Bologna: Cooperative Libraria Universitaria Editrice Bologna 1978.

Part I covers the material and texture of sediments. Part II has five chapters on processes, with many interesting illustrations: mechanics (complete and includes equations), sedimentary structures (two chapters) plus one each on chemical and organic sedimentation. Part III has 11 chapters that cover all the major environments and the first chapter provides the general methodology. Three tables at the back summarize all the facies. Many references (most to the English-language literature) and refreshing schematic illustrations. Excellent insights into the Italian experience in sedimentology.

Selley, R.C.: An introduction to sedimentology, 408 pp. London, New York, and San Francisco: Academic Press 1976.

Ten well-written chapters with Chaps. 6 to 10 especially relevant. Transportation and Sedimentation, Sedimentary Structures, Environments and Facies, Sedimentary Basins, and Applied Sedimentology. Very well written and illustrated.

Suguio, Kenitiro: Rochas sedimentares, 500 pp. São Paulo: Editors Edgard Blucher Ltda., Ed. da Universidade de São Paulo, 1980.

Properties, genesis, and economic significance of sedimentary rocks are set forth in ten chapters which include mineralogy, methods, sediment transport and sedimentary structures, sedimentary tectonics, and a chapter on the practical importance of sedimentary rocks. Much on sandstone and many references to the rapidly growing Brazilian literature of sedimentology.

Sandy Environments and Depositional Systems

Today the recognition of the diverse environments of sandstones and their depositional systems is well advanced and is essential for basin analysis, diagenesis, and nearly all economic applications. The references below effectively introduce you to this important aspect of the study of sandstones.

Davis, R.A., Jr.: Depositional systems, 669 pp. Englewood Cliffs, New Jersey: Prentice-Hall, Inc. 1983.

Sixteen chapters that cover all the depositional systems after a brief initial review of sedimentary rocks, processes and structures. Glossary and over 800 references, almost all in English.

Fisher, W.L., and Brown, L.F., Jr.: Clastic depositional systems—a genetic approach to facies analysis, Rev. Ed., 105 pp. Austin: Bur. Econ. Geology, Univ. Texas (1984).

Ten depositional systems logically and completely outlined plus references. Excellent summary. No illustrations.

Fraser, G.S.: Evolution of clastic depositional sequences, 398 pp. Boston: IHRC Press, 1986.

"Event stratigraphy" in continental, coastal, shallow marine, slope and deep marine environments in five parts with 234 illustrations.

Fraser, Gordon S., and Suttner, L.J.: Alluvial fans and fan deltas: An exploration guide, 223 pp. Boston: IHRDC Press, 1986.

Processes, characteristics, subfacies and broad controls on alluvial fans and fan deltas plus wireline log characteristics and diagenesis. One hundred illustrations.

Galloway, W.E., and Hobday, D.K.: Terrigenous clastic depositional systems, 423 pp. New York, Berlin, Heidelberg, Tokyo: Springer-Verlag, 1983.

Fourteen chapters covering all the environments begin with "Genetics".

Gilreath, J.A.; Cox, J.W.; Fett, T.H.; and Grace, L.M.: Practical dipmeter interpretations, various paging. Houston, Tx.: Schlumberger Educational Services. 1985.

Comprehensive coverage with 24 headings. Exceptionally clear text and illustrations explain much about external and internal sandstone geometry and paleocurrents.

Gnaccolini, Mario: Sedimenti processi e ambienti sedimentari, 179 pp. Milano: Inst. di Geologia dell'Universita de Milano. Edizioni COP.T.E. (Quaderni Riccardo Assereto di Sciences della Terra), 1978.	Subtitled "Guide to a Sedimentology Course", this is an elementary account—in paperback.
Harms, J.C., Southard; J.B., Spearing, D.R.; and Walker, R.G.: Depositional environments as interpreted from primary sedimentary structures and stratified sequences. Soc. Econ. Paleontologists and Mineralogists Short Course No. 2, 161 pp., (1976).	Many illustrations help the reader better relate sedimentary structures to processes and environment.
Hayes, M.O., and Kana, T.W., (Eds.): Terrigenous clastic depositional environments, various paging. Columbia, S.C.: Coastal Research Division, Department of Geology, Univ. South Carolina 19108, 1976.	Excellent for strandline environments, but also includes others. Large bibliography.
Ivanovich, R.S.: Sedimentologi scheskie osnovy lithologii, 406 pp. Lenningrad: "NEDRA", 1977 (Russian)	"Sedimentological Principles of Lithology" contains five parts: scope, methods and laws of sedimentology; particle size distributions and how to use them; probability applied to bedding sequences; analysis of compaction and thickness; and searching for hidden periodicities in stratigraphic sections. Chief emphasis is on sandstones and shales and how mathematics helps us understand them.
Klein, George deVries: Sandstone depositional models for exploration of fossil fuels, 2nd Ed., 149 pp. Minneapolis, Minnesota: Burgess Pub. Co. 1980.	Six chapters cover all the terrigenous environments.
Reading, H.G. (Ed.): Sedimentary environments and facies, 2nd Ed., 615 pp. New York: Elsevier Publ. Co. 1986.	Fifteen outstanding chapters with over 600 beautiful illustrations. Much on sandstone and a fine overview of the European literature.
Reineck, Hans-Erich: Aktuogeologie klasticher Sedimente. Senckenberg. Buch 61, 348 pp. (1984).	Two parts: Stratification, trace fossils, and structures followed by a description of 11 sedimentary environments. Two hundred and fifty illustrations.
Reineck, H.-E., and Singh, I.B.: Depositional sedimentary environments. 2nd Ed., 549 pp. New York–Heidelberg–Berlin: Springer-Verlag, 1970.	Outstanding documentation of the sedimentary structures and trace fossils of modern environments, especially those of shallow water. Very well illustrated with 576 figures and many references, especially to the European literature.
Rigsby, J.K., and Hamblin, W.K. (Eds.): Recognition of ancient sedimentary environments. Soc. Econ. Paleon. Mineral. Spec. Pub. 16, 340 pp. (1972).	About 1200 references—and many different points of view on both terrigenous and carbonate environments. Some paleocurrent maps. Thirteen summary papers.
Scholle, Peter A., and Spearing, Darwin (Eds.): Sandstone and depositional environments. Am. Assoc. Petroleum Geologists Mem. 31, 410 pp. (1982).	Twelve chapters, all notable for their abundant colored pictures of outcrops, maps, and cores. Much on the subsurface including some seismic sections.
Selley, R.D.: Ancient sedimentary environments, 2nd Ed., 254 pp. Ithaca, New York: Cornell Univ. Press 1978.	Excellent, short, and very well written with emphasis on the subsurface as well as the outcrop.

Shelton, John W.: Models of sand and sandstone deposits: A methodology for determining sand genesis and trend. Oklahoma Geol. Survey Bull. 118, 122 pp. (1973).

Well-described case histories by an ex-Shell sedimentologist. Paleocurrent data emphasized. North American examples.

Spalletti, Luis A.: Paleoambientes sedimentarios (en siliclastic sequences). Assoc. Geologica Argentina, Ser. B, Didactica Complementaria 8, 175 pp. (1980).

Two parts: Eight sections on processes and the needed methodology and a second part on paleoenvironments with six sections (fluvial, eolian, sandy shorelines, diamictites, lakes, and deltas). Much Argentine literature is cited. Outstanding for the geophysicists and sedimentologists whose mother language is Spanish. Order from Assoc. Geological Argentina, MAIPU 645, piso 1, Buenos Aires 1006, Argentina.

Spearing, D.R., Compiler (Ed.): Summary sheets of sedimentary deposits with bibliographies. Geol. Soc. America Misc. Chart 8 (1974).

Seven sheets, one each for alluvial fans, alluvial, eolian, regressive shoreline, coastal barriers, tidal, and turbidity deposits. Useful and informative with many references, all about sand and sandstone. Excellent.

Visher, G.S.: Exploration stratigraphy, 334 pp. Tulsa, Oklahoma: Penn Well Books 1984.

Essentials by an experienced practitioner of the art.

Walker, R.G. (Ed.): Facies models, 2nd Ed., 312 pp. Geoscience Canada Geological Assoc. Canada (Bus. and Econ. Serv. Ltd., 111 Peter St., Suite 509, Toronto, Ontario M5V 2H1), 1984.

Nineteen well-written chapters of which eight are on sands and one on seismic models thereof. Highly recommended.

Specialized Studies

These are legion and here we include only a very few—the ones that we believe to have the widest general utility—concerning use of the dipmeter, compaction, paleocurrents, sedimentary structures, seismic stratigraphy, trace fossils, etc.

Allen, John R.L.: Sedimentary structures, their character and physical basis, I and II, 593 and 553 pp. (Developments in Sedimentology, Vols. 30A and 30B). Amsterdam–Oxford–New York: Elsevier Scientific Pub. Co. 1982.

Twenty-six chapters provide an exhaustive treatment of relevant fluid flow as seen by a famous sedimentologist. Extended discussion of all the primary and early secondary structures.

Allen, John R.L.: Principles of physical sedimentology, 400 pp. Winchester, Massachusetts, Allen and Unwin 1985.

The object of this book is to understand via elementary general physics the small-scale sedimentary structures of detrital sediments. Thirteen chapters devoted to fluid flow, sediment transport, and deposition. See also the companion volume "Experiments in Physical Sedimentology."

Allen, John R.L.: Experiments in physical sedimentology, 64 pp. Winchester, Massachusetts, Allen and Unwin 1985.

A wide range of experiments that can be done using simple and easily obtained materials are described. Subjects include desiccation cracks, obstacle marks, bedforms due to currents and waves, flow in current and straight channels, and much more. See also "Principles of Physical Sedimentology."

Anstey, Nigel A.: Seismic exploration for sandstone reservoirs. 138 pp. Boston: IHRC 1980.

Beautifully written and illustrated exposition of how seismic sections can provide greatly expanded insights into the depositional environments of sandstones and their later deformation. Compare to Sheriff below.

Berg, R.R.: Reservoir sandstones, 481 pp. Engelwood Cliffs, N.J.: Prentice Hall, 1986.

Nine chapters: the first three are devoted to sandstone properties, sedimentary processes and interpretation of reservoir morphology and are followed by six chapters on the major sandy environments. A first of its kind and notable for its uniformly drafted and informative illustrations.

Chilingarian, G.V., and Wolf, K.H. (Eds.): Compaction of coarse-grained sediments, 552 pp. (Developments in Sedimentology Vol. 18A). Amsterdam–Oxford–New York: Elsevier Scientific Publ. Co., 1975.	Eight chapters with much of general interest for the study of sandstones but most emphasis is on compaction including its mechanics and how it distorts the original geometry of a sand-shale sequence.
Collinson, J.D., and Thompson, D.B.: Sedimentary structures, 240 pp. Winchester, Massachusetts: Allen and Unwin 1982.	Comprehensive field and laboratory account of sedimentary structures with emphasis on origin and use. Many illustrations and two appendices.
Conybeare, C.E.B.: Geomorphology of oil and gas fields in sandstone bodies, 342 pp. Canberra: Australian National University Developments in Petroleum Science 4 1976.	Case histories of sandstone bodies in oil fields with much on paleogeomorphology.
Frey, Robert W. (Ed.): The study of trace fossils, 562 pp. New York–Heidelberg–Berlin: Springer-Verlag 1975.	Today, the recognition and use of trace fossils is an essential part of the study of sand and sandstone to which much of this book is addressed. Five parts and 23 papers.
Geological Society of London: State of the art meeting 1977 on sandstone diagenesis. Jour. Geol. Soc. London 135, Pt. 1, 1–186 (1978).	Fourteen papers mostly based on studies from the North Sea but nearly all rich in general significance. Strong emphasis on how the depositional environment of sandstones controls porosity, permeability and diagenetic history. Excellent petrology reinforced by SEM studies. Pioneer volume.
Ginsburg, R.N. (Ed.): Tidal deposits, 428 pp. New York–Heidelberg–Berlin: Springer-Verlag, 1975.	The first 197 pages are devoted to modern and ancient tidal deposits, many of which are sand-rich. Sedimentary structures and vertical sequences emphasized. Short, well-illustrated articles.
Harms, J.C.; Southard, J.B.; and Walker, R.G.: Structures and sequences in clastic rocks. Soc. Econ. Paleon. Mineral. Short Course 9, various paging (1982).	Inexpensive, readily available, soft-cover summary of formative processes as seen in the early 1980s by three experts from North America.
Komar, Paul D.: Beach processes and sedimentation, 429 pp. Englewood Cliffs, New Jersey: Prentice-Hall, Inc. 1976.	Sand is the dominant theme of this book—how it is transported and deposited by waves and currents along sandy coastlines. Wave theory, physiography, and sedimentary structures also included.
Miall, Andrew D. (Ed.): Fluvial sedimentology. Canadian Soc. Petroleum Geologists Mem. 5, 859 pp. (1978).	Almost everything that one needs to know about fluvial sandstones, all in eight parts—modern and ancient sandy deposits, paleohydraulics, bedforms and bars, and more.
Miall, Andrew D. (Ed.): Sedimentation and tectonics in alluvial basins. Geol. Assoc. Canada, Spec. Paper 23, 272 pp. (1981).	Sandstone predominates and eight papers show the many different ways it can be studied and its close dependency on tectonics. Molasse much discussed.
Miall, Andrew D.: Principles of sedimentary basin analysis, 490 pp. Springer Verlag: New York Berlin Heidelberg Tokyo, 1984.	Comprehensive treatment includes much on sedimentology, depositional systems, tectonics, facies and methods all in 10 chapters with 387 figures.
Middleton, Gerard V., and Southard, John B.: Mechanics of sediment movement. Soc. Econ. Paleon. Mineral. Short Course 3 (Binghamton), various paging (1977).	Eight chapters ranging from "Behavior of Fluids," "Settling of Spheres" through "Turbulent Flow" to "Bed Configurations" and "Sediment Gravity Flows," all by two leading experts.

Nowatzki, C.H.; Reich dos Santos, B.; Araújo dos Santos, M.A.; and de Silva, M.E.: Atlas de estructuras sedimentares Pré-Gondwânicas e Gondwânicas do Estado do Rio Grande do Sul, Brazil. Parte 1—Estructuras primarias. Acta Geol. Leopoldensia VI, 3–286 (1982).

Many field photographs of the primary sedimentary structures of sandstones in southwestern Brazil.

Nowatzki, C.H.; Reich dos Santos, B.; Araujo dos Santos, M.A.; and Gonzaga, T.D.: Atlas de estruturas sedimentares Pre-Gondwanicas e Gondwanicas do Estado do Rio Grande, Brazil–Parte II: Estruturas quimicas e organicas. Acta Geol. Lepoldensia 15, Anno 7, 5–132 (1983).

Forty-five full page plates of chemical structures and of trace fossils, plus table of stratigraphic units.

Picard, M. Dane, and High, L.R., Jr.: Sedimentary structures of ephemeral streams, 223 pp. (Developments in Sedimentology, Vol. 17). Amsterdam: Elsevier Scientific Pub. Co. 1973.

Seven chapters with 139 photographs, mostly from the Unita Basin of Utah and nearby Wyoming and Colorado.

Potter, Paul Edwin, and Pettijohn, F.J.: Paleocurrents and basin analysis, 2nd Ed., 420 pp. New York: Springer-Verlag, 1977.

Over one hundred pages of new material have been added to the original ten chapters, plus many new references and figures.

Ryder, R.T.; Lee, Myung, W.; and Smith, G.N.: Seismic models of sandstone stratigraphic traps in Rocky Mountain basins, 77 pp. Tulsa, Oklahoma: Am. Assoc. Petroleum Geologists (Methods in Exploration Series), 1981.

Fourteen oil fields, many of which are sand-rich, with seismic cross sections showing velocities in color, plus commentaries. The wave of the future for the subsurface study of sandstone?

Scholle, P.A., and Schluger, P.R. (Eds.): Aspects of diagenesis. Soc. Econ. Paleon. Mineral. Spec. Pub. 26, 443 pp. (1979).

Two parts: eight papers on how to determine paleotemperatures and fifteen papers on studies of specific sandstones. Rich source of photomicrographs, SEM pictures, graphs, and theory.

Serra, O.: Sedimentary environments from wireline logs, 211 pp. New York: Schlumberger Corp. 1985.

Careful review of the characteristics and processes of the depositional environments and how to recognize them on wireline logs including the dipmeter. Color pictures and graphs plus some log calculations and three appendices.

Sheriff, Robert E.: Seismic stratigraphy, 227 pp. Boston: IHRD 1980.

Nine chapters include processing, seismic sequences, velocity, and resolution and more. Broad, nonmathematical viewpoint provides good starting place.

Stride, A.H. (Ed.): Offshore tidal sands, 222 p. London New York: Chapman and Hall 1982

Seven chapters: tidal currents on shelves, bedforms, sand transport, modern tidal sands, faunas occurring on modern tidal shelves, and ancient tidal deposits.

Tucker, M.E.: Sedimentary rocks in the field, 252 pp. New York: Halsted Press 1981.

Elementary but with many illustrations, and sandstones are well represented.

Zanke, U.: Grundlagen der Sedimentbewegung, 402 pp. Berlin–Heidelberg–New York: Springer-Verlag 1982.

Eleven chapters all focusing on sand and written by a sedimentologist. Complete and thorough and not excessively complicated. Good for the advanced reader.

Sedimentary Petrology

The texts below all contain much on the petrology of sandstone and represent widely different styles and viewpoints. See also some of the general texts.

Berthois, L.: 1. Étude sédimentologique des roches meubles (techniques et méthodes), 178 pp. Paris: Doin Editeurs 1975.

Some material on fluid transport and more on methods, mineralogy, X-ray techniques, and clay mineral analysis.

Blatt, Harvey: Sedimentary petrology, 564 pp. San Francisco: W.H. Freeman Co. 1982.

Fifteen chapters starting with plate tectonics and weathering and hence into all the sedimentary rock types and ending on "The Development of a Research Project" and "The Practice of Sedimentary Petrology." Much on processes. Useful glossary has over 300 terms.

Ehlers, Ernest G., and Blatt, Harvey: Petrology: Igneous, sedimentary and metamorphic, 732 pp. San Francisco: W.H. Freeman and Co. 1982.

Seven chapters devoted to sedimentary rocks, (pp. 247–506) and includes much on processes of transport as well as the formation of sand and other sediments. Section on petrology of sandstones at convergent margins.

Fitzpatrick, E.A.: Micro-morphology of soils, 433 pp. London: Chapman and Hall 1984.

The study of soils in thin section has much relevance to the study of sandstones in thin section. Broadly speaking, this volume is an updated Krumbein and Pettijohn with many clear and informative illustrations and text. See especially Chap. 11, "Ancillary Techniques." Excellent for describing alteration of ferrigenous debris.

Folk, R.L.: Petrology of sedimentary rocks, 182 pp. Austin, Texas: Hemphill Publishing Co., 1980.

A classic landmark for the teaching and development of sandstone petrology. See pp. 101–146 for an integration of the Krynine view of tectonics with that of Dickinson.

Greensmith, J.T.: Petrology of the sedimentary rocks, 5th Ed., 502 pp. New York: Hafner Pub. Co. 1971.

Eighteen chapters cover all the sedimentary rocks plus one on lithologic associations. Classical line drawings (137 illustrations), tables (26), and much descriptive material.

Greensmith, J.T.: Petrology of the sedimentary rocks, 6th Ed. 241 pp. London: Allen and Unwin 1978.

One chapter on sands and sandstones follows classification, transport, and structures. Excellent pen and ink drawings.

Krinsley, D.H., and Doornkamp, John C.: Atlas of quartz sand surface textures, 91 pp. Cambridge: Cambridge Univ. Press, 1973.

Two parts: the first on methods and basic concepts, and the second with 122 SEM photographs and their annotations.

Lucas, G.; Cros, P.; and Lang, J.: 2. Étude microscopique des roches meubles et consolidées, 503 pp. Paris: Doin Editeurs 1976.

Chiefly about carbonates, but contains some photomicrographs and discussion of sand and sandstones. Over 400 photomicrographs.

Lützner, Harald; Falk, Fritz; Ellenberg, Jürgen; and Grumbt, Eberhard: Tabellarische Dokumentation klastischer Sedimente. Akad. Wissenschaften DDR, Zentral-Institut für Physik der Erde 20, 153 pp. (1974).

A manual on how to describe terrigenous sands and sandstones—almost all aspects—but with strong emphasis on bedding and sedimentary structures. Many tables and charts and good for German references.

Parfenoff, A.; Pomerol, C.; and Tourenq, J.: Les minéraux en graines, 578 pp. Paris: Masson 1970.

Three major parts: Interpretation and application of heavy mineral studies (paleogeography, tectonics, economic geology, and stratigraphy), methods of study (includes sampling and sample preparation, a review of relevant crystallography and optics, X-ray identification, radioactivity, spectrographic methods, qualitative chemical tests, counting, and representation), and a long section on mineral descriptions and determinative tables with some colored plates in which opaque minerals are given their due. Unusual French–Russian dictionary of mineral names. Very well done and deserving of wider attention. All in easy-to-read French.

Scholle, Peter A.: A color-illustrated guide to constituents, textures, cements, and porosities of sandstones and associated rocks. Am. Assoc. Petroleum Geologists Mem. 28, 200 pp. (1979).

The first color atlas of sandstones in thin section with brief summaries of detrital grains, classification, texture, diagenetic and deformation fabrics, porosity, and techniques. Outstandingly practical.

Tucker, M.E.: Sedimentary petrology, an introduction, 252 pp. New York: John Wiley and Sons 1981.

Composition, petrography, sedimentary structures, and diagenesis of major rock groups: sandstones (pp. 10–76) and volcaniclastics (pp. 219–228).

Williams, Howell; Turner, F.J.; and Gilbert, C.M.: Petrography, an introduction to the study of rocks in thin sections, 2nd Ed., 626 pp. San Francisco: W.H. Freeman and Co. 1982.

Pages 277 to 427 are devoted to sedimentary rocks. Drawings of photomicrographs are classic. Good for textures of rock fragments.

Manuals, Encyclopedias, and Special Tables

Adams, A.E.; MacKenzie, W.S.; and Guilford, C.: Atlas of sedimentary rocks under the microscope, 104 pp., New York: John Wiley and Sons (1984).

Two hundred and ten outstanding colored photographs, all of the highest quality plus brief description. See also MacKenzie and others (1980) and (1982).

Broche, Jacques; Cananova; Roger; and Loup, Gustave: Atlas des minéraux en grains identification par photographies en couleurs, 173 pp. Société pur le Dévelopement Minièr de la Côte d'Ivoire (SODEMI), République de Côte d'Ivoire, Ministère des Mines, 1977.

Devoted to prospecting for alluvial mineral deposits with a section on descriptive mineralogy plus colored plates (French, German, English and Russian).

Devismes, Pierre: Atlas photographique des mineraux d'alluvions. Mem. Bur. Rech. Geol. Minières *95*, 203 (1978).

Six hundred and forty-one beautiful colored photographs (10.5 to 14×) of loose grains of 173 mineral species that occur in alluvial deposits, mostly sands and gravels. Methods of preparation and captions in French and English. One of a kind and indispensable for alluvial prospectors. Examples selected from over 300,000 separates.

Fairbridge, R.W.; and Bourgeois, Joanne (Eds.): The encyclopedia of sedimentology 901 pp. (Encyclopedia of Earth Sciences, Stroudsburg, Pennsylvania: Dowden, Hutchison and Ross, Inc., 1978.

A good place to start your study of a wide range of topics with short articles, hundreds of which directly relate to sand and sandstones on the shelf, slope, rise and trough. Many illustrations.

Le Ribault, Löic: L'exoscopie des quartz, 150 pp. Paris: Masson 1977.

Comprehensive, with many pictures and tables. Main emphasis is the environmental history and identification of single grains. Easy-to-read French.

Netto, A. Sérgio T.: Manual de Sedimentologia, 194 pp. Salvador, Bahia Brazil: Salvador, Bahia: Petrobras, Sector de Ensino na Bahia 1980.

Terrigeneous sediments, chiefly the sands, are emphasized in six chapters starting with the cycle of deposition, followed by structures, textures, composition and a chapter entitled "Sedimentology Applied to the Study of Reservoir Rocks". Elementary and concise, but contains some statistics and has five short appendices.

Nowatzki, C.H.; Santos, M.A.A.; Leao, H.Z.; Schuster, V.L.L.; and Wacker, M.L.: Glossario de estructuras sedimentares. Estudos Tehnologicos, Acta Geologica Leopoldensia 18 and 19, Ano 8, pp. 7–432 (1984).

Comprehensive glossary of sedimentary structures in Portuguese (pp. 19–192) followed by a glossary of English-language structures defined in Portuguese (pp. 193–256) and 157 photographs.

MacKenzie, W.S., and Guilford, C.: Atlas of rock-forming minerals in thin section, 98 pp. New York: John Wiley and Sons 1980.

Over 200 beautiful colored photomicrographs in plane and polarized light, each with short descriptions and key optical properties. Essential microscope companion.

MacKenzie, W.S.; Donaldson, C.H.; and Guilford, C.: Atlas of igneous rocks and their textures, 148 pp. New York: John Wiley and Sons 1982.

Two parts—textures and varieties of igneous rocks—both with beautiful color photographs, each with clear, simple description. Keep it by your microscope when puzzling over rock fragments.

Ricci Lucchi, Franco: Sedimentografia, 288 pp. Bologna: Nico a Zanichelli S. p. A. 1970.

One hundred and seventy beautiful plates of outcrops many of which are Italian turbidites. English to Italian glossary of sedimentary structures.

Rodrigo, L.A., and Coumes, F.: Manual de sedimentologia, 149 pp. La Paz, Bolivia: Universidad Major de San Andres 1973.

Subtitled "Manual of Techniques," there are fourteen practically oriented chapters plus useful tables, 67 figures, and a preface by Albert V. Carozzi.

Teruggi, Mario E.: Diccionario sedimentologico, Vol. 1, Rocas clásticas y piroclásticas, 104 pp. Buenos Aires: Ediciones Científicas Argentinas Librart (ECAL) 1982.

This is the first of a four-volume effort in sedimentology. The dictionary defines some 4000 terms and includes diagrams plus a list of English terms and their Spanish equivalents. Some terms discussed at length.

Welton, Joann E., SEM petrology atlas, 237 pp. Tulsa, Amer. Assoc. Petroleum Geologists, Methods in Exploration Series, 1984.

SEM photographs, EDX analysis of minerals and some colored thin sections plus energy table of characteristic X-ray transitions and glossary.

Zuffa, G.G. (Ed.): Provenance of arenites, 408 pp. Dordrecht-Boston-Lancaster: D. Reidel Pub. Co., 1985 (NATO ASI series C: Mathematical and Physical Sciences, v. 148).

Eleven articles by experts on the petrography and provenance of sand and sandstone from secondary porosity to surface textures, detrital quartz, heavy minerals, and green particles to plate tectonics.

Part I. The Fundamental Properties of Sandstones

Mineralogy, texture and sedimentary structures are essential to the study of all sedimentary rocks, be they limestones, coal, evaporites or sandstones. In the broadest sense, this trilology represents what one needs to know to get started in the study of sediments. Hence we have grouped these three together and placed them first recognizing that the experienced reader may briefly pass them by whereas the less experienced may profit by a careful reading. We have not tried to be encyclopedic or comprehensive, for the many advances in our knowledge of these three kinds of properties have made that impossible, given the scope of this book. Rather we call attention to the main uses of these properties as the bases of the more interpretive, later sections of the book. Perhaps mineralogy, texture and sedimentary structures are not the most exciting aspects of sands and sandstones, but certainly they are the most fundamental.

CHAPTER 2

Mineral and Chemical Composition

Introduction

Sandstones are mixtures of mineral grains and rock fragments coming from naturally disaggregated products of erosion of rocks of all kinds. The total variety of rock types in any given eroding watershed may be represented in the sediment product. Theoretically, therefore, the number of mineral species to be found in all sandstones is as large as the total number of mineral species known. Even a given specific sandstone might be expected to contain a large variety of minerals, since a glance at any geologic map will show the average watershed to have rocks with many kinds of minerals present. In fact the expectation proves to be untrue, for the abundant minerals of sandstones belong to a few major groups; many varieties of heavy minerals (most present in trace amounts) may be found, but the list is by no means very large. Obviously, the processes determining mineral composition of sandstones are more complex than simple mixing ones from source areas of different kinds. The discrepancy is great between observed and theoretically possible combinations of minerals.

Minerals may be lost or modified by weathering in the source area, by transportation to the site of sedimentation, and by diagenesis (Fig. 2-1). Because the mineralogy of a sandstone is the inheritance from the source area as modified by sedimentary processes, a mineralogical examination is one of the most practical studies that can be made to obtain information on provenance, including tectonics and climate; effects of transportation, including distance and direction; and additions of chemically deposited minerals during sedimentation and diagenesis.

Weathering

Chemical and physical weathering processes operate selectively to alter the chemical and mineralogical composition of the source rock as it is disaggregated into mineral and rock fragment detritus (see Chap. 7 and surveys of weathering by Ollier, 1969; Loughnan, 1969; Blatt *et al.*, 1980). A detailed consideration of weathering is beyond the scope of this book; here we treat those matters that are of direct concern for sand and sandstone under the individual mineral groups—especially under the feldspars (p. 32) and clay minerals (p. 38). Minerals may be affected by weathering at the source in a variety of ways, chiefly by chemical decomposition or alteration. Thus, in the course of weathering, feldspars may alter to kaolinite or an intermediate product; pyroxenes and amphiboles may simply dissolve and be transported as dissolved ions. In contrast, some minerals, such as quartz, are very slightly soluble and will in general be transported unchanged in amount or character from the source rocks. Thus, very largely, all sandstones represent a residuum from surficial chemical weathering processes. The longer or more intense those processes, the less the residuum will resemble the original mixture of source rocks from which it came. We can, therefore, by noting the extent of loss of unstable minerals, deduce the amount of weathering in the source area.

Transport

The transportation of minerals from the source area to the sedimentary basin where they are finally deposited can also result in attrition of minerals. Most important is differential abrasion, whereby the softer minerals may be decreased in size relative to the harder minerals, or simply become more rounded than the harder ones. Many minerals may be so soft as to not survive rigorous transportation as sand-sized grains; these will then find their place in silts or muds rather than sands. Rock fragments

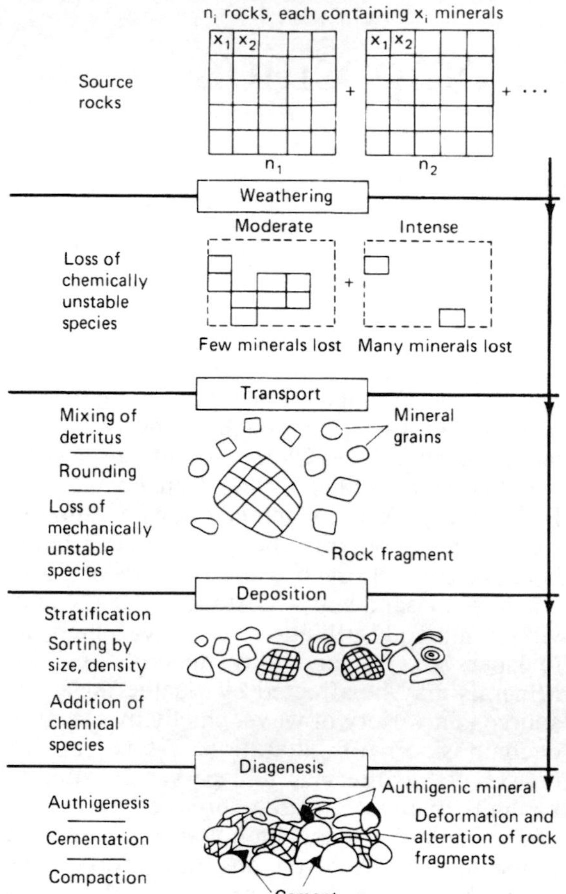

FIGURE 2-1. The sedimentary cycle of a sandstone. Though processes such as weathering and transport are shown as distinct, they overlap in nature.

may be broken into constituent finer mineral grains. The effect of transportation is also to sort on the basis of shape. To the extent that this factor operates, we may find sands derived from the same source materials that will be of varying mineralogical as well as textural composition in different parts of the sedimentary basin.

Diagenesis

Finally, the mineral composition of sandstones may be drastically altered by dissolution, precipitation, or alteration during diagenesis. In this process unstable minerals may be lost completely or partially. New minerals, carbonates in particular, may be added by precipitation from solution. If we can recognize these diagenetic effects and subtract them from the composition of the entire rock, we can arrive at an evaluation of what the original sand may have been.

Mixing of Sources

Aside from those processes which result in the disappearance or appearance of minerals in a sandstone, the simple effects of mixing may give rise to compositions which differ drastically from the source rocks. Many terranes from which sand grains are eroded are composed of a variety of igneous, metamorphic, and sedimentary rocks; the minerals from all these rocks become mixed. Certain minerals, such as olivines and pyroxenes, though they may be abundant in specific igneous or metamorphic rock types, are not abundant in the overall composition of an average terrane. Thus, where olivine may be a significant rock-forming mineral and constitute up to 30 percent of some basic igneous rocks, it is usually so diluted that it rarely makes up more than 1 to 2 percent of completely unweathered sandy detritus. In contrast to this, feldspar is such an abundant mineral in so many rock types, as is quartz, that it is unlikely that it will disappear by simple mixing and dilution.

Sedimentary Differentiation

When we consider that most sediments may contain materials that have been recycled from older sediments, we recognize that the processes of sedimentary differentiation—the segregation of different sizes and compositions by weathering and transportation—can extend through one or many sedimentary cycles. Some processes by which minerals are lost or gained may operate intensively during a single cycle. Others, such as the abrasion rounding of some minerals, may need a number of cycles before their effect can be noticed. The problem of interpretation of mineral composition of sediments lies in separating the effects of the last sedimentary cycle of weathering, transport, deposition, and diagenesis from the long-term effects of many cycles. Sedimentary differentiation is important in the general geochemical cycle of the crust and mantle and the growth of continental crust. The net result of the alteration of igneous rocks to sediments by weathering, transport, deposition, and diagenesis is to reinforce the transformation of basaltic materials of oceanic, lower crustal, and mantle origin to the sialic composition of continental crust. Sandstone mineral compositions reflect these changes and are a guide to continental evolution during geologic time (see Chap. 12).

Rock Fragments

The size of mineral crystals in source rocks determines the presence of monomineralic or rock fragments in the sandstone produced therefrom. Finely crystalline source rocks—igneous, metamorphic, or sedimentary—whose grain or crystal size is less than fine sand size, 1/16-mm diameter, contribute only rock fragments to sandstones. The coarser-grained source rocks will contribute mostly mineral grains, but, depending upon the ratio of grain size of source rock to grain size of sandstone, may also provide rock fragments. Thus the larger the grain of sandstone, the more likely the occurrence of rock fragments and the better the assessment of source terrane lithologies. We commonly find rock fragments made up of a cluster of many mineral grains, although some consist of but a few, say two, three or four. Just as minerals may be lost by decay or abraded to very fine grains and thus deposited with the muds during a sedimentary cycle, so may rock fragments. For example, if quartzite is the only rock fragment type in a coarse quartzose sandstone, we cannot simply interpret rock fragment composition to mean that the source lands consisted only of quartzite. Other lithologic types may have been eroded from the source terrane only to be subsequently eliminated as sand-sized material by mechanical or chemical processes. The chemical alteration of unstable minerals goes on within rock fragments just as it does for separate mineral grains and so influences the stability of the rock particles. Thus we face the same choice of interpretations as for mineral grains: the absence of a mineral or a rock type may be due either to (1) its absence at the source or (2) its disappearance during weathering and transport. The better the general knowledge of sedimentary processes and the specific knowledge of the particular geologic problem, the better one can make this choice, which is all-important to the reconstruction of geologic history.

Detrital and Chemical Minerals

The interpretation of the mineralogy of a sandstone is dependent on a proper assignment of the mineral species present to a meaningful genetic category. For example, a single list of minerals present is virtually meaningless; we need instead several lists by categories, such as primary detrital, precipitated cement, and post-depositional alteration products.

The distinction between "detrital" and "chemical" minerals has genetic significance by extension of the concept of detrital and chemical rocks. Objectively, the petrographer distinguishes rounded grains that indicate a detrital origin from interlocking crystal growth fabrics that indicate a chemical origin. We distinguish between minerals that are only slightly soluble in water, such as the detrital silicates, and minerals soluble in water, such as the chemical carbonates. The distinction cannot be made for any mineral species as such, for many minerals are both detrital and chemical in origin in the same rock. For example, rounded detrital quartz grains may be cemented by chemically precipitated secondary quartz. It is also possible to find carbonate and gypsum sands with all of the characteristics of detrital silicate sands. Clay minerals partake of some of the characteristics of both detrital and chemical groups. The distinctions between the detrital and chemical categories then are mainly based on the geometry of the particular grain. Figure 2-2 shows diagrammatically the relationships between the detrital and the chemical mineral groups. We may sum up the utility of the two groups by noting that detrital minerals reflect the source contribution of preexisting rocks, removal by weathering and transport, and diagenesis. The chemical group allows us to interpret the environment of deposition and of diagenesis.

Sandstone Mineral Ages

Until a decade ago sedimentary petrologists were only able to date the source detritus of sandstones by the rare diagnostic fossil in chert or other lithic fragments, though there were occasional attempts by geochronologists to determine radioactive age of zircons, detrital glauconites, or other datable minerals (Allen *et al.*, 1964; Tatsumoto and Patterson, 1964; Allen, 1969; Brookins and Voss, 1970; Aronson and Tilton, 1971; Gebauer and Grünenfelder, 1977). Since then, more sensitive mass spectrometric and mineral separation methods and the use of newer decay schemes ($^{40}A/^{39}A$, $^{147}Sm/^{143}Nd$) have allowed the dating of a wider range of detritus with more precision and resolution (Allen, 1972; Allen *et al.*, 1974; Lugmair, 1974; Lugmair and Scheinin, 1975; DePaolo and Wasserburg, 1976). Careful work now offers the

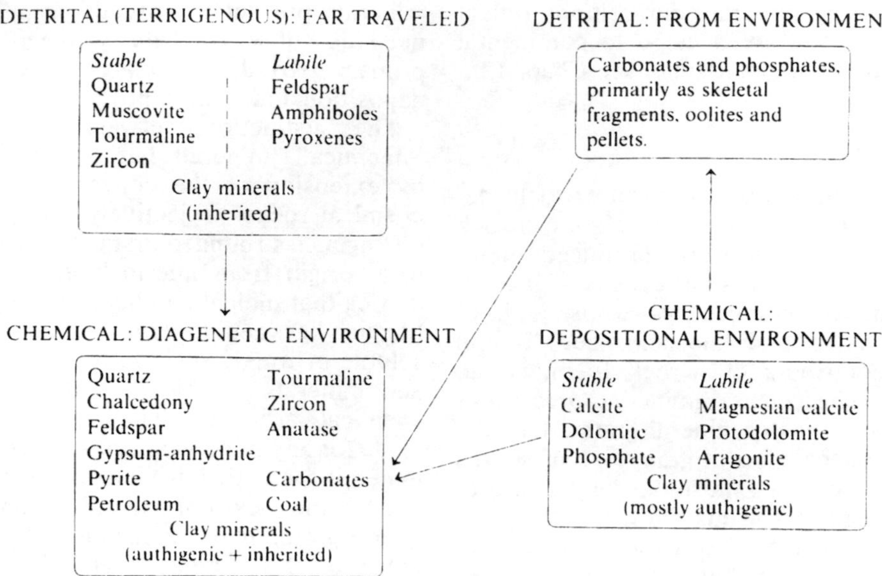

FIGURE 2-2. Derivation of minerals in sandstones.

prospect of determination of individual components with a range of ages and even the possibility of dating individual grains. Dating source rocks of sandstones allows the correlation of sedimentary sequences with stages in mantle and crustal evolution modeled from isotopic data (McCullough and Wasserburg, 1978; O'Nions *et al.*, 1979). Because the youngest detrital mineral in the sediment can be no younger than the age of the sediment itself, and also because many ages of preexisting rocks may be represented in the source area, an apparent radioactive age determination of a bulk sample of the detrital minerals will give us some sort of average age. The ages of source rocks that we refer to are in terms of "radioactive events," which may simply refer to metamorphic episodes rather than the first date of crystallization of the source rock. The dating of detrital minerals commonly involves many assumptions and difficult mineral separation problems. The major assumption that we know may be violated is that there is no diagenetic effect on the detrital mineral assemblage. Hurley *et al.* (1963) first showed that the apparent ages of micas and illites of Atlantic Ocean sediments are probably too young because of the presence of diagenetically formed illite. Such diagenetically formed materials will have some younger date than the date of the sediment, and will influence the total average date of detrital and diagenetic minerals in the younger-age direction (see Chap. 11). Autochthonous chemical minerals in sandstone, such as glauconites, and contemporaneous ash fall materials can be dated by $^{40}K/^{40}Ar$ or $^{40}Ar/^{39}Ar$ methods—with caution—to give the stratigraphic age of Tertiary and some Mesozoic rocks (Lipson, 1958; Evernden *et al.*, 1961; Polevaya *et al.*, 1961; Dodson *et al.*, 1964).

X-Ray and Other Methods

Although normally mineralogy of sandstones has been determined almost entirely by the polarizing microscope, the widespread availability of X-ray diffraction equipment makes possible its use as a supplementary tool. Precise analyses in short times for varieties of feldspars, carbonates, and heavy minerals identifiable under the microscope only with painstaking effort are possible in this way (Pryor and Hester, 1969). In addition, mineralogy based on X-ray data can include structural and compositional information not given by optical methods. For much work in sedimentary geochemistry, we need more precise mineralogical determinations and precise chemical compositions of the mineral phases present. For that work the electron microprobe is available. The data from the electron microprobe give a quantitative chemical analysis and from those data a mineralogic composition can be calculated using standard normative methods. The probe allows the accurate and precise determination of plagioclase and other mineral grains that have widely varying compositions. The electron probe may be integral with a scanning electron microscope (SEM) in a single instrument, in which case the petrographer can determine both texture and mineralogy of the same grain. The SEM, when linked to an energy-dispersive spectrometer in a more common configuration, can also give chemical composition as well as texture, though somewhat less precisely. Additional information comes from using SEM in the back-scattered electron mode (Pye and Krinsley, 1983). The ion microprobe, or secondary ion mass spectrometer (SIMS), can also be used for the analysis of elements as well as isotopes. SIMS and a growing number of other instrumental methods used for the study of chemistry and structure of surfaces, such as Auger spectroscopy, contribute to our growing knowledge of the chemical composition of sedimentary minerals. Instrumental neutron activation analysis and mass spectrometry give information on rare earth and other trace elements that may be useful in determining provenance or the effects of sedimentary differentiation (McLennan *et al.*, 1980; Bhatia and Taylor, 1981; Basu *et al.*, 1982). New knowledge of the structural states of clay minerals is coming from scanning transmission electron microscopy (STEM) and analytical electron microscopy (AEM), techniques that allow us to peer at the highest resolution of details of crystal structure, such as the interlayering of clay minerals (Lee and Peacor, 1983).

Surface Appearance

The surface appearance of sandstones is strongly influenced by mineral composition. The pure quartz arenites that contain virtually all quartz tend to be white or very light gray in color. The same types of sandstones, which may contain authigenic quartz overgrowths, many of them euhedral, have been called "sparkling" sandstones because of their glint in the light. On the other hand, sandstones rich in rock fragments and clay tend to be various shades of gray, greenish gray, and dark gray. The graywackes, again partly because of their fine grain, appear to be very dark. Geologists have long been familiar with reddish and brownish colors given to sandstones by the contained iron minerals, most frequently coatings of hematite or limonite on the silicate grains (see p. 50 and Chap. 11). Color may be attractive for the identification of certain sandstones, in particular for stratigraphic marker beds, and the field description of red and green to gray colors may be important in environmental or diagenetic analysis. The precise analysis of color in terms of mineral composition, iron oxidation state, texture, and other fundamental properties of the rock remains difficult.

In the following description, specific mineral groups are classed as detrital or nondetrital on the basis of their dominant mode of occurrence; thus all of the silica minerals are discussed under detrital minerals, although it is recognized that some of them are of secondary chemical origin.

The Detrital Minerals

The Silica Minerals

Only one crystalline polymorph of SiO_2, low quartz, is thermodynamically stable under sedimentary conditions, and it is one of the most common minerals in sandstones (Siever, 1957, p. 822; Frondel, 1962, p. 3). Other well-crystallized polymorphs, such as tridymite and cristobalite, are rarely found. Until the 1970s, poorly crystallized silicas in sediments were frequently identified as cristobalite on the basis of some similarities in X-ray diffraction pattern to that of crystallized cristobalite (Swineford and Franks, 1959). We now recognize all of the primary silica components of sediments, such as the tests of radiolarians and frustules of diatoms, as varieties of amorphous silica or opaline silica. This material, also called opal-A, transforms with diagenesis into a disordered modification of cristobalite, opal-CT (Jones and Segnit, 1971). Further diagenesis produces quartz. Because the opal-A to opal-CT to quartz transformation is largely complete under normal circumstances after a few tens of millions of years, only later Tertiary sands are likely to contain the unstable disordered forms.

Amorphous silica varieties, including opal-A, siliceous sinter, silica glass (lechatelierite), and others, are abundant in modern and very recent volcanic sands, but for older volcanic and most nonvolcanic sandstones we need consider only chert (chalcedony) in addition to quartz. In practically all ancient rocks older than Tertiary, chert is made up of microcrystalline quartz (Midgley, 1951; Folk and Weaver, 1952), as determined by X-ray diffraction. The anomalously low indices of refraction and birefringence of chalcedony noted under the microscope stem from the fibrous nature of the quartz and from water-filled holes in the structure that are readily seen by electron microscopy. Some Tertiary rocks contain opal-A and -CT, according to our reinterpretation of the data of Frye and Swineford (1946). Opaline silica may originate as a devitrification product of volcanic ash, as small irregular secretions of plant origin, opal phytoliths (Baker, 1959) , or as the hard parts of diatoms, radiolarians, and the siliceous sponges. Though Krynine (1941) and others have long inferred primary or penecontemporaneous authigenic quartz, there is only one report of the precipitation of crystalline quartz at the surface of the earth at low temperatures in any pre-burial modern environments. That occurrence was in a manganese nodule described by Harder and Menschel (1967). Mackenzie and Gees (1971) were able to experimentally precipitate quartz from seawater under low-temperature conditions but there is no clear-cut evidence that modern sands contain quartz overgrowths produced at the time of sedimentation by chemical processes. A summary of the sedimentary geochemistry of silica is given by Calvert (1983) and Wollast and Mackenzie (1983).

Quartz grains are the common detrital constituents of most sandstones, and, because quartz occurs in so many igneous and metamorphic rocks, attempts were made long ago to use them for source rock determination. The terms mono- and polycrystalline are often used to describe quartz varieties: monocrystalline quartz refers to grains consisting of single crystals and polycrystalline to aggregates of crystals. The terms unit and composite are also used. The optical characteristics of mono- and polycrystalline quartz, together with the nature of their inclusions, their shapes, and grain boundaries are important in provenance determinations, as discussed in Chap. 7.

The characters first used for differentiating quartz varieties were the nature of undulatory extinction (Mackie, 1896) and types of inclusions (Gilligan, 1920, pp. 259–60). There followed a history of refinement, culminating in Krynine's (1940, pp. 13–20) enumeration of ten different varieties, based on inclusions, shapes of grains, the nature of the extinction, and the form of the boundaries of the grains, as well as the polycrystalline nature of some grains. The varieties were assigned to igneous, metamorphic, or sedimentary source rocks.

Blatt and Christie (1963) restudied the problem of undulatory extinction and concluded that this criterion is not valid for source determinations in the way that it had been used in the past. They stressed the importance of using the universal stage to determine true undulatory extinction and, by a study of various source rocks and sediments, concluded that the only rocks that contained a high proportion of nonundulose quartz were volcanic extrusives and the quartz arenites of Paleozoic and Precambrian age. They concluded that undulose extinction was simply the optical expression of a strained crystal and that most rocks had been subjected to some sort of deformation, either in the course of crystallization or after formation. Basu *et al.* (1975) restudied undulose and polycrystalline quartz varieties, compared universal with flat-stage identifications, and concluded that careful petrography with a flat stage makes possible discrimination among plutonic, low- and high-grade metamorphic sources. The explanation for the quartz of older quartz arenites being unstrained was sought by Blatt and Christie (1963; p. 571) in the greater thermodynamic stability of the unstrained as opposed to the strained grains. This was correlated with the greater probability of strained grains wearing out during transportation in the course of multiple sedimentary cycles. Fractures in quartz derived from igneous and metamorphic rocks are common and are responsible for much of the decrease in size during transportation (Moss, 1972).

Polycrystalline quartz grains include those from igneous and metamorphic rocks, quartzite, sandstones, and cherts (Fig. 2-3), although chert, because of its fineness of grain, is commonly considered separately. Chert grains are recognizable by their very fine and relatively uniform crystal size and sometimes include fragments of fossils. Some early Precambrian cherts and those affected by metamorphism may have larger crystal sizes as a result of thermally induced grain growth and may not always be easily distinguished from polycrystalline grains derived from schists, granulites, and gneisses. Excluding many microcrystalline cherts, the genetic distinction between these possible origins is not fully established by optical criteria such as internal suturing, inclusions, grain shape, and internal crystal size. An example of this type of study is Blatt's (1967), in which he suggests that internal crystal size and morphology may help distinguish gneissic from granitic polycrystalline quartz. Overlapping characteristics are the principal problem. Polycrystalline quartz is commonly less abundant in well-rounded, well-sorted quartz-rich sandstones than in poorly rounded sandstones rich in feldspar and rock fragments. Apparently it is more easily eliminated by abrasion than monocrystalline quartz. Nonetheless, polycrystalline quartz is a most useful index for mapping petrographic provinces. Certainly it furnishes a good

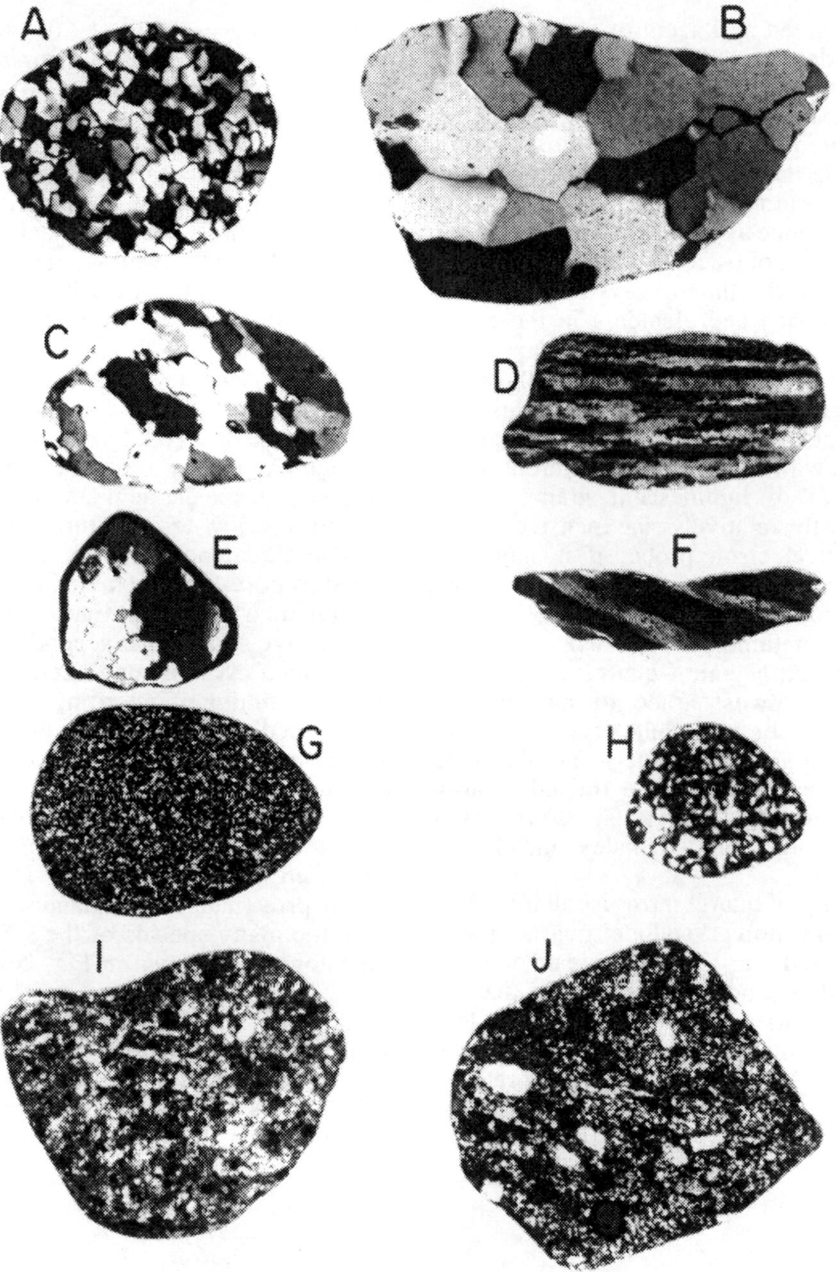

FIGURE 2-3. Composite of polycrystalline quartz grains and chert (various sources): A) and B) polycrystalline quartz with uniformly sized grains having mostly straight contacts (polygonized quartz), × 80 and 120; C) polycrystalline quartz grain consisting of elongate, slightly sutured grains, × 40; D) polycrystalline quartz with almost perfect orientation of elongate crystals, some of which have sutured contacts, × 40; E) polycrystalline grain with coarse interlocking mosaic of crystals, × 40; F) "pseudo"-polycrystalline grain that in reality is monocrystalline but has distinct zones of undulatory extinction, × 40; G) fine-grained chert, × 40; H) coarse-grained chert, × 80; I) spicular chert, × 120; and J) silty chert grain.

example of how a varietal type may be useful even if its ultimate origin is not fully clear.

There has also been some study of quartz types by chemical methods, although rarely have they been used to systematically map varietal types across a basin. In general, the chemical methods are slower, more difficult, and more expensive than optical ones.

The study of inclusions may be useful, but it is time-consuming and difficult. A review of quartz inclusions in various rock types is given by Clocchiatti (1975).

The trace elements of quartz varieties have

been investigated and seem to show some differences as determined by spectrography (Dennen, 1964). Cathodo-luminescence petrography (Smith and Stenstrom, 1965; Sippel, 1968; Sibley and Blatt, 1976) has become most useful in distinguishing the quality and quantity of trace elements in various quartz grains in sandstones. The luminescence is largely a function of the amount and kind of trace elements incorporated in the quartz at the time of crystallization. The incorporation of trace elements in the crystal structure is a strong function of temperature. The higher-temperature varieties are those that contain greater numbers and concentrations of trace elements. Because trace element composition varies with source rocks, different provenance classes of luminescent grains can be identified with relatively inexpensive equipment—or an electron probe, if it should be available. The most pronounced differences are between various dully luminescent colors of igneous and metamorphic quartz and nonluminescing authigenic quartz. Secondary quartz of the lowest grade of metamorphic rocks may also be nonluminescent or weakly luminescent in shades of blue to purple. Some cathodoluminescence may be related to abrasion or distortion of the quartz structure related to the transport agent (Krinsley and Tovey, 1978).

The varieties of quartz most useful for source evaluation are polycrystalline quartz grains, chert grains, and those grains made up of quartz plus rounded secondary overgrowths that can be recognized as second-cycle grains. Here, again, cathodo-luminescence helps greatly in identification. Though it may be theoretically possible to see multicycle grains—those with two or more generations of rounded overgrowths—the probabilities of proving such an origin are very low. SEM examination of detrital quartz grains can lead to recognition of weathering and fracture textures (Doornkamp, 1974; Smalley *et al.*, 1978) and textures indicative of specific transport agents such as eolian (Nieter and Krinsley, 1976; Krinsley and McCoy, 1977) or glacial (Rehmer and Hepburn, 1974; Bull *et al.*, 1980), though unambiguous criteria are elusive. Le Ribault (1977) gives a review with many illustrations.

Quartz is ubiquitous as a chemical cement, almost everywhere deposited in optical continuity with the original detrital grain. Such secondary quartz is relatively free of inclusions (though there may be some small fluid inclusions) and trace elements, and this is obviously related to a low-temperature origin. Chert and opaline silica, both opal-A and -CT, are also found as pore-filling cements or as rims around detrital quartz grains. Crystal growth experiments have shown that secondary overgrowths on strained crystals may become strained simply as a function of inheriting a crystal structure of the seed crystal. In view of this, it may not always be simply assumed that a strained overgrowth implies tectonic stress after the authigenic quartz was precipitated. Diagenetic processes resulting in silica precipitation are discussed in detail in Chap. 11. The ways in which provenance and diagenesis combine to form the many species of the silica minerals in sandstones are shown in Fig. 2-4.

Feldspars

All varieties of feldspar have been noted as detrital minerals. The feldspar minerals have been

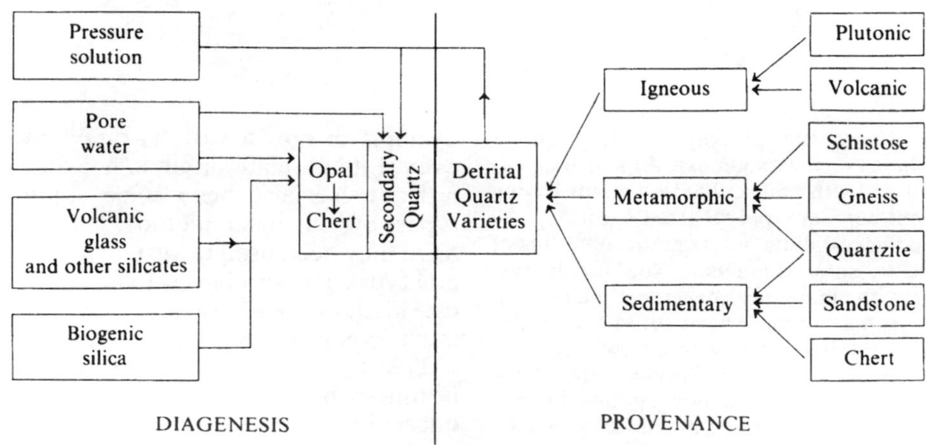

FIGURE 2-4. The origin of silica minerals in sandstones.

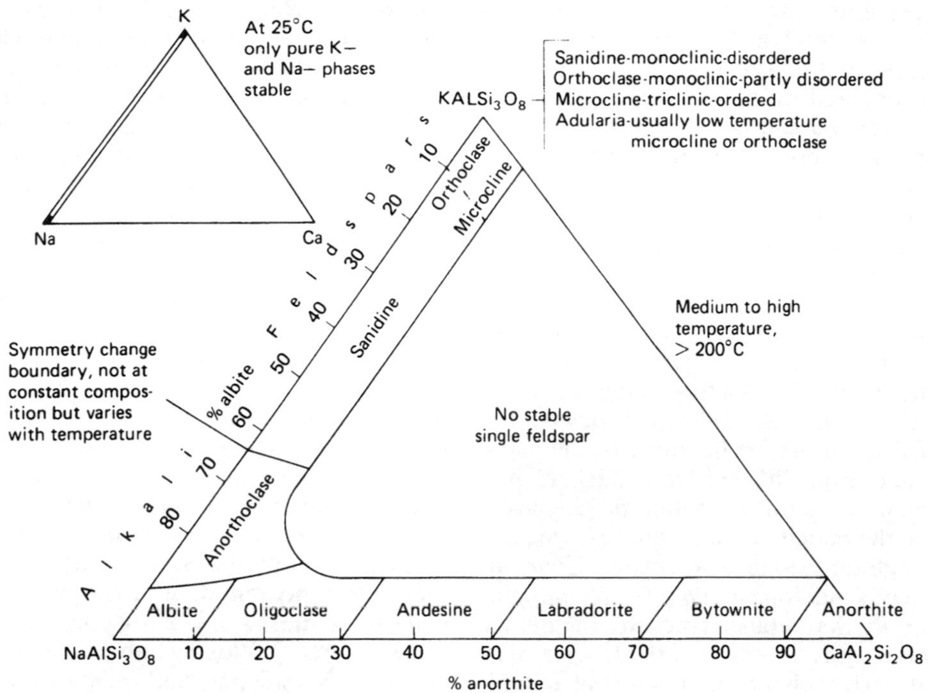

FIGURE 2-5. The ternary K–Na–Ca feldspar system. The crystallography of the phases varies with temperature, especially Al-Si disorder and Na–K–Ca disorder. At high temperatures the plagioclases are a solid-solution series. At intermediate temperatures the plagioclases consist of several unmixing series. Similarly, the anorthoclases are higher temperature, the perthites lower temperature.

studied intensively in the past two decades from mineralogical, crystallographic, and petrologic points of view; an extended treatment is given by Deer *et al.* (1963, pp. 1–178) and in a monograph by Smith (1974). Figure 2-5 shows the ternary K–Na–Ca feldspar system and terminology. In the past, relatively few petrographic analyses gave a breakdown of feldspar varieties; even now the sole distinction among the feldspars for most petrographers is between K-feldspar and plagioclase. That distinction and a growing number of more precise identifications have been stimulated by the recognition of the genetic relations between types of plutonic and volcanic feldspathic source rocks and plate late tectonic settings (see Chaps. 7 and 12 for a full discussion). In the average sandstone, K-feldspar (orthoclase, microcline) is most abundant and sodic plagioclases far outweigh calcic ones. Potassium-feldspar, especially microcline, is a characteristic feldspar of arkoses and feldspathic sandstones, especially those of continental block provenance. The few published analyses of all sandstones in which varieties of alkali feldspar are quantitatively distinguished show less microcline than orthoclase. This is not an identification problem, for microcline normally shows its characteristic twinning and so is easily recognized. Microcline is stated to be the most abundant species by many authors who do not report quantitative data. There is little doubt from the published literature that sodic varieties predominate among the plagioclases; they are the only feldspars in many graywackes. There are few sandstones with appreciable quantities of bytownite and anorthite.

In general, the amount of untwinned K-feldspar is probably underestimated relative to plagioclase, because it is that feldspar which is most easily not identified as such in thin sections, if staining techniques are not used. It is also well known that many different feldspar compositions may occur in the same thin section. This situation is far different from metamorphic or igneous rocks, which at equilibrium will normally contain two or, if perthite is included, three varieties, not counting zoned crystals. The fact that there are many different compositions present together thus is proof of detrital origin. Electron microprobe analysis of feldspar compositions is giving us much needed infomation on the composition of feldspars in sandstones. Not only do we get more accurate analyses of relative proportions of potassium, sodium, and calcium of individual feldspar grains in thin sections, but useful information is obtained on minor element distribution (e.g., Trevena and Nash, 1981).

The relative proportions of K-feldspar and albitic and anorthitic plagioclases may be controlled either by the relative abundance of those feldspars in the igneous and metamorphic source rocks or by differential stability in earth surface environments. The anorthitic component of plagioclase is far less frequently encountered in continental source rocks than is the albitic component; this is a function of the commonness of intermediate and silicic rocks at the surface of the continents. In contrast, the mafic rocks of the oceanic lithosphere and island arcs contain far more abundant calcic plagioclases. We believe the paucity of anorthitic plagioclase in the average sandstone is the result, in part, of dominance of sandstone derived from continental sources in the geologic record. In greater part, however, it is the result of the low stability of anorthite. The relative abundance of K-feldspar and albitic plagioclase in source rocks is highly variable, dependent on bulk· compositions that can be specified only with respect to the particular terrane that is being eroded. Any systematic greater abundance of K-feldspar requires recourse to stability arguments. Studies of groundwater compositions in relation to weathering of igneous rocks by Feth *et al.* (1964) and Garrels and Mackenzie (1967, 1971) confirm the impression gained from petrographic analysis that the order of stability in the weathering environment is K-feldspar—most stable, albite—less stable, and anorthite—least stable. There is a far greater difference in stability between albite and anorthite than between the alkali feldspars. These weathering studies confirm an extrapolation to low temperatures of the elevated-temperature experimental stability data of Orville (1963), which show the same order of stability. Experimental and observational studies of the dissolution kinetics of feldspars at low temperatures lend further support for this order of stability (Petrović, 1976; Petrović *et al.*, 1976; Busenberg and Clemency, 1976; Holdren and Berner, 1979; Berner and Holdren, 1979). All of these investigations as well as the calculations of Gibbs free energy changes during weathering reactions (Curtis, 1976), confirm the stability series proposed by Goldich (1938, pp. 36 and 54) based on soil and rock weathering studies. Compatible with stability relations is the composition of authigenic feldspars: pure K-feldspar and albite.

Feldspar in Recent and Modern Sands. Feldspar is an abundant constituent in modern sands of diverse origins. As can be seen from Table 2-1, the feldspar content varies from 1 percent or less to 77 percent. And although some of these sands contain 25 percent feldspar or more, few, if any, would be termed "arkose" or even "subarkose." In nearly all, the percentage of rock particles exceeds feldspar and the sands, if lithified, would more resemble lithic arenites. The reason few have the arkose composition is that these sands are far-traveled and have a mixed provenance. Arkoses, usually being derived almost wholly from a highly feldspathic igneous terrane, almost always form in close proximity to their sources and, in many cases, have a very limited distribution. These accumulations, though local, may be very thick.

The seven "grus" samples taken from the drainage area of the South Platte River, Colorado, are representative of the ultimate source material of most sand. They contain an average of 69.0 percent feldspar, about equally divided between K-feldspar and plagioclase (Hayes, 1962).

Potter (1978) studied the sands of the world's biggest rivers and found an unweighted average of all kinds of feldspar of 10.7 percent with a range from 0 to 53 percent. He found approximately equal amounts of K-feldspar and plagioclase. The feldspar abundances of big rivers vary in response to the source regions and are affected by weathering and transport with climate as an important variable (see Chaps. 5, and 12). River sands seem to be more feldspathic than either dune or beach sands. They average twice as much feldspar content as either of the latter. This may be related to some attrition of feldspar on beaches and dunes or to the fact that many of the latter are reworked or transported from less feldspathic relict Pleistocene continental shelf sands or have a large admixture of that origin. It is certainly worth noting that even some beaches and dunes have appreciable quantities of feldspar, a fact that militates against a universal origin of the pure quartz arenites as the products of disappearance of feldspar in such environments. Field and Pilkey (1969) have determined the abundance of feldspars in continental shelf sediments off the southeastern United States and concluded that the low amounts, averaging 3 to 5 percent, reflect provenance from Piedmont rivers draining deeply weathered source areas. This is in contrast to the much higher proportions, 11 to 30 percent, in New England shelf sediments reported by McMaster and Garrison (1966). The

TABLE 2-1. Feldspar content of Recent and Pleistocene North American sands.

Type	Locality	No. of samples	Range	Average	Reference
Gruss	Colorado	7	—	69	Hayes (1962)
Glacial, outwash	Illinois	24	—	14	Willman (1942)
Glacial, till (sand portion)	Illinois	5	15 to 20	17	Willman (1942)
River	Missouri River	26	26 to 45	32	Hayes (1962)
Rivers	South Platte and Platte, Colorado	11	—	50	Hayes (1962)
Rivers, small	Virginia	3	—	14	Giles and Pilkey (1965)
Rivers, small	North Carolina	12	—	7	Giles and Pilkey (1965)
Rivers, small	South Carolina	17	—	5	Giles and Pilkey (1965)
Rivers, small	Georgia	8	—	7	Giles and Pilkey (1965)
River	Illinois River	3	9 to 12	10	Willman (1942)
River	Ohio River	4	6 to 21	15	Willman (1942)
River	Wabash River	2	17 to 19	8	Willman (1942)
River	Mississippi River (Illinois)	13	16 to 34	25	Willman (19420
River	Mississippi (Illinois to Gulf)	62	15 to 26	21	Russell (1937)
River	Appalachicola River (Florida)	2	—	6	Hsu (1960)
River	Mobile River, Alabama	1	—	1	Hsu (1960)
River	Brazos River, Texas	2	—	11	Hsu (1960)
River	Colorado River, Texas	1	—	18	Hsu (1960)
River	Columbia River	4	—	23	Whetten (1966)
Rivers, small	Mexican streams	7	13 to 28	18	Webb and Potter (1969)
River, small	Jacalitos Creek, Calif.	1	—	25	Williams, Turner, and Gilbert (1954) (p. 285)
Beach, lake	Lake Erie	7	22 to 28	25	Pettijohn and Lundahl (1943)
Beach, lake	Lake Michigan	4	—	14	Willman (1942)
Beach, sea	Quebec, Labrador (Greenland)	9	27 to 77	49	Martens (1929)
Beach, sea	Massachusetts	17	1 to 26	8	Trowbridge and Shepard (1932)
Beach, sea	Gulf Coast, Louisiana	15	12 to 19	17	Hsu (1960)
Beach, sea	East Gulf	13	—	<1	Hsu (1960)
Beach, sea	Texas Gulf	14	8 to 9	8	Hsu (1960)
Beach, sea	North Carolina	17	—	5	Giles and Pilkey (1965)
Beach, sea	South Carolina	17	—	5	Giles and Pilkey (1965)
Beach, sea	Georgia	11	—	4	Giles and Pilkey (1965)
Beach, sea	Florida	18	—	2	Giles and Pilkey (1965)
Dunes	Illinois	47	8 to 29	18	Willman (1942)
Dunes	North Carolina	17	—	5	Giles and Pilkey (1965)
Dunes	South Carolina	7	—	3	Giles and Pilkey (1965)
Dunes	Georgia	11	—	2	Giles and Pilkey (1965)
Dunes	Florida	15	—	2	Giles and Pilkey (1965)
Continental shelf	Southeastern U.S.	80	—	3.3	Field and Pilkey (1969)[a]
Continental shelf	Southeastern U.S.	67	—	5.2	Field and Pilkey (1969)[b]
Summary					
	Average river	168	—	22.0	
	Average beach	142	—	10.1	
	Average dune	95	—	10.6	
	Grand average	404	1 to 77	15.3	

[a] 0.250–0.364 mm size fraction.
[b] 0.125–0.177 mm size fraction.

latter are derived from a much less weathered source overlain by feldspathic Pleistocene till.

Deep-sea sands from the western North Atlantic and Gulf of Mexico were investigated by Hubert and Neal (1967), who found an average of 12 percent feldspar, very close to Potter's average of 12.2 percent for modern big river sands other than those of tropical rivers of low to moderate relief (Potter, 1978). Valloni and Maynard (1980) and Maynard *et al.* (1982) studied a large assemblage of deep-sea sands in relation to plate tectonic position (see Chap. 12). They show feldspar averages ranging from 16 percent for fore-arc basins to 53 percent for subduction zones. The ratio of plagioclase to total feldspar ranged from 0.31 for trailing edge to 0.90 for fore-arc basins.

Inspection of Table 2-1 shows that, independent of their depositional environment, the sands of the northern United States are more feldspathic than those of the southern or southeastern states. This difference is probably due to the fact that the sands of the northern areas were formed by reworking of the glacial drift, the sand fraction of which is quite feldspathic.

The average of 404 river, beach, and dune samples is 15.3 percent feldspar, which is exactly the same as the average of 435 sandstones from the Russian platform (see Table 2-3). No claims can be made for the adequacy of either sample; that of Recent and Pleistocene sands is probably overweighted with sands of glacial derivation and hence more feldspathic than would otherwise be the case.

TABLE 2-3. Feldspar content of sandstones of Russian platform (Ronov *et al.*, 1963).

Age	No. of samples	Feldspar
Precambrian	65	30.5
Cambrian	18	16.6
Silurian	14	9.6
Devonian	177	8.9
Carboniferous	95	4.8
Triassic	5	61.6
Jurassic	23	42.8
Cretaceous	20	15.0
Tertiary	10	31.1
Quaternary	8	22.6
	Average	15.3

Feldspar in Ancient Sandstones. The average sandstone is calculated (from the average bulk chemical composition) to contain 11.5 percent feldspar (Clarke, 1924, p. 33). Leith and Mead (1915, p. 76) estimate by calculation 8.4 percent. Krynine (1948, Figure 11) gives a figure of 11 to 12 percent. Actual modal analyses show that some sandstones are essentially feldspar-free whereas in others feldspar constitutes over 90 percent of the framework fraction (Tables 2-2 and 2-3). The average feldspar content of North American sands and sandstones determined from a random selection of published analyses is about 14 percent; the average sandstone of the Russian platform is estimated to be about 15 percent feldspar.

The cause of the variations shown in the two tables is not readily apparent. In the North American data the highest values are for sandstones in the orogenic belts, the lowest in the sandstones of the continental interior. Most of the Mesozoic and Tertiary sandstones are from the Coast Ranges and the Rocky Mountain areas whereas the Paleozoic sandstones are from the continental interior and are generally texturally and mineralogically more mature and hence feldspar-poor. However, even the Paleozoic sandstones of the Appalachians and Ouachita orogenic belts are poor in feldspar though rich in rock fragments. The high feldspar content of some of the Russian sandstones is attributed to tectonism, i.e., rapid uplift and erosion of incompletely weathered debris; the low feldspar content of others is correlated with the end-stages of erosion and low relief and crustal stability (Ronov *et al.*, 1963, p. 225).

TABLE 2-2. Feldspar content of North American sands and sandstones.

Age	Number of formations	Percent feldspar
Pre-Devonian	35	5.1
Devonian–Permian	29	5.8
Mesozoic	12	25.0
Tertiary	22	21.0
Pleistocene–Recent		15.3
Unweighted mean		14.4

The Origin of Feldspar in Sandstones. Because feldspars, as quartz, are so ubiquitous in metamorphic and igneous rocks, the presence of undifferentiated feldspar as such has little interpretive value for precise source rock assessment unless the feldspar composition is specified. The exact composition can best be determined by electron probe or single-crystal X-ray diffraction, somewhat less accurately by optical methods and powder X-ray diffraction. Staining techniques are rapid and convenient for distinguishing plagioclase from K-feldspar (van der Plas, 1966, pp. 49–52). The value of feldspars of different types for provenance is partly based on the limited distribution of certain types, as, for example, sanidine, which is associated with high-temperature contact metamorphic or volcanic rocks. Microcline, in contrast, is widely distributed in metamorphic and plutonic igneous rocks but not in volcanics. Suttner and Basu (1977) have used order–disorder relations of single sand grains of alkali feldspars to distinguish source rocks with different cooling histories that imposed varying degrees of order. Plagioclase compositions correlate with the chemical compositions of metamorphic and igneous rocks but are not diagnostic of specific rock types. Optical heterogeneity of feldspars was studied by Rimsaite (1967), who related such characters as oscillatory or normal zoning and various kinds of intergrowths to composition and origin of source rock types. In spite of the complexities of their distribution in source rocks, feldspars are so much more characteristic of and abundant in metamorphic and igneous than in sedimentary terranes that their presence in a sandstone is most useful in a general way. Their abundance is related to (1) source rock composition (see Chap. 7), (2) chemical weathering in the source area, (3) abrasion and solution during transportation, and (4) solution and precipitation during diagenesis (Fig. 2-6).

We may relate chemical weathering in the source area to the concept of sedimentary tectonics in the following way. The survival of unstable minerals in the weathering environment, and thus their transportation to the sedimentary basin, is a function of the ratio of chemical to mechanical weathering in the source area. The ratio of chemical to mechanical weathering is related to the topography of the source area in the main. The topography, of course, is largely a function of tectonics. Thus we may go from the quartz–feldspar ratio to an evaluation of the tectonic state of the source area, having corrected for other things such as source rock composition and changes during transportation and diagenesis. The argument of sedimentary tectonics tends to underplay the role of climate or at least climate that is not directly controlled by topography. There has been continuing discussion of the relative importance of climate and tectonics (that is, topography) in the chemical weathering of crystalline rocks and the formation of soils (Strakov, 1967, pp. 3–23) and the formation of river sands (Basu, 1976; Potter, 1978; James *et al.*, 1981; Franzinelli and Potter, 1983) and applications to ancient sandstones (Todd, 1968). It is true that both may be important in particular cases. It appears yet to be demonstrated, however, that one or the other is always of overriding importance. To the extent that this is so, there is a certain indeterminacy in the interpretation of the ratio of feldspar to quartz or generally of stable to unstable minerals. A somewhat similar discussion has gone on with respect to the disappearance of feldspar and other unstable minerals as a result of abrasion and solution in the course of transportation, as discussed on p. 53 of this chapter.

Authigenic feldspars in sandstones have been reported so frequently that there is little doubt of their common occurrence (Baskin, 1956; Pettijohn, 1975, p. 204; Boles, 1982). The chemical conditions for the precipitation of feldspars are determined by the relative amounts of their components in solution, that is, K^+, Na^+, Ca^{++}, and SiO_2, as well as by pH. These variables as related to feldspar precipitation in general have been discussed by Garrels and Christ (1965, pp. 359–63) and with specific application to authi-

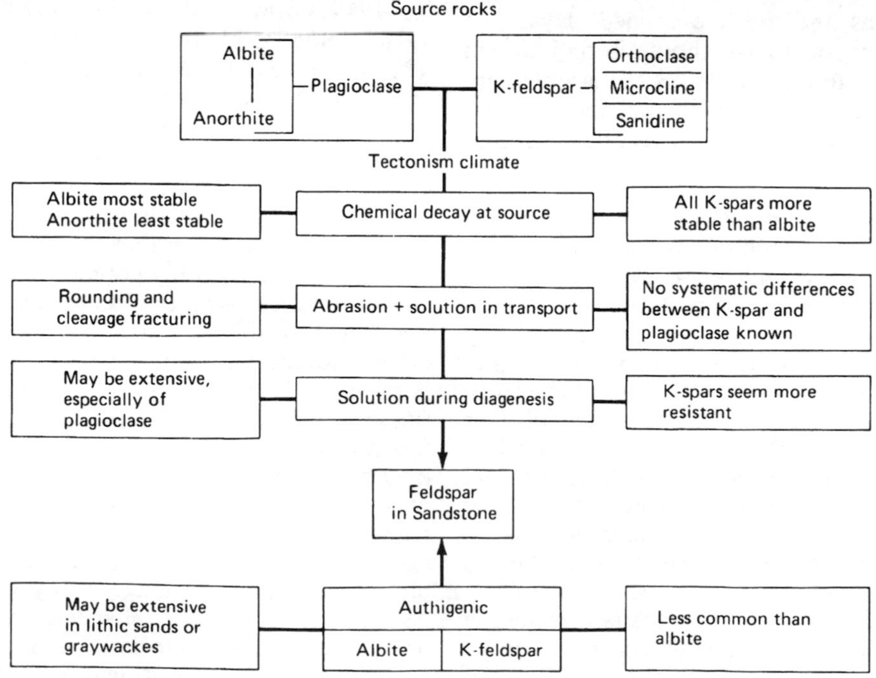

FIGURE 2-6. The origin of feldspars in sandstones.

genic feldspars in sediments by Kastner and Siever (1979). A measure of the stability of feldspar with respect to a solution and thus of the probability of its precipitation may be given by the appropriate alkali metal–hydrogen ion ratio, given a minimum amount of silica present in solution. A consideration of the temperature coefficients of the reaction as given by Hemley and Jones (1964) as well as the geological facts of the occurrence of authigenic feldspars suggests that slightly elevated temperatures, presumably associated with moderate to deep burial, are an important factor in their precipitation. A fuller discussion of feldspar authigenesis, including the sometimes extensive dissolution of detrital feldspars, is given in Chap. 11.

Micas, Chlorites, and Clay Minerals

These minerals can be considered together because they are all closely related in chemical composition (hydrous aluminosilicates) and crystal structure (sheet structures). They also are related to each other by virtue of their occurrence in sandstones. It is frequently impossible to note any discontinuity in size between large flakes, easily identifiable as detrital micas, and the very fine-grained interstitial clay characteristic of many sandstones. Nevertheless, there does seem to be some virtue in considering separately the two end-members, the large detrital grains and the fine-grained clays. The two can occur together without discontinuity in the size distribution or as two size varieties separated by a discontinuity in the size distribution, or each size can occur separately without the other size present.

Large Detrital Grains. Large detrital grains of muscovite, biotite, and chlorite are common in sandstones. They may be either fresh or altered, and species can be readily identified under the microscope. In most sandstones they are a minor constituent, except along fine-grained partings and in some shaly sandstones, where they may be abundant. Because of their thin sheet shape and consequent lower settling velocity, they are associated with quartz and feldspar grains of smaller sand or silt size (Doyle *et al.*, 1968, 1983). Submarine fans on the continental rise below submarine canyons show a greater abundance of micas in the finer-grain fractions farther out on the fan, while the more proximal parts of the fan, coarser-grained, have less mica. This is in general true of silts in comparison with sands. Much the same pattern is shown for turbidite sandstones, where the micas are associated with the more distal portions of subsea fans or interchannel deposits. Here too the coarser upper fan and channel deposits have less mica in them. The well-crystallized detrital micas can be interpreted in much the same fashion as feldspars, but may serve the purpose most usefully in the study of very fine-grained sandstones and siltstones. Muscovite is much more resistant to chemical weathering than biotite or chlorite, but there are not enough data to permit an evaluation of the stability of biotite relative to the several varieties of chlorite during weathering. There appears to be a greater tendency of chlorite to degrade to finer particles than biotite. So it is reasonable that chlorite is frequently lumped with the fine clay fraction in sandstones. The alteration of biotite to chlorite in place is also common. Frequently one can see the pattern of alteration in a single grain. The alteration of grains of biotite to glauconite was observed by Galliher (1939).

Fine-grained Clays. The much finer-grained clay found in sandstone as the essential constituent of matrix and argillaceous rock fragments includes all major groups of clays: the kaolin group (kaolinite, dickite, halloysite), the micas (muscovite, illite, glauconite), the smectite group (montmorillonite, nontronite, saponite, and many others), the chlorite group, and the mixed-layer group (corrensite and others) (Millot, 1964; Grim, 1968, pp. 558, 564–65; Carroll, 1970; Velde, 1977; Brindley and Brown, 1980). The structures of these groups are made up of sheets of alumina octahedra and silica tetrahedra (Fig. 2-7). The division of clay mineral groups is primarily on a structural rather than a compositional basis. In the kaolin group one octahedral layer is linked to one tetrahedral layer; the packet is a 1 : 1 group. Kaolinite, dickite, and nacrite are stacking polymorphs. In the micas an octahedral layer is sandwiched between two tetrahedral layers, a 2 : 1 group. The packets are joined with interlayer cations, such as potassium, in biotite and muscovite. The chlorite group has much the same kind of packet but the interlayer position between packets is occupied by a brucite, $Mg(OH)_2$, layer. The packets of smectites are similar to those of micas but contain water with cations in interlayer positions. Many compositional varieties are possible within groups. Though many of these are named, the compositional spectrum, especially of the smectites, chlorites, and mixed-layer minerals (a group transitional between smectite and others), is broad and continuous.

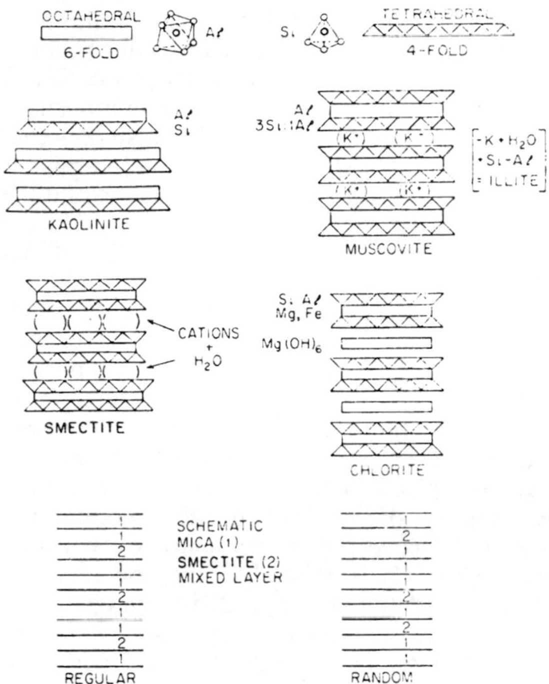

FIGURE 2-7. Terminology and structure of clay minerals.

The structure of the clay minerals is basic to the concept of ion exchange—an exchange of interlayer, and to a lesser extent octahedral, cations with the external chemical environment. Thus a calcium-rich montmorillonite can be put in a sodium-rich solution and become a sodium-montmorillonite, the only change being in the interlayer position. Equation 2-1 represents this exchange and Eq. 2-2 defines the exchange equilibrium constant, where a is the activity of the substance and K the equilibrium constant.

$$\text{Ca-montmorillonite} + 2\text{Na}^{+} \rightarrow \text{Na-montmorillonite} + \text{Ca}^{++} \quad (2\text{-}1)$$

$$K_{\text{Ca-Na}} = \frac{(a_{\text{Na-mont}})(a_{\text{Ca}^{++}})}{(a_{\text{Ca-mont}})(a_{\text{Na}^{+}})^2} \cong \frac{(a_{\text{Ca}^{++}})}{(a_{\text{Na}^{+}})^2} \quad (2\text{-}2)$$

The traditional measure of exchange, rather than using individual exchange equilibrium constants, lumps them in a cation exchange capacity (CEC) for all cations as milliequivalents (meq) per 100 grams. The CEC is by far the greatest for the smectites (in the vicinity of 100 meq/100 g), intermediate for the mixed-layer, illite, and chlorite groups, and lowest for the kaolinite group. Cations in octahedral positions are not generally exchangeable in short-term laboratory experiments, and so for relatively short periods of time, clays will act as stable detrital minerals insofar as octahedral and tetrahedral sheets are concerned, but as labile chemical components insofar as interlayer positions are concerned. Thus we may expect that over the short period of time it takes a clay mineral to be transported from river water to sea water and buried, it will exchange mainly interlayer positions as well as all those ions available on edges and corners on the outside of crystallites. This is typified by the change in clays from Ca-form in rivers to Na-form in marine deposits, shown in Eq. 2-1. Over the time scale of millions of years, characteristic of diagenesis, all parts of the clay mineral structure may react with the environment.

Laboratory work confirms thin-section studies that have shown the probability of direct precipitation of clay minerals from aqueous solutions at low temperatures and pressures (see, for example, Siffert, 1962). In addition, laboratory experiments have shown that clay minerals will partly dissolve and alter in hours or days (Mackenzie and Garrels, 1965; Siever, 1968). Clays will also absorb silica from solution in conjunction with cations to transform to somewhat different compositions; these changes are strong functions of pH as well as of the activity of dissolved cations (Siever and Woodford, 1973). Over periods of up to six months, clay minerals come to some quasi-equilibrium with respect to the solutions that bathe them. Work on the clay minerals in river and sea waters has likewise shown that the data are compatible with relatively quick equilibration of some sort, probably at the clay surface only, with the surrounding waters. There is some question as to whether the clays that react with their surrounding waters are those well-crystallized clays that are identified by X-ray diffraction procedures or rather are that proportion of the clay mineral fraction which is in the very fine-grained and relatively amorphous state characteristic of many soil clays. If the latter, they are likely not to be identified by X-ray diffraction and at the same time are likely to be more chemically reactive. The long-term diagenetic transformations of clays are treated in Chap. 11.

Precise identification of clay minerals is primarily by X-ray diffraction methods though rough approximate identification by optical methods is necessary for thin-section study. But precise identification by optical methods in thin section is often difficult (Grim, 1968, p. 432) though there are extensive treatments available of optical identification procedures in immersion media (Correns and Piller, 1955; Grim, 1968, pp. 412–33). Procedures for quantitative estimation of the relative abundance of the major clay mineral groups (mica, kaolinite, chlorite, smectite) include separation and size fractionation of clay from sandstone, making diffraction patterns with and without addition of ethylene glycol (for determination of montmorillonite or other expandable clays), measuring X-ray peak areas, and applying correction factors for X-ray scattering efficiency (Johns *et al.*, 1954; Carroll, 1970). The difficulties are such that analyses do not have a precision better than 10 percent unless extraordinary care is taken to insure reproducibility.

Transmission electron microscopy (Beutelspacher and van der Marel, 1968), electron diffraction (Zvyagin, 1967), and infrared spectroscopy (Farmer and Russell, 1964) have found much use for detailed studies of clay minerals. The SEM is now almost an everyday tool for the petrographic study of clays. SEM in the back-scattered electron mode combined with energy-dispersive X-ray microanalysis on the same instrument can be used to study size, shape, and composition of clays (Pye and Krinsley, 1983). Analytical electron microscopy is used for detailed investigations of clay mineral structures, such as the variation of cation populations of single layers (Lee and Peacor, 1983). The electron probe is very useful for determining chemical composition of clays but care must be taken to avoid volatilization, particularly of the alkali metals.

The major origin of clay minerals is as subaerial weathering products of silicate minerals. All of the major clay groups are found in weathered residues and soils. Which clays will be found depends on the interaction of climate, geomorphology, and parent rock. Humid climates and well-drained topographies lead to extensive weathering of feldspars and their silicates to kaolinite. Lower rainfall and poorer drainage may result in the formation of smectite from the same parent materials. Mafic silicates in many climates will go to smectite. Illite products may form in intermediate to humid climates. Tardy *et al.* (1973), for example, give a simplified scheme for the weathering of different silicates from granitic rocks (Fig. 2-8a) and for the variations with climate of these same alterations in sandy residual soils (Fig. 2-8b). These generalizations are fraught with exceptions, for the process of clay mineral formation is most complex. Thermodynamic calculations of the stability of clay minerals during weathering are limited in application to multiphase weathering products in many situations because the fundamental thermochemical properties of poorly crystallized and variable composition clays are not well known and because the reactions are kinetically controlled and rarely at equilibrium. As this large subject is beyond the scope of this book, the reader is referred to extensive reviews given by Millot (1964), Grim (1968), and Velde (1977).

Interpretation of clay mineralogy in sandstones proceeds from a knowledge of (1) the ways in which weathering processes form clays from primary silicate minerals and alter them from preexisting clays, (2) the nature of ion exchange and other short-term chemical changes that take place in the environment of deposition, and (3) the kinds of overall structural and chemical changes that take place over a long time. Thus (1) is related to the source rock, climate, and topography; (2) is related to the environment of deposition and later (groundwater) environments; and (3) is a function of long-term diagenesis. This might make clay mineralogy a powerful tool were it not for the fact that each type of change partially obliterates the record of earlier changes. Nonetheless, in comparative petrologic studies, clay minerals may find some utility for mapping petrographic provinces, for evaluating environmental differences, and for assessing the effects of diagenesis.

Studies by Biscaye (1965), Griffin *et al.* (1968), and others as well as the many studies of DSDP cores have shown that the clay mineral distributions of Atlantic and Pacific sediments can be understood most easily in terms of detrital petrographic provinces, where the transport is by a combination of wind and ocean currents. Though near-shore differential flocculation has some effect on clay mineral distributions in shallow waters, oceanic province boundaries do not fit such an origin. There is a clear correspondence between the climate, weathering, and source rock compositions of the adjacent continents and the distribution of clay minerals on the sea floor. This does not rule out certain early diagenetic changes that can be found in oceanic sediments, such as the alteration of volcanic ash to smectite and/or zeolite, and the formation of glauconite (see p. 49). Nevertheless, the major factor determining the clay mineral composition of oceanic sediments is the source contribution rather than the sedimentary environment. That is not necessarily so for other environments, such as marine and nonmarine evaporites, where mixed-layer, smec-

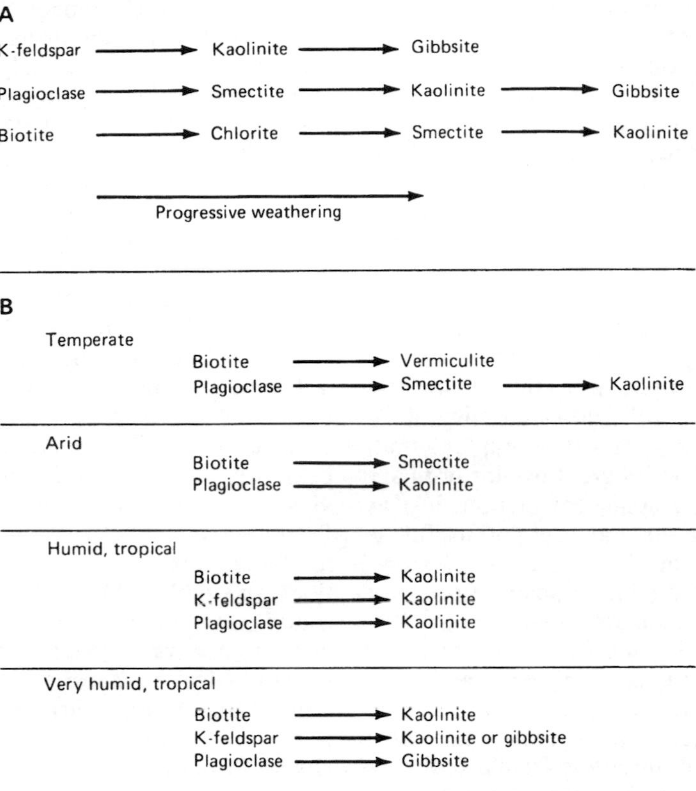

FIGURE 2-8. Formation of clay minerals by the weathering of granites. A) Progressive weathering of feldspars and biotite; order of appearance is the same regardless of climate. B) Weathering paths in different climates. Transformations simplified and modified from Tardy *et al.*, 1973.

tite, and zeolite formation may be strongly influenced by the environment during and immediately after sedimentation. Studies of clays in ancient sandstones have proceeded with this background of interaction of provenance, environment, and diagenesis. Petrographic provinces related to provenance are mapped partly on the basis of clay minerals and special imprints of environments are recognized by clay mineralogy. The effects of diagenesis are particularly important and have been related to depth of burial and formation waters. Authigenic clays are common. Kaolinite and dickite have been noted frequently in arenites. One study (Glass *et al.*, 1957) has definitely shown that the kaolinite is authigenic and could not be a different source contribution because the adjacent shales do not contain kaolinite. Much of the clay matrix of graywackes is apparently secondary, a diagenetic product. Volcaniclastic sands that contain a high fraction of labile components, such as glass and calcic plagioclase, together with detrital clays derived from land weathering of volcanics, tend to be rich in smectites and later alter to other clays (see Chap. 11).

The conditions for the formation and chemical stability of the various clay minerals, at least in idealized form, are considered by Garrels and Christ (1965, pp. 352–62) and Krauskopf (1979). As in the case of the feldspars, the pertinent parameters are the concentrations of alkali metals, alkaline earths, hydrogen ion, silica, and alumina in the environment (see Chap. 11). Thus it appears that the clue to understanding the origin of clay minerals in a particular rock will come from areal mapping of the clay mineral composition rather than an attempt only to infer origin from a particular composition in a particular place.

Heavy Minerals

In this group we include various silicates and oxides that are found in small quantities in sandstones, the total quantity of such constituents rarely making up more than one percent of the rock. They range from tourmaline and zircon, which do not occur in large amounts in any source rock, but are resistant to mechanical and chemical attack, to the amphiboles and pyroxenes, which may be abundant constituents of some source rocks, but show little resistance to decay. To the extent that the heavy minerals survive the hazards of weathering, transport, and diagenesis and to the degree that they occur in a restricted range of provenance types, they are most useful for source rock interpretation (see Chap. 7). Because it is most unusual for a source area–sedimentary basin couple to change appreciably and synchronously over short periods of geologic time, thus giving rise to a different and laterally constant mineral assemblage for each time–stratigraphic unit, heavy minerals are not of great use for ordinary detailed time–stratigraphic correlation, just as lithologic facies in general are not useful as time–stratigraphic markers. To the extent, however, that heavy minerals can be used for facies correlation indicating sedimentary dispersal from particular source areas which are undergoing tectonic evolution, they can be extraordinarily useful. Thus one can map the progress of an orogenic episode in a source area which led to gradual unroofing from a sedimentary to a metamorphic to an igneous terrane by noting the change in heavy mineral suites going upwards in the sandstones derived from that source. The stratigraphy of the heavy mineral zones of the sandstones will be the reverse of the sequence in the source area.

When used in conjunction with textural analysis and light mineralogy, heavy mineralogy becomes most valuable for establishing petrographic provinces and source rock types. Again it is the *mapping* that proves the utility of the heavy minerals. Because the size distributions of each mineral variety may be different, one must be careful to either include the entire size range to cover the entire heavy mineral assemblage, or to count the heavy minerals within a particular group. Hydraulic ratios (Rittenhouse, 1943; Briggs *et al.*, 1962) can be used with a proper understanding of the density–size relationships of heavy minerals to evaluate more carefully the total heavy mineral fraction.

Heavy minerals may be concentrated to form placer deposits (Zimmerle, 1973), typified by the monazite sands of India, an economic deposit. Garnet and other sand placers are found on many beaches. The heavy mineral that has received much attention in the search for uranium ores is uraninite, which has been considered to be detrital in a number of Precambrian ores such as those found in the Elliot Lake district of Ontario (Robinson and Spooner, 1982). The presence of uraninite as a detrital is linked to lack of oxygen in the early atmosphere (Holland, 1984). The literature on uraninite and sandstone-type uranium deposits is extensive and is summarized by Adler in an entire volume devoted to "Formation of Ore Deposits" (1974).

As covered in Chap. 3, the roundness of heavy minerals, as of any detrital grains, may indicate petrographic provinces by showing different abrasion histories. The stability and persistence of heavy minerals with age is covered in Chap. 11. The use of ore microscopic methods for the study of opaque heavy minerals such as magnetite, ilmenite, and hematite has been reviewed by Stumpfl (1958). Major reviews of heavy minerals have been given by Parfenoff *et al.* (1970), who included study methods, interpretation, and mineral descriptions and tables; Mitchell (1975), who concentrated on heavy minerals in soils; and in a group of three papers by Nickel (experimental dissolution), Dietz (experimental transport textural modification), and Grimm (weathering in a gravel formation) in a volume on "Stability of Heavy Minerals" (1973). The descriptive tables of heavy minerals given by Milner (1962) remain a standby for identification.

Hand (1967) compared theoretical and actual hydraulic equivalents of heavy minerals in beach and dune sands and found a general correspondence but sufficient discrepancy in detail to indicate that more than just settling from suspension was involved. The difference in quartz and garnet settling velocities was confirmed as a criterion for beach versus dune sand origin. Hubert (1960, pp. 208–21) proposed a zircon–tourmaline–rutile (ZTR) index of heavy mineral assemblages that he related to a source contribution modified by sedimentary differentiation in the environment. Force (1980) noted that the use of rutile in the ZTR index is limited because most rutile is derived from high-grade metamorphic rocks and is rare in most igneous rocks. Kelling *et al.* (1975) showed that the ZTR index was not strongly affected by heavy mineral size, even though individual minerals such as amphibole and garnet were.

Special attention has been given to two of the commonest stable heavy minerals, tourmaline and zircon, in much the same way that quartz varieties have been studied in the hope of gaining greater effectiveness in provenance determinations. Tomita (1954) surveyed the literature on zircons and concluded that purple or rose-pink zircon comes only from Archean gneisses or granites; the color was considered radiogenic, its depth or intensity increasing with age. Poldervaart (1955) studied zircons from sedimentary rocks in comparison with their distribution in igneous rocks and found that shape in relation to size was a more important indication of provenance than other variables: the larger the zircon, the more rounded, if sedimentary; the smaller, the more rounded in the rare occurrences of resorbed zircons in granites. Callender and Folk (1958) measured a variety of properties of zircons in Tertiary sand of central Texas, including color, elongation, and inclusions, but found only the degree of idiomorphism to be correlated with another petrologic variable, the relative amount of volcanogenic detritus. Krynine (1946) attempted to distinguish color varieties of tourmaline as a basis for provenance, but there is no simple relationship between color and composition or between composition and the type of igneous or metamorphic rock in which tourmaline is found (Deer *et al.*, 1962, p. 313).

As work on the tectonic significance of light mineral fraction mineralogy of sandstones gained dominance in the past generation, the use of heavy minerals declined. There is the prospect now, however, that there may be some resurgence of interest in heavy mineral provenance in relation to plate tectonics. An example is Zimmerle's (1984) study of spinels that are inferred to originate in ophiolite belts and subduction zones.

Rock Fragments

The major types of rock fragments in sandstones are (1) the argillaceous group including shale, slate, phyllite, and schist, (2) volcanic rocks including glass, and (3) the silica group consisting of quartz and cherts (Fig. 2-9). Of lesser importance but locally important are the carbonate rock fragments. Primarily used for provenance studies, the rock fragments have been grouped in a number of ways depending on the emphasis of the investigator (see Chap. 7). Recognition of different kinds of volcanic and metamorphic rock fragments has been spurred by attempts to relate provenance to the igneous rocks characteristic of different plate tectonic settings; especially of island arcs and other subduction zone environments. Feldspar, glass, and other components of rock fragments can be analyzed under the microscope, if sufficiently coarse-grained, or by electron probe, if fine-grained, allowing firmer identification of rock type and thus provenance. Individual mineral components of rock fragments may even be counted as separate components in a modal analysis (Ingersoll *et al.*, 1984).

Rock fragments may become deformed, be altered, and lose their identity by diagenetic change, leading to problems of recognition. Rock fragments are clearly recognized and identifiable in thin sections of modern sands. The argillaceous rock fragments are probably not identified as such in many older sandstones because they may be squashed, deformed, and molded about the more competent grains so that they blend into or appear as clay matrix (Allen, 1962, p. 678). All types of such squashing can be seen in suitable thin sections. It is likely that in a great many of the oldest sandstones the squashing is almost complete and the former rock fragments are confused with a fine-grained clay matrix. It is almost impossible to give a nonarbitrary and completely reproducible way of recognizing squashed rock fragments when all outlines are gone. Clay aggregates formed during diagenesis may also be confused with rock fragments. The tougher varieties of argillaceous material (more highly metamorphosed) are more likely to be identified as rock fragments; soft shales rarely survive as detrital fragments of sand size and, even if they do, are more likely to have lost their identity after deposition. There is some question as to the viability of argillaceous rock fragments during stream transport; most seem to be quickly degraded to silt or clay size in modern streams. Some tumbler barrel experiments also indicate a lack of durability of shale fragments with transport. In consequence, it may be possible that abundant argillaceous rock fragments in a sandstone indicate a nearby source of that material. Thus one might speak of a "cannibalism" index in terms of the proportion of soft argillaceous rock fragments preserved in a sandstone.

Volcanic rock fragments, including pyroclastic debris, are abundant in some sandstones. In fact, they may be the major or only constituent of some. The fragments may be of older volcanic terranes or of contemporaneous origin from volcanic activity within or near the depositional environment. When altered diagenetically, such rocks are likely to contain zeolite minerals. Devitrification of glasses within the sandstones can lead to the production of opal cement. Some formations can equally well be considered as volcanic rocks or as sandstones made up of volcanic rock fragments. There is probably only a semantic difference between the consideration of a welded tuff as a volcanic

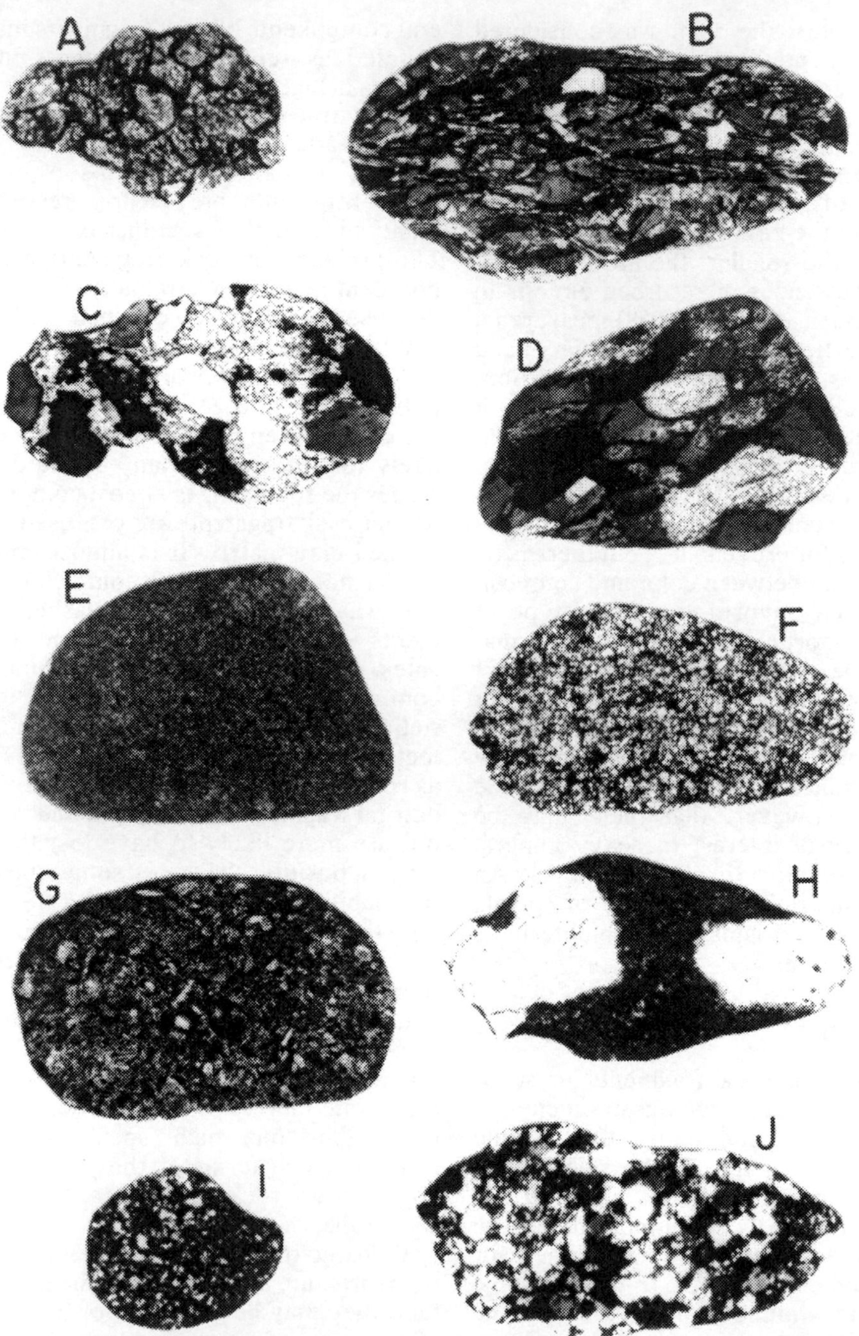

FIGURE 2-9. Rock fragments from modern sands of the Ohio River Basin, U.S.A., × 80: A) detrital carbonate; B) muscovite schist; C) sandy carbonate; D) hornblende schist; E) micritic limestone; F) microcrystalline carbonate; G) silty shale; H) sandstone cemented by chert; I) chert; and J) coarse siltstone. Photographs courtesy of James F. Friberg.

rock or as a sediment in which the conditions for cementation are especially related to the temperature of formation.

Rock Particles in Recent and Ancient Sands. Rock fragments are widely distributed in modern sands but the kinds and even the total quantity present were seldom specified until recently. Such data on their occurrence as could be obtained by a search of the literature are given in Table 2-4. These data are, in some cases, incomplete and for certain size grades only. To the extent that only the finer grades

TABLE 2-4. Rock fragment content of North American Recent and Pleistocene sands.

Type	Locality	No. of samples	Range	Average	Reference
River	Mississippi River	62	9 to 19	13	Russell (1937)
River	Ohio River	1	—	11	Hunter (1967)[a]
River	Mississippi River (below junction of Missouri)	2	22 to 27	24	Hunter (1967)[a]
River	Columbia River	11	—	31	Whetten (1966)[b]
Rivers, small	Mexican streams	7	45 to 69	57	Webb and Potter (1969)
River, small	Jacalitos Creek, California	1	—	40	Gilbert *in* Williams *et al.* (1954, p. 285)
River	Rio Grande, Texas	1	—	40	Nanz (1954)[c]
Average river		85		20	
Beach	Lake Michigan	2	—	17	Hunter (1967)[a]
Beach	Texas	2	—	15	Nanz (1954)[c]

[a] Data from 1.25–2.75ϕ range only.
[b] Includes some coarse silt grades also.
[c] Estimated from published plot.

were examined, the rock fragment content is too low. The original analyses in such sands show that, unlike feldspar, the rock fragment content tends to be markedly size dependent, the proportion of such fragments increasing with increasing size (Okada, 1966, Table I; Hunter, 1967). There is also some ambiguity about what is reported as a "rock fragment." Polycrystalline quartz, chert, and ordinary rock fragments may be placed together or reported separately. In Table 2-4 chert is included with the rock fragments.

As is the case with feldspar, some small streams draining areas of restricted lithology are apt to be highest in rock fragments. In the sands of large rivers, coming from more diverse sources, the arithmetic average of rock fragments is 30 percent, as determined by Potter (1978). He has pointed out the importance of specific provenance as well as tectonism and climate in controlling the proportion of rock fragments. Data on the rock fragments of shoreline and shelf environments are too sparse to allow any far-reaching conclusions to be drawn, but Davies and Ethridge (1975) have shown the dependence of rock fragments, among other constituents, of sands of these environments on abrasion, transportation, and sedimentation processes. The data of Nanz (1954, Figure 19), for example, suggest that rock fragment content is less on beaches and dunes than in rivers of the same geographic province. The rock fragment content of deep-sea sands was studied by Valloni and Maynard (1981), who correlated volcanic and other rock fragments with plate tectonic position. They found the highest proportion of rock fragments to be those of volcanic fore- and back-arcs with lesser amounts for leading edge environments and the least amounts for training edges. The proportion of volcanic to other fragments was highest, as would be expected, in fore- and back-arc environments.

It is obvious from this review of the meager literature on the rock fragment content of modern sands that more and better data are needed. Petrographers have tended to neglect the rock fragments. Moreover, study of the constituents as grains is not adequate. Most rock fragments are opaque to transmitted light and, unless sectioned, are unidentifiable. What is needed are studies such as those of Potter (1978) and Franzinelli and Potter (1983) in which thin sections of the grains are made. Though there have been many studies of the sedimentary structures of modern sands artifically indurated in place, in few have thin sections of those beds been made to see how rock fragment composition varies with light and heavy mineral abundances, including of placers.

The abundance and kind of rock fragments in ancient sands is much better known than in modern sands due primarily to the prevalence of thin-section analysis of these indurated sands.

Several factors determine the rock fragment content, namely, (1) grain size, (2) provenance, (3) maturity, and (4) age. In general, the rock fragment content is a function of grain size—the coarser the sand, the higher the rock fragment content, other things being equal (Shiki, 1959; Allen, 1962, p. 673).

As noted by Potter and Pryor (1961, pp. 1224–26), the sands derived from the Canadian Shield contain more feldspar than rock particles; those of Appalachian derivation show the reverse. The Shield area is more largely "granitic" whereas the Appalachians, including the crystalline Piedmont, is a sedimentary-metamorphic terrane. The texturally mature sands—those better rounded and better sorted—tend to have a lower rock fragment content (as well as a lower feldspar content).

The Chemical Minerals

Carbonates

The carbonate minerals in sandstones are both detrital and chemically precipitated, with the latter group by far predominant.

Detrital. Detrital carbonate is abundant in calcareous sands as skeletal fragments, oolites, and faecal pellets, none of which are traveled from a distant source. Such debris may also be admixed in varying, but usually minor, amounts in terrigenous sandstones. Majewske (1971), Horowitz and Potter (1971), and Scholle (1978) provide guides to the recognition of skeletal debris in thin section.

Detrital carbonate grains derived from source regions outside the basin are also present in some sandstones (Fig. 2-9a). Such grains have rounded, abraded boundaries and are significant in some molasse sandstones (Füchtbauer, 1964, Fig. 16). Potter (1978) found the greatest abundances of carbonate rock fragments in two sand types: modern big river sands of semi-arid and arid climates and sands derived from Pleistocene glacial drift eroded from carbonate bedrock. Both high relief with rapid erosion—and glacial erosion is comparable—and arid climate favor terrigenous carbonate. Of the two, high relief with a resulting high ratio of mechanical erosion to chemical weathering may be somewhat more important. But the dryer the climate, the lower relief need be. In any case, the solubility of carbonate plus its softness and cleavage all decrease the probability of its survival by stream transport. Thus some correction factor is probably needed to reconstruct the relative amount of carbonate rock in the source area. Acid etching and/or staining are useful when studying thin sections rich in carbonate rock fragments.

Calcite and Dolomite. Calcite and dolomite are abundant in sandstones as pore-filling and replacement cements of post-depositional origin. Grains of detrital and primary dolomite have been identified by (e.g., Sabins, 1962). Some pore-filling cement may be recrystallized from originally detrital carbonate grains or may be a precipitate from an aqueous solution in an originally empty pore space. The primary and early diagenetic cement carbonate of many modern sediments may include magnesian calcites. In contrast, the calcite and dolomite of later diagenetic cements of sandstones are relatively pure, with no excess magnesium in the calcite and no excess calcium in the dolomite. Aragonite has not been reported as a precipitated cement in ancient sandstones, though it is well known in beach-rock in modern carbonate environments and aragonite-cemented modern sands (Allen *et al.*, 1969). Presumably the aragonite of ancient sandstones inverted to calcite during early or middle diagenetic recrystallization.

Calcite may be a primary cement or recrystallized from earlier detrital origin but dolomite appears almost always as a replacement of a calcite precursor. The clue to this difference lies in replacement textures (see Chap. 11) and crystal habit in particular. Calcite is commonly anhedral, made up of an interlocking crystal mosaic whose individual crystals may vary from a few microns to centimeters (as in sand crystals). In contrast, dolomite and ferroandolomite most commonly assume a rhombohedral form. Though this is not exclusively so, the presence of rhombohedra is one of the quickest clues to the identification of dolomite, as the distinction between calcite and dolomite by ordinary optical criteria is difficult. Differentiation between calcite and dolomite is most quickly achieved by one of a variety of staining tests, the stains usually taking advantage of the presence of small amounts of iron in all dolomites (Friedman, 1959; Warne, 1962).

Fe–Mn Carbonates. The iron-rich carbonates are less common than calcite and dolomite in most sandstones, but are abundant constituents of some. Investigations by Goldsmith and Graf (1960) and Goldsmith *et al.* (1962) have elucidated the phase relations in the system Ca–Mg–Fe–Mn–CO_2. Fig. 2-10 shows the phase diagram for the Ca–Mg–Fe part of this system.

Of the various minerals in this chemical system, the ones that are important in sandstones, in addition to calcite and dolomite, are siderite and the iron-rich dolomites or ankerites. Rhodochrosite is known in some concretionary forms in sandstones, but has rarely been reported as cement. Compositional banding of iron and manganese cements is revealed by cathodo-luminescence; these bands are the result of changes in the cation ratios of the solu-

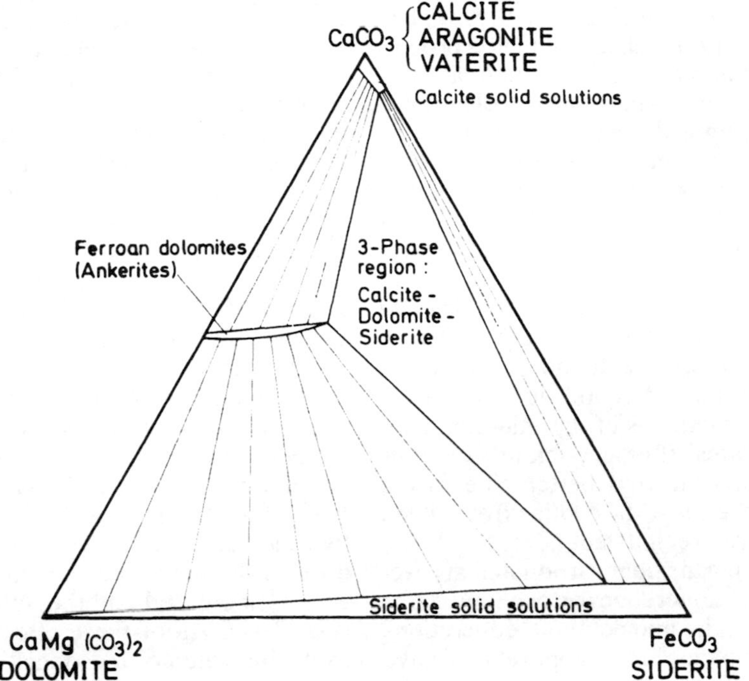

FIGURE 2-10. Schematic phase diagram of the Ca–Mg–Fe–CO_2 system, estimated, at 25°C. Estimate based on data in Goldsmith *et al.* (1962).

tions from which the carbonates were precipitated during diagenesis.

The precipitation of calcite and dolomite is primarily influenced by calcium and magnesium activities and pH, whereas that of the iron and manganese carbonates has an additional dependence on the oxidation–reduction balance of the environment. Because iron and manganese can be in carbonate minerals only in their reduced state, a redox potential is immediately established. The redox potential and carbon dioxide pressure necessary for the precipitation of siderite have been discussed by Garrels and Christ (1965, pp. 201–11). For any given amount of dissolved carbonate these relations can be shown on an Eh–pH diagram (Fig. 2-11). The most probable reducing agent responsible for the reduction of iron is organic matter. Thus we may look for clues to the formation of siderite and other reduced metal carbonates in terms of a linkage to environments in which organic matter, either in soluble or solid form, are abundant; such places are stagnant basins, or tidal or estuarine environments where organic productivity is high. Hallam (1966, 1967) suggested that sideritic concretions in the Jurassic of England are tied to shoreline conditions and indirectly to the presence of organic matter

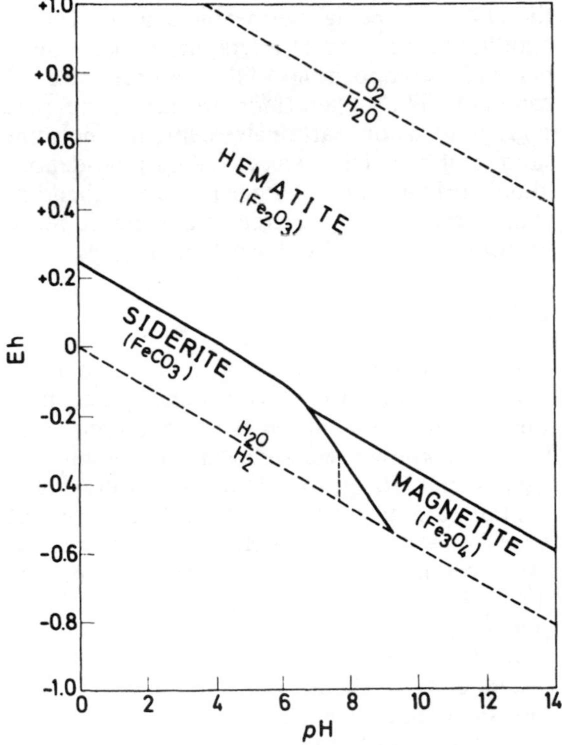

FIGURE 2-11. Eh–pH diagram of stability field of siderite in relation to hematite and magnetite in water at 25°C., 1 atm total pressure, and total dissolved carbonate of 10^{-2} moles per liter. Dotted line is for activity of Fe^{++} in solution = 10^{-6}. (Modified from Garrels and Christ, 1965, Fig. 7.13).

which supplied the reducing capacity. Matsumoto and Iijima (1981) related changing proportions of siderite and calcite cements in a series of Japanese Mesozoic sandstones to changes in diagenetic conditions during deep burial. Commonly, different carbonate cements are seen as successive cements precipitated under varying conditions but these need not be so. As is shown in the phase diagram of the carbonates (Fig. 2-10), it is possible at low temperatures to have a coexisting equilibrium assemblage of calcite, an iron-rich dolomite, and siderite. These three cements are found in what appears to be this kind of equilibrium in certain Carboniferous sandstones of the mid-continent of the United States. If other such associations are noted, we may have a better clue to the composition of the subsurface fluids from which the cements were precipitated.

The iron-rich magnesium carbonates are frequently found as concretionary accumulations in sandstones; calcite and dolomite concretions are also well known. These concretions have not been studied a great deal, but do seem to be related to bedding planes, which may be either the result of special permeability along certain bedding planes or a stratigraphic expression of particular compositions of the water in the environment. The concretions appear to be pure segregations of carbonate cement enclosing sand grains, which speaks for their post-depositional origin. The alignment along bedding planes may also be related to diagenetic redistribution of originally detrital carbonate.

Sulfates

Gypsum, anhydrite, and barite are the three common sulfates found as cementing agents in sandstones. Though modern gypsum sand dunes are known, we are aware of few ancient gypsum-rich rocks that have been interpreted as lithified gypsum sand dunes. The sulfate of these cements is ultimately derived from that dissolved in seawater or in subsurface brines. Much of our understanding of the origin of gypsum and anhydrite cement has come from work of the 1960s and 1970s on the sediments of the Trucial coast of the Persian Gulf, where evaporites have been mapped on the sabkha flats, broad reaches of supratidal flats formed in an arid area periodically inundated by marine waters at storm tidal surges and by windblown salt spray (Kinsman, 1964, 1969; Shearman, 1978). Beneath the sabkha a combination of surface evaporation and downward migration of evaporated seawater from the flats is responsible for much of the gypsum and anhydrite that has been deposited in pore spaces of preexisting sediments below the flats. The water migrates in response to a density gradient induced by the evaporation of the brines at the surface. In contrast to the open marine evaporite environment, the sabkha is an environment in which sand may accumulate, both by shoreline processes and as eolian deposits related to an arid hinterland. Barrier islands in arid regions, such as Laguna Madre, south Texas, may be made up of sands that at depth are cemented by gypsum and/or anhydrite.

The conditions for the presence of gypsum as opposed to anhydrite as functions of pressure, temperature, and salinity have been studied by a number of workers, the last two most definitively by Hardie (1967). Gypsum, containing two molecules of water, will convert to anhydrite as the temperature is increased, the pressure is increased, or the salinity is increased (Fig. 2-12). All of these effects are predictable from the volume and density relationships of the two minerals. It now appears that the depth of burial is minimal, 2000 to 3000 ft (645 to 970 m), for the pressure and temperature to be reached at which gypsum converts to anhydrite, depending on the salinity of the brine. At high surface temperatures and atmospheric pressure, as the salinity increases greatly by evaporation (corresponding to decrease in activity of H_2O in Fig. 2-12) anhydrite becomes the stable precipitate. Yet because of the kinetic difficulty in the crystallization of anhydrite, gypsum is the mineral found as the first crystallized mineral in almost all natural environments. The typical picture emerging from studies of sabkhas is one where gypsum forms first as crystals growing in pore space and then by displacement of detrital grains. Under conditions of very high brine salinity, the gypsum may dehydrate to anhydrite. If the sediment is then buried, increasing temperature and pressure will keep anhydrite stable. Or the shallowly buried anhydrite may rehydrate to gypsum if the pore water salinity decreases by invasion by less saline surface waters. A full description of these and other aspects of evaporite deposition is given by Borchert (1959) and Braitsch (1971), and this topic is summarized by Blatt *et al.* (1980 p. 541–547).

Barite occurs as concretions as well as cement. The origin of barite in modern sediments seems to be related to volcanism on the sea floor according to Arrhenius and Bonatti (1965). This has been deduced from studies of barite formation in modern sediments in the deep

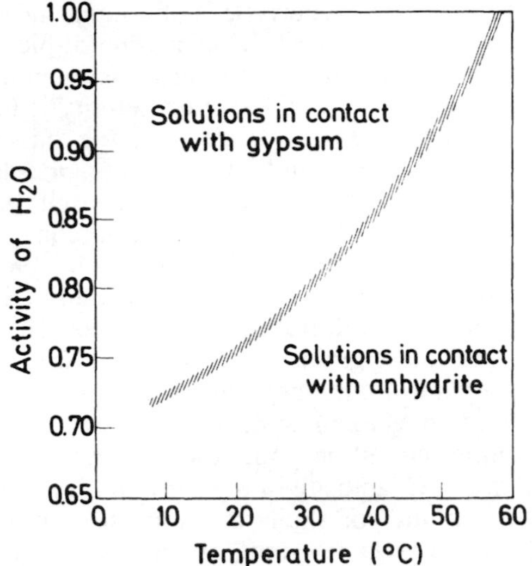

FIGURE 2-12. Temperature vs. activity of H_2O plotted for brines in contact with gypsum and anhydrite from experimental data and natural occurrences. (Modified from Hardie, 1967, Fig. 6). Activity of H_2O is a thermodynamic measure related to salinity; the activity of pure water = 1.

ocean; barite appears to be localized with respect to volcanic provinces in the sea. There is also evidence that much of the barite formed in economically workable deposits as well as sandstone cements and concretions is the result of barite precipitation from ascending brines from deeper-seated barium-containing basement rocks, according to Hanor (1967). Füchtbauer (1967, 1974) has related barite, among other cements, to marine depositional environments. Because of the very low solubility product of barium sulfate, the presence of barite is more to be ascribed to the presence of barium than to that of sulfate, because sulfate is almost always present either from seawater or the typical subsurface brine.

Sulfides

Pyrite is the principal sulfide found in sandstones. Lesser amounts of marcasite have been reported, and in the occasionally economically valuable deposit, other sulfides as well. The sulfide appears to be almost entirely diagenetic in origin, though certain Precambrian deposits such as those of the Mississagi Quartzite of the Blind River area in Ontario and the Witwatersrand uranium deposits in South Africa have large amounts of apparently detrital rounded pyrite grains. These detrital rounded grains in the Precambrian deposits have been considered to indicate a nonoxygenic atmosphere roughly two billion years ago (Rutten, 1962, p. 100).

Pyrite is genetically related to a more or less amorphous common mineral, FeS, "hydrotroilite," now properly called mackinawite, that is found in modern sediments under reducing conditions. FeS forms first by the bacterial reduction of sulfate in seawater and during early diagenesis is converted to FeS_2, pyrite (Berner, 1964, 1980). Thus pyrite in a sandstone, if it is not detrital, is presumptive evidence of at least locally reducing conditions. The reducing conditions have often been tied to a silled anoxic basin, but in the case of sandstones it is more likely that it is the presence of large amounts of organic matter on tidal flats that is responsible for the reduction. There is some evidence from modern sediment studies and ancient sediments that pyrite directly precipitated by bacteria can also be found, though no examples from sandstones have been quoted.

Because the pyrite that is formed diagenetically may commonly be found as rounded aggregates, it is not always easy to distinguish such diagenetic pyrite from possible rounded detrital grains of pyrite. It is presumed that in an oxygen-rich environment, such as the average weathering environment of the earth today, the oxidation of the sulfide will proceed so rapidly that pyrite, though a hard, abrasion-resistant mineral, will be destroyed in large quantities. Thus we find pyrite as a common but very minor constituent of the heavy mineral suite of most modern sands. But pyrite may persist for many miles downstream of areas where mine wastes containing pyrite have been dumped into a river. The persistence of pyrite as a detrital mineral is a subject that needs much more investigation.

Other Minerals

A variety of other minerals are of chemical origin, the most important being phosphates ("collophane" or carbonate–apatite), iron silicates (such as chamosite, greenalite, and glauconite), the zeolites, and the iron and titanium oxides.

Phosphate. The precipitation of phosphate in some sandstones appears to be related to phosphatic bones or shells. In other sands it appears as an authigenic precipitate or replacement of carbonate formed during early diagenesis in anoxic or suboxic interstitial water environments. Phosphate also occurs as surficial or soil segregations and concretions. The mineralogy of these phosphates is complex, most of the minerals being some mixture of hydroxy-fluor-carbonate-apatite. Carbonate fluorapatite, $Ca_5(PO_4)_3(F, CO_3)$, now widely termed francolite, is highly variable in composition, with substitution of Mg, Sr, and Na for Ca and some SO_4 for PO_4. It makes up the bulk of sedimentary phosphates. Carbonate hydroxylapatite, most frequently called dahllite, is similarly variable in composition. Both detrital bone and shell fragments and cements and nodules appear to be poorly crystallized and have only vaguely defined optical properties. In both modern and ancient sediments, phosphate abundance has been related to organic productivity and, in particular, to upwelling nutrients on a relatively shallow marine shelf. Phosphate-rich mudstones and carbonates grade into muddy phosphatic sandstones in formations such as the well-known Phosphoria (Permian) of the western United States. Environmental indicators reflect a shallow-water marine setting. In many of these deposits there is evidence of early diagenetic origin of nodules and pellets. An extensive review of phosphate deposits has been given by Cook (1976), a series of petrologic, mineralogic, and chemical studies of pelletal phosphates have been reported by Lucas *et al.* (1979), and a summary of modern ocean deposits has been provided by Kolodny (1981).

Iron Silicates. The iron silicates can be considered to be varieties of clay minerals that are iron-rich. Glauconite appears to be the most common such mineral throughout the geologic column. Simple, spheroidal, and ovoidal glauconite pellets occur in quartz arenites and lithic arenites and to a lesser extent in feldspathic and other types of sandstones. Glauconite is a mica related to illite in structure, but with a great deal of both ferrous and ferric iron (Burst, 1958). The glauconites appear to be associated with mildly reducing conditions, perhaps locally induced by the presence of abundant organic matter, related to relatively near-shore and shallow-water environments. Pryor (1975) relates concentrations of glauconite along unconformities to slow sedimentation in which the only deposition is biogenic—pelletization of suspended muds along quiet water surfaces of nondeposition. It is probable that there are a variety of conditions suitable for the formation of glauconite, as for the other iron silicates that are found in sediments, all of them relating to conditions which allow the entrance of ferrous iron into the clay structure. Almost always glauconite appears to be of marine origin, though some few occurrences of nonmarine origin have been reported. Thus a general picture emerges of glauconite forming in sedimentary environments in which biogenic activity is high and sedimentation rates are low, the mineral forming as a penecontemporaneous or early diagenetic mineral (Seed, 1965; Bell and Goodell, 1967; Velde, 1977). Glauconite may be reworked and transported (Triplehorn, 1966); some sands, such as those of the Weald of England, contain both transported and autochthonous glauconite grains (Allen *et al.*, 1964).

The chamosite minerals are a variety of chlorite, and in general appear to need some of the same conditions for their formation as glauconite. Chamosite in the form of faecal pellets and fillings of shells is found in relatively shallow water, a few tens of meters on warm continental shelves, such as that of the Orinoco off the coast of South America (Porrenga, 1967). Phanerozoic oolitic chamosite ironstones in sandy deposits have been reviewed by Van Houten and Battacharyya (1982). Examples of chamosite beds in sandy formations are the Jurassic Northampton formation of England (Taylor, 1949) and the Devonian Shatti formation of Libya (Van Houten and Karasek, 1981). It is not clear whether the minerals greenalite and stilpnomelane are primary or diagenetic sedimentary minerals; most may be metamorphic equivalents of iron-rich clay minerals. The most likely origin of the iron-rich clays and iron silicates in sedimentary environments is the diagenetic transformation of ferric hydroxide rims or accumulations that come down from rivers into estuarine or near-shore environments. The ferric hydroxide, when intimately mixed with detrital clay minerals such as chlorites or illites, may spontaneously reorganize and, if sufficient organic matter is in the environment to provide a slightly reducing environment, may enter the clay mineral structure and form the iron silicates. The sandstones that contain these minerals are most frequently associated with the ironstone shales or carbonates which have been used as iron ores.

Iron and Titanium Oxides. The same general scheme of origin inferred for the iron silicates has been suggested for the hematitic and limonitic iron oxide coatings of red sandstones. The iron in these formations is contained either in hematite or in the mixture of goethite and lepidocrocite (normally goethite is by far the dominant mineral) referred to by the field name limonite. Evidence from modern stream loads makes it doubtful that much of this iron was transported as detrital hematite or limonite. Some sand grains are coated with hematite pigment in rivers and in continental shelf sands derived from them (Judd *et al.*, 1970). The load of ferric hydroxide colloid in many streams suggests origination in soils, with transport in rivers and the near-shore ocean as a dilute suspension or adsorbed on the surfaces of clay minerals. After deposition, the ferric hydroxides spontaneously dewater to form limonite or, if complete dehydration occurs, hematite. Van Houten (1973) has reviewed the origin of red beds in this light.

Walker (1967, 1974) and Walker *et al.* (1978) have argued convincingly that many red sandstones of desert and more humid environments owe their hematite to precipitation by soil or groundwaters that get their iron from the weathering of iron-rich primary silicates, such as pyroxenes and amphiboles. Both processes may be at work in one or another situation. Both involve diagenetic precipitation of the iron oxide.

Red sands may occur interbedded with green, purple, or other variegated colors; the varied colors have been linked to post-depositional processes operating in different alluvial and deltaic environments under differing climatic conditions (McBride, 1974).

Much work on the remanent magnetism of sandstones is compatible with magnetic reversal stratigraphy based on other kinds of rocks, and supports the general view that the hematite–limonite is of immediately post-depositional origin (Walker *et al.*, 1981). Other reports indicate a detrital magnetic signature coming from larger grains of hematite (Steiner, 1983). There is some evidence of chemical remagnetization during later diagenesis (Turner and Ixer, 1977; Wisniowiecki *et al.*, 1983).

The titanium and iron–titanium oxides are quantitatively unimportant in sandstones, found mainly as minor components of the heavy mineral suites. These oxides are resistant to chemical and mechanical decomposition and so tend to persist for long times throughout the sedimentary cycle.

Zeolites. The zeolite minerals have assumed steadily increasing importance in mineralogic study of sandstones. These minerals, formerly considered as curiosities and rarely identified accurately, have now been recognized to be widespread and abundant in sandstones, especially those associated with volcaniclastics in subduction zone environments. Volcaniclastic sands containing zeolites may have been deposited in sedimentary environments ranging from deep-sea, slope, or shelf marine to deltaic, fluviatile, lacustrine, and freshwater evaporite continental. Zeolites are an important mineral facies occurring in hydrothermal alteration zones and hot spring deposits that affect surrounding sandstones. Sands associated with alkaline and/or saline lakes in closed basins may contain zeolites formed by reaction of detritus and interstitial waters. Zeolites of weathering origin are found on alkaline, saline soils formed on sandstones in dry environments.

Though there are more than 30 naturally occurring zeolite minerals and 20 of these are found in sediments, only eight are found in major abundance in sediments: analcime, chabazite, clinoptilolite–heulandite, erionite, ferrierite, laumontite, mordenite, and phillipsite (Munson and Sheppard, 1974). Most common in sandstones are analcime, laumontite, clinoptilolite, and mordenite. Table 2-5 lists some common zeolites, their idealized formulas, and the geological ages of their host rocks. From the ages shown it is obvious that zeolites must disappear by diagenesis or metamorphism over long geological times. They are most common in Cenozoic rocks, much less common in Meso-

TABLE 2-5. Common zeolite minerals found in sediments (modified from Iijima, 1978, and Deer *et al.*, 1963).

Zeolite	Composition	Geologic age
Analcime	$Na[AlSi_2O_6] \cdot H_2O$	Quaternary–Carboniferous
Chabazite	$(Ca, Na_2)[Al_2Si_4O_{12}] \cdot 6H_2O$	Quaternary–Miocene
*Clinoptilolite	$(Ca, Na_2)[Al_2Si_7O_{18}] \cdot 6H_2O$	Quaternary–Carboniferous
Erionite	$(Na_2, K_2, Ca, Mg)_{4.5}[Al_9Si_{27}O_{72}] \cdot 27H_2O$	Quaternary–Eocene
Ferrierite	$(Na, K)_4Mg_2Al_6Si_{30}O_{73} \cdot 20H_2O$	Miocene
*Heulandite	$(Ca, Na_2)[Al_2Si_7O_{18}] \cdot 6H_2O$	Pliocene–Carboniferous
Laumontite	$Ca[Al_2Si_4O_{12}] \cdot 4\text{-}3\frac{1}{2}H_2O$	Pliocene–Devonian
Mordenite	$(Na_2K_2Ca)[Al_2Si_{10}O_{24}] \cdot 7H_2O$	Quaternary–Carboniferous
Phillipsite	$(\frac{1}{2}Ca, Na, K)_3[Al_3Si_5O_{16}] \cdot 6H_2O$	Quaternary–Carboniferous

* Clinoptilotite is considered to be the high-silica member of the heulandite group.

zoic ones, and virtually absent from the early Paleozoic and Precambrian.

Analcime is found in and associated with some sandstones in continental rift valleys such as the Triassic Lockatong formation of the Newark Series of the Appalachians (Van Houten, 1962). The Lockatong rocks are thought to have been deposited in a lacustrine environment as groups of detrital and chemical cycles related to climate. Analcime is also common as an alteration product of volcaniclastic detritus in the diagenesis of both marine and continental sandstones, together with clinoptilolite, laumontite, and mordenite (Iijima, 1978). Though phillipsite is abundant in some marine sediments, it is not so common in sandstones. The formation of zeolites in sandstones is exclusively diagenetic; no significant amount is found as detrital grains, partly because of their generally fine grain size and partly because they are quickly broken down during weathering to clay minerals, most frequently to mixed-layer and smectite clays.

In many sandstones, zeolite minerals are closely linked to the presence of volcanic glass. It appears that the zeolites form as early decomposition products of volcanic glasses in weathering and continental and marine sedimentary environments. Other zeolites form even in the absence of glass in response to high-pH, high-silica environments such as the alkaline lakes of the western United States. Sequences of zeolite and other silicate minerals have been found with increasing depth in fore- and back-arc accumulations (Coombs *et al.*, 1959; Packham and Crook, 1960). Based on depth, translated to pressure and temperature zonations, these sequences were designated the zeolite facies, a true metamorphic facies, by Turner and Verhoogen (1960, p. 532). A discussion of the diagenesis of zeolites is covered in the section on volcaniclastics in Chap. 11. A complete review of the mineralogy and diagenetic origin of zeolites in sediments was given by Hay (1966). Reviews of current work in mineralogy, geochemistry, and sedimentary petrology, as well as practical applications, are in Sand and Mumpton (1978).

Organic Matter

Organic matter in sandstones is a biochemical precipitate characteristic of the environment. The organic matter always has been profoundly altered by diagenesis. The varieties of organic matter in sandstones range from oil and gas to solid particles of high carbon content material. None are as easily characterized as minerals; they are identified and classified according to their chemical composition. They can also be partly characterized by their physical properties just as are minerals, but analytical chemical descriptions are much too difficult and time-consuming to permit quick and easy identification. Much work in organic geochemistry has been devoted to specifying the kinds and amounts of organic compounds in recent and ancient sediments and their analytical chemistry, now largely dominated by gas chromatography–mass spectrometry (Eglinton and Murphy, 1969). Two important works concentrating largely on organic geochemistry related to the origin of oil and gas are those by Tissot and Welte (1978) and Hunt (1979).

We can easily distinguish the black opaque matter from the brownish translucent matter and the yellowish and brownish transparent materials that are noted in sandstones. The darkness of color is related to increasing carbon content. Most petrographic descriptions have simply lumped all of these materials into organic matter—if they are mentioned at all. We can identify woody and leafy material as precursors of coaly fragments, but such fossils cannot be used to specify the origin of kerogen, oil and gas. Reflectance of polished surfaces of coaly materials has long been used in coal petrology and is now being applied to carbonaceous grains in sandstones (Bostick, 1979). The clearest and simplest relationship among carbon content, reflectance, degree of coal metamorphism—rank—and temperature is shown by one component of coal, vitrinite, and hence vitrinite reflectance has become a widespread technique for assessing the degree of burial diagenesis.

The organic compounds of sandstones are derived from the breakdown of plant and animal tissues in the sedimentary environment; later the decomposition products may be solubilized, transported, and reprecipitated during stages of diagenesis. The ultimate chemical product of all these processes is apparently more linked to a complex diagenetic history than to the particular pathways linked to specific environments. But, as in the metamorphism of coal, from lignite to anthracite, the chemical changes are the result of a kinetic process that is primarily a function of temperature. The existence of organic matter is obviously the result of either high organic productivity in comparison to the oxidative breakdown of such organic material or to the presence of particular biologic entities that are highly resistant to oxidation, such as some of the plant cutins, spores, and other epidermal materials. Mineral charcoal, fusain, is also a prominent constituent of sandstones and is, as in coal, most likely of forest fire origin.

Unlike shales and limestones, sandstones in general are characterized by extremely small amounts of organic matter, less than 0.1 percent. Some graywackes are an exception and may contain up to several percent organic matter. The organic matter that is native to sandstones is dominantly detrital. Even tidal flat sandstones, where organic productivity is high, have a relatively small residue of organic matter left after the early biological diagenetic processes by which organisms, micro- and macro-, oxidize the original organic component. Modern sands, whether alluvial, deltaic, near-shore marine, or deep-sea turbidite, tend to be associated with aerated environments; the currents that bring in the sand also tend to mix the water and keep it oxygenated. The presence of oxygen and the high permeability of the sand promote oxidation of much of the organic matter.

Most of the fluid organic matter, oil and gas, has been introduced into the sandstone after deposition by migration from the beds in which the organic matter was indigenous. The fluids are hydrocarbons derived from a complex series of organic reactions during early and later diagenesis, first in the source beds and later in the reservoir. The relationship of source beds to reservoirs, their geologic history and geometry, and the geology and geochemistry of the transformation of sedimentary organic matter to oil and gas is a large subject beyond the scope of this book, and the reader is referred to works on petroleum geology and geochemistry.

Relation of Mineralogy to Texture

Mineralogy and Size

Because mineral and rock fragments break into smaller pieces and become reduced in size by abrasion as a function of cleavage, fracture, brittleness, and hardness, they may be expected to have size distributions in sandstones that are related to the relevant mineralogical properties. Most work has been done on the effects of abrasion on mineral composition (Pettijohn, 1975, p. 493), which could be used to predict the differential rate of size decrease of minerals, but few analyses of size distributions of different minerals have been made that reflect abrasion resistance. One such analysis is that of Füchtbauer (1964, Fig. 25), who compared quartz, rock fragments, and feldspar, and showed that the quartz was appreciably larger in modal and median size than the others. The importance of this effect was further demonstrated by Heim (1974) and by Odom *et al.* (1976), who found very different proportions of detrital feldspar in different size fractions (medium, 4%; fine, 8%; very fine, 34%). Charles and Blatt (1978) have found size variations among quartz, chert, and feldspars and recommended use of the 4.0–4.5 phi size class to eliminate the effects of size. Long known is the tendency for finer-grained and silty sands to contain more clay than coarser ones. But it is not possible to generalize yet on the data of Füchtbauer, who pointed out that a coarse-grained sand might be classified as a different sandstone type from a fine-grained sand from the same stratigraphic unit, just because of the differential size distribution of minerals. Ingersoll *et al.* (1984) have urged the use of point-counting methods in which sand-sized crystals and grains of rock fragments are counted as such, thus minimizing grain size–mineralogy variations.

The size of heavy minerals has been investigated extensively by van Andel (1955, pp. 530–35), who showed how the different size distributions were related to hydraulic ratios and thus to the densities of minerals. It appears that the size effect on heavy minerals is a complex function of both the hydraulic ratio and the differential resistance to abrasion, for the analyses are not easily predictable on the basis of either alone. A further complicating factor in the interpretation of size distributions of heavy minerals is the original size distribution in source rocks, which might be vastly different for different minerals, as, for example, between small accessory zircons in a granite and large tourmaline crystals from pegmatites. Faced with the possibility of significant differences in mineralogy with size, the petrologist is well advised to sample both coarser and finer beds to establish the range of compositions.

Mineralogy and Resistance to Abrasion

The possibility that selective abrasion would drastically affect mineral composition by loss of finer and softer minerals has been argued for strongly by many who believe that quartz arenites may be derived by abrasion from lithic arenites rich in feldspar and rock fragments. Some of the best evidence for this possibility has been provided by Mack (1978). Feldspar with its good cleavage and many of the softer and more cleavable rock fragments are most likely to abrade quickly. The harder metamorphic and igneous rock fragments appear to persist for longer distances of transport. Ethridge (1977) has determined that destruction of rock fragments in river channels has been significant in increasing sediment maturity but that one has to take into account other factors such as weathering, dilution, winnowing, and selective sorting. Quartz pebble abundance increases downstream relative to less resistant lithologies such as schist, shale, volcanics, and slate. The initial size decrease in unstables is rapid on headwaters followed by slower decrease in lower reaches (Adams, 1979).

The disappearance of soft minerals that are minor components of source contributions accounts in part for the general absence of minerals like talc, gypsum, and most carbonates, though there is no doubt that solution plays an important role for the carbonates and sulfates. What complicates this problem is the association of chemical alteration with mechanical degradation, so that when minerals such as talc become extremely fine-grained, they become more prone to sorb cations and transform to clay minerals. If mechanical degradation were the only important effect, then we should see the same overall mineral composition but softer minerals should be more abundant in fine silt and clay sizes. Thus the feldspar that supposedly is worn away from a quartz sand should appear as an important constituent of the associated shales and this apparently does not happen frequently, though Odom *et al.* (1976) correlated feldspar abundance and size distributions with environmental differences.

Chemical Composition

It is obvious that the chemical composition of sandstones is a different way of designating the mineralogical composition and vice versa. It is rarely possible, if ever, to calculate the bulk chemical composition from modal mineralogy because of the wide range of compositions in individual mineral species, particularly the clay minerals. Such compositional variations cannot easily be analyzed by optical petrography or X-ray diffraction, and electron probe analysis is time-consuming. It is even harder to calculate the abundance of minor or trace elements. To study the distribution of the elements that otherwise might not be revealed, we use bulk chemical analysis. Bulk chemical compositions of metamorphic and igneous rocks are generally more simply interpreted than those of sandstones, both because the crystalline rocks are usually treated as equilibrium mineral assemblages and because they do not involve the complex mechanical mixing and segregation processes inherent in the making of a sandstone. Bulk chemical composition, for example, does not distinguish between a detrital component and cement, a most important distinction. Nonetheless, chemical composition is useful in establishing a baseline for the study of metamorphic derivatives of sandstones and may shed additional light on provenance and environment.

Ordinarily the chemical analysis is given in terms of the oxides, with a precision of three or four significant figures. It is doubtful that this level of precision is warranted for much work on sandstones, for there is little reason to believe that such specimens are chosen randomly from a homogeneous population of rocks in the field. In addition to haphazard sampling, there is doubt that very small differences in major element composition will prove to be geologically significant, for sandstones are highly variable in composition, reflecting derivation from diverse sources (Pettijohn, 1963).

Chemical Composition as a Function of Mineral Constituents

Alkalies(Group I). The abundant elements of this group in sandstones, as in most other crustal rocks, are Na and K, though in sandstones neither exceeds a few percent of the total. In the non-clayey sandstones the bulk of the Na and K is in alkali feldspars and muscovite, the only igneous or metamorphic alkali metal minerals that survive as detritals to any significant extent. The Na and K in the argillaceous sandstones may be mainly in the clay minerals illite and smectite. K commonly exceeds Na in both argillaceous and non-argillaceous sandstones, a reflection of dominance of mica over feldspar in most of the latter and illite over smectite in the former. The rocks in which Na is more abundant are primarily the graywackes, in which Na-rich feldspars are the main source of the high Na content. It is not certain if this is related to a dominance of detrital albitic feldspar over K-feldspar or calcic plagioclases; some, perhaps much, of the Na content is related to the albitization of feldspars that is common in the diagenesis or low-grade metamorphism of these rocks (Boles, 1982). It may in some graywackes that come from Na-rich volcanics. Little is known of the distribution of Cs and Rb in sandstones; it is likely that they are held primarily in the clays. Li, Na, and K are all found in zeolite minerals, either as major components or as impurities.

Alkaline Earths (*Group II*). Carbonates and clay minerals account for most of the elements of this group. The only other mineral containing a Group II element in significant amounts is anorthitic (calcic) plagioclase. Ca and Mg are the only two elements of quantitative importance. Little is known of the distribution of Sr and Ba and practically nothing is known about Be in sandstones.

Ca is characteristically present as carbonate, either in calcite or dolomite, rarely in aragonite, or gypsum or anhydrite. Minor contributions to total Ca content may come from smectitic clays. Sandstones that contain anorthite plagioclase may owe their Ca to this mineral. Mg is partitioned between dolomite and chloritic and smectitic clays. Ca is more abundant than Mg in most sandstone types, a reflection of the generally greater abundance of calcite over dolomite. In graywackes the amounts of the two elements are more nearly equal as a result of the large amount of chloritic clay that is typical of the matrix of many of these rocks.

Ba and Sr, when present in more than trace amounts, are found as sulfates in the barite–celestite solid-solution system. In mixed carbonate–silicate sands, particularly the very young ones, Ba and Sr are in the carbonate, proxying for Ca or Mg.

The elements of Group II originate as chemical precipitates in sandstones, primarily as the carbonates, and much may be diagenetic. The exception is the Mg in chloritic clays and this too may be of the same origin as discussed in the section on the origin of graywacke matrix in Chap. 11.

Aluminum (*Group III*). Al is the only abundant element of Group III present in sandstones. It is present almost exclusively in aluminosilicates, though gibbsite and boehmite may be significant in some sandstones and minor amounts of Al can be attributed to heavy minerals such as spinels, pyroxenes, epidote, and micas. Abundance of Al is in general, then, related to the abundance of feldspars, micas, and clays, regardless of whether the clay is in argillaceous rock fragments or in interstitial clay, either detrital matrix or diagenetic. The variability of Al content in the different clay minerals is probably less important in affecting the bulk composition than is the quantity of clay. This is illustrated by the fact that the more aluminous clays are associated with quartz arenites but the quartz arenites as a group are low in Al, because they contain so little clay. At the other end of the sandstone compositional spectrum, the graywackes, Al is high not because the clays are aluminous, for they are not, but because both clays and feldspars are abundant.

Boron is present in trace quantities in the clay fraction of sandstones, although little is known about it. For the shales and muds, where it has been chiefly studied, there is a large literature concerning its possible significance (Potter *et al.*, 1980). Complicating the interpretation of boron in the clay fraction of sandstones is good permeability, which favors post-depositional diagenesis of the clay minerals.

Silica. Silica abundance is rather simply related to the ratio of silicate to nonsilicate minerals. Clearly, those sandstones with large amounts of carbonate, sulfate, or oxide, all of which are most commonly present as cements, will be low in silica. Next in importance are the relative amounts of quartz and chert in relation to other silicates or aluminosilicates. In this way silica content matches roughly with maturity factors in mineralogical composition. But to the extent that relative amounts of feldspar and clay minerals have some effect, feldspar being more siliceous than most clays, silica content would not correlate well with maturity, except for very mature quartz arenites.

Others. Both ferrous and ferric iron are present in sandstones as components of many minerals. Fe^{++} is in clays, mainly chlorites, with lesser amounts in smectites and illites; in carbonates as siderite and in ferroan dolomites (ankerites); in sulfides, primarily pyrite; and as a minor constituent of feldspars as other silicates. Fe^{3+} is normally present as the hydrous and anhydrous oxides, hematite, goethite, and lepidocrocite; and in the glauconites. Volcanic rock fragments may be rich in Fe^{++}; to the extent that they or Fe-rich glasses dominate the volcaniclastic sands, those sands will have a high Fe^{++} content. The Fe-oxides, though they may give the sandstone a pronounced red appearance, are normally dispersed as fine-grained pigment and normally are not sufficiently abundant to strongly affect total iron composition.

Titanium is present in small amounts, mostly in clays, but some is in the heavy minerals rutile, ilmenite, brookite, and anatase. Though these minerals may be authigenic, it is doubtful that the whole rock will be enriched in Ti by this means, except in some placers. Sulfur is present as sulfate, mainly as gypsum and anhydrite, and sulfide, as pyrite and marcasite. Thus, with the exception of some few sandstones in which gypsum or pyrite can be demonstrated to be detrital, S is restricted to the chemical precipitates, and, within that group, largely to those of diagenetic origin. Phosphorus is present as detrital apatite and as biochemically precipitated carbonate–apatite (francolite) as well as minor impurities in a host of other minerals. There is a great deal of scattered information on trace elements in sediments (see, for example, Wedepohl, 1969) but little systematic application to problems of sandstone petrology. Bhatia and Taylor (1981) have investigated the abundances of hafnium, lanthanum, thorium, and uranium in Paleozoic graywackes of the Tasman geosyncline, Australia, as provenance indicators. They found significant differences related to plate tectonic settings and source region igneous rocks (see discussion of chemical composition and plate tectonics below). Basu *et al.* (1982) have studied rare earth elements of Holocene fluvial sands and concluded that major rock type is the dominant control of first-cycle sands.

Isotopic Composition

Though much work has been done on the isotopic composition of sedimentary rocks and minerals (Faure, 1977), relatively little effort has been devoted to sandstones. Some studies of $^{12}C/^{13}C$ and $^{18}O/^{16}O$ ratios of carbonate minerals, such as those of Land and Dutton (1978) and Stanley and Faure (1979) on cements of sandstones, have been used to determine the origin and temperature of formation of diagenetic materials. Little has been done on $^{32}S/^{34}S$ except in conjunction with ore deposits, mostly hydrothermal origin (Maynard, 1983). The study of the stable isotopes of oxygen, carbon, and sulfur in the cements or other chemical components of sandstones can be petrologically useful for the determination of temperature–salinity relationships that may be related to depth of burial, replacement versus void-filling origin of carbonate cements, biogenic versus abiogenic formation, primary versus diagenetic formation, and probably many other factors. Savin and Epstein (1970) have reported oxygen isotope ratios of seven modern sands and five Precambrian sandstones and concluded that $^{18}O/^{16}O$ ratios may indicate high-temperature rock provenance of detrital quartz and feldspar as opposed to low-temperature origin of authigenic quartz and feldspar. They also suggested how the temperature of authigenic feldspar formations might be calculated. Clayton *et al.* (1978) studied oxygen isotopes of quartz silts from buried sediments and ancient soils, and concluded that isotopic exchange during weathering and diagenesis was small enough to use quartz oxygen isotopes for provenance. Rb–Sr studies also seem to confirm the validity of quartz isotopes as provenance indicators (Powers *et al.*, 1979). Peterman *et al.* (1981) used Sr isotope ratios as well as U–Th

Structural frame work: Anion Groups	Exchangeable cations		Associated Petrographic Type
High SiO_2/Al_2O_3 (mature) (little clay or detrital Al-silicate)	Alkaline-earth-rich (carbonate cement) Alkaline-earth-poor (silica cement)		Quartz arenites
Low SiO_2/Al_2O_3 (immature) (clay + detrital Al-silicate)	Alkali-metal-rich (feldspar and clays)	$Na_2O > K_2O$	Feldspathic graywackes
		$Na_2O < K_2O$	Arkoses; Lithic graywackes
	Alkali-metal-poor (aluminous clays)		Lithic arenites

FIGURE 2-13. A chemical classification of sandstones.

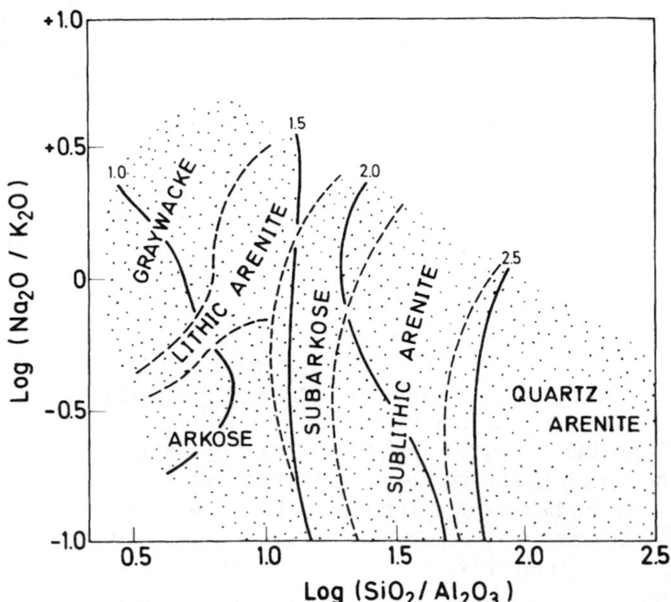

FIGURE 2-14. SiO_2/Al_2O_3 vs. Na_2O/K_2O ratios of sandstones. Stipple outlines area of most analyses. Contours (heavy lines) show values of log $\left(\frac{SiO_2 + Al_2O_3}{Na_2O + K_2O}\right)$. Analyses drawn largely from Pettijohn (1963); modified from original plot by J.S. Hanor.

contents to determine provenance of an Eocene graywacke of Oregon as a mixture of a mafic source—probably subjacent basalts—and a felsic source—Mesozoic graywackes. Radioactive dating of sandstone detritals is discussed in the introduction to this chapter.

Chemical Classification of Sandstones

Probably the simplest way to categorize sandstones by chemical composition rather than by mineralogical–textural composition is one that takes into account the general geochemical behavior of the elements by following the Periodic Table of the elements in association with their crystal chemistry, as shown in Fig. 2-13. This classification is not designed for and is not particularly useful for naming purposes but simply shows some of the relationships between elemental composition, mineralogy, and rock type.

The classification differentiates mature and immature as those terms are used in both mineralogical and textural classifications (see Chaps. 3 and 5). It does so by using the ratio of SiO_2 to Al_2O_3 rather than by using either alone. The mature quartz-rich sandstones have high SiO_2/Al_2O_3 ratios by virtue of the absence of aluminosilicates, either clays (a textural attribute) or primary silicates (a mineralogical attribute). This kind of division should not be confused with weathering maturity in the geochemical sense, in which low SiO_2/Al_2O_3 ratios are characteristic of the most maturely weathered soils in which SiO_2 is leached before Al_2O_3.

The differences between alkali metal-rich and alkali metal-poor sandstones can almost equally well be taken as corresponding to a maturity index defined in chemical compositional terms (Fig. 2-14), where the contours are drawn for the ratio of $SiO_2 + Al_2O_3$ to $Na_2O + K_2O$ and the individual analyses are plotted by the ratio of Na_2O to K_2O and the ratio of SiO_2 to Al_2O_3. Either way of looking at the chemical analysis is bound to reveal the same maturation concept that is expressed by modal analyses; the longer sand has been in the sedimentary "mill," the more it tends to be enriched in quartz at the expense of everything else.

The differences between immature sandstones, all of which have low SiO_2/Al_2O_3 ratios, can be expressed as further subdivisions of that ratio into "low" and "very low" or, as is done here, into alkali metal-rich and alkali metal-poor. Though not a neat division, it does separate most lithic sandstones from other immature sandstones, primarily reflecting the fact that many of the lithic sandstones tend to have more aluminous, and therefore less alkali metal-rich, clays in their rock fragments and matrices. The exceptions are those lithic sandstones close in nature to lithic graywackes, in which rock fragments may be from volcanic or other alkali metal-rich rocks.

Alkali metal-rich immature sandstones are here divided into two classes on the basis of a long recognized difference in Na_2O/K_2O ratios. The high Na_2O/K_2O ratios of graywackes are attributable to a dominance of albitic plagioclase feldspars over K-feldspars and K-micas. To the extent that some, primarily the younger, graywackes contain abundant smectite, the ratio is further increased. The frequency with which albitic rims, either diagenetic or low-grade metamorphic, have been observed in many graywackes is petrographic evidence of a secondary source of Na that is not related to provenance. Thus the ratio of feldspars is not purely nor may not even be largely due to distinctive source areas with more plagioclase, but in part comes from the typical secondary growth of Na-feldspars in strongly deformed and metamorphic belts of which graywackes are so typical.

The division of the mature sandstones takes recognition of the fact that the chemical analysis is strongly affected by the type of cementing agent. All gradations from pure silica to pure carbonate cement are known, sometimes in a single stratigraphic unit. One might erect another pigeonhole for analyses that reflect other cements, such as Ba-rich for barite-cemented sandstones and P-rich for phosphate-cemented sandstones. Regardless of the scheme, it will reflect the diagenetic additions to the mature sandstones.

Thus, in contrast to the igneous and metamorphic rocks, where bulk chemical analyses are all-important in any classification scheme and can be more or less neatly harmonized with modal mineralogical analyses, chemical analyses of sandstones do not fit so neatly with mineralogical analyses. The bulk chemistry of sandstones reflects the sometimes divergent and sometimes reinforcing effects of mineralogical versus textural sorting and differentiation, and the widely differing effects of source, environment, and diagenesis in shaping the rock's present composition.

Chemical Composition as a Function of Sandstone Type

The range of chemical composition of the common sandstone types and the average sandstone was summarized by Pettijohn (1963, Tables 12 and 13), from which Table 2-6 has been taken. The quartz arenites are nearly pure SiO_2, the Al_2O_3 coming from clay and the CaO from calcite cement. The lithic arenites show much more Al_2O_3, Fe-oxides, CaO, and K_2O, most deriving from argillaceous rock fragments. Graywackes have lower SiO_2 than most sandstones, more Al_2O_3, and dominance of Na_2O over K_2O and MgO over CaO. Arkoses have high Al_2O_3, K_2O, and Na_2O contents, predictable from the high proportion of feldspars.

Chemical Composition and Tectonic Setting

The foregoing discussion of chemical composition of sandstones is genetic only in the sense that it centers around the multiplicity of chemical and mineralogical processes affecting sandstones of all types. Another way of looking at chemical composition is to relate it to the larger geologic patterns of tectonics and sedimentation, to organize analyses to reflect genetic groupings more strongly controlled by those geologic processes and history that impose recurring distinct associations. The organizing principle of the attempts made thus far is the control of provenance—hence detrital chemical composition—by plate tectonics (Crook, 1974; Potter, 1978; Maynard *et al.*, 1982; Bhatia, 1983). A full discussion of plate tectonics and sandstone origin is given in Chap. 12; here we cover only some aspects of chemical composition.

The point of departure is the control of source-area rock types and their relative abundance by the dominance of one or another kind of igneous composition. Thus, for example, young oceanic island arcs are dominated by basalt and associated mafic volcaniclastics, giving rise to detrital materials rich in Fe, Mg, and Ca and poor in Si, Al, and K. Potter (1978) related high SiO_2/Al_2O_3 and high K_2O/Na_2O ratios to Atlantic-type passive margin big-river sands and distinguished them from collision-type coasts that have lower such ratios. Maynard *et al.* (1982) found that deep-sea sands on passive margins are distinct, high in SiO_2 and K_2O/Na_2O, and contrast with the extremely low K_2O/Na_2O ratios and very high Fe_2O_3 + Mg abundances of fore-arc basins of island arcs. On the other hand, deep-sea sands of strike-slip, continental margin-arc, and intra-oceanic back-arc basins showed intermediate and similar compositions. Bhatia and Taylor (1981) found arc-derived graywackes are lower in Hf, La, Th, and U than graywackes from rifted conti-

TABLE 2-6. Mean composition of principal sandstone classes and average sandstone (Pettijohn, 1963, p. 15).

	Quartz arenite	Lithic arenite	Gray-Wacke	Arkose	Average sandstone		
					A	B	C
SiO_2	95.4	66.1	66.7	77.1	78.66	84.86	77.6
Al_2O_3	1.1	8.1	13.5	8.7	4.78	5.96	7.1
Fe_2O_3	0.4	3.8	1.6	1.5	1.08	1.39	1.7
FeO	0.2	1.4	3.5	0.7	0.30	0.84	1.5
MgO	0.1	2.4	2.1	0.5	1.17	0.52	1.2
CaO	1.6	6.2	2.5	2.7	5.52	1.05	3.1
Na_2O	0.1	0.9	2.9	1.5	0.45	0.76	1.2
K_2O	0.2	1.3	2.0	2.8	1.32	1.16	1.3
H_2O+	0.3	3.6	2.4	0.9	1.33	1.47	1.7
H_2O-		0.7	0.6		0.31	0.27	0.4
TiO_2	0.2	0.3	0.6	0.3	0.25	0.41	0.4
P_2O_5		0.1	0.2	0.1	0.08	0.06	0.1
MnO		0.1	0.1	0.2	trace	trace	0.1
CO_2	1.1	5.0	1.2	3.0	5.04	1.01	2.5
SO_3			0.3		0.07	0.09	0.1
Cl					trace	trace	trace
F							trace
S			0.1				trace
BaO					0.05	0.01	trace
SrO					trace	none	trace
C			0.1				trace
Ignition Loss							
Total	100.7	100.0	100.4	100.0	100.41	99.86	100.0

A. Composite analysis of 253 sandstones.
B. Composite analysis of 371 sandstones used for building purposes.
C. Computed by taking 26 parts average graywacke, 25 parts average lithic arenite, 15 parts average arkose, and 34 parts average quartz arenite.

nental margins and marginal basins. Bhatia (1983) in a study of Paleozoic sandstone suites of eastern Australia compared four plate tectonic settings: oceanic island arc, continental island arc, active continental margins, and passive margins. From his own and published analyses he computed average chemical compositions for sandstones of each of these settings (Table 2-7). The most discriminating parameters proved to be: total Fe (as Fe_2O_3) + MgO; TiO_2; Al_2O_3/SiO_2; K_2O/Na_2O; and $Al_2O_3/CaO + Na_2O$.

TABLE 2-7. Average chemical composition of sandstones of various tectonic settings (after Bhatia, 1983, Table 10).

	Oceanic island arc	Continental island arc	Active continental margin	Passive margin
SiO_2	58.83	70.69	73.86	81.95
TiO_2	1.06	0.64	0.46	0.49
Al_2O_3	17.11	14.04	12.89	8.41
Fe_2O_3	1.95	1.43	1.30	1.32
FeO	5.52	3.05	1.58	1.76
MnO	0.15	0.10	0.10	0.05
MgO	3.65	1.97	1.23	1.39
CaO	5.83	2.68	2.48	1.89
Na_2O	4.10	3.12	2.77	1.07
K_2O	1.60	1.89	2.90	1.71
P_2O_5	0.26	0.16	0.09	0.12
Fe_2O_3* + MgO	11.73	6.79	4.63	2.89
Al_2O_3/SiO_2	0.29	0.20	0.18	0.10
K_2O/Na_2O	0.39	0.61	0.99	1.60
$Al_2O_3/(CaO + Na_2O)$	1.72	2.42	2.56	4.15

All averages on volatile-free basis.
* Total iron as Fe_2O_3.

Though one would not want to use these more or less strong correlations of chemical composition versus plate tectonic setting as the major basis for interpretation of sandstone origin, the relationships give insight into the ways tectonics and geochemical processes interact in sandstone formation. This is important in studies of metamorphosed sandstones, in which ordinary methods of sedimentary petrology are limited. More important, these correlations are central to ideas of continental evolution and changes in sandstone compositions through geologic time (see Chap. 12).

References

Adams, J.: Wear of unsound pebbles in river headwaters. Science 203, 171–172 (1979).

Adler, H.H. (Ed.): Formation of uranium ore deposits. Proc. Symposium Intern. Atomic Energy Agency, Vienna, 747 pp. (1974).

Allen, J.R.L.: Petrology, origin and deposition of the highest Lower Old Red sandstone of Shropshire, England. Jour. Sed. Petrology 32, 657–697 (1962).

Allen, P.: Lower Cretaceous sourcelands and the North Atlantic. Nature 222, 657–658 (1969).

Allen, P.: Wealden detrital tourmaline: Implications for northwestern Europe. Geol. Soc. London Jour. 128, 273–294 (1972).

Allen, P.; Dodson, M.H.; and Rex, D.C.: Potassium-argon dates and the origin of Wealden glauconites. Nature 202, 585–586 (1964).

Allen, P.; Sutton, J.; and Watson, J.V.: Torridonian tourmaline-quartz pebbles and the Precambrian crust northwest of Britain. Geol. Soc. London Jour. 130, 85–91 (1974).

Allen, R.C.; Gavish, E.; Friedman, G.M.; and Sanders, J.E.: Aragonite-cemented sandstone from outer continental shelf off Delaware Bay: submarine lithification mechanism yields product resembling beachrock. Jour. Sed. Petrology 39, 136–149 (1969).

Andel, Tj.H. van: Sediments of the Rhone delta, pt. II, Sources and deposition of heavy minerals. Koninkl. Nederlandsch. Geol. Mijnb. Genoot. Geol. Ser. 15, 357–556 (1955).

Aronson, J.L., and Tilton, G.R.: Probable Precambrian detrital zircons in New Caledonia and Southwest Pacific continental structure. Geol. Soc. America Bull. 82, 3449–3456 (1971).

Arrhenius, G., and Bonatti, E.: Neptunism and volcanism in the ocean. In: Sears, M. (Ed.): Progress in oceanography, Vol. 3, pp. 7–22. London: Pergamon 1965.

Baker, G.: Opal phytoliths in some Victorian soils and "red rain" residues. Australian Jour. Botany 7, 64–87 (1959).

Baskin, Yehuda: A study of authigenic feldspars. Jour. Geology 64, 132–155 (1956).

Basu, A.: Petrology of Holocene fluvial sand derived from plutonic source rocks: Implications to paleoclimatic interpretation. Jour. Sed. Petrology 46, 694–709 (1976).

Basu, A.; Blanchard, D.P.; and Brannon, J.C.: Rare earth elements in the sedimentary cycle: A pilot study of the first leg. Sedimentology 29, 737–742 (1982).

Basu, A.; Young, S.W.; Suttner, L.J.; James, W.C.; and Mack, G.H.: Re-evaluation of the use of undulatory extinction and polycrystallinity in detrital quartz for provenance interpretation. Jour. Sed. Petrology 45, 873–882 (1975).

Bates, T.F., and Comer, J.J.: Electron microscopy of clay surfaces. Clays and clay minerals: 3rd Nat'l. Conf. on Clays and Clay Minerals. Proc. U.S. Natl. Acad. Sci. Pub. 395, 1–25 (1955).

Bell, D.L., and Goodell, H.G.: A comparative study of glauconite and the associated clay fraction in modern marine sediments. Sedimentology 9, 169–202 (1967).

Berner, R.A.: Iron sulfides formed from aqueous solution at low temperatures and atmospheric pressure. Jour. Geology 72, 293–306 (1964).

Berner, R.A.: Early diagenesis, 241 pp. Princeton, New Jersey: Princeton Univ. Press 1980.

Berner, R.A., and Holdren, G.R., Jr.: Mechanism of feldspar weathering, pt. II, Observation of feldspars from soils. Geochim. et Cosmochim. Acta 43, 1173–1186 (1979).

Beutelspacher, H., and van der Marel, H.W.: Atlas of electron microscopy of clay minerals and their admixtures, 333 pp. Amsterdam: Elsevier 1968.

Bhatia, M.R.: Plate tectonics and geochemical composition of sandstones. Jour. Geology 91, 611–627 (1983).

Bhatia, M.R., and Taylor, S.R.: Trace-element geochemistry and sedimentary provinces: A study from the Tasman Geosyncline, Australia. Chem. Geology 33, 115–125(1981).

Birks, L.S.: Electron probe microanalysis, 253 pp. New York: Interscience 1963.

Biscaye, Pierre: Mineralogy and sedimentation of recent deep-sea clay in the Atlantic Ocean and adjacent seas and ocean. Geol. Soc. America Bull. 76, 803–832 (1965).

Blatt, H.: Original characters of clastic quartz. Jour. Sed. Petrology 37, 401–424 (1967).

Blatt, H., and Christie, J.M.: Undulatory extinction in quartz of igneous and metamorphic rocks and its significance in provenance studies of sedimentary rocks. Jour. Sed. Petrology 33, 559–579 (1963).

Blatt, H.; Middleton, G.; and Murray, R.: Origin of sedimentary rocks, 2nd. Ed., 782 pp. Englewood Cliffs, New Jersey: Prentice-Hall 1980.

Boles, J.R.: Active albitization of plagioclase, Gulf Coast Tertiary. Am. Jour. Sci. 282, 165–180 (1982).

Borchert, H.: Ozeane Salzlagerstätten, 237 pp. Berlin: Gebr. Bornträger 1959.

Bostick, N.: Microscopic measurement of the level of catagenesis of solid organic matter in sedimentary rocks to aid exploration for petroleum and to determine former burial temperatures—a review.

In: Scholle, P.A., and Schluger, P.R. (Eds.): Aspects of diagenesis. Soc. Econ. Paleon. Mineral. Spec. Pub. 26, 17–44 (1979).

Braitsch, O.: Salt deposits, their origin and composition, 297 pp. Berlin: Springer-Verlag 1971.

Breger, I.A. (Ed.): Organic geochemistry, 658 pp. New York: Macmillan 1963.

Briggs, L.I.; McCulloch, D.S.; and Moser, Frank: The hydraulic shape of sand particles. Jour. Sed. Petrology 32, 645–656 (1962).

Brindley, G.W., and Brown, G. (Eds.): Crystal structures of clay minerals and their X-ray identification, 490 pp. London: Mineralog. Soc. Great Britain 1980.

Brookins, D.G., and Voss, J.D.: Age dating of muscovites from Pennsylvanian sandstones near Wamego, Kansas. Am. Assoc. Petroleum Geologists Bull. 54, 353–356 (1970).

Bull, P.A.; Culver, S.J.; and Gardner, R.: Chattermark trails as paleoenvironmental indicators. Geology 8, 318–322 (1980).

Burst, J.F.: "Glauconite" pellets: Their mineral nature and application to stratigraphic problems. Am. Assoc. Petroleum Geologists Bull. 42, 310–327 (1958).

Busenberg, E., and Clemency, C.V.: The dissolution kinetics of feldspars at 25°C and 1 atm. CO_2 partial pressure. Geochim. et Cosmochim. Acta 40, 41–50 (1976).

Callender, D.L., and Folk, R.L.: Idiomorphic zircon, key to volcanism in the lower Tertiary sands of central Texas. Am. Jour. Sci. 256, 257–269 (1958).

Calvert, S.E.: Sedimentary geochemistry of silicon. In: Aston, S.R. (Ed.): Silicon geochemistry and biogeochemistry, pp. 143–186. New York: Academic Press 1983.

Carroll, D.: Clay minerals: A guide to their X-ray identification, Geol. Soc. America Spec. Pub. 126, 80 pp. (1970).

Charles, R.G., and Blatt, H.: Quartz, chert, and feldspars in modern fluvial muds and sands. Jour. Sed. Petrology 48, 427–432 (1978).

Clarke, F.W.: The data of geochemistry, 5th Ed. U.S. Geol. Survey Bull. 770, 841 pp. (1924).

Clayton, R.N.; Jackson, M.L.; and Sridhar, K.: Resistance of quartz silt to isotopic exchange under burial and intense weathering conditions. Geochim. et Cosmochim. Acta 42, 1517–1522 (1978).

Clocchiatti, R.: Les inclusions vitreuses des cristaux de quartz. Soc. Géol. France Mem. 122, 95 pp. (1975).

Cloud, P.E.: Physical limits of glauconite formation. Am. Assoc. Petroleum Geologists Bull. 39, 484–492 (1955).

Columbo, Umberto, and Hobson, G.C. (Eds.): Advances in organic geochemistry, 488 pp. New York: Macmillan 1964.

Cook, P.J.: Sedimentary phosphate deposits. In: Wolf, K.H. (Ed.): Handbook of strata-bound and stratiform ore deposits, Vol. 7, pp. 505–535. Amsterdam: Elsevier 1976.

Coombs, D.S.; Ellis, A.J.; Fyfe, W.S.; and Gaylor, A.M.: The zeolite facies, with comments on the interpretation of hydrothermal syntheses. Geochim. et Cosmochim. Acta 17, 53–107 (1959).

Correns, C.W., and Piller, H.: Mikroskopie der feinkörnigen Silikatminerale. In: Mikroskopie der Silikate, pt. I, Handbuch der Mikroskopie in der Technik, Vol. IV, pp. 699–780. Mikroskopie der Gesteine, 796 pp. Frankfurt: Umschau Verlag 1955.

Crook, K.A.W.: Lithogenesis and geotectonics: the significance of compositional variations in flysch arenites (graywackes). In: Dott, R.H., and Shaver, R.H. (Eds.): Modern and ancient geosynclinal sedimentation. Soc. Econ. Paleon. Mineral. Spec. Pub. 19, 304–310 (1974).

Curtis, C.D.: Stability of minerals in surface weathering reactions: a general thermochemical approach. Earth Surface Processes 1, 63–70 (1976).

Davies, D.K., and Ethridge, F.G.: Sandstone composition and depositional environment. Am. Assoc. Petroleum Geologists Bull. 59, 239–264 (1975).

Deer, W.A.; Howie, R.A.; and Zussman, J.: Rock-forming minerals, Vol. 1, Ortho- and ring silicates, 333 pp. New York: Wiley 1962.

Deer, W.A.; Howie, R.A.; and Zussman, J.: Rock-forming minerals, Vol. 4, Framework silicates, 435 pp. New York: Wiley 1963.

Degens, Egon: Geochemistry of sediments: a brief survey, 342 pp. Englewood Cliffs, New Jersey: Prentice-Hall 1965.

Degens, E.T.; Williams, E.G.; and Keith M.L.: Environmental studies of Carboniferous sediments, pt. I, Geochemical criteria for differentiating marine and freshwater shales. Am. Assoc. Petroleum Geologists Bull. 41, 2427–2455 (1957).

Degens, E.T.; Williams, E.G.; and Keith, M.L.: Environmental studies of Carboniferous sediments, pt. II, Application of geochemical criteria. Am. Assoc. Petroleum Geologists Bull. 42, 981–997 (1958).

Dennen, W.H.: Impurities in quartz. Geol. Soc. America Bull. 75, 241–246 (1964).

DePaolo, D.J., and Wasserburg, G.J.: Nd isotopic variations and petrogenetic models. Geophys. Research Letters 3, 249–252 (1976).

Dietz, V.: Experiments on the influence of transport on shape and roundness of heavy minerals. In: Stability of heavy minerals, 44 pp. Stuttgart: Schweizerbart'sche Verlagsbuchhandlung 1973.

Dodson, M.H.; Rex, D.C.; Casey, R.; and Allen, P.: Glauconite dates from the Upper Jurassic and Lower Cretaceous. Geol. Soc. London Jour. 1205, 145–158 (1964).

Doornkamp, J.C.: Tropical weathering and the ultramicroscopic characteristics of regolith quartz on Dartmoor. Geog. Annaler 56, 73–82 (1974).

Doyle, L.J.; Carder, K.L.; and Steward, R.G.: The hydraulic equivalence of mica. Jour. Sed. Petrology 53, 643–648 (1983).

Doyle, L.J.; Cleary, W.J.; and Pilkey, O.H.: Mica:

Its use in determining shelf-depositional regimes. Marine Geol. 6, 381–389 (1968).

Eglinton, G., and Murphy, M.T.J. (Eds.): Organic geochemistry, 720 pp. Berlin–Heidelberg–New York: Springer-Verlag 1969.

Ethridge, F.G.: Petrology, transport, and environment in isochronous upper Devonian sandstone and siltstone units, New York. Jour. Sed. Petrology 47, 53–65 (1977).

Evernden, J.F.; Curtis, G.H.; Obradovich, J.; and Kistler, R.W.: On the evaluation of glauconite and illite for dating sedimentary rocks. Geochim. et Cosmochim. Acta 23, 78–99 (1961).

Farmer, V.C., and Russell, J.D.: The infra-red spectra of layer silicates. Spectrochim. Acta 20, 1149–1173 (1964).

Faure, G.: Principles of isotope geology, 464 pp. New York: Wiley 1977.

Feth, J.H.; Roberson, C.E.; and Polzer, W.L.: Sources of mineral constituents in water from granitic rocks, Sierra Nevada, California and Nevada, 70 pp. U.S. Geol. Survey Water-Supply Paper 1535-I (1964).

Field, M.E., and Pilkey, O.H.: Feldspar in Atlantic continental margin sands off the southeastern United States. Geol. Soc. America Bull. 80, 2097–2102 (1969).

Folk, R.L., and Weaver, C.E.: A study of the texture and composition of chert. Am. Jour. Sci. 250, 498–510 (1952).

Force, E.R.: The provenance of rutile. Jour. Sed. Petrology 50, 485–488 (1980).

Franzinelli, E., and Potter, P.E.: Petrology, chemistry and texture of modern river sands, Amazon River system. Jour. Geology 91, 23–39 (1983).

Friedman, G.M.: Identification of carbonate minerals by staining methods. Jour. Sed. Petrology 29, 87–97 (1959).

Frondel, C.: Dana's system of mineralogy, 7th Ed., 334 pp. New York: Wiley 1962.

Frye, J.C., and Swineford, A.: Silicified rock in the Ogallala formation. Kansas Geol. Survey Bull. 64, pt. II, 33–76 (1946).

Füchtbauer, H.: Sedimentpetrographische Untersuchungen in der älteren Molasse nördlich der Alpen. Eclogae geol. Helvetiae 57, 157–298 (1964).

Füchtbauer, H.: Influence of different types of diagenesis on sandstone porosity. Proc. 7th World Petroleum Cong., 353–369 (1967).

Füchtbauer, H.: Sediments and sedimentary rocks I. Pt. II, 464 pp. New York: Wiley-Halsted Press 1974.

Galliher, E.W.: Biotite–glauconite transformation and associated minerals. In: Recent marine sediments, pp. 513–515. Tulsa: Am. Assoc. Petroleum Geologists 1939.

Garrels, R.M., Christ, C.L.: Solutions, minerals, and equilibria, 435 pp. New York: Harper and Row 1965.

Garrels, R.M., and Mackenzie, F.T.: Origin of the chemical compositions of some springs and lakes. In: Equilibrium concepts in natural water systems. American Chemical Society Advances in Chemistry Ser. 67, 222–242 (1967).

Garrels, R.M., and Mackenzie, F.T.: Evolution of Sedimentary rocks, 397 pp. New York: W.W. Norton and Co. 1971.

Garrels, R.M.; Mackenzie, F.T.; and Siever, R.: Sedimentary cycling in relation to the history of the continents and oceans. In: Robertson, E.D. (Ed.): The nature of the solid earth, pp. 93–121. New York: McGraw-Hill 1971.

Gebauer, D., and Grünenfelder, M.: U–Pb systematics of detrital zircons from some unmetamorphosed to slightly metamorphosed sediments of Central Europe. Contrib. Mineral. Petrol. 65, 29–37 (1977).

Giles, R.T., and Pilkey, O.H.: Atlantic beach and dune sediments of the southern United States. Jour. Sed. Petrology 35, 900–910 (1965).

Gilligan, A.: The petrography of the Millstone Grit of Yorkshire. Geol. Soc. London Quart. Jour. 75, 251–294 (1920).

Glass, H.D.; Potter, P.E.; and Siever, R.: Clay mineralogy of some basal Pennsylvanian sandstones, clays and shales. Am. Assoc. Petroleum Geologists Bull. 40, 750–754 (1957).

Goldberg, E.D., and Griffin, J.J.: Sedimentation rates and mineralogy in the South Atlantic. Jour. Geophys. Research 69, 4293–4309 (1964).

Goldich, S.S.: A study in rock weathering. Jour. Geology 46, 17–58 (1938).

Goldschmidt, V.M.: In: Muir, A. (Ed.): Geochemistry, 730 pp. Oxford: Oxford Univ. Press 1954.

Goldsmith, J.R., and Graf, D.L.: Subsolidus relations in the system $CaCO_3$–$MgCO_3$–$MnCO_3$. Jour. Geology 68, 324–335 (1960).

Goldsmith, J.R.; Graf, D.L.; Witters, Juanita; and Northrop, D.A.: Studies in the system $CaCO_3$–$MgCO_3$–$MnCO_3$–$FeCO_3$. Jour. Geology 70, 659–688 (1962).

Graf, D.L.: Carbonate mineralogy, carbonate sediments, pt. I. Geochemistry of carbonate sediments and sedimentary carbonate rocks. Illinois Geol. Survey Circ. 297, 39 pp. (1960).

Griffin, J.J.; Windom, H.; and Goldberg, E.D.: The distribution of clay minerals in the world ocean. Deep-Sea Res. 15, 433–459 (1968).

Grim, R.E.: Clay mineralogy, 2nd Ed., 596 pp. New York: McGraw-Hill 1968.

Grimm, W.D.: Stepwise heavy mineral weathering in the residual quartz gravel, Bavarian Molasse (Germany). In: Stability of heavy minerals, 125 pp. Stuttgart: Schweizerbart'sche Verlagsbuchhandlung 1973.

Hallam, A.: Depositional environment of British Liassic ironstones considered in the context of their facies relationships. Nature 209, 1306–1307 (1966).

Hallam, A.: Siderite- and calcite-bearing concretionary nodules in the Lias of Yorkshire. Geol. Mag. [Great Britain] 104, 222–227 (1967).

Hand, B.M.: Differentiation of beach and dune sands, using settling velocities of light and heavy minerals. Jour. Sed. Petrology 37, 514–520 (1967).

Hanor, J.S.: Regional control and zoning of barite in eastern North America. Econ. Geology 62, 870 (1967).

Harder, H., and Menschel, G.: Quarzbildungen am Meeresboden. Die Naturwissenschaften 54, 561 (1967).

Hardie, L.A.: The gypsum–anhydrite equilibrium at one atmosphere pressure. Am. Mineralogist 52, 171–200 (1967).

Hay, R.L.: Zeolites and zeolite reactions in sedimentary rocks. Geol. Soc. America Spec. Paper 85, 130 pp. (1966).

Hayes, J.R.: Quartz and feldspar content in South Platte, Platte and Missouri river sands. Jour. Sed. Petrology 32, 793–800 (1962).

Heim, D.: Über die Feldspäte im Germanischen Buntsandstein, ihre Korngrössenabhängigkeit, Verbreitung und paläogeographische Bedeutung. Geol. Rundschau 63, 943–970 (1974).

Hemley, J.J., and Jones, W.R.: Chemical aspects of hydrothermal alteration with emphasis on hydrogen metasomatism. Econ. Geology 59, 538–569 (1964).

Hirst, D.M., and Nicholls, G.D.: Techniques in sedimentary geochemistry, pt. I, Separation of the detrital and non-detrital fractions of limestones. Jour. Sed. Petrology 28, 468–481 (1958).

Holdren, G.R., Jr., and Berner, R.A.: Mechanism of feldspar weathering—I. Experimental studies. Geochim. et Cosmochim. Acta 43, 1161–1171 (1979).

Holland, H.D.: The chemical evolution of the atmosphere and oceans, 582 pp. New York: Wiley-Interscience 1984.

Horowitz, Alan, and Potter, P.E.: Introductory petrography of fossils, 325 pp. Berlin–Heidelberg–New York: Springer 1971.

Hsu, K.Jinghwa: Texture and mineralogy of the Recent sands of the Gulf Coast. Jour. Sed. Petrology 30, 380–403 (1960).

Hubert, J.F.: Petrology of the Fountain and Lyons Formations, Front Range, Colorado. Colorado School Mines Quart. 55, no. 1, 1–242 (1960).

Hubert, J.F., and Neal, W.F.: Mineral composition and dispersal patterns of deep-sea sands in the western North Atlantic petrologic province. Geol. Soc. America Bull. 78, 749–772 (1967).

Hunt, J.M.: Petroleum geochemistry and geology, 617 pp. San Francisco: W.H. Freeman 1979.

Hunter, R.E.: The petrography of some Illinois Pleistocene and Recent sands. Sedimentary Geology 1, 57–75 (1967).

Hurley, P.M.; Hunt, J.M.; Pinson, W.H., Jr.; and Fairbairn, H.W.: K–Ar age values on the clay fractions in dated shales. Geochim. et Cosmochim Acta 27, 279–284 (1963).

Iijima, A.: Geological occurrences of zeolite in marine environments. In: Sand, L.B., and Mumpton, F.J. (Eds.): Natural zeolites: Occurrence, properties, use, pp. 175–198. New York: Pergamon Press 1978.

Ingersoll, R.V.; Bullard, T.F.; Ford, R.L.; Grimm, J.P.; Pickle, J.D.; and Sares, S.W.: The effect of grain size on detrital modes: a test of the Gazzi–Dickinson point-counting method. Jour. Sed. Petrology 54, 103–116 (1984).

James, W.C.; Mack, G.H.; and Suttner, L.J.: Relative alteration of microcline and sodic plagioclase in semi-arid and humid climates. Jour. Sed. Petrology 51, 151–164 (1981) .

Johns, W.D.; Grim, R.E.; and Bradley, W.F.: Quantitative estimations of clay minerals. Jour. Sed. Petrology 24, 242–251 (1954).

Jones, J.B., and Segnit, E.R.: The nature of opal—I. Nomenclature and constituent phases. Geol. Soc. Australia Jour. 18, 56–68 (1971).

Judd, J.B.; Smith, W.C.; and Pilkey, O.H.: The environmental significance of iron-stained quartz grains on the southeastern United States Atlantic shelf. Marine Geol. 8, 355–362 (1970).

Kastner, M., and Siever, R.: Low temperature feldspars in sedimentary rocks. Am. Jour. Sci. 279, 435–479 (1979).

Kelling, G.; Sheng, H.; and Stanley, D.J.: Mineralogic composition of sand-sized sediment on the outer margin off the Mid-Atlantic States: assessment of the influence of the ancestral Hudson and other fluvial systems. Geol. Soc. America Bull. 86, 853–862 (1975).

Kinsman, D.J.J.: The Recent carbonate sediments near Halat el Bahrani Trucial Coast, Persian Gulf. In: Deltaic and shallow marine deposits. Developments in Sedimentology 1, 185–192 (1964).

Kinsman, D.J.J.: Modes of formation, sedimentary associations, and diagnostic features of shallow-water and supratidal evaporites. Am. Assoc. Petroleum Geologists Bull. 53, 830–840 (1969).

Kolodny, Y.: Phosphorites. In: Emiliani, C. (Ed.): The sea, Vol. 7: The oceanic lithosphere, pp. 981–1023. New York: John Wiley and Sons 1981.

Krauskopf, K.: Introduction to geochemistry, 2nd Ed., 617 pp. New York: McGraw-Hill 1979.

Krinsley, D.H., and McCoy, F.W.: Significance and origin of surface textures on broken sand grains in deep-sea sediments. Sedimentology 24, 857–862 (1977).

Krinsley, D.H., and Tovey, N.K.: Cathodoluminescence in quartz and grains. Scanning Electron Microscopy 1, 887–894 (1978).

Krynine, P.D.: Petrology and genesis of the Third Bradford Sand. Pennsylvania State College Bull. 29, 134 pp. (1940).

Krynine, P.D.: Petrographic studies of variations in cementing material in the Oriskany sand. Proc. 10th Pennsylvania Min. Ind. Conf., Pennsylvania State College Bull. 33, 108–116 (1941).

Krynine, P.D.: The tourmaline group in sediments. Jour. Geology 54, 65–87 (1946).

Krynine, P.D.: The megascopic study and field classification of sedimentary rocks. Jour. Geology 56, 130–165 (1948).

Land, L.S., and Dutton, S.P.: Cementation of a Pennsylvanian deltaic sandstone: isotope data. Jour. Sed. Petrology 48, 1167–1176 (1978).

Lee, H.L., and Peacor, D.R.: Intralayer transitions in phyllosilicates of Martinsburg shale. Nature 303, 608–609 (1983).

Leith, C.K., and Mead, W.J.: Metamorphic geology, 337 pp. New York: Henry Holt 1915.

Le Ribault, L.: L'exoscopie des quartz, 150 pp. Paris: Masson 1977.

Lerman, A.: Boron in clays and estimation of paleosalinities. Sedimentology 6, 267–286 (1966).

Lipson, J.: Potassium–argon dating of sedimentary rocks. Geol. Soc. America Bull. 69, 137–150 (1958).

Loughnan, F.C.: Chemical weathering of the silicate minerals, 154 pp. New York: Elsevier (1969).

Lovell, J.P.B.: Tyee formation: a study of proximality in turbidites. Jour. Sed. Petrology 39, 935–953 (1969).

Lucas, J.; Chaabani, F.; Prevot, L.; El Mountassir, M.; Menor, E.; Staman, G.; Gündogdu, N.; and Doubinger, J.: Studies on phosphorite deposits. Sci. Géol. Bull. (Strasbourg) 32, 105 pp. (1979).

Lugmair, G.W.: Sm–Nd ages: A new dating method (abs.). Meteoritics 9, pp. 369 (1974).

Lugmair, G.W., and Scheinin, N.B.: Sm–Nd systematics of the Stannern meteorite. Meteoritics 10, 447–448 (1975).

Mack, G.H.: The survivability of labile light mineral grains in fluvial, aeolien, and littoral environments; the Permian Cutler-Cedar Mesa formations Moab Utah. Sedimentology 25, 587–603 (1978).

Mackenzie, F.T., and Garrels, R.M.: Silicates: reactivity with sea water. Science 150, 57–58 (1965).

Mackenzie, F.T., and Gees, R.: Quartz synthesis at Earth surface conditions. Science 173, 533–534 (1971).

Mackie, W.: The sands and sandstones of Eastern Moray. Edinburgh Geol. Soc. Trans. 7, 148–172 (1896).

Majewske, Otto P.: Recognition of invertebrate fossil fragments in rocks and thin sections, 101 pp. Leiden: Brill 1971.

Manskaya, S.M.; and Drozdova, T.V.: Shapiro, L., and Breger, I.A. (Trans. and Eds.): Geochemistry of organic substances, 345 pp. London: Pergamon 1968.

Martens, J.H.C.: Beach sands of Quebec and Labrador. Field Museum Nat. History Pub. 260 (Geol. Series) 1929.

Matsumoto, R., and Iijima, A.: Origin and diagenetic evolution of Ca–Mg–Fe carbonates in some coalfields of Japan. Sedimentology 28, 239–259 (1981).

Maynard, J.B.: Geochemistry of sedimentary ore deposits, 305 pp. Berlin–Heidelberg–New York: Springer-Verlag 1983.

Maynard, J.B.; Valloni, R.; and Yu, H.S.: Composition of modern deep-sea sands from arc-related basins. In: Leggett, J. (Ed.): Trench and fore-arc sedimentation. Geol. Soc. London, 551–561 (1982).

McBride, E.F.: Significance of color in red, green, purple, olive, brown, and gray beds of Difunta Group, northeastern Mexico. Jour. Sed. Petrology 44, 760–773 (1974).

McCullough, M.T., and Wasserburg, G.J.: Sm–Nd and Rb–Sr chronology of continental crust formation. Science 200, 1003–1011 (1978).

McLennon, S.M.; Nance, W.B.; and Taylor, S.R.: Rare earth element–thorium correlations in sedimentary rocks and the composition of the continental crust. Geochim. et Cosmochim. Acta 44, 1833–1839 (1980).

McMaster, R.L., and Garrison, L.: Mineralogy and origin of southern New England shelf sediments. Jour. Sed. Petrology 36, 1131–1142 (1966).

Midgley, H.G.: Chalcedony and flint. Geol. Mag. [Great Britain] 88, 179–184 (1951).

Millot, G.: Geologie des argiles, 499 pp. Paris: Masson 1964.

Milner, H.B.: Sedimentary petrography, Vol. II, Principles and applications, 715 pp. New York: Macmillan 1962.

Mitchell, W.A.: Heavy minerals. In: Gieseking, J.E. (Ed.): Soil components, Vol. 2: Inorganic components, pp. 450–469. New York: Springer-Verlag 1975.

Moss, A.J.: Initial fluviatile fragmentation of granitic quartz. Jour. Sed. Petrology 42, 905–916 (1972).

Munson, R.A., and Sheppard, R.A.: National zeolites: Their properties, occurrences, and uses. Minerals Sci. Eng. 6, 19–30 (1974).

Nanz, R.H., Jr.: Genesis of Oligocene sandstone reservoir, Seeligson field, Jim Wells and Kleberg Counties, Texas. Am. Assoc. Petroleum Geologists Bull. 38, 96–117 (1954).

Nickel, E.: Experimental dissolution of light and heavy minerals in comparison with weathering and intrastratal solution. In: Stability of heavy minerals, 68 pp. Stuttgart: Schweizerbart'sche Verlagsbuchhandlung 1973.

Nieter, W.M., and Krinsley, D.H.: The production and recognition of aeolian features on sand grains by silt abrasion. Sedimentology 23, 713–720 (1976).

Odom, I.E.; Doe, T.W.; and Dott, R.H., Jr.: Nature of feldspar–grain size relations in some quartz-rich sandstones. Jour. Sed. Petrology 46, 862–870 (1976).

Okada, H.: Non-graywacke "turbidite" sandstones in the Welsh geosyncline. Sedimentology 7, 211–232 (1966).

Ollier, C.: Weathering, 304 p. New York: Elsevier 1969.

O'Nions, R.K.; Evensen, N.M.; and Hamilton, P.J.: Geochemical modeling of mantle differentiation and crustal growth. Jour. Geophys. Research 84, 6091–6101 (1979).

Orville, P.: Alkali ion exchange between vapor and feldspar phases. Am. Jour. Sci. 261, 201–237 (1963).

Packham, G.H., and Crook, K.A.W.: The principle of diagenetic facies and some of its implications. Jour. Geology 68, 392–407 (1960).

Parfenoff, A.; Pomerol, C.; and Tourenq, J.: Les minéraux en grains, 571 pp. Paris: Masson 1970.

Peterman, Z.E., Coleman, R.G., and Bunker, C.M.: Provenance of Eocene graywackes of the Fluornoy Formation near Agness, Oregon—a geochemical approach. Geology 9, 81–86, (1981).

Petrović, R.: Rate control in feldspar dissolution—II. The protective effect of precipitates. Geochim. et Cosmochim. Acta 40, 1509–1521 (1976).

Petrović, R.; Berner, R.A.; and Goldhaber, M.B.: Rate control in dissolution of alkali feldspars—I. Study of residual feldspar grains by X-ray photoelectron spectroscopy. Geochim. et Cosmochim. Acta 40, 537–548 (1976).

Pettijohn, F.J.: Chemical composition of sandstones—excluding carbonate and volcanic sands. U.S. Geol. Survey Prof. Paper 400-S, 19 pp. (1963).

Pettijohn, F.J. Sedimentary rocks, 3rd Ed., 628 pp. New York: Harper and Row 1975.

Pettijohn, F.J., and Lundahl, A.C.: Shape and roundness of Lake Erie beach sands. Jour. Sed. Petrology 13, 69–78 (1943).

Plas, L. van der: The identification of detrital feldspars, 305 pp. Amsterdam: Elsevier 1966.

Poldervaart, A.: Zircons in rocks, pt. I, Sedimentary rocks. Am. Jour. Sci. 253, 433–461 (1955).

Polevaya, N.I.; Murina, G.A.; and Kozakov, G.A.: Utilization of glauconite in absolute dating. N.Y. Acad. Sci. Ann. 91, 298–310 (1961).

Porrenga, D.H.: Glauconite and chamosite as depth indicators in the marine environment. Marine Geol. 5, 495–501 (1967).

Potter, P.E.: Petrology and chemistry of modern big river sands. Jour. Geology 86, 423–449 (1978).

Potter, P.E.; Maynard, J.B.; and Pryor, W.A.: Sedimentology of shale, 306 pp. Berlin–Heidelberg–New York: Springer-Verlag 1980.

Potter, P.E., and Pryor, W.A.: Dispersal centers of Paleozoic and later clastics of the upper Mississippi valley and adjacent areas. Geol. Soc. America Bull. 72, 1195–1250 (1961).

Powers, L.S.; Brueckner, H.K.; and Krinsley, D.H.: Rb–Sr provenance ages from weathered and stream transported quartz grains from the Harney Peak granite, Black Hills, South Dakota. Geochim. et Cosmochim. Acta 43, 137–146 (1979).

Pryor, W.A.: Biogenic sedimentation and alteration of argillaceous sediments in shallow-marine environments. Geol. Soc. America Bull. 86, 1244–1254 (1975).

Pryor, W.A., and Hester, N.C.: X-ray diffraction analysis of heavy minerals. Jour. Sed. Petrology 39, 1384–1389 (1969).

Pye, K., and Krinsley, D.H.: Interlayered clay stacks in Jurassic shales. Nature 304, 618–620 (1983).

Rehmer, J.A., and Hepburn, J.C.: Quartz sand surface textural evidence for a glacial origin of the Squantum "Tillite," Boston Basin, Massachusetts. Geology 2, 413–415 (1974).

Rimsaite, J.: Optical heterogeneity of feldspars observed in diverse Canadian rocks. Schweiz. Min. Petrogr. Mitt. 47, 61–76 (1967).

Rittenhouse, G.A.: Transportation and deposition of heavy minerals. Geol. Soc. America Bull. 54, 1725–1780 (1943).

Robinson, A., and Spooner, E.T.C.: Source of the detrital components of uraniferous conglomerates, Quirke Ore Zone, Elliot Lake, Ontario, Canada. Nature 299, 622–624 (1982).

Ronov, A.B.; Mikhailovskaya, M.S.; and Solodkova, I.I.: Evolution of the chemical and mineralogical composition of arenaceous rocks. In: Chemistry of the earth's crust, Vol. 1. U.S.S.R. Acad. Sci., Israel Progr. Sci., Translations, 1966, 212-262 (1963).

Russell, R.D.: Mineral composition of Mississippi River sands. Geol. Soc. America Bull. 48, 1307–1348 (1937).

Rutten, M.G.: The geological aspects of the origin of life on earth, 146 pp. Amsterdam: Elsevier 1962.

Sabins, F.F., Jr.: Grains of detrital, secondary, and primary dolomite from Cretaceous strata of the western interior. Geol. Soc. America Bull. 73, 1183–1196 (1962).

Sand, L.B., and Mumpton, F.J. (Eds.): Natural zeolites: occurrence, properties, use, 576 pp. New York: Pergamon Press 1978.

Savin, S.M., and Epstein, Samuel: The oxygen isotopic compositions of coarse grained sedimentary rocks and minerals. Geochim. et Cosmochim. Acta 34, 323–329 (1970).

Scholle, P.A.: A color illustrated guide to carbonate rock constituents, textures, cements, and porosities. Am. Assoc. Petroleum Geologists Mem. 27, 241 pp. (1978).

Seed, D.P.: The formation of vermicular pellets in New Zealand glauconites. Am. Mineralogist 50, 1097–1106 (1965).

Shearman, D.J.: Evaporites of coastal sabkhas. In: Dean, W.E., and Schreiber, B.C. (Eds.): Marine evaporites. S.E.P.M. Short Course 4, 6–42 (1978).

Shiki, T.: Studies on sandstone in the Maizuru Zone, Southwest Japan, pt. I, Importance of some relations between mineral composition and grain size. Mem. Coll. Sci., Univ. Kyoto, Ser. B, 29, 291–324 (1959).

Shimp, N.F.; Witter, J.; Potter, P.E.; and Schleicher, J.A.: Distinguishing marine and freshwater muds. Jour. Geology 77, 566–580 (1969).

Sibley, D.F., and Blatt, H.: Intergranular pressure solution and cementation of the Tuscarora orthoquartzite. Jour. Sed. Petrology 46, 881–896 (1976).

Siever, R.: The silica budget in the sedimentary cycle. Am. Mineralogist 42, 821–841 (1957).

Siever, R.: Establishment of equilibrium between clays and sea water. Earth Planet. Sci. Letters 5, 106–110 (1968).

Siever, R., and Kastner, Miriam: Mineralogy and

petrology of some Mid-Atlantic Ridge sediments. Jour. Marine Research 25, 263–278 (1967).

Siever, R., and Woodford, N.: Sorption of silica by clay minerals. Geochim. et Cosmochim. Acta 37, 1851–1880 (1973).

Siffert, Bernard: Quelques réactions de la silice en solution: la formation des argiles. Service Carte Géol. Alsace Lorraine Mémoires No. 21, 86 pp. (1962).

Sippel, R.F.: Sandstone petrology, evidence from luminescence petrography. Jour. Sed. Petrology 38, 530–554 (1968).

Smalley, I.J.; Krinsley, D.H.; Moon, C.F.; and Bentley, S.P.: Processes of quartz fracture in nature and the formation of clastic sediments. In: Pusch, R.; Easterling, K.; Lundberg, B.; and Stephansson, O. (Eds.): Mechanisms of deformation and fracture, pp. 112–121. Sweden: Luleå 1978.

Smith, J.V.: Feldspar minerals, Vols. 1–3. Berlin-New York: Springer-Verlag 1974.

Smith, J.V., and Stenstrom, R.C.: Electron excited luminescence as a petrologic tool. Jour. Geology 73, 627–635 (1965).

Stanley, K.O., and Faure, G.: Isotopic composition and sources of strontium in sandstone cements: the high plains sequence of Wyoming and Nebraska. Jour. Sed. Petrology 49, 45–54 (1979).

Steiner, M.B.: Detrital remanent magnetization in hematite. Jour. Geophys. Research 88, 6523–6539 (1983).

Strakov, N.M.: Principles of lithogenesis, Vol. 1, 245 pp. New York: Consultants Bureau 1967.

Stumpfl, E.: Erzmikroskopische Untersuchungen an Schwermineralien in Sanden. Geol. Jahrbuch 73, 685–724 (1958).

Suttner, L.J., and Basu, A.: Structural state of detrital alkali feldspars. Sedimentology 24, 63–74 (1977).

Sweatman, T.R., and Long, J.V.P.: Quantitative electron-probe microanalysis of rock-forming minerals. Jour. Petrology 10, 332–379 (1969).

Swineford, A., and Franks, P.O.: Opal in the Ogallala formation, Kansas. In: Silica in sediments. Soc. Econ. Paleon. Mineral. Spec. Pub. 7, 111–120 (1959).

Tardy, Y.; Bocquier, G.; Paquet, H.; and Millot, G.: Formation of clay from granite and its distribution in relation to climate and topography. Geoderma 10, 271–284 (1973).

Tatsumoto, M., and Patterson, C.: Age studies of zircon and feldspar concentrates from the Franconia sandstone. Jour. Geology 72, 232–242 (1964).

Taylor, J.H.: Petrology of the Northampton sand ironstone fomation. Geol. Survey Great Britain Mem., 111 pp. (1949).

Tissot, B.P., and Welte, D.H.: Petroleum formation and occurrence, 538 pp. Berlin–Heidelberg–New York: Springer-Verlag 1978.

Todd, T.W.: Paleoclimatology and the relative stability of feldspar minerals under atmospheric conditions. Jour. Sed. Petrology 38, 832–844 (1968).

Tomita, Toru: Geologic significance of the color of granite zircon, and the discovery of the Precambrian in Japan. Kyushu Univ. Fac. Sci. Mem. 4, 135–161 (1954).

Trevena, A.S., and Nash, W.P.: An electron microprobe study of detrital feldspar. Jour. Sed. Petrology 51, 137–150 (1981).

Triplehorn, D.M.: Morphology, internal structure, and origin of glauconite pellets. Sedimentology 6, 247–266 (1966).

Trowbridge, A.C., and Shepard, F.P.: Sedimentation in Massachusetts Bay. Jour. Sed. Petrology 2, 3–37 (1932).

Turner, F.J., and Verhoogen, John: Igneous and metamorphic petrology, 694 pp. New York: McGraw Hill 1960.

Turner, P., and Ixer, R.A.: Diagenetic development of unstable and stable magnetization in the St. Bees sandstone (Triassic) of northern England. Earth Planet. Sci. Letters 34, 113–124 (1977).

Valloni, R., and Maynard, J.B.: Detrital modes of recent deep-sea sands and their relation to tectonic setting: A first approximation. Sedimentology 28, 75–83 (1981).

Van Houten, F.B.: Cyclic sedimentation and the origin of analcime-rich Upper Triassic Lockatong formation, west-central New Jersey and adjacent Pennsylvania. Am. Jour. Sci. 260, 561–576 (1962).

Van Houten, F.B.: Origin of red beds—a review: 1961–72. Ann. Rev. Earth Planet. Sci. 1, 39–61 (1973).

Van Houten, F.B., and Battacharyya, D.P.: Phanerozoic oolitic ironstones—geologic record and facies model. Ann. Rev. Earth Planet. Sci. 10, 441–457 (1982).

Van Houten, F.B., and Karasek, R.M.: Sedimentologic framework of late Devonian oolitic iron formation, Shatti Valley, west-central Libya. Jour. Sed. Petrology 51, 415–427 (1981).

Velde, B.: Clays and clay minerals in natural and synthetic systems, 218 pp. New York: Elsevier 1977.

Walker, T.R.: Formation of red beds in modern and ancient deserts. Geol. Soc. America Bull. 78, 353–368 (1967).

Walker, T.R.: Formation of red beds in moist tropical climates: A hypothesis. Geol. Soc. America Bull. 85, 633–638 (1974).

Walker, T.R.; Larson, E.E.; and Hoblitt, R.P.: Nature and origin of hematite in the Moenkopi formation (Triassic), Colorado Plateau: A contribution to the origin of magnetism in red beds. Jour. Geophys. Research 86, 317–333 (1981).

Walker, T.R.; Waugh, B.; and Crone, A.J.: Diagenesis in first-cycle desert alluvium of Cenozoic Age, southwestern United States and northwestern Mexico. Geol. Soc. America Bull. 89, 19–32 (1978).

Warne, S.St.J.: A quick field or laboratory staining scheme for the differentiation of the major carbonate minerals. Jour. Sed. Petrology 32, 29–38 (1962).

Webb, W.M., and Potter, P.E.: Petrology and geochemistry of modern sands derived from a volcanic terrain, western Chihuahua. Bol. Soc. Mexicana 32, 45–61 (1969).

Wedepohl, K.H. (Ed.): Handbook of geochemistry, Vols. 1, 2. Berlin–Heidelberg–New York: Springer-Verlag 1969.

Whetten, J.T.: Sediments from the lower Columbia River and origin of graywacke. Science 152, 1057–1058 (1966).

Williams, Howell; Turner, F.J.; and Gilbert, C.M.: Petrography, 406 pp. San Francisco: W.H. Freeman and Co. 1954.

Willman, H.B.: Feldspar in Illinois sands: a study in resources. Illinois Geol. Surv. Rept. Inv. 79, 87 pp. (1942).

Wisniowiecki, M.J.; Van der Voo, R.; McCabe, C.; and Kelly, W.C.: A Pennsylvanian paleomagnetic pole from the mineralized late Cambrian Bonneterre formation, southeast Missouri. Jour. Geophys. Research 88, 6540–6548 (1983).

Wollast, R., and Mackenzie, F.T.: The global cycle of silica. In: Aston, S.R. (Ed.): Silicon geochemistry and biogeochemistry, pp. 39–76. New York: Academic Press 1983.

Zimmerle, W.: Fossil heavy mineral concentrations. Geol. Rundschau 62, 536–548 (1973).

Zimmerle, W.: The geotectonic significance of detrital brown spinel in sediments. Mitt. Geol.-Paläont. Inst. Univ. Hamburg 56, 337–360 (1984).

Zvyagin, B.B., and Lyse, S. (Trans.): Electrondiffraction analysis of clay mineral structure (Translated from the Russian by S. Lyse), 364 pp. New York: Plenum Press 1967.

CHAPTER 3
Texture

Introduction

One of the major objectives of sedimentology is to infer from the properties of a bed of sand or sandstone how it was transported and deposited by either water or air and in what depositional environment. The sedimentologist primarily uses texture, sedimentary structures, the vertical sequence of texture and structures within the sandstone body as well as both the geometry of the body and its associated lithologies to help answer this question. Textural studies were among the very first of these to receive attention.

In this chapter we first review textural concepts and definitions and then examine the utility of texture as a guide to the natural history of sand and sandstone.

Texture includes the shape, roundness, surface features, grain size, and fabric of the components—principally the detrital ones—plus the pore system of a sandstone. Also commonly included are such texturally dependent properties as bulk density, sonic transmissibility, and permeability. In short, texture includes all the descriptors of the geometry, size, and shape of particles and pores that form a sand or sandstones—and all their interrelationships—so clearly it is a complex subject.

The object of research on texture has been chiefly identification of the environment of deposition, the underlying assumption being that the physical processes at the site of deposition impart a distinctive textural "fingerprint" to the sand. In other words, it is assumed that currents, wave height, and water depth, to name but a few, are the chief determinants of grain size, surface texture, roundness, and fabric. The fundamental problem of identifying the ancient environments of deposition of sand and sandstone by textural measurements is to determine the extent to which grain size, rounding, and surface texture are inherited from a previous transport history—or are really only the response to processes that acted immediately before burial in the last environment. With few exceptions, environmental discrimination by texture has had more success with modern than with ancient consolidated sands. But even with modern sands, discrimination is far from consistent so that the search for universal texture discriminators still continues.

Other objectives of the study of textures are to relate them to such physical properties as porosity, permeability and crushing strength, to distinguish stratigraphic units, and to map dispersal patterns by textural parameters, using chiefly variation in grain size. The last is discussed in Chap. 9.

Grain Size

Geologists have long been fascinated by the problem of extracting geologic information from a grain size (granulometric) analysis of sand or sandstone. As a result, there is a larger literature on the techniques and interpretation of grain size than on any other aspect of texture. There is certainly no sign of slackening of the rate of publication in this area.

Meaning of Size

Although it is one of the most widely used terms in sedimentology, the "size" of a particle is not uniquely defined except perhaps for only the most simple of geometric objects such as a sphere (diameter) or a cube (length of an edge). But for irregular particles such as sand grains, size commonly depends on the method of measurement, which in turn depends on the object of study. Moreover, as the particle becomes

TABLE 3-1. Differing definitions of particle size (simplified from Allen, 1981, Table 4.1)

Volume diameter—diameter of a sphere having the same volume as the particle

Surface diameter—diameter of a sphere having the same surface as the particle

Surface volume diameter—diameter of a sphere having the same external surface to volume ratio as the particle

Drag diameter—diameter of a sphere having the same resistance to motion as the particle in a fluid of the same viscosity and at the same velocity (d_d approximates to d_s when R, Reynolds' number, is small)

Free-falling diameter—diameter of a sphere having the same density and the same free-falling speed as the particle in a fluid of the same density and viscosity

Stokes' diameter—the diameter of a free-falling particle in the laminar flow region ($R < 0.2$)

Projected area diameter—diameter of a circle having the same area as the projected area of the particle resting in a *stable position*

Projected area diameter—diameter of a circle having the same area as the projected area of the particle in *random orientation*

Perimeter diameter—diameter of a circle having the same perimeter as the projected outline of the particle

Sieve diameter—the width of the minimum square aperture through which the particle will pass

Feret's diameter—the mean value of the distance between pairs of parallel tangents to the projected outline of the particle

Martin's diameter—the mean chord length of the projected outline of the particle

Unrolled diameter—the mean chord length through the center of gravity of the particle

less equant, discrepancy between the measures becomes progressively greater. Table 3-1 summarizes the different definitions of particle size. A sedimentologist studying behavior of grains in a fluid might find their fall velocity, drag, or Stokes' diameter, measures related to fluid flow, much more meaningful than volume diameter. On the other hand, a petrographer studying the downcurrent gradient of abundance of carbonate rock fragments in an alluvial sand might consider their post-depositional solubility to be important and, consequently, specific surface diameter might be most useful. Thus the measure chosen will reflect the object of study as well as the technique used. In general, one measure of size converts to another by expressing it in terms of a corresponding equivalent diameter. Unless otherwise specified, the term "size" as here used is the diameter of the grain (in mm), as defined by sieving inasmuch as we are here concerned only with sand—not gravels, silts, or clays.

Techniques

Summaries of methods of particle size measurement, written from a nongeologic point of view, are given by A.S.T.M. (1959) , Herdan (1960), Irani and Callis (1963), Allen (1981), and Orr (1985). Brewer (1964, pp. 18–40), Folk (1966), and Riviere (1977) provide excellent geologically oriented summaries of techniques, and Gibbs (1974) gives a very complete discussion of suspension, much of which has significance for sands.

The different techniques of size measurement apply to widely different size ranges (Fig. 3-1). With the exception of the settling tube and use of the thin section, measurement techniques in sedimentology have been stabilized for more than two decades. Sieving, which measures the width of the minimum square aperture through which a grain will pass, still remains the preferred method for unconsolidated sands and friable sandstones. It is relatively rapid, trouble-

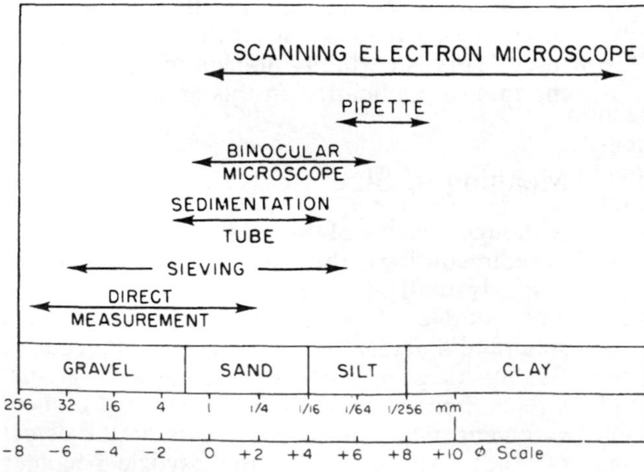

FIGURE 3-1. Range of applicability of different techniques of size analysis.

free, cheap, and the oldest, most tested method. But because the problem of correlating thin section and sieve size estimates is now better understood (Harrell and Eriksson, 1979), direct measurement in thin section is more widely used and is the only possible choice for completely cemented sandstones, even though only the mean and sorting can be obtained. Conversion equations are necessary when size data generated from sieving and thin section are to be combined. The binocular microscope can also be used to estimate mean and sorting quickly and easily using standard comparison sets (Swann *et al.*, 1959; Emrich and Wobber, 1963; Beard and Weyl, 1973). The settling tube, formerly used only for clay-sized sediment, is now commonly used for unconsolidated sands and employs modern instrumentation to measure settling velocities rapidly and accurately. It yields an automatically recorded result, based on weight percent or cross-sectional area of individual grains, that can be empirically related to a standard size distribution curve (Anderson and Kurtz, 1979; McCave and Jarvis, 1973; Taira and Scholle, 1977, Table 1). Like measurement in thin section, the mean and sorting are normally obtained, although Reed *et al.* (1975) and Taira and Scholle (1979) have also computed skewness and kurtosis.

Choice of method depends both on the degree of consolidation and the objectives of the study. If the sand is unconsolidated and only the mean and sorting are of interest and their range of variability is large, the binocular microscope may suffice. Or, if available, the settling tube should be used for unconsolidated sands because it is rapid and sensitive and only small samples of a few grams are needed. Indurated sands require thin sections or very careful disaggregation (Suczek, 1983) and subsequent sieving.

Because size is a continuous variable and many particles are needed for a good estimate, a grade scale is always necessary. A *grade scale* consists of a series of class intervals having some constant relationship to one another. Udden (1898, pp. 6–7) recognized early the need for a grade scale and based his on powers of 2. A convenient scale to use is a logarithmic one, because a difference of a millimeter between sand grains is significant but between boulders is trivial. In addition, samples from single populations of grains tend to plot as straight lines on arithmetic probability graph paper. Kolmogorov (1941) has shown that the lognormal distribution is the theoretically expected one under certain conditions of grinding and crushing, if one assumes homogeneous material. Subsequent reviews of theoretical grain size distributions are found in Middleton (1970), Dapples (1975), and Shultz (1975). Wentworth (1922) and later Krumbein (1934) modified Udden's scale to obtain the now widely used *phi* scale defined as

$$\phi = -\log_2 S$$

where S is the diameter in millimeters and is commonly determined either by sieving or by measurement in thin section. Conversion of millimeters to phi units can be made from tables (Page, 1955) or by slide rule (Folk, 1964); reconversion can be made the same way or the log itself can be used, keeping in mind that the base is 2. Table 3-2 shows the relationships between size in millimeters, mesh, and phi units and also gives terminology. Quarter phi units should be used for the best estimates of size parameters and shape of the cumulative curve.

Settling velocity is increasingly used to define the distribution of particles in a sand, because more effective settling tubes (Gibbs, 1974) are becoming available, and more and more sedimentologists recognize that settling velocity is probably the single most meaningful hydraulic property of a grain. Very small samples can be used and automation is impressive (Taira and Lienert, 1979). However, one should remember that only unconsolidated sands can be analyzed with settling tubes.

Middleton (1967, p. 484) proposed the psi distribution for settling velocities, $\psi = -\log_2 w$, which differs from the phi distribution in that fall velocity, w, is used rather than grain diameter. Taira and Scholle (1977) proposed another distribution based on settling velocities, the tau distribution defined as $\tau = -\log_{10} w$, which differs from Middleton's psi distribution only in that the particles are "counted" by photocell (cross-sectional area) rather than weighed as they accumulate at the bottom of the settling column and that the log transformation is to the base 10 rather than 2. Taira and Scholle note that a very wide range of particles from 10^0 to 10^4 microns can be analyzed by settling column and claim that the method has much greater significance for environmental discrimination. The frequency distribution of settling velocities is, from the standpoint of hydraulics, more logical than a frequency distribution of grain diameters and its greater use could be a significant forward step because much sand, especially the finer sizes, travels by saltation and temporary suspension. All settling velocities can be converted to an equivalent diameter (Table 3-3).

TABLE 3-2. Terminology and class intervals for grade scales.

	U.S. Standard sieve mesh	Millimeters		Phi (ϕ) units	Wentworth size class
GRAVEL	Use wire squares	4096		−12	
		1024		−10	Boulder
		256	256	−8	
		64	64	−6.	Cobble
		16		−4	Pebble
	5	4	4	−2	
	6	3.36		−1.75	
	7	2.83		−1.5	Granule
	8	2.38		−1.25	
	10	2.00	2	−1.0	
SAND	12	1.68		−0.75	
	14	1.41		−0.5	Very coarse sand
	16	1.19		−0.25	
	18	1.00	1	0.0	
	20	0.84		0.25	
	25	0.71		0.5	Coarse sand
	30	0.59		0.75	
	35	0.50	1/2	1.0	
	40	0.42		1.25	
	45	0.35		1.5	Medium sand
	50	0.30		1.75	
	60	0.25	1/4	2.0	
	70	0.210		2.25	
	80	0.177		2.5	Fine sand
	100	0.149		2.75	
	120	0.125	1/8	3.0	
	140	0.105		3.25	
	170	0.088		3.5	Very fine sand
	200	0.074		3.75	
	230	0.0625	1/16	4.0	
SILT	270	0.053		4.25	
	325	0.044		4.5	Coarse silt
		0.037		4.75	
		0.031	1/32	5.0	
		0.0156	1/64	6.0	Medium silt
	Use pipette or hydrometer	0.0078	1/128	7.0	Fine silt
		0.0039	1/256	8.0	Very fine silt
MUD		0.0020		9.0	
		0.00098		10.0	Clay
		0.00049		11.0	
		0.00024		12.0	
		0.00012		13.0	
		0.00006		14.0	

Statistical Measures

A size distribution is usually described either by employing some form of the cumulative frequency curve or by numerically specifying its central tendency by mean, mode, or median and its form or shape by sorting, skewness, and kurtosis. These parameters can be either obtained from the cumulative curve (graphic measures) or calculated directly from the sieve analysis (moment measures). Because a size analysis is usually open-ended—it is rarely practical to use sieves smaller than 30 microns—most graphic measures are superior to moment measures, even though the latter can be calculated rapidly by computer. Computer programs also generate graphic plots (Slatt and Press, 1976; Burger, 1976). Moreover, visual inspection of the cumulative curve may show whether two or more populations are present and reveal possible errors in the analytical method. Hence it always pays to see the cumulative curve which is determined by using either half or quarter phi class intervals, preferably the latter (Fig. 3-2). Sedimentation tubes directly yield a continuous cumulative curve of fall velocities

An ideal lognormal distribution plots as a sin-

TABLE 3-3. Common logarithms of settling velocity and grain diameters at 20°C (Taira and Scholle, 1977, Table 1).

τ	Settling velocity (cm/s)	Microns	Phi
2.0	100.0	15,284.70	−3.934
1.9	79.4	10,136.62	−3.342
1.8	63.0	6,848.72	−2.776
1.7	50.1	4.728.46	−2.241
1.6	39.8	3,340.97	−1.740
1.5	31.6	2,416.17	−1.273
1.4	25.1	1,786.74	−0.837
1.3	20.0	1,348.65	−0.432
1.2	15.8	1,037.21	−0.053
1.1	12.6	811.15	0.302
1.0	10.0	644.02	0.635
0.9	7.94	518.52	0.948
0.8	6.30	422.74	1.242
0.7	5.01	348.68	1.520
0.6	3.98	290.67	1.783
0.5	3.16	244.67	2.031
0.4	2.51	207.75	2.267
0.3	2.00	177.79	2.492
0.2	1.58	153.17	2.707
0.1	1.26	132.74	2.913
0.0	1.00	115.62	3.113
−0.1	0.794	101.14	3.306
−0.2	0.630	88.78	3.494
−0.3	0.501	78.15	3.678
−0.4	0.398	68.97	3.858
−0.5	0.316	60.98	4.035
−0.6	0.251	54.01	4.211
−0.7	0.200	47.89	4.384
−0.8	0.158	42.52	4.556
−0.9	0.126	37.77	4.727
−1.0	0.100	33.58	4.896

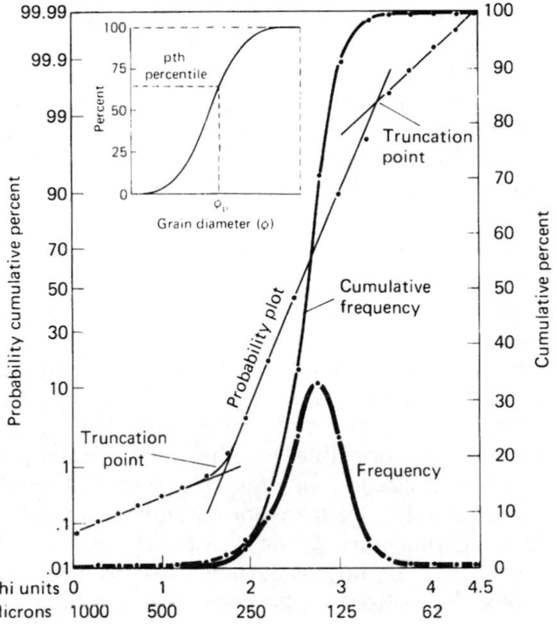

FIGURE 3-2. Segmented cumulative curve plotted on lognormal probability paper. (Redrawn from Visher, 1969, Fig. 4). Small inset defines a percentile.

gle straight line on arithmetic probability paper (when size is expressed in phi units) and on logarithmitic probability paper (when size is expressed in millimeters). Although some grain-size distributions of sands do plot as straight lines on the above graph papers, most do not (Fig. 3-2), so that the cumulative curve is commonly segmented and generally has at least one or perhaps as many as three or more inflection points. Visher (1969) first proposed that the inflection point between the two coarsest fractions separates grains that were always in contact with the bed—grains transported by sliding and rolling (surface creep)—from grains transported by intermittent suspension (saltation), whereas the second inflection point is believed to separate salting grains from those carried totally in suspension. Initial studies of breaks in the cumulative distribution curve were made by Doeglas (1946) and Sindowski (1957), followed by Glaister and Nelson (1974). Visher (1969), Middleton (1976), and others proposed that grains move by traction as bedload, when settling velocity is greater than stream velocity (Chap. 8), and in intermittent suspension when settling velocity is greater than shear velocity (Chap. 8), shear velocity being a measure of flow intensity along the bottom. Viard and Breyer (1979) tested Middleton's hypothesis on sands of the Platte River and had some, but not complete, success. Where comparative studies have been made, significant differences are claimed between the cumulative curves based on sieving (Reed *et al.*, 1975, p. 1324)—sieves separate particles on the basis of minimal cross-sectional area regardless of grain density, whereas a settling tube directly measures the velocity of the falling grains, which depends on size, density, and shape. As we will see, much effort has been expended on correlating the shape of the cumulative curve with the depositional environment and attempts have been made to give a hydraulic interpretation to the graphic size plot, recognizing, of course, that the availability of different size fractions for local transport also plays a role. Another factor, at least for ancient sandstones, is diagenesis, which may greatly enhance the proportion of fines, chiefly by alteration of argillaceous and volcanic rock fragments.

Other methods of description are essentially graphic and include plots of sand–silt–clay ratios, which are very popular with modern sediment workers, *CM* diagrams (Passega, 1957, 1964), the arithmetic-cumulative probability plots of the Dutch school (Doeglas, 1964; van Andel and Postma, 1954, pp. 80–94), and logarithmic probability plots (Sindowski, 1957). The above methods are examples of what has been called "textural fingerprinting."

Central Tendency. The grain size of a sand is commonly clustered around some average value such as the mean, mode, or median (Table 3-4), which is controlled by a combination of two factors: the average competence of the depositing medium and the initial size of the source materials. Contrasting methods of transport are responsible for the almost universal sharp segregation of clay from silt and sand. Clay particles are transported entirely as turbulent suspensions while most silt, almost all sand, and all larger particles are transported along the bottom as bedload by a combination of sliding, saltation, and rolling. The lateral sequence sand→silt→clay found over distances of tens to hundreds of kilometers in many basins is almost always related to declining current strength rather than abrasion of sand (Kuenen, 1959).

To specify both central tendency and shape of the curve, linear combinations of percentiles are used, a *linear combination* being nothing more than the addition or subtraction of a series of terms such as, for example, the estimate of the mean $(\phi_{10} + \phi_{30} + \phi_{50} + \phi_{70} + \phi_{90})/5$, where the ϕ's are percentiles. The more percentiles that are read from the cumulative curve, the more accurately one can specify the desired parameter, such as the mean, mode, or sorting. Thus, initially, only two percentiles, ϕ_{25} and ϕ_{75}, called the first and third quartiles, were used to determine sorting, but subsequently others seeking more information from the cumulative curve have used as many as five or more (Folk and Ward, 1957, pp. 13–14; McCammon, 1962). Hand calculators now reduce

TABLE 3-4. Graphic and moment measures.

Name	Graphic formula[a]	Moment formula[b]	Remarks
Mean	$Me_\phi = \dfrac{(\phi_{16} + \phi_{50} + \phi_{84})}{3}$	First moment $\bar{x}_\phi = \sum_{i=1}^{n} f_i m_{i\phi}/100$	All three measures of central tendency reflect average kinetic energy of depositing medium plus size distribution of available sediment.
Median	$Md_\phi = \phi_{50}$		
Mode	M_ϕ = Midpoint of most abundant class interval		
Bimodality Index	$Mi_\phi = 1 + \dfrac{(\phi_f - \phi_c)}{2\phi}$		Measure of bimodality, if present: ϕ_f is midpoint of finest mode; ϕ_c is midpoint of coarsest.
Sorting	Inclusive graphic standard deviation $s_I = \dfrac{\phi_{84} - \phi_{16}}{4} + \dfrac{\phi_{95} - \phi_5}{6.6}$	Second moment $s_\phi = \left[\sum_{i=1}^{n} f_i(M_{i\phi} - \bar{x}_\phi)^2/100\right]^{1/2}$	Measures dispersion, which is dependent upon velocity variations plus bimodality.
Skewness	Inclusive graphic skewness[c] $SK_I = \dfrac{\phi_{84} + \phi_{16} - 2\phi_{50}}{2(\phi_{84} - \phi_{16})} + \dfrac{\phi_{95} + \phi_5 - 2\phi_{50}}{2(\phi_{95} - \phi_5)}$	Third moment $3_\phi = \sum_{i=1}^{n} \dfrac{f_i(M_{i\phi} - \bar{x}_\phi)^3}{100s_\phi^3}$	Measures asymmetry, the direction of "tails," which are widely believed to have environmental significance. Skewness varies from +1.0 (positive) to 0.0 (symmetrical) to −1.0 (negative).
Kurtosis	$K_G = \dfrac{(\phi_{95} - \phi_5)}{2.44(\phi_{75} - \phi_{25})}$	Fourth moment $4_\phi = \sum_{i=1}^{n} \dfrac{f_i(M_{i\phi} - \bar{x}_\phi)^4}{100s_\phi^4}$	Measures peakedness. Graphic measure is ratio of sorting of centered 90 percent to centered 50 percent. K_G = 1.0 for normal curve, $K_G > 1.0$ for peaked curve, and $K_G < 1.0$ for flattened curve.

[a] All after Folk and Ward (1957) except bimodality index after Sahu (1964).

[b] f_i = fraction of total weight in each class interval; $m_{i\phi}$ = midpoint of each class interval in phi units.

[c] Also $SK_I = \dfrac{\phi_{84} - \phi_{50}}{\phi_{84} - \phi_{16}} - \dfrac{\phi_{50} - \phi_5}{\phi_{95} - \phi_5}$ according to Warren (1974)

much tedious work (Benson, 1981) and personal computers do even more. Table 3-4 summarizes the most commonly used graphic and moment measures.

Because the mean of a distribution depends on the entire size curve, it is a better estimate of central tendency than the median. In a statistically normal distribution, mean, mode, and median all coincide. One of the most effective uses of central tendency is, as we will see in Chap. 9, to record its vertical variation in single sandstone bodies and, in so doing, correlate it with the vertical sequence of sedimentary structures to see if the sandstone body fines or coarsens upward, a major key to the reconstruction of its depositional environment. Simply plotting some measure of size in a thick sandy section can also be very informative (Fig. 3-3). Leggewie *et al.* (1977, Fig. 2) have correlated median sizes estimated by hand lens with those based on a complete sieve analysis.

Maximum Size. This is commonly estimated either by using the 95th or 99th percentiles or by using the average diameter of the five to ten largest grains and is a rough measure of maximum current competency. Because the average diameter of the largest grains can be quickly estimated, usually by simply examining the thin section for its largest grains and then selecting the largest five or ten and averaging their diameter, mapping maximum size, either in a long vertical section or laterally, is very attractive.

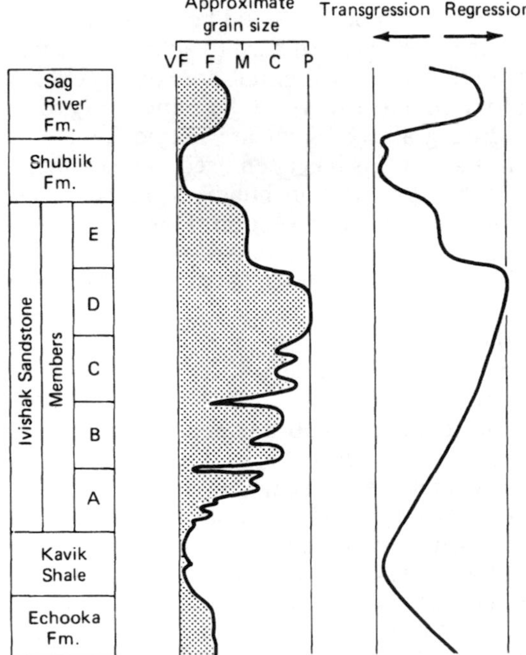

FIGURE 3-3. Diagrammatic illustration of approximate grain size and its relation to transgression and regression in a 400-ft section of Permian-Triassic sandstones of the Prudhoe Bay Field, North Slope, Alaska. (Modified from Jones and Speers, 1977, Fig. 15).

Bimodality. The formula for the index proposed by Sahu (1964, p. 77) to measure bimodality is given in Table 3-4. Bimodality may result from a variety of causes: a combination of bed and suspension load transport, infiltration of fines, post-depositional diagenesis, or lack of certain size grades in some source materials. The presence of two or more modes complicates interpretation of all statistical measures. Bimodality is thought to be a unique characteristic of some eolian sands (Folk, 1968; Warren, 1972).

Sorting. Dispersion around central tendency determines sorting. Because the tails of a distribution have been thought to be environmentally sensitive, estimates of sorting have been designed to reflect them. Thus an estimate of sorting based on the 84th and 16th percentiles should be superior to one based on the 75th and 25th percentiles, that is, one closer to the central tendency. Folk and Ward (1957, p. 13) combined each of these measures in their inclusive graphic standard deviation, σ_I, which is the average of sorting based on both the central and exterior parts of the size distribution (Table 3-4). One can also estimate sorting visually or measure it semiquantitatively in thin section.

Fluctuations in velocity, contributions from suspension as well as traction transport to the same bed, sampling more than one sedimentation unit, and source sands with more than one mode all contribute to sorting variation. The effect of inheritance on sorting has been emphasized by Folk and Robles (1964, pp. 290–91). Their point of view may be stated as follows: commonly sands contain several grain populations, each with its own size distribution, so that sorting is best when the different modes are close together and worse when they are far apart. Thus in polymodal sediments, sorting, as well as other measures, simply reflects the relative magnitude and separation of the different modes.

What are the limiting values of sorting that have been observed? Folk and Robles (1964, p. 290) have suggested that sands deposited in the surf zone have values of σ_I of 0.3–0.6ϕ. Little other information is available on limiting values of sorting in other environments. Even though

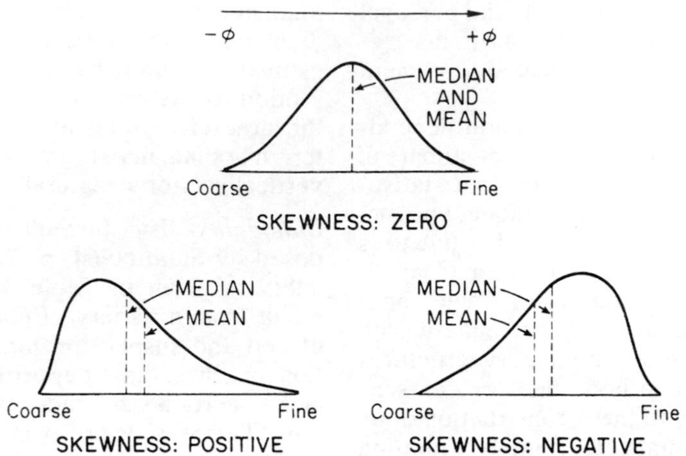

FIGURE 3-4. Positively and negatively skewed size curves. (Modified from Friedman, 1961, Fig. 7). Skewness is thought by some to be an environmental discriminator but it is very sensitive to the method of analysis.

it is hard to estimate limiting values, such an inquiry strikes us as worthwhile.

Skewness. The asymmetry of a distribution is measured by skewness and is determined by the relative importance of the tails of the distribution (Fig. 3-4). As with sorting, a number of skewness formulas have been proposed. Folk and Ward (1957, p. 14) provide the most comprehensive graphic formula, one that uses six percentiles (Table 3-4). Those who have attempted to draw conclusions about either transport processes or the environmental significance of the size distribution have focused attention on the tails of the distribution and paid particular attention to skewness.

Kurtosis. Roughly speaking, kurtosis (Table 3-4) is a measure of the peakedness of the distribution. If a distribution is flatter than a normal one, it is called platykurtic; if more peaked, it is called leptokurtic. Few geologic conclusions have been deduced from kurtosis values alone.

The moment measure formulas for skewness and kurtosis both illustrate why, if meaningful conclusions are to be drawn about them, much care must be used in obtaining the original size data. Because the deviations from the mean as well as the inclusive sample standard deviation are raised to the third and fourth powers, errors in the original data can produce widely misleading estimates of both skewness and kurtosis.

Sampling. Proper sampling is vital for grain size studies. Determination of average grain size of a deposit at a given point requires either a channel sample or a composite obtained from spot samples taken across the deposit. Values of sorting, skewness, and kurtosis obtained from such samples may be used in determining regional variations. If geographic size variations of the pebbles in a pebbly sandstone are to be established, one should record from outcrop only the largest pebble; a more statistically stable estimate can be obtained by recording the average of a small number, such as the largest ten. Rigorously, the number of pebbles observed at each point should be the same, but in practice this has been hard to achieve. To minimize effects of differential abrasion, pebble lithology should always be the same. Fifty to one hundred grams of sand are ample for a sieve analysis and less has been used in a sedimentation tube. Two to four hundred grains are usually measured in thin section. However, Davies *et al.* (1971, Fig. 3) have used only 50 grains to estimate the mean diameter of quartz grains in thin sections. Only 50 grains were used because of the need to estimate rapidly grain size in many samples from small side wall cores and cuttings.

A different procedure is needed for environmental studies. Here it is best to obtain samples from a single bed or sedimentation unit, defined by Otto (1938, p. 575) as "that thickness of sediment which was deposited under essentially constant physical conditions." Attention to sedimentary structures now assumes importance because ripple mark, crossbedding, parting lamination, graded beds, and other structures constitute natural units of sampling. In other words, the type of sedimentary structure, mean grain size, and sorting as well as possibly even the shape of the cumulative curve, especially when plotted on logarithmic probability paper, are related to intensity of fluid flow. When sampled in this way, the size parameters should yield a maximum of environmental information. As before, either single spot samples or composites from the same structure may be analyzed, the former probably being superior.

Shape and Roundness

Shape and roundness are the properties of sand grains that have significance for the study of the effect of the transport process on the debris furnished by the source area. These properties reveal the modification of angular grains of many shapes by abrasion, solution, and current sorting. Though they are fundamental, shape and roundness of sand grains have received but a fraction of the effort that has been devoted to size studies, possibly because most methods involve individual grain measurements that are tedious as well as imprecise. Roundness in particular reflects abrasion history, which in turn depends on such diverse geologic controls as relief, kinds of source rocks, transport process, and the mineralogy of the grain. Shape has hydraulic importance because it determines the ratio of surface area to volume of a grain—the greater this ratio, the more easily will the grain be entrained by hydraulic forces acting on the surface and the slower will it settle. Photographs by the scanning electron microscope (SEM) (Fig. 3-5) beautifully illustrate the diverse shapes and roundness of modern sands—a diversity still inviting full explanation.

Shape is defined by various ratios of a particle's long (L), intermediate (I), and short (S) axes. There are two aspects of shape: sphericity and form. *Sphericity* is a quantitative parameter measuring the departure of a body from equidimensionality. Sneed and Folk (1958, pp. 117–25) review the various measures of sphericity and propose a new one, *maximum projection sphericity* (also called effective settling sphericity), which is the ratio of the cross-sectional area of a sphere of the same volume as the particle divided by its maximum projection area; this is quantitatively defined as $\psi_p = \sqrt[3]{S^2/LI}$. They believe this quantity to be a better measure of a particle's behavior in a fluid medium than previous ones. One can calculate maximum projection sphericity graphically (Fig. 3-6). In a comprehensive study of the shape of sand particles Briggs *et al.* (1962, p. 655) noted that shape can be as important as density in affecting fall velocity.

Grain shape and size also affect the electrical formation factor (ratio of resistivity of a porous media to the resistivity of the pore fluid) of sands (Jackson *et al.*, 1978); i.e., different shapes of framework grains alter the geometry of the pore system and thus alter electrical resistance.

Roundness is geometrically distinct from shape and is concerned with the curvature of corners. It is quantitatively defined as $\Sigma_{i=1}^{n}(r_i/R)/n$, where r_i is the radius of a circle inscribed in the ith corner of a grain, n is the number of corners, and R is the smallest radius that will circumscribe the grain. One uses the silhouette of the maximum projection plane of the particle to determine r_i and R. Standard sets of images are commonly used to estimate roundness and differing degrees of roundness from angular to well-rounded as quantitatively defined. Wadell (1935) and Krumbein (1934) are largely responsible for our present-day concepts of roundness and the chart used to estimate it. Later Powers (1953) proposed a scale based on two sets of images for grains of different sphericity (see Appendix, Fig. A-2). The rho scale of Folk (1955, p. 297) is based on logarithmic intervals of the Powers Scale.

Ehrlich and Weinberg (1970) and Ehrlich *et al.* (1974) have approached grain roundness in a different manner by describing individual grain outlines with polar coordinates and then using closed-form Fourier series to obtain a power spectrum—the magnitude of the different harmonics that describe the individual grain outlines. Loosely speaking, this is a type of mathematical fingerprinting that specifies the complexity of the outline of the grain and its roundness. Conversion of a grain outline described by polar coordinates into its Fourier harmonics is similar to the process wherein a geophysicist converts a seismic wave in the time domain to the frequency domain where it is more easily analyzed and processed by computer (Dobrin, 1976, pp. 156–57 and 163–64). The Fourier grain shape method, although slow, has utility for studies of provenance and has also been applied to the stratigraphic zoning of an 11,000-ft (3600-m) Pliocene sandstone section in the Gulf Coast (Mrakovich *et al.*, 1976). The method also provides a mathematical equation that regenerates grain outline as precisely as required. Serra (1982) provides a lengthy mathematical treatment of image analysis including porosity as well as grain shape and roundness.

Briggs *et al.* (1962, p. 654) found that visual estimates of grain roundness had negligible correlation with the fluid dynamic behavior of sand grains. Boggs (1967) describes in detail the use of an electronic particle size analyzer to determine roundness and sphericity of sand grains. A

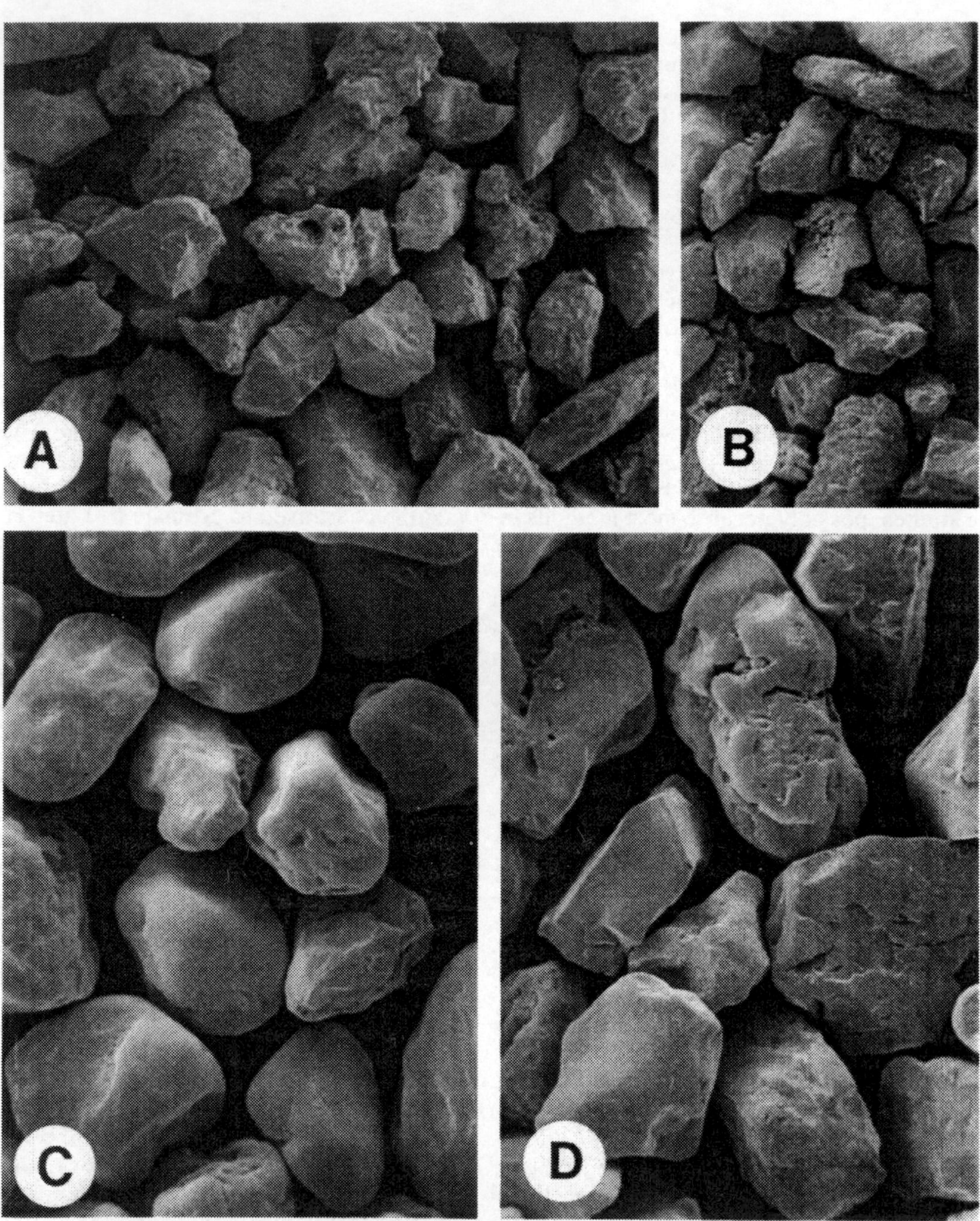

FIGURE 3-5. SEM photographs of selected modern river sands in the Amazon Basin: A, Rio Amazonas, X100 has Andean source; B, Rio Solimoes, X100 has Andean course; C, Rio Tocantins, X50 has a mixed Paleozoic-Precambrian source with some round grains; and D, Rio Negro, X100 has a Precambrian source. From Franzinelli and Potter, 1983, Fig. 6. Reproduced by permission of the University of Chicago Press.

minor use of grain roundness is to distinguish between stratigraphic units and different sources of sand in sedimentary basins—perhaps between a cratonic source that supplied a large proportion of well-rounded grains and a marginal geosynclinal source that yielded more angular grains.

Winkelmolen (1969) and Winkelmolen and Veenstra (1974) have made careful studies of another aspect of particle morphology called *rollability*—the minimum angle of slope which

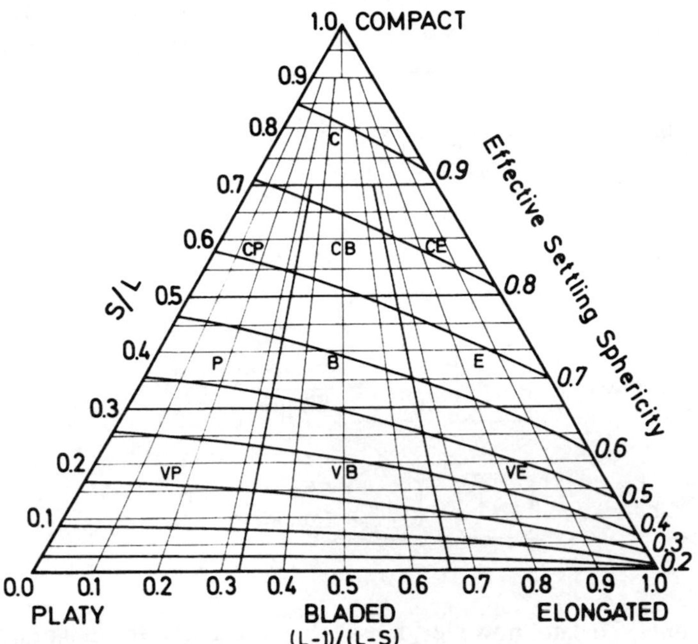

FIGURE 3-6. Graphic calculation of maximum projection sphericity. To determine sphericity, first compute S/L and then $(L - I)/(L - S)$. The intersection of the ratios S/L and $(L - I)/(L - S)$ determines a point in the triangles, the sphericity of which is obtained by interpolating between the curved lines. L = long diameter; I = intermediate diameter; S = short diameter. The ten form classes are: *C*, compact; *CP*, compact platy; *CB*, compact-bladed; *CE*, compact-elongate; *P*, platy; *B*, bladed; *E*, elongate; *VP*, very platy; *VB*, very bladed; and *VE*, very elongate. (Modified from Sneed and Folk, 1958, Fig. 2).

will cause a grain to roll or fall. Rollability is a property derived from roundness and shape, as shown in Fig. 3-7. It is in essence a *functional* shape description—in contrast to a purely geometric one—and is determined for loose sand in a machine. Winkelmolen used such a device and concluded that plots of rollability against grain size were helpful in discriminating between some beach and dune sands. His work is one of the most complete studies on record of the sedimentological aspects of grain shape.

Harrell (1984) describes a roller micrometer and its use in studies of grain shape. His micrometer uses the smallest and intermediate axes of sand grains.

What significance do studies of the shape and roundness of sand grains have? The degree of rounding of a detrital particle depends on its size, physical characteristics, and history of abrasion. Laboratory studies and measurement of sand roundness in modern sediments have shown grain rounding to be a very slow process, one that rapidly becomes much slower as size decreases. Hence after a given amount of transport, larger grains are better rounded than smaller ones. Kuenen's experiment (1959, p. 186) showed that 20,000 km of transport would cause no more than a 1 percent loss in the weight of an angular, medium-grained quartz sand, thus confirming earlier experimental studies. Although one might wish to estimate quantitatively the distance of travel for sand from its average roundness or percentage of angular grains, this is as yet impossible because the processes of rounding are still poorly understood. One of the difficulties is that the rate of rounding may vary greatly with depositional environment or rate of sedimentation. Chemical solution appears to play a secondary role for most silicates but is no doubt of some importance for more soluble detrital carbonate grains.

Well-rounded grains are the result of either many cycles of transport, each contributing its small share of rounding, or of intensive abrasion in a special environment where rounding was accomplished very rapidly. Petrographers appear to be about equally divided between these two, some favoring the "inheritance" viewpoint and others the "environment" viewpoint. The beach and, to some extent, the dune are the environments that have been suspected as producing well-rounded grains, especially in the geologic past where greater tectonic stability on ancient cratons may have been responsible for greater reworking of sand than in

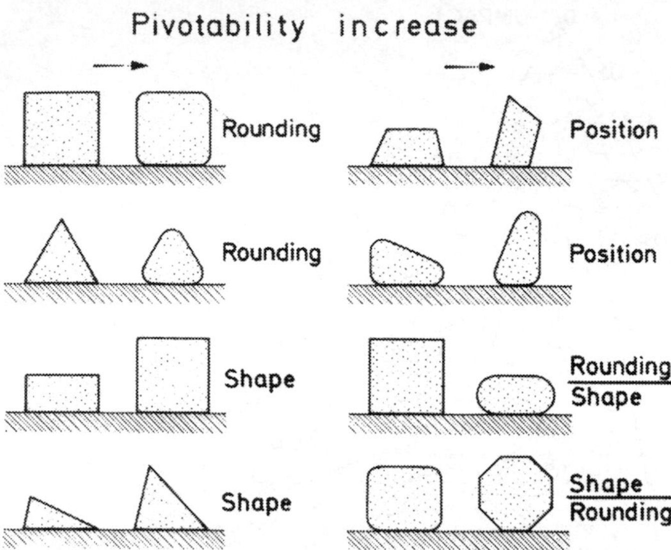

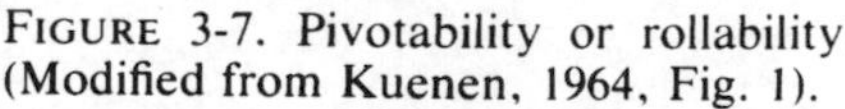
FIGURE 3-7. Pivotability or rollability. (Modified from Kuenen, 1964, Fig. 1).

present-day sediments. To date, however, most studies of modern beach deposits generally do not show strikingly more rounded grains than their presumed continental source sand.

Because of the slow rate of rounding of sand-size particles, roundness trends in most sands are difficult to establish. Some investigators have simply used roundness to help define mineral associations or dispersal patterns based on the presence or absence of an arbitrary abundance of angular grains. Inasmuch as roundness is a function of size, determination of the percentage of angular or round grains should be made only with the same size fraction and mineral. Roundness provinces may also be defined by the angularity of heavy minerals, generally either zircon or tourmaline, as well as quartz. Although rounded overgrowths are decisive evidence of multicycle grains, they are usually so small in number as to be impractical for petrographic mapping. Multicycle grains are easiest to recognize with cathodo-luminescence (Sippel, 1968).

After we develop some insight into fluid flow (Chap. 8), we can examine the relative effectiveness of wind versus water in the rounding of sand grains.

Surface Textures

Surface textures of sand grains were first studied by the binocular and polarizing microscope; later, replicates of grain surfaces were studied by the electron microscope, and more recently the scanning electron microscope (SEM) has been used with magnifications in excess of 30,000 times (Krinsley and Doornkamp, 1973; Margolis and Krinsley, 1974; Krinsley, 1978; Le Ribault, 1975, 1977, 1978) to great advantage (Fig. 3-8). The SEM is superior to the electron microscope in three ways—no replicates of the surface need be made, 40 or more grains can be placed on the same mount, and a simulated three-dimensional view—one that can be rotated—is obtained. As a result of this technology, a single grain 1 mm in diameter, when magnified 20,000 times, has a surface area equivalent to 1 km^2 and becomes a study in itself (cf. Wellendorf and Krinsley, 1980, Fig. 1)! The literature on surface textures of quartz grains is very large and rapidly growing. Quartz is studied because it is ubiquitous and both mechanically and chemically stable.

The SEM has revealed the great variety of microstructures that occur on sand grains—conchoidal fractures, friction-created striae of many kinds, isolated depressions, upturned cleavage plates, chattermarks, and solution pits as well as flowers, globules, and pellicules of quartz (which give rise to terms such as "silica plastering" and "capping layer") plus euhedra of quartz, calcite, dolomite, and clay minerals to name but a very few (Fig. 3-9). Different magnifications are required to see these different microstructures. All of these microstructures may be polished, weathered, or fresh. Another variant, and an important one according to Le Ribault, is *where* on the surface does the microstructure occur—on summits, depressions, or between them? In *L'exoscopie des quartz*, Le Ribault (1977) recognized 73 morphologic features, which are well illustrated by

FIGURE 3-8. SEM mosaic of a single quartz grain, 300 microns in diameter, based on over 1100 SEM photographs. Photograph by David Krinsley. Reprinted by permisstion from *Nature*, Vol. 283, No. 5745, ©1980—Macmillan Journals Limited.

145 photographs on 24 plates. These features are matched, in a complex matrix, against six major environments with a total of 28 subenvironments. Recognition of these features permitted Le Ribault (1975, pp. 59–61) to infer complete environmental histories on uncemented sand grains, as shown by the following example:

Torrential sedimentation → Delta → Eolian → Soil
Glacial → Torrential → Marine → Soil weathered quartz

Earlier approaches used the number of impact pits per μm^2 as well as 21 other surface features (Krinsley and Doornkamp, 1973, Fig. 1; Margolis and Krinsley, 1974, Fig. 1) and were essentially based on the morphology and density of mechanical traces. Consequently, they do not provide as complete a data base for establishing the evolving "natural surface history" of a grain as does a system that additionally emphasizes "where" such features occur relative to surface relief—on grain surface summits, depressions, or in between. Nonetheless, the summary tables and figures of Krinsley and Doornkamp (1973, pp. 25–27), Margolis and Krinsley (1974, Fig. 1), Le Ribault (1977, p. 145), Higgs (1979, Fig. 2), and Giresse and Le Ribault (1981) are exceptionally well organized and easy to read and use (Fig. 3-9).

What is to be said of this research? If Le

Legend

	% grains showing feature
■ Abundant	> 75
▪ Common	25-75
△ Sparse	5-25
□ Rare	< 5

Environment	Small irregular pits (< 10 μ)	Medium irregular pits	Large irregular pits (> 100 μ)	Small conchoidal fracture	Medium conchoidal fracture	Large conchoidal fracture	Straight steps	Arcuate steps	Fracture plates/planes	Parallel striations	Imbricated grinding features	Adhering particles	Meandering ridges	Straight scratches	Curved scratches	Impact 'v's	Angular outline	Rounded outline	Low relief (< 0.5 μ)	Medium relief	High relief (> 1.0 μ)	Oriented etch pits	Anastomosing etch pattern	Solution pits	Solution crevasses	Scaling	Silica globules	Silica flowers	Silica pellicle	Crystalline overgrowth
Crystalline source rock				■	■	■	■	■	▪			■					■		■											
Subaqueous – Fluviatile – Low energy	■	■	■											△	△	△	▪	▪		■	▪						▪		▪	△
Subaqueous – Fluviatile – Medium energy	■	■	■											△	△	△	▪	▪		■	▪						▪		▪	
Subaqueous – Fluviatile – High energy	■	■	■											▪	▪	▪	▪	▪		■	▪									
Subaqueous – Fluviatile – Torrential	■	■	■	▪	▪	▪	▪	▪						▪	▪	▪	■	△	▪	■	▪									
Subaqueous – Deltaic – Subaerial	■	■	■											△	△	△	▪	▪		■	■							▪		▪
Subaqueous – Deltaic – Marsh	■	■	■											△	△	△	▪	▪		■	■								▪	△
Subaqueous – Deltaic – Channel – Landward	■	■	■											△	△	△	▪	▪		■	■						▪		△	▪
Subaqueous – Deltaic – Channel – Medial	■	■	■											△	△	△	▪	▪		■	■						△			△
Subaqueous – Deltaic – Channel – Seaward	■	■	■											▪	▪	▪	▪	▪		■	■	▪	△							
Subaqueous – Marine – Intertidal	■	■	▪	△	△		△	△						■	■	■	△	■	△	■	■	△					△			
Subaqueous – Marine – Subtidal	■	■	▪											▪	▪	▪	△	■		■	■	▪	△							
Aeolian – Coastal	■	■	▪	△	△	△	△	△				▪	▪	▪	▪	▪	△	■	△	■	■									
Aeolian – Hot desert	■	■	▪		△	△						▪	△				△	■		■	■		△	△	△	△				
Glacial				■	■	■	■	■	■	▪	▪	■	△	■	▪		■		■											
Pedologic – Silica Dissolved – Temperate	■	■	■														▪	▪		■	■	▪	△			△				
Pedologic – Silica Dissolved – Tropical	■	■	■														▪	▪		■	■	■	▪	■	■	■				
Pedologic – Silica Precipitated – Temperate	■	■	■														▪	▪		■	■						■	△	■	▪
Pedologic – Silica Precipitated – Tropical	■	■	■														▪	▪		■	■						■	▪	■	▪
Subsurface diagenetic	■	■	■									▪					▪	▪		■	■		▪	▪	▪		■		■	▪

FIGURE 3-9. Environmental variation and terminology employed to describe surface textures of quartz as seen by the SEM on modern sands, and possible environmental significance. (Redrawn from Higgs, 1979, Fig. 2).

Ribault and Krinsley are correct, an effective means of studying successive depositional environments is at hand for unconsolidated sands. Needed, however, are similar studies in modern sands with widely differing provenances—with great contrasts in inheritance—because environmental discrimination by texture is rich in techniques that had initial local success but later failed as universal environmental discriminators. At best, the study of surface textures of sand grains is mostly applicable to environmental discrimination of modern sands.

Textural Maturity

Sedimentologists, recognizing the wide range of sandstone textures from clay-rich and angular to well-sorted and rounded, have extended the concept of maturity from the mineralogy of sands to the description of textural variation. Plumley (1948, pp. 573–574) defined a mature gravel or sand as one consisting of a well-sorted, mineralogically mature framework of grains. Folk (1951) defined *textural maturity* as the degree to which a sand is free of interstitial clay and is well sorted and well rounded. He gives a clear procedure for determination of three stages of textural maturity (see Appendix, Table A-1). These three stages are based on the idea that in transport clay is first removed, then framework grains are sorted, and much later they are finally rounded. In short, proponents of the idea of textural maturity attempt to relate winnowing and abrasion by currents to sorting and rounding. The *final* depositional environment, however, controls textural maturity. Thus during a storm, well-rounded and well-sorted marine shelf sand might be mixed with much clay in deeper water. Or bioturbation can produce a like mixture.

It is widely believed that sorting is best when sand is repeatedly exposed to reworking by currents of moderate intensity. Probably the worst sorting occurs in sands subject to one brief episode of mass transport, such as a submarine slide, which is deposited below wave base in deep water. Kuenen (1964, pp. 213–15) suggested the term "repository" for an environment where grains become unavailable for transport once deposited. The sand of a barchan dune moving over bedrock is a good example of an environment lacking repositories and thus one with excellent sorting. A similar environment is the berm of a beach.

The late stage of maturity—grain roundness—has never been quantified as to rate in natural environments, although Kuenen's (1959, p. 186) experiments showed it to be exceedingly slow.

Appreciable clay in either a well-sorted or well-rounded framework is called a *textural inversion*. A poorly sorted framework of very well-rounded grains provides another example of a textural inversion. The relationships between textural and mineralogical maturity are not fully established, largely because they have not been systematically and quantitatively studied.

The concept of textural maturity has been applied primarily to ancient sandstones where it is used for much the same purposes as a size analysis. Both rounding and sorting of framework grains are easily determined in thin section. But clay content is much more troublesome, for it is commonly difficult to distinguish detrital from authigenic clay in many ancient sandstones, a distinction that is vital, if textural maturity is to be a meaningful term. As a means of easy and rapid textural description in consolidated sandstones, however, the concept of textural maturity is useful.

Evaluation

In the vast literature of textural studies, the major question is the nature of the size distribution and the utility of textural studies as environmental discriminators—how effective are cumulative curves, textural parameters derived from them, graphic plots, and the surface features of quartz grains at discriminating different sedimentary environments? The basic problem is distinguishing inherited characteristics from those of the depositional environment and, in ancient sandstones, minimizing the effects of diagenesis (Fig. 3-10).

Textural fingerprinting is, as we have seen, an empirical technique based on sampling of modern sands from known different environments using either graphic plots or statistical tests for discrimination. Underlying this work is the idea that successful discrimination of modern environments of sand deposition will permit us to identify and discriminate between the environments of deposition of ancient, consolidated sandstones. Grain size as well as grain shape and roundness plus surface texture have all been used. Examples of such studies for grain size include Visher (1969), Glaister and Nelson (1974), Stapor and Tanner (1975), and Sagoe and Visher (1977) for breaks in the size curve, and Friedman (1961, 1979) for bivariable plots. Discriminate functions have been used by Sahu (1964), Moiola and Weiser (1968), Taira and Scholle (1979), and Reed *et al.* (1975) and factor analysis by Klovan (1966) and Solohub and Klovan (1970). Factor analysis is thought by Glaister and Nelson (1974, p. 237) to be more useful in *mapping* the size distribution of modern sands rather than classifying the distributions into different groups. For additional approaches, see the list of papers compiled by Socci and Tanner (1980).

The key problem seems to be twofold—the grain size distribution reflects provenance as well as the hydraulics of the depositional site, and the hydraulics of the depositional site do not always correlate with the depositional environment. In one test of graphic methods, for example, Sedimentation Seminar (1981) found only 13 percent of 57 samples from the sands of the Rio Solimões of Brazil to indicate a fluvial environment. Hence the search for universal discriminators seems unlikely to be successful, even though individual studies such as those by Reed *et al.* (1975) and Taira and Scholle (1979) yield successful local results.

We suggest that grain size analysis will always be most useful (1) when closely integrated with the type of sedimentary structure which itself depends on grain size and flow intensity, (2) when grain size is plotted against vertical profile and related to log form, and (3) when settling velocities rather than grain diameters are used.

Hydraulic ratios have had some limited use in distinguishing between water- and air-deposited sands—an idea first proposed by Ruhkin (1937), developed by Hand (1967), and applied by Stiedtmann (1974) and Bart (1977), who both sampled laminae of foresets of crossbeds and inferred an eolian origin, because the cumulative distributions of velocities of quartz in water were less than that of the heavy minerals in the same laminae.

In sum, it is difficult today to write with conviction on behalf of textural studies of modern sands as effective environmental discriminations—especially when the great scope of past efforts is recognized. In addition, many of the techniques used for modern sands cannot easily be applied to indurated ancient sandstones.

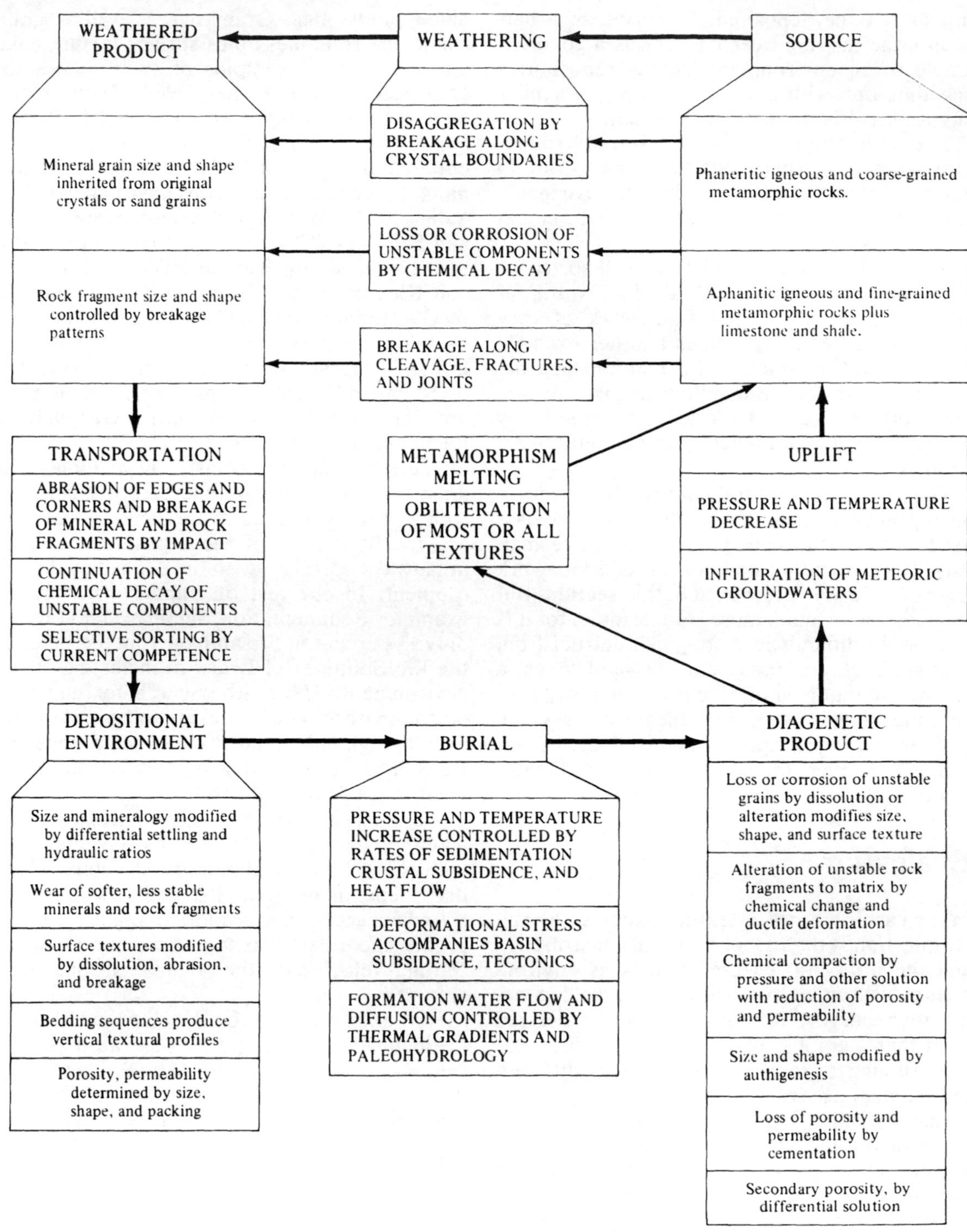

FIGURE 3-10. Rock cycle and texture. In all textural studies directed toward environmental recognition, it is necessary to consider—or at least recognize—effects other than those produced by the last depositional environment.

There has been some discussion of the lognormal curve as the appropriate curve to represent a textural mature sandstone. An alternative is Rosin's law of crushing (Rosin and Rammler, 1933). Dapples (1975) reviewed this question and concluded that transport in a long river leads to a distribution that follows Rosin's law whereas repeated reworking and resultant winnowing of beaches and/or marine shelves with very slow sedimentation rates and subsidence—as is typical of the supermature sandstones of stable cratons—follows the lognormal distribution (Fig. 3-2). If Dapples' conclusion is generally true, which distribution prevails in dune sands exposed to repeated winnowing? The Rosin distribution is also believed to best describe sands produced by glacial breakage and grinding and grains resulting from explosive volcanism.

Control of Physical Properties

Many attempts have been made to relate physical properties, such as permeability, sonic transmissibility, tensile strength, and thermal conductivity, to the size distribution. These efforts have usually proved to be moderately successful. It is necessary to have a carefully designed experiment because, in general, any aggregate physical property is a complex function of size, grain shape, fabric, and composition, with the result that three physical properties may have to be sorted out via regression analysis (Krumbein and Graybill, 1965, Chaps. 10 and 12; Davis, 1986, pp. 469–477; Till, 1974, pp. 83–103). This is especially true when the property is a "sensitive" one, such as permeability. A stepwise regression procedure permits one to test the effect of adding or subtracting successive independent variables, x_1, x_2, x_3, . . . , x_n to the dependent variable. How porosity and grain size are related to bulk density and sound velocity was well shown by Hamilton (1970).

Fabric

The fabric of sand and sandstones—that is, the ways in which the grains are put together to make an aggregate—is dependent on the currents that deposit large numbers of grains of diverse size, shape, and roundness and on how the aggregate is later compacted by physical and chemical processes.

Grain Contacts. Grain-to-grain relations as seen in thin section are described by a combination of qualitative terms and quantitative indices, both of which attempt to infer three-dimensional relationships in the plane of the thin section. While we have the terminology and a little of the methodology, we lack systematic knowledge of the spatial distribution of fabric types in a sandstone body. Such knowledge is needed for a better understanding of the origin and relative ages of their cementing agents and diagenetic processes. One impediment to progress is lack of reliable instrumentation that could determine grain-to-grain relations more quickly than grain-by-grain microscopic counting methods and perhaps do so in three dimensions as well.

Another difficulty in mapping fabric type is that fabric is highly variable—much like permeability—as can be seen in many single thin sections. Some of this variation is linked to primary deposition, subtle differences in original grain-to-grain relations being characteristic of each lamination and bed. These differences influence, of course, later diagenetic processes. What is needed is a better way of increasing the "signal-to-noise ratio" of fabric studies.

Table 3-5 defines the qualitative and quantitative terms used in studies of grain-to-grain relations in sandstones and Fig. 3-11 illustrates their application. An overview of fabric is given in Table 3-6.

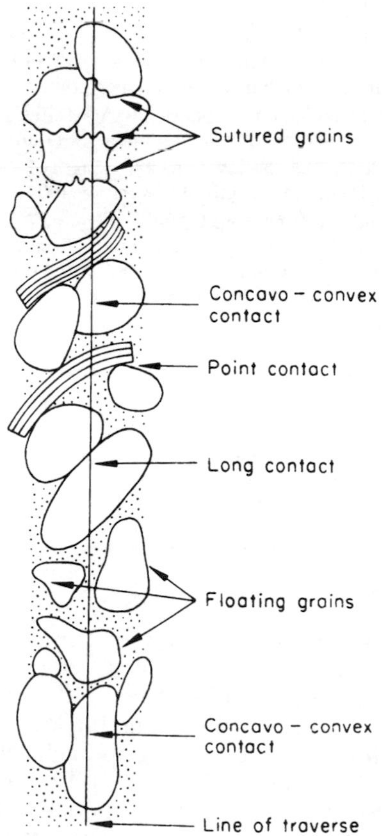

FIGURE 3-11. Definition sketch of fabric terminology: quartz (white), mica (lined), and matrix (stippled).

TABLE 3-5. Terms used to specify sandstone fabric.

QUALITATIVE

Concavo-convex contact (Taylor, 1950, p. 707): one that appears as a curved line in the plane of section.

Contact strength (Füchtbauer, 1968, p. 365): a linear combination of contact types yields a single number (see also weighted contact packing).

Fixed margin (Allen, 1962, p. 678): that part of a grain in contact with another in the plane of section.

Fixed grain (Allen, 1962, p. 678): fixed margin exceeds free margin.

Floating grain: no contacts with other grains in the plane of section.

Framework fraction: the stress-transmitting portion of a sand.

Free margin (Allen, 1962, p. 678): that part of grain not in contact with other grains in the plane of section.

Free grain (Allen, 1962, p. 678): free margin exceeds fixed margin.

Long contact (Taylor, 1950, p. 707): a contact that appears as a straight line in the plane of section.

Packing (Kahn, 1956, p. 390): mutual spatial relationships among grains.

Sutured contact: mutual stylolitic interpenetration of two or more grains.

Tangential contact (Taylor, 1950, p. 707): one that appears as a point in the plane of section.

QUANTITATIVE

Condensation index (Allen, 1962, p. 678): ratio of percentage of fixed rock fragments to percentage of free grains.

Contact index: number of contacts per grain.

Horizontal packing intercept (Mellon, 1964, Fig. 7): average horizontal distance between framework grains.

Packing density (Kahn, 1956, p. 390): length of grains intercepted divided by length of traverse × 100.

Packing index (Emery and Griffiths, 1954, p. 71): the product of the number of quartz to quartz contacts per traverse and the average quartz diameter, the product being divided by the total length of traverse.

Packing proximity (Kahn, 1956, p. 390): number of grain to grain contacts divided by total number of contacts of all kinds (grain to matrix and grain to cement) × 100.

Vertical packing intercept (Mellon, 1964, Fig. 7): average vertical distance between framework grains.

Weighted contact packing (Hoholick *et al.*, 1982): the basic contact types are weighted by powers of two to provide more discrimination between fabrics than contact strength.

The proportions of the four contact types—point, long, concavo-convex, and sutured—have been used by Füchtbauer (1967, p. 365) to define a single number he called "contact strength." Füchtbauer's approach has been modified by Hoholick *et al.* (1982) to also include floating grains and, additionally, these authors weighted the different contact types by powers of 2 rather than linearly as did Füchtbauer. Cathodo-luminescence is more appropriate for the study of grain contact types than is the petrographic microscope (Sippel, 1968; Nickel, 1978).

TABLE 3-6. Sandstone fabric (modified from Adams, 1964, Table 1).

A. Grain to grain relations specified chiefly by qualitative observation supplemented by some quantitative indices on nonoriented samples. Chiefly used to interpret and predict reservoir porosity.
   1. Much pressure solution.
      Many sutured contacts, grain to grain contacts large, packing density high, and no porosity and cement.
   2. Moderate pressure solution.
      Some sutured contacts, principally equidimensional grains, chiefly with concavo-convex contacts. Cement is mostly quartz overgrowths with or without minor amounts of carbonate or clay. Little porosity.
   3. Minor pressure solution.
      Mostly original grain outlines with long and tangential contacts. Low to moderate number of grain to grain contacts and moderate packing density. May be either well or poorly cemented by quartz overgrowths, clay, and carbonate. May have moderate porosity, if poorly cemented.
   4. No pressure solution.
      Chiefly original grain outlines with either tangential contacts or floating grains. Number of grain to grain contacts is low as is packing density. Cements are mostly carbonate or clay. Porosity is high when cementation is limited.

B. Orientation of framework specified by quantitative measurement of oriented samples. Porosity and cementation are independent of orientation. Chiefly used to determine current direction in undeformed sediments.
   1. Particulate methods.
      Visual, direct measurement of either long axes or apparent long axes of framework grains usually in thin section.
   2. Aggregate methods.
      Measurement by an appropriate black box of a bulk geophysical property that can be correlated with the orientation of the framework grains.

Most of the study of contact types in sandstones has been made as part of investigations of their diagenetic history, either using samples from vertical profiles of deep wells (Taylor, 1950; Zimmerle, 1976, Fig. 72) or more rarely sampling a single sandstone as it is traced down dip into a basin. However, there has been some effort to relate contact types to strength (Bell, 1978). An exceptional example of the relation between grain fabrics and rock behavior is the flexible sandstones called *itacolumites*, whose fabrics have been analyzed by Dusseault (1980). Contact types in a sandstone clearly need attention when porosity—be it primary or dissolution—and permeability are studied (see Chap. 11).

Orientation. Orientation of framework grains has been related to permeability, used for paleoecologic studies, and correlated with sedimentary structures—to better understand how structures are formed. Grain imbrication may also be useful in the study of seemingly massive sandstones—for example, is grain orientation random and hence the product of bioturbation or perhaps some type of mass flow (Hiscott and Middleton, 1980) or does it represent, in fact, normal depositional fabric such as that shown in Fig. 3-12? Regional fabric studies are rare, chiefly because paleocurrent data based on sedimentary structures can be obtained more rapidly and thus provide better regional information.

Shape fabrics can be measured directly by apparent long axes in oriented thin sections and (or from photographs of them) by photometric methods using thin sections, by magnetic susceptibility, and by anisotropic electrical conductivity. The first two methods are based on summations of the orientation of individual grains and the last two on a bulk property which is related to the aggregate orientation of the framework. Taira and Lienert (1979) compared direct measurement of long axes in thin sections to photometric and magnetic methods with good results (Fig. 3-13). The latter is by far the most rapid, however, and also gives a three-dimensional estimate of orientation—one that thus includes imbrication. Shelton *et al.* (1974) give a good demonstration of electrical anistropy as a measure of bulk fabric, using plugs of modern sand in a braided river. Stained acetate peels have also been used (Hiscott and Middleton, 1980).

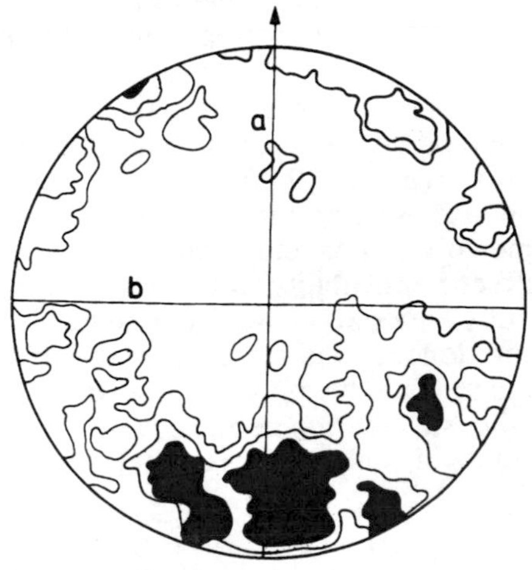

FIGURE 3-12. Orientation of sand grains in flume. Long axis of quartz grains plotted on equal area net. Note imbrication. Arrow shows current. (Modified from Rusnak, 1957, Fig. 9d).

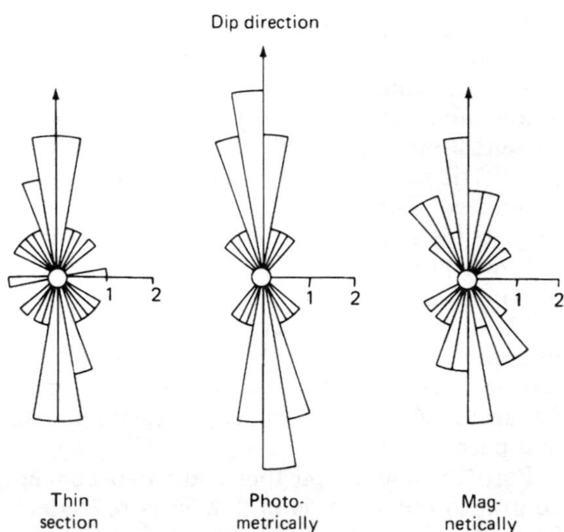

FIGURE 3-13. Orientation of sand grains in plane of bedding as measured individually in thin section, photometrically, and magnetically. Data from Taira and Scholle (1979, Fig. 9).

Potter and Pettijohn (1977, pp. 70–72) summarize other aspects of the fabric literature of sand and sandstone.

Porosity and Permeability

A sandstone consists of a framework of grains, interstitial detrital silt and clay, chemical cements, and an interconnecting network of void space or pores. The study of the pore system of rocks is included in the area of petrophysics. Contributions to this field have been made by petroleum engineers, groundwater hydrologists, soil scientists, and geologists. Very much still remains to be learned.

There are four types of porosity in sandstones: *interparticulate* (intergranular), the voids between framework grains, small detrital grains, and/or dense authigenic minerals; *intraparticulate* (intragranular) within grains such as in most authigenic clays; *fracture porosity*, micro- and macro-; and *solution porosity* related to the solution of framework or cements. The latter is also included in secondary porosity. Mechanical compaction of the framework (the rotation and wedging of rigid grains), pressure solution at points of contact, and cementation all combine to reduce porosity (in some quartzites almost to zero), but solution at depth and in outcrop—*decementation*—enhance it (see Chapter 11).

It is the pore system that permits a sandstone to store and transmit fluid—groundwater, oil, gas, and mineralizing solutions. The size, shape, and pattern of this pore system in a sand or sandstone is very difficult to specify. Lack of geometric regularity and small size are the chief obstacles. The scanning electron microscope with its high depth of field beautifully illustrates the intricate surfaces of pores in sandstone (Fig. 3-14). SEM photographs and energy-dispersive EDX analyses help determine the geometry and mineralogy of secondary pore fillings that form during burial and the actual size of the pores (Welton, 1984), and thus better explain porosity and permeability.

Porosity and permeability are two concepts central to the analysis of flow in pore systems. Like density, both are intensive properties. *Void space* is that portion of a sandstone not occupied by its solid components. Voids may either be connected or isolated. Porosity, a scalar quantity, is expressed as a percent. *Absolute* or *total porosity* is defined as

$$P_t = \frac{\text{bulk volume} - \text{solid volume}}{\text{bulk volume}} \times 100,$$

and *effective porosity* as

$$P_e = \frac{\text{interconnected pore volume}}{\text{bulk volume}} \times 100.$$

Absolute or total porosity is the percentage of all void space in a sand, whether the pores are connected or not, whereas effective porosity is the percentage of interconnected pore space. Effective porosity, commonly less than total porosity, is what is ordinarily measured in most reservoir studies. Pumice and foam plastic are two substances with very large absolute porosity but very small effective porosity. See Pirson (1958, pp. 31–40) and Monicard (1980, pp. 7–42) for a summary of the many different laboratory measurement techniques used to determine porosity. Different methods are used to determine total and effective porosities. Semiquantitative estimates can also be obtained from electric logs, micro-logs, and sonic logs. Sneider *et al.* (1983) give criteria to identify and describe porosity types in cuttings.

The packing of equal-sized spheres has been studied in order to approximate porosities found in granular materials, since the early work of Graton and Fraser (1935). See Scheidegger (1974, pp. 19–23) for a review. For a given mode of packing of uniformly sized spheres, porosity is independent of size of the spheres: tightest packing (rhombohedral) has a porosity of 25.9 percent and loosest packing has a porosity of 47.6 percent. The abundance of grains of mixed sizes and nonspherical framework grains, and the filling of pores by cementation and interstitial clay minimizes the significance of these studies for sandstone, however. In addition, it has been virtually impossible to specify the irregular packing of framework grains of sandstones by crystallographic terms such as rhombohedral, isometric, etc.

Most Phanerozoic sandstones that form petroleum reservoirs commonly have porosities of 5 to 20 percent. As cementation increases, particularly silica cementation, fracture porosity is likely to increase with respect to intergranular porosity, because the sandstone is more brittle and thus fractures more readily.

Specific permeability, k, measures the ability of a medium to transmit a fluid and its unit of measurement is the darcy, which has the dimensions of length squared. For the special case of the horizontal flow of an incompressible fluid, *Darcy's law* can also be written as

$$Q = \frac{k}{\mu} A \frac{\Delta P}{L}$$

where Q is the flow rate, k is the permeability constant of the medium, μ is the fluid viscosity measured in centipoises (both k and μ are, in effect, simply constants of proportionality), A is the cross-sectional area, ΔP is the change in pressure, and L is the length over which the pressure drop occurs. Pirson (1958, pp. 62–67) and Monicard (1980, pp. 45–84) describe laboratory methods for the measurement of permeability, and Scheidegger (1974) provides an extended mathematical treatment; Matheron (1967) has a very different approach to the study of pore systems, one based largely on mathematical set theory, which has found some adherents (Watson, 1975).

Grain size, sorting, orientation, and packing of framework grains and cementation and bedding affect permeability in a complex way that is not fully understood even though much study has been done. It has been found that the finer the grain size and the poorer the sorting of loose sand, the smaller its permeability, as shown by Krumbein and Monk (1943, p. 10) in the formula

$$k = Cd^2e^{-1.35\sigma}$$

where C is a constant, d is geometric mean grain diameter, e is the base of the natural logarithms, and σ is sorting of the sand. Lerman (1979, Table 2.1, Fig. 2.5) gives eight other

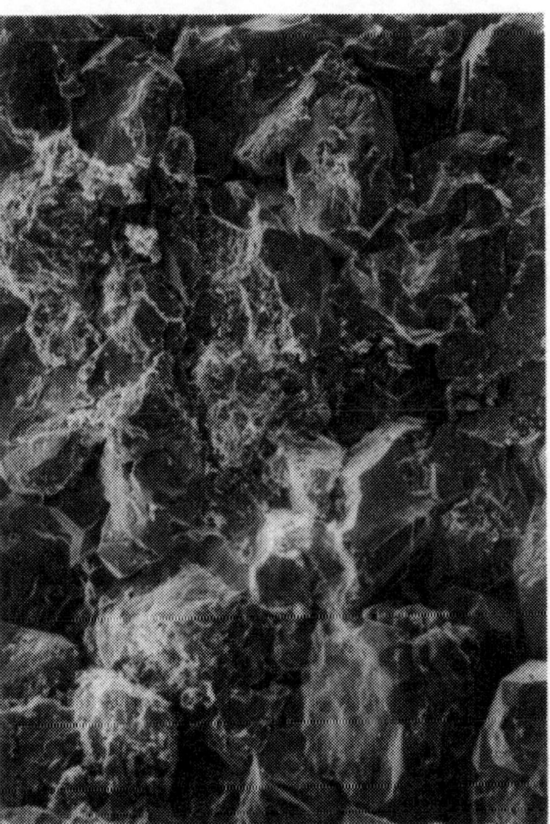

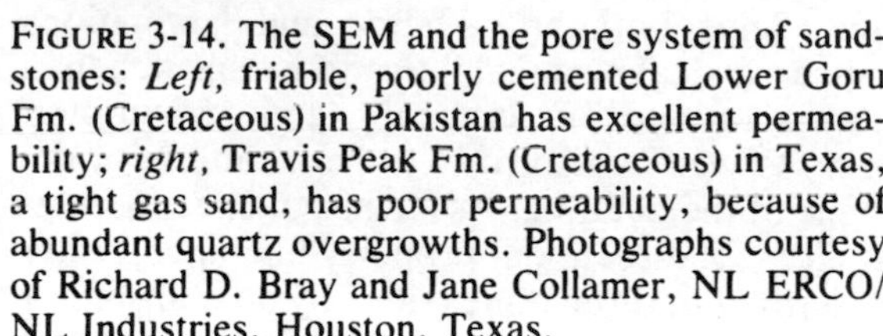

FIGURE 3-14. The SEM and the pore system of sandstones: *Left,* friable, poorly cemented Lower Goru Fm. (Cretaceous) in Pakistan has excellent permeability; *right,* Travis Peak Fm. (Cretaceous) in Texas, a tight gas sand, has poor permeability, because of abundant quartz overgrowths. Photographs courtesy of Richard D. Bray and Jane Collamer, NL ERCO/NL Industries, Houston, Texas.

equations relating permeability to porosity and grain size and shows a graphic plot for each.

Beard and Weyl (1973) studied artificial mixtures of sands and related permeability and porosity to size and sorting. A useful set of photomicrographs of thin sections is included in their report. They observed that permeability decreased with grain size in artificial sand mixtures and that porosity increased with sorting—from 28 to 48 percent. Permeability decreases with finer grain size because of increasing internal surface area, which increases the surface drag on the flowing fluid. Porosity increases with better sorting because fewer small grains occur between the larger grains. However, some differences have been observed in the sands of different environments. Pryor (1973) measured the porosity and permeability of 922 modern river, dune, and beach sands and related both to grain size, sorting, and lamination and bedding. He found permeability to increase with sorting and porosity to increase with grain size in river sands—but not in dune or beach sands. When permeability and porosity are segregated by environment in ancient sandstones, plots commonly show subfields, some of which may overlap (Fig. 3-15). Thus there is evidence that depositional processes affect the porosity and permeability distribution of ancient sandstones. More studies of porosity, permeability, size, sorting, and lamination of modern sands in different environments clearly would be helpful, especially if carefully related to lamination, bedding, and structures. Minor structures can have dramatic effect on permeability and porosity (Fig. 3-16). Such studies provide baselines for initial porosities and permeabilities that are useful in understanding their decrease during later burial and diagenesis.

What can be said of grain orientation, packing, and permeability? Orientation and packing of the framework grains of a sandstone appear to have a weak control on permeability in the plane of the bedding but a strong control in sections perpendicular to bedding (Mast and Potter, 1963, pp. 558–59). Framework sand grains

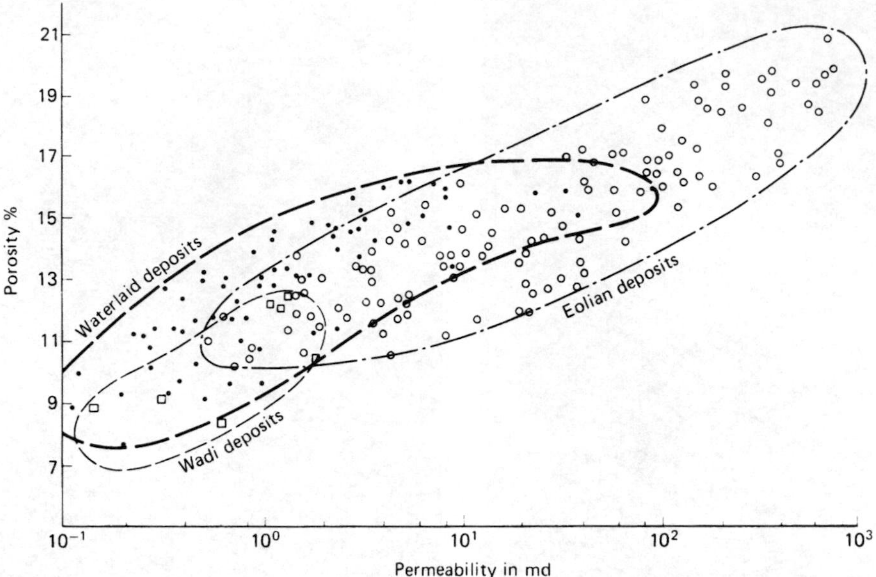

FIGURE 3-15. Porosity, permeability, and depositional environment in the Leman gas field of the North Sea. (Redrawn from Van Veen, 1975, Fig. 11).

accumulate with their long axes parallel to current flow and imbricate upcurrent 15 to 18 degrees from the depositional interface. As a consequence, an anisotropy is imposed on the pore system. Bedding too has an important effect on permeability—one that probably appreciably exceeds that of fabric—for slight pauses in sand

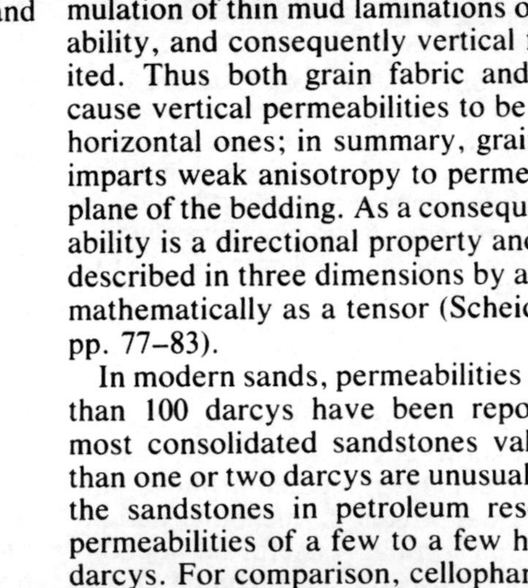

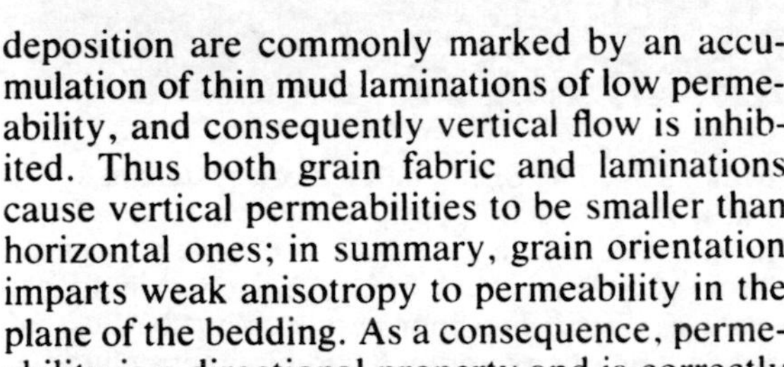

deposition are commonly marked by an accumulation of thin mud laminations of low permeability, and consequently vertical flow is inhibited. Thus both grain fabric and laminations cause vertical permeabilities to be smaller than horizontal ones; in summary, grain orientation imparts weak anisotropy to permeability in the plane of the bedding. As a consequence, permeability is a directional property and is correctly described in three dimensions by an ellipsoid or mathematically as a tensor (Scheidegger, 1974, pp. 77–83).

In modern sands, permeabilities of 10 to more than 100 darcys have been reported. But in most consolidated sandstones values of more than one or two darcys are unusual and many of the sandstones in petroleum reservoirs have permeabilities of a few to a few hundred millidarcys. For comparison, cellophane commonly has a permeability of less than one millidarcy. Permeability in reservoirs, especially from the same bedding facies in a reservoir, tends to be lognormally distributed, whereas porosities typically have a normal distribution. Moreover, variance of permeability is much greater than that of porosity (Fig. 3-15) and can vary markedly vertically, centimeter by centimeter.

Nearly always there is good correlation between effective porosity and permeability in a sandstone reservoir (Fig. 3-15). The Kozeny–Carman equation (Scheidegger, 1974, p. 141) re-

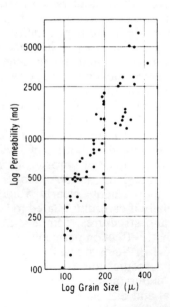

FIGURE 3-16. Dependence of permeability on grain size in the Bentheimer Sandstone of the Scherrhorn oil field near Lingen, Germany. Regression equation is $\log y = 2.1007 + 2.221 \log x$, where y is permeability in millidarcys and x is grain size in millimeters. Diagram based on random selection of data from Fig. 49 of von Engelhardt (1960).

lates permeability to internal surface area, S_v,

$$K = \frac{P_e^3}{5\,S_v^2(1 - P_e)^2}$$

where S_v is the specific surface exposed to the fluid per unit volume of solid (not porous material) and P_e is effective porosity. Specific surface area per solid volume is closely related to grain size—the finer the grain size, the greater are values of S_v (Fig. 3-16).

The Kozeny–Carman equation gives insight into the dependence of permeability on effective porosity, especially in unconsolidated sands. As a first approximation, permeability is a complicated power function of porosity but inversely proportional to the second power of specific surface. The Kozeny–Carman equation is useful in that it explains why a fine-grained sand or silt with effective porosity identical to that of a coarser-grained sand has smaller permeability: as grain size decreases, specific surface increases and consequently so does resistance to flow (Fig. 3-17).

And why is the permeability of sandstones so much more variable than their porosity? There appear to be two principal reasons. A given porosity defines neither the drag imparted to a fluid by the small-scale roughness of the walls of the pore system (Fig. 3-14) nor the path length that the fluid must flow to go between any two points in the sandstone. Change in either or both can alter permeability without change of porosity. The Kozeny–Carman equation may also shed some light on the much greater spatial variability of permeability than porosity. For example, while progressive cementation decreases both specific surface and porosity, perhaps they decrease at variable rates so that permeability, being inversely proportional to the *square* of specific surface, is much more variable than porosity. Von Engelhardt and Pitter (1951) give much quantitative data on the relationships between porosity, permeability, and grain size. Von Engelhardt (1960) gives a thorough treatment of the porosity, transmissive characteristics, and chemical composition of the pore system in sediments.

Porosity and permeability in ancient sandstones are best studied jointly, because they are both related to pore space. But no matter how studied, it is always best to tabulate the petrographic observations on thin sections from core plugs by age, mineralogy, grain size, and sedimentary facies—or perhaps even by geothermal gradient—to help sort out and isolate the variables, which, acting together, commonly combine to give the data an appreciable scatter.

Selley (1975, 1978) did this in his study of porosity of sandstones in the North Sea, where he related porosity to sandstone facies and to quartz content (Fig. 3-17). He concluded that in the North Sea porosity–depth curves are linear and the gradient increases with increasing quartz content, abnormal pressures, and presence of hydrocarbons. He also formulated the concept of a "porosity window" (Fig. 3-18)—a parallelepiped on a porosity–depth plot which shows ranges of porosity likely to be found in sandstones of different environments at differ-

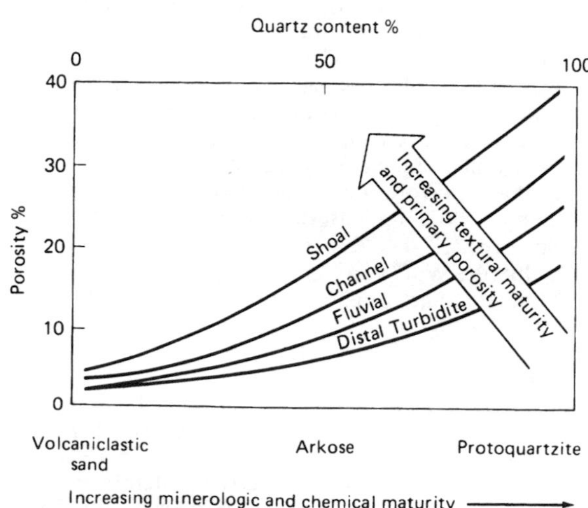

FIGURE 3-17. Porosity and mineralogical maturity. (Redrawn from Selley, 1975, Fig. 9).

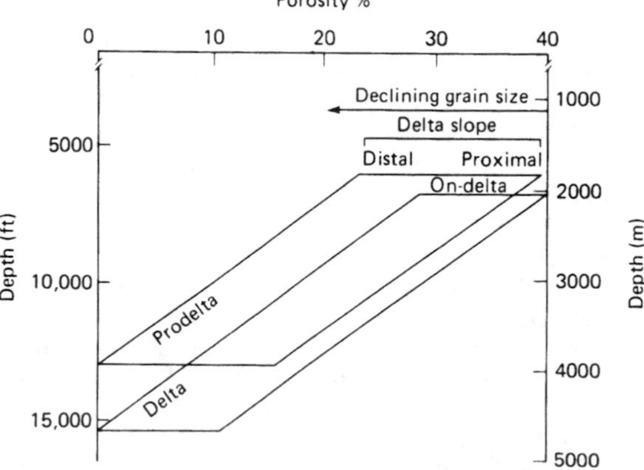

FIGURE 3-18. Estimated changes of porosity with depth and "porosity windows," in the North Sea. (Modified from Selley, 1978, Fig. 12). Windows such as these are calculated from both porosity gradients and the porosity ranges of different types of sandstone.

ent depths. Selley emphasizes that, when porosities are compared, estimated total depth of burial (Chap. 12) rather than present depth of burial should be used. The starting point for all studies is the porosity and permeability of modern sands.

Nagtegaal (1980) has recognized nine different curves which represent how porosity decreases with depth. Porosity–depth curves can be nonlinear, negative exponentials or be linear and discontinuous (Hoholick *et al.*, 1984)—the discontinuity and greater porosity at depth being related to the solution of framework grains and cements, called *solution porosity*. Pittman (1980, pp. 50–54) gives a general review of how solution porosity has been studied, as do Schmidt and McDonald (1979). The full recognition of solution porosity in quartz-rich sandstones waited many years—long after it was well known in carbonates—and is discussed at length in Chap. 11.

References

Adams, W.L.: Diagenetic aspects of Lower Morrowan, Pennsylvanian sandstones, northwestern Oklahoma. Am. Assoc. Petroleum Geologists Bull. 48, 1568–1580 (1964).

Allen, J.R.L.: Petrology, origin and deposition of the highest Lower Old Red Sandstone of Shropshire, England. Jour. Sed. Petrology 32, 657–697 (1962).

Allen, Terence: Particle size measurement, 3rd Ed., 678 pp. London: Chapman and Hall 1981.

Andel, Tj. H. van, and Postma, H.: Recent sediments of the Gulf of Paria, Vol. 1. Koninkl. Nederlandse Akad. Wetensch. Verh. 20, 245 pp. (1954).

Anderson, John B., and Kurtz, Dennis, D.: "RUASA": An automated rapid sediment analyzer. Jour. Sed. Petrology 49, 625–627 (1979).

A.S.T.M.: Symposium on particle size measurement. Am. Soc. Testing Materials, Spec. Tech. Publ. 234, 303 pp. (1959).

Bart, H.A.: Sedimentology of cross-stratified sandstones in Arkaree Group, Miocene, southern Wyoming. Sedimentary Geology 19, 165–184 (1977).

Beard, D.C., and Weyl, P.K.: Influence of texture on porosity and permeability of unconsolidated sand. Am. Assoc. Petroleum Geologists Bull. 57, 349–369 (1973).

Bell, F.G.: The physical and mechanical properties of the Fell Sandstones, Northumberland, England. Engineering Geology 12, 1–29 (1978).

Benson, D.J.: Textural analysis with Texas Instruments 59 Programmable Calculator. Jour. Sed. Petrology 51, 641–642 (1981).

Boggs, Sam., Jr.: Measurement of roundness and sphericity parameters using an electronic particle size analyzer. Jour. Sed. Petrology 37, 908–913 (1967).

Brewer, Roy: Fabric and mineral analysis of soils, 470 pp. New York: Wiley 1964.

Briggs, L.I.; McCulloch, D.S.; and Moser, Frank: The hydraulic shape of sand particles. Jour. Sed. Petrology 32, 645–656 (1962).

Burger, Heinz: Log-normal interpolation in grain size analysis. Sedimentology 23, 395–405 (1976).

Dapples, E.C.: Laws of distribution applied to sand sizes. In: Whitten, E.H. Timothy (Ed.): Quantitative studies in the geological sciences. Geol. Soc. America Mem. 142, 37–61 (1975).

Davies, David K.; Ethridge, Frank G.; and Berg, Robert R.: Recognition of barrier environments. Am. Assoc. Petroleum Geologists Bull. 55, 550–565 (1971).

Davis, John C.: Statistics and data analysis, 2nd ed., 646 pp. New York, John Wiley and Sons 1986.

Dobrin, M.B.: Introduction to geophysical prospecting, 3rd. Ed., 630 pp. New York: McGraw-Hill 1976.

Doeglas, D.J.: Interpretation of the results of mechanical analysis. Jour. Sed. Petrology 16, 19–40 (1946).

Dusseault, M.B.: Itacolumites: the flexible sandstones. Quart. Jour. Eng. Geol. London 13, 119–128 (1980).

Ehrlich, R., and Weinberg, B.: An exact method for characterization of grain shape: Jour. Sed. Petrology 40, 205–212 (1970).

Ehrlich, Robert; Orzeck, John J.; and Weinberg, Bernhard: Detrital quartz as a natural tracer—Fourier grain shape analysis. Jour. Sed. Petrology 44, 145–150 (1974).

Emery, J.R., and Griffiths, J.C.: Reconaissance investigation into relationships between behavior and petrographic properties of some Mississippian sediments. Pennsylvania State Univ., Min. Ind. Expt. Sta. Bull. 62, 67–80 (1954).

Emrich, Grover, and Wobber, F.J.: A rapid method for estimating sedimentary parameters: Jour. Sed. Petrology 33, 831–841 (1963).

Engelhardt, Wolf von: Die Unterscheidung wasser- und windsortieter Sande auf Grund der Korngrößenverteilung ihrer leichten und schweren Gemengteile. Chemie Erde 12, 451–465 (1940).

Engelhardt, Wolf von: Der Porenraum der Sedimente, 207 pp. Berlin–Göttingen–Heidelberg: Springer, 1960.

Engelhardt, Wolf von, and Pitter, H.: Über die Zusammenhänge zwischen Porosität, Permeabilität und Korngröße bei Sand und Sandsteinen. Heidelberger Beitr. Min. Petrogr. 2, 477–491 (1951).

Folk, R.L.: Stages of textural maturity in sedimentary rocks. Jour. Sed. Petrology 21, 127–130 (1951).

Folk, R.L.: Student operator error in determination of roundness, sphericity, and grain size. Jour. Sed. Petrology 25, 297–301 (1955).

Folk, R.L.: Rapid phi-millimeter conversion by ordinary slide rule. Jour. Sed. Petrology 44, 658–659 (1964).

Folk, R.L.: A review of grain-size parameters. Sedimentology 6, 73–93 (1966).

Folk, R.L.: Bi-modal supermature sandstones: Product of the desert floor. Proc. 23rd Intern. Geol. Cong., Prague 8, 9–32 (1968).

Folk, R.L., and Robles, Rogelio: Carbonate sands of Isla Perez, Alacran Reef complex, Yucatan. Jour. Geology 72, 255–292 (1964).

Folk, R.L., and Ward, W.C.: Brazos River bar: A study in the significance of grain size parameters. Jour. Sed. Petrology 27, 3–26 (1957).

Franzinell, E., and Potter, P.E.: Petrology, chemistry and texture of modern river sands, Amazon River System. Jour. Geology 91, 23–39(1983).

Friedman, G.M.: Distinction between dune, beach, and river sands from their textural characteristics. Jour. Sed. Petrology 31, 514–529 (1961).

Friedman, G.M.: Address of the retiring President of the International Association of Sedimentologists: Differences in size distributions of populations of particles among sands of various origins. Sedimentology 26, 3–32 (1979).

Füchtbauer, Hans: Influence of different types of diagenesis on sandstone porosity. 7th World Petroleum Cong., Mexico, Proc. 2, 353–369 (1967).

Gibbs, R.J.: A settling tube system for sand-size analysis. Jour. Sed. Petrology 44, 583–588 (1974).

Giresse, P., and Le Ribault, Lôic: Contribution de l'étude exoscopique des quartz à la reconstitution paléogéographique des derniers episodes des Quaternaire littoral du Congo. Quaternary Research 15, 86–100 (1981).

Glaister, R.P., and Nelson, H.W.: Grain-size distributions: An aid in facies identification. Bull. Canadian Petroleum Geology 22, 203–240 (1974).

Graton, L.C., and Fraser, H.J.: Experimental study of permeability and porosity of clastic sediments. Jour. Geology 43, 785–909 (1935).

Hamilton, Edwin L.: Sound velocity and related properties of marine sediments, North Pacific. Jour. Geophys. Research 75, 4423–4446 (1970).

Hand, B.M.: Differentiation of beach and dune sands, using settling velocities of light and heavy minerals. Jour. Sed. Petrology 37, 514–520 (1967).

Harrell, J.: Roller micrometer analysis of grain shape. Jour. Sed. Petrology 54, 643–645 (1984).

Harrell, J.A., and Eriksson, K.A.: Empirical conversion equations for thin-section and sieve derived size distribution parameters. Jour. Sed. Petrology 49, 273–280 (1979).

Herdan, G.: Small particle statistics, 418 pp. New York: Academic Press 1960.

Higgs, Roger.: Quartz-grain surface features of Mesozoic–Cenozoic sands from the Labrador and western Greenland continental margins. Jour. Sed. Petrology 49, 599–610 (1979).

Hiscott, R.N., and Middleton, G.V.: Fabric of coarse deep-water sandstones, Turelle Formation, Quebec, Canada. Jour. Sed. Petrology 50, 703–722 (1980).

Hoholick, John; Metarko, Thomas, A.; and Potter, Paul Edwin: Weighted contact packing—improved formula for grain packing. Mountain Geologist 19, 79–82 (1982).

Hoholick, J., Metarko, T.A., and Potter, P.E.: Regional variations of porosity and cement: St. Peter and Mount Simon Sandstones in Illinois Basin: Amer. Assoc. Petroleum Geologists Bull. 68, 733–764 (1984).

Irani, R.R., and Callis, C.F.: Particle size: measurement, interpretation and application, 165 pp. New York: Wiley 1963.

Jackson, P.D.; Smith, D.T.; and Stanford, P.N.: Resistivity–porosity–particle shape relationships for marine sands. Geophysics 43, 1250–1258 (1978).

Jones, H.P., and Speers, R.G.: Permo-Triassic reservoirs of Prudhoe Bay Field, North Slope Alaska. In: Braunstein, Jules (Ed.): North American oil and gas fields. Am. Assoc. Petroleum Geologists Mem. 24, 23–50 (1977).

Kahn, J.S.: The analysis and distribution of the properties of packing in sand size sediments. 1. On the measurement of packing in sandstones. Jour. Geology 64, 385–395 (1956).

Klovan, J.E.: The use of factor analysis in determining depositional environments from grain-size distributions. Jour. Sed. Petrology 36, 115–125 (1966).

Kolmogorov, A.N.: Über das logarithmische Verteilungsgesetz der Teilchen bei Zerstückelung. Dokl. Akad. Nauk SSSR 31, 99–101 (1941).

Krinsley, D.H., and Doornkamp, J.: Atlas of quartz sand grain textures, 91 pp. Cambridge: Cambridge Univ. Press, 1973.

Krinsley, David H.: The present state and future prospects of environmental discrimination by scanning electron microscopy. In: Whalley, Brian (Ed.): Scanning electron microscopy in the study of sediments, pp. 196–180. Norwich, England: Geo. Abstracts, 1978.

Krumbein, W.C.: Size frequency distribution of sediments: Jour. Sed. Petrology 4, 65–77 (1934).

Krumbein, W.C., and Graybill, F.A.: An introduction to statistical models in geology, 475 pp. New York: McGraw-Hill 1965.

Krumbein, W.C., and Monk, G.D.: Permeability as a function of the size parameters of unconsolidated sands. Am. Inst. Mining Metall. Engineers. Tech. Pub. 1492, 11 pp. (1943).

Kuenen, Ph.H.: Experimental abrasion: 3. Fluviatile action on sand. Am. Jour. Sci. 275, 172–190 (1959).

Kuenen, Ph.H.: Pivotability studies of sand by a shapesorter. In: van Straaten, L.M.J.U. (Ed.): Developments in Sedimentology, vol. 1, pp. 208–215. Amsterdam: Elsevier 1964.

Lambasi, J., Jr.: Hydraulic control of grain-size distributions in a macrotidal estuary. Sedimentology 27, 433–446 (1980).

Leggewie, Rüdiger von; Füchtbauer, Hans; and El-Najjar, Ramdam: Zur Bilanz des Buntsandsteinbeckens (Korngrössenverteilung und Gesteinsbruchstücke). Geol. Rundschau 66, 551–557 (1977).

Le Ribault, Löic.: L'exoscopie, méthodes et applications. Compagnie Française des Pétroles, Notes et Mémoires 12, 232 pp. (1976).

Le Ribault, Löic.: L'exoscopie des quartz, 150 pp. Paris: Masson 1977.

Le Ribault, Löic.: The exoscopy of quartz sand grains. In: Walley, Brian (Ed.): Scanning electron microscopy in the study of sediments, pp. 319–328. Norwich, England: Geo. Abstracts 1978.

Lerman, A.: Geochemical processes and sediment environments, 481 pp. New York: Wiley 1979.

Margolis, S.V., and Krinsley, D.H.: Processes of formation and environmental occurrence of micro features on detrital quartz grains. Am. Jour. Science 274, 449–464 (1974).

Mast, R.F., and Potter, P.E.: Sedimentary structures, sand shape fabrics, and permeability, pt. 2. Jour. Geology 71, 548–565 (1963).

Matheron, G.: Eléments pour une théorie des milieux poreux, 166 pp. Paris: Masson 1967.

McCammon, R.B.: Efficiencies of percentile measures for describing the mean size and sorting of sedimentary particles. Jour. Geology 70, 453–465 (1962).

McCave, I.N., and Jarvis, J.: Use of the Model T Coulter Counter in size analysis of fine to coarse sand. Sedimentology 20, 305–315 (1973).

Mellon, G.B.: Discriminatory analysis of calcite- and silicate-cemented phases of the Mountain Park Sandstone. Jour. Geology 72, 786–809 (1964).

Middleton, G.V.: Experiments on density and turbidity currents, 3. The deposition of sediment. Canadian Jour. Earth Sci. 41, 475–505 (1967).

Middleton, G.V.: Generation of the log-normal frequency distribution in sediments. In: Romanova, M.A., and Sarmanov, O.V. (Eds.): Topics in mathematical geology, pp. 34–42. New York: Consultants Bureau 1970.

Middleton, G.V.: Hydraulic interpretation of sand size distributions: Jour. Geology 84, 405–426 (1976).

Moiola, R.J., and Weiser, D.: Textural parameters: An evaluation. Jour. Sed. Petrology 38, 45–53 (1968).

Monicard, R.P.: Properties of reservoir rocks: Core analyses, 168 pp. Houston: Gulf Publishing Co. (Paris: Editions Technip) 1980.

Mrakovich, John V.; Ehrlick, Robert; and Weinberg, Bernard: New techniques for stratigraphic analysis and correlation—Fourier grain shape analysis Louisiana offshore Pliocene. Jour. Sed. Petrology 46, 226–233 (1976).

Nagtegaal, P.J.C.: Sandstone—framework instability as a function of burial diagenesis. Jour. Geol. Soc. London 135, 101–105 (1978).

Nagtegaal, P.J.C.: Diagenetic models for predicting clastic reservoir quality: Revista Instituto Investigaciones Geologicas (Barcelona), 34, 5–19 (1980).

Nickel, E.: The present status of cathodoluminescence as a tool in sedimentology. Minerals Sci. Eng., 73–100 (1978).

Oda, M.: The mechanism of fabric changes during compressional deformation of sand. Soils and Foundations 12, No. 2, 1–18 (1972a).

Oda, M.: Initial fabrics and their relations to mechanical properties of granular material. Soils and Foundations 12, No. 1, 17–36 (1972b).

Orr, C.: Recent advances in particle size measurement. In: Shahinpoor, M. (Ed.): Advances in the mechanics and flow of granular materials, Vol. 1. Rockport, Massachusetts: Trans Tech Pub. 1985.

Otto, George: The sedimentation unit and its use in field sampling. Jour. Geology 46, 569–582 (1938).

Page, H.G.: Phi-millimeter conversion table. Jour. Sed. Petrology 25, 285–292 (1955).

Passega, Renato: Texture as characteristic of clastic deposition. Am. Assoc. Petroleum Geologists Bull. 41, 1952–1984 (1957).

Passega, Renato: Grain size representation by CM patterns as a geological tool. Jour. Sed. Petrology 34, 830–847 (1964).

Pirson, S.J.: Oil reservoir engineering, 2nd Ed., 735 pp. New York: McGraw-Hill 1958.

Pittman, E.D.: Recent advances in sandstone diagenesis, Ann. Rev. Earth Planet. Sci. 7, 39–62 (1979).

Plumley, W.J.: Black Hills terrace gravels: A study in sediment transport. Jour. Geology 48, 526–577 (1948).

Potter, P.E., and Pettijohn, F.J.: Paleocurrents and basin analysis, 2nd Ed., 425 pp. Berlin–Heidelberg–New York: Springer 1977.

Powers, M.C.: A new roundness scale for sedimentary particles. Jour. Sed. Petrology 23, 117–119 (1953).

Pryor, Wayne, A.: Permeability–porosity patterns and variations in some Holocene sand bodies. Jour. Sed. Petrology 57, 162–189 (1973).

Reed, W.E.; Le Fever, R.; and Moir, G.J.: Depositional environment interpretation from settling-velocity (Psi) distributions. Geol. Soc. America Bull. 86, 1321–1328 (1975).

Riviere, A.: Methodes granulometriques; techniques et interpretaions, 170 pp. Paris: Masson, 1977.

Rosin, P., and Rammler, E.: The laws governing the fineness of powdered coal. Jour. Inst. Fuel 7, 29–36 (1933).

Ruhkin, L.B.. A new method for the determination of the conditions of deposition of ancient sands. Proc. Soviet Geol. 7, 953–957 (1937).

Rusnak, G.A.: The orientation of sand grains under conditions of unidirectional fluid flow, 1. Theory and experiment. Jour. Geology 65, 384–409 (1957).

Sagoe, Kweku-Mensah O., and Visher, Glenn S.: Population breaks in grain-size distributions of sand—a theoretical model. Jour. Sed. Petrology 47, 285–310 (1977).

Sahu, Basanta K.: Depositional mechanisms from the size analysis of clastic sediments. Jour. Sed. Petrology 34, 73–84 (1964).

Scheidegger, Adrian E.: The physics of flow through porous media, 3rd Ed., 353 pp. Toronto: Univ. Toronto Press 1974.

Schmidt, Victor, and McDonald, D.A.: The role of

secondary porosity in the course of sandstone diagenesis. In: Scholle, Peter A., and Schluger, Paul R. (Eds.): Aspects of diagenesis. Soc. Econ. Paleont. Mineral. Spec. Pub. 26, 175–207 (1979).

Sedimentation Seminar: Comparison of methods of size analysis for sands of the Amazon–Solimões Rivers, Brazil and Peru. Sedimentology 28, 123–128 (1981).

Selley, R.C.: Genesis, migration and entrapment of the Jurassic oil in the North Sea. Proc. Soc. Petroleum Engineers, Spring Meeting, London, Paper 5269, 7 pp. (1975).

Selley, R.C.: Porosity gradients in North Sea oil-bearing sandstones. Jour. Geol. Soc. London 135, 119–132 (1978).

Serra, J.: Image analysis and mathematical morphology, 628 pp. New York: Academic Press 1982.

Shelton, J.W.; Burman, H.R.; and Noble, R.L.: Directional features in braided-meandering-stream deposits, Cimarron River, north-central Oklahoma. Jour. Sed. Petrology 44, 1114–1117 (1974).

Shultz, D.M.: Mass-size distributions—a review and a proposed model. In: Patil, G.P., Kotz, S., and Ord, J.K. (Eds.): Statistical distributions in scientific work 2, D. Reidel Pub. Co.: Dördrecht, Holland, 275–288 (1975).

Sindowski, Karl-Heinz: Die synoptische Methode des Kornkurven-Vergleiches zur Ausdeutung fossiler Sedimentationsräume. Geol. Jahrbuch 73, 235–275 (1957).

Sippel, R.F.: Sandstone petrology, evidence from luminescence petrography. Jour. Sed. Petrology 38, 530–554 (1968).

Slatt, R.M., and Press, D.E.: Computer program for presentation of grain-size data by the graphic method. Sedimentology 23, 121–131 (1976).

Sneed, E.D., and Folk, R.L.: Pebbles in the lower Colorado River, Texas, a study in particle morphogenesis. Jour. Geology 66, 114–150 (1958).

Sneider, R.M.; King, H.R.; Hawkes, H.E.; and Davis, T.B.: Methods for detection and characterization of reservoir rock, Deep Basin Gas Area, western Canada. Jour. Petrol. Tech. 35, 1725–1734 (1983).

Socci, A., and Tanner, W.F.: Little-known but important papers on grain size analysis. Sedimentology 27, 231–232 (1980).

Solohub, J.T., and Klovan, J.E.: Evaluation of grain-size parameters in lacustrine environments. Jour. Sed. Petrology 40, 81–101 (1970).

Stapor, F.W., and Tanner, W.F.: Hydrodynamic implications of beach, beach ridge and dune grain study: Jour. Sed. Petrology 45, 925–931 (1975).

Stiedtmann, James R.: Evidence for eolian origin of cross-stratification in sandstone of the Casper Formation, southernmost Laramie Basin, Wyoming. Geol. Soc. America Bull. 85, 1835–1842 (1974).

Suczek, C.A.: Disaggregation of quartzites. Jour. Sed. Petrology 53, 672–673 (1983).

Swann, D.H., Fisher, R.W., and Walters, M.J.: Visual estimates of grain size distribution in some Chester sandstones. Illinois Geol. Survey Circ. 280, 43 pp. (1959).

Taira, A., and Lienert, Barry R.: The comparative reliability of magnetic, photometric, and microscopic methods of determining the orientations of sedimentary grains. Jour. Sed. Petrology 49, 759–772 (1979).

Taira, A., and Scholle, P.A.: Design and calibration of a photo-extinction settling tube for grain size analysis. Jour. Sed. Petrology 47, 1347–1360 (1977).

Taira, A., and Scholle, P.A.: Discrimination of depositional environments using settling tube data. Jour. Sed. Petrology 49, 787–800 (1979).

Taylor, J.M.: Pore-space reduction in sandstone: Am. Assoc. Petroleum Geologists Bull. 34, 701–716 (1950).

Till, Roger: Statistical methods for the earth scientist, 154 pp. New York: Wiley 1974.

Udden, J.A.: Mechanical composition of wind deposits. Augustana Library Pub. 1, 69 pp. (1898).

Udden, J.A.: Mechanical composition of clastic sediments. Geol. Soc. America Bull. 25, 655–744 (1914).

Van Veen, F.R.: Geology of the Leman Gas Field. In: Woodland, A.W. (Ed.): Petroleum and the continental shelf of northwest Europe, pp. 223–231. New York: Wiley 1975.

Viard, J.P., and Breyer, John A.: Description and hydraulic interpretation of grain size cumulative curves from the Platte River System. Sedimentology 26, 427–439 (1979).

Visher, G.A.: Grain size distributions and depositional processes: Jour. Sed. Petrology 39, 1074–1106 (1969).

Wadell, Hakon: Volume, shape and roundness of rock particles. Jour. Geology 40, 443–451 (1935).

Warren, A.: Observations on dunes and bimodal sands in the Tenere desert. Sedimentology 19, 37–44 (1972).

Warren, Guyon: Simplified form of the Folk–Ward skewness parameter. Jour. Sed. Petrology 44, 259 (1974).

Watson, Geoffrey S.: Texture analysis. In: Whitten, E.H. Timothy (Ed.): Quantitative studies in the geological sciences. Geol. Soc. America Mem. 142, 367–391 (1975).

Wellendorf, William, and Krinsley, David: The relations between the crystallography of quartz and upturned aeolian cleavage plates. Sedimentology 27, 447–454 (1980).

Welton, J.E.: SEM petrology atlas, 237 pp. Tulsa: Am. Assoc. Petroleum Geologists 1984.

Wentworth, C.K.: A scale of grade and class terms for clastic sediments. Jour. Geology 30, 377–392 (1922).

Winkelmolen, A.M.: The rollability apparatus. Sedimentology 13, 291–305 (1969).

Winkelmolen, A.M., and Veenstra, H.J.: Size and shape sorting in a Dutch tidal inlet. Sedimentology 21, 107–126 (1974).

Zimmerle, Winfried: Petrographische Beschreibung und Deutung der erbohrten Schichten. Geol. Jahrbuch 427, 91–305 (1976).

CHAPTER 4
Sedimentary Structures and Bedding

Introduction

Like texture and composition, sedimentary structures and bedding are inherent in sedimentation. Both are made visible by variations in grain size and to a lesser extent by mineralogy (Fig. 4-1). Because the great majority of structures can be seen with the naked eye, their study is as old as geology itself and, therefore, most of what we know has arisen from observation of ancient sediments. However, modern sediments and flume experiments have also contributed significantly to the study and understanding of structures, as we will see in Chap. 8. Structures have been used (1) as guides to determine the agent or environment of deposition, (2) as guides to stratigraphic order, by determination of top and bottom, (3) to map paleocurrent systems, (4) as indices of flow conditions, and (5) to assess chemical changes after deposition.

Obviously then, sedimentary structures are as important to the study of sand and sandstones as are texture and mineralogy. But unlike texture and mineralogy, most sedimentary structures can be studied only in outcrops and cores so that microscopic study is largely inappropriate. Sedimentary structures are most informative when they are closely tied to lithology, facies, and vertical sequences—in short, when the kind and abundance of sedimentary structures are carefully mapped either vertically or laterally across a sandstone body, a facies, or even a sedimentary basin. When a sedimentary structure provides information about paleocurrents, measurements and maps are obligatory.

The sedimentary structures of volcaniclastic sandstones are essentially the same as those of the common subaqueous sandstones, but those of the ignimbrites are very different, as covered in Chap. 6.

In this chapter we emphasize the recognition and occurrence of sedimentary structures and include brief summaries of their significance for environmental recognition and paleocurrent analysis. Hydraulic interpretation is deferred to Chap. 8 where we will find that the origin of primary sedimentary structures depends on fluid velocity and density and depth of water and is closely correlated with grain size—but only loosely with many sedimentary environments.

It is worth noting that some structures, such as ripple marks and crossbedding, are readily observable in both ancient and modern sands. Others, such as the sole marks of which flute casts are an example, are seen only in those ancient sandstones which separate along bedding planes; the unconsolidated nature of modern sands precludes examination of the underside of sand layers.

A structure cannot be defined in the same precise manner as a geometric object, such as a cube or a cylinder. As with fossils and organic forms in general, a picture is essential for its description.

There are four broad types of sedimentary structures (Table 4-1): (1) *current structures* formed by currents of water, air, and even ice as sediment is transported and deposited, (2) *deformational structures* formed shortly after deposition and before consolidation, mostly by slumping and foundering but also by escaping fluid and gas, (3) *biogenic structures* of organic origin such as tracks, trails, and burrows plus a few formed by plants, and (4) *chemical structures* formed by chemical processes during and after lithification of the sand. Current, deformational, and most organic structures are all made very early in a sand's history before effective consolidation, but only the structures of current origin are strictly "primary." It is chiefly these structures that have been used to help interpret

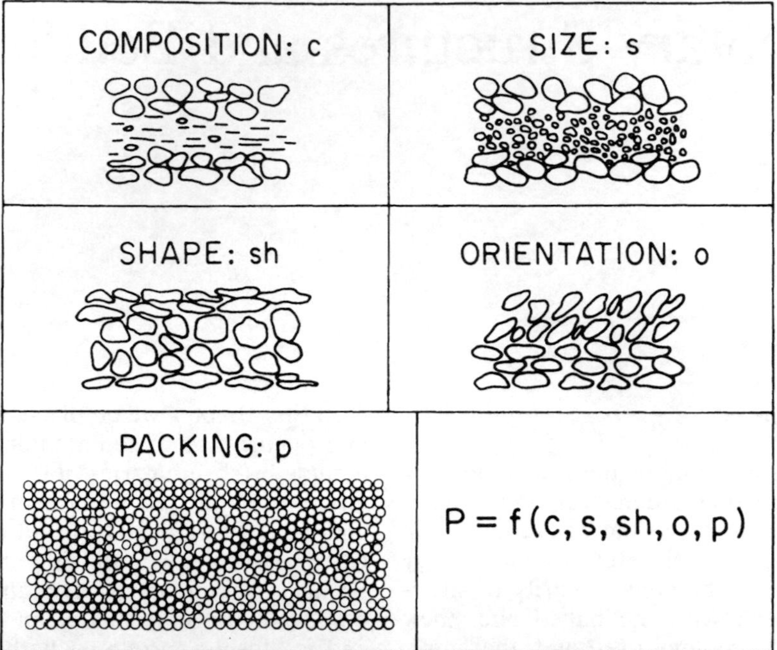

FIGURE 4-1. Bedding as the product of different combinations of composition, size, shape, orientation, and packing. (Modified from Griffiths, 1961, Fig. 3).

ancient environments and map paleocurrent systems. Practically all the current structures and many of the deformational and biogenic structures can form in a matter of hours, some even in a matter of minutes. In contrast, the chemical structures develop over much longer intervals—perhaps hundreds or thousands of years. Some structures, of course, are a mixture of more than one origin and hence they can

TABLE 4-1. Sedimentary structures: process and structure.

CURRENT	DEFORMATIONAL
Depositional	desiccation (mud crack casts)
beach cusps	eruption (sand volcanoes and spring pits)
graded bedding	founder and load structures
parallel lamination (parting lineation)	impact (spray, hail, and rain pits)
sand waves (ripple mark and crossbedding)	injection (neptunian dikes and sills)
wave and swash marks	slump (folds, faults, and breccia)
CURRENT	BIOGENIC
Erosional	Animal
channels	crawling trails
obstacle scours	feeding trails
rill marks	grazing trails
scour marks	residence structures
Tool marks	resting trails
bounce, brush, prod, and skip marks	Plant
roll marks	impressions
slide and groove marks	rootlets
striations and grooves	CHEMICAL
	cementation (sand crystals)
	crystallization (salt and ice)
	diffusion (color banding)
	pressure solution (stylolites)
	replacement (nodules)

be and have been classified in many different ways.

The literature of sedimentary structures is vast and goes far back in geology. Collinson and Thompson (1982) provide a comprehensive reference to most aspects of sedimentary structures, including their history of study, and Allen (1982a, b) emphasizes their sediment mechanics.

TABLE 4-2. Classification of primary sedimentary structures (Pettijohn and Potter, 1964, p.5).

BEDDING, EXTERNAL FORM

1. Beds *equal* or *subequal* in thickness; beds laterally uniform in thickness; beds continuous
2. Beds *unequal* in thickness; beds laterally uniform in thickness; beds continuous
3. Beds *unequal* in thickness; beds laterally variable in thickness; beds continuous
4. Beds *unequal* in thickness; beds laterally variable in thickness; beds discontinuous

BEDDING, INTERNAL ORGANIZATION AND STRUCTURE

1. Massive (structureless)
2. Laminated (horizontally laminated; cross-laminated)
3. Graded
4. Imbricated and other oriented internal fabrics
5. Growth structures (stromatolites, etc.)

BEDDING PLANE MARKINGS AND IRREGULARITIES

1. On base of bed
   (a) Load structures (load casts)
   (b) Current structures (scour marks and tool marks)
   (c) Organic markings (ichnofossils)
2. Within the bed
   (a) Parting lineation
   (b) Organic markings
3. On top of the bed
   (a) Ripple marks
   (b) Erosional marks (rill marks; current crescents)
   (c) Pits and small impressions (bubble and rain prints)
   (d) Mud cracks, mud-crack casts, ice-crystal casts, salt-crystal casts
   (e) Organic markings (ichnofossils)

BEDDING DEFORMED BY PENECONTEMPORANEOUS PROCESSES

1. Founder and load structures (ball-and-pillow structures, load casts)
2. Convolute bedding
3. Slump structures (folds, faults, and breccias)
4. Injection structures (sandstone dikes, etc.)
5. Organic structures (burrows, "churned" beds, etc.)

Current and Deformational Structures

Bedding

Most sedimentary structures can be classified and defined as some aspect of *bedding* (Table 4-2). Our approach [here] is mainly descriptive, is based mainly on form and geometry, and is in terms of four types of bedding. The genetic aspects of bedding and related structures are covered in Chap. 8. A *set* is a lithologic unit composed of two or more consecutive beds of the same lithology.

What are the basic properties of a bed? They are its thickness and lateral continuity. Several attempts have been made to describe thickness and redefine the terms which denote thickness (Grumbt, 1969, Fig. 7). Lateral continuity is a variant of the thickness problem. Some beds persist virtually without change in thickness even in large outcrops whereas others, such as ripples, pinch and swell (wavy and flaser bedding) and some pinch out (Reineck and Wunderlich, 1968).

Bedding, like other rock properties, is susceptible to measurement, particularly in vertical sections. Measurement of bedding thickness is an elemental part of all stratigraphic studies. Bed thickness is related to current competence, stronger currents producing thicker and commonly coarser-grained beds. Many bedding thicknesses seem to be lognormal (Fig. 4-2) and practically all are strongly skewed toward the thinner beds. Kelley (1956, p. 299) proposed a *stratification index*, defined as the number of beds $\times$ 100 divided by the thickness of the section measured, which, although not widely used, is essentially the reciprocal of the average thickness of the beds. Bokman (1957) proposed a geometric *theta scale* to facilitate statistical computation. A geometric scale such as Bokman's tends to normalize originally skewed thickness distributions much the same as the phi scale does for size distributions. As with any statistical distribution, mean, mode, and some measure of dispersion are needed to specify any particular bedding sequence. Plots of vertical profiles of bed thickness (rhythmograms) have been used for precise correlation or to search for cycles (Dean and Anderson, 1967). The most striking success has been with glacial varves and evaporites. Rhythmograms may also be useful in attempting correlation in facies, such as turbidites, that lack good marker beds.

The four bedding types of Table 4-2 represent a progression from maximum order (uniform thickness within and between beds) to minimum order (variable thickness within and between beds plus discontinuous beds). This is a progression from uniform to very variable flow

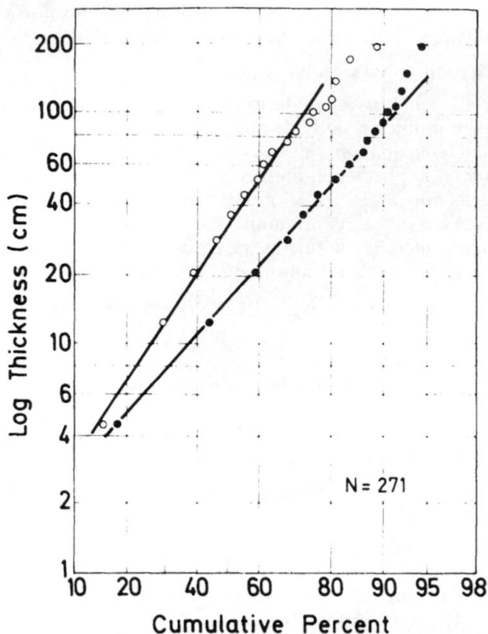

FIGURE 4-2. Approach to lognormality of thickness of turbidite sandstone beds (Scott, 1966, Fig. 13). Whether lognormal or not, virtually all such distributions are strongly skewed toward the thinner beds.

conditions—from deposition virtually without erosion to deposition with appreciable erosional scour and fill.

Internal Organization and Structure. Internally, beds may be (1) massive or structureless, (2) may be either "horizontally" laminated or show diagonal or cross-laminations, (3) may be graded, (4) may display internal imbrication, or (5) may exhibit "growth" bedding produced by rhythmic precipitation or by organisms such as stromatolites. All occur in sands and sandstones, although algal stromatolites are very rare.

Massive bedding is bedding seemingly without internal structure. Such lack of internal structure can be deceptive, however, as Hamblin (1965) and workers on the cores from the Deep Sea Drilling Project have shown (Fig. 4-3). Truly massive beds of sand appear to be very rare which is indeed fortunate, for if they were common, we would be hard pressed to explain them. Some, but not all, pebbly sandstones are massively bedded and contain isolated pebbles and even cobbles "floating" in massive sandstone and are loosely referred to as "sandstone mudflows" but are properly termed *debris flows*.

Laminated bedding, characterized by sedimentation units less than 10 mm thick, is moderately common and many such laminated beds yield excellent flagstone (Fig. 4-4). Laminated bedding forms a small part of many sands and occurs in virtually every major environment. Its origin—whether the product of weak or strong currents—is as yet not well established.

Crossbedding is one of the most characteristic structures of sands (Fig. 4-5). This structure, also known as current bedding, cross-lamination, cross-stratification, diagonal bedding, and inclined bedding, may be seen in both modern and ancient sands. As here defined, it is a struc-

FIGURE 4-3. X-ray and normal (inset) photograph of fine-grained, laminated sandstone (Hamblin, 1965, Fig. 14).

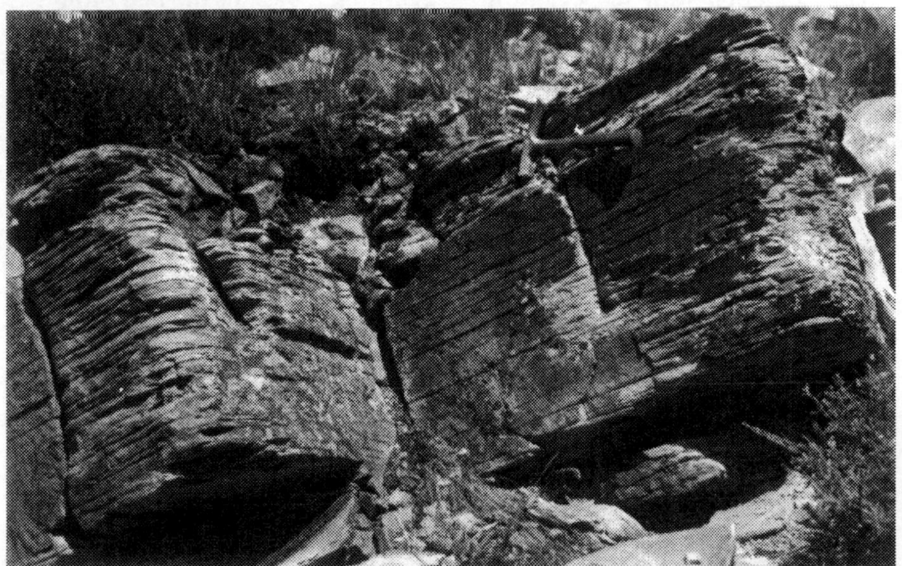

FIGURE 4-4. Laminated bedding of fine-grained Sappington Formation (Mississippian–Devonian). Bedding planes have weak parting lineation. NW 1/4 SW 1/4 sec. 31, T. 2 N., R. 1 E., Broadwater County, Montana, U.S.A.

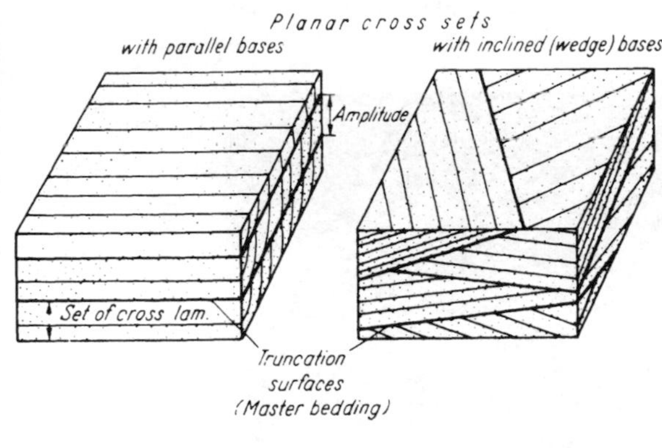

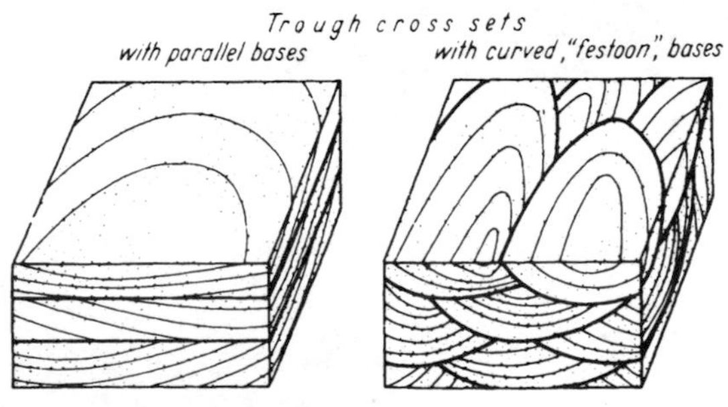

FIGURE 4-5. Four major types of crossbedding. (Redrawn from Dalziel and Dott, 1970, Fig. 10). See also hummocky crossbedding (Fig. 4-9).

ture confined to a single sedimentation unit characterized by internal bedding or laminations, called foreset bedding, inclined to the principal surface of accumulation. This definition is restrictive and excludes the inclined bedding of talus and of lateral accretion deposits such as the slip slope of a point bar as well as other stratification with a high initial dip such as that formed by the basinward growth of a delta front as well as the lower-angle stratification related to a prograding beach. We distinguish *crossbedding* from *cross-lamination* by the thickness of the set—does the set have a maximum thickness greater or less than 10 mm thick? But see Campbell (1967) for a different usage.

Several classifications of crossbedding, dependent upon the geometry of the structure, have been proposed of which the most widely followed is that of McKee and Wier (1953). Allen (1963) recognized some fifteen varieties of which only his epsilon type—inclined bedding formed by lateral accretion—is widely used. Allen's pi-type crossbedding is also called "sidefill" bedding. In actual practice, it is difficult to apply these classifications owing to the fact that exposures are seldom adequate or complete enough to determine to which class a given crossbed set belongs. It can even be difficult, in small outcrops, to distinguish between planar and trough crossbedding (see Fig. 4-5); as a rule, only successive sections cut in unconsolidated sands will correctly reveal true geometry. Normally one is restricted to two situations: a vertical cross section or a bedding-plane exposure. The vertical section may have any orientation with respect to the crossbedding. The most useful, perhaps, is a longitudinal section, that is, one parallel to the direction of current flow. In such a section, one can see whether or not the traces of the planes which bound the set are *parallel* or *convergent*, whether the traces of the foreset surfaces are straight or curved, and whether they are *tangent* to the base of the bed or not. One can also determine the *scale* of the crossbedding, that is, the thickness of the set (Figs. 4-6 and 4-7). Crossbedding varies from sets but a centimeter or two thick (small scale) to those 30 or more meters thick (large scale). In most sandstones the average is 15 to 60 cm. The observer can also measure the dihedral angle between the up-

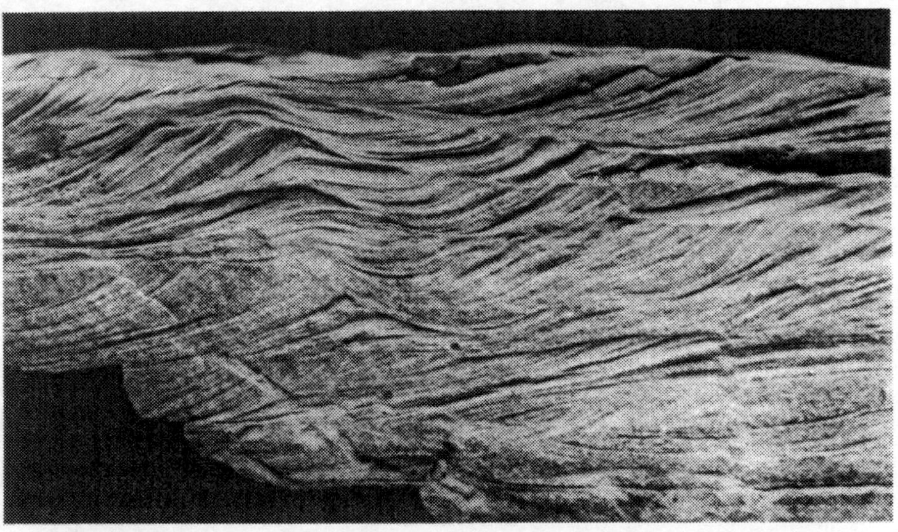

FIGURE 4-6. Cross-lamination near top of a fine-grained turbidite bed. Approximately one-half natural size. Paleocene flysch near east end of San Telmo beach, Zumaja, Vascongadas Province, Spain.

FIGURE 4-7. Crossbedding in Lower Devonian sandstone, transport from left to right. Oued Tairi, Tassili Externe, Department of Oasis, Algeria.

per bounding surface of the set and the foreset planes. This angle is often referred to as the *angle of repose* or *inclination*, a designation approximately but not strictly correct as the measured angle may be affected by deformation. Most important is to measure the *azimuth* of the dip of the foreset planes which is the presumed downcurrent direction. DeCelles *et al.* (1983) review methods of measurement for trough crossbedding and suggest two new ones. In tectonically tilted strata, the dip of the foreset planes and that of the true bedding are both recorded and the downcurrent azimuth determined after correction for tectonic tilt—a correction made by use of the stereographic projection or by computation.

Vertical sections normal to the current flow yield very little information except in the case of trough crossbedding. If a complete section of a trough is exposed, one can measure the depth of the trough (maximum thickness of the set) and its width to compute a width/depth ratio. In general, there seems to be a fixed width/depth ratio which is independent of scale.

Most useful are bedding-plane sections, especially of the trough crossbedded units. Such sections enable one to determine the bisectrix of the trough—the downcurrent direction. Bedding-plane sections are the most useful for distinguishing between trough crossbedding and the ordinary planar-tabular sets. The horizontal traces of the foreset surfaces are sharply curved in the trough crossbedding; they are straight or nearly so in the planar crossbedding.

It is important to observe the relation between crossbedded sets. They may be superposed one on the other. Commonly one set is separated from the next by ordinary bedded strata. The latter may be, in fact, "top-set" or "back-set" beds which, though appearing horizontal in outcrop, in fact dip upcurrent at very low angles. Back-set bedding—gently inclined foresets that dip upstream and are found on antidunes—is present in some rapidly deposited sands, but is not common. Ripple cross-lamination presents special situations. Ripples that are superimposed slightly out of phase and regularly displaced form what has been termed "climbing ripples" or "ripple-drift lamination."

Sets of crossbedding or cross-lamination contain low-angle inclined surfaces which separate adjacent foresets, usually with similar orientation, and are called *reactivation surfaces* (Fig. 4-8). Reactivation surfaces are figured in many studies of crossbedding and appear to have several origins: erosion in the lee of an advancing

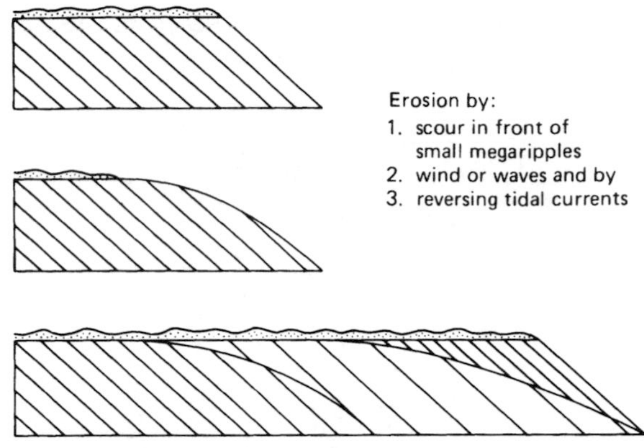

Erosion by:
1. scour in front of small megaripples
2. wind or waves and by
3. reversing tidal currents

FIGURE 4-8. Origin of reactivation surfaces in crossbed.

sand wave at constant water depth, plus erosion resulting from either decrease in water depth or partial change in flow direction (McCabe and Jones, 1977, pp. 713–14).

"Hummocky" crossbedding (Fig. 4-9) is somewhat ill defined but refers to low mounds and hollows formed by wave action, according to Harms *et al.* (1982, Fig. 3-15). Hummocky crossbedding commonly consists of fine to medium, well-laminated, gently inclined layers and has great variance of paleocurrents (see p. 334).

Cross-laminations may be deformed—either at the time of deposition by some kind of "soft-sediment deformation" or by shear during tectonic movements. Normal penecontemporaneous deformation results in oversteepening of the foresets, in some cases even overturning; tectonic deformation leads to either oversteepening or flattening and other distortions (Ramsey, 1961). Penecontemporaneously de-

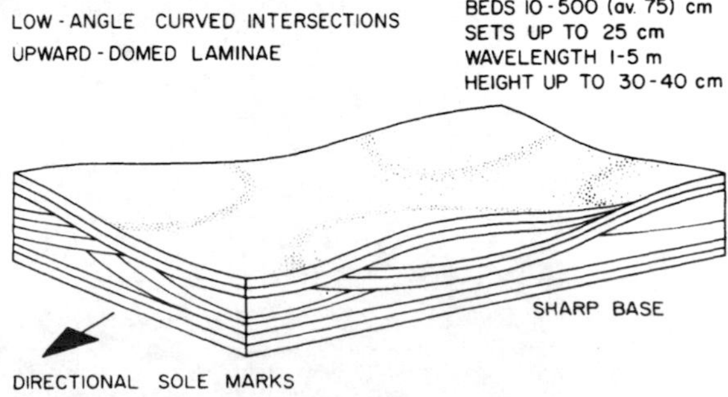

FIGURE 4-9. Idealized block diagram of hummocky crossbedding in the Fernie-Kootenay Formation (Jurassic) in the southern Rocky Mountains of Canada (Hamblin and Walker, 1979, Fig. 7).

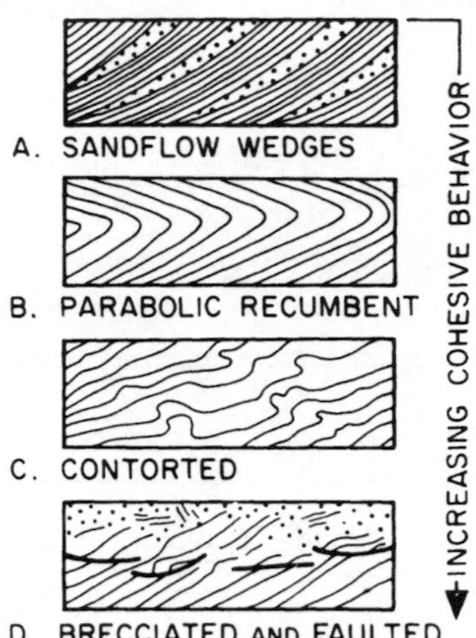

FIGURE 4-10. Four major types of penecontemporaneously deformed crossbedding (Doe and Dott, 1980, Fig. 3). Published by permission of the Journal of Sedimentary Petrology.

formed crossbedding has been classified into four types (Fig. 4-10).

Crossbedding, as here defined, is the product of downcurrent migration of a sand wave of some kind. The very smallest-scale cross-lamination is the product of ripple migration. That of most fluvial sandstones is formed by migration of subaqueous "dunes" forming medium-scale structures. Large-scale crossbedding is the product of migration of large dunes, either subaqueous or eolian. Such factors as scale, angle of inclination, tangency or lack of it, planar or trough structure, or trough dimensions and depth/width ratios have been in part investigated (Harms *et al.*, 1982, Chap. 3) and are discussed in Chap. 8, but there is still much to learn from the mechanics of sediment transport—how bedforms vary with grain size, flow intensity, water depth, and other factors. Wide experience suggests that most types of crossbedding are more closely linked to physical processes than to sedimentary environments—hummocky crossbedding being the most probable exception. Maximum interpretive value is always obtained, however, by closely relating the kind and scale of crossbedding and cross-lamination to vertical sequence and facies and by always carefully measuring paleocurrent direction.

Graded bedding is defined by an upward decline of grain size *within* a bed and is of several varieties (Figs. 4-11 and 4-12). Graded beds form by deposition from a decaying current and may range from a centimeter or less to several meters in thickness. The graded materials may be silt, sand, or even gravels. In a general way, the coarser the material, the thicker the graded unit. In general, most graded sandstones, com-

FIGURE 4-11. Gıaded bedding, Archean turbidite, Minnitaki Lake, Ontario, Canada. Photograph by R. G. Walker.

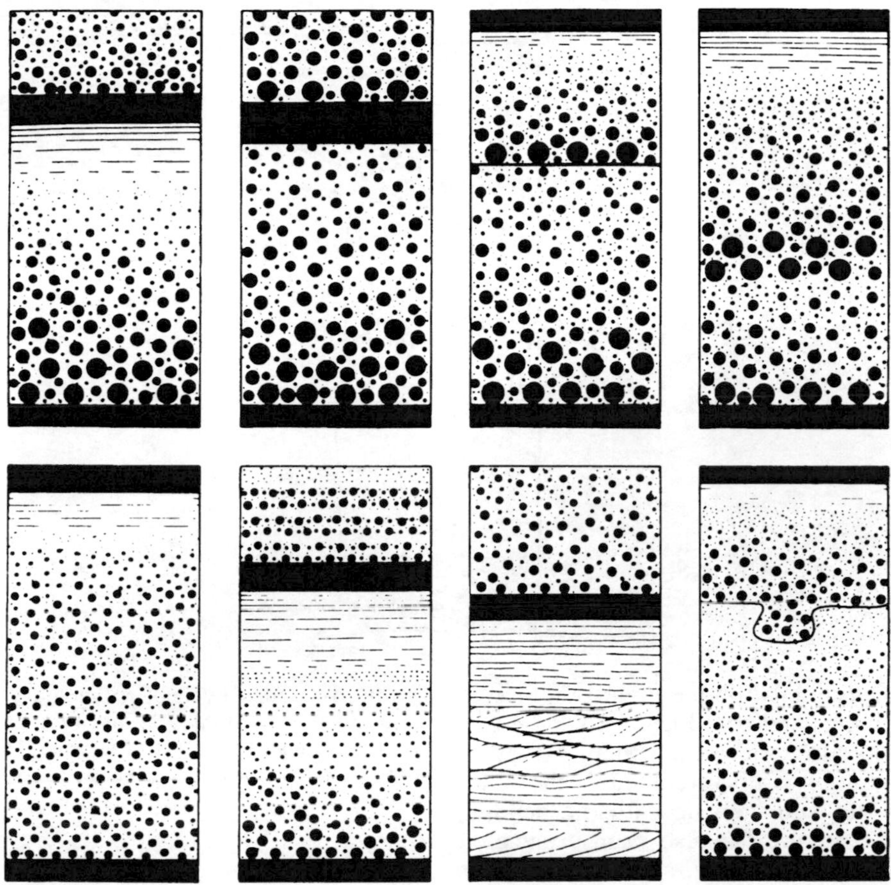

FIGURE 4-12. Varieties of graded bedding (Kuenen, 1953, Fig. 1).

monly the graywackes, range from a few centimeters to a meter in thickness. Much thicker beds are likely to be composites. Graded sequences display a lognormal thickness distribution. Graded beds commonly have a distinct internal, vertical sequence of structures, which like grain size, is the response to a decaying current (Fig. 4-13). Typically, graded bedding is found in thick sections of immature sandstones—the graywackes—of geosynclines. Here density or turbidity currents—turbid mixtures of mud and sand and water—are believed by most sedimentologists to flow periodically downslope and transport sand into deep water, where normally only muds would accumulate. Unlike crossbedding, the graded beds of turbidites can have wide lateral extent, some individual ones having been correlated for many kilometers.

Bedding Plane Markings and Irregularities. Many bedding planes of sandstones, if examined closely, show a variety of structures. They may be divided into those on the *base* or sole of the bed, those on *top* of the bed, and those on planes *within* the bed.

Sole markings are features characteristic of the underside of sandstone and some limestone beds that rest on siltstone and shales (Table 4-3). These features have been known for many years but only recently studied intensively. Most of these structures are the "negatives" or "casts" of depressions or markings originally produced in the mud over which the sand was spread. The structures originate by (1) the action of currents on the mud surface, (2) unequal loading of the soft hydroplastic mud, or (3) the activity of organisms on this surface. Although they occur in almost all sands, sole marks are particularly abundant in turbidites where they provide the best means of determining current flow.

Of the various structures produced by the action of currents, the most common is the *flute* formed on a somewhat firm mud surface and filled with sand and hence preserved as a raised

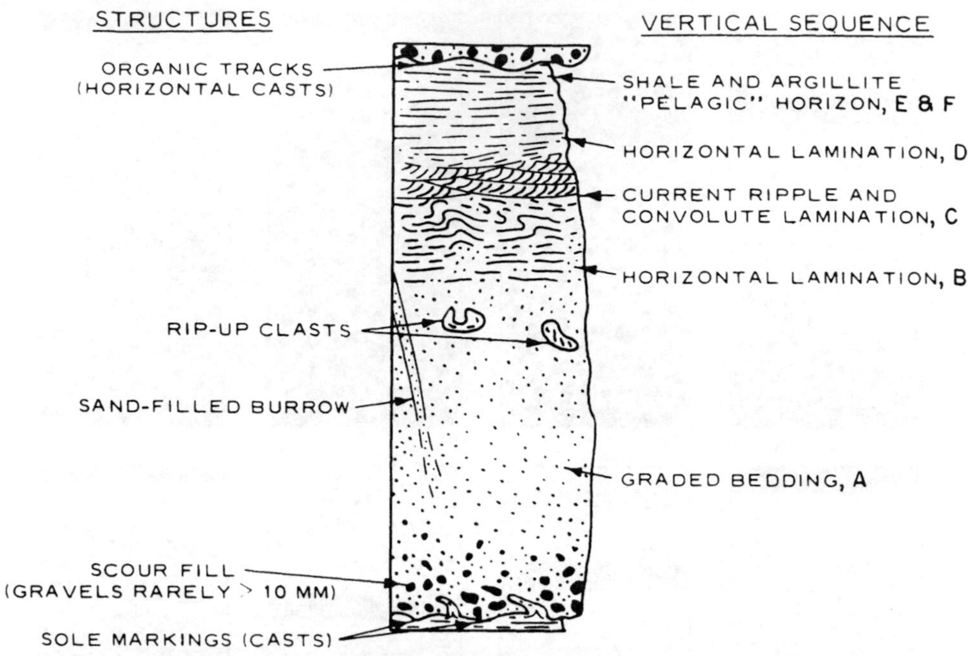

FIGURE 4-13. Ideal cycle of sedimentary structures in a turbidite bed (Stanley, 1963, Fig. 2) and Bouma units A through F; see also Fig. 8-36. Dish structure (Fig. 4-24) can be associated with convolute lamination.

structure or *flute cast* on the underside or sole of the overlying sand bed (Figs. 4-14 and 4-15).

Flute casts are subconical structures with a rounded or bulbous upcurrent nose, the other end flaring out and merging with the bedding plane. The structure has also been designated as flute molds, flow marks, scour casts, scour finger, vortex cast, and turboglyph. Flute casts vary from a few centimeters in length to giant structures a meter or even two meters in length. Solitary flutes are rare; most commonly they occur in swarms within which the individual flute casts may be widely spaced, close-spaced, or even overlapping. It is common for successive sandstone beds to show flute casts. In other words, when conditions were right for the production of a flute swarm during the deposition of one sand bed, these conditions persisted during the formation of subsequent beds.

Flute casts vary in shape, those of a given swarm being more or less alike. Some are elon-

TABLE 4-3. Inorganic sole markings and deformation structures (modified from Potter and Pettijohn, 1977, Table 5-1).

Agent	Process	Name of structure
Produced by current	Current scour	Flute (casts)
	Engraved by moving objects (tools)	Tool marks
	(a) Drag	Drag marks (groove and striation casts)
	(b) Saltation	Bounce, brush, and prod marks (and casts)
	(c) Rolling	Roll marks
	Unknown	Channels
Produced by gravity	Unequal loading	Load pockets (load casts)
	Foundering (thixotropic transformation)	Ball-and-pillow structure
	Slump or slide	Slide marks (and casts); slump folds and faults and breccias
Produced by "liquidation"	Injection and other processes	Sandstone dikes and sills; sand volcanoes; mud volcanoes
Complex current and gravitational interaction		Convolute laminations

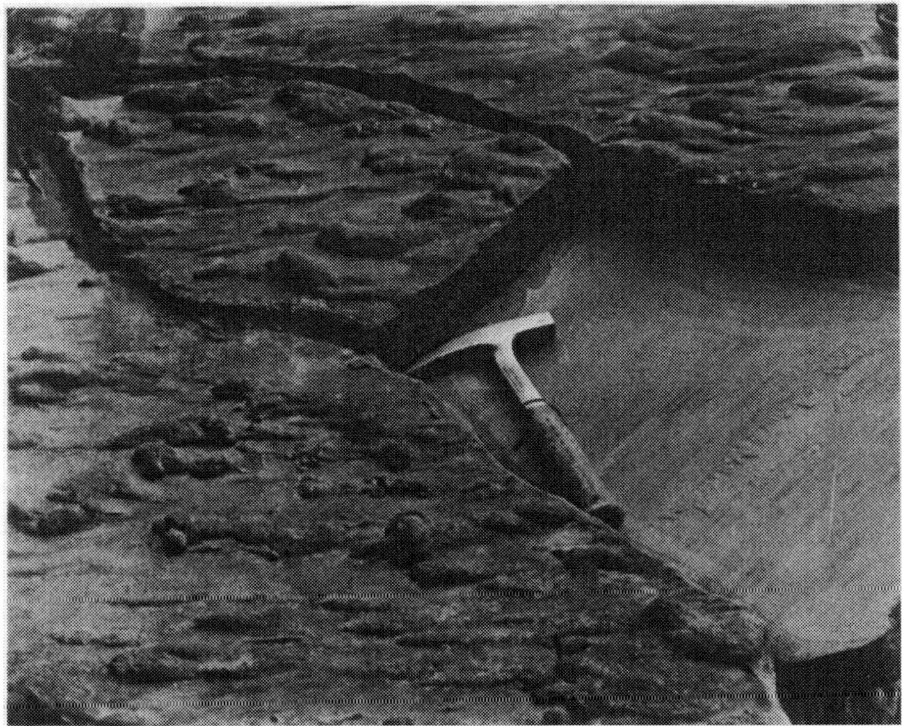

FIGURE 4-14. Partial section through a turbidite sandstone: flute casts and parting lineation in overturned bed of Marnoso Arenacea (Miocene). Current from right to left. Note divergence in orientation between direction indicated by flutes and parting lineation. Near Marradi, Tuscany, Italy.

gate, relatively narrow structures; others have a broader deltoid form. Some have good bilateral symmetry; others show less regular form, commonly with a twisted "beak." There seems to be a transition from well-defined flute casts to more irregular transverse scour casts and to relatively weak, greatly elongate furrow casts.

The less regular forms resemble load casts but have a more regular shape and also show clear evidence of their erosional origin, such as the transection of laminations in the subjacent mud or silt. The laminations may be differentially eroded so that the sand filling shows "terraces" forming a sculptured flute cast. In contrast, the laminations associated with load casts are not transected by the structure but are, instead, distorted or deformed by it. Some flute fillings were the cause of loading and deformation. Such "load-casted" flute marks are good examples of hybrid structure—the type that are difficult to classify.

Although flutes may be associated with grooves, such is not the rule. In general, they are mutually exclusive.

Flutes appear to be the product of local eddies. Their size is dependent on the size of the eddy, the latter being perhaps a function of the current strength. When the flow conditions are right for the production of one eddy, they are right for the generation of a field of such vortices. There are many unknown factors which govern the size, shape, and spacing of flutes. Allen (1982b, Fig. 7-2) provides a comprehensive classification for the very diverse assemblages of flutes.

Flutes are the most common and most useful of the current-produced sole marks. Their form is a sure guide to the direction of current flow and, although not exclusively the product of turbidity flows, they are most characteristic of the flysch facies.

A structure produced by current scour, and hence akin to flutes, is the *current crescent* (Hufeisenwülste). It is indeed a horseshoe-shaped depression developed by current scour around an obstacle lying on the sand surface. These commonly form around shale intraclasts in fluvial sandstones. The clast and moat are buried with additional sand. Ultimately, the shale weathers out, leaving a hole. The moat appears as a raised ridge on the underside of the sandstone bed surrounding the hole.

FIGURE 4-15. Flute casts from Denbigh Grits, Salopian (Middle Silurian) at Penstrowed Quarry, Wales. Photograph by Norman McIver.

Groove casts (Shrock, 1948, p. 162) are raised, rectilinear, rounded to sharp-crested features found on the underside of some sandstone beds. They are particularly characteristic of turbidite sands. They are presumed to originate by the filling of a corresponding furrow in the underlying shale (mud) and were in fact called "mud furrows" by Hall who observed them in the Devonian of New York State over 100 years ago (Hall, 1843, p. 424). They have also been called "drag marks" and "drag casts" from their presumed formation by objects being dragged over the mud bottom (Kuenen, 1957, p. 243).

Groove casts seldom appear singly—they generally occur in sets, commonly as two sets intersecting at an acute angle on the same surface (Fig. 4-16). The individual ridges usually display a relief of only a millimeter or two, rarely more than a centimeter. They are remarkably straight and in most exposures show neither a beginning nor an end. Some are multiple, and are ornamented with a second-order series of microgrooves or ridges. Within a given set there is little or no deviation in azimuths. Groove casts may be few or many; the latter sets partially effacing the earlier ones. Groove marks should be distinguished from *slide marks* or casts (Fig. 4-17) formed by the movement of a large object—such as a shale raft—over the bottom. Such sliding objects tend to pivot or rotate so that the marks they produce all curve alike. Normal grooves, on the other hand, being produced by many individual objects show no such coordinated behavior. Groove casts may be associated with prod casts, skip and bounce casts, and brush marks but rarely with flute casts.

Terminations are seldom seen; rarer still are terminations marked by a shell fragment or other recognizable tool presumed to be responsible for the original groove.

Groove casts are observable only on the base of the indurated sands resting on shale which has weathered away, exposing the bottom surface of the sandstone bed. Like flute casts, they

FIGURE 4-16. Bottom markings in Denbigh Grits, Penstrowed Quarry, Wales. Note two sets of groove casts. Photograph by N. L. McIver.

FIGURE 4-17. Large, multiple subparallel slide marks on base of near-vertical turbidite. *Arenisca numulitica* flysch (Eocene) along the Zumaja–Guetaria Highway (Km 34), Vascongadas Province, Spain.

are most abundant on the base of turbidite sands and are the most common hieroglyphic structure of the flysch facies.

Their origin was long unsuspected. It now seems clear that they are the product of tools swept along by the current which engrave the surface of a relatively firm mud bottom. This view is supported by the finding of such tools—shells, large sand grains, mud lumps—at the downcurrent end of the groove and the parallelism of the grooves with the direction of current flow as shown by other criteria. The exact dynamics is not yet clear. Most objects moved by a current proceed by rolling and saltatory leaps and are constantly rotating or twisting about. To form a groove requires continuous contact, even pressure, and nonrotary movement. Eddy motion produces flutes, not grooves. What conditions control each and which of the two is the more proximal?

Groove casts, by reason of their abundance, are among the most useful paleocurrent indicators, especially if used in conjunction with structures which yield the sense of the motion. If the currents responsible for them were indeed turbidity or density currents moving downslope, then some interpretive problems arise. The divergence in direction shown by intersecting sets or by aberrant directions on a particular bed within a more regular sequence raises questions about the paleoslope. Clearly not all grooves are cut by currents moving downslope.

As may be seen in Table 4-3, there are other marks or structures in addition to the more common flute and groove casts. Included here are those produced by objects which touch bottom intermittently and those which are generated by rolling objects. The first group includes bounce, brush, and prod casts. *Bounce casts*, also *skip casts*, are marks spaced at rather regular intervals and are impact structures made by an object pursuing a saltatory path. The *brush cast* differs in that the bottom contacts are accidental and not regularly repeated and moreover are prolonged enough for the construction of a slight mound of material pushed up by the forward-moving object. The *prod cast* is, as the name implies, generated by an object, such as a partially water-logged stick, impinging on the bottom, being forced downward into it, and then being rotated in the forward direction and lifted free of it. The more prominent downcurrent terminal point of the mark lies at the end of a short groove.

Roll marks are diverse. Common in some flysch sequences are those formed by rolling, in wheel or hoop fashion, of planar coiled shells, especially cephalopods. These leave a characteristic "signature" or track (Seilacher, 1963).

Mudcracks develop in cohesive materials which undergo shrinkage upon loss of water. Common muds illustrate this very well. Noncohesive granular materials, such as sand, cannot be expected to show mudcracks. Nevertheless, a polygonal pattern of raised ridges is present on the *underside* of some sandstones. This structure is in reality produced by an input of sand over a mudcracked surface, the sand filling of the cracks becoming a part of the overlying sand layer itself. The shale ultimately weathers away, leaving the "casts" of the crack fillings welded to the sandstone layer itself.

Structures characteristic of the top of sandstone beds include ripple marks, rill marks, pits and prints and, in the finer sands and silts, ice-crystal casts and molds. Biogenic structures are also common on some beds. All of the above marks can also occur as casts on the bottom of the bed.

Historically, one of the earlier observed structures of sand and sandstone and one about which the most has been written are *ripples* and *ripple marks*. There is an enormous literature on the subject—not only by geologists but also by those interested in the physics of grain movement and ripple or wave phenomena. Early classic papers by geologists are those of Kindle (1917) and Bucher (1919). One of the most recent comprehensive reviews of the subject is that of Allen (1969).

Most of the earlier work dealt with the form or morphology of ripples as seen in modern sands or as displayed on bedding planes of the older sandstones. During the past decade interest has shifted to the structure of ripples and to features, such as ripple-drift, best observed in cross section.

Ripple marks are sand waves of the smallest scale and thus form from weaker currents than those that produce dunes which generate the common and large-scale crossbedding. Ripple marks are characteristic of noncohesive granular materials of sand size. They may develop in either siliceous or carbonate sands but do not form in either coarse materials such as gravel or the finer silts and muds.

Classification has been particularly troublesome because of the great variety of ripples and the gradation from one type to another. Figure 4-18 illustrates but one of the many different varieties of ripple mark. Very broadly, ripples are of two types: those with symmetrical cross section generally attributed to wave-generated oscillatory currents and those with asymmetrical cross sections produced by unidirectional currents of wind or water. Small-scale current ripples are of many varieties and forms and belong to a sequence such that they either grade or intermingle with one another (Fig. 4-19). Ripple marks display patterns termed *rectilinear*, if continuous and normal to the current, or *sinuous*, if continuous but with some sinuousity. Many current ripples do not form a continuous pattern but are instead broken up into laterally compressed crescentic structures of U-shape in plan. They are termed *crescentic* or barchanoid if the extremities point downcurrent and *lunate* if they point upstream. Several indices based on such parameters as length and height have been thought to discriminate between eolian and

FIGURE 4-18. Fossil ripple mark. Horton Group (Mississippian), Minas Basin, Walton, Nova Scotia. Note two sets of ripples. Photograph by H. P. Eugster.

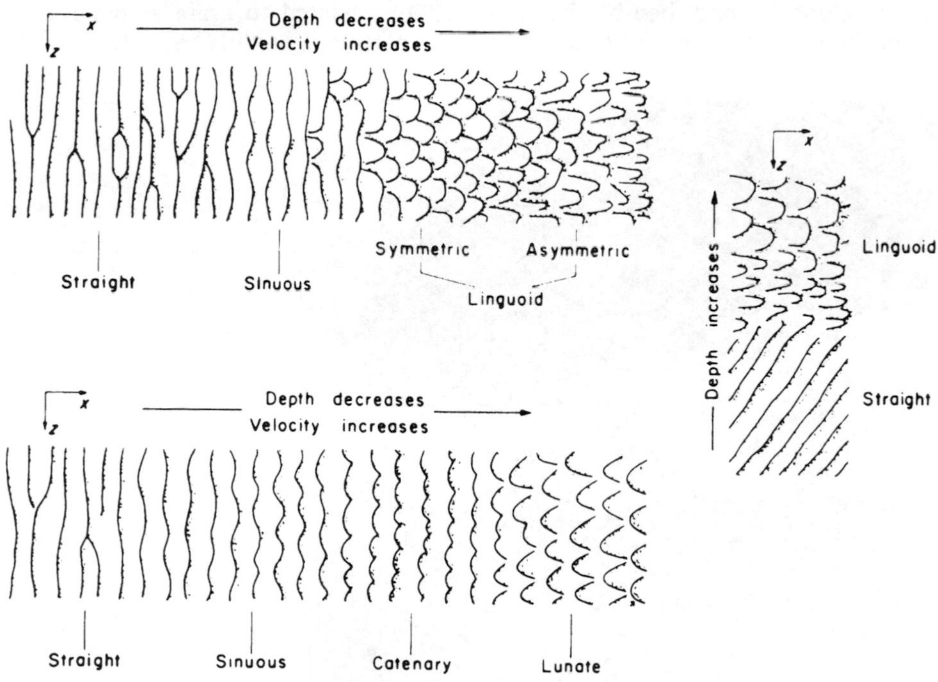

FIGURE 4-19. Nomenclature and relationship between different patterns of current ripples (Allen, 1969, Fig. 4-61).

aqueous ripples. Since undoubted eolian ripples are rarely, if ever, found in the geologic record, this distinction has little meaning.

Because ripples form wherever sand is in motion, they occur in a wide range of environments from the intertidal zone to the deep sea. In shallow seas they tend to be parallel to the shoreline and this tendency has useful paleogeographic significance. Their orientation should be systematically recorded.

In many ripple systems sand is transported up the stoss side and avalanches down the lee side, causing the ripple to migrate downstream without any change in the level of sedimentation. If, however, sand is *added* to the system, the rate of accumulation of the lee side increases and each ripple climbs the backslope of the one in front of it to form *climbing* ripples or ripple-drift cross-lamination (Sorby, 1859, 1908, p. 184; Walker, 1963, 1969; Jopling and Walker, 1968; Hunter, 1977).

In detail the morphology is variable and several subtypes of ripple-drift structure are recognizable. In one, all the deposition is on the lee side; another is characterized by deposition on both stoss and lee sides with individual laminations continuous across the system. A third type is characterized by a sort of grading, mud being collected in the ripple trough and silt and sand being segregated on the stoss side. Although laminations are continuous, in this case they change composition from the stoss to the lee sides. This type of ripple-drift seems to be characteristic of turbidite sedimentation (Walker, 1963).

The angle of climb is controlled by the rates of deposition from suspension relative to the rate of bedload movement. At high-flow regimes (Chap. 8), the sediment is swept to the lee side and the angle of climb is small (Walker, 1969).

The "rib-and-furrow" structure of some sandstones is simply the bedding-plane expression of micro-cross-lamination generated by migration of crescentic ripples. It is trough cross-stratification in miniature.

The ripple structure, as seen in cross section, may be deformed. Local oversteepening and overturning of the foresets are common, especially in rippled silts and fine sand. There seems to be some relation between these deformed ripples and convolute bedding—the latter perhaps an extreme end-product of ripple deformation. Isolated or "starved" ripples which migrate over a mud bottom may show subsidence into the mud, that is, initiate load casts.

Some sands separate along bedding planes forming very regular flags. These parting surfaces may display a faint lineation, *parting lineation*, which is also called primary current lineation, related to grain alignment. Commonly the parting is slightly imperfect so that the sur-

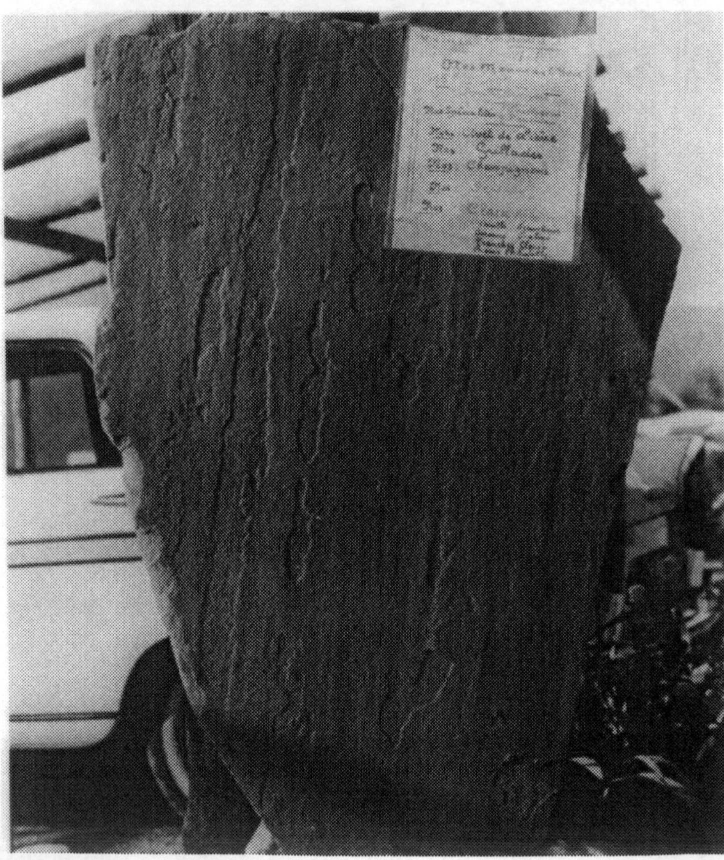

FIGURE 4-20. Parting lineation in Annot Sandstone (Tertiary), a turbidite. Current parallels parting. Menu gives scale. Near Piera-Cava, Maritime Alps, France.

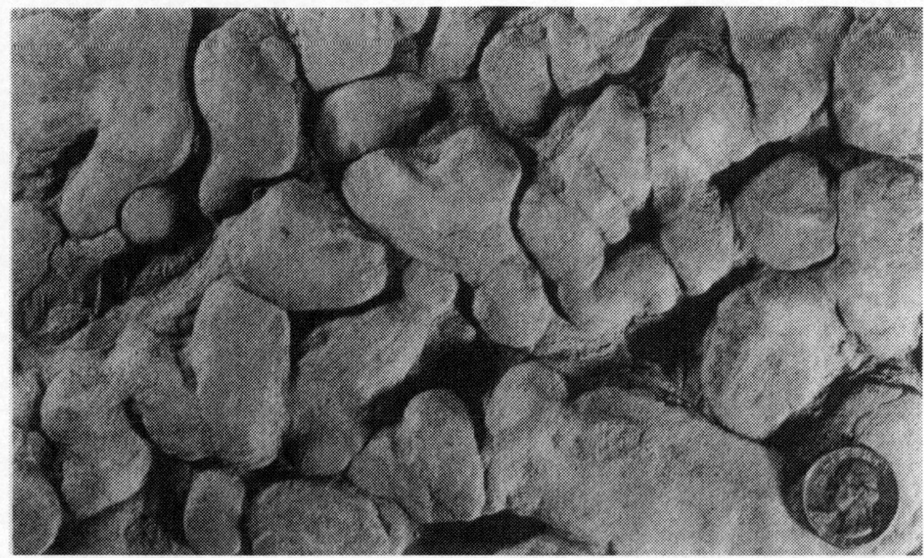

FIGURE 4-21. Bottom view of small load casts at base of fine-grained, thin-bedded Sample Sandstone (Mississippian). NW 1/4 SE 1/4 NW 1/4 sec. 25, T. 8 N., R. 4 W., Owen County, Indiana, U.S.A.

face has plasterlike remnants clinging to the bedding surface. These irregular patches are elongate in the direction of the lineation and are best seen in intense low-angle illumination. The term *parting step lineation* (Fig. 4-20) has been applied to this structure (McBride and Yeakel, 1963).

Contemporaneously Deformed Bedding

Following deposition, or concurrent with it and before consolidation, a sand deposit may undergo deformation. Various structures are produced. These may be attributed to one or another of three different processes. One is a convective-like pattern of motion which results in a vertical transfer of material. Such motion is initiated by an unstable density stratification such as occurs where a bed of sand is deposited on a less dense, water-saturated finer silt or clay. If the underlying material undergoes a thixotropic transformation (Chap. 8), with loss of strength, a series of convective cells may be set up with downward movement of sand and compensatory upward movement of silt or clay. The motion may be slow or it may be rapid and catastrophic. Comprehensive theory and description are given by Allen (1982b, Chaps. 8 and 9).

A second process of deformation is the result of an instability due to oversteepened depositional slopes. The resulting movement has a large lateral component and hence a near-horizontal transfer of material. This may be a slow creep or rapid slump or slide.

A third process may involve a pseudoliquefaction of the sand—development of a "quicksand" capable of injection as sills, dikes, or diapirs.

A fourth process, one very common in sandy glacial outwash, is collapse around a melted ice block.

Of the various structures due to vertical readjustments, *load casts* or load pockets are the most common (Fig. 4-21). These are bulbous, mammillary or papilliform downward protrusions of sand produced by unequal loading of the underlying hydroplastic mud. These structures may become sacklike, attached to the mother bed by a constricted neck; in rare cases they become detached and sink downward in the underlying materials. These are *load pouches*, and if detached, *load balls*.

As indicated by their process of formation, load casts are not "casts" at all. Unlike flute casts, they did not fill a preexisting cavity or "mold." Laminations in the underlying mud are deformed and the mud is forced upward in a thin tonguelike fashion between the downward-protruding load casts.

Load casts generally lack symmetry or plan of arrangement and may be very irregular in size and form. They appear as swellings on the underside of a sand bed, varying from slight bulges to deep or shallow rounded sacks. Load structures may have been initiated as flutes, or even grooves, and by "starved" or isolated sand ripples. Usually with the load-casted ripples there is a pattern of arrangement and an internal structure inherited from the parent structure.

Load casts are indicative of no particular environment. The only requirement is deposition of sand on a water-saturated hydroplastic layer. They seem to be more common in turbidite sequences but even here some beds are devoid of load casts—others will display them. Their growth in one case and not in the other seems to indicate the condition of the underlying bed. If one turbidite flow follows on the heels of another, conditions are more favorable for load-casting whereas if there is a significant time lapse and dewatering of the subjacent bed, no load-casting will take place.

Some sandstone beds, like some subaqueous lava flows, display an ellipsoidal or *pillow structure* (Fig. 4-22). The sand is broken up into numerous and generally closely packed, ball- or pillow-form masses. These have been called "hassocks" and "pseudo-nodules," which they resemble. The less satisfactory term "flow rolls" has also been used.

The closely packed balls or pillows vary from objects a few centimeters to bodies several meters in diameter. They are rarely spherical, are more commonly ellipsoidal or kidney-shaped. If the sandstone were laminated, such laminations within the balls or pillows are deformed, generally conforming to the rounded lower half of the ball or pillow, perhaps contorted in the central part. In structure, the pillows are like basins with rims slightly recurved inward. Or they may be likened in form to an inverted mushroom without a stem. The cup-like or basin structure generally is convex downward, concave upward, in some cases tilted but not overturned.

Despite earlier views to the contrary, the pillows are not concretions nor the products of spheroidal weathering, both of which are known in sandstones. Nor are they the products of slump as commonly stated. Their symmetry and orientation is a response to downward not lateral movement. That such saucer- or kidney-shaped structures can be produced by foundering of a sand bed was shown experimentally by Kuenen (1958, p. 18). With complete foundering, isolated detached balls of sandstone are formed (Fig. 4-23). The conviction has grown that some such mechanism is responsible for their formation. Perhaps this action was sudden or catastrophic. There is clearly an affinity between these structures and load pockets, and especially with the detached load balls. All are related to vertical transfer of materials. Perhaps the differences are related only to rates—one being slow, the other rapid. Just as for load

FIGURE 4-22. Ball-and-pillow structure. Annot Sandstone (Oligocene), Peira-Cava, Maritime Alps, France.

FIGURE 4-23. Small detached load balls of sandstone become smaller upward in bed of Berea Mississippian Sandstone near Camp Whitewood, near Windsor Mills, Ashtabula County, Ohio, U.S.A. (Potter *et al.*, 1984, Fig. WM-6).

casts, an underlying hydroplastic layer—a layer of mud that has not had time to dewater—is necessary.

Dish structure appears in cross section as a weakly concave-upward series of dish-like cross sections (Fig. 4-24) and was early recognized by Stauffer (1967) and more recently discussed by Stanley (1974) and Nagahama *et al.* (1975). It is most common in the sandstones of turbidites, where it may be associated with convolute lamination (Fig. 4-12). Escaping water from rapidly deposited sands is believed to be its origin. Rautman and Dott (1977) suggest that dewatering occurs in all environments.

Under some conditions, sands are deformed by gravity-induced movements with a large lateral component while the sediment is still unconsolidated and still in the environment of deposition (we thus exclude tectonic and other later movements). The terms "slump" and "slide" have been applied to such movements and "slump structures" to the resulting features produced. In general, "slump" conveys the idea of a local phenomenon; slides denote movement of greater horizontal displacement—kilometers in some cases. "Creep" also denotes downslope movement of a slow imperceptible nature. Slides and slumps are significant for paleocurrent and basinwide studies.

All sediments may be involved in these movements. They are not peculiar to sands, although sand deposits may be much deformed by them. One structure, involving only sand, is deformed crossbedding (Jones, 1962). This is made manifest by oversteepening of the foreset planes—which may display inclinations up to 90 degrees or even be overturned, the overturning being always in the downcurrent direction.

In rare cases the laminations are crumpled. Clearly there has been creep or slumping or some kind of shear imposed on the sand by drag of the depositing current or by lowering of water level, or earthquake shock. Tectonic folding will also distort crossbedding (Ramsey, 1961), producing either steepened foresets or very much flattened foresets. The deformation described here, however, is penecontemporaneous as the overlying or underlying strata may be unaffected.

Large-scale slumps and slides involving many beds, including interbedded shales and other lithologies, are known (Potter and Pettijohn, 1977, pp. 230–31). The effects vary from a chaotic mixture of the blocks of the more resistant beds caught up in a matrix formed from the least competent materials to a well-defined décollement above which the strata are distorted, folded, or brecciated. Where the movement involved interbedded but unconsolidated sand

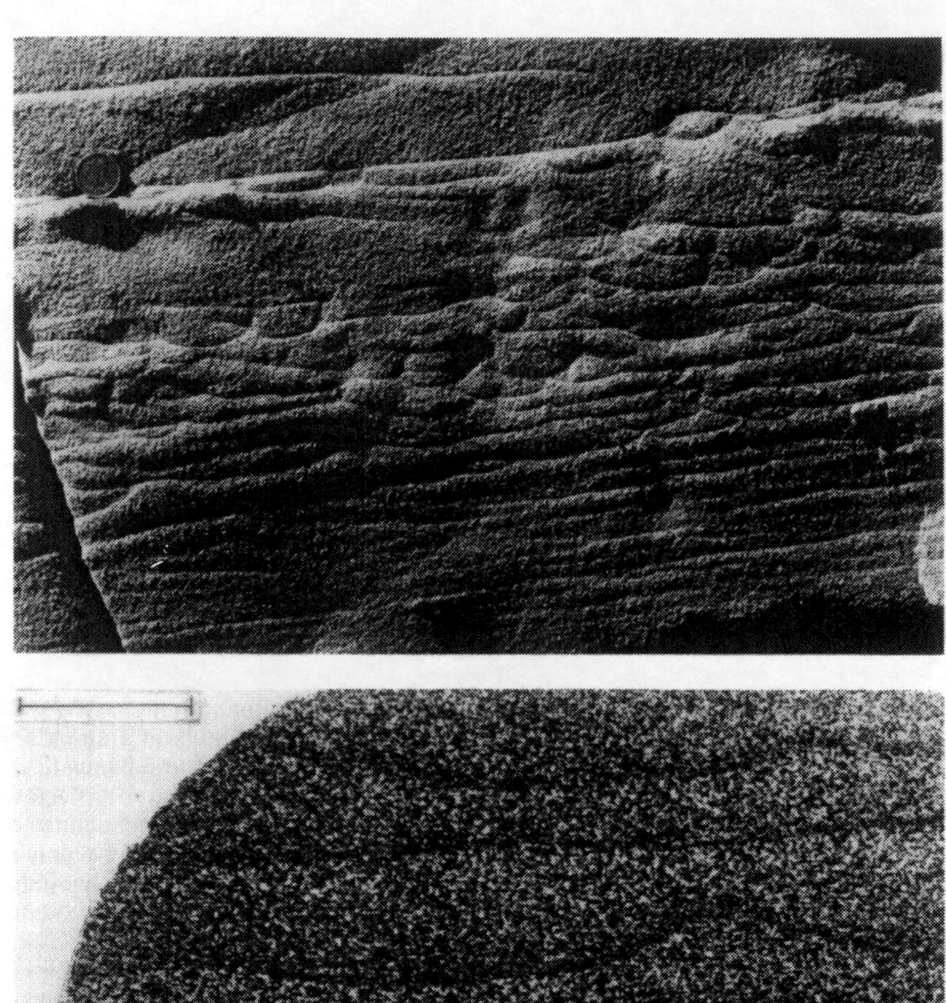

FIGURE 4-24. Dish structure as seen on outcrop (above) and on a polished slab (below) in Tertiary turbidites from Japan (Nagahama *et al.*, 1975, Pl. 6). Structure is formed by dewatering and is a guide to "the way up" in deformed sediments. Scale of polished slab is one centimeter.

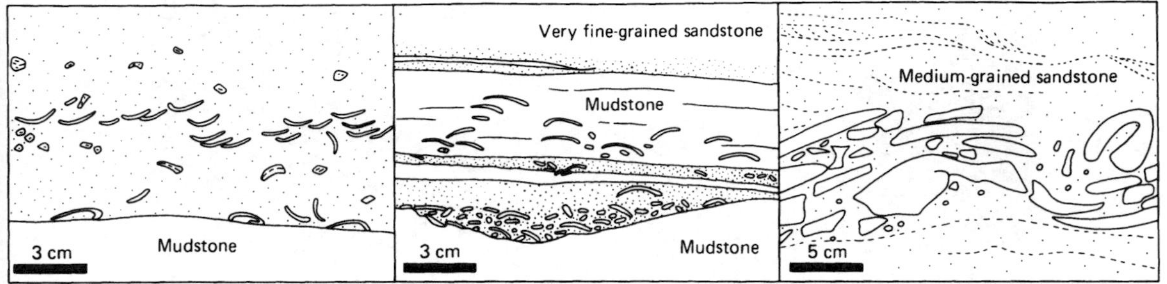

FIGURE 4-25. Shale clasts in turbidite sandstones of Pico Formation (Pleistocene), Santa Paula Creek, Ventura Basin, California (Crowell *et al.*, 1966, Figs. 17D, 17E, and 17F).

and clay, the latter, being tenacious, yields fragments which become enveloped in a sand matrix—the unconsolidated sand yielding by grain flow. The general term *debris flow* is used for mixtures of poorly sorted and isolated pebbles and cobbles resting in a sandy matrix or less commonly in a muddy matrix. See Johnson and Rodine (1984) for a comprehensive summary. The shale fragments may be bent or otherwise distorted (Fig. 4-25). If the muds are hydroplastic, then both sand and mud flow, and a streaked "migmatitic" mixture results. If there is a great deal of clay and relatively little sand or silt, the strata may retain integrity and be thrown into a series of disharmonic folds above a glide surface. Various types of penecontemporaneous slump have been described by Kuenen (1949) and Gregory (1969).

The circumstances which initiate slumps are many and varied. Apparently slides, once initiated, can move over a very flat surface, even a level area, and can, if they have enough momentum, move upslope. Some subaerial slides have been deduced to ride on a "cushion" composed of air.

Slides, slumps, olistostromes, and debris flow deposits are common associates of turbidites in fore-arc basin and accretionary terranes.

Convolute bedding, also known as convolute lamination, slip bedding, intrastratal contortions, and crinkled bedding, is one of the most difficult structures to define or explain (Fig. 4-26). Rich's expression, "intrastratal contortions," is perhaps the best as it emphasizes the intrastratal nature, i.e., the observation that these folds or convolutions affect the laminations *within* a bed but *not* the bed itself.

Convolute bedding seems to characterize relatively thin beds—2 to 25 cm thick—or coarse silt or fine sand. Within such a bed—either siliceous or carbonate—one observes a complex set of folds. The individual laminations are continuous and traceable from fold to fold, although micro-unconformities occur within the structure. The synclines tend to be broad and open; the anticlines tight and peaked. These structures tend to die out toward the base of the bed, and commonly toward the top of the bed also. In some beds the anticlines appear truncated as if eroded.

Clearly the distorted laminations are not involved in folds of the ordinary kind; that is, they do not record lateral compression or telescoping as is true of slump folds nor are they elongated as are ordinary folds. The bedding-plane section shows them to be a series of sharp domes and basins. These observations suggest that they are due to a complex system of vertical motions.

All evidence—confinement to very fine sand or coarse silt, restriction to a single bed, symmetry indicating vertical transfer of material—indicates some kind of internal readjustment of material in a quick or near-quick condition. Various theories have been put forward (Potter and Pettijohn, 1977, p. 228). None are wholly satisfactory.

Under some conditions sands become "quick," that is, have such a loose structure and contain so much water that they are capable of injection into fissures to form *sandstone dikes* (Fig. 4-27), or to be injected along bedding planes to form *sandstone sills*. Although some dikes are simple fillings, grain by grain, of widened joints from above, most bear evidence of instantaneous, forcible injection inasmuch as many follow the bedding for a short distance or even pass into sills.

Sandstone dikes and the problems of their origin have been reviewed elsewhere (Potter and Pettijohn, 1977, p. 229). They vary in size from a centimeter or two to several meters in width. They may be injected after consolidation or they may be injected early—prior to compaction. The former are sharp-edged, straight-

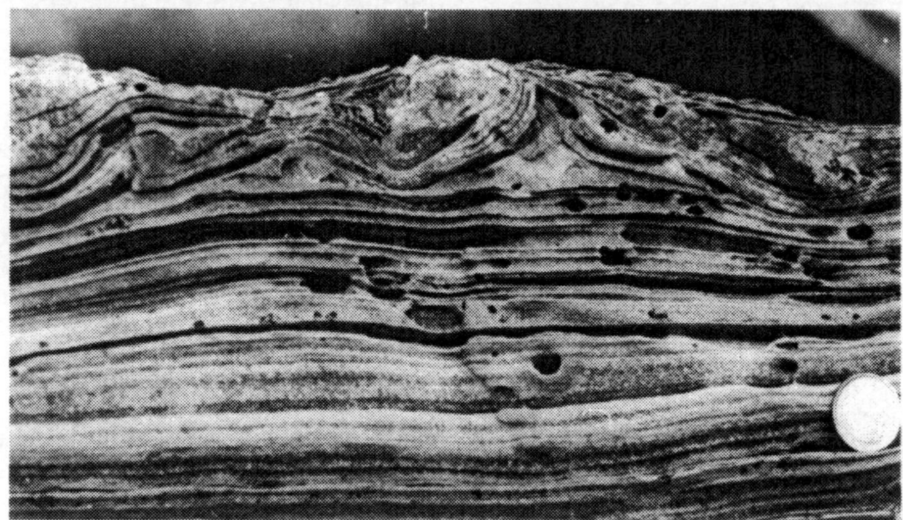

walled; the latter are contorted. They have recently been used as an index of tectonic state of tension in overthrust sheets; the tensional cracks are filled subsequently with sand and become sandstone dikes.

FIGURE 4-26. Convolute bedding at top of a fine-grained turbidite bed: cross section at bottom shows convoluted bed following lamination (see Fig. 4-13) whereas top is a view of convoluted upper surface of the same bed. Paleocene flysch, San Telmo Beach, near Zumaja, Vascongadas Province, Spain.

Biogenic Structures

As early as 1850, James Hall commented on tracks and trails in the Silurian sandstones of New York State, but for many years afterward biogenic structures were regarded by nearly all geologists as little more than exotic curiosities. Although their study began in ancient sandstones, modern sediment study early played an important role in the better understanding of biogenic structures, with the Germans being foremost in the investigation of these structures. Important early progress was made by Walther in the late 19th century at a marine station on the Bay of Naples and in the early 20th century by Rudolph Richter in the North Sea region at Senckenberg-am-Meer. Other noted pre-World War II German contributors were Abel, Krejci-Graf, and Häntzschel. But it was only in the early fifties that the interpretive value of biogenic structures began to be more fully exploited by Seilacher and others. Today, interest in biogenic structures is greater than ever.

FIGURE 4-27. Small sandstone dike extends downward from base of load cast to underlying, slightly load-casted bed in metagraywackes of Archean Yellowknife Supergroup, District of MacKenzie, Canada (Henderson, 1975, Fig. 6).

Terrigenous silts and sands are rarely rich in fossil remains, but they can contain good records of animal activity called *trace fossils, ichnofossils*, or *Lebenspuren*. Trace fossils are in fact best preserved in silts and sands, because they require textural contrast to be evident—burrowers, crawlers, and creepers in silt and sand-free muds leave little obvious evidence of their existence other than the disruption of lamination and distinctive textures (Fig. 4-28). Trace fossils may be found on the top of the bed, within it, on its base, or outside of it. It should be kept in mind, however, that trails on the bottom of a bed can be made by planned mining as well as by free grazing with later infilling of sand. Seilacher (1964) recognized five functional classes of trace fossils (Fig. 4-29).

What general statements can we make about trace fossils? According to Seilacher and others, trace fossils have a long time span but a restricted facies distribution, are practically never reworked, and are, unlike body fossils, greatly enhanced by diagenesis so that like alcoholic beverages they improve with age.

Table 4-4 gives a functional classification. Seilacher (1964, Fig. 7) recognized three broad facies, each defined by different proportions of the five trace fossil types shown in Table 4-4;

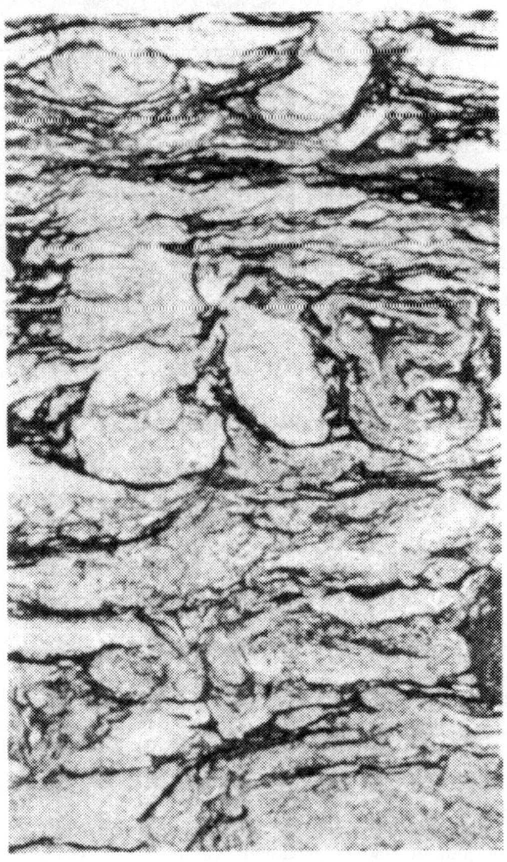

FIGURE 4-28. Bedding of siltstone and shale disrupted by burrowing. Tradewater Formation (Pennsylvanian) near Dawson Springs, Hopkins County, Kentucky, U.S.A.

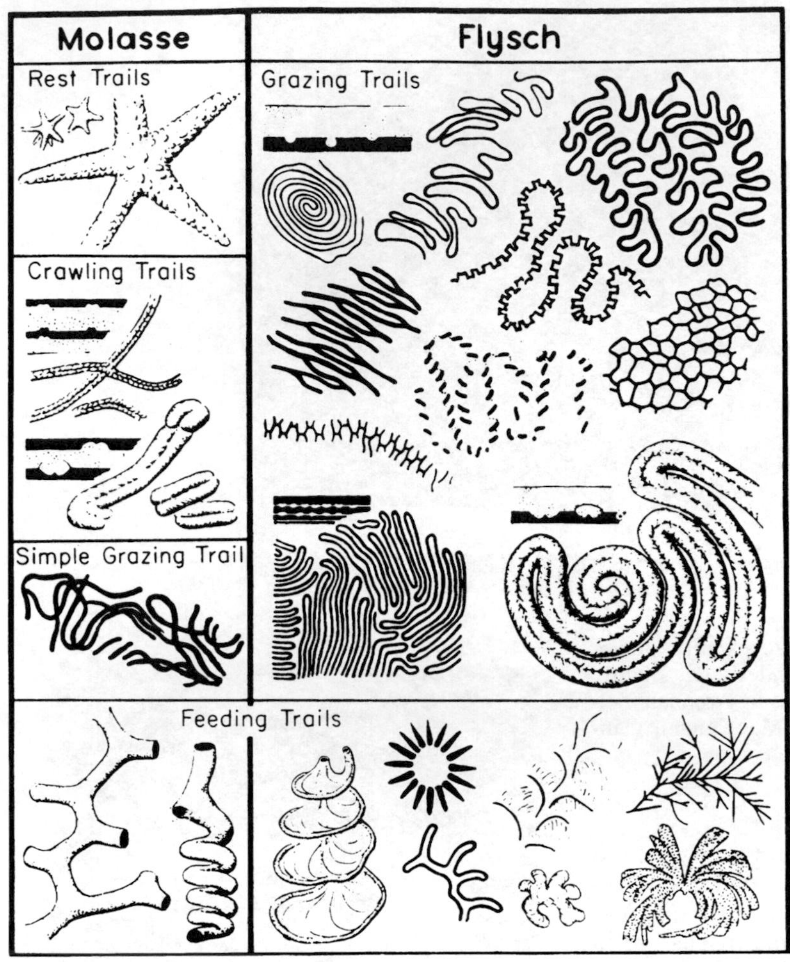

FIGURE 4-29. Five functional classes of trace fossils and their suggested occurrence in flysch and molasse deposits according to Seilacher (1955, Fig. 2). See Table 4-4 for companion terminology.

TABLE 4-4. Functional classification of ichnofossils (modified from Seilacher, 1953, Fig. 6)

CRAWLING TRAILS (Kriechspuren; *Repichnia*)
Trails or burrows produced by mobile benthonic organisms while moving.

FEEDING STRUCTURES (Fressbauten; *Fodinchia*)
Burrows made by semi-sessile bottom feeders; structure radiates from a place of origin.

GRAZING TRAILS (Weidespuren; *Pasichnia*)
Sinuous trails or burrows of mobile mud-eating organisms, "grazing trails" at or below the interface.

RESIDENCE STRUCTURES (Wohnbauten; *Domichnia*)
Permanent shelters made by either mobile or semi-attached organisms.

RESTING MARKS (Ruhespuren; *Cubichnia*)
Shallow resting tracks made by mobile animals while resting.

one facies, *Nereites*, is associated with flysch deposition and the other two, *Zoophycos* and *Cruziana*, with shelf deposition. Figure 4-30 is typical of a grazing trail in flysch, Fig. 4-28 represents an intensively burrowed 40-cm zone where molasse deposition was at a near standstill, and Fig. 4-31 shows a spectacularly bioturbated sandstone in Antarctica.

Of what use are trace fossils? Their narrow facies distribution may enable one to zone a shelf as it passes into deeper water (Farrow, 1966, Fig. 11), help establish progressive changes in water depth in vertical profiles and thus sharpen our perception of tectonic events (Seilacher, 1963a; Farrow, 1966, Fig. 11; Ksiaszkiewicz, 1977, p. 37–44) or perhaps help pinpoint a strand line (Weimer and Hoyt, 1964). However, Kitchell *et al.* (1978) have analyzed trace fossils from the deep sea and cast some doubt on the simple interpretation of water depth from these structures. Trace fossils also

FIGURE 4-30. Systematic grazing trail on underside of a flysch bed. Pietraforte formation (Cretaceous) near Florence, Tuscany, Italy.

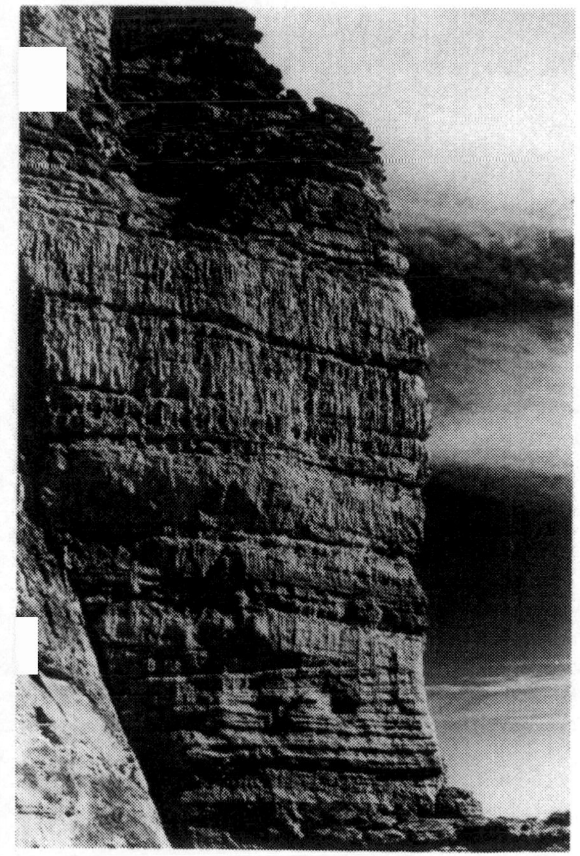

give good information on relative rates of sedimentation, slow deposition exposing the interface to intensive attack by burrowing organisms forming a bioturbate in which all laminations are destroyed. Faecal pellets may be especially abundant in such zones. In contrast, rapid sedimentation forms well-laminated sands. In the absence of other structures, Lebensspuren can be used to determine top and bottom of a sequence. A very minor use is their contribution to paleocurrent analysis (Crimes and Crossley, 1980).

Today, to thoroughly document the environment of deposition of an ancient sandstone, the study of its trace fossils is fully as important as the study of its sedimentary structures. To obtain maximum information from trace fossils, one should systematically make comparison between different known environments and *map* trace fossil distribution in the field. Four excellent general references to trace fossils are Crimes and Harper (1970), Frey (1975), Gall

◊FIGURE 4-31. *Skolithos linearis* overlies *Heimdallia-Diplichnites* in the NPW Mountain Sandstone (Devonian) on Table Mountain, South Victoria Land, Antarctica (Bradshaw, 1981, Fig. 1). Reproduced by permission of The New Zealand Jour. Geol. Geophysics.

(1976), and Crimes (1976). The short summary by Frey and Pemberton (1985) is well recommended. An informative example of the use of trace fossils is found in Bradshaw (1981).

Plant impressions and rootlets have value for identification of fresh and brackish water as opposed to marine environments. They are especially helpful if their occurrence is correlated with other characters, such as color, the rootlets being associated with red sediments, the marine fossils with green sands, etc.

Chemical Structures

Sedimentary structures of secondary origin are the result of both precipitation and solution. By and large, such structures have received but little attention since World War II. In sandstones, one finds structures due to solution and structures due to precipitation. Stylolites are an example of the first class; crystal aggregates such as sand crystals and barite rosettes are peculiar to sandstones and belong to the second class. Concretions, especially carbonate concretions, characterize both sands and silts.

Carbonate is the most common material that is precipitated to form chemical structures. Calcium carbonate may produce euhedral *sand crystals*. The luster mottling of some sandstones reflects such poikiloblastic calcite. Barite *rosettes* are of similar origin and character. Silica, pyrite, marcasite, rhodochrosite, phosphate, limonite–hematite, and other minerals may form nodules, concretions, and other structures in sandstones.

The *concretions* of sandstones are commonly larger than those of silts and clays, the greater permeability of the sand permitting greater transport of material in solution (Fig. 4-32). Laminations may pass undisturbed through the concretions. Such concretions may form quite early so that they may be the principal component of some intraformational conglomerates. Iron carbonate concretions may replace fine sands and silts; they are normally spherical and ellipsoidal, being flattened parallel to bedding, and may or may not have a nucleus. They vary in size from a few centimeters to bodies as much as 3 meters in diameter. Preferred shapes and orientation have been noted and may well reflect the anisotropic permeability of the host. Voidal concretions are hollow, variously shaped bodies with a hard rim of iron oxide or limonite. Extended tubelike shapes are common and even long ironstone pipes occur. These form above the water table, principally in friable sands.

Although it has but rarely been done, systematic mapping of concretionary facies should be

FIGURE 4-32. Large concretions weathered out of crossbedded sandstone, "Rock City" southwest of Minneapolis, Ottawa County, Kansas, U.S.A. Photograph by Alvin Hornbaker, Kansas Geological Survey.

FIGURE 4-33. Stylolites in sandstone, Allegheny Formation (Pennsylvanian), Garrett County, Maryland. Note two stylolite seams; core has parted or separated along the upper stylolitic surface. Photograph by H. P. Eugster.

FIGURE 4-34. Liesegang banding in Pounds Sandstone (Pennsylvanian). West side of northbound I-24, sec. 8, T. 12 S., R. 3 E., Union County, Illinois, U.S.A.

most rewarding, primarily because it would show relationship to associated facies, thus facilitating their integration with the entire section. If elongate, the direction of elongation may reflect paleogroundwater flow or even be related to anisotropy of permeability.

Replacement features such as chert and phosphate nodules—irregular tuberous bodies of matter unlike the host—are rare in sandstone.

Stylolites resulting from later pressure solution (Fig. 4-33) are rather common in clean sandstones but become less abundant as clay content increases. They are best observed in cores. These are discussed more fully in connection with pressure solution in Chap. 11.

Color-banding is a rhythmic precipitation of iron oxide in thin, closely spaced, generally curved layers (Fig. 4-34). It closely mimics bedding laminations, for which it may be mistaken.

Obtaining Maximum Value from Sedimentary Structures

Maximum information is usually obtained by relating the kind and abundance of sedimentary structures to position within the sand accumulation—be it a sand bank in a tidal estuary, a beach, or a single bed. Is a particular structure more common at the base, at the top, or near the margins of a sand? What are the associations of structures? How do such associations correlate with bedding thickness, grain size, and trace and body fossils? What is the vertical sequence of structures? How many bedding facies does a sand body have? *Field mapping* or careful logging of cores will nearly always show a systematic distribution of kinds, abundance, and magnitude of sedimentary structures in all sands so that one can speak of the cross-bedded, ripple-bedded or horizontally bedded facies of a sand body. Commonly, the relationship of such facies to each other is the key to the origin of the sand. Why? Because a sand deposit is the integrated response to a unified, overall process such that the whole is the ordered sum of its parts. Only by systematic mapping can the geologist effectively see this unified response—defined by the different bedding facies—and so best identify the ancient environment.

Collectively, current-formed structures, trace-fossils, and post-depositional structures contribute to improved knowledge of sedimentary environments, paleocurrents, and provenance (Fig. 4-35) and explain why sedimentary structures play a major role in today's study of sand and sandstone—especially when this information is combined with vertical variation in grain size to define fining- or coarsening-upward cycles (Fig. 4-36).

Because current structures are not restricted to single environments—practically every such structure can be found wherever sand is transported in volume—it is very rare indeed that one can deduce a specific environment from the *single* occurrence of a particular current structure. Hence, attempts to uniquely relate specific structures to a particular environment such

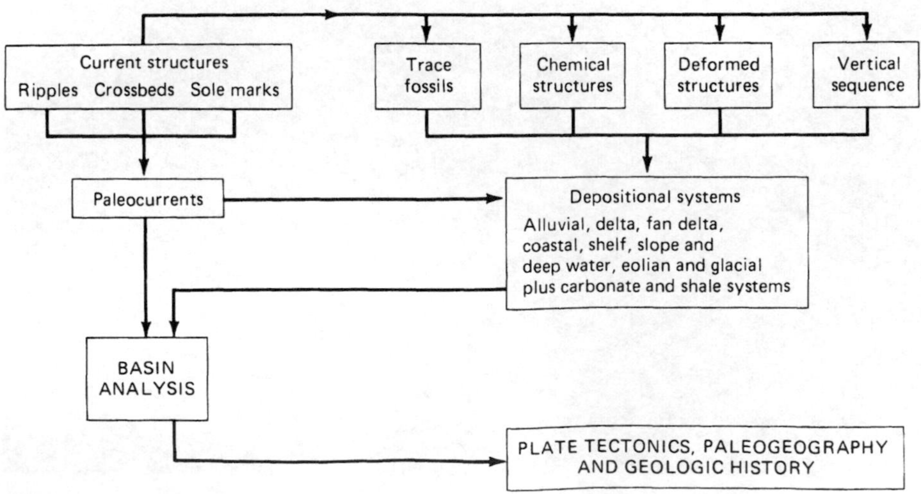

FIGURE 4-35. Relationships of sedimentary structures of all types to basin analysis, plate tectonics, and paleogeography.

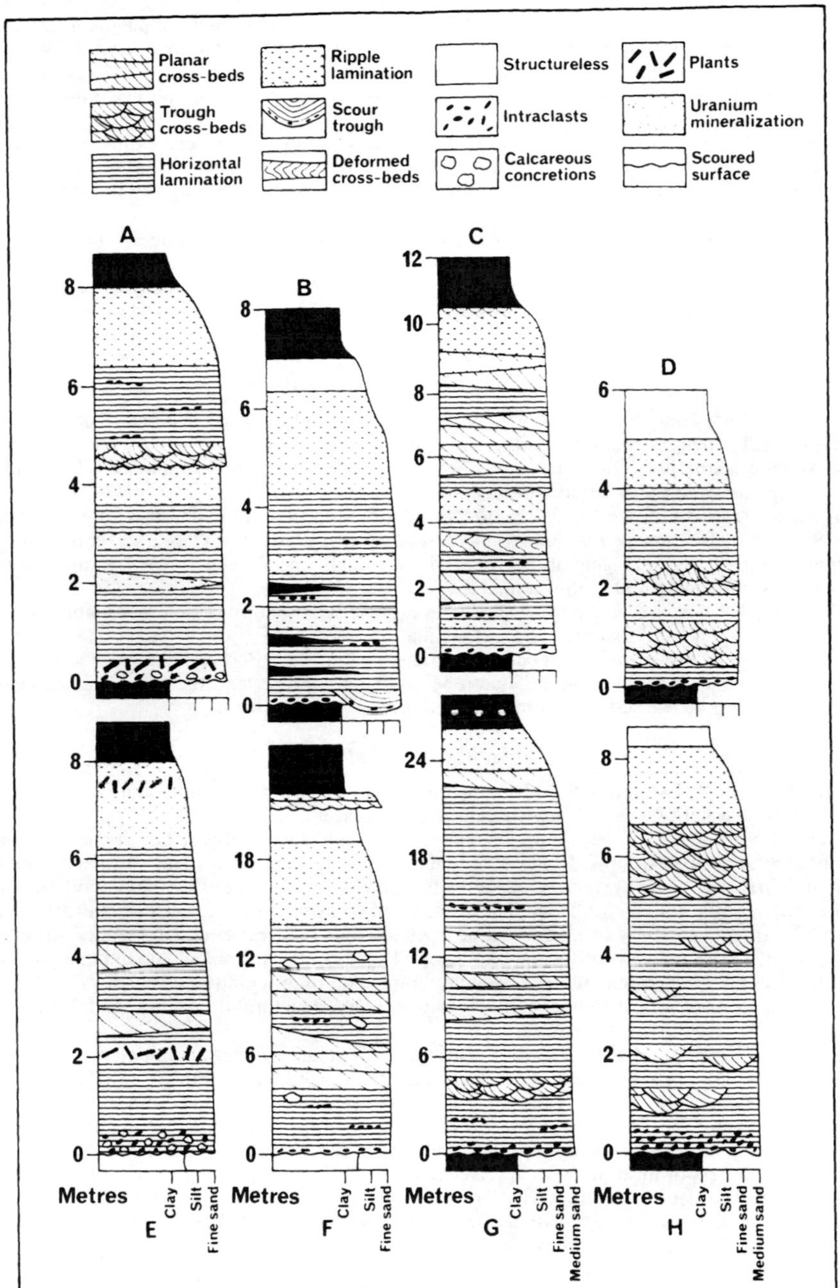

FIGURE 4-36. Changing proportions of sedimentary structures and sand–shale ratio in down-dip direction of Triassic deposits of Karoo Basin, South Africa (Turner, 1983, Fig. 4). Published by permission of the International Association of Sedimentologists.

as fore-beach, fluvial point bar, etc. have, to our knowledge rarely, if ever, been successful.

Good examples of sedimentary structures and trace fossils and their occurrence in modern sands are provided in Table 4-5. Studies of sedimentary structures in ancient sandstones are very numerous and only a very few are cited here: Khabakov (1962), Botvinkina (1965), Dzulynski and Walton (1965), Dimitrijević *et al.* (1967), Ricci-Lucchi (1970) for turbidites; Allen (1969), Picard and High (1973), and Miall (1978) for alluvial and deltaic sandstones; and Glennie (1970) for eolianites. General references for field use include Pettijohn and Potter (1964), Gubler (1966), Conybeare and Crook (1968), and Reineck and Singh (1980). Five texts that show how to use sedimentary structures of all kinds to help identify the environment of depo-

TABLE 4-5. Informative examples of studies of sedimentary structures and trace fossils in modern sands.

Streams, rivers, and fan deltas
Coleman, 1969: Excellent description and analysis of kinds of stratification and stability of paleocurrent indicators in sands of braided Brahmaputra River (pp. 208–229), which ranks fourth in the world in terms of discharge.
McGowen, 1970: Inventory of sedimentary structures in an artificial fan delta along the Texas Gulf Coast. Kinds of sedimentary structures, their origin, and paleocurrent significance based on surface observations, trenches, and radiographs are covered.
Picard and High, 1973: Describes and illustrates (139 figures) the sedimentary structures in small, ephemeral streams, mostly in Utah and nearby states. Table XIV lists associated features and gives interpretation. Field studies such as this are of general interest because they form a middle ground between flume experiments and the structures of large rivers. Compare to Williams (1971).
Singh, 1978: Studies of trace fossils in modern alluvial deposits are not common and we need more such as this one, which found lamellibrachs, crabs, insects, and earthworms to be the principal producers of trace fossils in an Indian river.
Williams, 1971: Good line drawings and photographs are combined with geometric and textural characteristics of large and small sand waves and flat bedding to provide improved hydraulic understanding of sands deposited by floods in an arid environment. See also Picard and High (1973).
Estuarine and tidal deposits
Howard and Frey, 1975: Very comprehensive report on animals and sandy–muddy sediments of Georgia estuaries. Key ideas: trace fossils, sedimentary structures, radiography, and bioturbation.
Dalrymple, 1984: Much quantitative data on bedforms and their relation to grain size and tidal hydraulics in the Bay of Fundy plus good documentation of internal stratification.
Dörjes, 1978: Organisms related to trace fossils on modern tidal flat, up to 4 km wide, in western Taiwan.
Howard, *et al.*, 1975: Animal–sediment relations in fluvial-marine point bars in a Georgia estuary studied by cores and surface sampling. Good data on changing abundances of individuals and species downstream as river becomes more saline and on changes of bedding and bioturbation (Fig. 6).
Jago, 1980: Small estuary has a well-sorted marine sand with profuse sedimentary structures. Figures 12 and 13 document sedimentary structures as seen in box cores and at low tide.
Klein, 1970: Pioneering paleocurrent study of modern intertidal bars in Nova Scotia demonstrated prevalence of bimodal orientation of sand waves and provides much data on bedforms, grain size, and their internal sedimentary structures. Good example of what can be learned from a modern sedimentary environment.
Beaches and shelves
Chowdhuri and Reineck, 1978: Estimates the changing abundance of different sedimentary structures, lithology, and bioturbation into deeper water of a North Sea shoreface. Excellent model to follow.
Davidson-Arnott and Greenwood, 1976: Facies model relates sedimentary structures to beach and near-shore topography. Twenty-seven photographs of box cores. Concise and informative. Compare with Hill and Hunter below.
Farrow *et al.*, 1979: A first-of-its-kind systematic, underwater television survey (250 km^2) of a modern marine sand→mud transition of the inner continental shelf off Scotland recognized 11 facies (trace fossils, organisms, and structures).
Hill and Hunter, 1976: Integrated study of macrobenthic zonation plus biogenic and sedimentary structures found in the beach of Padre Island, Texas, a coast with a low tidal range. Prime emphasis on ecology and many illustrations of trace fossils. One of the few fully integrated studies available. Some data on orientation of burrows.
Thompson, 1937: The first major article to document the sedimentary structure of beaches. Lamination emphasized, but almost all structures are inventoried. Classic.
Reineck, 1967: Excellent summary of the kinds and distribution of shallow-water sedimentary structures (physical and biogenetic) in the tidal flats, beaches, and shelves of the east side of the North Sea. Structures were studied by impregnating sediment in the field. Figure 27 is classic and shows seaward variation of infauna; Table 1 gives abundance of "Structures on Mud and Sand Flats and in Intermixed Areas." Outstanding model to emulate.
Reineck *et al.*, 1968: Very good documentation of trace fossils and sedimentary structures from shore to muddy shelf along the German coast into waters of more than 40 m. Excellent illustrations, especially of trace fossils. Classic.
Reineck, 1976: Sedimentary structures and bioturbation as sampled by box cores from waters more than 20 m deep define four zones seaward of an island along the North Sea bordering Germany. See author's Table 1 for excellent summary and compare to Howard and Frey (1975) and Howard *et al.* (1975).
Slope and deep sea
Hesse, 1977: Thin slices (8 mm thick) were cut from piston cores taken of muddy turbidites deposited in the Kurile Trench off Japan show convolute and parallel lamination, minor bioturbation, and some mottling. Structures and facies such as these usually are found more distal to sand-rich medial and proximal turbidites. Lack of bioturbation indicates deposition in oxygen-deficient bottom waters. Compare to Shepard *et al.* (1969).
Shepard *et al.*, 1969: Excellent photographs and X-radiographs of sands in box cores taken from natural submarine levees, channels, splays, etc., on the La Jolla subsea fan valley off California. Structures photographed include: graded beds, rip-up clasts, parallel lamination, massive beds, convoluted lamination and drag folds, cross-lamination, and more.
Eolian
McKee and Bigarella, 1979: The best single reference to the sedimentary structures and morphology of modern eolian sand dunes by two life-long students—plus hydrodynamic discussion. Forty-nine figures.

sition of ancient sandstones are Rigby and Hamblin (1972), Shawa (1979), Reading (1985), Collinson and Thompson (1982), and Harms *et al.* (1982). Chapters 9 and 10 "Paleocurrents and Dispersal" and "Sandy Depositional Systems") make ample use of this methodology.

A set of symbols for sedimentary structures is useful (Fig. 4-37); core or outcrop description is also facilitated by the use of a standard format—which even could be a punch card.

The paleocurrent significance of sedimentary structures cannot be underestimated for both paleogeography, paleogeology, and facies analysis. The study of paleocurrents in sands and sandstones is both exceptionally rewarding and easy to do. Paleocurrent studies span almost the entire geologic column—from the Recent to far back into the Precambrian—and now include metamorphic and volcaniclastic sandstones.

There are nine basic paleocurrent systems of which six are fairly common—parallel, conver-

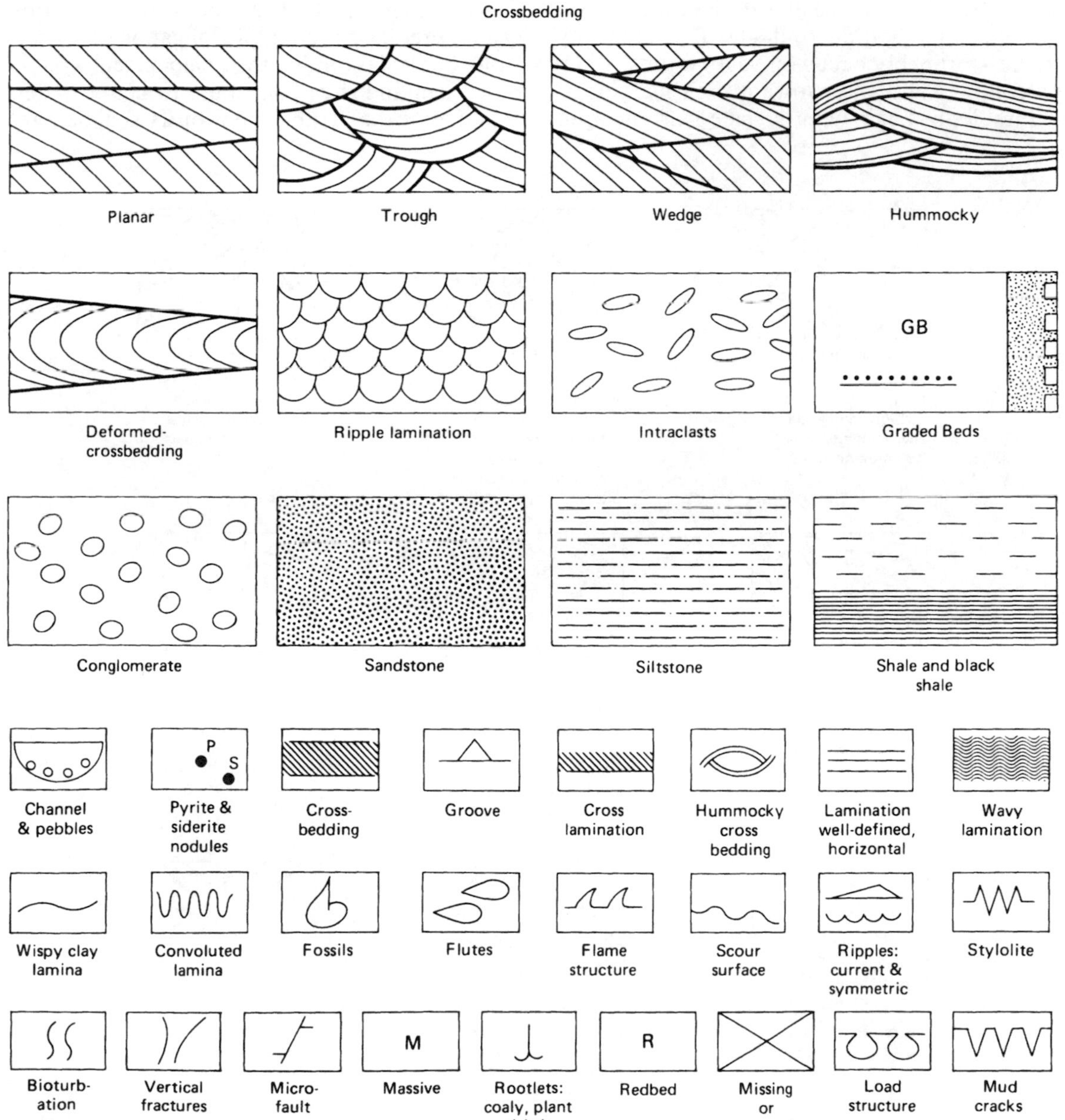

FIGURE 4-37. Symbols for the common sedimentary structures are slowly becoming standardized.

gent, divergent, curvilinear, reversing, and radial (Fig. 4-38). To a large degree parallel, convergent, divergent, curvilinear, and reversing patterns are scale independent—they have been mapped in single sandstone bodies covering a few tens of kilometers to entire basins covering thousands of square kilometers. The rotary pattern, on the other hand, has never actually been mapped, but only inferred for very large basins that might cover much of an entire continent or perhaps would be inferred in the marine current system of a developing rift or a proto-ocean. Small radial patterns are characteristic of volcaniclastics derived from a single volcanic source and—at the other extreme—probably occur on vast cratons, where paleoflow was always toward their margins.

First let us begin with the paleocurrent structures found in sands and sandstones and learn how to interpret, measure, and map them. Maps can be very diverse (Figs. 4-39, 4-40, and 4-41).

For a directional structure to be useful, it must be easy to measure, widespread, and closely correlated with the direction of the current that formed it (Table 4-6). The vast majority of maps have been made with crossbedding in either fluvial-deltaic, eolian, or marine shelf sands, or with sole marks in turbidites. Ripple mark has played a minor role.

What benefits can result from paleocurrent measurement? Knowledge of current direction can (1) predict direction of elongation of a sandstone body, (2) outline the paleocurrent system of a basin and thus contribute to better understanding of the arrangement of its sedimentary

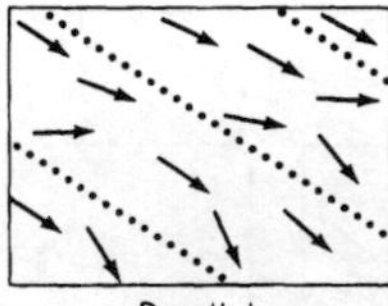

Parallel

Uniform paleoslope in fluvial, marine shelf turbidite and eolian environments. Common pattern.

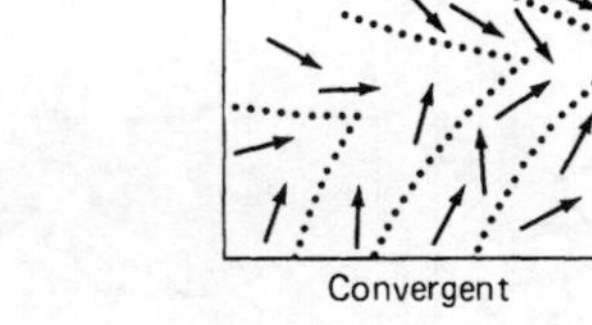

Convergent

Junction of two contrasting paleoslopes or of fans, either submarine or subaerial. Fluvial and turbidite environments most likely. Common pattern.

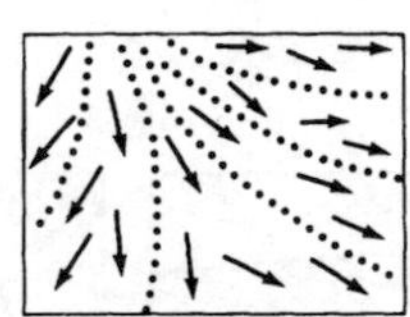

Divergent

Point source such as a submarine or subaerial fan or a glacial lobe. Common pattern.

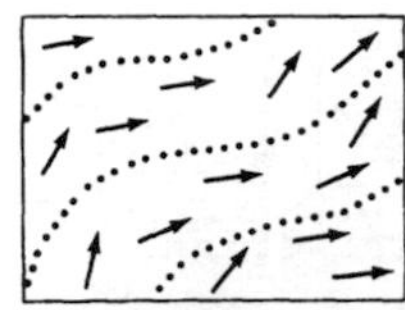

Curvilinear

Changeable slope possibly influenced by local syndepositional tectonic activity. Uncommon.

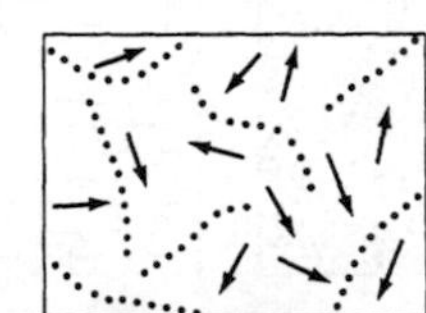

Random

Very rare, but can arise when sampling a sandstone with a very complex history. Also possible in sands formed by very high-sinuosity rivers.

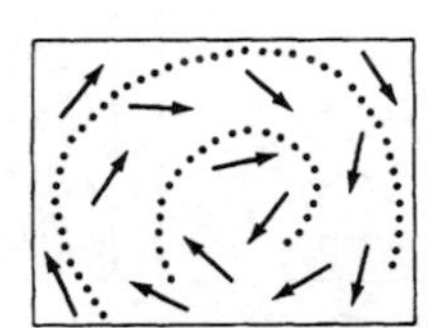

Rotary

Never found in a single sandstone but pattern might exist in a very wide marine basin of oceanic or continental dimensions.

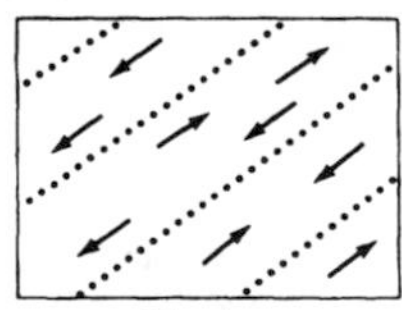

Bipolar

Characteristic of a tidal deposit where ebb and flow currents are of equal magnitude.

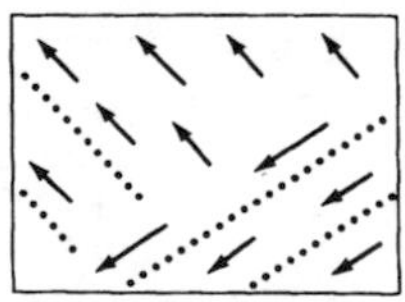

Orthogonal or oblique

Most likely to occur in the mixed environments of a coastal sandy barrier - dunes versus longshore currents and swash.

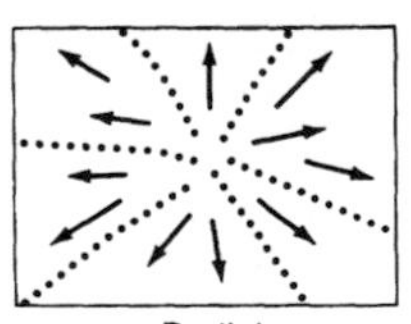

Radial

Characteristic of volcaniclastics and ignimbrites derived from a caldera and of broad cratons where paleoflow was toward continental margins.

FIGURE 4-38. The nine fundamental paleocurrent patterns.

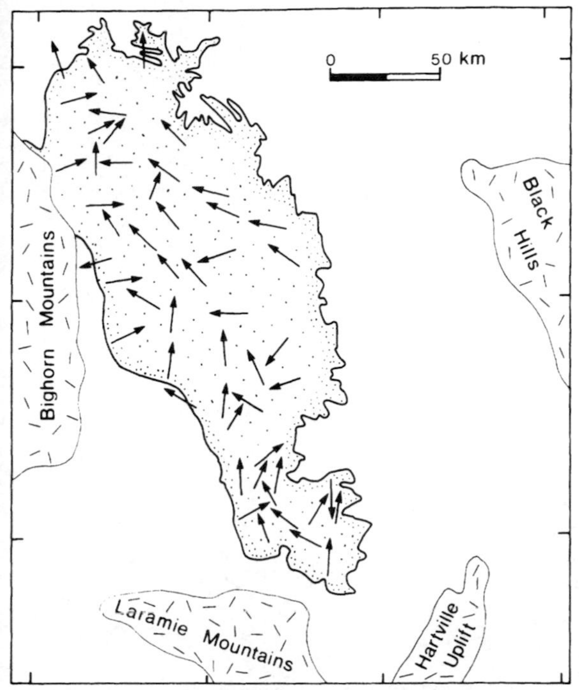

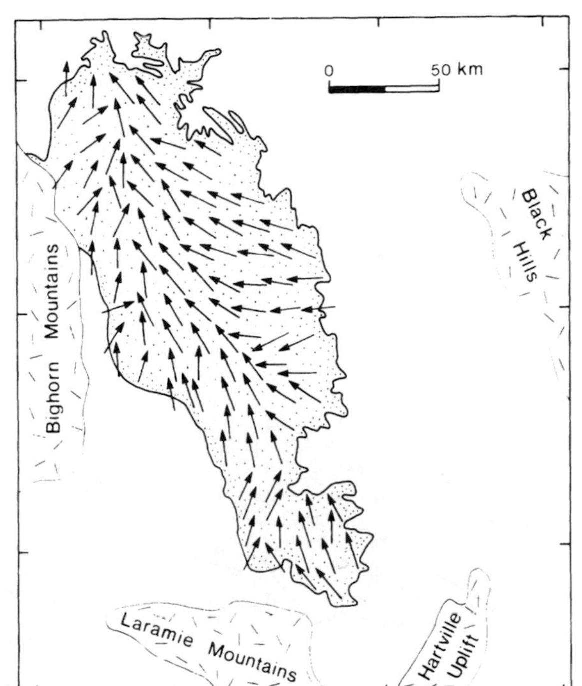

fill, (3) help to determine if a structural feature in a basin was active or not during deposition, (4) help locate source regions that lie beyond a basin margin, (5) aid paleoecology by establishing the direction of supply of nutrient-carrying currents, (6) contribute to some types of stratigraphic correlation problems, and (7) can exercise the dominant control, through grain fabric, on bulk geophysical properties such as electrical and thermal conductivity. As we have already seen, current systems determine the distribution of mineral associations in a basin and some size patterns as well. For effort expended, measurement of directional structures is probably more rewarding to the field geologist than anything else.

Directional structures are usually measured "as they come" in an outcrop as one moves either vertically up section or horizontally along it as in a road cut or along a creek bed. Usually half a dozen or so measurements of crossbedding per outcrop will suffice; but if orientation is very variable, as in some tidally influenced sands, 20 or more readings may be necessary to establish a pattern of bimodality. Vertical profiles of current direction are informative in assessing current stability; pronounced back and forth alternation of ripple mark and crossbedding suggests a strong tidal influence. Also by plotting paleocurrent direction against type of directional structure, one may perhaps better understand the variability of the paleocurrent system and how it is related to specific structures

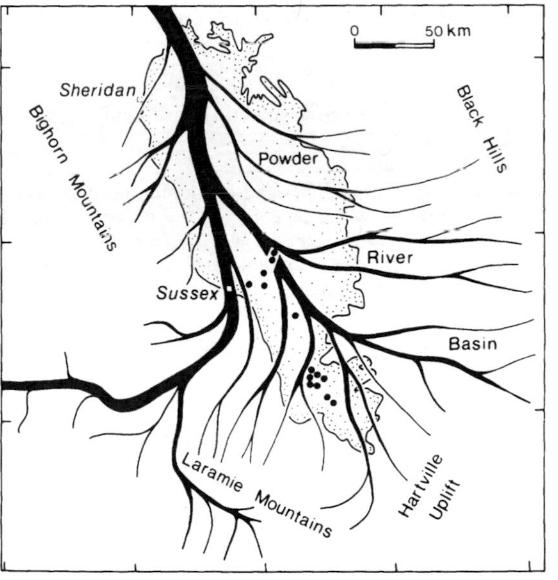

FIGURE 4-39. Crossbedding, paleodrainage, and uranium in the fluvial Eocene sandstones of the Powder River Basin, Montana, U.S.A. A) outcrop averages; B) smoothed vector field; and C) inferred paleodrainage system. (Redrawn from Seeland, 1976, Figs. 6, 7 and 8).

(see Cant, 1978, Fig. 6). In order to specify the convergence, divergence, or uniformity of paleocurrents most effectively, it is generally best to have more outcrops with fewer readings than a smaller number of outcrops each with numer-

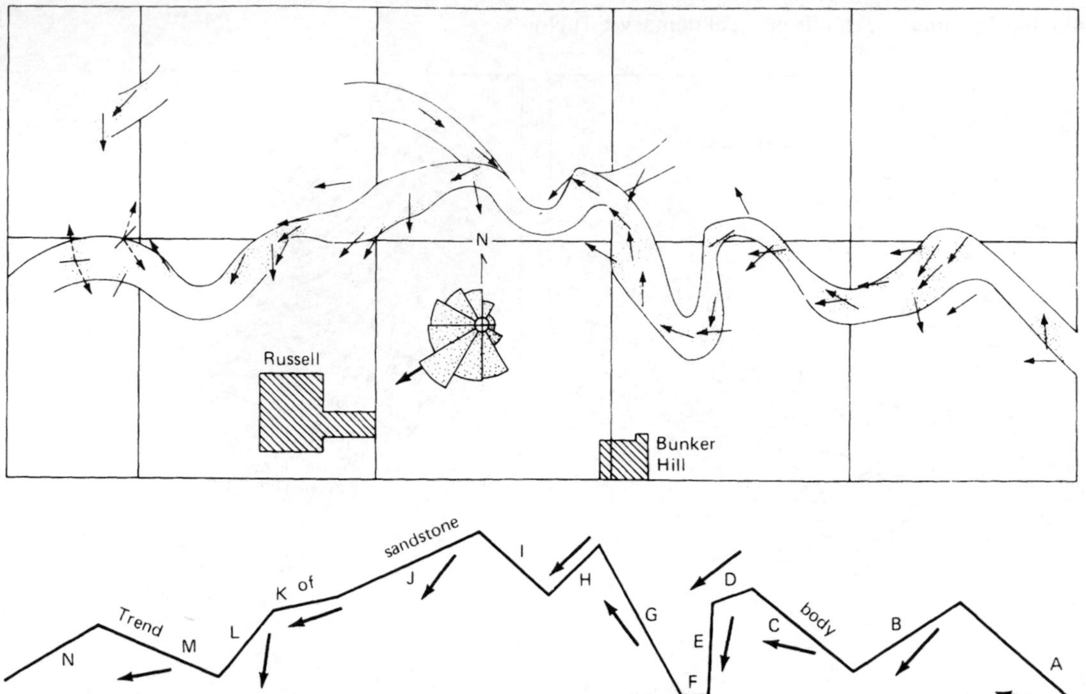

FIGURE 4-40. Crossbedding orientation and trend of Rocktown Channel Sandstone (Cretaceous) in Kansas. (Redrawn from Siemers, 1976, Fig. 6). Original data (above) and average orientation per straight line segment (below). Township grid is 6 × 6 miles (9.6 kilometers).

TABLE 4-6. Directional structures in sands and sandstones—measurement, occurrence and origin.

Structure and measurement	Occurrence and origin
Crossbedding: Trough axis and maximum dip direction of foreset provide local paleoflow. One measurement per set. Hummocky crossbedding has much more variable orientation. Back-set (antidune) bedding rare but present in some turbidites and in ash flows.	Present in almost all traction-transported sands and sandstones and generally thicker than 20 cm. Distinguish crossbedding formed by sand waves from the inclined bedding formed by lateral accretion (sidefill bedding) and from gently inclined beach bedding. Bipolar orientation indicates reversing tidal currents.
Cross-lamination: As above, but perhaps more variable in orientation.	Especially common in siltstones, fine sandstones, and turbidites. Cross-lamination is the internal structure of most ripple marks.
Flutes and grooves: Long dimension parallels current and blunt end of flute indicates upcurrent. Excellent local correlation with paleoflow. Measure average orientation per bed.	Only abundant in turbidites, but present in all sands except the eolianites.
Ripple marks: Direction of steep side indicates sense of flow for asymmetrical ripples, and strike of crest of symmetrical ripples generally parallels shoreline.	Many different names reflect great variation in morphology and hydraulics. Occur everywhere—from the deep sea to eolian dunes—but are perhaps most abundant along strandline. Most common in the fine to medium sands. Rarely measured.
Parting lineation: Trend of parting parallels flow. One measurement per each bed.	Found in almost all environments as a minor accessory structure and therefore rarely systematically measured.
Slides and slumps: Measure fold axes, direction of overturning, or inclinations of slide plane. Structural reorientation of fold axes can be time-consuming.	Most commonly studied in turbidites and slope deposits where failure occurs in downslope direction which may be different from orientation of sole marks.
Trace fossils and fossils: Not measured as much as they could be in sands and sandstone, but some studies exist.	Trace fossils are present in almost all sands but kind, abundance, and relation to current varies with trace fossil. Wood and plant debris have been measured and there are a few studies of the orientation of body fossils.
Clast orientation: Long axes are measured or dip and strike of maximum cross-sectional area.	Rarely used because other paleocurrent structures provide more information more rapidly.

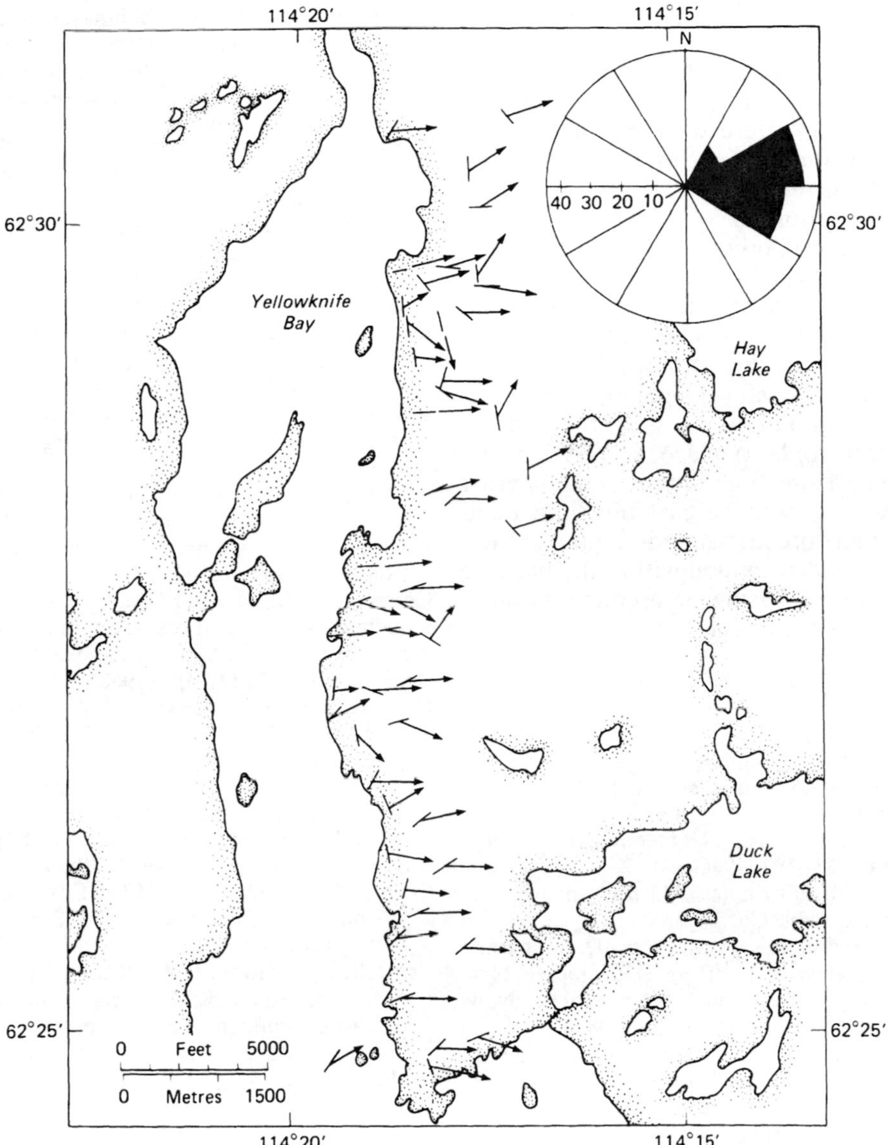

FIGURE 4-41. Well-oriented foresets of ripple cross-lamination from the C division of the Bouma cycle in Burwash Formation (Yellowknife Supergroup, Archean, along east side of Yellowknife Bay, District of Mackenzie, Canada) (Henderson, 1975, Fig. 31). Light line at base of arrow represents structural strike. Because sandstones and their interbedded shales are so well indurated, it was not possible to routinely measure sole marks in this thick turbidite sequence.

ous readings. It is best to compute and plot an average for each outcrop. The vector mean is a satisfactory average (Steinmetz, 1962; Jones, 1967) for all except markedly bimodal distributions. For the latter, either midpoints of the two modes should be plotted or special computations are needed (Jones and James, 1969). Weighting crossbedding orientation has been proposed using the cube of maximum observed thickness of the set (Miall, 1974, p. 1180). The resulting weighted vector mean can be very different from the unweighted vector mean.

Averages and variances are always calculated for the grand total of all observations. Sedimentologists often calculate the mean and variance of paleoflow for the different subfacies of a sandstone—for example, they may compare the vector means and variances of the ripple-bedded, the flat-bedded, and the crossbedded facies. Pocket calculators are used more and more (Lindholm, 1979; Freeman and Pierce, 1979) as are computer programs that include structural rotation (Park, 1974). Trend surface analysis has also been used (Shakesby, 1981).

Potter and Pettijohn (1977) discuss all the major paleocurrent structures and their sampling and interpretation. Here we have only briefly elaborated on some aspects of Table 4-6.

The paleocurrent significance of slides, slumps, and overturned folds has continued to receive attention (Corbett, 1973; Stone, 1976; Woodcock, 1976, 1979) and paleocurrent studies of trace fossils are becoming more commonplace (Beutner, 1975; Crimes and Crossley, 1980). A fairly complex structural analysis is needed to correctly infer the paleoslope significance of slump structures and their many smaller internal folds (Woodcock, 1979, pp. 96–97). Evidence from both slumps and internal folds always needs to be carefully integrated with that from directional sedimentary structures and facies distribution within the basin to obtain the correct, basinwide reconstruction of an ancient current system.

References

Allen, J.R.L.: The classification of cross-stratified units with notes on their origin. Sedimentology 2, 93–114 (1963).

Allen, J.R.L.: On bed forms and paleocurrents. Sedimentology 6, 153–190 (1966).

Allen, J.R.L.: Current ripples, 433 pp. Amsterdam: North Holland Publ. Co. 1969.

Allen, J.R.L.: Sedimentary structures: Their character and physical basis, 1, 593 pp. (Developments in Sedimentology, Vol. 30A). Amsterdam–New York: Elsevier Scientific Pub. Co. 1982a.

Allen, J.R.L.: Sedimentary structures: Their character and physical basis, 2, 663 pp. (Developments in Sedimentology, Vol. 30B). Amsterdam–New York: Elsevier Scientific Pub. Co. 1982b.

Beutner, E.C.: Did the worm turn? Deformed burrow as a slump indicator. Jour. Sed. Petrology 45, 212–214 (1975).

Bokman, J.W.: Suggested use of bed-thickness measurements in stratigraphic descriptions. Jour. Sed. Petrology 27, 333–335 (1957).

Botvinkina, L.N.: Methodicheskoe rukovodstvo po i zucheniiu stoistotis (Manual on the methods of studying bedding). Akad. Nauk, USSR Geol. Inst. Trans. 119, 253 pp. (1965).

Bradshaw, M.A.: Paleoenvironmental interpretations and systematics of Devonian trace fossils from the Taylor Group (Lower Beacon Supergroup), Antarctica. New Zealand Jour. Geol. Geophysics 24, 615–652 (1981).

Bucher, W.H.: On ripples and related sedimentary surface forms and their paleogeographic interpretation. Am. Jour. Sci. (ser. 4) 47, 149–210, 241–269 (1919).

Campbell, C.V.: Lamina, laminaset bed, and bedset. Sedimentology 8, 7–26 (1967).

Cant, Douglas J.: Development of a facies model for sandy braided river sedimentation: Comparison of the South Saskatchewan River and the Battery Point Formation. In: Miall, A.D. (Ed.): Fluvial sedimentology. Canadian Soc. Petroleum Geologists, Mem. 5, 627–639, 1978.

Chowdhuri, K.R., and Reineck, H.-E.: Primary sedimentary structures and their sequence in the shoreface barrier island Wangerooge (North Sea). Senckenbergiana Marit. 10, 15–29 (1978).

Coleman, James M.: Brahmaputra River: Channel processes and sedimentation. Sedimentology, Spec. Issue 3, 131–239 (1969).

Collinson, J.D., and Thompson, D.B.: Sedimentary structures, 194 pp. London: George Allen and Unwin, Ltd. 1982.

Corbett, K.D.: Open-cast slump sheets and their relationship to sandstone beds in an Upper Cambrian flysch sequence, Tasmania. Jour. Sed. Petrology 43, 147–159 (1973).

Conybeare, C.E.B., and Crook, K.A.W.: Manual of sedimentary structures. Australian Dept. Natl. Development. Bur. Min. Res., Geol., and Geophysics, Bull. 102, 327 pp. (1968).

Crimes, T.P. (Ed.): Trace fossils, Vol. 2, 351 pp. Liverpool: Seel House Press 1976 (Geol. Jour. Spec. Issue 9).

Crimes, T.P., and Crossley, J.D.: Inter-turbidite bottom current orientation from trace fossils with an example from the Silurian flysch of Wales. Jour. Sed. Petrology 50, 821–830 (1980).

Crimes, T.P., and Harper, J.C. (Eds.): Trace fossils, 547 pp. Liverpool: Seel House Press 1970 (Geol. Jour. Spec. Issue 3).

Crowell, J.C.; Hope, R.A.; Kahle, J.E.; Ovenshine, A.T.; and Sams, R.H.: Deep-water sedimentary structures Pliocene Pico Formation, Santa Paula Creek, Ventura Basin, California. California Div. Mines and Geology Spec. Rept. 89, 40 pp. (1966).

Dalrymple, R.W.: Morphology and internal structure of sandwaves in the Bay of Fundy. Sedimentology 31, 365–382 (1984).

Davidson-Arnott, R.G.D., and Greenwood, B.: Facies relationships on a barred coast, Kovchiboyquac Bay, New Brunswick, Canada. In: Davis, R.A., Jr., and Ethington, R.L. (Eds.): Beach and nearshore sedimentation. Soc. Econ. Paleon. Mineral. Spec. Pub. 24, 149–168 (1976).

Dalziel, I.W.D., and Dott, R.H., Jr.: Geology of the Baraboo district. Wisconsin Geol. Nat. History Survey Inf. Circ. 14, 164 pp. (1970).

Dean, W.E., Jr., and Anderson, R.Y.: Correlation of turbidite strata in the Pennsylvanian Haymond Formation, Marathon Region, Texas. Jour. Geology, 75, 59–75 (1967).

DeCelles, P.G.; Lanford, R.P.; and Schwartz, R.K.: Two new methods of paleocurrent determination from trough cross-stratification. Jour. Sed. Petrology 53, 629–642 (1983).

Dimitrijević, M.N.; Dimitrijević, M.D.; and Radošević, B.: Sedimentne teksture u turbiditima. Zavod Geološka Geofizička Istraživanja 16, 70 pp. (1967).

Doe, T.W., and Dott, R.H., Jr.: Genetic significance of deformed cross bedding—with examples from the Navajo and Weber Sandstones of Utah. Jour. Sed. Petrology 50, 793–812 (1980).

Dörjes, J.: Sedimentologische und faunistische Untersuchungen an Watten in Taiwan. II. Faunistische und akutopaläontologische Studien. Senckenbergiana Marit. 10, 117–143 (1978).

Dzulynski, S., and Sanders, J.E.: Sedimentary features of flysch and greywackes, 300 pp. Developments in Sedimentology, Vol. 7. Amsterdam: Elsevier Scientific Pub. Co. 1965.

Dzulynski, S., and Walton, E.K.: Current marks on firm mud bottems. Connecticut Acad. Art and Sci. Trans. 42, 57–96 (1962).

Farrow, G.F.: Bathymetric zonation of Jurassic trace fossils from the coast of Yorkshire, England: Palaeogeog. Paleoclimatol. Palaeoecol. 2, 103–151 (1966).

Farrow, G.; Scoffin, T.; Brown, B.; and Cucci, M.: An underwater television survey of facies variation on the inner Scottish shelf between Colonsay, Islay and Jura. Scottish Jour. Geology 15, 13–29 (1979).

Freeman, T., and Pierce, K.: Field statistical assessment of cross-bed data. Jour. Sed. Petrology 49, 624–625 (1979).

Frey, Robert W. (Ed.): The study of trace fossils, 562 pp. New York: Springer-Verlag, 1975.

Frey, R.W., and Pemberton, S.G.: Biogenic structures in outcrops and cores 1. Approaches to ichnology. Bull. Canadian Petroleum Geology 33, 72–115 (1985).

Gall, J.C.: Environments sédimentaires anciens et milieux de vie, 219 pp. Paris: Doin Editeurs 1976.

Glennie, K.: Desert sedimentary environments. Developments in Sedimentology, Vol. 14, 222 pp. Amsterdam–New York: Elsevier Pub. Co. 1970.

Gregory, M.R.: Sedimentary features and penecontemporaneous slumping in the Waitemata Group, Whangaparaoa Peninsula, North Auckland, New Zealand. New Zealand Jour. Geol. Geophysics 12, 248–282 (1969).

Griffiths, J.C.: Measurement of the properties of sediments. Jour. Geology 69, 487–498 (1961).

Grumbt, Eberhard: Beziehungen zwischen Korngrösse, Schichtung, Materialbestand und anderen sedimentologischen Merkmalen in feinklastischen Sedimenten. Geologie (Berlin) 18, 151–167 (1969).

Gubler, Y.; Bugnicourt, D.; Faber, S.; Kubler, B.; and Nyssen, R.: Essai de nomenclature et caractérisation des principales structures sédimentaires, 291 pp. Paris: Editions Technip 1966.

Hall, James: Remarks upon casts of mud furrows, wave lines, and other markings upon rocks of the New York System. Assoc. Am. Geol. Rept. [1941], 422–432 (1843).

Hamblin, W.K.: Internal structures of "homogeneous" sandstones. Kansas Geol. Survey Bull. 175, 568–582 (1965).

Hamblin, A.P., and Walker, R.G.: Storm-dominated shallow marine deposits: The Fernie–Kootney (Jurassic) transition, southern Rocky Mountains. Canadian Jour. Earth Sci. 16, 1673–1690 (1979).

Harms, J.C.; Southard, J.B.; and Walker, R.G.: Structures and sequences in clastic rocks. Soc. Econ. Paleon. Mineral. Short Course no. 9 (1982).

Henderson, John B.: Sedimentology of the Archean Yellowknife Supergroup at Yellowknife, District of Mackenzie. Geol. Survey Canada Bull. 246, 62 pp. (1975).

Hesse, Reinhard: X. Soft x-radiographs of sliced piston cores from the Japan and southern Kurile Trench and slope areas. In: Honza, E. (Ed.): Geological investigations of Japan and southern Kurile Trench and slope areas GH 76-2 Cruise, April–June 1976, pp. 86–108. Geol. Survey Japan Cruise Rept. 7 (1977).

Hill, G.W., and Hunter, R.E.: Interaction of biological and geological processes in the beach and nearshore environments, northern Padre Island, Texas. In: Davis, R.A., Jr., and Ethington, R.L. (Eds.): Beach and nearshore sedimentation, pp. 169–187. Soc. Econ. Paleon. Mineral. Spec. Pub. 24 (1976).

Howard, J.D.: Patterns of sediment dispersal in the Fountain Formation of Colorado. Mountain Geologist 3, 147–153 (1966).

Howard, J.D., and Frey, R.W.: Estuaries of the Georgia Coast, U.S.A.: Sedimentology and biology. II. Regional animal–sediment characteristics of Georgia estuaries. Senckenbergiana Marit. 7, 33–103 (1975).

Howard, J.D.; Elders, C.A.; and Heinbokel, J.F.: Estuaries of the Georgia coast, U.S.A.: Sedimentology and biology. V. Animal–sediment relationships in estuarine point bar deposits, Ogeechee River-Ossabaw Sound, Georgia. Senckenbergiana Marit. 7, 181–203 (1975).

Hunter, R.E.: Terminology of cross-stratified sedimentary layers and climbing ripple structures. Jour. Sed. Petrology 47, 697–706 (1977).

Jago, C.F.: Contemporary accumulation of marine sand in a macrotidal estuary, southwest Wales. Sedimentary Geology 26, 21–49 (1980).

Johnson, A.M., and Rodine, J.R.: Debris flow. In: Brunsden, D., and Prior, D.B. (Eds.): Slope instability, pp. 257–361. New York: John Wiley and Sons 1984.

Jones, G.P.: Deformed cross-stratification in Cretaceous Bima sandstone, Nigeria. Jour. Sed. Petrology 32, 231–239 (1962).

Jones, T.A.: Estimation and testing procedures for circular normally distributed data. Office Naval Research, ONR Task No. 388-078, Contract Nonr-1228 (36), 61 pp. (1967).

Jones, T.A., and James, W.R.: Analysis of bimodal orientation data. Mathematical Geology 1, 129–135 (1969).

Jopling, A.V., and Walker, R.G.: Morphology and origin of ripple-drift lamination, with examples from the Pleistocene of Massachusetts. Jour. Sed. Petrology 38, 971–984 (1968).

Kelley, V.C.: Thickness of strata. Jour. Sed. Petrology 26, 289–300 (1956).

Khabakov, A.V. (Ed.): Atlas tekstur: struktur osadochyhk gornykh porod (An atlas of textures and structures of sedimentary rocks, pt. 1, clastic and argillaceous rocks), 578 pp. Moscow: VSEGEI 1962.

Kindle, E.M.: Recent and fossil ripple mark. Canada Geol. Survey Mus. Bull. 25, 1–56 (1917).

Kitchell, J.A.; Kitchell, J.K.; Johnson, G.L.; and Hunkins, K.L.: Abyssal traces and megafauna: Comparison of productivity, diversity and density in the Arctic and Antarctic. Paleobiology 4, 171–180 (1978).

Klein, George deVries: Depositional and dispersal dynamics of intertidal sand bars. Jour. Sed. Petrology 40, 1095–1127 (1970).

Ksiażkiewicz, Marian: Ichnoskamienałości z osadów fliszowych Karpat Polskich (Trace fossils in the flysch of the Polish Carpathians). Palaeontologia Polonica 36, 208 pp. (1977).

Kuenen, Ph.H.: Slumping in the Carboniferous rocks of Pembrokeshire: Geol. Soc. London Quart. Jour. 104, 365–385 (1949).

Kuenen Ph.H.: Graded bedding, with observations on Lower Paleozoic rocks of Britain. Koninkl. Nederlandse Akad. Wetensch. Afd. Nat. Verh., 1st Ser., 20, 1–47 (1953).

Kuenen, Ph.H.: Sole markings of graded graywacke beds. Jour. Geology 65, 231–258 (1957).

Kuenen, Ph.H.: Experiments in geology. Glasgow Geol. Soc. Trans. 23, 1–28 (1958).

Lindholm, R.C.: Utilization of programmable calculators in sedimentology. Jour. Sed. Petrology 49, 615–620 (1979).

McBride, E.F., and Yeakel, L.S.: Relationship between parting lineation and rock fabric. Jour. Sed. Petrology 33, 779–782 (1963).

McCabe, P.J., and Jones, C.M.: Formation of reactivation surfaces within superimposed deltas and bedforms. Jour. Sed. Petrology 47, 707–715 (1977).

McGowen, J.H.: Gum Hollow fan delta Nueces Bay, Texas. Texas Bur. Econ. Geol., Rept. Invs. 69, 91 pp. (1970).

McKee, E.D., and Bigarella, J.J.: Sedimentary structures in dunes, U.S. Geol. Survey Prof. Paper 1052E, 83–136 (1979).

McKee, E.D., and Wier, G.W.: Terminology of stratification and cross-stratification. Geol. Soc. America Bull. 64, 381–390 (1953).

Miall, A.D.: Paleocurrent analysis in alluvial sediments: A discussion of directional variance and vector magnitude: Jour. Sed. Petrology 44, 1174–1185 (1974).

Miall, A.D. (Ed.): Fluvial sedimentology. Canadian Soc. Petroleum Geologists Mem. 5, 857 pp. (1978).

Nagahama, H.; Ota, Ryokei; and Aoyama, H.: Dish structure newly found in the Nichinan Group, Kyushu, Japan. Bull. Geol. Survey Japan 26, 217–225 (1975).

Park, J.M.: Paleocurrent analysis of sedimentary crossbed data with graphic output using three integrated computer programs. Mathematical Geology 6, 353–372 (1974).

Pettijohn, F.J., and Potter, P.E.: Atlas and glossary of primary sedimentary structures, 117 pl., 370 pp. New York: Springer 1964.

Picard, M. Dane, and High, L.R., Jr.: Sedimentary structures of ephemeral streams, 233 pp. Amsterdam–New York: Elsevier Scientific Pub. Co. 1973. (Developments in Sedimentology *17*).

Potter, P.E., and Pettijohn, F.J.: Paleocurrents and basin analysis, revised and up-dated 2nd ed., 425 pp. Berlin–Göttingen–Heidelberg: Springer 1977.

Potter, P.E.; DeReamer, J.; Jackson, D.; and Maynard, J.B.: Lithologic and environmental atlas of Berea Sandstone (Mississippian) in the Appalachian Basin. Appalachian Geol. Soc. Spec. Pub. 1, 158 pp. (1984).

Ramsey, J.G.: The effects of folding upon the orientation of sedimentation structures. Jour. Geology 69, 84–100 (1961).

Rautman, C.A., and Dott, R.H., Jr.: Dish structures formed by fluid escape in Jurassic shallow marine sandstones. Jour. Sed. Petrology 47, 101–106 (1977).

Reading, H.G. (Ed.): Sedimentary environments and facies, 2nd Ed., 680 pp. Oxford: Oxford Univ. Press 1985.

Reineck, Hans-Erich: Primargefüge, Bioturbation und Makrofauna als Indikatoren des Sandversatzes in Seegebeit vor Norderney (Nordsee). I. Zonierung von Primärgefügen und Bioturbation. Senckenbergiana Marit. 8, 155–169 (1976).

Reineck, H.E.: Layered sediments of tidal flats, beaches and shelf bottoms of the North Sea. In: Lauff, G.H. (Ed.): Estuaries. Am. Assoc. Adv. Sci. Spec. Pub. 83, 191–206 (1967).

Reineck, H.E., and Singh, I.B.: Depositional sedimentary environments, 2nd Ed., 549 pp. Berlin–Heidelberg–New York: Springer-Verlag, 1980.

Reineck, H.E., and Wunderlich, F.: Classification and origin of flaser and lenticular bedding. Sedimentology 11, 99–104 (1968).

Reineck, H.E.; Dörjes, J.; Gadow, S.; and Hertweck, G.: Sedimentologie, Faunenzonierung und Faziesabfolge vor der Ostküste der inneren Deutschen Bucht. Senckenbergiana Lethaea 49, 261–309 (1968).

Ricci-Lucchi, F.: Sedimentografia, 288 pp. Bologna: Zanichelli, 1970.

Rigby, J. Keith, and Hamblin, Wm. Kenneth (Eds.): Recognition of ancient sedimentary environments. Soc. Econ. Paleon. Mineral. Spec. Pub. 16, 340 pp. (1972).

Scott, K.M.: Sedimentology and dispersal pattern of a Cretaceous flysch sequence, Patagonian Andes,

Southern Chile. Am. Assoc. Petroleum Geologists Bull. 50, 72–107 (1966).
Seeland, D.A.: Relationships between early Tertiary sedimentation patterns and uranium mineralization in the Powder River Basin, Wyoming. In: Laudon, R.B. (Ed.): Geology and energy resources of the Powder River Basin. 28th Ann. Field Conference Guidebook, pp. 53–64. Casper: Wyoming Geol. Assoc. 1976.
Seilacher, Adolf: Die geologische Bedeutung fossiler Lebensspuren. Zeitschr. deutsch. geol. Gesell. 10, 214–227 (1953).
Seilacher, Adolf: Studien zur Palichnologie I. Uber die Methoden der Palichnologie. Neues Jahrb. Geologie u. Palaontologie Abh. 96, 421–452 (1955).
Seilacher, Adolf: Kaledonischer Unterbau der Irakiden. Neues Jahrb. Geologie u. Palaontologie Mh. 196, 527–542 (1963).
Seilacher, Adolf: Biogenic sedimentary structures. In: Imbrie, John, and Newell, Norman (Eds.): Approaches to paleoecology, pp. 296–316. New York: John Wiley and Sons 1964.
Shakesby, R.A.: The application of trend surface analysis to directional data. Geol. Mag. 118, 39–48 (1981).
Shawa, Monzer S.: Use of sedimentary structures for recognition of clastic environments, 2nd Ed., 66 pp. Calgary: Canadian Soc. Petroleum Geologists 1979.
Shepard, F.P.; Dill, R.F.; and von Rad, Ulrich: Physiography and sedimentary processes of La Jolla Submarine Fan and Fan-Valley, California. Am. Assoc. Petroleum Geologists Bull. 53, 390–420 (1969).
Shrock, R.R.: Sequence in layered rocks, 507 pp. New York: McGraw-Hill Book Co. 1948.
Siemers, Charles T.: Sedimentology of the Rocktown channel sandstone, upper part of the Dakota Formation (Cretaceous), Central Kansas. Jour. Sed. Petrology 46, 97–123 (1976).
Singh, Irdra Bir: On some Lebensspuren in the Ganges River sediments, India. Senckenbergiana Marit. 10, 67–73 (1978).
Sorby, H.C.: On the structures produced by the currents present during the deposition of stratified rocks: Geologist 2, 137–147 (1859).
Sorby, H.C.: On the application of quantitative methods to the study of the structure and history of rocks. Geol. Soc. London Quart. Jour. 64, 171–232 (1908).
Stanley, D.J.: Vertical petrographic variability in Annot sandstone turbidites: some preliminary observations and generalizations. Jour. Sed. Petrology 33, 783–788 (1963).
Stanley, D.J.: Dish structures and sand flow in ancient submarine valleys, French Maritime Alps. Bull. Centre Rech. Pau-S.N.P.A. 8, 351–371 (1974).
Steinmetz, Richard: Analyis of vectorial data. Jour. Sed. Petrology 32, 801–812 (1962).
Stauffer, P.H.: Grain-flow deposits and their implications, Santa Ynez Mountains, California. Jour. Sed. Petrology 37, 487–508 (1967).
Stone, B.D.: Analysis of slump slip lines and deformation fabric in slumped Pleistocene lake beds. Jour. Sed. Petrology 46, 313–325 (1976).
Thompson, W.O.: Original structures of beaches, bars and dunes. Geol. Soc. America Bull. 48, 723–752 (1937).
Turner, B.R.: Braid plain deposition of the Upper Triassic Molteno Formation in the main Karoo (Gondwana) Basin, South Africa. Sedimentology 30, 77–89 (1983).
Walker, R.G.: Distinctive types of ripple drift cross-lamination. Sedimentology 2, 173–188 (1963).
Walker, R.G.: Geometrical analysis of ripple-drift cross-lamination. Canadian Jour. Earth Sci. 6, 383–392 (1969).
Weimer, R.H., and Hoyt, J.H.: Burrows of *Callianassa major* Say, geologic indicators of littoral and shallow neritic environments. Jour. Paleontology 38, 761–767 (1964).
Williams, George E.: Flood deposits of the sand-bed ephemeral streams of central Australia. Sedimentology 17, 1–40 (1971).
Woodcock, N.H.: Ludlow Series slumps and turbidites and the form of the Montgomery Trough, Powys, Wales. Proc. Geol. Assoc. 87, 169–182 (1976).
Woodcock, N.H.: The use of slump structures as paleoslope orientation estimators. Sedimentology 26, 83–99 (1979).

Part II. The Petrography of Sandstones

Utilizing mineralogy and texture, petrography is the first major step toward putting the parts of a sandstone together to form an integrated, meaningful description which is the basis for almost all subsequent inferences about a sandstone's origin and economic potential. Classification is a natural consequence of the systematic petrographic description of sandstones for, like all natural materials, sandstones differ from one another and one needs a convenient practical shorthand to quickly and conveniently label and classify them. The many classification schemes hitherto available have been further multiplied in the service of plate tectonic interpretations but the general bases of classification remain the same. The two petrographic chapters that follow describe major petrographic types, their occurrence, and origin. Again we do not give encyclopedic coverage, for sedimentary petrologists, more than ever in recent years, have been describing an enormous diversity of sandstone types. We focus on some typical representatives and how interpretations follow from detailed study of the components of the sandstone. Together these three chapters form the basis for the serious study of sandstones, a work that was petrographically begun by Henry Clifton Sorby in 1877 and 1880 in his papers "The Application of the Microscope to Geology" and "On the Structure and Origin of Non-calcareous Stratified Rocks." The appendix offers guides to and gives an example of a detailed petrographic description and its interpretation.

CHAPTER 5
Petrography of Common Sands and Sandstones

There are three major types of sand: terrigenous, carbonate, and pyroclastic. *Terrigenous* sands are most abundant and are all ultimately derived from outside the basin of deposition by erosion of preexisting crystalline, volcanic, and sedimentary rocks and, except for eolianites, are all deposited by water. Silicates predominate. *Carbonate* sands are virtually all deposited in marine waters and consist primarily of skeletal grains, oolites plus other coated grains, some locally derived detrital carbonate called *intraclasts*, and *terrigenous* carbonate. The latter is really a terrigenous sand, and not abundant except where there is very rapid erosion of thick carbonate sections in orogenic belts or in some glacial situations. *Pyroclastic* sands are those derived directly from volcanic explosion as, for example, ash, lapilli, and bombs. They may be deposited on either land or in water. Pyroclastic sands are less abundant than either terrigenous or carbonate sands. The more inclusive term *volcaniclastic* refers to clastic sedimentary materials rich in volcanic debris which may be either of pyroclastic origin or a normal terrigenous (epiclastic) sand derived from an older volcanic terrane (discussed in detail in Chap. 6).

It is noteworthy that the above classification, although simple and fitting readily into our ordinary geologic thinking, is not internally consistent as it is based on a variety of concepts: terrigenous sands are those classified by their source, carbonate sands are defined by composition, and pyroclastic sands by their agent of formation, vulcanism. Nonetheless, the threefold division is a valid one because of the unlike modes of sand production represented. This classification is based only on the detrital fraction and hence minimizes the complications of diagenesis.

In nature there are gradations between all three types. Pyroclastic and carbonate sand may occur together, for example, in an oceanic setting where skeletal debris is accumulating on a shallow shelf marginal to islands of volcanic origin. Or terrigenous sand may be carried into the area by longshore currents to become a minor constituent of the indigenous oolitic and skeletal sands.

We are primarily concerned with the terrigenous sands. But before considering their classification, let us first take a look at the more general problems of petrographic nomenclature and classification.

Nomenclature and Classification

Nomenclature and classification are parts of the same problem. To give something a name is to set it apart from all other things—to put it in a class by itself. The need for classification and nomenclature is both to facilitate communication and to organize our thinking. Our problem is to consider the petrographic classification of sandstone—itself a name designating a class of sedimentary rocks.

To classify is to define, to draw limits based on some tangible property. Of all the many properties of sandstones, which shall be the defining parameters? We think the answer depends on the point of view and the use to be made of the classification. An engineer might select properties meaningful to engineering problems. A geologist might choose a different set of parameters related to pertinent geological questions. The ultimate geological question is origin and, therefore, the defining parameters should be as genetically significant as possible. A simple inventory of the minerals in a sand contributes little—what is really required is that they be arranged in meaningful genetic groups. One needs to discriminate, for example, between the original detrital components, those introduced as cements, and those produced by post-depositional metamorphism or weathering.

Even if it were desirable, no wholly descriptive classification is, in fact, possible. Such a classification, were it possible, would be largely meaningless and hence useless. Virtually all geologic "facts" are, in truth, interpretations based on origin. "Sandstone" is a good illustration. It is difficult to define independently of its origin. Even a simple term such as "pebble" has strong genetic connotations (which distinguish it from other round objects such as concretions).

Some rock names are derived from common speech and have been redefined. These include such terms as "sand" and "clay." Others, like "litharenite," have been coined to fill a need. Unfortunately, usage tends to vary with time and geography. Some older terms are abandoned and new ones invented. A few are tied to a particular place and first usage—a "type" locality. The term "graywacke" is an example. Abandonment of the term, as advocated by some, would not solve our problem—the problem of defining and naming the rock.

To summarize, classification to be geologically meaningful should be based on easily observed, genetically significant variables that are susceptible to quantitative analysis; the names applied to the classes defined are more acceptable if they correspond to current usage and do not depart greatly from original meaning.

Defining Parameters

All of the three types of sands—terrigenous, carbonate, and pyroclastic—can be considered in terms of their relative proportions of framework grains, detrital matrix (called *micrite* if carbonate rather than argillaceous mud), and chemical cement. Framework grains are those that support the sand, whereas matrix and chemical cement either wholly or partially fill the pore space.

The framework grains of noncarbonate sands are mainly quartz, feldspar, and fine-grained rock fragments (chert, limestone, siltstone, argillite, slate, glass, rhyolite and other volcanic fragments, etc.). Micas commonly play but a subordinate role as framework elements. Others such as heavy minerals and glauconite are negligible except locally (as in placers and greensands). The three basic framework components of terrigenous and pyroclastic sands are thus quartz, feldspar, and rock fragments.

Should a classification be based on the framework grains alone, or on framework grains plus matrix, or should the chemically precipitated cements also be included? The independence of mineralogy and texture bears upon this question. In igneous petrology both are used to define rock clans and are treated as independent variables. Can we do the same for sands and sandstones? To state the question differently, is mineralogical composition dependent or independent of grain size? As we show in Chap. 2, the two are not independent.

Regardless of the defining parameters finally selected, there is always the problem of subdivision and the naming of mixtures. How many subdivisions, what limits, and what names? The possibilities are really endless and consequently there is plenty of opportunity for individual expression.

A major sandstone type is distinguished by a dominant constituent. For example, if more than 50 percent of a sand consists of detritus derived from preexisting rocks, it is a terrigenous sand; or if 50 or more percent is derived directly from volcanic explosion, it is a pyroclastic sand.

But, as subdivisions within these major types are made, disagreement develops as where to place exact limits, even if the basic parameters are similarly chosen. It has been suggested that such problems can be avoided by not using a name at all but relying instead on only a few relevant numbers (Rodgers, 1950, p. 308; Füchtbauer, 1964, pp. 165–66; Boggs, 1967; Dickinson, 1970) or perhaps simply reporting the entire composition. This suggestion presumes that one has made a quantitative petrographic modal analysis. The latter is, of course, necessary for many statistical purposes and no quantitative classification of sandstones makes sense without the modal analyses to support it. Petrologic interpretation of a given sandstone is often based on its quantitative analysis—an analysis that also allows it to be classified. Though we might envision a literature that quoted only modal analyses and used no classification or nomenclature, such a procedure is all right up to the point of either discussion or final report writing when a name does become necessary for effective communication. Finally, regardless of the name, the petrologic interpretation rests on the modal analysis, not on the system of classification.

Major Trends in Sandstone Classification

The last generation saw an almost continuous flow of papers on sandstone classification and we are not through yet. Much of this literature has been summarized by Klein (1963) and Okada (1971). Without reviewing the history of the classification controversy in detail, it is worthwhile to note the major trends in the discussion. Some of the classification schemes have received wide usage; others have been proposed and hardly used except by the proposer. In a sense a good many petrographers have voted on their preference by how such classifications work when actually using the microscope rather than on abstract or theoretical grounds. We have been mindful of that kind of choice, for a classification is truly of value only if it can be and is used.

The use of the microscope is central. Though careful field observation, including the use of a hand lens, makes a large number of distinctions, it is doubtful that any of thc proposed classification schemes can be applied in any precise fashion without a careful and quantitative microscopical analysis. Thus a field classification is bound to be limited to general types, without quantitative limits, and is likely to be affected by variables of lesser interest, such as color or friability. There should be no difficulty in recognizing the pure quartz sandstones, the highly feldspathic arkoses, or the dark graywackes that are the end-members of the classification; there is less point in trying to discriminate precisely among rocks of mixed composition.

The major impetus to sandstone classification came from the proposals of P. D. Krynine in the years 1940–1948 and F. J. Pettijohn from 1943–1957. Both recognized the importance of mineralogy as a clue to source-rock composition and source-area tectonism but put different emphasis on the role of texture. It is clear that all of the other schemes proposed derive from one or the other or both of them. The composition triangle was used for representation of modal analyses. The relative amounts of feldspar (or other igneous minerals) and quartz (or other siliceous sedimentary minerals) were explicitly recognized clues to the composition of source rocks, the tectonics of the source terrane, and the weathering processes that produced detritus.

Krynine (1948) produced a composition triangle that considered only the detrital fraction and divided the rocks into classes based on the three-component system: *quartz* (+chert) – *feldspar* (+kaolin) – *phyllosilicates* (micas + chlorite). It was the mineral composition rather than the texture (mica–chlorite matrix or quartz–carbonate cement) that was the important criterion. Central to Krynine's ideas was the quantitative and interpretive significance of the graywackes, those sandstones with high proportions of the mica + chlorite component and varying amounts of the feldspar + kaolin component. He divided "low" and "high" rank graywackes on the relative amounts of the feldspar + kaolin and mica + chlorite components.

A somewhat different emphasis was given by Pettijohn to the classification of the rock fragment – mica + chlorite + clay matrix complex that Krynine had treated as only a mineralogical problem. In the fullest statement of his ideas (1954) Pettijohn erected a four-component system in which the matrix, a fine-grained mixture of various kinds of clay (commonly altered to mica–chlorite by diagenesis) was a component by itself, differentiated from the framework rock-fragment components. The rock-fragment component included the same micaceous low-rank metamorphic rocks that Krynine assigned to his mica + chlorite component; the difference was that Pettijohn essentially took that component of Krynine and divided it into two on the basis of texture. The importance of the detrital matrix was recognized by both Krynine and Pettijohn, but it was elevated to a classification variable by Pettijohn because he reasoned that it was a fluidity index of the transporting medium and closely related to the mechanics of turbidity flows of dense clay suspensions—a view we now know is only partially correct.

Many other attempts were made to resolve the ambiguities in the treatment of both texture and mineralogy in the same classification. Folk (1954, 1956) redefined Krynine's classification by introducing grain size end-members with all gradations between gravel, sand, silt, and mud as well as using the abundance of clay as an index of what he called "textural maturity," an extension of the idea of mineralogical maturity. Gilbert (Williams, Turner, and Gilbert, 1954, p. 290) used a textural criterion, the degree of sorting and detrital matrix, to divide all sandstones into suites, the wackes and the arenites. Each of these was classified according to the same general three-component system used by Pettijohn. Many other proposals have been made that amplify, amend, extend, or modify the earlier classifications (Table 5-1).

Dickinson (1970) advocated reserving such

TABLE 5-1. Summary of classifications of terrigenous sandstone (modified from McBride, 1963, Table 1).

Reference	Basis of classification	End-members of classification			Comments
Fischer (1933)	Mineralogy	Quartz	Feldspar	Rock fragments	First use of triangular diagram for sandstone composition?
Krynine (1948)		Quartz	Feldspar and kaolin	Micas and chlorite	Ignores rock fragments, Graywacke based solely on mica and chlorite.
Folk (1954)		Quartz and chert	Feldspar and volcanic rock fragments	Metamorphic rock fragments, micas, metamorphic quartz	Graywacke based solely on metamorphic constituents. Sedimentary rock fragments ignored.
van Andel (1958)		Quartz	Feldspar	Rock fragments and chert	Graywacke based solely on rock fragments and chert.
Füchtbauer (1959)		Quartz	Feldspar	Rock fragments and chert	Recognizes clay-rich and clay-poor sandstone types.
Hubert (1960)		Quartz, chert, and metaquartzite	Feldspar and feldspathic crystalline rock fragments	Micas and micaceous rock fragments	Non-micaceous rock fragments are not treated as a major constituent. Classification designed originally for feldspathic rocks.
Fujii (1962)		Quartz plus chert	Feldspar	Rock fragments	Divides triangle into five fields.
McBride (1963)		Quartz, chert and quartzite	Feldspar	Rock fragments	Ignores large micas. Has 8 classes.
Shutov (1967)		Quartz	Feldspar	Rock fragments	Divides triangle into 12 fields forming three major groups; graywacke based solely on rock fragment content.
Teodorovich (1967)		Quartz	Feldspar and mica/chlorite	Rock fragments	Eighteen subdivisions of triangle; addition of pyroclastic mineral requires tetrahedral representation and 11 additional subclasses.
Tallman (1949)	Texture and mineralogy	Quartz	Feldspar		Graywacke based solely on matrix content. Rock fragments ignored.
Dapples *et al.* (1953)		Quartz and chert	K- and Na-feldspar	Rock fragments and matrix	Graywacke based on sum of rock fragments and

Williams *et al.* (1954), Dott (1964)		Quartz, chert, quartzite	Feldspar	Unstable fine-grained rock fragments	Recognizes two suites on basis of > or < 10% matrix. Graywacke used as special rock type and not part of classification.
Bokman (1955)		Quartz	Feldspar and rock fragments	Clay	Graywacke based solely on clay content. Feldspar and rock fragments not differentiated.
Pettijohn (1957)		Quartz and chert	Feldspar	Rock fragments	Clay matrix is most important property of graywacke.
Sahu (1965)		Stable grains (quartz, chert, quartzite, plus tourmaline, etc.)	Unstable grains (feldspar, rock fragments, and micas)	Matrix	Eight clans of sandstone.
Krumbein and Sloss (1966)		Quartz	Feldspar	Clay, sericite, and chlorite	Ignores rock fragments. Graywacke based on clay, sericite, chlorite, and feldspar content.
Boggs (1967)		Siliceous resistates	Feldspar	Labile grains	Ten principal and 10 subclasses.
Okada (1971)		Quartz	Feldspar	Rock fragments including chert	Recognizes two suites: arenites and wackes (15% matrix).
Packham (1954)	Structure, texture, and mineralogy	Quartz and chert	Unstable minerals and rock fragments	Matrix	Recognizes two suites. Graywacke based on deposition by turbidity current.
Crook (1960)		Quartz and chert	Unstable minerals and rock fragments	Matrix	Recognizes three suites. Graywacke based on deposition by turbidity current.
Dickinson (1970)	Modal analysis	Quartz (mono-, poly-, crypto-)	Feldspar	Lithic (fine-grained rock fragments)	Rocks defined solely by modal analysis expressed by formula $Q_xF_yL_z$.
Zuffa (1980)	Grain composition		Carbonates, extrabasinal and intrabasinal	Noncarbonates, extrabasinal and intrabasinal	Provenance-oriented classification of "hybrid" sandstones.
		Terrigenous grains plus carbonate and noncarbonate grains from inside and outside the basin.			

Other papers dealing with sandstone classifications, not represented in this table, include those of Michot (1958), Kossovskaia (1962), Shutov (1965), Chab (1967), Chang (1967), Wang (1967), Chen (1968), Konta (1968), Peikh (1969), and Travis (1970).

terms as arkose and graywacke for imprecise field descriptions. He would prefer to characterize sandstones in terms of their modal composition and use simple percentage subscripts for Q, F, and L (quartz, feldspar, and lithic fragments). Such information permits plotting of the composition on the standard triangular diagram. An ambiguity arises, however, in the case of polycrystalline quartz (quartzite, chert). It could be plotted either as a lithic fragment or as quartz. If the emphasis is on the maturity of the sand, such quartz goes at the Q-pole; if provenance, then at the L-pole. For finer discrimination one may distinguish plagioclase (P) from K-feldspar (K), and volcanic (Lv) from sedimentary (Ls) particles. See Table 5-2. Dickinson has further elaborated this classification in relation to plate tectonics as a prime determinant of provenance (see Chap. 12).

Though there has been no overwhelming acceptance of any one classification, most petrographers have tended to clump about the classification schemes of Krynine as modified by Folk, Pettijohn, Gilbert, or Dickinson. The impression from the literature of the past decade is that most practicing microscopists prefer to recognize texture explicitly, and favor dividing sandstones into two great groups—the poorly sorted, matrix-rich, and the well-sorted, matrix-poor. Much of the matrix is recognized as squashed or altered rock fragments. Within that conceptual framework, there are minor variations in such matters as the choice between classifying a metamorphic quartzite grain as a particle of quartz or a metamorphic rock fragment. Such choices are probably always going to vary, for different sandstones may need "tailormade" decisions on how to classify most meaningfully.

We will always need the approximate, field-defined terms such as arkose, subarkose, lithic sandstone, and others for general converse. On the other hand, we regret an overemphasis on names and a tendency to subdivide or split classes. We restrict ourselves in the following sections to those names designating only the larger categories but for the readers' benefit have assembled all published terms at the end of this chapter.

Making a Choice

The classification we use for terrigenous sands and sandstones (Fig. 5-1) is a very simple one and one that, as indicated above, is generally consistent with current usage. It is appropriate for both ancient and modern sands. Basically, it uses only framework grains of quartz, feldspar, and rock fragments of sand size. As a secondary criterion, the classification distinguishes between the "clean" sands or *arenites*—sands with less than 15 percent matrix—and the "dirty" sands or *wackes*—those with more than 15 percent matrix. We recognize that matrix may be largely diagenetic and thus this distinction introduces diagenesis as well as transport and provenance. Among the matrix-poor sands, those with no more than five percent of

TABLE 5-2. Definition of grain populations for triangular compositional diagrams (after Dickinson, 1982, Table 1).

Triangular diagram	Uppermost pole	Lower left pole	Lower right pole
QFL	Q Quartzose grains (= Qm + Qp)	F Feldspar grains (= P + K)	L Unstable aphanitic lithic fragments (= Lv + Ls)
QmFLt	Qm Monocrystalline quartz grains	F (same as above)	Lt Total aphanitic lithic fragments (=L + Qp)
QmPK	Qm (same as above)	P Plagioclase grains	K K-feldspar grains
QpLvLs	Qp Polycrystalline quartzose lithic fragments (chert, quartzite, etc)	Lv Volcanic and metavolcanic lithic fragments (including hypabyssal)	Ls Sedimentary and metasedimentary lithic fragments (argillite, hornfels, shale, slate, etc)

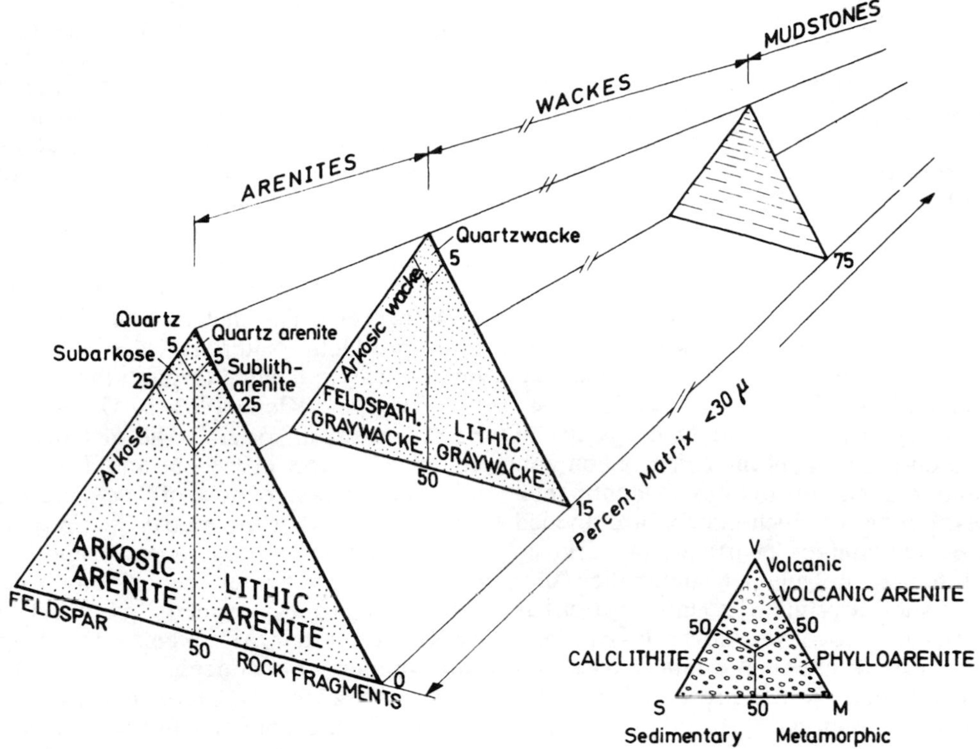

FIGURE 5-1. Classification of terrigenous and sandstones. (Modified from Dott, 1964, Fig. 3)

either feldspar or rock particles are called *quartz arenite;* commonly in the past we have called these sandstones *orthoquartzites*. Those with 25 percent or more feldspar and a smaller percent of rock fragments, are the *arkosic arenites*. *Arkoses* belong to this clan. Those with 25 or more percent of rock fragments, but a lesser amount of feldspar, are the *lithic arenites* (some may choose to contract this to *lithic-arenites*). Transitional classes, subarkose and sublithwacke (protoquartzite), may be recognized (Fig. 5-1).

The term *feldspathic arenite* is a loosely used, all-encompassing term. Many arenites contain some feldspar. Some contain a great deal. As here used, any sandstone with five or more percent feldspar is "feldspathic." As thus defined, feldspathic sandstone includes subarkose, arkose, some litharenites, and many graywackes. The reader will note that "arkose" is here defined in a narrower sense. Commonly it is said that any sandstone with 25 or more percent feldspar is an arkose. This is usually true but as here defined an arkose must also have more feldspar than rock fragments. If it does not, it is a *feldspathic lithic arenite*. Also we have followed Gilbert in dividing the arkoses into *arkosic arenite* (a "clean" arkose) and *arkosic wacke* (a "dirty" arkose)—the latter having a significant matrix content.

The *litharenites* (a contraction of lithic arenite) are those arenites with 25 or more percent of rock particles and a minimal matrix content. Most commonly these rock particles are pelitic in character (shale, siltstone, slate, phyllite, and mica schist). Consequently the litharenites with such phylloid fragments have been termed *phyllarenites* (Folk, 1968a, p. 131). Most lithic arenites are phyllarenites. Other important types include those in which the dominant rock particles are limestone or dolomite. The term *calclithite* (Folk, 1968a, p. 141) has been proposed to distinguish this carbonate sand from *calcarenite*—a carbonate sand produced by biochemical or chemical precipitation. Other lithic arenites include *chert arenite,* if chert is the dominant detrital rock particle. Some lithic arenites may be *volcanic arenites* if the rock detritus is derived from disintegration of extrusive or flow rocks.

The transitional class *sublitharenite* has a lesser rock particle content. The term *protoquartzite* has been used by Krynine (Payne, 1952) for these sandstones. Likewise the term *subarkose* is used for a transitional class of arenites with less feldspar than an arkose and with few or no rock particles.

For those sandstones with a significant matrix content (15 percent) the general term *wacke* (Fischer, 1933, p. 366; Williams, Turner, and Gilbert, 1954, p. 290) has been used. These are dominantly the *graywackes,* a term which has engendered considerable controversy. We retain the term and recognize two main classes: *feldspathic graywacke* and *lithic graywacke* depending on whether detrital feldspar or detrital rock particles dominate the rock particle fraction. *Quartz wackes* constitute a minor and relatively small group within the wacke clan.

In summary, we have two main groups of sandstone on the basis of their matrix content. The common sands or arenites of continental platforms, devoid of much matrix, are divided into three main families (quartz arenite, arkosic arenite, lithic arenite) and two subfamilies (subarkose and sublitharenite). The investigator has the option of recognizing some lesser sand types within the lithic arenite family based on the dominant rock particle present. It is noteworthy that the mineralogy or composition of the cement does not enter into the classification. The petrographer can, and should, indicate the character of the cement by a proper adjectival designation, for example, calcareous subarkose.

The above classification is a classification based on mineral composition and generally has minimal dependence on the environment of deposition. For example, a quartz arenite may have been deposited as a subaerial dune, in a stream or on a beach. Or an arkosic arenite may have been deposited on an alluvial piedmont fan or as a marine shelf sand. Because the character of the source rocks very largely determines mineral composition, this classification is largely relatable to source area composition and ultimately to tectonics.

We have used detrital matrix as a classifying criterion despite the fact that, as in the case of "micrite" in the carbonate rocks, there is little agreement among petrographers as to the upper size limit of matrix. Should it be 4 microns (1/250 mm) as in the Wentworth scale? Most have set it higher, some using 20 microns, some 30, and some even 62.5. We have selected 30 microns as the upper limit of matrix. This seems to us most satisfactory, although we recognize that others may alter this limit.

It is becoming increasingly clear that the matrix is not all of primary origin. As Kuenen (1966) has shown, the trapped pore waters of turbidity currents will contain only a small amount of suspended clay. The maximum percentage of clay in modern coarse turbidites is below ten percent though that amount increases with finer-grained sands and silts such as those of oceanic abyssal plains. Hence, contrary to earlier opinion, the matrix of the turbidite sands is not necessarily entirely or even largely the consequence of deposition from mud-laden waters. Other mechanisms are needed to account for those sandstones with matrix content of 20 to 40 percent. The matrix could be an infiltration product—materials filtered out from muddy waters flowing through the pore system (Emery, 1964; Klein, 1963, p. 571)—or it could be diagenetic in origin. The absence of such abundant matrix in Recent and Tertiary turbidites, in contrast to those of Paleozoic and older age, suggests that the matrix was produced by transformation of argillaceous and volcanic rock fragments after deposition (Cummins, 1962a)—a process termed "graywackization" by Kuenen (1966). Although some petrographic data support this concept (Greenly, 1899, p. 256) and transitions between rock fragments and matrix are common in these rocks, some conclusive evidence is experimental. Modern sands, containing feldspar and labile rock particles, subjected to water pressure of 1 kilobar and slight elevation of temperature, were converted to synthetic graywackes with matrix (Hawkins and Whetten, 1969). These experiments, coupled with Brenchley's observation (1969) that the matrix of some Ordovician graywackes, normally 40 to 60 percent of the whole rock, was absent where a calcite cement was present, strongly suggest a diagenetic origin. In part, the matrix may also be the result of mechanical action—squashing the weaker shale particles and pellets as well as similar grains which are molded around the more durable quartz grains and appear to be a matrix-filled pore. Dickinson (1970, p. 702) has formally recognized the diverse origins of matrix by giving names to the several kinds: protomatrix (primary), orthomatrix (recrystallized primary matrix), epimatrix (produced by diagenetic alteration of framework grains), and pseudomatrix (deformed original grains). He also outlined the criteria for recognition of the several types. Following Fischer (1933), Gilbert (Williams, Turner, and Gilbert, 1954, p. 290), and Dott (1964), we think the term *wacke* is a useful one to substitute for arenite when the sandstone has more than 15 percent matrix, as long as one remembers that the term itself is descriptive and implies nothing about the mechanism of transport or origin of the matrix. This usage of wacke differs from that of some who equate it with a sandstone presumed to have been deposited by a turbidity current. Wacke, therefore, is not a synonym for muddy or clayey sandstones. In our judgment, much of the matrix is generally not mud or clay but is derived in some manner from mechanically or chemically unstable framework grains.

If, as pointed out above, some or much of this matrix is diagenetic or postdepositional in origin, this utilization of a diagenetic feature is at variance with our disregard of the precipitated mineral cements. These do not enter into our classification except as adjectival modifiers. We are aware of this inconsistency but we justify our decision by the prevailing treatment of the graywackes—that class of sandstones with a significant matrix content. The graywacke problem and the significance of matrix is discussed at length later in the chapter.

Our approach to classification recognizes the influence of mineralogical composition on grain size and the role that diagenesis can play in the origin of matrix. Because of these factors, we believe that the conservative scheme of Fig. 5-1 represents about as many petrographic types as can be meaningfully used. With appropriate adjectival modifiers this classification can also be used for sands rich in unusual constituents. Should one encounter, for example, an arenite in which olivine is the dominant mineral, the rock can be called an olivine arenite. Or should the framework be glass shards, the sandstone can be described as a vitric arenite.

Complete description should, of course, include textural terms (Table 3-2) so that the final rock name might be, "medium-grained, well-sorted lithic arenite," the order being size, sorting, and rock name. To this one might add degree of cementation, porosity, and possibly color, if deemed necessary.

And what of carbonate or calcarenite sands? Like their terrigenous counterparts, the basic components are framework grains, micrite, and chemical cement. Framework grains that are derived from within the basin of deposition, and commonly have experienced some transportation, are called *allochems*. Because these allochems, which include skeletal debris, are capable of extensive subdivision, there are many more components to use in defining and mapping the calcarenites than are commonly available for terrigenous sands. Carbonate classifications are many and varied. Among the more widely used classifications are those of Folk (1959), Dunham (1962) and Wilson (1975). The student of sandstone is commonly involved with "hybrid" sands—mixtures of carbonate detritus with the ordinary constituents of sands. Zuffa (1980) has addressed the problem of classification of these materials.

In carbonate rocks, the problems of microcrystalline calcite are comparable to those of matrix in the terrigenous sandstone. Distinguishing between detrital micrite and later secondary recrystallization that produces microspar can be just as difficult as unraveling the transformations experienced by matrix in a lithic arenite. Folk (1965) emphasizes this complexity of the carbonate rocks.

Introduction to Petrography

There comes a time in the study of a sandstone body or formation when it is necessary to look at the rocks of which it is made. It is not enough to measure a stratigraphic section, to trace out the limits of a given sandstone, or even to study its structures in the field and map its paleocurrents. The rocks which constitute the formation cannot be ignored. A close look at these cannot be made without study of thin sections under the microscope. Such study is not to classify or name the rock but rather to understand it. It is necessary to know of what kinds and classes of minerals it is made and how they are put together. Petrographic studies can greatly expand our understanding of the geologic history—can contribute to the questions of provenance by supplementing or confirming paleocurrent studies, indicate probable source rocks as well as source areas, can assist, perhaps, in making environmental discriminations, provide insight into the nature of the rock fabric and pore system, and shed light on its diagenetic history—including grain alterations and cementation, changes that profoundly alter the porosity and permeability of the rock.

Our objectives are (1) to *describe* the principal families of sandstone and some of the more important species in each group and to point out some of the problems of origin of each of them, (2) to summarize what is known about their *relative abundance*, and (3) to summarize the principal theories of *sandstone petrogenesis*.

We have grouped sandstones into genetically meaningful classes namely, those based on maturity as expressed by composition and sorting. These are (1) the immature sandstones, that is, the least modified residues—those closest in composition to the parent rock, (2) the mature sandstones, those nearest the theoretical end-product, and (3) various hybrid types such as tuffaceous sandstone, and other hybrids with a significant fraction of sand formed by chemical or biochemical processes, and (4) a few rare types of sand which do not fit readily in any of the above categories.

The petrography of a sandstone is determined in large measure by provenance (Fig. 5-2). The character of the immature sands, in particular, is controlled primarily by the nature of the source rock. The ultimate source of a great many sands is, of course, the quartz-bearing plutonic rocks—typically the granites and quartz monzonites, and the feldspar-rich metamorphic rocks such as the gneisses and coarse schists and granulites. A large class of highly feldspathic sands—the *arkoses*—are a direct product of the disintegration of these rocks. Those sands rich in rock particles—the *lithic arenites*—are derived from supracrustal rocks including preexisting sediments (sandstones, shales, and limestones), their metamorphic equivalents, and the effusive volcanic rocks. The effect of provenance is greatly diminished in the mature sands—the *quartz arenites* (orthoquartzites) which tend to converge toward a common end-type. A great deal of skill is required to determine either the immediate or ultimate source of these sands.

Finally, all sands undergo diagenetic changes. These lead to a diversity of cementing agents. In some sands the changes lead to a breakdown of some of the framework elements and the production of a matrix. Matrix formation is accompanied by albitization of the feldspar and other changes profound enough to produce a separate class of rocks—*the graywackes*. Diagenetic changes also can leach framework grains and thus alter our perception of their provenance.

Feldspathic Sands and Arkose

Definitions

The term *arkose* was apparently first used by Brongniart who, in 1826, applied this term to some sandstones in the Auvergne district of France, the essential constituents being quartz and feldspar (Oriel, 1949). Incidental components include mica and clay, usually kaolinitic. The term has since been redefined as "sandstone containing 25 or more percent of feldspars usually derived from the disintegration of acid igneous rocks of granitoid texture" (Allen, 1936, p. 44). Most of the earlier definitions emphasized the coarseness of grain and the resemblance to granite, from which these sands were presumably derived.

The above definitions are all in some degree unsatisfactory. Graywackes may likewise con-

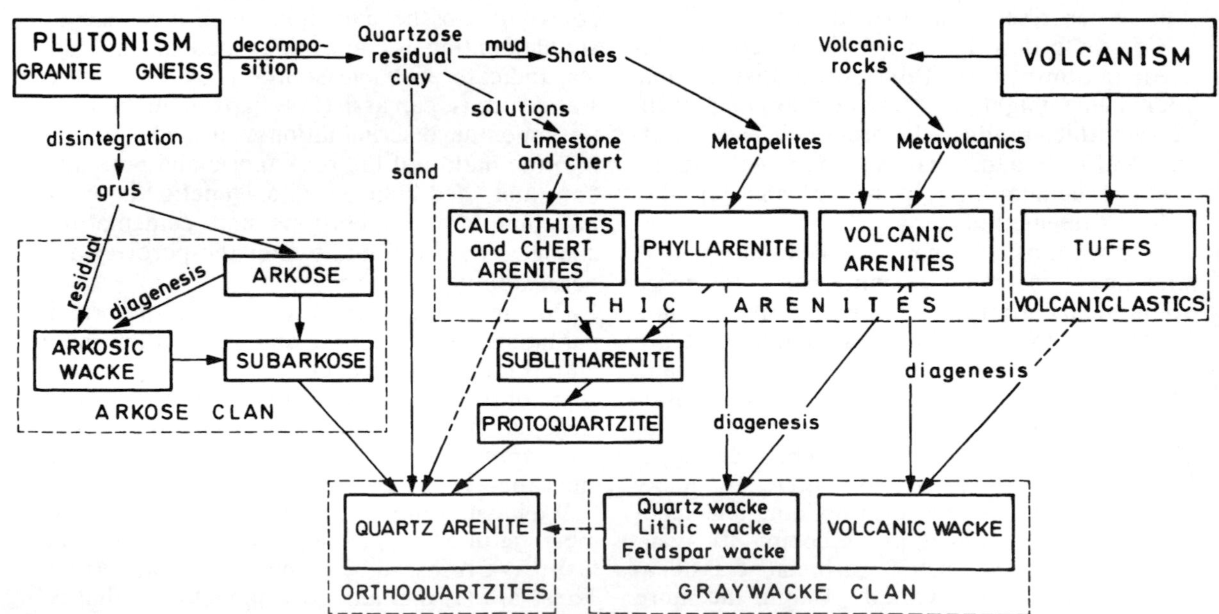

FIGURE 5-2. Provenance and evolution of the noncarbonate sands.

tain 25 or more percent of feldspar. Some authorities have applied the term arkose if the feldspar was conspicuous enough to be seen easily in hand specimen. Such rocks may contain 20 percent or even less feldspar (Krynine, 1950, p. 101). Furthermore, definitions based on inferred provenance are always difficult to apply but inasmuch as the feldspar of arkose, in many cases forming 40 to 50 percent of the rock, is characteristically potassium-feldspar (microcline), the granitic derivation seems well established, although there are arkoses with substantial, and in rare cases dominant, plagioclase.

Although there is no general agreement on how little feldspar a sandstone can have and still be called an arkose, the 25 percent figure is most often cited. Krynine (1940, p. 50) suggested 30 percent but later (1948, Fig. 11) indicated 25 percent to be the *average* feldspar content and designated some rocks with less than 20 percent feldspar arkose. There is no natural discontinuity in the abundance of feldspar. But because sandstones with about 25 percent or more have a field appearance that is distinctive, this figure may be the best basis for the choice. The term *feldspathic sandstone* merely means any sandstone with appreciable feldspar—arkoses and graywackes included. It has also been used to designate a class of sandstones less feldspathic than arkose but more so than normal sandstones. For this group of sands (10 to 25 percent feldspar) generally lacking in rock fragments, the term *subarkose* has been suggested (Folk, 1954; Pettijohn, 1954, p. 364). Arkose has also been redefined as a sand containing 25 percent or more of labile constituents (rock fragments and feldspar) of which feldspar forms half or more (Pettijohn, 1975, p. 214). By this definition, arkose might contain as little as 12.5 percent feldspar. Subarkose similarly defined, containing 10 to 25 percent labile components of which feldspar is dominant, might have as little as 5 percent feldspar.

General Description

Typical arkose is a coarse-grained rock consisting of quartz and feldspar. The feldspar imparts a pink color to the rock. Though normal arkose is pinkish to reddish, some may be derived from granitic or gneissic rocks containing gray or white feldspar, and such arkoses may themselves be gray or white—normally becoming lighter colored in outcrop. Some arkoses are associated with red beds but the two should not be confused as being necessarily the same.

In many cases the arkosic beds are massive and, being of coarse grain and pink color, they resemble granite, from which they were presumably derived. Some arkose is indeed no more than *in situ* disintegrated and weathered granite. Transported arkoses, on the other hand, usually display stratification, in some cases prominent crossbedding, somewhat better-rounded grains, and a higher proportion of quartz.

The dominant mineral of arkoses, as of most sandstones, is quartz though, exceptionally, feldspar may exceed quartz in volume (Table 5-3). Because of coarseness of grain, considerable polycrystalline quartz is present; also present may be composite granules consisting of both

TABLE 5-3. Mineral composition of arkose and subarkose (percent).

	A	B[1]	C	D	E	F[1]	G	H	I	J
Quartz	60	57	57	71	60	35	37.7	57	51	53.1
Microcline	34 }	35[2]	{ 27 }	25	13	59[2]	{ 0.7	24	30	18.5
Plagioclase	— }		{ 1 }				{ 45.4	6	11	0.4
Micas	—	—	—	—	T	—	4.2	3	1	6.9
Clay	—	—	—	—	5	—	12.0	9	7	17.0
Carbonate	—	P[3]	—	—	—	2	—	P[3]	P[3]	—
Other	6[4]	8[5]	14	4	8	4[5]	—	1	—	4.1

(1) Normative or calculated composition; (2) modal feldspar, given by Mackie as 55 and 60, respectively; (3) present in amounts under 1 percent; (4) chlorite; (5) iron oxide (hematite) and kaolin.

A. Sparagmite (Precambrian), Norway (Barth, 1938, p. 60).
B. Torridonian (Precambrian), Scotland (Mackie, 1905, p. 58).
C. Jotnian (Precambrian), Satakunta, Finland (Simonen and Kuovo, 1955, Table 2, No. 5).
D. Subarkose, Potsdam Sandstone (Cambrian), New York, U.S.A. (Wiesnet, 1961, p. 9). A subarkose.
E. Subarkose, Lamotte Sandstone (Cambrian), Missouri, U.S.A. (Ojakangas, 1963, p. 863). A subarkose.
F. Lower Old Red (Devonian), Scotland (Mackie, 1905, p. 58).
G. Arkose (Permian), Auvergne, France (Huckenholtz, 1963, p. 917).
H. Pale arkose (Triassic), Connecticut, U.S.A. (Krynine, 1950, p. 85).
I. Red arkose (Triassic), Connecticut, U.S.A. (Krynine, 1950, p. 85).
J. Arkose (Oligocene), Auvergne, France (Huckenholtz, 1963, p. 917).

quartz and feldspar. The grains are generally irregular and poorly rounded.

The feldspar is, with few exceptions, dominantly K-feldspar, usually microcline. It varies from extremely fresh to weathered (kaolinized) to a mixture of both fresh and weathered. In arkoses with a carbonate cement, the feldspar may show varying degrees of replacement—from grains with corroded borders to isolated but oriented residuals to completely replaced grains. Kaolinization, either before or after deposition, is common. In some porous arkoses the feldspars display limpid secondary overgrowths. If attached to clouded detrital cores, the alteration of the latter is clearly pre-depositional.

Large detrital micas characterize arkoses; both muscovite and biotite (and chloritized biotite) are common. The mica flakes, commonly considerably larger than the associated quartz and feldspar, tend to lie parallel to the bedding and hence to one another. The flakes may be bent or deformed by pressure of adjacent grains. The biotite may show chloritization or, more commonly, alteration and oxidation. Mafic minerals, other than biotite, are absent, implying that they were the chief loss by chemical weathering.

Arkosic sandstones of mixed provenance may contain rock fragments and pass by degrees into coarse lithic arenites.

Some arkoses contain matrix clay—commonly kaolinitic and ironstained. Other varieties have but little such matrix and are generally carbonate cemented. The oldest arkoses show secondary overgrowth of the quartz and feldspar and if such enlargements are carried to completion, the resulting rock may so much resemble a granite or granitic gneiss as to be mistaken for such in small outcrops. This is especially true in some Precambrian terranes.

Because they are usually derived from K-rich granitoid rocks, arkoses form a chemically homogeneous group as representative analyses show (Table 5-4). Arkoses, like their source rocks, are rich in Al_2O_3 and K_2O, the latter, unlike in graywackes, generally exceeding Na_2O (Fig. 5-3). Most arkoses, perhaps because they are generally subaerial rather than submarine, have, unlike graywackes, an excess of Fe_2O_3 over FeO. Those with a carbonate cement run high in CaO and CO_2.

Varieties and Types of Arkose

We have followed the usage suggested by Gilbert (Williams, Turner, and Gilbert, 1954, p. 310) and distinguished between those arkosic

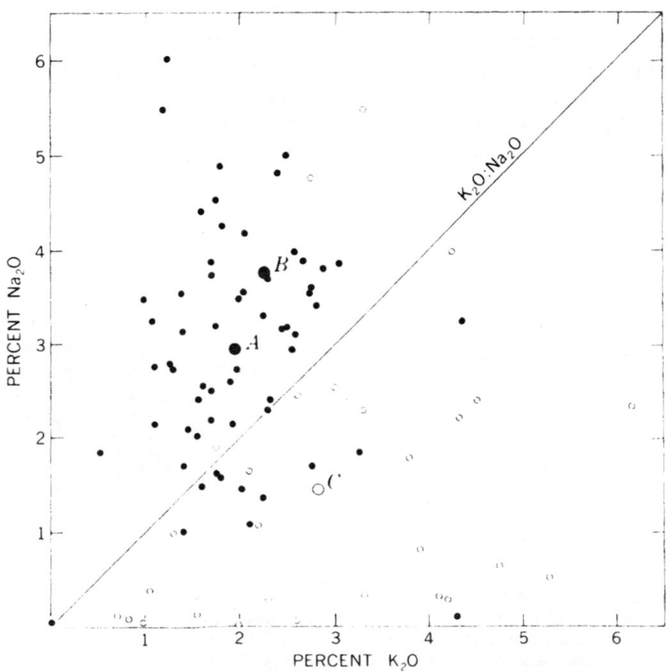

FIGURE 5-3. Na_2O/K_2O ratio in arkoses and graywackes. Solid black circles, graywackes; open circles, arkoses. A, average graywacke; B, composite New Zealand graywacke (Reed, 1957, p. 16); C, average arkose (from Pettijohn, 1963, Fig. 2; data from same source).

TABLE 5-4. Chemical analyses of arkose and subarkose (from Pettijohn, 1963, Table 8, with additions).

	A	B	C	D	E	F	G	H	I	J	K	L
SiO_2	79.30	75.80	80.89	87.02	92.60	73.32	59.24	92.13	85.74	69.94	72.21	76.6
Al_2O_3	9.94	11.74	7.57	2.86	3.52	11.31	6.65	4.42	6.84[1]	13.15	10.69	12.4
Fe_2O_3	1.00	0.59	2.90	0.49	0.44	3.54	2.02	0.37	0.79[2]	2.48	0.80	0.7
FeO	0.72	1.31	1.30	0.28		0.72	0.31	0.33	—		0.72	0.2
MgO	0.56	0.54	0.04	0.20	0.04	0.24	0.12	0.14	1.11	Trace	1.47	0.3
CaO	0.38	1.41	0.04	3.41	0.06	0.75	16.04	1.27	0.49	3.09	3.85	0.4
Na_2O	2.21	2.40	0.63	0.00	2.93	2.34	0.19	0.11	1.16	5.43	2.30	0.3
K_2O	4.32	4.51	4.75	1.98		6.16	2.30	0.72	2.19	3.30	3.32	3.8
H_2O+	0.55	0.86	1.11	—	0.17	0.30	1.26	—	—	—	1.46	2.7
H_2O-	0.41	0.03		—			—	—	—	—	0.08	
TiO_2	0.22	0.15	0.40	—	—	—	—	—	0.38[3]	—	0.22	0.6
P_2O_5	0.05	0.60	—	—	0.02	—	—	—	0.01	—	0.10	0.2
MnO	0.02	0.05	—	—	—	—	0.50[4]	0.24[4]	—	0.70	0.22	—
CO_2	—	Trace	—	—	0.06	0.92	12.16	None	—	—	2.66	—
Ign. loss	—	—	—	3.35	—	—	—	0.42	1.12	1.01	—	—
Total	99.68	99.99	99.63	99.65[5]	99.84	99.60	100.79	100.15	99.83	99.10	100.10[6]	100.6

(1) Contains MnO_2; (2) total iron; (3) contains ZrO_2 and V_2O_5; (4) reported as MnO_2; (5) includes 0.06 percent S; (6) sum given in original as 99.90.

A. Jotnian (Precambrian) Köyliö, Muurunmäki, Finland. II B. Wiik, analyst (Simonen and Kouvo, 1955, p. 63). 44 percent normative feldspar.
B. Torridonian (Precambrian), Kinlock, Skye, M. H. Kerr, analyst (Kennedy, 1951, p. 258). 53 percent normative feldspar.
C. Sparagmite (Lower Cambrian), Engerdalen, Norway (Barth, 1938, p. 58). 33.5 percent normative feldspar.
D. Calcareous subarkose (Cambrian or Ordovician), Bastard Township, Ontario, Canada (Keith, 1949, p. 21). About 12 percent feldspar and 7 percent calcite.
E. Subarkose, Potsdam sandstone (Cambrian), New York, U.S.A., P. L. D. Elmore and K. E. White, analysts. 17 percent normative feldspar (Wiesnet, 1961, p. 9).
F. Lower Old Red Sandstone (Devonian), Foyers, Loch Ness, Scotland (Mackie, 1905, p. 58). 52 percent normative feldspar.
G. Calcareous arkose, Old Red Sandstone (Devonian), Red Crags, Fochabers-on-Spey, Scotland (Mackie, 1905, p. 58). 16 percent normative feldspar and 28 percent normative calcite.
H. Subarkose, Rosebrae Sandstone (Devonian), Rosebrae, Elgin, Scotland (Mackie, 1950, p. 59). About 12 percent normative feldspar.
I. Subarkose of Whitehorse Group (Permian), Kansas, U.S.A. (Swineford, 1955, p. 122).
J. Portland Stone (Newark Group, Triassic), Portland, Connecticut, U.S.A. (Merrill, 1891, p. 420). 74 percent normative feldspar.
K. Molasse arkose (Oligocene, Zugertypus), Unterägeri, Kt. Zug. Switzerland, F. de Quervain, analyst (Niggli *et al.*, 1930, p. 262).
L. Arkose (Oligocene), Auvergne, France (Huckenholtz, 1963, p. 917). 19 percent feldspar.

sands with little matrix—the *arkosic arenites*—and those with a significant clay content—the *arkosic wackes*. The latter might be confused with the feldspathic graywackes except that, in general, the latter have sodic rather than potassic feldspar and a chlorite-rich rather than clay-rich (or sericitic) matrix and hence are dark rather than light colored.

Some of the arkosic wackes are produced by the *in situ* disintegration of granite and related rocks. Such material has been termed "grus." Grus has a mineral composition similar to that of the parent rock. Although not sorted by a transporting agent, it has a characteristic Rosin's Law size-distribution (McEwen *et al.*, 1959)—essentially the size-distribution which characterizes crushed materials. Such sedentary or *residual arkose* commonly marks an unconformity between a granitic basement and an overlying sedimentary sequence. The residual arkose may contain more quartz than normally found in granite; it grades upward to an arkose which is faintly stratified and which may contain scattered pebbles of granite and grades downward into fresh unweathered granite. When such a "graded unconformity" has been imprinted by metamorphism the relations become less clear and the relative age of the granite and the metasedimentary sequence may be misinterpreted. One of the best illustrations of this problem is the contact between the Archean Knife Lake sediments and the Saganaga Granite on Cache Bay on the Canada-Minnesota boundary (Gruner, 1941, p. 1599).

Residual arkosic materials may be shifted downslope and deposited as fans or aprons of waste materials which extend into the basin and are intercalated with more normal, better stratified and sorted sediments. Such tongues of "granite wash" encountered in drill holes may be mistaken for "basement" and lead to abandonment of drilling. Many of these arkoses contain an abundant clay-rich matrix. They are the product of very limited transportation, perhaps by mass movement, with very imperfect or incomplete sorting. Rocks of this type having a deep red-stained matrix have been designated *redstones* by Krynine (1950, p. 103) from their occurrence on Redstone Hill in Connecticut. This term has also been applied to phases of the Fountain Formation in Colorado (Hubert, 1960, p. 65). These rocks have a minimum of 20 percent of matrix some of which may be secondary.

Although arkoses with a prominent matrix are probably sedentary or residual, or if "granite wash," a product of limited transport, others may be the result of post-depositional alteration that can be an important part of the diagenesis of continental beds. Arkosic sands so altered will show pseudomorphs of kaolin after feldspar.

Arkosic materials which have undergone considerable reworking by rivers, or by the sea, are fairly well sorted and matrix-free. These are the *arkosic arenites* of Gilbert (Williams, Turner, and Gilbert, 1954, p. 292). Older deposits of this sort may have a normal mineral cement precipitated in the pore system. Such cement is commonly a carbonate—generally calcite. In these arkoses the grains may show incipient to good rounding. The proportion of quartz rises as the badly weathered feldspars are eliminated by abrasion. Stratification becomes noticeable and crossbedding may be conspicuous. As is common in carbonate-cemented sands, the cement may marginally replace or embay the framework grains. Embayment and partial to complete replacement of the feldspar is especially noteworthy.

In addition to the textural variations which distinguish the arkosic wackes from the arkosic arenites, there are mineralogical variations in the composition of the framework components. Arkose normally is, and by tradition is defined as, the product of disintegration of a granite. By "granite" is usually meant a coarse-grained plutonic rock, either igneous or metasomatic, of which K-feldspar is the dominant, or at least a major, constituent. What about the comparable sandstones derived from rocks in which plagioclase is the dominant or even the sole feldspar? These have been called "plagioclase arkoses" and although they are, in general, not very common, some notable examples have been described. The Paleocene Swauk Formation in the state of Washington is predominantly a plagioclase arkose, presumably derived from a quartz dioritic source.

Some volcaniclastic sands resemble arkose in external appearance although they contain little feldspar. The role of feldspar is taken by light-colored, acid volcanic rock fragments—in many cases reddish rhyolitic or related materials. These rocks also contain quartz, some of volcanic origin, and feldspar, usually zoned and volcanic in origin. These sandstones are derived from volcanic terranes—the more acidic flows providing an undue share of the debris from such regions. These sands, like most arkoses, have had limited transport, are coarse, only moderately well sorted, and generally deposited in continental basins associated with coarser clastics. They are described more fully in Chap. 6.

Field Occurrence and Examples

Arkoses, as here defined, form no more than 15 percent of all sandstones (Pettijohn, 1963, p. 515) though some estimates are significantly higher.

Arkoses occur in all major geologic systems and range in age from Precambrian to Recent. Well-known Precambrian examples include the Sparagmites of Norway and Sweden (Hadding, 1929, p. 151; Barth, 1938, p. 60) and their presumed equivalent in the Northwest Highlands of Scotland, the Torridon sandstones (Peach *et al.*, 1907, p. 278; Kennedy, 1951). The late Precambrian Jotnian sandstone of Finland (Simonen and Kuovo, 1955, p. 60) belongs here as does the lower part of the Huronian Lorrain Formation of the Canadian Shield which in places carries as much as 40 percent feldspar. All of these units form coarse, strongly crossbedded sequences and are associated with pebbly sands of conglomerates.

Well-known and well-described are the arkoses of the Old Red Sandstone of Scotland containing up to 60 percent feldspar (Mackie, 1899a), the arkoses of the Triassic Newark Group (Fig. 5-4) of Connecticut (Krynine, 1950) and the central and southern Appalachians (Fig. 5-5), and the arkosic beds of the Fountain (Fig. 5-6) and Lyons Formations (Pennsylvanian) of the Front Range of Colorado (Hubert, 1960). Another well-known arkose is that of the Swauk, over 5000 ft thick, of Paleocene age in Washington. The feldspar of this formation, unlike that of most arkoses, is predominantly plagioclase (Foster, 1960, p. 105). Where the Tertiary Molasse of Switzerland had a granitic source, it is an arkose commonly containing 50 to 60 percent of feldspar (Gasser, 1968, Table 10). For an extended tabulation of ancient arkoses, the reader is referred to Barton (1916).

The sandstones of the Precambrian Keweenawan Series of the Lake Superior region are in part arkosic (Irving, 1883, p. 128). For the most part, however, they are volcaniclastic are-

FIGURE 5-4. Sugarloaf arkose, Newark Series (Triassic), Mt. Tom, Massachusetts, U.S.A. Crossed nicols, ×20. A typical coarse red arkose consisting of a very poorly sorted mixture of angular quartz, feldspar, together with a little mica set in a red, ferruginous clayey matrix.

FIGURE 5-5. Arkose from Newark Series (Triassic). Deep River Basin, North Carolina, U.S.A. Crossed nicols, ×20. A very coarse, immature sandstone, with abundant rock fragments as well as quartz and feldspar. Note composite character of some of the quartz, the alteration of the finely twinned feldspar, and rock fragments.

FIGURE 5-6. Fountain Sandstone (Pennsylvanian). About 100 ft (30 m) above Precambrian basement, Red Mountain, 6 mi (10 km) south of Wyoming border, Laramie County, Colorado, U.S.A. Crossed nicols, ×55. An arkose marked by an abundance of detrital (?) dolomite grains, quartz, and feldspar cemented by calcite.

nites rather than true arkoses, rocks in which the quartz and feldspar is diluted by a large volume of acidic volcanic rock particles all cemented together by calcite.

Formations which themselves are not normally arkoses may have an arkosic basal facies at or near their contact with an underlying granitic basement. Such is the case with the Cambrian Lamotte Sandstone of Missouri (Ojakangas, 1963) and the Cambrian Potsdam Formation of New York (Wiesnet, 1961). These formations, which are primarily quartz arenites, locally contain as much as 30 percent and 52 percent feldspar, respectively. Rocks such as the Lamotte show a commingling of little worn and partially weathered feldspar and quartz derived from the subjacent terrane and smaller but well-rounded quartz derived from more distant sources.

From a review of the known occurrences of arkose, it is clear that, leaving aside the basal arkoses related to marine transgression over granites, significant arkose accumulation can be expected only when sharp uplifts bring a granitic or gneissic basement up into the zone of erosion. If subsidence occurs adjacent to such elevated basement blocks, a local but thick arkose accumulation can be expected.

Provenance and Tectonics of Arkose

Having noted the requirement for uplift of granitic or gneissic terranes to produce arkoses, we may further inquire where such situations occur. The rifting of continents along incipient or developed plate boundaries provides one of the most common tectonic associations of arkoses. Some of the best known, such as the Triassic Newark series of the eastern United States and the New Red Sandstones of England, are alluvial deposits along the rifts associated with the breakup of Pangaea. Many of the sands of the East African rift system are arkoses. These associations are discussed in detail in Chap. 12.

Significance and Origin

The question of presence or absence of feldspar in sandstones is at the heart of the arkose problem—its origin and significance. Obviously provenance is a major factor and arkose implies a richly feldspathic source area and one usually characterized by K-feldspar bearing plutonic granites and gneisses.

But beyond provenance the question is: under what conditions is feldspar released to the sediments rather than decomposed to clay in the source region? Normally, or certainly ultimately, the feldspar is so decomposed and the clay formed is separated from the quartz so that the resulting sand is quartz-rich and feldspar-free. Years ago, Mackie (1899b, p. 444) concluded that under conditions of extreme aridity or extreme cold the processes of weathering would be inhibited or retarded so that incompletely weathered materials would escape and become part of sediments produced. Arkoses thus became indicators of desert or glacial conditions. Many glacial sands are indeed highly feldspathic. But this concept has been challenged. It has been observed, for example, that many Eocene sandstones in California which contain up to 50 percent feldspar yield a flora which could only have lived under warm, humid conditions (Reed, 1928). The Tertiary Catahoula Sandstone of Texas also contains a similar quantity of feldspar (Goldman, 1915, p. 273) and likewise contains a flora characteristic of a tropical coastal region. Krynine (1935) observed modern arkose accumulation in tropical Mexico where the mean temperature is 80°F (27°C) and the annual precipitation 120 inches (305 cm).

These observations led Krynine to the view that the necessary condition for production of arkosic sands was high relief with consequent rapid erosion rather than adverse climatic conditions. Under conditions of high relief and rapid erosion producing deeply incised canyons, both fresh and partly weathered feldspars are incorporated in the sediments. If relief is low, slopes are more stable and covered by vegetation. Weathering is then more likely to go to completion and yield only a quartz sand. Detrital feldspar, therefore is *the result of a balance between the rate of decomposition and the rate of erosion*. Arkose may be, therefore, a product of rigorous climate *if* decomposition is inhibited but it may be a product of high relief *if* erosion is accelerated. Distinguishing between these possibilities in an ancient arkose is difficult but, on the whole, it seems more likely that high relief is more important than rigorous climate in arkose formation. The orogenic sediments, those related to orogenesis, are more feldspathic than those related to erosion and deposition in tectonically stable areas. Gibbs (1967) in his study of the Amazon River sediments has shown that high relief is the dominant factor. But Strakhov (1967, pp. 48, 95) argues strongly for the effect of climate.

It has been presumed that feldspar is more subject to abrasion and destruction during transport than quartz. This conclusion, set forth by Mackie (1899a, p. 149) as a result of his studies of Scottish streams, seems to be confirmed in some measure by similar studies of the mineralogy of the steep-gradient streams draining the Black Hills of South Dakota (Plumley, 1948, p. 562). However, larger, low-gradient streams seem capable of transporting feldspar long distances without significant loss. The Mississippi sands, for example, near Cairo, Illinois, contain about 25 percent feldspar; at the delta, some 1100 miles (1760 km) downstream, these sands still contain 20 percent feldspar (Russell, 1937, p. 1334). The presence of feldspar, therefore, does not imply short transport. Similarly, though reworking of sands may be extensive on some beaches, feldspar does not generally seem to disappear as a result of wave and longshore transport although some reduction has been reported by Mack (1978).

Lithic Arenites and Related Rocks

Definitions and Nomenclature

Sandstones which contain a substantial quantity of rock particles and which have little or no matrix materials but have instead an empty pore system or one filled with a precipitated mineral cement were early recognized and described but not specifically named. The term *subgraywacke* was applied to this group (Pettijohn, 1949, p. 255) because of the superficial resemblance of many of these sandstones to graywacke. In fact, the term "graywacke" or "low rank graywacke" itself has been applied to this class of rocks (Krynine, 1945, Table 1) but this usage is being gradually abandoned in favor of the term *lithic arenite*—a term proposed by Gilbert (Williams, Turner, and Gilbert, 1954, p. 304). The term has been shortened to *litharenite* (McBride, 1963, p. 669). Pettijohn (1954, Fig. 1) has used the term *lithic sandstones* to encompass both subgraywackes with 25 percent or more labile components, rock particles exceeding feldspar, and *protoquartzites,* a term suggested by Krynine (Payne, 1952), with 5 to 25 percent labile constituents. The latter have also been designated *sublitharenites* (McBride, 1963, p. 667).

General Description

Lithic arenites are generally light gray, "salt and pepper" sands with an abundance of rock particles, especially of sedimentary, low-rank metamorphic and volcanic rocks, subangular to rounded quartz, and with a chemical cement, either quartz or calcite. Mica flakes are common; feldspar is not. Mineral charcoal (fusain) and carbonaceous plant fragments are common in Devonian and younger lithic arenites. Shale pebbles may also be present. Detrital matrix is generally absent though a pseudomatrix (squashed shale particles) or authigenic precipitated clay may be present.

Of all sandstones, probably not even excepting graywackes, lithic arenites show the greatest diversity of both mineralogical and chemical composition. This variability reflects the importance and relative abundance of the diverse rock particles which these sands may contain. If the rock particle content is small, these sands pass over into the quartz arenites or orthoquartzites. On the other hand, rock particles may form over half of the framework and in a few rare cases, all of it.

The rock particles in lithic arenites are themselves diverse—their only common character is their fine grain. The number of rock types which have been described in these sandstones is formidable—even in a single sand as many as twenty have been identified (Mattiat, 1960). In general, however, only a few kinds are prominent. They fall into three main classes: (1) *volcanic,* that is, particles of aphanitic flow rocks, (2) low-grade *metamorphic* rock particles such as slate, phyllite, and mica schist, and (3) *sedimentary* rock particles which include various kinds of shale, siltstone, argillite, and related pelitic material but generally also some detrital chert and, in some cases, micritic limestone and dolomite. There are sandstones in which these latter constituents, chert and micritic limestone, become dominant.

The arenites with volcanic rock particles are a special class and are a product of a clearly defined provenance. But because they constitute an abundant and important class of sands apart from the normal lithic sandstones, they are treated in the following chapter. But as many lithic arenites, as well as graywackes, are of mixed provenance, volcanic rock particles are to be looked for as minor components in many of these sandstones.

The unique character of the lithic sandstones—the volcanic arenites excluded—is the abundance and variety of sand-sized particles of pelitic derivation: shale, siltstone, slate, phyllite, and mica schist. These may form a substantial part of the sand as a whole—up to 50 percent or more. Their uniqueness lies in the fact that clay-sized materials have been built up—either through diagenesis or metamorphism—into rocks that yield sand-sized materials so that what were once two classes of material, sand and mud, sorted and deposited separately from one another, are now deposited in one and the same place as well-sorted sand. The lithic arenites with the pelitic materials of sedimentary and metasedimentary origin are the most common. Representative modal analyses and averages of these phyllarenites, together with a few aberrant or special types are given in Table 5-5.

Although the rock particles are the definitive component of lithic arenites, quartz is a prominent, and generally dominant, detrital component. In those sands with rock particles of sedimentary origin (chert, limestone and dolomite, pelitic rock particles, etc.), it is probable much of the detrital quartz, perhaps most of it, is derived from sedimentary sources—mainly preexisting sandstones. Quartz, released by disintegration of older sandstones, is apt to be better rounded than the quartz of most arkoses and graywackes. Inasmuch as many lithic arenites have a quartz cement, such original rounded detrital quartz grains may show good secondary overgrowths. In those sands rich in metamorphic rock particles, the quartz also is presumably largely of metamorphic origin, although this is difficult to demonstrate. Such arenites will perhaps contain a higher proportion of undulatory and polycrystalline quartz than those of sedimentary or volcanic provenance. The quartz is apt to be angular to subangular.

As might be expected, sands derived principally from sedimentary and low-rank metamorphic sources will contain little feldspar. Such feldspar as is present is apt to be better rounded than the feldspar of arkose. Though one might expect a second-cycle arkose, the labile character of the feldspar makes this unlikely.

A notable constituent of many lithic arenites is detrital mica which tends to concentrate on certain bedding planes and impart a sheen to such surfaces. Both biotite and muscovite are present, the latter being the more common. The mica flakes are deposited parallel to the bedding and hence parallel to one another. They may show deformation due to compaction and appear bent and wrapped around adjacent quartz grains.

The lithic arenites may be cemented either with carbonates or with silica or both. Little or no matrix material is present though some of the weaker argillaceous rock particles may, especially in the older, more compacted varieties, be deformed between adjacent more durable quartz grains in such a manner as to resemble a matrix-filled pore. Such squashed rock particles are recognized only with difficulty (Allen, 1962, p. 669). They appear to fill some pores—not others. True matrix would be distributed among *all* pores. The pseudomatrix may also show relict bedding, albeit somewhat deformed, characteristic of shales and siltstones. Moreover, the fragments, though deformed by compaction, are not all alike either in color or in texture. True matrix materials should exhibit greater uniformity.

TABLE 5-5. Modal analyses of lithic sandstones (subgraywackes) and protoquartzites (sublitharenites) (from Pettijohn, 1963, Table 3, with additions).

	A	B	C	D	E	F	G	H
Quartz	50	60	78 }	65.4	{ 32.0	71	30.9	27
Feldspar	3–5	3	3 }		{ 2.2	8	10.0	2
Mica	—	1	—	—	0.2	tr	0.5	—
Rock fragments	40	35	15	10.6	43.0[2]	22[4]	33.0[1]	46[5]
"Clay" or matrix	10	2	4	6.8	6.9	2	5.5	5
Silica cement	—[3]	—	—	11.9	trace	—	—	—
Calcite cement	—	present	—	8.5	13.0	—	19.2	20

(1) Includes 15.0 percent chert; (2) includes 28.0 percent chert; (3) 5–10 percent, author's observation; (4) includes 5.0 percent chert; (5) includes 3.0 percent chert, 12.0 percent limestone, 27.0 percent dolomite.

A. Oswego Sandstone (Ordovician), Pennsylvania, U.S.A. (Krynine and Tuttle, 1941).
B. Bradford Sand (Devonian), Pennsylvania, U.S.A. (Krynine, 1940, C-1, Table 3).
C. Deese Formation (Pennsylvanian), Oklahoma, U.S.A. (Jacobsen, 1959, Table 4, Analysis D-112).
D. Salt Wash Member of Morrison Formation (Jurassic), Colorado Plateau, U.S.A. Mean of 25 thin sections (Griffiths, 1956, p. 25).
E. "Calcareous graywacke" (Cretaceous), Torok, Alaska, Average of 3 samples (Krynine in Payne and others, 1952).
F. Basal Claiborne Sand (Eocene), Texas, U.S.A. (Todd and Folk, 1957).
G. "Frio" Sandstone (Oligocene), Seeligson field, Jim Wells and Kleberg Counties, Texas, U.S.A. Average of 22 samples (Nanz, 1954, p. 112).
H. Molassesandstein (Tertiary), Germany (USM No. 186, Füchtbauer, 1964, p. 256).

Representative chemical analyses of lithic arenites are given in Table 5-6. Most of the analyses in this table are of phyllarenites. These are characterized by the high Al_2O_3 and in some, particularly the older ones, relatively high K_2O contents, reflecting the pelitic nature of the rock particles. They are relatively low in Na_2O and MgO, unlike graywackes, and high in CaO and CO_2 in the carbonate-cemented varieties. As might be expected, those lithic arenites in which chert or micritic limestone particles are dominant show an unusual quantity of SiO_2 or CaO and CO_2, respectively. A high $CaCO_3$ content, however, can denote a calcite cement rather than limestone detritus. The high MgO content of some of these sandstones is due to detrital dolomite.

The SiO_2 content is depressed by addition of carbonate cement or by the abundance of detrital limestone and dolomite particles, and is augmented by added quartz cement or by the abundance of detrital chert.

Special Types

Folk (1968a, p. 131) apparently coined the term *phyllarenite* for those lithic sandstones in which the dominant rock particles are of metamorphosed pelites: slate, phyllite, and mica schist. Those in which the rock particles are phyllite or schist had previously been termed *schist arenites* (Krynine, 1937, p. 427). The phyllarenites contain considerable mica and usually a little feldspar. With a diminishing proportion of such rock particles, these sandstones grade into a *subphyllarenite* (or protoquartzite) and ultimately into a *quartz arenite*. If a significant proportion of other rock particles is present, the term *polylitharenite* has been proposed (Folk, 1968a, p. 135).

TABLE 5-6. Chemical composition of lithic sandstones (subgraywackes) and protoquartzites (modified from Pettijohn, 1963, Table 4).

	A	B	C	D	E	F	G	H
SiO_2	92.91	74.45	40.35	84.01	65.00	56.80	51.52	47.75
Al_2O_3	3.78	10.83	7.43	2.57	9.57	8.48	5.77	6.41
Fe_2O_3	trace	4.62	3.27	0.17	1.59	1.67	2.43	2.39
FeO	0.91	—	—	0.26	1.08	—	—	—
MgO	trace	1.30	10.28	0.67	0.40	1.24	0.95	4.48
CaO	0.31	0.35	12.00	5.41	10.10	15.25	16.96	18.75
Na_2O	0.34	1.07	0.54	0.17	2.14	1.31	1.32	1.20
K_2O	0.61	1.51	0.93	0.86	1.43	1.46	1.90	1.02
H_2O+	1.19 (H₂O+ and H₂O−)	4.95	6.75	0.54	0.82	0.50	2.25	1.32
H_2O-		—	—	0.19	0.23	—	2.54	—
CO_2	—	trace	17.80	4.65	6.90	12.95	13.30	17.78
TiO_2	—	0.50	0.30	0.05	—	0.10	0.32	0.20
P_2O_5	—	trace	—	0.04	—	trace	0.10	0.10
SO_3	—	—	—	—	0.04	—	0.52	—
Cl	—	—	—	0.02	—	—	—	—
F	—	—	—	0.01	—	—	—	—
S	—	—	—	0.02	0.16	—	—	—
MnO	—	—	—	0.04	—	—	0.14	—
BaO	—	—	—	0.05	—	—	—	—
C	—	—	—	—	0.06	—	—	—
Ign.Loss	—	—	—	—	—	—	—	—
Total	100.05	99.58	99.65	99.73[1]	99.54[2]	99.76	100.06	101.40[3]

(1) Includes Cl, 0.02; F, 0.01, BaO, 0.05; (2) includes C, 0.06; Cu, 0.002; V, 0.017; Zn, <0.03; Cr, 0.003; (3) sum given as 99.40 in original.

A. Protoquartzite, Berea sandstone (Mississippian), Berea, Ohio, U.S.A., L. G. Eakins, analyst (Clarke, 1890, p. 159).

B. Coal measure sandstone (subgraywacke?) (Carboniferous), Westphalian coal basin, France-Belgium (Hornu and Wasmes), (Cayeux, 1929, p. 227).

C. Coal measure sandstone (calcareous subgraywacke?) (Carboniferous), Westphalian coal basin, France-Belgium (Hornu and Wasmes), (Cayeux, 1929, p. 227).

D. Protoquartzite, Salt Wash Member of Morrison Formation (Jurassic). Composite of 96 samples, Colorado Plateau, U.S.A. Unmineralized. V. C. Smith, analyst (Pettijohn, 1963, Table 4).

E. Calcareous subgraywacke (lithic arenite), (Oligocene, "Frio" Formation), Seeligson field, Jim Wells and Kleberg Counties, Texas, U.S.A. (Nanz, 1954, p. 114). Composite of 10 samples. Also included Cu 0.002, V 0.017, Zn less than 0.03, and Cr 0.003.

F. Calcareous subgraywacke (Tertiary Molasse Aquitanian), Lausanne, Switzerland (Cayeux, 1929, p. 161).

G. Calcareous subgraywacke (Tertiary Molasse), Grönchen, Burghalde, Kt. Aargan, Switzerland, J. Jakob, analyst (Niggli *et al.*, 1942, p. 263).

H. Calcareous subgraywacke (?) (Molasse Burdigalian), Voreppe (Isère), France (Cayeux, 1929, p. 163).

Many sands contain detrital chert grains; in a few the detrital chert forms a considerable or even major part of the sand detritus. These are the *chert arenites*. Chert forms, for example, from 20 to 90 percent of the grains in the Lower Cretaceous Cut Bank Sandstone of Montana (Sloss and Feray, 1948, p. 6). Similarly some of the sands in the Jurassic Morrison Formation of Montana are extremely chert-rich (Suttner, 1969, Fig. 11). In these, as in other sandstones, there is difficulty distinguishing chert particles from particles of devitrified rhyolite materials. Some chert grains will preserve fossils that in rare instances can be identified and stratigraphically date the source rock. It is not uncommon for ordinary lithic sands to contain 5 to 15 percent detrital chert (Table 5-5).

Chert-rich sands are more prone to diagenetic change than are quartz sands inasmuch as chert is more soluble than the quartz. Consequently, microstylolitic contacts are apt to form between adjacent grains. Chert is probably more susceptible to carbonate replacement than is quartz.

The chert-rich sands probably have a very local provenance, being the less soluble, sand-sized residue from a limestone terrane or being derived from a region with a significant amount of bedded chert. Chert-rich sands of volcanic affinities are probably far more common, however, the chert being an alteration product of volcanic glass.

Detrital carbonate grains are those derived from preexisting carbonate rocks by the ordinary processes of weathering, erosion, and transportation. They are, in other words, the clastic products of the wastage of a land mass on which limestones and dolomites are exposed. Most such detrital grains appear as polycrystalline aggregates. The term "extraclast" has been suggested for these particles in contrast to "intraclast" (Chanda, 1969).

Detrital carbonates occur in modern sands; locally they form a significant part of the sand. They are comparatively rare in ancient sandstones but there are important exceptions. Sandstones in which detrital carbonate forms a large and significant part of the rock have been called *calclithite* to distinguish them from *calcarenites* in which the carbonate detritus is intrabasinal (Folk, 1968a, p. 141). Carbonate detritus may exceed 50 percent.

Detrital dolomite is reported in Upper Cretaceous sandstones in the Uinta Basin of Utah (Sabins, 1962, p. 1185). Sabins (1962, p. 1186) also recognized "primary dolomite" in sandstones—the dolomite being single crystals of primary origin modified somewhat by abrasion prior to inclusion in the framework of the sand. Dolomite of this kind is said to be common in Cretaceous sands of the Western Interior of the United States (Sabins, 1962, p. 1188). The rounded corners of some dolomite rhombohedra and the close correlation between the size of the dolomite grains and the associated clastic quartz are cited as evidence of the concurrent sedimentation of the dolomite and quartz. Because dolomite grains are confined to marine sandstones and absent in associated nonmarine sands, they are considered "primary" rather than detrital. Dolomite grains of a similar nature, however, have been observed in the arkose of the Fountain Formation of Pennsylvanian age in Colorado. This occurrence in presumed fluvial sandstones suggests a detrital rather than a primary origin.

Although dolomite particles are common in some sandstones, calclithites as such are relatively rare. They are the analogue of and found in association with limestone conglomerates. The Oakville Sandstone (Miocene) of Texas consists of grains of Cretaceous limestones (Folk, 1968a, p. 141). Many of the Molasse sandstones north of the Alps are over half detrital limestone and dolomite (Füchtbauer, 1967, Fig. 3). The reduction of most limestone terranes of low relief in humid regions is by solution; the only residues moved by surface streams are red clays and cherts. Hence the calclithites probably record rapid erosion and therefore signify high relief whereas chert arenites indicate low relief and removal of the limestones by solution.

Calclithites are generally cemented with carbonate—probably self-cemented.

Field Occurrence and Examples

Lithic sandstones are very common, are widespread, and are of all ages.

Well-known and well-described examples include various Paleozoic sandstones of the central Appalachians: the Ordovician Juniata Formation (Yeakel, 1962) and Oswego Sandstone (Krynine and Tuttle, 1941), Fig. 5-7; the Devonian Third Bradford Sand (Krynine, 1940); the Mississippian Pocono Formation (Pelletier, 1958) and Mauch Chunk Formation (Meckel, 1967; Hoque, 1968); and the Pennsylvanian Pottsville Formation (Meckel, 1967), Fig. 5-8. All these are quartz-rich, feldspar-poor sandstones with a large component of rock particles of sedimentary and low-rank metamorphic provenances. Excepting the Third Bradford Sand, all are alluvial.

FIGURE 5-7. Oswego Sandstone (Ordovician). U.S. 322, Bald Eagle Mountain, near State College, Pennsylvania, U.S.A. Crossed nicols, ×20. The formation consists of 50 percent or more of quartz, and 30-40 percent of rock particles. Original outlines in quartz are difficult to see; the secondary quartz results in interlocking, sutured grain boundaries. Rock particles are mainly siltstone, fine-grained quartzite, and phyllitic low-grade metamorphic grains. Feldspar rare—up to 2 or 3 percent. A lithic arenite.

FIGURE 5-8. Pottsville Formation (Pennsylvanian). U.S. 61 at Pottsville, Pennsylvania, U.S.A. Crossed nicols, ×20. A coarse, ill-sorted lithic arenite composed of quartz and rock particles, the latter being in part weak shale and siltstones, forms a pseudomatrix. The rock particles are pelitic, sedimentary, and metamorphic. Quartz generally subangular.

Many of the sandstones of the Lower Old Red Sandstone of England are lithic arenites (Allen, 1962, p. 671).

Most sandstones associated with coal measures throughout the world are lithic arenites; perhaps more are protoquartzites. The sandstones of the Illinois Basin perhaps belong here, although their rock fragment content is low. They are, nevertheless, characterized by a significant proportion—some 8 to 10 percent—of feldspar, mica, and rock particles (Siever, 1957, p. 240). The coal measure sandstones of the Westphalian coal basin of France and Belgium are lithic arenites (Cayeux, 1929, p. 227).

Many Jurassic and Cretaceous sandstones in the Western United States are excellent lithic arenites (Figs. 5-9 and 5-10). Included here are the chert-rich arenites of Montana, of Jurassic age (Suttner, 1969), the Cretaceous Cut Bank Sand (Sloss and Feray, 1948), the Cretaceous Belly River Sandstone from Alberta, Canada (Lerbekmo, 1963), and the sandstones of the Cretaceous Chico Formation of California (Williams, Turner, and Gilbert, 1954, p. 306). There are many other examples.

Many of the Tertiary sands of the Gulf Coast turn out to be lithic arenites. Some of the better-known examples include the Oligocene "Frio" (Nanz, 1954), the Miocene Oakville Sandstone (Folk, 1968a, p. 141) of Texas, and the Eocene Wilcox Formation of Louisiana (Williams, Turner, and Gilbert, 1954, p. 305).

The Tertiary Molasse sandstones (Fig. 5-11) north of the Alps are mostly lithic arenites. They contain a large proportion of rock fragments and a relatively low content of quartz and feldspar (Füchtbauer, 1964, pp. 256–57; Gasser, 1968). Quartz ranges from 17 to 75 percent, averages 20 to 30 percent. Feldspar ranges from 1 percent or less to a maximum of 27 percent and averages between 2 and 18 percent. Rock particles, on the other hand, run as high as 72 percent which in some sands are largely limestones and dolomite grains. The carbonate grains form 18 to 33 percent of these sands. The rock particles in the more feldspathic sands are of metamorphic origin rather than from carbonate rocks. The carbonate-rich sands also carry considerable chert.

A large number of sandstones are best characterized as protoquartzites or sublitharenites. Good examples of these include the Precambrian Serpent Quartzite of the North Shore of Lake Huron, some parts of the Silurian Tus-

FIGURE 5-9. Frontier Formation (Cretaceous). SW$\frac{1}{4}$, SE$\frac{1}{4}$, Sec. 12, T. 42 N., R. 107 W., Fremont County, near Dubois, Wyoming, U.S.A. Crossed nicols, ×55. A moderately well-sorted lithic arenite consisting of about equal parts of subangular quartz and chert. The chert varies from dense isotropic to well-crystallized aggregates. See Fig. 5-10.

FIGURE 5-10. Viking or Cardium "B" sandstone (Cretaceous). Garrington Field, T. 35 N., R. 3 W., Alberta, Canada. Crossed nicols, ×55. A lithic arenite, similar to that shown in Fig. 5-9, consisting of a moderately well-sorted mixture of subangular to subrounded quartz and chert, the latter forming nearly one-third of the framework of the sand.

FIGURE 5-11. Altern molasse (Tertiary). Near Bregenz, Vorarlberg, Austria. Ordinary light ×100. An excellent example of a calclithite. Note relatively low content of quartz and great abundance of equigranular, microcrystalline rock particles—micritic limestone and dolomite. Some specimens of the sandstones contain as much as one-third to one-half detrital carbonate. Carbonate cement.

carora Quartzite of the Appalachians, and the Pennsylvania Anvil Rock Sandstone of the Illinois Basin. These may be lithic arenites which have been somewhat "cleaned up" in the neritic zone; true lithic arenites are perhaps more characteristic of alluvial deposits

Origin and Significance

It seems probable that most modern sands are lithic arenites or protoquartzites in composition. Although rock particles are commonly ignored in modern sands, perhaps largely because they cannot easily be identified from grain mounts but must be thin-sectioned, they generally exceed feldspar (Table 2-4). The sands of many large rivers contain an abundance of rock particles. Rock particles are present also on beaches but to a lesser degree than in the streams, perhaps because they are destroyed by the more vigorous abrasion in the surf zone.

Ancient lithic arenites are very common. They are estimated to constitute some 20 (Pettijohn, 1963) to 26 percent ("low rank graywacke"; Middleton, 1960, p. 1021) of all sandstones and thus outrank arkose by a good margin. They are the most common sandstones in Cretaceous and Tertiary flysch sequences, playing the same role that graywacke does in the older flysch facies (Cummins, 1962a), but unlike graywacke, they occur outside the geosyncline as well as in it. The typical molasse sandstone probably is a lithic arenite. See Chap. 12 for definitions of "flysch" and "molasse."

Lithic sandstones are, for the most part, immature sands and hence like other such sands require conditions favoring the production and deposition of relatively unstable materials. The mechanism of production of large volumes of sand from fine-grained rocks is not well understood. Thorough decomposition would yield silt and clay-sized materials. Disintegration without decomposition of coarse-grained rocks would yield sand, but the finer-grained rocks might be expected to yield either silt or block-sized pieces. Nonetheless, a large volume of sand of the finer-grained rocks is produced. The mechanically weak character of much of cleavable pelitic and metapelitic materials precludes prolonged transport or survival, particularly in the surf environment. For this reason, some of the lithic arenites may be indicators of relatively local provenance, perhaps even within uplifted parts of the same sedimentary basin. The predominance of lithic arenites in modern big river sands, especially in their head waters and middle reaches, shows, however, that this is not universally true.

The chemically unstable rock particles, limestone and volcanic rock fragments, require erosion of incompletely weathered materials. Such erosion is promoted by high relief and/or aridity. The chert arenites are an exception to this generalization. These, like the quartz arenites, are mature sandstones and imply prolonged and thorough weathering.

The lithic arenites reflect provenance. In the same way that the arkoses denote a plutonic provenance—granites and gneisses—lithic arenites denote a supracrustal provenance—volcanic, low-grade metamorphic, and sedimentary. The composition of the lithic arenites reflects each of these major provenances. Most lithic arenites arise from the denudation of a mixed sedimentary and metasedimentary terrane. With deeper dissection, the plutonic and higher-grade metamorphic sources should be unroofed with the result that the derived sands become increasingly feldspathic.

From the brief survey given above, lithic arenites seem to be (1) alluvial sandstones deposited on the flanks of marked uplifts as thick accumulations in closely associated molasse basins or (2) alluvial sandstones deposited on cratons by large rivers that derived much of their detritus from marginal and distant uplifts or (3) marine turbidite sands both in Tertiary geosynclinal depressions and the present deep sea. The alluvial sands show the usual structures of such deposits: crossbedding, basal position in an upward-fining cycle, and the like. The turbidite sands are part of a flysch-like sequence and have the usual graded bedding and sole markings characteristic of this facies. The alluvial and turbidite lithic arenites that contain abundant rock particles of continental derivation are associated with two principal tectonic regimes, the continental collision orogenic belt and the passive continental margins; some rift valley sands include lithic arenites as well as arkoses (see Chap. 12).

Graywackes and Related Rocks: The Wackes

Definitions and History of Term

The term "graywacke" has perhaps engendered more controversy than any other term in sedimentary petrography. There has been an outpouring of notes and papers dealing with this class of rocks and upon the proper definition of the term. Among the latter are those by Boswell (1960), Cummins (1962a), Krynine (1941a), McBride (1962a), McElroy (1954), Pettijohn (1960), Shiki and Mizutani (1965), Wieseneder (1961), and Dott (1964). Perhaps the most complete review of the nomenclatural problem is that of Dott.

Dott states that the term was originally a field term, attributed to Lasius (1789) who applied it to Upper Devonian-Lower Carboniferous Kulm strata of the Harz Mountains in Germany. It was, however, used by Werner in 1787 (Crook, 1970a). With the advent of microscopical methods of study, we now have a better understanding of these rocks and are capable of framing a sharper definition of the term. Helmbold and Van Houten (1958) and Mattiat (1960) restudied the type graywackes of the Harz and provided us with petrological "bench marks" with which to compare graywackes elsewhere.

The authors are inclined to follow Cummins (1962a, p. 52) who accepts Jamieson's definition of 1808 in which graywacke is defined as a ". . . kind of sandstone . . . composed of grains of sand, which are of various sizes . . . connected together by a basis of clay-slate, and hence this rock derives its gray color and solidity." This definition clearly fits an early description of the original graywackes of the Harz Mountains by Naumann (1858, p. 663) as well as the recent descriptions of the same rocks by Helmbold and Mattiat. It is also consistent with the usage of the term graywacke by most geologists of the last one hundred and fifty years.

The most marked deviations from traditional usage are those definitions which broaden the term to encompass sands rich in rock fragments, and in some cases well-sorted, and exclude those rich in feldspar. The latter restriction has largely been abandoned inasmuch as the classic Harz graywackes are themselves notably feldspathic. The designation "graywacke" for the clean sands rich in rock particles has also now been generally replaced by the term "lithic sandstone" (lithic arenite or litharenite) or "subgraywacke."

Although difficulty has been encountered in giving a precise definition, there exists a fairly homogeneous group of rocks similar in their essential characteristics to the type Harz sandstone and which by tradition and usage have been called "graywackes." The dark, fine-grained matrix in which the sand grains are set is an essential characteristic of these rocks.

Although "matrix" is a relative term and different arbitrary upper size limits as well as differing percentages of matrix have been utilized in defining graywacke, these difficulties are second-order and do not affect the identity of the bulk of the graywackes that have been studied and described. Graywackes are not just muddy or silty sandstones nor arkoses with a clayey pore filling. Unlike graywackes, the clay content of the arkose is a separate fraction—a much finer-grained component. Arkoses seem to have a clear bimodal distribution whereas in graywacke there is a continuum from the coarsest to the finest particle (Huckenholtz, 1963). Moreover, the matrix of arkoses is different from that of graywackes. It is apt to be kaolinitic and may also be red-stained rather than a dark chloritic "paste."

Muddy, quartz-rich sands are apt to be sands produced by organic burrowing, *bioturbates*, which are readily distinguished from graywackes. Neither they nor the clayey arkoses will resemble graywackes in color, induration, or chemical composition. Moreover, their manner of occurrence and usual structure will be quite unlike that of the typical graywacke. It is with the water-laid tuffs and submarine ash flows with their zeolitic products of altered glass that the sedimentary petrographer will have the most difficulties.

General Description

Graywackes are, as noted by most investigators, dark gray or black, generally tough, well-indurated rocks. Many exhibit graded bedding, convolute and small-scale current laminations, and various sole markings such as flute, groove, and load casts. They are generally rhythmically interbedded with mudstones or slates. These features are neither universal nor restricted to graywackes and, although the occurrence of such features is of interest, they are not diagnostic of themselves and hence are not a part of the definition.

The sand fraction is generally rich in quartz, has a varying proportion of feldspar and rock particles, and generally contains a little detrital mica. The quartz is varied in size and shape though it is generally quite angular and commonly shows pronounced undulatory extinction. It usually constitutes half or less of the sand fraction, the balance being feldspar and rock particles. The feldspar is largely plagioclase, showing the usual multiple twinning. It is more commonly sodic rather than calcic. The high Na_2O content of graywackes as a class suggests that the feldspar is nearly pure albite. K-feldspar may be wholly absent though it is present in small amounts in some (Bailey and Irwin, 1959). Although the feldspars are generally fresh, some contain inclusions of sericite, chlorite, and epidote—the last being, perhaps, a product of decalcification of the plagioclase. As with the quartz grains, the margins of the feldspars may be hazy as though encroached on by the matrix.

Rock particles are dominantly mudstone, shale, siltstone, slate and argillite, phyllite, and mica schist. Chert and micritic limestone, especially the former, and polycrystalline quartz and fine-grained quartzite may be plentiful. Many graywackes, however, contain particles of fine-grained igneous rocks, some with microlites of feldspar. Particularly common are acid igneous flow of rocks; less common is andesitic debris. In rare cases serpentine is present (Zimmerle, 1968). With an increasing proportion of volcanic rock and with the appearance of zoned plagioclase and broken crystal euhedra, the graywackes pass insensibly into water-laid tuffs and tuffaceous sandstone. The range in character and composition of graywackes is illustrated by Table 5-7.

Detrital micas, both biotite and chloritized biotite as well a muscovite, are common though not abundant constituents. Minor accessories include a carbonate, probably iron-rich ankerite, which occurs in irregular patches replacing matrix and also some rock particles and feldspar grains.

Other minor accessories include sulfides, probably mostly pyrite, which occur, like the carbonate, in irregular areas.

Unlike most other sandstones which are held together by a pore-filling mineral cement, graywackes are bound by a fine-grained matrix consisting of an intimate intergrowth of chlorite, sericite, and minute, silt-sized particles of quartz and feldspar. The matrix content seems also to be something of a function of the size of the sand fraction—the finer the grain, the higher the matrix content (Okada, 1966). Huckenholtz (1963, p. 914) states that in a strict sense "matrix" requires two distinct maxima in the grain size distribution curve—one in the sand and the other in the clay-size region. By this view, most graywackes would not have matrix. This restrictive definition of matrix is generally not adhered to. The matrix is, to some extent, intergrown with the larger sand grains. In many graywackes one has great difficulty in distinguishing between the matrix as such and ill-defined, fine-grained pelitic rock particles. This difficulty may lead to an overestimation of the abundance of matrix (de Booy, 1966).

The graywackes are a surprisingly homogeneous group chemically (Table 5-8). Despite the uncertainty about the definition, the bulk chemical composition of graywackes is much alike the world over, regardless of geologic age. Moreover, their compositions are quite unlike those of arkoses (Table 5-4 and Fig. 5-3).

As can be seen by reference to the table, graywackes are rich in Al_2O_3, FeO + Fe_2O_3, MgO, and Na_2O. The high Na_2O content no doubt reflects the albitic nature of the feldspar; the MgO is related to the chloritic matrix as is the high FeO content. The distinguishing chemical attributes are dominance of FeO over Fe_2O_3, usually dominance of MgO over CaO,

TABLE 5-7. Mineralogical composition of graywackes (based on modal analysis) (from Pettijohn, 1963, Table 5, with additions).

	A	B	C	D	E	F	G	H	I	J
Quartz	33	22	37	4	24	27	33	56	9	trace
Feldspar	15	5	12	10	32	19	21	37	43	30
Rock fragments	3	26	15	50	19	30	7	7	10	13
"Matrix"	45	47	32	32	P[1]	21	33	P[2]	25	45
Mica and chlorite	—	—	—	—	16	—	6	—	4	—
Miscellaneous	—	—	3	2	8	3	—	—	4[3]	10[3]

(1) Not separately reported; 38 percent of rock is "clay and silt"; (2) not separately reported; (3) hornblende and pyroxene.

A. Feldspathic graywacke (Precambrian), Ontario, Canada; average of 3 analyses (Pettijohn, 1943, p. 946).
B. Lithic graywacke (Martinsburg Shale), (Ordovician), Pennsylvania, U.S.A. (McBride, 1962b, p. 62).
C. Aberystwyth Grit (Silurian), Wales (Okada, 1967, Table 1, Analysis 70A).
D. Lithic graywacke (Devonian), Australia; average of 5 (Crook, 1955, p. 100).
E. Feldspathic graywacke (Devonian-Mississippian, Tanner), Harz Mountains, Germany (Helmbold, 1952, p. 256).
F. Graywacke (Kulm), Harz, Germany (Mattiat, 1960).
G. Graywacke (Lower Mesozoic), Porirua district, New Zealand (Webby, 1959, p. 472).
H. Feldspathic graywacke (Jurassic? Franciscan Formation), Calif., U.S.A.; average of 17 analyses (Taliaferro, 1943, p. 135).
I. Purari graywacke (Cretaceous), Papua; average of 4 (Edwards, 1950b, p. 164).
J. Tuffaceous Aure graywacke (Miocene), Papua; average of 2 (Edwards, 1950a, p. 129).

TABLE 5-8. Representative chemical analyses of graywackes (modified from Pettijohn, 1963, Table 6, with modifications).

	A	B	C	D	E	F	G	H	I	J
SiO_2	60.51	61.39	76.84	69.11	68.85	74.43	73.04	71.10	68.84	65.05
TiO_2	0.87	0.62	—	0.60	0.74	0.83	0.15	0.50	0.25	0.46
Al_2O_3	15.36	16.97	11.76	11.38	12.05	11.32	10.17	13.90	14.54	13.89
Fe_2O_3	0.76	0.39	0.55	1.41	2.72	0.81	0.56	trace	0.62	0.74
FeO	7.63	5.32	2.88	4.64	2.03	3.88	4.15	2.70	2.47	2.60
MnO	0.16	0.12	trace	0.17	0.05	0.04	0.18	0.05	nil	0.11
MgO	3.39	3.84	1.39	2.06	2.96	1.30	1.43	1.30	1.94	1.22
CaO	2.14	3.21	0.70	1.15	0.50	1.17	1.49	1.80	2.23	5.62
Na_2O	2.50	2.78	2.57	3.20	4.87	1.62	3.56	3.70	3.88	3.13
K_2O	1.69	1.25	1.62	1.76	1.81	1.74	1.37	2.30	2.68	1.41
H_2O+	3.38	2.44	1.87[1]	4.13	2.30	2.15	2.36[1]	1.90	1.60	2.30
H_2O-	1.15	0.06		0.05	0.77	0.20		0.26	0.35	0.28
P_2O_5	0.27	0.19	—	0.03	0.06	0.18	0.23	0.10	0.15	0.08
ZrO_2	—	0.07	—	—	—	—	—	—	0.05	—
CO_2	1.01	0.88	—	—	0.08	0.48	0.84	0.12	0.14	2.83
SO_3	—	—	—	—	—	—	—	—	0.15	—
S	0.42	0.15	—	—	0.08	0.12	0.10	trace	—	0.05
Cr_2O_3	—	0.01	—	—	—	—	—	—	—	—
BaO	—	0.06	—	—	trace	—	—	—	0.04	—
C	—	—	—	—	0.07	0.17	0.17	0.09	—	—
	100.24	99.75	100.18	99.69	99.94	100.45	99.80	99.80	99.93	99.77

(1) Loss on ignition.
A. Graywacke (Archean), Manitou Lake, Ontario, Canada, B. Brunn, analyst (Pettijohn, 1975b, p. 228).
B. Graywacke (Archean), Knife Lake, Minnesota, U.S.A., F. F. Grout, analyst (Grout, 1933, p. 997).
C. Graywacke Tyler slate (Precambrian Animikian), Hurley, Wisconsin, H. N. Stokes, analyst (Diller, 1898, p. 87).
D. Rensselaer Graywacke (Ordovician?), near Spencertown, N.Y., H. B. Wiik, analyst (Balk, 1953, p. 824).
E. Tanner Graywacke (Upper Devonian–Lower Carboniferous), Scharzfeld, Germany, R. Helmbold, analyst (Helmbold, 1952, p. 256).
F. Graywacke from Stanley Shale (Carboniferous), near Mena, Arkansas, B. Brunn, analyst (Pettijohn, 1972, p. 228).
G. Kulm (Carboniferous), Steinback, Frankenwald, Germany (Eigenfeld, 1933, p. 58).
H. Composite sample (Lower Mesozoic), prepared by using equal parts of 20 graywackes exposed along shoreline between Palmer Head and Hue-Te-Taka, Wellington, N.Z., J. A. Richie, analyst (Reed, 1957, p. 16).
I. Franciscan Formation (Jurassic?), Quarry Oakland Paving Co., Piedmont, California, Jas. W. Howson, analyst (Davis, 1918, p. 22).
J. Graywacke (Cretaceous?), Olympic Mountains, Washington, near Solduc, B. Brunn. analyst (Pettijohn, 1975b, p. 228).

and the dominance of Na_2O over K_2O. It is in these respects that graywackes differ notably from arkoses.

Many graywackes have a bulk composition closely resembling a particular source rock—commonly a granodiorite and not, as some have thought, a mafic igneous rock. Sands are normally chemically differentiated from their source rocks by weathering and sorting. The chemically undifferentiated nature of graywacke suggests that one or both processes have been incomplete. The sands of the present Columbia River (Whetten, 1966) have a bulk chemical composition surprisingly similar to that of graywacke. Did ancient graywackes have a provenance similar to the Columbia River sands?

Varieties and Types of Graywacke

As noted elsewhere, we distinguish between the matrix-free arkoses, that is, the arkosic arenites, and those with a kaolinitic clay matrix—the arkosic wackes. Is there a comparable distinction in the graywacke clan between matrix-free graywacke and matrix-rich graywacke? By our usage, matrix-free graywacke would be a contradiction of terms: A rock of graywacke composition without matrix would be, in most cases, a lithic arenite. Although the term "mature graywacke" was once applied to such sandstones, this usage has been replaced by "lithic arenite" even by Folk himself (1968a, p. 125).

It is possible, however, to recognize a number of varieties of graywacke. Those rich in rock fragments have been called *lithic graywackes;* those rich in feldspar are the *feldspathic graywackes* (Fig. 5–12).

Some graywackes are very quartz-poor and contain evident volcanic contributions such as volcanic quartz, euhedra or broken euhedra of feldspar, and an appreciable content of ferromagnesian minerals. These are the volcanic graywackes. They pass by degrees into true subaqueous tuffs. There are other graywackes abnormally rich in quartz and lacking feldspar. These are the "quarzwackes" of Fischer

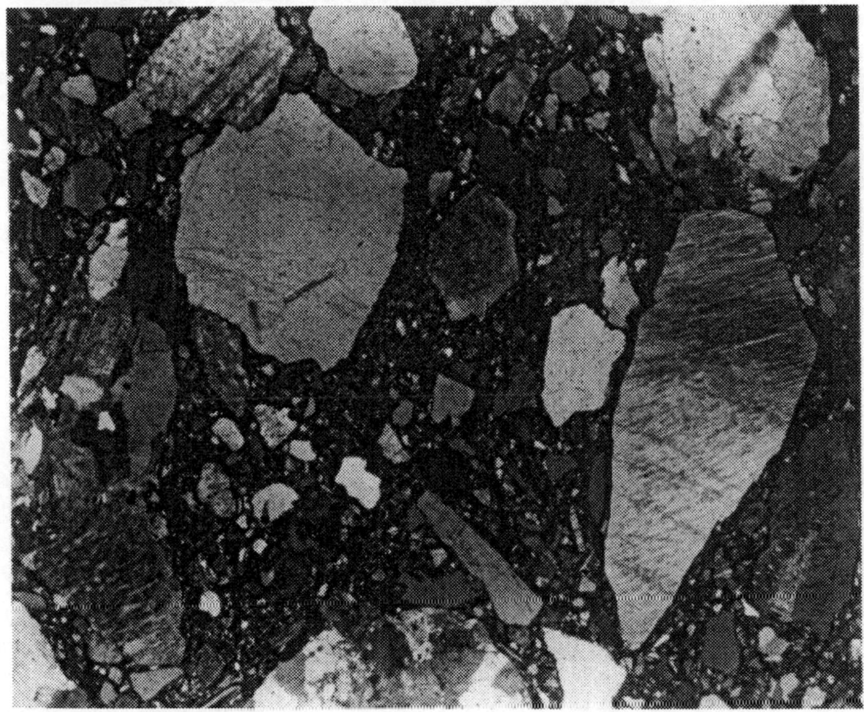

FIGURE 5-12. Recluse Formation, Epworth Group (Lower Proterozoic). Coronation Gulf, N.W.T., Canada. Crossed nicols, ×20. A feldspathic graywacke consisting of a poorly sorted mixture of angular debris in a fine-grained matrix. Mainly quartz and feldspar but includes also some rock particles.

(1933). In between the quartz-poor and quartz-rich varieties are the common and most abundant graywackes of the geologic record. Analyses D, I, and J, Table 5-7, are representative of the quartz-poor class; most of the other analyses in this table belong to the intermediate group.

Crook (1970b) has recognized three major classes of graywacke and has attributed their differences to provenance: the quartz-poor (under 15 percent quartz) being of volcanic provenance, the quartz-rich (over 65 percent and commonly 80 percent quartz) being of sedimentary provenance, and the intermediate class (15 to 65 percent quartz) being of a mixed provenance. Crook attributed the quartz-poor types to the island arc environment, the quartz-rich class to tectonically inactive continental margins, and the intermediate type to tectonically active margins of continents or microcontinents. Dickinson (1970) and others have similarly discussed provenance of graywackes in terms of plate tectonics.

Field Occurrence and Examples

Graywackes typically occur in geosynclinal tracts, with or without associated volcanics. Certainly they are the characteristic sandstones of the eugeosynclinal belts—active continental margins—where they are associated with pillow lavas or "greenstones" (Tyrrell, 1933; Bailey, 1936; Turner and Verhoogen, 1960, p. 270). Graywackes of this habit range widely in age from the oldest Precambrian to the Tertiary. Examples of the former are the many Archean sequences in the Canadian Shield (Pettijohn, 1943; Donaldson and Jackson, 1965), the Fennoscandian Shield (Simonen and Kuovo, 1951), and South Africa (Anhaeusser *et al.*, 1969, p. 2184). Included here also are the Precambrian Dalradian graywacke grits of Scotland (Sutton and Watson, 1955). Similar rocks occur in the older Paleozoic regions of Wales (Woodland, 1938; Okada, 1967), the Southern Uplands of Scotland (Walton, 1955) (Fig. 5-13), the Scandinavian Highlands, and Spitzbergen (Tyrrell, 1933, p. 25). The Ordovician of Newfoundland is characterized by a similar assemblage (Espenshade, 1937). This association is found also in the younger Hercynian fold belt of Cornwall and Devon as well as in the corresponding areas of Europe, particularly in the Harz (Fischer, 1933; Helmbold, 1952; Mattiat, 1960) (Fig. 5-14), and in the Rheinische Schiefergebirge

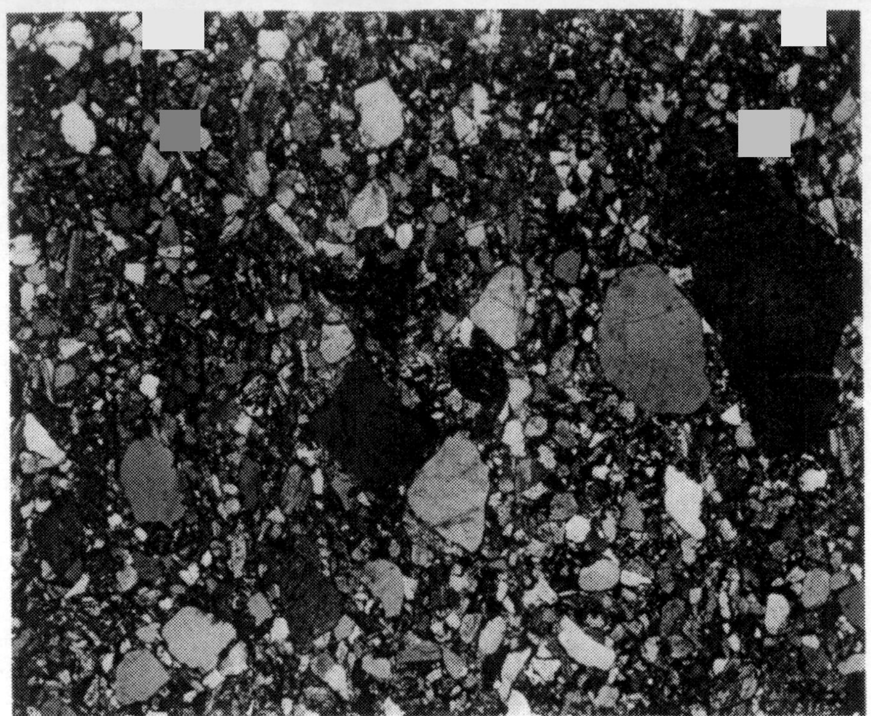

FIGURE 5-13. Caradocian (Ordovician). 370 meters east by north of Garvald, Scotland. Crossed nicols, ×20. A coarse, poorly sorted graywacke consisting of subangular to subrounded quartz and feldspar with a few rock fragments, including aphanitic volcanic rocks.

FIGURE 5-14. Kulm (Carboniferous). Quarry near Sösetalsperre, about 7 km from Osterode, Harz Mountains, Germany. Crossed nicols, ×20. The type or "classical" graywacke consisting mainly of angular, ill-sorted quartz (av. 26 percent), feldspar (av. 19 percent), and rock fragments (av. 38 percent) in fine-grained, chlorite-rich matrix.

(Henningsen, 1961). Well-described examples occur also in the Lower Mesozoic of the Wellington district of New Zealand (Reed, 1957) and in the Franciscan Formation (Jurassic) (Fig. 5-15), of the Coast Ranges of California (Davis, 1918; Taliaferro, 1943). Graywackes with the greenstone and chert associations are found in both the Mesozoic and Paleozoic of Alaska (Fig. 5-16) (Cady *et al.*, 1955; Loney, 1964), in the Olympic Mountains of Washington (Park, 1942), in the Paleozoic of New South Wales of Australia (Fig. 5-17) (Crook, 1955, 1960), and in the Cretaceous of the Northwest Borneo Geosyncline (Haile, 1963).

Graywackes also occur in geosynclinal tracts in which volcanism, if any, is minimal. The Animikian Tyler Slate, the Michigammee Slate, and the Thomson Slate near Duluth (Schwartz, 1942) of the Penokean fold belt south of Lake Superior contain good examples (Irving and Van Hise, 1892, p. 206f) as does the Chelmsford "sandstone" (Fig 5-18) of the Sudbury Basin. The sandstones of the Ordovician Martinsburg Shale (McBride, 1962b) (Fig 5-19) and the correlative Rensselaer Graywacke (Balk, 1953) and the Normanskill Shale (Weber and Middleton, 1961) of the Appalachian Mountains belong in this category as do those of the Carboniferous Stanley Shale and Jackfork Sandstone of the Ouachita Mountains of Arkansas and Oklahoma (Bokman, 1953) and the correlative Haymond Formation of the Marathon district of Texas (McBride, 1966). Well-described examples of Cretaceous age occur in Japan (Shiki, 1962; Okada, 1960, 1961). Many of the graded sandstones of the Apennines are graywackes (Fig. 5-20). All of these examples of graywackes in miogeosynclinal tracts are apt to be characterized by their abundance of rock particles rather than by feldspar. Although a few tuffaceous beds are found in some of these geosynclinal sections, no extensive submarine volcanism is recorded.

Clearly graywackes are characteristically found in Alpine-type orogenic belts and are absent in sequences deposited on the undeformed stable or platform areas. They are generally marine and, in the examples listed above, are largely believed to be turbidite sands and part of what we have called the "flysch" facies. Not all flysch sandstones, even though of turbidite origin, are graywackes. Neither are all sandstones

FIGURE 5-15. Franciscan Formation (Upper Jurassic?). U.S. 101, about 0.4 km north of Golden Gate Bridge, near San Francisco, California, U.S.A. Crossed nicols, ×20. A typical feldspathic graywacke, chiefly quartz, feldspar, and rock fragments in a fine-grained chloritic and sericitic matrix. Note poor rounding and lack of assortment of the material.

FIGURE 5-16. Kuskokwim, (Cretaceous ?). Cribby Creek—George River divide, Kuskokwim region, Alaska. Crossed nicols, ×20. A graywacke marked by its low quartz content and abundance of feldspar, some of which is subhedral, and many aphanitic flow rock fragments. Very poorly sorted. Clearly a graywacke with a considerable volcanic input.

FIGURE 5-17. Graywacke, Neranleigh Group, probably Silurian. Arundale District, New South Wales, Australia. Ordinary light, ×20. A graywacke characterized by its low quartz content and abundance of feldspar and rock particles, including those of volcanic rocks.

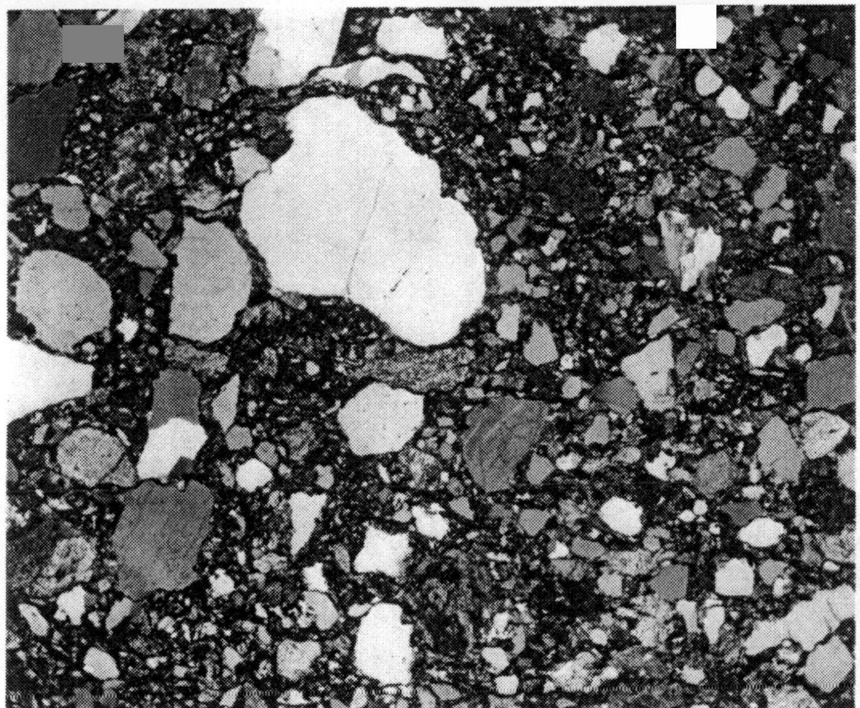

FIGURE 5-18. Chelmsford Sandstone (Precambrian). Sudbury district, near Chelmsford, Ontario, Canada. Crossed nicols, ×20. A graywacke with the usual unassorted, poorly rounded quartz, feldspar, and rock particles in a fine matrix.

FIGURE 5-19. Sandstone from Martinsburg Shale (Ordovician). Broadfording, Washington County, Maryland, U.S.A. Crossed nicols, ×20. A lithic graywacke consisting of poorly sorted, subangular sand-sized debris in a fine matrix. Martinsburg graywackes contain 20-60 percent quartz, 1–12 percent feldspar, and 3–44 (av. 24) percent rock fragments, mostly sedimentary in origin. Matrix forms 30–48 percent of the rock.

FIGURE 5-20. Pietraforte Sandstone (Cretaceous). Fiesole, near Florence, Tuscany, Italy. Crossed nicols, ×20. A good graywacke consisting of poorly sorted angular debris—mainly quartz, feldspar, and rock particles. Some samples of this sandstone contain up to 17 percent detrital dolomite.

in the orogenic belts graywackes; but graywackes do not seem to occur outside such belts.

Graywackes have been estimated to constitute about one-fifth to one-fourth of all sandstones (Middleton, 1960; Pettijohn, 1963). It has been presumed that they were of greater importance in the earlier periods of earth history. The bases for these statements will be considered in Chap. 12.

The Matrix Problem

The matrix is the essential characteristic of graywacke and is, as Cummins (1962a) noted, "the essence of the graywacke problem." By definition the matrix content exceeds 15 percent and is reported in some cases to exceed 50 percent. These estimates, however, were questioned by de Booy (1966) who argued that immature detrital sediments, especially graywackes, are often wrongly analyzed. The detrital grain boundaries are in most cases camouflaged by recrystallization phenomena. As a result, the matrix content is greatly overestimated. De Booy contends that the matrix content of many typical graywackes is no more than a few percent. Certainly under crossed nicols one has great difficulty in distinguishing between somewhat altered rock fragments and the recrystallized sericitic and chloritic matrix. In many cases, too, it is difficult to distinguish between the softer pelitic rock fragments, which may become deformed and squeezed into pores between more durable quartz grains, and true matrix. Such fragments give rise to a false or pseudomatrix.

The matrix has been variously interpreted, as discussed on pp. 146 and 431. It was thought by some to be the product of recrystallization of an original detrital material presumed to be mud. The problem became one of explaining the simultaneous deposition of mud and sand. Normally, as a result of current action, the two part company and are separately accumulated. Even the point bars in the lower reaches of the Mississippi River—a dominantly mud-carrying stream—are mud free. It has been suggested that flocculation by the electrolytes of seawater would have precipitated the clay-sized materials (Woodland, 1938). Evidence to support this view is generally lacking. Few, if any, modern near-shore, shallow-marine sands have the requisite character, that is, interstitial mud, to become graywackes on lithification.

The matrix of graywackes has been attributed to their manner of transport and deposition. Many graywackes in the geologic record bear the earmarks of turbidity current action and are in fact turbidites. As Kuenen and Migliorini (1950, p. 123) stated, ". . . all the exceptional and puzzling features of typical series of graded graywackes can be readily accounted for by the activity of turbidity currents of high density." They reasoned that when ponded the turbidity current would cease and deposit both its load of sand and suspended mud, producing a graded graywacke. The deposition in deep marine basins would explain our general failure to find the modern analogue of the ancient graywackes. But a consideration of settling velocities and analysis of flume studies do not substantiate the view that much mud matrix can be deposited by simple settling of clay together with sand (Chap. 8). In addition, graywackes, that is, matrix-rich sands, do not appear to be common among present deep-sea sands of presumed turbidite origin (Hollister and Heezen, 1964). Kuenen (1966, p. 267) subsequently reviewed the problem and concluded on both theoretical and experimental grounds that the original matrix, if any, was well below ten percent in the coarser rocks.

Emery (1964) and Klein (1963, p. 571) believed that the clay might have been mechanically introduced—by movement of interstitial water from overlying or underlying beds—and that initially clean and well-sorted turbidites were thus converted to matrix-rich graywackes. Kuenen rejected this suggestion and concluded that the matrix of the coarser graywackes is very largely of secondary origin and that there is no need to presume such graywackes were carried and deposited by some different kind of turbidity current or by some different kind of mechanism than that assumed for Recent deep-sea sands or Tertiary turbidites.

It is evident that the matrix is indeed recrystallized material. It consists of chlorite, sericite, and quartz in the older graywackes and zeolites and montmorillonite in the younger ones. That some matrix material is truly authigenic was noted by Krynine (1940, p. 22), who observed sericite replacing detrital quartz, a relationship also observed between chlorite and quartz. The original water-worn boundaries of the quartz have wholly disappeared, the existing boundary being a kind of "chevaux-de-frise" of green chlorite crystals projecting into clear quartz (Greenly, 1897, p. 256). As discussed earlier, there is wide consensus that much, if not all matrix, is diagenetic (Cummins, 1962a; Brenchley, 1969; Hawkins and Whetten, 1969).

Cummins (1962a) noted that most Tertiary and Recent turbidites do not have the abundant matrix found in the comparable turbidites of the early flysch sequences. The latter have presumably undergone deep burial and incipient matamorphism involving pressure solution at grain contacts and especially alteration of the labile constituents—mainly unstable rock fragments but also including feldspar (Fig. 5-21). Although some deep-sea turbidites cored by the DSDP program disclosed lithification, much of that proved to be due to silicification by dissolution and reprecipitation of biogenic silica. We may conclude, therefore, that the modern equivalents of the matrix-rich ancient graywackes are the immature sands with high amounts of argillaceous and/or volcanic rock fragments that were deposited in basins where they are des-

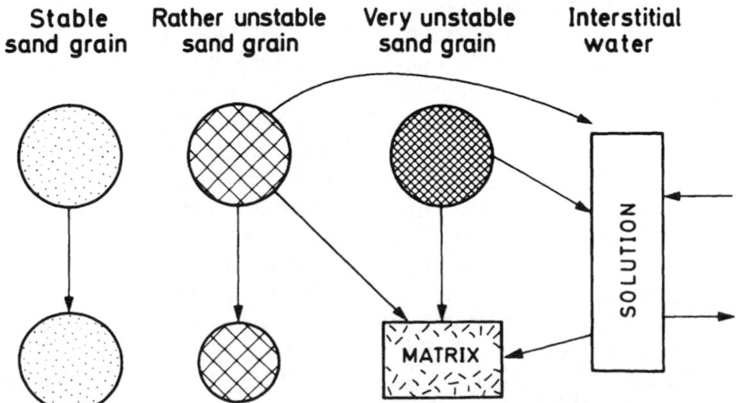

FIGURE 5-21. Diagram showing presumed post-depositional origin of matrix of graywackes. (Modified from Cummins, 1962a, Fig. 4)

tined for deep burial and low-grade metamorphism or high-grade diagenesis—termed "graywackisation" by Kuenen.

In conclusion, it can be said that matrix, the fine interstitial material characteristic of graywackes (and some arkoses), is most probably of more than one origin, most of it being secondary.

The Problem of Na_2O

Various explanations have been offered to account for the high Na_2O content of the graywackes (Engel and Engel, 1953, pp. 1086–91). The high Na_2O content seems to be due to the high content of albitic feldspar. The Tanner graywacke, for example, contains 3.5 percent Na_2O. This quantity of Na_2O would require 30 percent pure albitic feldspar, an estimate consistent with the observation that the feldspar actually forms from 30 to 40 percent of this rock, of which some 85 to 90 percent has a composition An_{3-10}.

Clearly the feldspar is a detrital component. But was the original material albitic or is its albitic nature a post-depositional feature? The close association of many (but not all) graywackes with oceanic volcanic rocks suggests that the problem is related to the alteration of basalts, ophiolites, and processes associated with subduction zones. There is some evidence of albitization *in situ*. The irregular patches of carbonate, neither detrital nor a pore-filling cement and which replaces both detrital grains and matrix, are clearly a post-depositional product. The albitization of intermediate plagioclase feldspar would have provided the CaO which was deposited elsewhere as a carbonate. One is reminded of the innumerable vugs and veinlets of calcite associated with the albitized greenstones. But if the alteration of the feldspar is post-depositional, why are not the associated and interbedded slates also Na_2O-rich? The pelitic strata interbedded with the graywackes of New Zealand, for example, have a normal Na_2O/K_2O ratio in contrast to that of the cogenetic graywackes (Reed, 1957, p. 28). See Fig. 5-22. Garrels *et al.* (1971) suggest that the high K_2O/Na_2O content of shales and slates is also a diagenetic characteristic, and is a reflection of the growth of illite at the expense of montmorillonite, mixed layer, and kaolinite assemblage. Thus the Na_2O ends up in the feldspar and the K_2O in the mica. It has been suggested that the absence of K-feldspar is due to post-depositional solution and is not a primary characteristic (Gluskoter, 1964, p. 341). This would fit the compatibility of albite and muscovite at low to moderate temperatures (see Chap. 11).

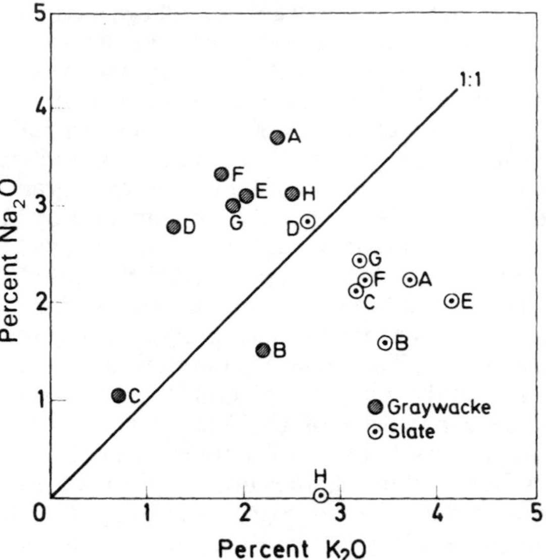

FIGURE 5-22. Diagram showing Na_2O/K_2O content of graywackes and co-genetic shales. A) New Zealand Mesozoic, composite sample (Reed, 1957, Tables 1 and 3). B) New Zealand, Mesozoic (Reed, 1957, Tables 1 and 3, Analyses 17 and 6). C) North Spirit Lake, Ontario, Archean (Donaldson and Jackson, 1965, Table 3, cols. 2 and 5). D) Knife Lake Minnesota, Archean (Grout, 1933, Table 1, cols. 1 and 2). E, F, and G) Yellowknife, Great Slave Lake, N.W.T., Archean (Henderson, pers. comm.). H) Casummit Lake, Ontario, Archean (Horwood, 1938).

Albitization of detrital feldspars in deeply buried sandstones of the Great Valley Sequence in California has been described by Dickinson *et al.* (1969, p. 519). Such alteration began at 7500 ft and was complete at 20,000 ft. It is associated with chloritization of the detrital biotite. However, it is absent or slight in those sandstones with a calcareous cement. Dickinson *et al.* also noted that the normal calcareous sandstones had 10 to 25 percent interstitial matrix whereas the calcareous sands has 5 percent or less. They concluded the matrix was a diagenetic product.

Albitization of the feldspars has also been reported in the buried Eocene sands of the Gulf Coast (Boles, 1982).

It seems probable, therefore, that the high albite content and chloritic matrix is a post-burial product. Perhaps graywackes owe their distinctive properties to such "burial metamorphism."

Significance of Graywackes

Many, if not most, graywackes are redeposited sands. In general, such redeposited sediments are not good guides to the climate either of the source area or the place of deposition. Insofar as the graywackes are marine, as most of them are, they tell us little about the climate of the depositional site. On the other hand, the detritus of which graywackes are composed must, like that of arkose, require an environment in which erosion, transportation, and deposition are so rapid that complete chemical weathering of the materials does not take place. Inasmuch as arkosic sands can form under humid tropical conditions as well as under an arid or arctic climate, so also can the detritus of graywackes. Krynine (1937, p. 427 and Fig. 4) reports on graywackes of the Siwalik Series (late Tertiary) in northwestern India which he believed, on the basis of both lithologic and faunal evidence, were deposited under tropical conditions varying from savanna through prairie to steppe. Fischer (1933, p. 339) also concluded, from a study of graywackes in Germany, that they could form in several different climatic environments. Taliaferro (1943, p. 139) believed the graywackes of the Franciscan Formation (Upper Jurassic) to have been derived ". . . from a high, rugged, recently uplifted landmass under rigorous climatic conditions, high rainfall, and possibly a cold climate in the highlands with well wooded lower slopes." This inference was based on the abundance of fresh feldspar and of carbonized fragments of wood. Fine woody debris is common in many post Devonian graywackes.

Because graywacke is not confined to the Precambrian, or even the pre-Devonian, the absence of a vascular land plant cover does not seem to be a requirement for its formation as was suggested by Kaiser (1932).

Graywacke, however, does indicate a special tectonic environment. Fischer called it a "poured-in" type of sediment. To the extent that graywackes are turbidites, each bed is essentially an instantaneous event. Its restriction to geosynclinal tracts and its dominance in the flysch of the older systems suggests that much or most is a marine turbidite. Although graywackes are not by definition turbidites, the fact remains that most of those in the geologic record are turbidites. It may be, as Cummins suggests, that the turbidites of the younger orogenic flysch sequences are precursors of graywacke and will be converted to graywacke by diagenetic alteration or low-grade metamorphism. Such alteration accompanied by Na_2O-enrichment related to their marine environment could produce the distinctive characters of graywacke. By this interpretation, graywacke is a high-grade diagenetic or low-grade metamorphic product. Some authors do not, however, consider this a closed question (Shiki and Mizutani, 1965, p. 32).

It is uncertain how much, if any, graywacke is nonmarine. As noted above, Krynine (1937) described some presumed nonmarine graywackes (Cummins, 1962b) but an overwhelming proportion of the graywackes of the past seem to have been marine. Some have thought that graywackes reflect reducing conditions, perhaps even a reducing atmosphere (MacGregor, 1927), presumably because of the high FeO/Fe_2O_3 ratio. The existence of graywackes of Paleozoic and later age precludes a reducing atmosphere. The associated shales generally lack the abnormally high content of organic carbon or early diagenetic pyrite indicative of strongly reducing conditions.

To a certain extent graywacke and arkose do reflect a different provenance. Arkose records a granitic provenance. And although granite or related rocks are very common and widespread in some areas, they give rise to arkose only under relatively restricted circumstances. On the other hand, graywacke seems to have had a less restricted provenance—the abundant quartz and feldspars of some, mixed with low-rank metamorphic rock particles and even volcanic detritus, denote a more varied provenance than do arkoses. Is this because it is commonly a redeposited sediment, the detritus being supplied by large rivers, with an extensive drainage basin of varied lithology, and by longshore drift? It may be, as Shiki (1962, p. 316) suggests, that there are at least two types of graywackes which differ in their provenance, the first being rich in feldspar and therefore of crystalline plutonic derivation, the second being a lithic graywacke and hence of a supracrustal source. Moreover, those characteristic of the eugeosynclines may have a significant volcanic input. These views are similar to those of Crook (1970b), who attributes the low-quartz graywackes to a volcanic provenance and the quartz-rich varieties to a sedimentary provenance. Certainly graywacke is *not* the ferromagnesian equivalent of an arkose. (Twenhofel,

1932, p. 231). A fuller discussion is given in Chap. 12.

Quartz Arenites (Orthoquartzites)

Definitions and Nomenclature

Almost all sands contain detrital quartz; most are quartz-rich, that is, have over 25 percent quartz, and a very large number consist almost wholly of detrital quartz. The term "quartzose sandstone" is therefore more or less redundant. But a name is required for those end-products of sand evolution which contain an overwhelming abundance of quartz. The term *orthoquartzite* has been used for this class of sands. Although apparently first used by Tieje (1921, p. 655), it was popularized by Krynine (1945) who applied this term to sands made up entirely of quartz grains cemented by silica. These are the sedimentary quartzites as opposed to the quartzites of metamorphic origin (metaquartzites). The term "quartzite" in the traditional sense applied to a rock so thoroughly cemented that the rock breaks across the grains rather than around them. However, as Krynine (1948, p. 152) pointed out, many orthoquartzites contain some carbonate cement as an additional end-member and as the proportion of this nonsiliceous material increases, they begin to show less cohesiveness. Although the term "orthoquartzite" has come to be accepted and applied to these less cohesive sands and even to friable or modern quartz sands, it engenders some confusion because of the long-standing concept of quartzite as a durable, non-friable rock (see, for example, Mathur, 1958). Hence other terms have been suggested. Most acceptable is the term *quartz arenite* (or *quartzarenite*) proposed by Gilbert (Williams, Turner, and Gilbert, 1954, p. 294) and McBride (1963). Although we have used the term "orthoquartzite" in the past, we consider "quartz arenite" the better term for this class of sandstones. The trend in the recent literature seems to indicate concurrence in this view.

General Description

The quartz arenites are very common—perhaps the best-known type of sandstone. They may be defined simply as sands of which the detrital fraction is 95 or more percent quartz. In general, they are cemented with quartz deposited in crystallographic and optical continuity with the detrital quartz, and in many cases so well cemented that they are indeed quartzites.

Quartz arenites are generally white rocks, some are tinged with pink, a few are deeper red. The pink or reddish coloration is due to a film or coating of hematite on the grains. The iron oxide content, however, may be only a percent or less of the whole rock. Many quartz arenites are relatively thin, widespread blanket sands with a minimum of interbedded shale. In other cases, the sand accumulation may be thick—several thousand feet (1000 m) or more as in the case of some Precambrian formations. Ripple-marking and/or crossbedding are characteristic structures.

As noted, many are indeed quartzites and hence durable and erosion resistant and form high ridges or hills. Others, especially those cemented by carbonates, are less quartzitic and some, perhaps due to leaching of the cement, are friable. In many of the friable sands, however, the quartz grains show silica overgrowths leading to the reconstruction of the quartz crystal form with brilliant, sharp-edged facets that reflect light so that the sand has a prominent "sparkle" in bright sunlight.

These sands are, as noted above, those in which quartz forms 95 or more percent of the detrital fraction. In many, especially the quartz-cemented varieties, the bulk chemical analyses reflect this dominance of quartz, the silica content being 99 percent or even more (Table 5-9). Such deposits are the largest and purest concentrations of silica in the earth's crust and constitute a commercial source of silica for the manufacture of glass and other needs.

The quartz of these sands is largely monocrystalline quartz—the polycrystalline grains are less stable and tend to be eliminated (Blatt, 1967, p. 422). The proportion of quartz with undulatory extinction tends to be lower in these sands for similar reasons (Blatt and Christie, 1963, p. 571). The quartz grains of most orthoquartzites are highly rounded—some approaching perfect roundness. Many exhibit a frosted or "matte" surface. Inasmuch as sorting is excellent, these sands are both the most texturally and compositionally mature of all sands. Some approach the theoretical end point in sand evolution.

Other constituents, if any, are likely to include a few well-worn grains of chert or other equally durable rock particles. Though insignificant in volume, these few rock particles may be the clue to the provenance and "line of descent" of the particular quartz arenite in question. The heavy mineral fraction is generally

TABLE 5-9. Chemical composition of representative orthoquartzites (Pettijohn, 1963, Table 2).

	A	B	C	D	E	F	G	H	I	J
SiO_2	98.87	95.32	97.58	97.36	98.91	83.79	99.54	99.40	97.30	93.13
Al_2O_3	0.41	2.85	0.31	0.73	0.62	0.48	0.35	0.20	1.40	3.86
Fe_2O_3	0.08	0.05	1.20	0.63	0.09	0.063	0.09	0.01	0.30	0.11
FeO	0.11	—	—	0.14	—	—	—	—	—	0.54
MgO	0.04	0.04	0.10	0.01	0.02	0.05	0.06	0.01	0.03	0.25
CaO	—	trace	0.14	0.04		8.81	0.19	<0.01	<0.05	0.19
Na_2O	0.08	0.30	0.10	0.08	0.01	—	—	0.08	<0.05	—
K_2O	0.15		0.03	0.19	0.02	—	—	trace	0.20	—
H_2O+	0.17	—	—	0.54	—	—	0.25	0.04	—	—
H_2O-		—	—	0.14	—	—	—	0.01	—	—
TiO_2	—	—	—	0.05	0.05	—	0.03	0.02	0.28	—
P_2O_5	—	—	—	0.02	—	—	—	none	—	—
MnO	—	—	—	0.01	—	—	—	trace	0.003	—
ZrO_2	—	—	—	—	—	—	—	<0.01	0.06	—
CO_2	—	—	—	—	—	6.93[2]	—	—	—	—
Ign. loss	—	1.44	0.03	—	0.27	—	—	0.28	—	1.43
Total	99.91	100.00	99.62[1]	99.94	99.99	100.13[3]	100.51	100.05[4]	99.57	99.51

(1) Including SO_3, 0.13; (2) calculated; (3) includes organic matter, 0.006; (4) includes Cr_2O_3, 0.00008; BaO and SrO, none; NiO less than 0.001; CuO less than 0.00027; CoO less than 0.0002.

A. Mesnard Quartzite (Precambrian), Marquette County, Michigan, U.S.A., R. D. Hall, analyst (Leith and Van Hise, 1911, p. 256).
B. Lorrain Quartzite (Precambrian), Plummer Township, Ontario, Canada, M. F. Connor, analyst (Collins, 1925, p. 68).
C. Sioux Quartzite (Precambrian), Sioux Falls, S. Dakota, U.S.A. (Rothrock, 1944, p. 151).
D. Lauhavouri Sandstone (Cambrian?), Tiiliharju, Finland, Pentii Ojanperä, analyst (Simonen and Kuovo, 1955, p. 79). Quartz 70–75; feldspar 0.1–1.4; rock fragments 0.1–5.6; silica cement 18–20.
E. St. Peter Sandstone (Ordovician), Mendota, Minnesota, U.S.A., A. William, analyst (Thiel, 1935, p. 601).
F. Simpson Sand (Ordovician), Cool Creek, Oklahoma, U.S.A. (Buttram, 1913, p. 50).
G. Tuscarora Quartzite (Silurian), Hyndman, Pennsylvania, U.S.A. (Fettke, 1918, p. 263).
H. Oriskany Sandstone (Devonian), Berkeley Springs Quarry, Berkeley Springs, West Virginia, U.S.A., Pennsylvania Glass Sand Corp., Sharp-Schurtz Co., analysts. Analysis supplied courtesy of Pennsylvania Glass Sand Corp.
I. Mansfield Formation (basal Pennsylvanian), Crawford County, Indiana, U.S.A., M. E. Coller, R. K. Leininger, R. F. Blakely, analysts. Computed mineral composition: Quartz 95.3; orthoclase 1.2; kaolin 3.0; ilmenite 0.3 (Murray and Patton, 1953, p. 28).
J. Berea Sandstone (Mississippian), Berea, Ohio, U.S.A., N. W. Lord, analyst. A protoquartzite (Cushing *et al.*, 1931, p. 110).

highly restricted and consists usually only of very well-rounded tourmaline and zircon plus some ilmenite or leucoxene derived from the ilmenite.

The bulk chemical composition may be drastically altered by the addition of the cement. The silica-cemented sands, of course, are not so altered, but those with calcite or anhydrite will show a silica percentage depressed by the addition of CaO and CO_2 or SO_3.

Silica is the most common cementing agent and even in those sands in which another cement dominates, some silica cement is present (see Chap. 11). The silica is almost always quartz, deposited in optical and crystallographic continuity with the rounded detrital quartz. In sands incompletely cemented the overgrowths exhibit well-formed crystal facets—which in thin section give the remaining pores straight-edged boundaries. In the more friable sands the regenerated quartz euhedra show well-formed pyramidal terminations. Casual inspection shows that the crystal axis tends to coincide approximately with the long dimension of the detrital core. As cementation proceeds, the overgrowths meet along irregular boundaries so that the end result of the enlargement process is an interlocking quartz mosaic within the elements of which the rounded outlines of the detrital grains may be visible, being delineated by a thin ring of dustlike inclusions. In some quartz arenites this line of demarcation is faint or altogether absent.

Pressure-solution phenomena are well-known and especially notable and clearly displayed in many orthoquartzites. Clear and well-developed stylolitic seams are present in some cases (Heald, 1955). Microstylolites involving detrital chert grains are known in other cases. Pressure-solution along grain boundaries can presumably lead to transformation of a quartz sand into a quartzite. This conversion by pressure-solution at grain contacts has been described in detail by Skolnick (1965). In many grains, it is difficult to know whether or not the interlocking sutured boundaries of the quartz mosaic are the product of pressure-solution or whether they are nothing more than the result of mutual intereference of the respective overgrowths. Examination by the method of cathodoluminesence microscopy (Sippel, 1968) suggests that many presumed cases of pressure-solution are nothing more than the end-product of quartz enlargement. Under cathodoluminescence the detrital grains are clearly visible and display uninterrupted detrital outlines surrounded by secondary quartz.

A few quartz arenites are cemented with other forms of silica—opal or chalcedony, some with both. Opal may form concretionary coatings on the grains which extend into the pores, partially or wholly filling them. That the opaline coatings are post- rather than pre-depositional is demonstrated by their absence at the points of contact of adjacent quartz grains. Chalcedony plays a similar role and differs from opal in that it may show a microfibrous fan-shaped structure and, unlike opal, is birefringent. In general, opaline cements are characteristic only of the younger sandstones. As in the cherts, the older rocks are chalcedonic or quartzose—presumed devitrification products of opal.

Carbonates, usually calcite, are a common cement in other quartz sands. They take several forms. In most cases each individual pore is filled by a single crystal of calcite; in others, the calcite crystallized in large "poikilitic" patches which enclose many sand grains. These are the so-called "luster-mottled" sandstones (Cayeux, 1929, p. 154). Such sandstones display striking reflections from the calcite cleavage of these coarse-textured cements. In others, the calcite forms drusy coatings on the individual quartz grains. Although calcite is the usual carbonate cement, a few sands contain other carbonate species. Siderite is present in some sands and gives rise, on oxidation, to iron oxide which then plays the role of cement. It is our impression, however, that the siderite-cemented quartz arenites are rare; siderite is more commonly found as cement in lithic sandstones.

Sands cemented with dolomite are common but not as abundant as those cemented with calcite (see, for example, Swineford, 1947, p. 79). Dolomite occurs as small rhombic crystals, in some cases about the same size as the sand grains.

Other cements are much less common. Sands cemented with anhydrite, barite, and celestite have been described. The barite-cemented sands are restricted to small nodular bodies consisting of tabular barite crystals arranged in a rosette-like manner (Shead, 1923; Tarr, 1933). Celestite cementation is likewise a very local and restricted phenomenon (Swineford, 1947, p. 82). The quartz grains enclosed in the barite and celestite cements show markedly less corrosion by the cement than those cemented with carbonates. Unlike barite and celestite, anhydrite is a more common cement. It is generally associated with other cementing minerals and is usually the last to have been precipitated (Waldschmidt, 1941, p. 1866). Anhydrite is readily identified by its rectilinear cleavage.

Varieties

Because quartz arenites are 95 or more percent quartz, there is not much variation between one and the next except in the kind of cement which they contain. The most marked contrast is between the quartz-cemented and carbonate-cemented arenites.

On the other hand, it is important to make a distinction, if possible, between the quartz arenites derived directly from a plutonic source rock and those derived from preexisting sandstones. It is very difficult, however, to make this distinction. First-cycle sands are apt to be less well rounded and to contain a greater proportion of polycrystalline and undulatory quartz, and are more likely to retain a little feldspar and to show a greater diversity of heavy minerals—or at least less well-rounded zircons and tourmaline. Multicycle sands, on the other hand, might contain grains of older quartzites and detrital chert, especially in the coarser size grades. Worn overgrowths would be proof of second-cycle origin. Unfortunately, such features are uncommon.

Multicycle quartz arenites are commonly associated with transitional types—with subarkose and especially with protoquartzites (sublitharenites). They are presumed to be the better "cleaned-up" sands. Not uncommonly, the sands in a sedimentary sequence show a progressive decrease in feldspar or rock-particle content with decreasing age. The "evolutionary" succession that terminates in orthoquartzites suggests that the latter are indeed multicycle sands.

A peculiar feature of some quartz arenites is their bimodal texture. Attention has been called to this feature especially by Folk (1968b) who attributed it to the action of the wind in desert areas. It is presumed to be due to a selective removal of the fine sand fraction (0.1-0.3 mm), leaving behind both the coarser and finer sizes. Folk described present-day bimodal sands from the Simpson Desert of Australia. A similar bimodal texture is reported to be common in many Cambro-Ordovician and some Precambrian sands. These formations may be marine or fluvial as well as eolian but the final size distribution was created in the deflationary eolian environment. Folk (1968b, p. 12) tabulates some 21 North American examples of bimodal sands. A good Precambrian example is the Odjick Formation of the Coronation Gulf Geosyncline.

Field Occurrence and Examples

Quartz arenites have a worldwide distribution. Well-known North American examples include the thick, widespread Precambrian quartzites: the Sioux Quartzite (Fig. 5-23) of Minnesota, Iowa, and South Dakota (Rothrock, 1944), the Baraboo and Waterloo Quartzites of Wisconsin (Brett, 1955), the Sturgeon Quartzite of northern Michigan and the correlative Mesnard Quartzite of the Marquette district on Lake Superior, the Palms Quartzite of the Gogebic district, the uppermost Lorrain Quartzite (Fig. 5-24) on the north shore of Lake Huron in Ontario, and the Odjick Formation (Fig. 5-25) in the Northwest Territories. The Wishart Formation of the Labrador Trough contains excellent quartz arenites (Donaldson, 1966, p. 34). The Hinckley Sandstone of Minnesota (Thiel and Dutton, 1935) and the Sibley Sandstone of the Thunder Bay area of Lake Superior are orthoquartzites of Keweenawan age. The Athabaska Formation of northern Saskatchewan is fairly typical of many of the Precambrian quartz arenites (Fahrig, 1961). It appears to have covered a very large area and even today has an areal extent of 40,000 mi^2 (104,000 km^2), contains some quartz pebbles and granules, has a uniform transport direction, is largely free of shale, and has a maximum thickness of about 5000 ft (1510 m). The Thelon Formation of the Northwest Territories of Canada is largely sandstone and is similar in character and extent, and perhaps in age, to the Athabaska (Donaldson, 1967). The Precambrian Uinta Mountain Group of Utah contains some quartz arenites said by Krynine to be first cycle (Krynine, 1941b).

The sandstones of the Cambro-Ordovician of the Upper Mississippi Valley are primarily quartz arenites. These include the Cambrian sandstones—the Dresbach Sandstone, Franconia Sandstone, and Jordan Sandstone (Graham, 1930)—and the Ordovician New Richmond and St. Peter Sandstones (Fig. 5-26) (Dake, 1921; Thiel, 1935). Quartz arenites of equivalent age and similar character occur in Missouri—the

FIGURE 5-23. Sioux Quartzite (Precambrian). Dell Rapids, South Dakota, U.S.A. Crossed nicols, ×55. A typical orthoquartzite—a truly quartzitic rock produced by thorough cementation of a quartz arenite. Note well-preserved boundaries of highly rounded detrital quartz separating detrital core from secondary quartz overgrowth. Essentially 100 percent quartz.

FIGURE 5-24. Lorrain Quartzite (Huronian, Precambrian). Near Bruce Mines, Ontario, Canada. Crossed nicols, ×55. An orthoquartzite (near 97 percent quartz), well-sorted and well-rounded quartz. Original grain boundaries well shown.

FIGURE 5-25. Odjick Formation, Epworth Group, Lower Proterozoic. Coronation Gulf, Northwest Territories, Canada. Crossed nicols, ×20. A quartz arenite with a bimodal size distribution presumed to be indicative of an eolian origin. Note extraordinary rounding of the large quartz grains, the interstices between which contain smaller (and less well-rounded) grains. Large microcrystalline grains are chert.

FIGURE 5-26. St. Peter Sandstone (Ordovician). Twin City Brick Yard, St. Paul, Minnesota, U.S.A. Crossed nicols, ×55. A famous quartz arenite consisting wholly of very well-sorted and highly rounded monocrystalline quartz. A friable sand with little or no cement in most places. Specimen photographed was impregnated with resin.

Cambrian Lamotte Sandstone (Ojakangas, 1963) and the sandstones of the Ordovician Roubidoux Formation—and in Oklahoma—the sands of the Simpson Group (Ordovician), including, for example, the sandstones of the Tulip Creek Formation of the Arbuckle region, and the Blakely and Crystal Mountain Sandstones of the Ouachitas. In the New York region, the Cambrian Potsdam Sandstone is in most places a quartz arenite (Fig. 5-27) as are also the sandstones in the Cambrian Gatesburg Formation and the Chickies Quartzite of Pennsylvania and the Antietam Sandstone of Maryland and West Virginia (Schwab, 1970). In the western United States the Cambrian is characterized by quartz arenites including the Flathead Quartzite of Wyoming and the Lower Cambrian quartzites of British Columbia. The Ordovician Eureka Quartzite of Nevada and its equivalent, the Swan Peak Quartzite of Idaho are high-purity deposits consisting of over 99 percent highly rounded and frosted quartz (Ketner, 1966). The Pennsylvanian Tensleep Sandstone of Wyoming and its facies equivalents in the Casper Formation and the Weber Quartzite of Utah are good quartz arenites, the latter being interpreted by Krynine (1941b) as second cycle. The Shawangunk Conglomerate and the Tuscarora Quartzite (Silurian) of Pennsylvania and New Jersey are similarly interpreted. The Devonian Oriskany Sandstone of Pennsylvania (Fig. 5-28), in places a glass sand, is a good example of a marine quartz arenite. The Dakota Sandstone (Cretaceous) of the Great Plains is in many places a quartz arenite (Fig. 5-29). Some of the Miocene sands of New Jersey are very clean quartz sands, containing 97 to 99 percent quartz. These are locally well enough cemented with opal, chalcedony, or quartz to deserve the name quartzite (Friedman, 1954). See Fig. 5-30.

In Europe there are a number of Cambrian quartz arenites. Among these are the Hardeberga Sandstone of Sweden (Hadding, 1929, p. 77), the sandstone of Lauhavuori, Finland (Simonen and Kuovo, 1955, p. 75), and the Cambrian Malvern Quartzite of England. The Cretaceous seems also to have been a time of quartz arenite development. The Lower Senonian Quadersandstein of the Harz of Germany is an excellent example (Rinne, 1923, p. 280) as is the Ashdown Sandstone (Lower Wealden) of Sussex (Allen, 1949). Other European quartz arenites include some sands from the Lias near Celle and the Dogger near Braunschweig

FIGURE 5-27. Red Sandstone. Potsdam, New York, U.S.A. Crossed nicols, ×55. Unusually well-displayed example of secondary enlargement of well-rounded quartz. The boundaries between detrital core and overgrowth made very distinct by coating of iron oxide on the original detrital grains.

FIGURE 5-28. Oriskany Sandstone (Devonian). Near Berkeley Springs, West Virginia, U.S.A. Crossed nicols, ×55. A glass sand consisting solely of quartz. A well-sorted sand with a good marine fauna. Original grain outlines very indistinct.

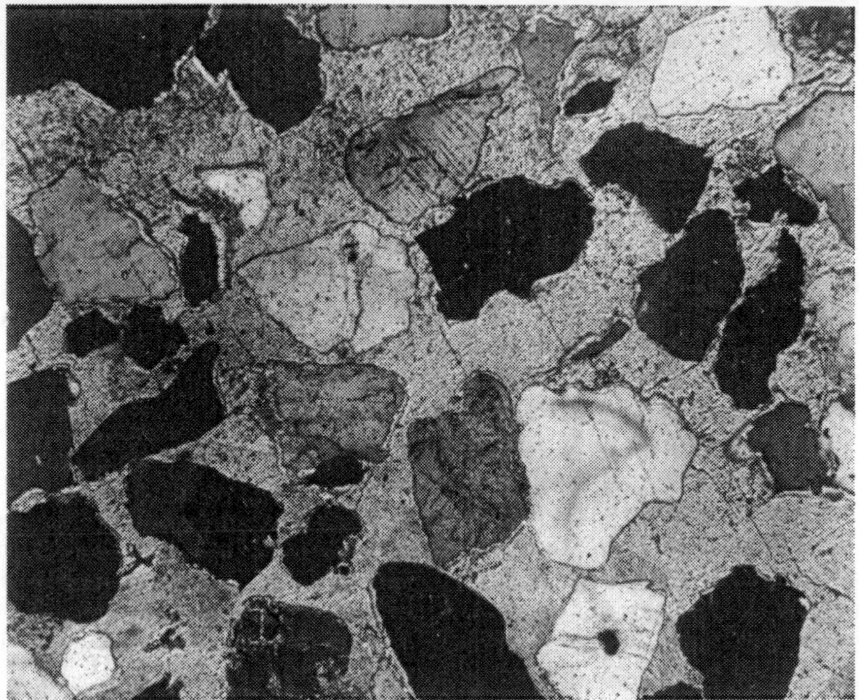

FIGURE 5-29. Dakota Sandstone (Cretaceous). Kansas, U.S.A. Crossed nicols, ×55. An excellent example of a calcite-cemented quartz sand showing local etching or embayments in quartz indicating solution and replacement. The cement is carbonate which has same crystallographic (and optical) orientation over entire field of view.

FIGURE 5-30. Kirkwood Formation (Miocene). Near Greenwich, New Jersey, U.S.A. Crossed nicols, ×20. A chalcedony-cemented quartz arenite containing silicified shell fragments. Chalcedonic coatings, once opal, coat the quartz grains. Remaining interstices filled with fibrous chalcedonic material. Original quartz well-sorted and subangular to well-rounded.

(Füchtbauer, 1959, Table 2) and the Tertiary Fontainebleau Sandstone of France (Cayeux, 1929, p. 154). The Nubian Sandstone of Egypt is at least in part a good quartz arenite as is the Precambrian Scarp Sandstone of the Vindhyan System along the Son River of India.

Distribution in Space and Time

Because quartz arenites are by definition essentially quartz sands with less than five percent other constituents, they are exceptional sands. They are exceptional in the sense that modern quartz arenites are absent in temperate and northern latitudes or if they exist at all, are small accumulations derived from or formed by redeposition of an older quartz arenite. Some of the sands on the Gulf Coast of northwestern Florida are nearly pure quartz, being over 99 percent quartz (Hsu, 1960, Table 9). These sands may have been derived from the Coastal Plain sediments of Cretaceous or Tertiary age. The pure quartz sands of the Libyan Desert in Egypt that contain 95 or more percent quartz (Mizutani and Suwa 1966) were probably derived from disintegration of the Nubian Sandstone. However, sands with over 95 percent quartz have been reported from some tropical rivers including the Congo, Niger, San Francisco, and Paraña (Potter, 1978).

Although quartz arenites are estimated to constitute about one third of all sandstones (Pettijohn, 1963), they are not equally abundant everywhere or throughout geologic time. It is probably no accident that of the 44 examples cited in the preceding section, all but five are Precambrian or Paleozoic and about one-third are Cambrian. There seem to be but few examples of Mesozoic and Cenozoic quartz arenites although the Nubian Sandstone of the Middle East approaches a true quartz arenite as do several Cretaceous sandstones such as the Dakota Sandstone of Kansas (Swineford, 1947), the Woodbine sand of Texas, and the Quadersandstein of the Harz region of Germany. It would seem that the older sandstones were more largely derived from the craton whereas those of later times were derived from orogenic belts or lands and are, therefore, less mature. It has even been suggested that the abundance of Late Precambrian and Cambrian quartz arenites is related to more vigorous and effective tidal action due to a presumed lesser distance between the earth and moon (Merifield and Lamar, 1968).

Although quartz arenites seem to be the characteristic sandstone of the stable cratonic areas and of little importance in the geosynclinal or orogenic belts, there are some notable exceptions. Craton-derived quartz sands do extend into the miogeosynclinal belts, being most prominent along the cratonic margin of these tracts. Ketner (1968), for example, has described an Ordovician quartzite, over 1000 ft (300 m) thick, of high purity, in the Cordilleran miogeosyncline. It appears to have been derived by erosion of Cambrian sandstones in northern Alberta and transported over 1000 mi (1600 km) southward along the axis of the geosyncline. In general, eugeosynclines are without the true quartz arenites—the typical arenite of these tracts being graywacke. In rare cases, however, clean quartz arenites are reported from eugeosynclines. The Silurian Clough Quartzite (Billings, 1956, p. 21), a metamorphic quartzite, clearly derived from a quartz arenite and, containing marine fossils, lies in the midst of the eugeosynclinal sequence in New England. Ketner (1966) has described a high-purity Ordovician quartzite (Valmy Formation) in the eugeosynclinal tract in Nevada. This quartzite is associated with bedded cherts and volcanic greenstones.

Significance and Origin of Quartz Arenites

What is the geologic significance of these sands? The salient facts to be accounted for are: (1) their supermature petrography, (2) their sheetlike geometry, (3) their cratogenic derivation, and (4) their distribution, that is, primarily on the craton or its margins rather than in the geosynclines.

The quartz arenites exhibit the best sorting, the best rounding, the highest concentration of quartz, and the most restricted heavy mineral suit of any of the sands. On the basis of these characters, most investigators have concluded that such sands could not have been derived directly from a weathered granite, that instead they were derived from preexisting sandstones. In short, they are multicycle. On the basis of experimental work, especially that of Kuenen (1959), and on field observation of some large modern streams (Russell, 1937, p. 1346), it seems improbable that a quartz arenite sand of the kind found in the geologic record could ever be produced by river transport, no matter how prolonged.

The problem of the quartz arenites is reviewed by Suttner *et al.* (1981). Based on a study of modern sands formed under both arid and humid conditions and on certain theoretical considerations, they conclude that first-cycle quartz arenites cannot be formed under "average" conditions and that most of the quartz arenites of the geologic record are of multicycle origin. Perhaps, however, a combination of a warm humid climate, low relief, and slow rate of sedimentation could yield a first-cycle quartz arenite.

According to Kuenen (1960), only in the desert dune environment is sand effectively rounded. This conclusion would imply that most of the sand of the quartz arenites had an eolian episode at some time in its history—not necessarily the last stage represented by the present accumulation. The frosting observed on the grains of many of these sands has been taken as confirmatory evidence of eolian action, as has the bimodal texture described by Folk (1968b). The absence of shale has been thought to be related to the desert eolian episode in the history of the sand. Wind action effectively separates or winnows the clay-sized dust from the quartz sand. The deflation process is also almost always fatal to the micas (Krynine, 1942, p. 550).

Kuenen dismisses beach action also as too restricted to be quantitatively important, even if it could be shown that such action is capable of concentrating and rounding quartz to the degree required. One cannot wholly dismiss the possibility, however, that intense weathering followed by prolonged action of the surf might produce an quartz arenite. Most modern beach sands are not quartz arenites although there are places where the sand is nearly pure quartz, such as the Gulf Coast of northwestern Florida, the sand being 99.65 percent silica (Burchard, 1907, p. 382). These sands, however, may have been derived from coastal plain sediments of Cretaceous or Tertiary age. On the other hand, there is some support for the notion of a progressive "cleaning-up" of sands by beach action. Not a few sedimentary sequences progress upward from lithic arenites or arkoses to quartz arenite. The lowest member of the 7000-ft (2100 m) Lorrain Quartzite (Precambrian) in the Bruce Mines area of Ontario is an arkose with 25-30 percent feldspar. The feldspar shows an upward decline so that the upper-most Lorrain is a clean orthoquartzite with less than one percent feldspar. The Tuscarora Quartzite (Silurian) of West Virginia is a further illustration (Folk, 1960). The lowest beds contain 50 to 70 percent quartz; the highest have over 95 percent quartz. This upward increase in quartz is accompanied by a compensatory decrease in the quantity of rock particles (15 to 35 percent) and chert (3 to 6 percent) to less than five percent. Similarly, the lowest Pennsylvanian sandstones in the Anthracite Basin of Pennsylvania have 15 to 20 percent rock particles; the highest sands have less than 3 percent (Meckel, 1967). It seems unlikely in all the above examples that the provenance of the lower beds was appreciably different from that of the higher ones. If the change in character of the sands as one proceeds upward is not due to changing character of the sand supplied, it must be due to action in the environment of deposition. Folk interpreted the change as due to a shift from an essentially fluvial environment to that of the beach. He believed that it was to surf action that the highest sands owe their quartz enrichment.

The Lower Carboniferous Kellerwald Quartzite, a pure quartz arenite in the Variscan geosyncline west of the Rhine, is thought to be a shelf-edge bar, produced by reworking and winnowing of the graywacke materials that characterize most of the section (Meischner, 1971, p. 19). If so, it is, like parts of the Tuscarora, a product of the environment.

The Cambrian Eriboll Sandstone of Scotland (Swett *et al.*, 1971) is an example of a first-cycle quartz arenite attributed to prolonged tidal action.

Some of the quartz arenites are clearly marine—as the contained marine fauna testifies (Seilacher, 1968). Others are presumed to be marine even though fossils are absent, in part because these sands are interbedded with or pass laterally into marine limestones and dolomites. Because the shallow-water sands of today are mainly restricted to a narrow littoral zone, it has been supposed that these orthoquartzitic sand sheets record the advance of a transgressive shoreline. Thiel (1935), for example, thought the Ordovician St. Peter Sandstone of the Upper Mississippi Valley to be due to marine deposition related to several retreats and advances with intervals of wind action to account for the rounding and frosting. The size and character of the crossbedding and ripple-marking and, in some cases, association with limestones or dolomites seem to indicate a marine origin, or at least deposition in a standing body of water, for many of the Precambrian quartz arenites.

A problem which arises with these and other quartz arenites is the paucity of shale associated with them. Unlike the lithic and arkosic sandstones and the graywackes, little or no shale is interbedded with the quartz arenites nor, indeed, are there significant shaly formations in the sequence of which the quartz arenites are a part. Although this absence of clays can be explained by eolian action, it seems more probable to us that the reason is that most quartz arenites of the past were deposited on shallow marine shelves where, if the modern world can be taken as a guide, much if not most of the clay is in suspension and is carried over the shelf edge (see Chap. 8).

In summary, the necessary conditions for the formation of quartz arenites are extraordinary weathering to eliminate the feldspar coupled with the most effective removal of clay and rounding of the quartz. Whether the latter was achieved by desert eolian action or by action of the surf is uncertain.

The known quartz arenites, certainly the best North American examples, seem to have been deposited on the stable cratonic portions of the continents and such thin quartz arenites as occur in the marginal geosynclines are of cratonic or shield derivation. The best examples—the Cambro-Ordovician sands of the North American continental interior—are illustrative of the deposition on the craton. They are also of cratonic derivation (Potter and Pryor, 1961). Such sands as those of the Cambrian Chickies Quartzite and Antietam Sandstone which occur not on the craton but in a marginal geosyncline are probably shield-derived. The great Precambrian orthoquartzites of the Lake Superior region seem also to have been derived from the interior of the Canadian Shield (Pettijohn, 1957). All of the Paleozoic and some of the Precambrian quartz arenites seem to have been deposited by a marine transgression of this stable interior platform.

Miscellaneous Sandstones

A number of sandstones do not fall into the major groups reviewed in the foregoing portion of this chapter. Some of these are of rare occurrence or of small volume, such as serpentine and placer sands. Others are hybrid in nature, in reality intermediate in composition between the epiclastic sands and those formed by endogenetic processes of precipitation; these are the subject of the next section.

Placer sands (and sandstones) are of relatively small volume though they are commonly of great economic value. They may be a source of rutile, titanium, zirconium, and thorium. Placers are local accumulations, most commonly products of fluvial and beach action. Such deposits are of slight areal extent and thickness, the latter being rarely more than a meter or two.

We distinguish here between those deposits in which one or several "heavy" minerals, such as magnetite, become abundant enough to form a large part of the sand and those deposits exploited for such minerals as gold and diamonds, which are a minor constituent albeit a valuable one. We discuss only the former deposits. Modern beach placers are common. Those of Oregon have local concentrations of chromite (Griggs, 1945); beach placers in Florida have been worked for ilmenite, zircon, and rutile (Martens, 1928). Ancient placers are also known and exploited. Examples include the rutile-bearing sands of the Pliocene Cohansey formation of New Jersey (Markewicz, 1969) and the Upper Cretaceous magnetite sands of Montana (Stebinger, 1914, p. 329).

A very unusual placer-type sandstone is that from the Precambrian Rand Basin of South Africa which contains thin layers of detrital pyrite (Ramdohr, 1958, p. 16).

A peculiar and rare type of sandstone is *serpentine sandstone*. It has been described by Zimmerle (1968) and Okada (1964). Although serpentinite is a minor constituent of some sandstones, it is rarely the dominant mineral. It is of local derivation, of small volume, in beds one or two meters thick, commonly intercalated with other sands. Serpentine may form as much as 50 to 85 percent of the sand (Okada, 1964, p. 28). Much of the rock is a calcite cement. The deposits described by Zimmerle were part of a graded Flysch-type sequence and were not as closely associated with serpentinite source rocks; their serpentine content was less—3 to 56 percent.

Hybrid Sands and Sandstones

The sands we have so far dealt with are *epiclastic,* that is, sands in which the framework components, usually quartz, feldspar, and rock particles, are the debris of a wasting landmass—that is, are derived from preexisting rocks or sediment. We shall now briefly discuss those sands in which a significant portion of the framework is of another origin—formed *in* the basin or deposited by chemical or biochemical precipitation or produced elsewhere by volcanic action. These sands are thus neither wholly epiclastic nor pyroclastic nor endogenetic but are instead a cross between several of these and are thus *hybrid* sands.

Examples of sands containing both intrabasinal and extrabasinal framework elements include the greensands in which glauconite is a significant constitutent, phosphatic sands in which the phosphate mineral is a framework component (rather than cement) and, most common of all, those sands consisting of a mixture of detrital quartz and shell or other calcareous debris indigenous to the depositional basin.

Greensands

Glauconite occurs as granules which may be mixed in all proportions with ordinary sand. Some greensands contain over 50 percent glauconite. The glauconite may be concentrated in certain laminations or scattered throughout the sand. If abundant enough, it imparts a speckled appearance to the rock.

Under the binocular microscope the granules appear as very dark, almost black, grains with a smooth rounded surface, some being mammilated, and of diverse shapes, mostly ovoid. In thin section the glauconite varies from very pale bluish green to greenish yellow to a dark grass green. It may also be oxidized to a brownish yellow or reddish brown. The grains are rounded, are of about the same size as the associated quartz, exhibit a polylobate outline, are generally structureless, and may show some shrinkage cracks which taper inward. Some grains are deformed and are molded about the quartz or squeezed into a pore. Under crossed nicols, the glauconite appears as a microcrystalline aggregate. Birefringence is high so that the grain appears much the same under crossed nicols as in ordinary light. In the standard thin section it varies from a pale green to a deep grass green. Oxidized grains are yellowish or brown.

Glauconite is not confined to a particular type of sand. It occurs in some nearly pure quartz arenites; it also occurs in less mature feldspathic and micaceous sands. Commonly it is associated with shell materials, the glauconitic sand filling the shells, and, in the younger sands, the glauconite alone may fill foraminiferal tests. Very commonly the shell materials of the Cambrian sandstones are phosphatic. Clearly most glauconitic sandstones are marine. In some sands the glauconite is associated with pellets—many of which show partial conversion to glauconite—or with biotite flakes which likewise exhibit all degrees of glauconization. Some glauconite forms a coating on other mineral grains—even on heavy minerals, especially ilmenite (Grim, 1936, p. 201).

There is a large literature on the mineralogy of glauconite and hence many analyses of the mineral itself. There are few analyses of greensands. Glauconitic sands are represented by three analyses (A, B, and C) in Table 5-10. As might be expected, the composition of greensands is highly variable depending on the proportion of clastic material, interstitial calcite, and secondary siderite.

Greensands are characterized by their high total-iron content, principally Fe_2O_3, and by the high content of K_2O. Quite commonly also, as shown by the New Jersey greensands, they are high in phosphorus. As would be expected, they also contain a good deal of combined water. Some greensands are high enough in iron to be classed as an iron-bearing formation or ironstone (over 15 percent Fe).

Although glauconitic sandstones have a wide range in both space and time and perhaps occur in the rocks of every geologic system, one gains the impression that they are more common in some systems than others. They are particularly common, for example, in the rocks of the Cambrian. They occur in the sands of the Bright Angel Shale in the Grand Canyon, in the Reagan Sandstone of Oklahoma, in the Franconia Sandstone of Wisconsin, all of Cambrian age, and in the Lower and Middle Cambrian rocks of Wales and of Scandinavia (Hadding, 1932). Likewise the Cretaceous seems to have been a period of glauconite formation (Fig. 5-31). Greensands of this age occur in the east of England (Cambridge Greensand) and in Europe. This period of greensand formation continued in the Paleocene and Eocene of the Atlantic Coastal Plain (Aquia Formation; Drobnyk, 1965). Greensand is relatively rare in rocks of Precambrian age although glauconite sandstones are known from the Semri Series of the Vindhyan System of presumed Precambrian age in peninsular India (Auden, 1933, p. 160).

Glauconite has a wide distribution on the present sea floor. A good example of its occurrence and association with sand is that in Monterey Bay, California (Galliher, 1935).

There is still considerable uncertainty about the origin and significance of glauconite (Goldman, 1919; Hadding, 1932; Allen, 1937; Takihashi, 1939; Cloud, 1955; Burst, 1958). It apparently requires marine conditions for its formation though it may be redeposited, usually oxidized, in a nonmarine sand. It seems to be more marine than chamosite inasmuch as the Clinton (Silurian) oolitic hematite-chamosite ironstone of the Appalachians grades to the east into semicontinental hematitic sandstone and to

TABLE 5-10. Chemical analyses of miscellaneous sandstones (after Pettijohn, 1963, Table 9, with modifications).

	A	B	C	D	E	F
SiO_2	57.40	50.74	75.95	45.43	48.85	51.32
Al_2O_3	6.89	1.93	2.91	0.03	11.82	2.92
Fe_2O_3	11.98	17.36	10.29	2.92	1.83	0.72
FeO	3.04	3.34	—	—	1.22	—
MgO	2.41	3.76	1.37	0.61	0.45	0.58
CaO	1.78	2.86	0.10	26.21	12.85	24.70
Na_2O	1.11	1.53	0.35	0.34	0.47	—
K_2O	4.85	6.68	2.99	0.16	0.64	—
H_2O+	5.36	9.08	5.40	2.78	2.75	—
H_2O-	4.46	—	—	—	—	—
TiO_2	0.29	—	0.20	0.11	trace	0.30
P_2O_5	0.22	1.79	—	16.05	10.70	—
CO_2	—	0.88	—	3.12	3.40	20.00[4]
MnO	0.03	—	—	0.02	—	—
SO_3	0.45	—	—	0.86	—	trace
F	—	—	—	1.87	2.86	—
Total	100.29[1]	99.95	99.56	101.25[2]	97.84	100.54
Less 0				−0.79	−1.20	
				100.46[3]	96.64	

(1) Includes BaO, 0.02; (2) includes C, 0.45, FeS_2, 0.29; (3) given as 99.01 in original; (4) by calculation.

A. "Greensand" (Middle Eocene), Pahi Peninsula, New Zealand (Ferrar, 1934, p. 47).
B. "Greensand marl" (Upper Cretaceous), New Jersey, U.S.A., R. K. Bailey, analyst (Mansfield, 1920, p. 553).
C. Greensand, opal-centered (Thanetien), Angre, Belgium (Cayeux, 1929, p. 130).
D. Phosphate sandstone, "Upper phosphorite stratum" (Cenomanian), Kursk, Schchigri, U.S.S.R. (Bushinsky, 1935, p. 90). About 38 percent quartz, 45 percent phosphorite, 5 percent glauconite.
E. Phosphatic sandstone, Saint Pôt, Boulannais, France (Cayeux, 1929, p. 191).
F. Calcarenaceous sandstone, Loyalhanna Formation (Mississippian), Pennsylvania, U.S.A. (Hickok and Moyer, 1940, p. 464).

FIGURE 5-31. Greensand, Ft. Augustus Formation (Cretaceous). Ft. Augustus No. 1, T. 55 S., R. 21 W., west 4th Meridian, 2,883 ft, Alberta, Canada. Ordinary light, ×55. A sandstone rich in glauconite (large round pellets) together with subangular quartz cemented by carbonate. This formation also contains some detrital chert.

the west into fully marine glauconitic strata (Hunter, 1970, p. 118). Normal salinity, weakly reducing conditions, and a slow rate of deposition are required for its formation. The mineral appears to be a diagenetic product created by alteration of biotite (Galliher, 1935), or by reorganization and iron enrichment of mud pellets, replacement of shells and other debris (Takihashi, 1939), or even by precipitation in cavities or as coatings on detrital grains (Grim, 1936, p. 201).

Glauconite does occur in the Precambrian (see, for example Auden, 1933, p. 160; Gulbrandsen *et al.*, 1963)—despite some contrary opinions (Cloud, 1955): There is no good explanation for the apparently greater abundance in the Cambrian and Cretaceous-Eocene strata. Much yet remains to be learned about its origin and distribution.

Phosphatic Sandstones

Phosphatic sandstone is sand cemented with calcium phosphate (carbonate fluorapatite) or sandstone that contains an appreciable quantity of phosphatic debris or precipitated granules or oolites of phosphate.

Many sands contain a little phosphatic debris—largely that of phosphatic skeletal materials. The glauconitic sands are especially prone to be phosphatic. In most phosphatic sandstone, however, the phosphate forms a cement, or occurs as a drusy coating on the quartz grains or as a microcrystalline pore filling (Bushinsky, 1935). On the other hand, some phosphorites are themselves "sands" consisting of sand-sized phosphate granules or ooids which may be mixed in varying proportions with detrital quartz (Fig. 5-32).

Calcarenaceous Sandstones

By far the most common sands of a "mixed" origin are those which consist of a mixture of detrital quartz and sand-sized chemical or biochemical carbonate. These have been called *calcarenaceous sands* (Pettijohn, 1972, p. 237) to distinguish them from a calcareous sand, the latter having carbonate as a cement rather than as a framework element. The calcarenaceous sands grade, with increasing proportions of the carbonate materials, into the calcarenite class of limestones (Figs. 5-33 and 5-34). Care should be taken to differentiate between calcarena-

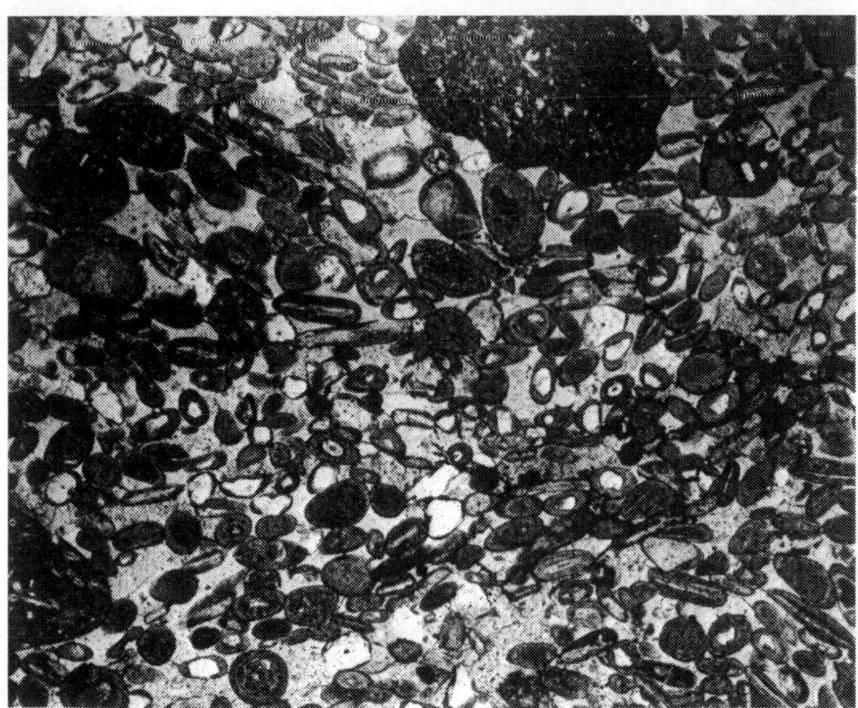

FIGURE 5-32. Phosphatic sandstone, Phosphoria Formation (Permian). SE$\frac{1}{4}$, NE$\frac{1}{4}$, Sec. 13, T. 1 N., R. 3 W., Jefferson County, Montana, U.S.A. Ordinary light, ×20. An oolitic phosphate rock. Many oolites have quartz sand grains as nuclei.

FIGURE 5-33. A well-sorted skeletal limestone "Bedford limestone", Salem Limestone (Mississippian). Quarry near Bedford, Lawrence County, Indiana. Ordinary light, ×20. A biocalcarenite cemented with clear crystalline calcite (biosparite). Skeletal debris includes foraminiferal tests, crinoidal debris, bryozoan, and brachiopod debris. Elongate fragments parallel to bedding. A good example of a skeletal carbonate sand.

ceous sandstone and calclithite, the latter being a lithic sandstone derived by the destruction of preexisting limestones or dolomites.

Under the microscope, the calcarenaceous sandstones are seen to consist of carbonate detritus such as foraminiferal tests, shell and other skeletal fragments, carbonate intraclasts and pellets, and carbonate oolites, mingled in all proportions with quartz and other epiclastic debris. The usual cement is calcite. Inasmuch as all the sand-sized material is current-deposited, the structures of the rock are those of a sandstone. Many are strongly crossbedded. As would be expected, sandstones of this type are rich in Ca and CO_2—richer than a calcareous sandstone—and relatively low in SiO_2 (analysis F of Table 5-10).

Modern examples of sands of this type are common, a notable one being the sands of the east coast of Florida where the siliceous components are carried by southward shore drift and mingled with shelly debris produced locally. In the northern section, quartz forms most of the sand; further south the carbonate detritus increases until it constitutes more than half of the sand (Martens, 1931, p. 82).

Ancient examples are common, one of the best described being the Mississippian Loyalhanna Limestone of Pennsylvania and Maryland (Adams, 1970, p. 83). The Loyalhanna is, perhaps, the most crossbedded formation in the central Appalachians. A typical sample contains 38 percent monocrystalline quartz, 5 percent polycrystalline quartz, 1 percent chert, 26 percent carbonate detritus (intraclasts, ooids, and skeletal grains), and 28 percent carbonate cement. The chemical composition of a rock of this type is given by analysis F in Table 5-10.

Portions of the Conococheague Limestone (Cambrian) and the Grove Limestone (Cambro-Ordovician) of Maryland contain calcarenaceous sandstone beds. The Cretaceous Cow Creek Formation (Fig. 5-35) of Burnet County, Texas, is likewise a calcarenaceous sand, probably a beach deposit. Almost any sequence of limestones interbedded with sandstones will contain some beds of mixed composition.

Tuffaceous Sandstones

A volcanic eruption very commonly generates a great volume of tuffaceous material which, on

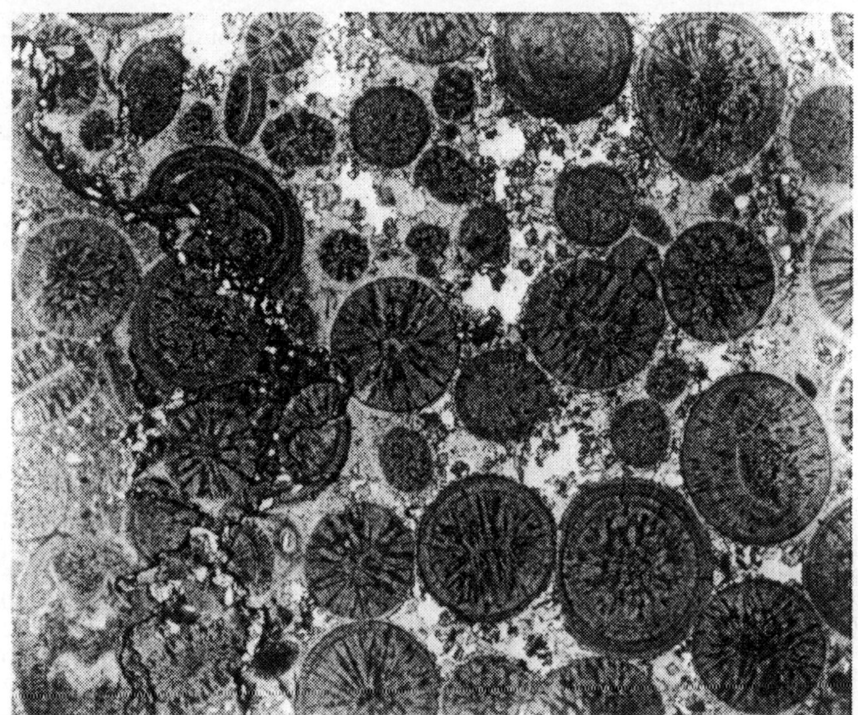

FIGURE 5-34. Warrior Formation (Cambrian). Tyrone, Pennsylvania, U.S.A. Ordinary light, ×20. An oolitic calcarenite (oosparite) composed of spherical oolites with both radial and concentric structure embedded in a carbonate cement. Contains small scattered quartz silt particles. Note stylolite seam which transects oolites and along which is a residuum rich in quartz particles. A "concentration" sand—originally an oolitic sand.

FIGURE 5-35. Cow Creek Limestone (Cretaceous). Cow Creek locality, Edwards Plateau, near Fredericksburg, Texas, U.S.A. Crossed nicols, ×55. Consists of subrounded detrital quartz, skeletal debris (fibrous structure), and some micritic carbonate grains cemented by calcite.

fall-out, overwhelms normal sedimentation and produces a tuff. Recognition of a fully developed tuff—especially one of recent origin—is no special problem. On the other hand, the detection of pyroclastic debris where such material is diluted by and mingled with sands of other origins is not easy. Such material is commonly overlooked. Moreover, it is difficult to distinguish between such newly formed pyroclastic materials and debris produced by disintegration of older volcanic rocks.

Reworking and redistribution of pyroclastic materials is common and such redeposited debris may be mixed with other sands in all proportions to produce a hybrid rock or tuffaceous sandstone (as opposed to a tuff). The alert petrographer will look for the telltale features which signify pyroclastic contamination. Criteria include euhedral feldspars, many of which are broken, are commonly zoned, generally oscillatory. Especially significant is volcanic quartz with its bipyramidal form which is commonly rounded or embayed due to magmatic resorption. Minerals rare in ordinary sands, such as olivine and pyroxene, usually denote tuffaceous origin. Glass in various forms—as envelopes surrounding crystals, as collapsed pumice, and as shards, all of which may be devitrified or altered but retain something of the pumice structure—is one of the best criteria of volcanic contamination. Highly tuffaceous sands have abnormally low quartz content (less than ten percent in some cases). Such a low quartz content is especially significant when associated sandstones contain a normal quantity of quartz. Likewise a high feldspar to quartz ratio is meaningful, particularly if the feldspar tends to be euhedral.

The discrimination between water-laid tuffs, tuffaceous sandstones, and graywackes is one of the most difficult tasks a sedimentary petrographer is called on to make. For an elaboration of this problem and for a more complete description of the volcaniclastic sands, the reader is referred to Chap. 6.

Relative Abundance of Sandstones and the "Average" Sandstone

Several estimates of the abundance of the common types of sandstones have been published (Krynine, 1948; Tallman, 1949; Pettijohn, 1972; Middleton, 1960; Pettijohn, 1963). Because some disagreement exists on the defining parameters and on their limits, the several estimates are somewhat unlike (Table 5-11). The estimates of Middleton and Pettijohn are in reasonable agreement and because their estimates were derived by very different means—one from published chemical analyses and the other from a university rock collection—they can be accepted with fewest reservations.

If graywackes are indeed a diagenetic derivative of lithic arenites, we should include them with the lithic sands. If we do so, then the proportions are quartz-arenite sand 34 percent, arkosic sand 15 to 16 percent, and lithic sands 46 to 50 percent. Lithic sand is certainly dominant—nearly half of all sands belong in this category. On this point Middleton, Krynine, and Pettijohn all agree.

Elsewhere we have given the mean chemical composition of the principal sandstone types and also the various estimates of the bulk chemical composition of the average sandstone (Chap. 2). But what is the composition of this sandstone in mineralogical terms? Pettijohn's average sandstone (Pettijohn, 1963, Table 13) may be recast approximately as follows: quartz, 59 percent; feldspar, 22 percent; kaolin, 6 percent; chlorite, 4 percent; calcite, 6 percent; and

TABLE 5-11. Relative abundance of sandstone classes.

Class	Krynine[a] (1948)	Tallman[b] (1949)	Middleton[c] (1960)	Pettijohn[d] (1963)
Quartz arenite (orthoquartzite)	22.5	45	34	34
Arkose	32.5	17	16	15
Lithic arenite ("low rank" graywacke or subgraywacke)	35.0	17	24	26
Graywacke ("high rank" graywacke)	10.0	21	26	20
Miscellaneous	—	—	—	5

[a] Basis of estimate not stated.
[b] Based on sample of 275 sandstones, Cambrian to Tertiary in age, from all parts of the United States.
[c] Based on 167 sandstones for which chemical analyses appear in the published literature.
[d] Based on 121 sandstones in the Johns Hopkins University collection for which thin sections were available. Age and distribution of samples given in Pettijohn, 1963, Table 11.

iron oxide 2 percent. Such a calculated mineral composition does not distinguish between grains and cement, nor does it permit an estimate of the proportion of rock particles. Probably the latter are hidden in the figures for kaolin, chlorite, and feldspar. If all of the kaolin and chlorite and a third of the feldspar are considered to be present in the rock particles, the average sandstone would be basically 65 percent quartz, 15 percent feldspar, and 18 percent rock particles on a cement-free basis (carbonate and iron oxide omitted). Clarke (1924, p. 3) has given the calculated mineral composition of the average sandstone as 66.8 percent quartz, 11.5 percent feldspar, 11.1 percent carbonates, and 10.6 percent "other."

How do these estimates based on calculations compare with actual modal analyses? The average sandstone of the Russian platform, according to Ronov *et al.* (1963), is 69.7 percent quartz, 15.3 percent feldspar, 2.6 percent rock particles, 2.9 percent mica, and 1.8 percent carbonate. See also discussion in Chapter 2.

Data are quite inadequate to get a very good estimate of what the average modern sand really is. Since an "educated guess" leads us to believe that alluvial sands are by far the most common types of sand in the geological record, especially in many thick geosynclinal sections, the sands of large rivers, therefore, should most closely resemble the average sandstone. The average river sand today consists of about 22 percent feldspar and about 20 percent rock particles and, by difference, 58 percent quartz. These figures agree reasonably well with the estimate made by recasting the chemical composition of the "average" sandstone, considering the uncertainties in the recasting process and the inadequacies of the data on modern river sands.

These calculations and estimates, however crude they may be, point to several major conclusions. As noted in the Introduction (Chap. 1), the bulk of the sand of the world is on the continents and most of that occurs in the folded belts rather than on the cratonic platform. If the Appalachian geosyncline is typical of miogeosynclinal belts, and we believe it is, most of its sands are alluvial. The average sandstone, as demonstrated above, most nearly resembles the sands of modern rivers. The differences may reflect the bias of the data on modern streams—the sample being overly weighted by glacial materials of Shield derivation and hence abnormally rich in feldspar and by the inclusion of near-shore marine sands in the average of ancient sands, which tends to reduce the feldspar content and enhances the proportion of quartz.

Sandstone Petrogenesis

Six factors have been cited as controls on or modifiers of petrographic types of sandstone—source rocks, the climate and relief of the source area, fluvial abrasion during transport to the depositional basin, abrasion and/or sorting on high-energy beaches, by tides, or in eolian dune fields, and changes after burial. What is the relative importance of these factors and how, if at all, does their importance change in different geologic settings?

This question, like most in sedimentology, is best answered by examining the occurrence of the petrographic types of sandstones in the ancient record and by marshaling evidence gathered from the petrographic study of modern sands that can be closely related to climate, tectonic setting, source rocks, and depositional environment. Like most of the major issues in geology, an overall appraisal is qualitative rather than quantitative, even though there is much more quantitative data on the composition of modern sands than was available at the time of our first edition.

Our judgment is summarized in the flow diagram of Fig. 5-36, where relief and rainfall are the key modifiers of the mineralogy of source rocks and also determine the *volume* of sand production. Relief is a factor because it determines the potential energy available for the transport of debris to base level and generally determines the length of time source materials can remain in the zone of weathering—with high relief, deep soils are exceedingly rare. Rainfall is also a factor, because most weathering reactions depend on water in two important ways—hydrolysis and to flush away the weathering products both the solids and solutes (cf. pp. 26–27, especially Fig. 2-1). Hence the more water that flushes through a soil horizon, the more will its minerals be attacked and dissolved. On the other hand, temperature is not a major factor because it does not influence the volume of sand produced and is less important for weathering than rainfall, even though high temperatures accelerate chemical reactions, especially by their effect on biologically mediated reactions. Hence temperature is not explictly considered in Figure 5-36. Nor are the various types of source rocks listed. Although *locally* specific source rocks may indeed dominate the framework composition of a sand and produce

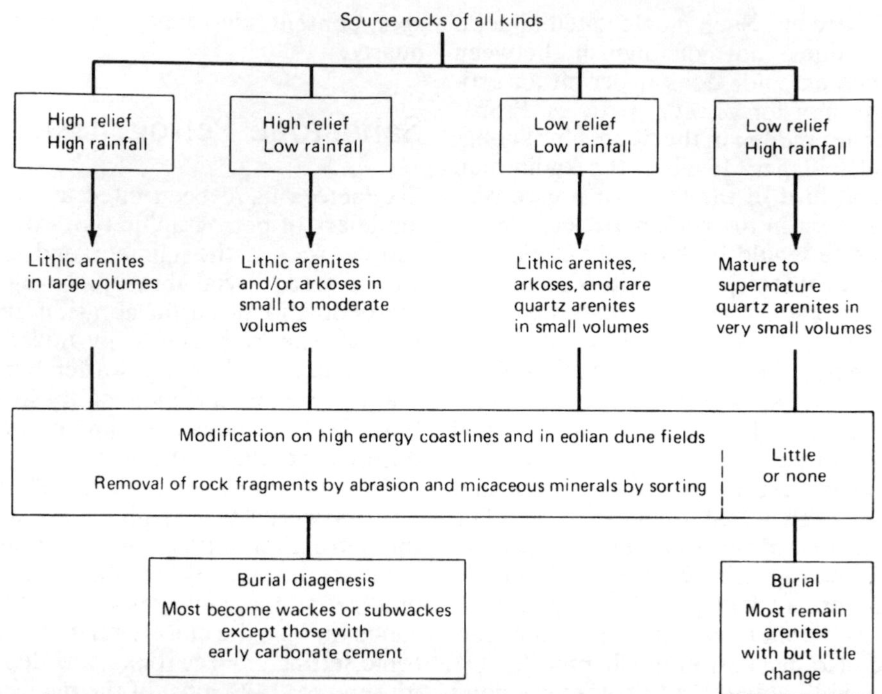

FIGURE 5-36. "Flow diagram" showing climatic and other environmental controls of sandstone petrography. (After Potter and Franzinelli, 1985 Fig. 4).

frameworks consisting of either serpentine, glaucophane, detrital carbonate, volcanic glass, pure monocrystalline quartz, greenschist grains, or mixtures of quartz and feldspar, most ancient sandstones and all modern big-river sands have a far smaller range of framework composition—a composition that results from the mixing of many sources all subject to potential modification by climate and weathering. Moreover, there appears to be little *compositional* modification by fluvial abrasion (Kuenen, 1959). More abrasion and/or better sorting occurs on beaches and probably much more in eolian dune systems and in tidal sands. We do assign a major role in sandstone petrogenesis to burial, for it is the diagenetic processes of burial that produce most wackes via the squashing, crushing, and chemical alteration of unstable rock fragments and of feldspar on the one hand, and cementation of arenites by precipitated cements on the other. Thus lithic arenites and arkoses are all the *potential precursors* of lithic and felspathic wackes—unless cemented by early calcium carbonate or unless diagenetic reactions are inhibited by the early emplacement of oil. Thus, with the exception of those sands actually deposited as pebbly mudstones, we view most of the wackes as of diagenetic origin. As pointed out by Walker *et al.* (1978), important enrichment of secondary clay may occur even in alluvial fans and in the nonmarine fill of arid intermontaine basins well prior to later burial in a thick marine basin. Framework composition can also effect the chemical cements in sandstones, especially volcanic ones (cf. Galloway, 1974, Fig. 8), although contrasts in cementing agents do not, of course, alter petrographic types based on framework composition and matrix content.

Another diagenetic factor is secondary porosity. Shanmugam (1984, p. 131–135) has discussed how secondary dissolution of grains can alter framework composition and gives estimates of its importance—in eleven petroleum reservoirs secondary porosity typically accounted for 30 to 50 percent of total porosity. Thus in an ancient sandstone diagenetic alteration and elimination of framework grains can be appreciable.

Consider now how relief and rainfall interact to produce major rock types. High relief means high potential energy which, when combined with high rainfall, yields a large volume of immature lithic arenite—because weathering in the source region is mostly physical rather than chemical. Judging by the composition of modern rivers draining mountain ranges, rock fragments overshadow feldspar by ratios of commonly about 3 to 6 times or more. On the other hand, high relief and low rainfall yield only small to moderate volumes of either lithic arenites or locally arkosic sands because of incomplete weathering and mixing. Low relief and low rainfall combine to produce small volumes of sand of highly variable types—pure quartz sands derived from the mature, preexisting sandstones veneering old cratons, local residual arkoses above granitic basements, or lithic arenites from either preexisting immature sandstones or from plateau basalts or beveled metamorphic terranes. In short, weathering in an arid climate will be minimal and framework composition will closely mimic source rocks. The combination of low relief and high rainfall yields only mature or supermature quartz arenites. Why? Because in a flat or low-lying region with high rainfall, sediment production is low, weathering is deep and intense, and streams yield only quartz, perhaps a few rare grains of potash feldspar, and some grains derived from ferricretes and laterites, if the climate is tropical. A rain forest developed on an old craton is the model and the prime example in most of the Amazon Basin (Potter and Franzinelli, 1985). Millot (1964, p. 433) perceptively called a deeply weathered, low lying tropical landscape *le paysage de tectonique null ou paysage geochimique*. Possible weathering on a tropical floodplain of immature sand derived from a distant mountain range is still another way in which climate can alter framework composition. Because the resulting sand will consist chiefly of monocrystalline quartz—polycrystalline quartz is more susceptible to weathering—little can happen to it during later transport by either abrasion or sorting. Nor is there an opportunity for a quartz arenite to become a quartz wacke during burial—there are no rock fragments or feldspar to be converted into "clay paste." Some studies that evaluate the role of climate on the composition of modern sands include those by Mann and Cavaroc (1973), Basu (1976), Potter (1978, p. 445–447), Potter (1986), and Suttner and Dutta (1986). There is also another argument to be made on behalf of climate as a major factor affecting sandstone composition—when and where it is not overwhelmed by high relief. Tropical to subtropical weathering appears to have extended to middle and even possibly high latitudes throughout much of earth history except for major periods of glaciation (Dury, 1971; Feininger, 1971). Thus perceptions of sand production based on observations from today's middle-latitude, temperate climates probably are not very applicable to most of earth history.

What then of abrasion and/or sorting in the surf zone of beaches and in eolian dune fields? Because both are high-energy environments, it would seem that weak rock fragments (soft argillites and shales, for example) might well be destroyed, as well as grains with good cleavage, such as feldspar. Kuenen's experiments on the eolian abrasion of sand demonstrated that weight loss of quartz per kilometer is 100 to 1000 times greater per kilometer by eolian than by stream abrasion (Kuenen, 1960, p. 442). In addition, hydraulic sorting on beaches and in the air is very effective in removing platy minerals such as the micas. There is some quantitative evidence from ancient sandstones for these possibilities (Davies and Ethridge, 1975; Ethridge *et al.*, 1975), although these studies included clay content and did not simply examine the mineralogy of the sand fraction itself. An early study that emphasized the role of depositional environment on sandstone type was that of Hubert (1960), who studied the Pennsylvanian Fountain and Lyons Formations where they overlap Precambrian basement in Colorado. He largely ascribed contrasting petrographic types to different environments. Later, Ojakangas (1963) also studied interbedded petrographic types in the Cambrian Lamotte Sandstone of Missouri where it lies directly above basement. He concluded that contamination by distally derived coastal sands was also a factor in producing different petrographic types (Ojakangas, 1963, p. 871). In any case, variations in petrographic type produced by contrasts in depositional environment will be maximal where different depositional environments directly overlie granitic basement or are directly derived from it (Mack, 1978) rather than where riverine detritus derived from afar is reworked along a coastline. Needed are a series of systematic paired comparisons between modern river sands and associated beaches and/or their coastal dunes. The role of tidal currents and sand abrasion needs more study (Swett *et al.*, 1971). An additional useful study would be to determine the framework composition of modern continental dune fields and follow possible mineralogical changes from upland sources downwind a long distance (cf. Fig. 10-58). Are any modern continental dune sands lithic arenites? McKee (1979, p. 191) has suggested that most ancient eolianites are quartz arenites. But again more systematic petrographic data is needed. In all such studies the initial weathering of the sand is a factor because weathered rock fragments and feldspar are more subject to differential abrasion and/or sorting than their fresh equivalents. Thus climate needs again to be considered. Another consideration for the role of depositional environment and framework modification is initial composition. Little but a change in mean size and possibly sorting can occur when a pure quartz sand is delivered by a river to either a high-energy beach or a dune field. Thus climate, degree of weathering of the detritus, and initial composition influence the degree to which a depositional environment may alter sand fraction mineralogy enough to produce a major change in petrographic type. The possibility of contamination by a different source should always be evaluated.

We recognize, of course, that the continental geography of a depositional basin—whether it was supplied by a terrane located on a leading continental margin, a trailing margin, or from an Andean versus a Himalayan mountain belt—dramatically affects framework composition (Dickinson, 1974 and 1984, Potter, 1978) so that tectonics is always a factor to consider, as we do at length in Chapter 12. Viewed in this manner, continental geography determined by plate tectonics is a broad and essential determinant of source rocks and their derivative sands. Still another way plate movement will affect sandstone type is by movement of a continent from one climatic zone to another—from the tropics to the arctic. Or simply lack of movement—possible plate stagnation—should be marked by quartz-rich sands in most of the geologic past. But in all these possibilities relief and rainfall are everywhere important modifiers of the sand supplied to a depositional basin.

Glossary of Rock Names Applied to Sands and Sandstones

The great interest in the related problems of classification and nomenclature of sandstones, which began with the 1948 issue of the *Journal of Geology* on classification of sediments, especially Krynine's paper which contained a rational classification of the sandstones, has led to a proliferation of new terms and many varied redefinitions of older ones. This proliferation is a burden to the average reader and a source of confusion and misunderstanding. We felt the need, therefore, for a glossary which contains both the new and old terms and which provides a reference to the original definition in the case of the new terms and references to the several variants in usage and redefinitions of the older terms.

The only other glossary restricted to petrographic terminology of sandstones that we know of is that prepared by V. T. Allen and published in the *Report of the Committee on Sedimentation* of the National Research Council in 1936. This has proved of great help with the older terms.

We have included every term known to us, including many which in our judgment might best be forgotten, as well as some old and now obsolete terms. We did not elect to pass judgment on the terms but felt rather that there was need for a glossary as complete as we could make it. The reader will observe, in the chapter on the petrography of sandstones, that we find need for relatively few rock names—no more than a dozen or fifteen of the one hundred or more contained in the glossary. The reader, however, will encounter many terms, good or bad, useful or not, and will need a definition.

We have included only rock names. Terms denoting textures or structures of sandstones have been omitted. These are covered to some extent in general dictionaries and, in the case of structures, in special glossaries (Pettijohn and Potter, 1964).

Where possible, we have given the reference to the first use of the term; such reference being placed immediately after the term. But inasmuch as many of the terms, such as grit, sand, brownstone, and the like were common language or trade terms appropriated by the petrologist, it is not always possible to cite an original definition.

It is probably not practical to give all the variants of many well-known terms. It would be difficult, for example, to run down and give a reference to all the size-definitions of "sand." The effort at quantification of many rock names has led to a wide variation in the definitions of these terms. Not only is there a great variation in the choice of the mineralogical parameters used in classification (whether, for example, chert should be included with quartz or put with rock particles), but there is a wide variation in the percentage limits of these components. The combinations and permutations that are possible—and that, in fact, exist—are very large. A complete tabulation would be tedious to make and probably rather cumbersome to use. Many of these variations have been shown in graphic form in Klein's 1963 review of the classification question.

The reader will note that the glossary is alphabetical. Moreover, the adjectival modifier, if there is one, such as in feldspathic graywacke, is listed first, not afterwards. This arrangement is simplest to construct; any other raises troublesome problems. For example, we believe it is easier to list and find "coral sand" or "lithic graywacke" than "sand, coral" or "graywacke, lithic."

We have restricted our glossary to terms found in the English-language literature.

ANEMOSILICARENITE (-YTE) (Grabau, 1904; 1913, p. 293) — An eolian sand of siliceous composition.

ARENITE (ARENYTE) (Grabau, 1904; 1913, p. 285) — 1. A term of Latin derivation (*arena:* sand) used to denote a consolidated or lithified sand without respect to composition. Composition may be indicated by suitable prefix, thus: calcarenite, silicarenite. Adj.: *arenaceous*.
2. A sandstone with less than ten percent matrix material, as opposed to "wacke" (Williams, Turner, and Gilbert, 1954, p. 290). Synonyms: Sandstone, psammite.

ARGILLACEOUS SANDSTONE — The term "argillaceous sandstone" may be loosely applied to impure sandstones containing an indefinite amount of fine silt and clay (Williams, Turner, and Gilbert, 1954, p. 290); considered by some as a synonym for "wacke." Used synonymously with the field term "dirty" sandstone.

ARKOSE (Brongiart, 1826) — 1. A sandstone, formed principally by mechanical disintegration of a granitic rock, consisting of coarse grains of quartz and feldspar (Brongiart).
2. Any feldspathic sandstone.
3. A highly feldspathic sandstone with 30 percent or more feldspar (Krynine, 1940, p. 50).
4. An arenite or sandstone consisting of 80 percent or more quartz and feldspar, feldspar (and rock particles) over 25 percent, and feldspar exceeding rock particles (Pettijohn, 1957, p. 332).
5. A general term including (a) "arkosic arenites," clean sands with 25 percent or more feldspar, and (b) "arkosic wackes," sands with over 10 percent matrix and 25 percent or more feldspar (Williams, Turner, and Gilbert, 1954, p. 294).

ARKOSIC ARENITE (Williams, Turner, and Gilbert, 1954, p. 294, Fig. 97) — A matrix-free (under 10 percent) arenite with feldspar exceeding rock fragments; an arkose.

ARKOSIC GRAYWACKE (Williams, Turner, and Gilbert, 1954, p. 294, 314) — A graywacke with abundance of feldspar; a hard, dark-colored arkosic wacke.

ARKOSIC SANDSTONE — An arkose; a feldspathic sandstone.

ARKOSIC WACKE (Williams, Turner, and Gilbert, 1954, p. 291) — A wacke containing more feldspar grains than rock fragments; a feldspathic graywacke.

ARKOSITE (Tieje, 1921, p. 655) — 1. Well-cemented arkose but without interlocked grains (see arkositite).
2. An arkose which is the lithified equivalent of a quartzite (Grout, 1932, p. 367).
Synonyms: Quartzitic arkose; arkosic quartzite.

ARKOSITITE (Tieje, 1921, p. 655) — An arkose cemented *with* grains interlocked; bears same relation to arkosite as quartz sandstone does to quartzite.

ARTICULITE — According to Holmes (1920, p. 36) the term articulite was used by Wetherell in 1867 for flexible sandstone. It does not seem to be in general use.

ASPHALTIC SAND — A natural mixture of asphalt with varying proportions of sand, a tar sand; a bituminous sand.

ATMOSILICARENITE (-YTE) (Grabau, 1904; 1913, p. 296)	A siliceous sand resulting from atmospheric action (weathering) leading to disintegration of parent rock; example: gruss. Results from passive action of atmosphere in contrast to atmosphere in motion: see anemosilicarenite.
AUTOARENITE (-YTE) (Grabau, 1904; 1913, p. 296)	A sand produced by crushing due to earth movements or tectonic pressures; the sand-size equivalent of an autoclastic breccia.
BASAL ARKOSE	An arkosic sandstone basal to a sedimentary sequence resting unconformably on a granitic terrane; the arkosic equivalent of a granitic basal conglomerate. May grade downward into sedentary or residual arkose.
BIOARENITE (-YTE) (Grabau, 1904; 1913, p. 296)	A sand produced by the activity of organisms, including man, for example, roofing granules prepared by crushing of quartzite, greenstones, and other rocks.
BITUMINOUS SAND	A sand-asphalt mixture. Synonyms: Asphaltic sand; tar sand.
BLACK SAND	Usually refers to a magnetite or ilmenite sand; a common placerlike accumulation especially on beaches very much enriched in magnetite or ilmenite.
BROWNSTONE	A ferruginous quartz sandstone in which the grains are generally coated with iron oxide (Grabau, 1920, p. 579). The term was once widely used for the reddish-brown sandstone from the Triassic of the eastern United States, especially the Portland stone of Connecticut, once extensively quarried as a building stone. It has been little used for other sandstones.
BURRSTONE (also BUHRSTONE)	Given as "Burrh-stone" in Humble (1840, p. 35); also "burystone." A term used for a porous siliceous sandstone of angular grain suitable for millstones. Many geologists consider that the term has no special geological significance and should be discarded (Allen, 1936, p. 24).
CALCARENACEOUS ORTHOQUARTZITE (Pettijohn, 1957, p. 405)	A term suggested for a sandstone consisting of subequal proportions of *detrital* carbonate and quartz.
CALCARENITE (-YTE) (Grabau, 1904; 1913, p. 290)	A term for a sandstone composed of grains of calcium carbonate with or without carbonate cement. Commonly considered to be a limestone and commonly held to be intrabasinal in origin (composed of oolites, skeletal debris, and intraclasts) rather than terrestrial. A terrestrial sand derived from preexisting carbonate rocks is *calclithite*.
CALCLITHITE (Folk, 1959, p. 36)	A terrigenous sandstone containing over 50 percent carbonate particles derived by weathering and erosion of preexisting limestones and dolomites. The 50 percent figure is probably not generally adhered to. Twenty five percent of rock particles defines a lithic arenite; of these 50 percent or more must be terrigenous carbonate detritus.
CARBONACEOUS SAND	Any sandstone that contains an appreciable amount of *detrital* carbonaceous or woody particles.
CHERT ARENITE (Folk, 1968, p. 125)	A lithic arenite in which chert particles are the dominant rock-particle constituent.
CLASMOSCHIST	A term attributed by Roberts (1839, p. 71) to Coneybeare to be used in place of graywacke. Obsolete.
CORAL SAND	A calcareous sand formed by break-down of reef rock. Coral detritus forms a significant but not necessarily dominant part of this debris.
CRYSTAL SANDSTONE	1. A sandstone in which calcite has been deposited in the pores in large patches or units having a single crystallographic orientation resulting in a "poikiloblastic" or "lustermottling" effect. In some rare sandstones with incomplete cementation the carbonate occurs as sand-filled scalenohedra of calcite—"sand crystals." 2. Term more commonly applied to sandstones in which the quartz

grains have been enlarged by deposition of silica so that the grains show regenerated crystal facets and in some cases nearly perfect quartz euhedra. Crystal sandstones of this nature sparkle in bright sunlight. Sometimes called "sparkling sandstone."

FELDSPATHIC ARENITE (Williams, Turner, and Gilbert, 1954, p. 316)	A feldspathic sandstone, with 10 to 25 percent detrital feldspar, usually also containing a smaller proportion of rock particles.
FELDSPATHIC GRAYWACKE	1. A graywacke containing appreciable feldspar. 2. A graywacke in which feldspar exceeds rock particles (Pettijohn, 1954, p. 363). Synonyms: "High-rank graywacke" (Krynine, 1945); "feldspathic wacke" and "arkosic wacke" are arkoses with a significant proportion of matrix (Williams, Turner, and Gilbert, 1954, p. 292), not graywackes.
FELDSPATHIC LITHARENITE (McBride, 1963, p. 667)	A lithic arenite containing appreciable (over 10 percent) feldspar.
FELDSPATHIC LITHWACKE (Casshyap, 1967)	Essentially a lithic graywacke (over 15 percent matrix) in which rock fragments exceed feldspar but the latter forms 10 percent or more of the sand fraction.
FELDSPATHIC POLY-LITHARENITE (Folk, 1968, p. 135)	A polylitharenite containing appreciable (over 10 percent) feldspar.
FELDSPATHIC SANDSTONE	1. Any sandstone containing appreciable feldspar; would include arkose, some graywackes, etc. 2. A sandstone with 10 to 25 percent feldspar but without appreciable matrix (Pettijohn, 1949, Table 55). Synonym: Subarkose.
FELDSPATHIC SUBLITH-ARENITE (McBride, 1963, p. 667)	A lithic subarkose.
FELDSPATHIC WACKE (Williams, Turner, and Gilbert, 1954, p. 292)	A wacke with 10 to 25 percent feldspar (a wacke with over 25 percent feldspar is an "arkosic wacke").
FLAGSTONE	A name applied to thin-bedded sandstones capable of being split or parted to form flags suitable for paving. A calcareous sandstone which splits readily along micaceous layers (Tyrrell, 1926, p. 210). Thin-bedded argillaceous sandstone used chiefly for paving (Ries, 1910, p. 117).
FLEXIBLE SANDSTONE	See *Itacolumite*.
FLYSCH SANDSTONE	A sandstone characteristic of the flysch facies; generally a turbidite, commonly graywacke in the older systems.
FRANGITE (from Latin: *frango*—to break up; Bastin, 1909, p. 450)	All sedimentary rocks formed by the disintegration but without decomposition of igneous rocks, and without extensive mechanical sorting; includes arkoses, graywackes, grits and their metamorphic derivatives.
FREESTONE	1. Any kind of stone, the texture of which is so free or loose that it may be easily worked (Humble, 1840, p. 100). 2. Any stone, such as sandstone, which can be freely worked or quarried, especially one that cuts well in all directions. 3. A uniform, thick-bedded sandstone with few divisional planes; rarely a similar limestone (Tyrrell, 1926, p. 210).
GANISTER (Also gannister)	A hard, compact, highly siliceous sedimentary rock, with fine, uniform granular texture sand composed essentially of angular quartz grains cemented with secondary silica. A fine-grained quartzose sandstone consisting of angular grains, cemented with silica (Williams, Turner, and Gilbert, 1954, p. 283; see also Thomas, 1918, p. 3). Originally a local name for a variety of sandstone found in the Yorkshire and Derbyshire coal fields (Roberts, 1839, p. 66).

GLAUCONARENITE (-YTE) (Grabau, 1904, p. 245)	A glauconitic sand.
GLAUCONITIC SANDSTONE	See *Greensand.*
GRANITE WASH	1. A term loosely applied to tongues of poorly sorted, little-rounded arkose intercalated with ordinary sediments derived from nearby buried hills of granite. 2. May also be applied to modern alluvium derived locally from exposed granite hills.
GRANULITE (-YTE) (Grabau, 1904; 1913, p. 283)	A sand-sized deposit of constructional but nonclastic origin (ex. oolitic sand); corresponds to the term *arenite,* a sand of clastic origin (Commonly now used for a large number of coarse-grained metamorphic rocks such as "pyroxene granulite").
GRAYWACKE (also GREYWACKE) (Lasius, 1789)	Term was originally applied to the dark, tough Paleozoic Kulm sandstones in the Harz Mountains of Germany by Lasius. From the German "grauewacke." 1. Defined by Geike (1885, p. 162) as "a compact aggregate of rounded or subangular grains of quartz, feldspar, slate, or other minerals or rocks cemented by a paste . . . gray, as its name denotes, is the prevailing color. . . . The rock is distinguished from ordinary sandstone by its darker hue, its hardness, the variety of its component grains, and above all, by the compact cement in which the grains are embedded." Various attempts have been made to redefine the term more precisely. These have involved (1) the proportion of matrix ("paste") and (2) the proportions of the labile components (feldspars and rock particles). Pettijohn (1954), for example, required 15 or more percent matrix and 25 or more percent labile constituents. 2. Other workers have departed from the classical definition and expanded it to emcompass sandstones differing materially from the type graywacke. Some ignore the matrix; others exclude feldspar. For example: Krynine (1940, p. 51) defined graywacke simply as "a clastic rock containing a substantial amount (20 percent and over) of dark rock fragments or dark-colored ferromagnesian minerals." Later he (1945) stressed only the rock particles—primarily "chert, slate, schists, phyllites, etc.", that is, largely low-grade metapelites. This usage was followed by Folk (1954). Graywacke as thus defined is essentially a lithic arenite (var. phyllarenite). This is also the usage of Shutov (1967, Fig. 6). Folk (1968, p. 125) has dropped the term "graywacke" and replaced it by "lithic arenite." 3. Graywacke has also been defined as "a variety of sandstone composed of material derived from the disintegration of basic igneous rocks of granular texture. . . . Thus defined it is the ferromagnesian equivalent of arkose" (Twenhofel, 1932, p. 231). This usage has gained little or no support. For a review of the problem of defining graywacke see Dott (1964).
GREENSAND	Sands containing appreciable glauconite are referred to as green sands, or greensand. They are usually mixtures of quartz and glauconite in all possible proportions. Synonyms: Glauconarenite, glauconitic sandstone.
GRIT	First used as a provincial term for a coarse-grained sand or sandstone (Humble, 1840, p. 35), for example: Millstone Grit. Later sharpness of grain became a part of the definition (Holmes, 1920, p. 50).
GRUSS (also GRUS)	The fragmental products of *in situ* disintegration (with little or no decomposition) of granite. Synonyms: Residual arkose, sedentary arkose. Also used for *any* rock that is finely granulated but not decomposed by weathering.
HIGH-RANK GRAYWACKE (Krynine, 1945)	Essentially a lithic arenite or sandstone with appreciable detrital feldspar; may also be a classical graywacke rich in feldspar. Synonym:

"Feldspathic graywacke" (Folk, 1954); "Feldspathic litharenite" (Folk, 1968, p. 124).

HYDRARENITE (-YTE) (Grabau, 1904; 1913, p. 294) — A water-laid sandstone of variable composition.

HYDROSILICARENITE (-YTE) (Grabau, 1904; 1913, p. 294) — A water-laid siliceous sandstone, that is, one consisting of siliceous detritus (quartz, feldspar, etc.).

ITACOLUMITE — A peculiar quartz schist named for Itacolumi, a mountain in Brazil, composed of interlocking quartz grains with some mica, which exhibits some flexibility (Page, 1865, p. 204). Also known in North Carolina and elsewhere. Is considered by some to be a metamorphic rock rather than a sandstone. A term attributed to Humboldt by Holmes (1928, p. 126).
Synonym: Flexible sandstone.

LITHARENITE (McBride, 1963) — A contraction of the term "lithic arenite." Defined as a sandstone with over 25 percent rock particles and less than 10 percent feldspar.

LITHIC ARENITE (Williams, Turner, and Gilbert, 1954, p. 304) — Sandstones, with less than 10 percent matrix in which rock particles (unstable, fine-grained) are an important constituent.
Synonym: "Low-rank graywacke" of Krynine and "lithic sandstone" of Pettijohn.

LITHIC ARKOSE (Folk, 1968, p. 124; McBride, 1963) — An arkose with a feldspar/rock fragment ratio between 1 : 1 and 3 : 1 (Folk, p. 124); an arkose with over 10 percent rock fragments (McBride, 1963, Fig. 1).

LITHIC ARKOSIC WACKE (Casshyap, 1967) — A graywacke in which feldspar exceeds rock particles.
Synonym: "Feldspathic graywacke" (Pettijohn, 1954).

LITHIC GRAYWACKE (Pettijohn, 1954) — A graywacke in which rock particles exceed feldspar.

LITHIC SANDSTONE (Pettijohn, 1954) — 1. A sandstone, with less than 15 percent matrix, and 5 percent or more feldspar and rock particles but with rock particles exceeding feldspar. Subdivided into subgraywackes and protoquartzites.
2. Now (1970) used interchangeably with lithic arenite and defined as having 25 or more percent labile constituents but with rock particles exceeding feldspar.
Synonym: Lithic arenite of Gilbert (Williams, Turner, and Gilbert, 1954, p. 304).

LITHIC SUBARKOSE (McBride, 1963) — A sandstone or arenite composed of abundant subequal amounts of detrital feldspar and rock fragments (more than 10 percent but less than 25 percent). The term "lithic subarkose" is preferred to the more cumbersome but equally appropriate "feldspathic sublitharenite."

LITHIC SUBARKOSIC WACKE (Casshyap, 1967) — A wacke with subequal proportion of feldspar and rock fragments but no more than 25 percent of either.

LITHIC WACKE (Williams, Turner, and Gilbert, 1954, p. 301) — A sandstone with 10 percent or more of matrix and a significant proportion of detrital rock particles—rock particles exceeding feldspar; "lithic graywacke" of Gilbert is a lithic wacke characterized by great hardness and gray color.
Synonym: Essentially same, in most cases, as lithic graywacke of Pettijohn (1954).

LOW-RANK GRAYWACKE (Krynine, 1945) — A lithic arenite in most cases; also a graywacke as defined in the classical sense with little or no feldspar.

MENGWACKE (Fischer, 1933, p. 336) — A wacke with 33–90 percent unstable mineral constituents.

METAQUARTZITE (Krynine, 1945) — A quartzite of metamorphic origin in contrast to orthoquartzite which is a primary sedimentary quartzite.
Synonym: Paraquartzite (Tieje, 1921).

MICROBRECCIA — Ill-sorted sandy rocks in which the grains are sharply angular (Williams, Turner, and Gilbert, 1954, p. 283).

MOLASSE SANDSTONE (Cayeux, 1929, p. 164)	A sandstone of the molasse facies; characterized by Cayeux as poorly rounded, poorly sorted, coarse sand rich in rock fragments and generally calcareous. Probably generally a lithic arenite, in places arkosic. The product formed by the demolition of a newly elevated orogenic belt.
OIL SAND	Oil-saturated sand.
ORTHOARENITE (Marchese and Garrasino, 1969, p. 283)	An arenite with detrital matrix under 15 percent.
ORTHOQUARTZITE (Tieje, 1921, p. 655)	1. A sandstone converted to a quartzite with interlocking grains "cemented only through infiltration and pressure" as opposed to "paraquartzite," a quartzite originating mainly through contact metamorphism (Tieje, 1921, p. 655). 2. A term revived by Krynine (1941, 1945, 1948) for sandstones consisting almost exclusively of detrital quartz. Although commonly cemented by quartz and truly "quartzitic" in character, most contain some carbonate cement and as the proportion of this nonsiliceous material increases, the orthoquartzites become less cohesive. The term has, by usage, been extended to all sandstones, indurated or friable, with 95 or more percent detrital quartz (Pettijohn, 1957, p. 295). Synonyms: *Quartz arenite, quartzitic sandstone* (Krynine, 1940, p. 51) a sedimentary quartzite, *quartzose sandstone* (Krynine, 1940, p. 51) 95 percent or more quartz but *not* cemented by silica.
PARAQUARTZITE (Tieje, 1921, p. 655)	A truly metamorphic quartzite derived from a sandstone by action of heat and pressure. Synonym: Metaquartzite.
PARARENITE (Marchese and Garrasino, 1969, p. 283)	An arenite with detrital matrix between 15 and 70 percent; prefix "para-" may be used with other terms, such as *paralithite*. We would suggest that the prefix be used for any matrix-rich sandstones whether the matrix is detrital or diagenetic.
PHYLLARENITE (Folk, 1968, p. 131)	Essentially a lithic arenite characterized by an abundance of detrital rock particles of low-grade metamorphic pelitic rocks: slate, phyllite, mica schist. Synonym: Most so-called "low-rank graywackes" of Krynine and Folk.
PLAGIOCLASE ARKOSE (Folk, 1968, p. 130)	An arkose, the chief feldspar of which is plagioclase.
POLYLITHARENITE (Folk, 1968, p. 135)	A lithic arenite with a diversity of sand-sized rock particles—volcanic, sedimentary, and metamorphic.
PROTOQUARTZITE (Krynine, 1952)	A term first used by Krynine (Payne and others, 1952) but never really defined. A member of the series "low-rank graywacke-quartzose graywacke-protoquartzite-orthoquartzite" and hence by implication a sandstone with an appreciable but not excessive quantity of detrital rock particles and little or no feldspar. Defined by Pettijohn (1954) as a sandstone with 5 to 25 percent rock particles, a variety of lithic sandstone. Detrital feldspar less than detrital rock particles. Synonym: Sublitharenite (McBride, 1963).
PSAMMITE (also PSAMMYTE)	From the Greek *psammos:* sand; adj. *psammitic*. 1. Formerly used in Europe for fine-grained clayey sandstone in which the component grains are scarcely distinguishable with naked eye (Oldham, 1879, p. 44). 2. Generally used to denote sandstones without reference to composition. Synonym: Arenite, sandstone. Note: Tyrrell (1921) would apply "psammitic" only to metamorphic rocks; the Latin term "arenaceous" would be applied to sedimentary rocks; the Greek terms would be used when the rocks were hardened and altered beyond the limits implied by the Latin terms.

QUARTZ ARENITE (Williams, Turner, and Gilbert, 1954, p. 292) — Sandstones consisting essentially of quartz without appreciable matrix (under 10 percent), containing no more than 10 percent of either feldspar or rock particles.
Most authors would restrict allowable "contaminants" to 5 percent or less, or to 5 percent or less of either feldspar or rock particles.
Synonym: Quartzose sandstone, orthoquartzite, quartzarenite.

QUARTZ GRAYWACKE (Williams, Turner, and Gilbert, 1954, p. 294) — An exceptionally quartz-rich graywacke; quartz wacke.

QUARTZARENITE (McBride, 1963) — A contraction of quartz arenite.

QUARTZ-FREE WACKE (Quarz-frei wacke of Fischer, 1933) — A wacke with over 90 percent unstable mineral constituents.

QUARTZITE — The term "quartzite" is commonly used for a metamorphic rock produced by recrystallization of a sandstone. It has been defined by Holmes (1920, p. 194) as "a granulose metamorphic rock, representing a recrystallized sandstone, consisting predominantly of quartz." The term "quartzite" has come to include also sedimentary quartzites (orthoquartzites) or sandstones cemented with silica which has grown in optical continuity around each detrital quartz grain. The term "metaquartzite" or "paraquartzite" is reserved for the truly metamorphic quartzite.
Quartzites have been defined as rocks which fracture through rather than around the constituent grains.

QUARTZITIC GRIT — A grit with properties of a quartzite.

QUARTZITIC SANDSTONE — A sedimentary quartzite (Krynine, 1940, p. 51); a sandstone approaching the character of a quartzite; an orthoquartzite.

QUARTZOSE SANDSTONE — 1. A sandstone of 95 or more percent detrital quartz but not cemented with silica as opposed to quartzitic sandstone so cemented (Krynine, 1940, p. 51).
2. A sandstone of 95 or more percent detrital quartz, essentially synonymous with *quartz arenite* or *orthoquartzite*.

QUARZMENGWACKE (Fischer, 1933, p. 336) — A wacke with 10 to 33 percent unstable mineral constituents.

QUARZWACKE (Fischer, 1933, p. 336) — 1. A wacke with less than 10 percent unstable mineral constituents.
2. Redefined as a sandstone with 10 percent or more matrix and no more than 10 percent each detrital feldspar and rock particles (Williams, Turner, and Gilbert, 1954, p. 292).
3. Used by Krumbein and Sloss (1963, p. 173) as synonym for subgraywacke (Pettijohn, 1949, p. 255).

QUICKSAND — Medium to fine-grained sand containing much water which yields readily to pressure or weight and hence apt to engulf persons and animals coming upon it. Capable of injection into fissures.

REDSTONE (Krynine, 1950, p. 103) — A brick-red, clayey sandstone consisting of angular quartz, mica, and feldspar grains embedded in a matrix of red hematitic clay. Names for Redstone Hill, southwest of Southington, Connecticut, U.S.A. A ferruginous arkosic wacke.

RESIDUAL ARKOSE — An arkose formed *in situ* by disintegration of a granite; an untransported arkose.
Also called *sedentary arkose*, commonly grading into the underlying granite (Barton, 1916, p. 447); related to gruss.

SAND — 1. Noncohesive granular material of specified size (commonly 1/16 to 2 mm in diameter). May be of organic, chemical, volcanic, or clastic origin and of widely varying composition such as $CaCO_3$, SiO_2, $CaSO_4 \cdot 2H_2O$, etc.
2. *Detrital* material of specified size range.
3. A term applied to a siliceous detrital deposit composed mainly of quartz particles.

4. Material having a terminal fall velocity less than the upward eddy currents and an upper limit such that a grain resting on the surface ceases to be movable either by direct pressure of the fluid or by impact of other moving grains (Bagnold, 1941, p. 6).
5. A drilling term for an oil-bearing horizon.
Many definitions emphasize "water-worn," "fragments," "clastic," "siliceous," and "detrital." But many sands are not water-worn, for example, gruss, nor fragmental, for example, oolitic sand, nor siliceous, for example, gypsum, nor detrital, for example, coral sand.

SAND ROCK (Tieje, 1921, p. 655)	A weakly or poorly cemented sand may be called *sand rock*. Also rural terminology in Appalachia.
SANDSTONE	Defined by Lyell (1833, p. 79) as "any stone which is composed of an agglutionation of grains of sand, whether calcareous, siliceous, or any other mineral nature." In practice, only consolidated sands which are dominantly siliceous are designated sandstone. A consolidated calcareous sand would be called limestone.
SCHIST ARENITE	A term attributed to Adolph Knopf (Krynine, 1937, p. 427) for sandstones with a significant proportion of metamorphic rock particles (schists and phyllites): a particular kind of lithic arenite.
SCHIST WACKE (Williams, Turner, and Gilbert, 1954, p. 292)	A lithic wacke characterized by an abundance of metapelitic rock particles (slate, phyllite, schist).
SEDARENITE (Folk, 1968, fig. on p. 124)	A lithic arenite, the rock particles of which are of sedimentary origin in contrast to volcanic arenite and phyllarenite (of metamorphic rock particles).
SILICARENITE (-YTE) (Grabau, 1904; 1913, p. 290)	A purely siliceous sand or arenite such as the St. Peter Sandstone.
SILICEOUS SANDSTONE	1. A sandstone cemented with silica; a hard quartzitic sandstone. 2. A sandstone, the detrital components of which are siliceous (quartz, feldspar, etc.).
SILICINATE QUARTZOSE SANDSTONE (Allen, 1936, p. 40)	The adjective "silicinate" (also "calcarinate" and "ferruginate") is used to denote the composition of the *cement* of the sandstone.
SPARAGMITE (Blaas, 1898, p. 18)	A term applied to coarse Eocambrian arkoses in Norway and Sweden.
SUBARKOSE (Pettijohn, 1954)	A sandstone with 5 to 25 percent labile components of which feldspar exceeds rock particles; a feldspathic sandstone with less feldspar than that of a normal arkose.
SUBARKOSE WACKE (Casshyap, 1967)	Essentially a wacke (over 15 percent matrix) with 5 to 25 percent feldspar: a species of feldspathic graywacke. Synonym: Feldspathic wacke (Williams, Turner, and Gilbert, 1954, p. 292).
SUBFELDSPATHIC LITHIC ARENITE (Williams, Turner, and Gilbert, 1954, Fig. 97)	An arenite with 10 or less percent feldspar and a larger quantity of rock fragments.
SUBFELDSPATHIC LITHIC WACKE (Williams, Turner, and Gilbert, 1954, Fig. 96)	A lithic wacke containing less than 10 percent feldspar. A species of feldspathic graywacke. Synonym: Subarkose wacke (Casshyap, 1967).
SUBGRAYWACKE (Pettijohn, 1949, p. 255)	1. A sandstone (wacke) with over 20 percent matrix and less than 10 percent feldspar (Pettijohn, 1949, p. 255); essentially a quartzwacke. 2. A sandstone, with less than 15 percent matrix and over 25 percent labile grains in which rock particles exceed feldspar; essentially a lithic arenite (Pettijohn, 1954). Subgraywacke has a superficial resemblance to a graywacke especially in color and rock particle content, but as defined under No. 2 above, it is without matrix.

SUBLITHARENITE (McBride, 1963, p. 667)	A rock analogous to subarkose but containing rock fragments instead of feldspar; with 5 to 25 percent rock fragments and 0 to 10 percent feldspar, and 65 to 95 percent quartz. Synonym: Protoquartzite.
SUBLITHWACKE (Casshyap, 1967)	A wacke with 5 to 25 percent detrital rock particles; a sublitharenite with over 15 percent matrix.
SUBPHYLLARENITE (Folk, 1968, p. 132, 134)	A sublitharenite with metapelitic rock particles.
TUFFACEOUS SANDSTONE	A sandstone composed of or containing an appreciable proportion of material of pyroclastic origin.
VOLCANIC ARENITE	1. A lithic arenite consisting of terrigenous volcanic detritus of epiclastic origin. 2. An arenite of pyroclastic debris, including rock particles, crystal debris, and glass fragments: a tuff. Synonym: Volcanic sandstone.
VOLCANIC SANDSTONE	See *volcanic arenite* above.
VOLCANIC WACKE (Williams, Turner, and Gilbert, 1954, p. 303)	A volcanic sand with a matrix of fine volcanic debris; a species of graywacke of strictly volcanic origin.
VOLCANICLASTIC SANDSTONE (Fisher, 1961)	A sandstone consisting of either (a) pyroclastic debris or (b) terrigenous volcanic detritus of epiclastic origin.
WACKE	1. An unsorted mixture of sand, silt, and clay; a "loam." 2. Rocks in which the clastic grains are approximately evenly divided between the several size grades (Fischer, 1933). 3. A sandstone with 10 or more percent argillaceous matrix materials (Williams, Turner, and Gilbert, 1954, p. 290). 4. A dirty green to brownish-black clay arising as end-product of *in situ* decomposition of basalt (Geike, 1882, p. 161).

References to Glossary

Allen, V.T.: Terminology of medium-grained sediments. Rept. Comm. Sedimentation Natl. Research Council 1935-36, 18–47 (1936).

Bagnold, R.A.: The physics of blown sand and desert dunes. 265 pp. London: Methuen 1941.

Barton, D.C.: The geological significance and classification of arkose deposits. Jour. Geology 24, 417–449 (1916).

Bastin, E.S.: Chemical composition as a criterion in identifying metamorphosed sediments. Jour. Geology 17, 445–472 (1909).

* Blaas, J.: Katechismus der Petrographie, 2nd ed., 242 pp. Leipzig: J. J. Weber 1898.

* Brongiart, A.: L'arkose, caractères minéralogiques et histoire géognostique de cette roche. Paris: Ann. Sci. Naturelles 8, 113–163 (1826).

Casshyap, S.M.: On the classification of argillaceous sandstone. Ann. Geol. Dept., Aligarh Muslim Univ. 3, 48–50 (1967).

Cayeux, L.: Les roches sédimentaries de France—roches siliceuses. 774 pp. Paris: Impr. Nationale 1929.

Dott, R.L., Jr.: Wacke, graywacke and matrix—What approach to immature sandstone classification? Jour. Sed. Petrology 34, 625–632 (1964).

Fischer, Georg: Die Petrographie der Grauwacken. Jahr. preuss. geol. Landesanstalt 54, 320–343 (1933).

Fisher, R.V.: Proposed classification of volcaniclastic sediments and rocks. Geol. Soc. America. Bull. 72, 1409–1414 (1961).

Folk, R.L.: The distinction between grain size and mineral composition in sedimentary-rock nomenclature. Jour. Geology 62, 344–359 (1954).

— Practical petrographic classification of limestones. Am. Assoc. Petroleum Geologists Bull. 43, 1–38 (1959).

— Petrology of sedimentary rocks. 170 pp. Austin, Texas: Hemphill's, 1968.

Geike, A.: Textbook of geology, 971 pp. London: Macmillan 1882.

Grabau, A.W.: On the classification of sedimentary rocks. Am. Geologist 33, 228–247 (1904).

Grabau, A.W. Principles of stratigraphy, 1185 pp. New York: A. G. Seiler and Co. Reprinted 1960 by Dover Publications, New York. The paperback reprint has the same paging as the 1913 edition.

Grabau, A.W. Textbook of geology, Part I., 864 pp. Boston: Heath 1920.

Grout, F.F.: Petrography and petrology, 522 pp. New York: McGraw Hill 1932.

Holmes, A.: The nomenclature of petrology, 2nd ed. 284 pp. London: Murby and Co. 1928.

* Humble, William: Dictionary of geology and mineralogy, 279 pp. London: H. Washbourne 1840.

* Original reference not seen by authors.

Krumbein, W.C., and Sloss, L.L.: Stratigraphy and sedimentation. 2nd ed. 660 pp. San Francisco: Freeman 1963.

Krynine, P.D.: Petrography and genesis of the Siwalik series. Am. Jour. Sci., ser. 5., 34, 422–446 (1937).

Krynine, P.D. Petrology and genesis of the Third Bradford Sand. Pennsylvania State College Bull. 29, 134 pp. (1940).

Krynine, P.D. Paleogeographic and tectonic significance of sedimentary quartzites (abs.). Geol. Soc. America Bull. 52, 1915 (1941).

Krynine, P.D. Sediments and the search for oil. Producers Monthly 9, 12–22 (1945).

Krynine, P.D.: The megascopic study and field classification of sedimentary rocks. Jour. Geology 56, 130–165 (1948).

* Lasius, G.S. Otto: Beobachtungen über das Harzgebirge mit Karte. Hannover 1789.

Lyell, Charles: Principles of geology, 3, appendix and glossary, 398 pp. London: Murray 1833.

Marchese, H.G., Garrasino, C.A.F.: Clasificación descriptiva de areniscas. Rev. Assoc. Geol. Argentina 24, 281–286 (1969).

McBride, E.F.: A classification of common sandstones. Jour. Sed. Petrology 33, 664–669 (1963).

* Oldham, Thomas: Geological glossary, for the use of students, 62 pp. London: Edward Stanford 1879.

Page, David: Handbook of geological terms, geology and physical geography, 2nd ed., 506 pp. Edinburgh-London: Blackwood 1865.

Payne, T.G., and others: The arctic slope of Alaska. U.S. Geol. Survey, Oil and Gas Investig. Map OM 126, sheet 2, 1952.

Pettijohn, F.J.: Sedimentary rocks, 1st ed., 526 pp. New York: Harper 1949.

Pettijohn, F.J. Classification of sandstones. Jour. Geology 62, 360–365 (1954).

Pettijohn, F.J. Sedimentary rocks, 2nd ed., 718 pp. New York: Harper 1957.

Ries, Heinrich: Economic geology, 589 pp. New York: John Wiley and Sons 1910.

* Roberts, George: Etymological and explanatory dictionary of the terms and language of geology, 139 pp. London: Longmans 1839.

Shutov, V.D. Klassifikatsiia peschanikov (Classification of sandstones). Litologiya i Poieznye Iskopalmye 5, 86–103 (1967).

Thomas, H.H.: Refractory materials. Mem. Geol. Survey. Spec. Rept., Min. Resources Great Britain 6, 241 p. (1918) also 16, 159 p. (1920).

Tieje, A.J.: Description and naming of sedimentary rocks. Jour. Geology 29, 650–666 (1921).

Twenhofel, W.H.: Treatise on sedimentation, 2nd ed., 926 pp. Baltimore: Williams and Wilkens 1932.

Tyrrell, G.W. The principles of petrology. 349 pp. New York: Dutton 1926.

* Original reference not seen by authors.

Williams, H.; Turner, F.J.; and Gilbert, C.M.: Petrology. 406 pp. San Francisco: Freeman 1954.

References

Adams, R.W.: Loyalhanna Limestone—cross-bedding and provenance. In: Fisher, G.W.; Pettijohn, F.J.; Reed, J.C., Jr.; and Weaver, K.N. (Eds.): Studies of Appalachian geology—central and southern, pp. 83–100. New York: Interscience 1970.

Allen, J.R.L.: Petrology, origin, and deposition of the highest Lower Old Red Sandstone of Shropshire, England, Jour. Sed. Petrology 32, 657–697 (1962).

Allen, Percival: Wealden petrology: The Top Ashdown Pebble Bed and the Top Ashdown Sandstone. Geol. Soc. London Quart. Jour. 104, 257–321 (1949).

Allen, V.T.: Terminology of medium-grained sediments. Rept. Comm. Sedimentation 1935–1936. Natl. Research Council, pp. 18–47 (1936).

Allen, V.T.: A study of Missouri glauconite. Am. Mineralogist 22, 1180–1183 (1937).

van Andel, Tj.H.: Origin and classification of Cretaceous Paleocene and Eocene sandstones of western Venezuela. Am. Assoc. Petroleum Geologists Bull. 42, 734–763 (1958).

Anhaeusser, C.R.; Mason, Robert; Viljoen, M.J.; and Viljoen, R.P.: A reappraisal of some aspects of Precambrian shield geology. Geol. Soc. America Bull. 80, 2175–2200 (1969).

Auden, J.B.: Vindhyan sedimentation in the Son Valley, Mirzapur District. India Geol. Survey Mem. 62, pt. 2, 250 pp. (1933).

Bailey, E.B.: Sedimentation in relation to tectonics. Geol. Soc. America Bull. 47, 1716–1718 (1936).

Bailey, E.H., and Irwin, W.P.: K-feldspar content of Jurassic and Cretaceous graywackes of northern Coast Ranges and Sacramento Valley California. Am. Assoc. Petroleum Geologists Bull. 43, 2797–2809 (1959).

Balk, R.: The structure of graywacke areas and Taconic Range, east of Troy, New York. Geol. Soc. America Bull. 64, 811–864 (1953).

Barth, T.F.W.: Progressive metamorphism of sparagmite rocks of southern Norway. Norsk geol. tidsskr. 18, 54–65 (1938).

Barton, D.C.: The geological significance and genetic classification of arkose deposits. Jour. Geology 24, 417–449 (1916).

Basu, A.: Petrology of Holocene fluvial sand derived from plutonic source rocks: Implications to paleoclimatic interpretation. Jour. Sed. Petrology 46, 694–709 (1976).

Bathurst, R.G.C.: Carbonate sediments and their diagenesis, 620 pp.(Developments in sedimentology, Vol. 12). Amsterdam: Elsevier (1971).

Billings, M.P.: The geology of New Hampshire. New Hampshire State Planning and Development Commission, 203 pp. (1956).

Blatt, H.: Original characters of clastic quartz. Jour. Sed. Petrology 37, 401–424 (1967).

Blatt, H., and Christie, J.M.: Undulatory extinction in quartz of igneous and metamorphic rocks and its significance in provenance studies of sedimentary rocks. Jour. Sed. Petrology 33, 559–579 (1963).

Boggs, Sam, Jr.: A numerical method for sandstone classification. Jour. Sed. Petrology 37, 548–555 (1967).

Bokman, John: Lithology and petrology of the Stanley and Jackfork formations. Jour. Geology 61, 152–170 (1953).

Bokman, John: Sandstone classification in relation to composition and texture. Jour. Sed. Petrology 25, 201–206 (1955).

Boles, J.R.: Active albitization of plagioclase, Gulf Coast Tertiary. Amer. Jour. Sci. 282, 165–180 (1982).

de Booy, T.: Petrology of detritus in sediments, a valuable tool. Koninkl. Nederlandse Akad. Wetensch. Proc., Ser. B, 64, 277–282 (1966).

Boswell, P.G.H.: The term graywacke. Jour. Sed. Petrology 30, 154 (1960).

Brenchley, P.J.: Origin of matrix in Ordovician greywackes, Berwyn Hills, North Wales, Jour. Sed. Petrology 39, 1297–1301 (1969).

Brett, G.W.: Cross-bedding in the Baraboo Quartzite of Wisconsin. Jour. Geology 63, 143–148 (1955).

Burchard, E.F.: Notes on various glass sands mainly undeveloped. U.S. Geol. Survey Bull. 315, 377–382 (1907).

Burst, F.F.: "Glauconite" pellets; their mineral nature and applications to stratigraphic interpretations. Am. Assoc. Petroleum Geologists Bull. 42, 310–327 (1958).

Bushinsky, G.I.: Structure and origin of the phosphorites of the U.S.S.R. Jour. Sed. Petrology 5, 81–92 (1935).

Buttram, Frank: The glass sands of Oklahoma. Oklahoma Geol. Survey Bull. 10, 91 pp. (1913).

Cady, W.M.; Wallace, R.E.; Hoare, J.M.; and Webber, E.J.: The central Kuskokwim region, Alaska. U.S. Geol. Survey Prof. Paper 268, 132 pp. (1955).

Cayeux, Lucien: Les roches sédimentaires de France, Roches siliceuses, 774 pp. Paris: Impr. Nationale 1929.

Cháb, Jan: Poznámka ki klasifikaci psamitůn (Note on the classification of psammites). Věst.-Ůstred. ůst. Geol. Ceshost 42, 225–227 (1967).

Chanda, S.K.: Calclithite fragments vs. extraclasts: A discussion. Jour. Sed. Petrology 39, 1640–1641 (1969).

Chang, Shih-Chicao: A new sandstone classification scheme. Geol. Soc. China. Proc. No. 10. pp. 107–114 (1967).

Chen, P.Y.: A modification of sandstone classification. Jour. Sed. Petrology 38, 54–60 (1968).

Clarke, F.W.: Report of work done in the division of chemistry and physics. U.S. Geol. Survey Bull. 60, 174 pp. (1890).

Clarke, F.W.: Data of geochemistry. U.S. Geol. Survey Bull. 770, 841 pp. (1924).

Cloud, P.E., Jr.: Physical limits of glauconite formation. Am. Assoc. Petroleum Geologists Bull. 39, 484–492 (1955).

Collins, W.H.: North shore of Lake Huron. Canada Geol. Survey Mem. 143, 160 pp. (1925).

Crook, A.W.: Classification of arenites. Am. Jour. Sci. 258, 419–428 (1960).

Crook, K.A.W.: Petrology of graywacke suite sediments from the Turon River-Coolamigal Creek district, N.S.W. Jour. and Proc. Royal Soc. New South Wales 88, 97–105 (1955).

Crook, K.A.W.: Petrology of Tamworth Group, Lower and Middle Devonian Tamworth-Nundle District New South Wales. Jour. Sed. Petrology 30, 353–369 (1960).

Crook, K.A.W.: Graywackes. In: Encyclopaedia Britannica 10. Chicago: Encyclopaedia Britannica 1970a.

Crook, K.A.W.: Geotectonic significance of graywackes: Relevance of Recent sediments from Niugini. 42nd ANZAAS Congress, Port Moresby, Niugini, August, 1970b.

Cummins, W.A.: The greywacke problem. Liverpool Manchester Geol. Jour. 3, 51–72 (1962a).

Cummins, W.A.: Greywacke in Lower Siwaliks, Simla Hills. Nature 196, 1085 (1962b).

Cushing, H.P.; Leverett, Frank; and Van Horn. F.R.: Geology and mineral resources of the Cleveland district, Ohio. U.S. Geol. Survey Bull. 818, 138 pp. (1931).

Dake, C.L.: The problem of the St. Peter sandstone. Missouri Univ. School Mines and Metal., Bull. (Tech. Ser.) 6, 158 pp. (1921).

Dapples, E.C.; Krumbein, W.C.; and Sloss, L.L.: Petrographic and lithologic attributes of sandstones. Jour. Geology 61, 291–317 (1953).

Davies, D.K., and Ethridge, F.G.: Sandstone composition and depositional environments. Am. Assoc. Petroleum Geologists Bull. 59, 239–264 (1975).

Davis, E.F.: The Franciscan sandstone. California Univ. Publ., Bull. Dept. Geology 11, 6–16 (1918).

Dickinson, W.R.: Plate tectonics and sedimentation. In: Dickinson, W.R. (Ed.): Tectonics and sedimentation. Soc. Econ. Paleon. Mineral. Spec. Pub. 22, 1–27 (1974).

Dickinson, W.R.: Singatoka dune sands. Viti Levu (Fiji). Sedimentary Geology 2, 115–124 (1968).

Dickinson, W.R.: Interpreting detrital modes of graywacke and arkose. Jour. Sed. Petrology 40, 695–707 (1970).

Dickinson, W.R.: Compositions of sandstones in Circum-Pacific subduction complexes and fore-arc basins. Am. Assoc. Petroleum Geologists Bull. 66, 121–137 (1982).

Dickinson, W.R.: Interpreting provenance relations from detrital modes of sandstones, pp. 333–362. In: Zuffa, G. G. (Ed.): Provenance of arenites. Dordrecht: D. Reidel Pub. Co. (1984).

Dickinson, W.R., Ojakangas, R.W., and Stewart, R.J.: Burial metamorphism of the Late Mesozoic

Great Valley Sequence, Cache Creek, California. Geol. Soc. America, Bull. 50, 519–526 (1969).

Diller, J. S.: The educational series of rock specimens, etc. U.S. Geol. Survey Bull. 150, 84–87 (1898).

Donaldson, J.A.: Marion Lake Map-area. Quebec-Newfoundland. Canada Geol. Survey Mem. 338, 85 pp. (1966).

Donaldson, J.A.: Two Proterozoic clastic sequences. A sedimentological comparison. Geol. Assoc. Canada Proc. 18, 33–54 (1967).

Donaldson, J.A., and Jackson, G.D.: Archean sedimentary rocks of North Spirit Lake Area. northwestern Ontario. Canadian Jour. Earth Sci. 2, 622–647 (1965).

Dott, R.H., Jr.: Wacke, graywacke and matrix—what approach to immature sandstone classification? Jour. Sed. Petrology 34, 625–632 (1964).

Drobnyk, J.W.: Petrology of the Paleocene-Eocene Aquia formation of Virginia, Maryland and Delaware. Jour. Sed. Petrology 35, 626–642 (1965).

Dunham, R.J.: Classification of carbonate rocks according to depositional texture. In. Ham, W.E. (Ed.): Classification of carbonate rocks. Tulsa, OK, Amer. Assoc. Petrol. Geol., Mem. 1, 108–121 (1962).

Dury, G.H.: Relict deep weathering and duricrusting in relation to the paleoenvironments of middle latitudes. Geog. Jour. 137, 511–522 (1971).

Edwards, A.B.: The petrology of the Miocene sediments of the Aure Trough, Papua. Royal Soc. Victoria Proc. (N.S.) 60, 123–148 (1950a).

Edwards, A.B.: The petrology of the Cretaceous greywackes of the Purari Valley, Papua, Royal Soc. Victoria Proc. (N.S.) 60, 163–171 (1950b).

Eigenfeld, R.: Die Kulmconglomerat von Teuschnitz im Frankenwalde. Sächs. Akad. Wiss. Abh., math. phys. Kl. 42, 58 (1933).

Emery, K.O.: Turbidites—Precambrian to present. In: Studies on oceanography, 568 pp. Tokyo: Univ. Tokyo Press 1964.

Engel, A.E.J., and Engel, C.G.: Grenville series in the northwest Adirondack Mountains. New York. Geol. Soc. America Bull. 64, 1013–1097 (1953).

Espenshade, G.: Geology and mineral deposits of the Pilleys Island area. Dept. Natl. Resources Newfoundland, Geol. Survey Bull. 6, 56 pp. (1937).

Ethridge, F.G.; Gopinath, T.R.; and Davies, D.K.: Recognition of deltaic environments from small samples. In: M.L. Broussard (Ed.): Deltas, models for exploration. Houston Geol. Soc., 151–164 (1975).

Fahrig, W.F.: The geology of the Athabasca formation. Canada Geol. Survey Bull. 68, 41 pp. (1961).

Feininger, T.: Chemical weathering and glacial erosion of crystalline rocks and the origin of till. U.S. Geol. Survey Prof. Paper 750-C, 65–81 (1971).

Ferrar, H.T.: The geology of the Dargaville-Rodney Subdivision. New Zealand Geol. Survey Bull. 34, 78 pp. (1934).

Fettke, C.R.: Glass manufacture and the glass sand industry of Pennsylvania. Pennsylvania Topog. and Geol. Survey Rept. 12, 278 pp. (1918).

Fischer, Georg: Die Petrographie der Grauwacken. Jahrb. preuss. geol. Landesanstalt 54, 320–343 (1933).

Folk, R.L.: The distinction between grain size and mineral composition in sedimentary rock nomenclature. Jour. Geology 62, 344–359 (1954).

Folk, R.L.: The role of texture and composition in sandstone classification. Jour. Sed. Petrology 26, 166–171 (1956).

Folk, R.L.: Practical petrographic classification of limestones. Am. Assoc. Petroleum Geologists Bull. 43, 1–38 (1959).

Folk, R.L.: Petrography and origin of the Tuscarora, Rose Hill, and Keefer formations. Lower and Middle Silurian of eastern West Virginia. Jour. Sed. Petrology 30, 1–58 (1960).

Folk, R.L.: Some aspects of recrystallization in ancient limestones. In: Pray, L. C., and Murray, R. C., (Eds.): Dolomitization and limestone diagenesis. Soc. Econ. Paleon. Mineral. Spec. Pub. 13, 14–48 (1965).

Folk, R.L.: Petrology of sedimentary rocks, 170 pp. Austin, Texas: Hemphill's Bookstore 1968a.

Folk, R.L.: Bimodal supermature sandstones. Product of the desert floor. XXIII Intern. Geol. Congress Proc. 8, 9–32 (1968b).

Foster, R.J.: Tertiary geology of a portion of the central Cascade Mountains. Washington. Geol. Soc. America Bull. 71, 99–125 (1960).

Friedman, Melvin: Miocene orthoquartzite from New Jersey. Jour. Sed. Petrology 24, 235–241 (1954).

Füchtbauer, Hans: Zur Nomenklatur der Sedimentgesteine. Erdöl und Kohle 12, 605–613 (1959).

Füchtbauer, Hans: Sedimentpetrographische Untersuchungen in der älteren Molasse nördlich der Alpen. Eclogae geol. Helvetiae 57, 157–298 (1964).

Füchtbauer, Hans: Die Sandsteine in der Molasse nördlich der Alpen. Geol. Rundschau 56, 266–300 (1967).

Fujii, Koji: Petrography of the Upper Paleozoic sandstones from the Yatsushiro area. Kyushu Univ. Mem. Fac. Sci., Ser. D., 179–218 (1962).

Galliher, E.W.: Glauconite genesis. Geol. Soc. America Bull. 46, 1351–1366 (1935).

Galloway, J.F.: Deposition and diagenetic alteration of sandstone in northeast Pacific arc-related basins: Implications for graywacke diagenesis. Geol. Soc. America Bull. 85, 379–390 (1974).

Garner, H.F.: Stratigraphic-sedimentary significance of contemporary climate and relief in four regions of the Andes Mountains. Geol. Soc. America Bull. 70, 1327–1368 (1959).

Garrels, R.M.; Mackenzie, F.T.; and Siever, Raymond: Sedimentary cycling in relation to the history of the continents and oceans. In: Robertson, E.C. (Ed.): The nature of the solid earth, pp. 93–121. New York: McGraw-Hill 1971.

Gasser, Urs: Die innere Zone der subalpinen Molasse des Entlebuchs (Kt. Luzern). Geologie und Sedimentologie. Eclogae geol. Helvetiae 61, 229–319 (1968).

Gibbs, R.J.: The geochemistry of the Amazon River System. Part I. The factors that control the salinity and the composition and concentration of the suspended solids. Geol. Soc. America Bull. 78, 1203–1232 (1967).

Gluskoter, H.J.: Orthoclase distribution and authigenesis in the Franciscan Formation of a portion of western Marin County California. Jour. Sed. Petrology 34, 335–343 (1964).

Goldman, M.I.: Petrographic evidence on the origin of the Catahoula Sandstone. Am. Jour. Sci., Ser. 4, 39, 261–287 (1915).

Goldman, M.I.: General character, mode of occurrence and origin of glauconite. Jour. Washington Acad. Sci. 9, 501–502 (1919).

Graham, W.A.P.: A textural and petrographic study of the Cambrian sandstones of Minnesota. Jour. Geology 38, 696–716 (1930).

Greenly, E.: Incipient metamorphism in the Harlech Grits. Edinburgh Geol. Soc. Trans. 7, 254–258 (1899).

Griffiths, J.C.: Petrographical investigations of the Salt Wash sediments. U.S. Atomic Energy Comm., Tech. Rept. RME-3122 (Pts. I and II), 84 pp., 1956.

Griggs, A.B.: Chromite-bearing sands of the southern part of the coast of Oregon. Bull. U.S. Geol. Survey 945E, 113–150 (1945).

Grim, Ralph: The Eocene sediments of Mississippi. Mississippi State Geol. Survey Bull. 30, 240 pp. (1936).

Grout, F.F.: Contact metamorphism of the slates of Minnesota by granite and gabbro magmas. Geol. Soc. America Bull. 44, 989–1040 (1933).

Gruner, J.W.: Structural geology of the Knife Lake area of northeastern Minnesota. Geol. Soc. America Bull. 52, 1577–1642 (1941).

Gulbrandsen, R.A.; Goldich, S.S.; and Thomas, H.H.: Glauconite from the Precambrian Belt Series. Montana Science 140, 390–391 (1963).

Hadding, Assar: The pre-Quaternary sedimentary rocks of Sweden. III. The Paleozoic and Mesozoic sandstones of Sweden. Lunds Univ. Årsskr., N.F. Avd. 2, 25, Nr. 3, 287 pp. (1929).

Hadding, Assar: The pre-Quaternary sedimentary rocks of Sweden. IV. Glauconite and glauconitic rocks. Medd. Lunds Geol. Min. Inst., no. 51. 175 pp. (1932).

Haile, N.S.: The Cretaceous-Cenozoic Northwest Borneo Geosyncline. British Borneo Geol. Survey Bull. 4, 1–18 (1963).

Hawkins, J.W., and Whetten, J.T.: Graywacke matrix minerals. Hydrothermal reactions with Columbia River sediments. Science 166, 868–870 (1969).

Heald, M.T.: Stylolites in sandstone. Jour. Geology 63, 101–114 (1955).

Helmbold, Reinhard: Beitrag zur Petrographie der Tanner Grauwacken. Heidelberger Beitr. Min. Petrogr. 3, 253–288 (1952).

Helmbold, Reinhard, and Van Houten, F.B.: Contribution to the petrography of the Tanner graywacke. Geol. Soc. America Bull. 69, 301–314 (1958).

Henningsen, Dierk: Untersuchungen über Stoffbestand und Paläogeographie der Giessener Grauwacke. Geol. Rundschau 51, 600–626 (1961).

Hickok IV, W.O., and Moyer, F.T.: Geology and mineral resources of Fayette County. Pennsylvania Geol. Survey Bull. C-26, 530 pp. (1940).

Hollister, C.D., and Heezen, B.C.: Modern graywacke-type sands. Science 146, 1573–1574 (1964).

Hoque, Momin ul: Sedimentologic and paleocurrent study of Mauch Chunk sandstones (Mississippian) of south-central and western Pennsylvania. Am. Assoc. Petroleum Geologists 52, 246–263 (1968).

Horwood, H.C.: Geology of the Casummit Lake area and the Argosy Mine. Ontario Dept. Mines 46th Ann. Rept. 1937, 46, Pt. 7, 33 pp. (1938).

Hsu, K.J.: Texture and mineralogy of the Recent sands of the Gulf Coast. Jour. Sed. Petrology, 30, 380–403 (1960).

Hubert, J.F.: Petrology of the Fountain and Lyons Formations, Front Range, Colorado. Colorado School Mines Quart. 55, no. 1, 242 p. (1960).

Huckenholtz, H.G.: Mineral composition and texture in graywackes from the Harz Mountains (Germany) and in arkoses from the Auvergne (France). Jour. Sed. Petrology 33, 914–918 (1963).

Hunter, R.E.: Facies of iron sedimentation in the Clinton Group. In: Fisher, G.W., and others (Eds.): Studies of Appalachian geology; Central and southern. p. 101–121. New York: Interscience 1970.

Irving, R.D.: The copper-bearing rocks of Lake Superior. U.S. Geol. Survey Mono. 5, 464 pp. (1883).

Irving, R.D., and Van Hise, C.R.: The Penokee iron-bearing Series of Michigan and Wisconsin. U.S. Geol. Survey Mono. 19, 534 pp. (1892).

Jacobsen, Lynn: Petrology of Pennsylvanian sandstones and conglomerates of the Ardmore Basin. Oklahoma Geol. Survey Bull. 79, 144 pp. (1959).

Kaiser, E.: Der Grundsatz des Aktualismus in der Geologie. Zeitschr. deutsch. geol. Gesell. 83, 401–402 (1932).

Kay, Marshall: North American geosynclines. Geol. Soc. America Mem. 48, 143 p. (1951).

Keith, M.L.: Sandstone as a source of silica sands in southeastern Ontario. Ontario Dept. Mines Ann. Rept. 55, pt. 5, 36 pp. (1949).

Kennedy, W.Q.: Sedimentary differentiation as a factor in the Moine-Torridonian correlation. Geol. Mag. 88, 257–266 (1951).

Ketner, K.B.: Comparison of Ordovician eugeosynclinal and miogeosynclinal quartzites of the Cordilleran geosyncline. In: Geological Survey Research 1966. U.S. Geol. Survey Prof. Paper 550-C, C 54–C 60 (1966).

Ketner, K.B.: Origin of Ordovician quartzite in the Cordilleran miogeosyncline. In: Geological Survey Research 1968. U.S. Geol. Survey Prof. Paper 600 B, B 169–B 177 (1968).

Klein, G. de Vries: Triassic sedimentation, Maritime Provinces, Canada. Geol. Soc. America Bull. 73, 1127–1146 (1962).

Klein, G. de Vries: Analysis and review of sandstone classification in the North American geological literature. 1940–1960. Geol. Soc. America Bull. 74, 555–576 (1963).

Konta, Jiri: Problem of the quantitative petrological classification in the rock series arkose-graywacke-quartz sandstone-clay shale. Contr. Mineral. Petrol. 19, 125–132 (1968).

Kossovskaia, A.G.: K voprosu o klassifikatsii peschanykh porod po mineralogischeskomu sostavu (On the question of classification of sandstones according to mineralogical composition). Uchen. Zap. Leningr. Gos. Un. V., No. 310. Seriia geolog. Nauk 12, 201–211 (1962).

Krumbein, W.C., and Sloss, L.L.: Stratigraphy and sedimentation, 491 pp. San Francisco: Freeman 1951.

Krumbein, W.C., and Sloss, L.L.: Stratigraphy and sedimentation, 2nd ed., 660 pp. San Francisco: Freeman 1966.

Krynine, P.D.: Arkose deposits in the humid tropics, a study of sedimentation in southern Mexico. Am. Jour. Sci. 29, 353–363 (1935).

Krynine, P.D.: Petrography and genesis of the Siwalik Series. Am. Jour. Sci. 34, 422–446 (1937).

Krynine, P.D.: Petrology and genesis of the Third Bradford Sand. Pennsylvania State College Bull. 29, 134 p. (1940).

Krynine, P.D.: Graywackes and the petrology of Bradford Oil Field, Pennsylvania. Am. Assoc. Petroleum Geologists Bull. 25, 2071–2074 (1941a).

Krynine, P.D.: Paleogeographic and tectonic significance of sedimentary quartzites (abstr.). Geol. Soc. America Bull. 52, 1915 (1941b).

Krynine, P.D.: Differential sedimentation and its products during one complete geosynclinal cycle. Anales 1st Congreso Panamerican Ingenieria de Minas y Geologia Geology Part 1, 2, 537–561 (1942).

Krynine, P.D.: Sediments and the search for oil. Producers Monthly 9, 12–22 (1945).

Krynine, P.D.: The megascopic study and field classification of sedimentary rocks. Jour. Geology 56, 130–165 (1948).

Krynine, P.D.: Petrology, stratigraphy and origin of the Triassic sedimentary rocks of Connecticut. Connecticut State Geol. Natl. Hist. Survey Bull. 73, 247 p. (1950).

Krynine, P.D., and Tuttle, O.F.: Petrology of Ordovician-Silurian boundary in central Pennsylvania (abs.). Geol. Soc. America Bull. 52, 1917–1918 (1941).

Kuenen, Ph.H.: Experimental abrasion 3: Fluviatile action on sand. Am. Jour. Sci. 257, 172–190 (1959).

Kuenen, Ph.H.: Experimental abrasion 4: Eolian action. Jour. Geol. 68, 427–449 (1960).

Kuenen, Ph.H.: Matrix of turbidites. Experimental approach. Sedimentology 7, 267–297 (1966).

Kuenen, Ph.H., and Migliorini, C.I.: Turbidity currents as a cause of graded bedding. Jour. Geology 58, 91–127 (1950).

Lasius, Georg: Beobachtungen im Harzgebirge, pp. 132–152. Hannover: Helwing 1789.

Leith, C.K., and Van Hise, C.R.: The geology of the Lake Superior region. U.S. Geol. Survey Mono. 52, 641 pp. (1911).

Lerbekmo, J.F.: Petrology of the Belly River Formation, southern Alberta Foothills. Sedimentology 2, 54–86 (1963).

Loney, R.A.: Stratigraphy and petrography of the Pybus-Gambier area, Admiralty Island, Alaska. U.S. Geol. Survey Bull. 1178, 103 pp. (1964).

MacGregor, A.M.: The problem of the Precambrian atmosphere. South Africa Jour. Sci. 24, 155–172 (1927).

Mack, Greg H.: The survivability of labile light-mineral grains in fluvial, aeolian and littoral marine environments: the Permian Cutler and Cedar Mesa Formations, Moab Utah. Sedimentology 25, 587–604 (1978).

Mackie, W.: The sands and sandstones of eastern Moray. Edinburgh Geol. Soc. Trans. 7, 148–172 (1899a).

Mackie, W.: The feldspars present in sedimentary rocks as indication of the condition of contemporaneous climate. Edinburgh Geol. Soc. Proc. 7, 443–468 (1899b).

Mackie, W.: Seventy chemical analyses of rocks. Edinburgh Geol. Soc. Proc. 8, 33–60 (1905).

Mann, W.R., and Cavaroc, V.V.: Composition of sand released from three source areas under humid, low relief weathering in the North Carolina Piedmont. Jour. Sed. Petrology 43, 870–881 (1973).

Mansfield, G.R.: The physical and chemical character of New Jersey Greensands. Econ. Geology 15, 547–566 (1920).

Markewicz, F.J.: Ilmenite deposits of the New Jersey Coastal Plain. In: Sibitzky, S. (Ed.): Geology of selected areas in New Jersey and eastern Pennsylvania and guidebook of excursions, pp. 363–382. New Brunswick, New Jersey: Rutgers Univ. Press 1969.

Martens, J.H.C.: Beach deposits of ilmenite, zircon and rutile in Florida. Florida State Geol. Survey, 19th Ann. Rept., 124–154 (1928).

Martens, J.H.C.: Beaches of Florida. Florida State Geol. Survey, 22nd Ann. Rept., 67–119 (1931).

Mathur, S.M.: On the term "orthoquartzite". Eclogae geol. Helvetiae 51, 695–696 (1958).

Mattiat, B.: Beitrag zur Petrographie der Oberharzer Kulmgrauwacke. Beitr. Min. Petrogr. 7, 242–280 (1960).

McBride, E.F.: The term graywacke (discussion). Jour. Sed. Petrology 32, 614–615 (1962a).

McBride, E.F.: Flysch and associated beds of the

Martinsburg Formation (Ordovician), central Appalachians. Jour. Sed. Petrology 32, 39–91 (1962b).

McBride, E.F.: A classification of sandstones. Jour. Sed. Petrology 33, 664–669 (1963).

McBride, E.F.: Sedimentary petrology and history of the Haymond Formation (Pennsylvanian). Marathon Basin, Texas. Univ. Texas, Bur. Econ. Geology Rept. Inv. No. 57, 101 pp. (1966).

McElroy, C.T.: The use of the term "greywacke" in rock nomenclature in New South Wales. Australian Jour. Sci. 16, 150–151 (1954).

McEwen, M.C.; Fessenden, F.W.; and Rogers, J.J.W.: Texture and composition of some weathered granites and slightly transported arkosic sands. Jour. Sed. Petrology 29, 477–492 (1959).

McKee, E.D.: Ancient sandstones considered to be eolian. In: McKee, E.D. (Ed.): A study of global sand seas. U.S. Geol. Survey Prof. Paper 1052, 187–238 (1979).

Meckel, L.D.: Origin of Pottsville conglomerates (Pennsylvanian) in the central Appalachians. Geol. Soc. America Bull. 78, 223–258 (1967).

Meischner, D.: Clastic sedimentation in the Variscan Geocyncline east of the River Rhine. In. Miller, G. (Ed.): Sedimentology of parts of central Europe, pp. 9–43. Guidebook 8th Intern. Sed. Congr. Heidelberg, 1971.

Merifield, P.M.; and Lamar, D.L.: Sand waves and early earth-moon history. Jour. Geophys. Research 73, 4767–4474 (1968).

Merrill, G.P.: Stones for building and decoration, 3rd ed., 551 pp. New York: John Wiley and Sons 1891.

Michot, Paul: Classification et terminologie des roches lapidifiées de la série psammito-pelitique. Ann. Soc. Géol. Belgique 81, 311–342 (1958).

Middleton, G.V.: Chemical composition of sandstones. Geol. Soc. America Bull. 71, 1011–1026 (1960).

Millot, Georges: Geologie des argiles, 499 pp. Paris: Masson et Cie (1964).

Mizutani, S.; and Suwa, K.: Orthoquartzitic sand from the Libyan Desert, Egypt. Nagoya Univ. Jour. Earth Sci. 14, 137–150 (1966).

Murray, H.H.; and Patton, J.B.: Preliminary report on high-silica sand in Indiana. Indiana Dept. Conserv. Geol. Survey Rept. Prog. 5, 35 pp. (1953).

Nanz, R.H., Jr.: Genesis of Oligocene sandstone reservoir, Seeligson field, Jim Wells and Kleberg counties, Texas. Am. Assoc. Petroleum Geologists Bull. 38, 96–117 (1954).

Naumann, C.F.: Lehrbuch der Geognosie, Vol. 1, 960 pp. Leipzig: Engelman 1858.

Niggli P.; de Quervain, F.; and Winterhalter, R.U.: Chemismus schweizerischer Gesteine. Beitr. Geologie Schweiz, Geotechn. Ser. No. 14, 389 pp. (1930).

Ojakangas, R.W.: Petrology and sedimentation of the Upper Cambrian Lamotte Sandstone in Missouri. Jour. Sed. Petrology 33, 860–873 (1963).

Okada, H.: Sandstones of the Cretaceous Mifune Group. Kyushu. Japan. Kyushu Univ. Mem. Fac. Sci., Ser. D, Geology 10, 1–40 (1960).

Okada, H.: Cretaceous sandstones of Goshonoura Island, Kyushu, Kyushu Univ. Mem. Fac. Sci., Ser. D, Geology 11, 1–48 (1961).

Okada, H.: Serpentine sandstone from Hokkaido. Mem. Fac. Sci., Kyushu Univ., Ser. D. Geol. 15, 23–38 (1964).

Okada, H.: Non-graywacke "turbidite" sandstones in the Welsh geosynclines. Sedimentology 7, 211–232 (1966).

Okada, H.: Composition and cementation of some Lower Paleozoic grits in Wales. Kyushu Univ. Mem. Fac. Sci., Ser. D, Geology 18, 261–276 (1967).

Okada, H.: Classification of sandstone: analysis and proposal. Jour. Geol. 79, 509–525 (1971).

Oriel, S.S.: Definitions of arkose. Am. Jour. Sci. 247, 824–829 (1949).

Packham, G.H.: Sedimentary structures as an important feature in the classification of sandstones. Am. Jour. Sci. 252, 466–476 (1954).

Park, C.F., Jr.: Manganese resources of the Olympic Peninsula, Washington. U.S. Geol. Survey Bull. 931-R, 435–457 (1942).

Payne, T.G. and others: The arctic slope of Alaska. U.S. Geol. Survey, Oil and Gas Investig. Map OM 126, sheet 2, 1952.

Peach, B.N.; Horne, John; Gunn, W.; Clough, C. T.; and Hinxman, L. W.: The geological structure of the North-West Highlands of Scotland. Mem. Geol. Survey Great Britain, 668 pp. (1907).

Peikh, V.: Klassifikatsiya peschanikov po veshchestvennomu sostavu (Classification of sands according to their mineralogic composition). Vest. Moskov Univ. Geol. 24, 87–98 (1969).

Pelletier, B.R.: Pocono paleocurrents in Pennsylvania and Maryland. Geol. Soc. America Bull. 69, 1033–1064 (1958).

Pettijohn, F.J.: Archean sedimentation. Geol. Soc. America Bull. 54, 925–972 (1943).

Pettijohn, F.J.: Sedimentary rocks, 1st ed., 526 pp. New York: Harper 1949.

Pettijohn, F.J.: Classification of sandstones. Jour. Geology 62, 360–365 (1954).

Pettijohn, F.J.: Paleocurrents of Lake Superior Precambrian quartzites. Geol. Soc. America Bull. 68, 469–480 (1957).

Pettijohn, F.J.: Sedimentary rocks, 3rd ed., 628 pp. New York: Harper and Row 1975.

Pettijohn, F.J.: The term graywacke. Jour. Sed. Petrology 30, 627 (1960).

Pettijohn, F.J.: Chemical composition of sandstones—excluding carbonate and volcanic sands. In: Data of geochemistry, 6th ed. U.S. Geol. Survey Prof. Paper 440S. 19 pp. (1963).

Plumley, W.J.: Black Hills terrace gravels. A study in sediment transport. Jour. Geology 56, 526–577 (1948).

Potter, Paul Edwin: Petrology and chemistry of modern big river sands. Jour. Geology 86, 423–449 (1978).

Potter, Paul Edwin: South America and a few grains of sand: Part 1—Beach sands. Jour. Geology 94, 301–319 (1986).

Potter, P.E., and Pryor, W.A.: Dispersal centers of Paleozoic and later clastics of the Upper Mississippi Valley and adjacent areas. Geol. Soc. America Bull. 72, 1195–1250 (1961).

Potter, Paul Edwin, and Franzinelli, Elena: Fraction analysis of modern river sand of Rios Negro and Solimões, Brazil, implications for the origin of quartz-rich sandstones. Rev. Brasileira Geociencias 15, 31–35 (1985).

Ramdohr, Paul: New observations on the ores of the Witwatersrand in South Africa and their genetic significance. Trans. Geol. Soc. South Africa 61, Annexure, 50 pp. (1958).

Reed, J.J.: Petrology of the Lower Mesozoic rocks of the Wellington District. New Zealand Geol. Survey Bull. (N.S.) 57, 60 pp. (1957).

Reed, R.D.: The occurrence of feldspar in California sandstones. Am. Assoc. Petroleum Geologists Bull. 12, 1023–1024 (1928).

Rinne, Friedrich: Gesteinskunde, 374 pp. Leipzig: Dr. Max Jänecke 1923.

Rodgers, John: The nomenclature and classification of sedimentary rocks. Am. Jour. Sci. 248, 297–311 (1950).

Ronov, A.B.; Mikhailovskaya, M.S.; and Solodkova, I.I.: Evolution of the chemical and mineralogical composition of arenaceous rocks. In: Chemistry of the earth's crust. v. 1 (Israel Program Sci. Trans., 1966). pp. 212–262 (1963).

Rothrock, E.P: A geology of South Dakota. part 3, Mineral resources. South Dakota Geol. Survey Bull. 15, 255 pp. (1944).

Russell, R.D.: Mineral composition of Mississippi River sands. Geol. Soc. America Bull. 48, 1307–1348 (1937).

Sabins, F.F., Jr.: Grains of detrital, secondary, and primary dolomite from Cretaceous strata of the Western Interior. Geol. Soc. America Bull. 73, 1183–1196 (1962).

Sahu, B.K.: Classification of common sandstones. Punjab Univ. Research Bull. 16, N.S. 315–322 (1965).

Schwab, F.L.: Origin of the Antietam Formation (Late Precambrian?-Lower Cambrian) central Virginia. Jour. Sed. Petrology 40, 354–355 (1970).

Schwartz, G.M.: Correlation and metamorphism of the Thomson Formation, Minnesota. Geol. Soc. America Bull. 52, 1001–1020 (1942).

Seilacher, A.: Origin and diagenesis of the Oriskany sandstone (Lower Devonian, Appalachians) as reflected in its shell fossils. In: Müller G., and Friedman, G.M., (Eds.): Recent developments in carbonate sedimentology in Central Europe. pp. 175–185. Berlin-Heidelberg-New York: Springer 1968.

Shanmugam, G.: Types of porosity in sandstone and their significance in interpreting provenance, pp. 115–137. In: Zuffa, G.G. (Ed.): Provenance of arenites. Dordrecht: D. Reidel Pub. Co. (1984).

Shead, A.C.: Notes on barite in Oklahoma with chemical analysis of sand barite rosettes. Oklahoma Acad. Sci. Proc. 3, 102–106 (1923).

Shiki, T.: Studies on sandstones in the Maizuri Zone, southwest Japan. III. Graywackes and arkose sandstone in and out of the Maizuri Zone. Kyoto Univ., Mem. Coll. Sci., Ser. B. 29, 291–324 (1962).

Shiki, T., and Mizutani, S.: On "graywacke". Geoscience 81, 21–32 (1965) (in Japanese w. Engl. summ.).

Shutov, V.D.: Obzor i analiz mineralogicheskikh klassifikatsii peschanykh porod (Survey and analysis of mineralogical classifications of sandstones). Litol. i Polez. Iskop., I. 95–112 (1965).

Shutov, V.D.: Klassifikatsiia peschanikov (Classification of sandstones). Litol. i Polez. Iskop. 5, 86–103 (1967).

Siever, Raymond: Trivoli sandstone of Williamson County, Illinois, Jour. Geology 57, 614–617 (1949).

Siever, Raymond: Pennsylvanian sandstones of the Eastern Interior coal basin. Jour. Sed. Petrology 27, 227–250 (1957).

Simonen, Ahti, and Kuovo, Olavi: Archean varved schists north of Tampere in Finland. Soc. Géol. Finlande Compte Rendus. 24, 93–117 (1951).

Simonen, Ahti, and Kuovo, Olavi: Sandstones in Finland. Comm. Géol. Finlande, Bull. 168, 57–87 (1955).

Sippel, R.F.: Sandstone petrology, Evidence from luminescence petrography. Jour. Sed. Petrology 38, 530–554 (1968).

Skolnick, Herbert: The quartzite problem. Jour. Sed. Petrology 35, 12–21 (1965).

Sloss, L.L., Feray, D.E.: Microstylolites in sandstone. Jour. Sed. Petrology 18, 3–13 (1948).

Stebinger, E.: Titaniferous magnetite beds on the Blackfeet Indian Reservation, Montana. In: Contributions to economic geology, U.S. Geol. Survey Bull. 540, 329–337 (1914).

Strakhov, N.M.: Principles of lithogenesis. v. 2, First English ed. 1969, 609 pp. New York: Consultants Bureau 1967.

Suttner, L.J.: Stratigraphic and petrographic analysis of Upper Jurassic-Lower Cretaceous Morrison and Kootenai Formations, southwest Montana. Am. Assoc. Petroleum Geologists Bull. 53, 1391–1410 (1969).

Suttner, L.J., Basu, A., and Mack, G. H.: Climate and the origin of quartz arenites. Jour. Sed. Petrology 51, 1235–1246 (1981).

Suttner, L.J., and Dutta, P.K.: Alluvial sandstone composition and paleoclimate I. Framework mineralogy. Jour. Sed. Petrology 56, 329–345 (1986).

Sutton, J., and Watson, J.: The deposition of the upper Dalradian rocks of the Banffshire coast. Geol. Assoc. London Proc. 66, 101–133 (1955).

Swett, K.; Klein, G. deVries; and Smit, D.E.: A Cambrian tidal sand body—the Eriboll Sandstone of northwest Scotland: An ancient-recent analog. Jour. Geol. 79, 400–415 (1971).

Swineford, Ada: Cemented sandstones of the Dakota

and Kiowa Formations in Kansas. Kansas State Geol. Survey Bull. 70, 57–104 (1947).

Swineford, Ada: Petrography of Upper Permian rocks in south-central Kansas. Kansas State Geol. Survey Bull. 111, 179 pp. (1955).

Takihashi, J.: Synopsis of glauconization. In: Trask, P.D. (Ed.): Recent marine sediments, pp. 503–513. Tulsa, Oklahoma: Am. Assoc. Petroleum Geologists (1939).

Taliaferro, N.L.: Franciscan-Knoxville problem. Am. Assoc. Petroleum Geologists Bull. 27, 109–219 (1943).

Tallman, S.L.: Sandstone types, their abundance and cementing agents. Jour. Geology 57, 582–591 (1949).

Tarr, W.A.: The origin of the sand barites of the Lower Permian of Oklahoma. Am. Mineralogist 18, 260–272 (1933).

Teodorovich, G.I.: Rasshirennaya klassifikatsiya peschanikov po veshchestvennomu sostavu (Comprehensive classification of sandstones based on their composition). Izvest. Akad. Nauk SSSR, ser. Geol., 6, 75–95 (1967).

Thiel, G.A.: Sedimentary and petrographic analysis of the St. Peter Sandstone. Geol. Soc. America Bull. 46, 559–614 (1935).

Thiel, G.A., and Dutton, C.E.: The architectural, structural and monumental stones of Minnesota. Minnesota Geol. Survey Bull. 25, 160 pp. (1935).

Tieje, A.J.: Suggestions as to the description and naming of sedimentary rocks. Jour. Geology 29, 650–666 (1921).

Todd, T.W., and Folk, R.L.: Basal Claiborne of Texas, record of Appalachian tectonism during Eocene. Am. Assoc. Petroleum Geologists Bull. 41, 2545–2566 (1957).

Travis, R.B.: Nomenclature for sedimentary rocks. Am. Assoc. Petroleum Geologists Bull. 54, 1095–1107 (1970).

Turner, F.J., and Verhoogen, J.: Igneous and metamorphic petrology, 694 pp. New York: McGraw-Hill 1960.

Twenhofel, W.H.: Treatise on sedimentation, 926 pp. Baltimore: Williams and Wilkens 1932.

Tyrrell, G.W.: Greenstones and graywackes. Reunion Intern. pour l'Etude du Précambrien 1931, Comptes Rendus, 24–26 (1933).

Waldschmidt, W.A.: Cementing materials in sandstones and their influence on the migration of oil. Am. Assoc. Petroleum Geologists Bull. 25, 1859–1879 (1941).

Walker, T.R.; Waugh, B; and Grone, A.J.: Diagenesis in first-cycle desert alluvium of Cenozoic age, southwestern United States and northwestern Mexico. Geol. Soc. America Bull. 89, 19–32 (1978).

Walton, E.K.: Silurian greywackes in Peebleshire. Royal Soc. Edinburgh Proc. 65, 327–357 (1955).

Wang, Chao-Siang: On the occurrence of quartz wacke and its bearing on the problems of sandstone classification. Geol. Soc. China. Proc. 10, 99–106 (1967).

Webby, B.D.: Sedimentation of the alternating graywacke and argillite strata in the Porirua district. New Zealand Jour. Geol. Geophysics 2, 461–478 (1959).

Weber, J.N., and Middleton, G.V.: Geochemistry of turbidites of the Normanskill and Charny formations. Geochim. et Cosmochim. Acta 22, 200–288 (1961).

Whetten, J.T.: Sediments from the lower Columbia River and origin of graywacke. Science 152, 1057–1058 (1966).

Wieseneder, H.: Über die Gesteinsbezeichnung Grauwacke. Tschermaks min. petrog. Mitt. 7, 451–454 (1961).

Wiesnet, D.R.: Composition, grain size, roundness and sphericity of the Potsdam Sandstone (Cambrian) in northeastern New York. Jour. Sed. Petrology 31, 5–14 (1961).

Williams, Howel; Turner, F.J.; and Gilbert C.M.: Petrography, 406 pp. San Francisco: Freeman 1954.

Wilson, J.L.: Carbonate facies in geologic history, 471 pp. New York: Springer-Verlag 1975.

Woodland, A.W.: Petrological studies in the Harlech Grit series of Merionethshire II. Geol. Mag. 74, 440–454 (1938).

Yeakel, L.S., Jr.: Tuscarora, Juniata, and Bald Eagle paleocurrents and paleogeography in the central Appalachians. Geol. Soc. America Bull. 73, 1515–1540 (1962).

Zimmerle, W.: Serpentine graywackes from the North Coast basin, Colombia, and their tectonic significance. Neues Jahrb. Mineral. Abh. 109, 156–182 (1968).

Zuffa, G.G.: Hybrid arenites: Their composition and classification. Jour. Sed. Petrology 50, 21–29 (1980).

Chapter 6
Volcaniclastic Sandstones and Associated Rocks

Introduction

Sandstones, conglomerates, and breccias of volcanic origin provide exciting opportunities for expanded insight into earth history and improved mineral exploration.

Of greatest importance is the close relation of volcaniclastic deposits to plate tectonics. As a result, volcaniclastic sandstones and their associated rocks form significant deposits along active, subduction continental margins and of oceanic magmatic arcs. Volcanics are also characteristic of rifts and hot spots.

Secondly, they form an important part of the geologic record. Today, there are at least 26 well-defined active volcanic arcs in the world—mostly in the Pacific—plus many more inactive ones onshore in Tertiary deposits marginal to the Pacific basin, and each of these areas is rich in volcaniclastic deposits. Moreover, volcanics were even more abundant in the early history of the Earth (Ronov, 1964) and include many volcaniclastic sandstones. An enhanced understanding of their origin will tell us much about depositional systems, crustal composition, and tectonics in the Archean. Volcanoes were once active on the Moon and Mars, and apparently are still active on Io, one of the moons of Jupiter. Depending on the planetary atmosphere and the eruptive process, we can deduce much about presumed volcaniclastic deposits on other planets.

Thirdly, volcaniclastic sandstones are also closely associated with Precambrian greenstone belts and their sulfide and gold mineralization (Fox, 1984). Their unstable mineralogy also makes them reactive hosts to hydrothermal solutions of all ages. Pyroclastics and their associated sediments and flows are also useful guides to volcanic centers where hydrothermal activity was greatest. The sedimentological gradients of volcaniclastics help locate these centers (Fox, 1977). In addition, uranium occurs in acid volcanic rocks and sediments derived from them (Goodell and Waters, 1981).

Finally, the volcaniclastics, especially air falls and pyroclastic flows, have special significance for archeology and folklore and contemporary human activity. Consider, for example, Krakatoa in 1883 (Simkin and Fiske, 1983) and Mount St. Helens in 1980 (Lipman and Millineaux, 1981) or Pompeii and Mount Vesuvius in Roman times in Italy.

Sands rich in volcanic debris are of three dominant types: those produced by explosive ejection from a volcanic cone on land or underwater, those formed by lahars (debris flows), and those derived from the erosion of a volcanic terrane, one composed of extrusives, lahars, and/or pyroclastic deposits. The reworked volcaniclastic sands are comparable in sedimentary structures, geometry, and thickness to other terrigenous sandstones except for the abundance of framework grains of volcanic origin. In contrast, the true pyroclastics—hot, gassy, debris-charged flows and airfalls—originate as explosive, igneous rocks but are deposited as sedimentary ones. Thus they interest both the igneous and the sedimentary petrologist. Pyroclastic material may form local agglomerates or cinder cones. Other pyroclastics may be transported high in the atmosphere over much longer distances to form widespread beds of ash, which serve as useful time-equivalent stratigraphic markers in many sedimentary basins. Volcanic debris that is blown high in the air may be deposited on either land or in water and is called *air fall*; it contrasts with *pyroclastic flows*, material that accumulated from a hot, incandescent, turbulent mixture of debris and gas that explosively escaped from a fissure, cone, or vent and flowed downhill along the surface much in the same manner as a deep-water, marine turbidity current. Both air falls and pyroclastic flows may, of course, be interbedded with lava flows, lahars, or other sediments. Redeposited volcaniclastic sandstones commonly occur with other volcanic deposits or provide a clue to them. Hence our treatment is broad.

In part, lack of study by sedimentary petrologists may be a historical accident in that sedimentary petrology had its beginnings in regions remote from active volcanism. Fortunately, today excellent summaries of volcanism and the different types of volcaniclastic deposits are available and include those by Botvinkina (1974), Bullard (1976), Williams and McBirney (1979), and the informative summary by Clapperton (1977). Also recommended for general igneous petrology is the text by Best (1982, pp. 71–107). The early text by Rittmann (1962, pp. 62–111) is classic and *Volcanic Activity and Ecology* by Sheets and Grayson (1979) effectively illustrates how volcanoes have affected humanity. The sedimentologist can learn much from the atlas *Volcanic Landforms and Surface Features* by Green and Short (1971). The International Association of Volcanology publishes a catalog of the active volcanoes of the world. Simkin *et al.* (1981) give a worldwide summary of present volcanoes, Self and Sparks (1981) consider tephra (the Greek word used for volcanic debris), and Fisher and Schmincke (1984) have a comprehensive text entitled *Pyroclastic Rocks*. A short but effective review is that of Lajoie (1984).

Ancient volcaniclastic deposits are recognized by their petrographic composition, by their association with lavas, and by their distinctive sedimentary structures such as antidunes, bomb impact structures, stretched pumice, inverse grading, and others (see Shrock, 1948, Chap. 6). The vast majority of structures, however, are those which occur in fluvial and turbidite sandstones.

Characteristic Petrographic Features

Volcaniclastic sands are mineralogically more immature than any other group of sandstones, because they are either directly or indirectly the product of volcanism, generally with little modification by weathering and long distance of transport. Hence, their provenance is readily determined. Special earmarks of a volcaniclastic origin include glass, pumice and scoria, plagioclase and mafic mineral euhedra, characteristic rock fragments, zoned feldspars, cristobalite, basaltic hornblende, and much diagenetic mineralogical alteration.

Glass may occur as bubble-wall shards (Figs. 6-1 and 6-2), as pumice shards, or as binding matrix, where it plays the role of cement. The amount of glass seen in a thin section of a volcaniclastic sandstone can be deceptive until one uses crossed nicols (Fig. 6-3). Glass may be colorless, red, yellow, brown, or black depending on impurities and the oxidation state of iron. The approximate silica content of the glass can be determined from its index of refraction (Huber and Reinhart, 1966), although minor components, such as iron and water, affect the index

FIGURE 6-1. Undistorted bubble-wall glass shards (left) versus distorted and flattened (right).

FIGURE 6-2. An ash fall consists of an open framework of little abraded glass shards cemented by poikiloblastic calcite. Ellensburg Formation (Tertiary), Washington, U.S.A., × 100.

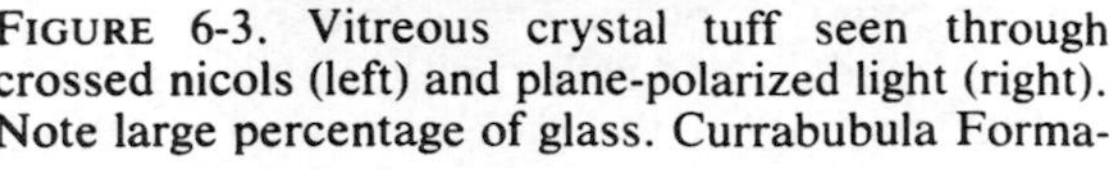

FIGURE 6-3. Vitreous crystal tuff seen through crossed nicols (left) and plane-polarized light (right). Note large percentage of glass. Currabubula Formation (Carboniferous), Werrie Basin, New South Wales, Australia.

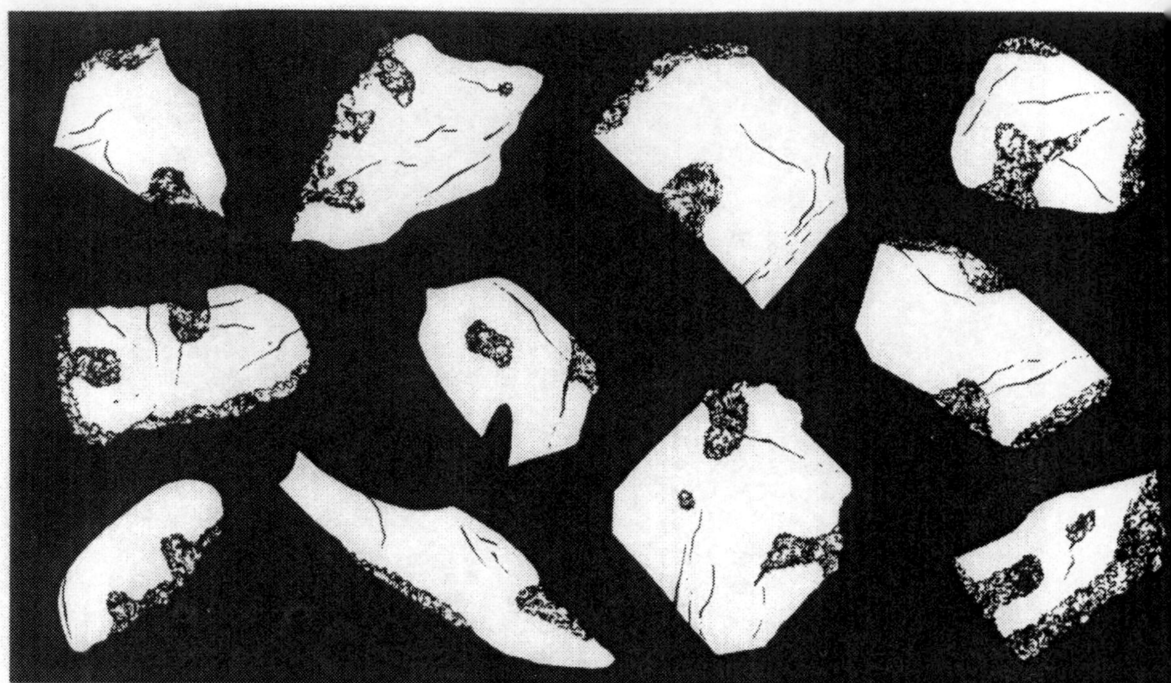

FIGURE 6-4. Embayed quartz with attached aphanitic and glassy blebs (Webb and Potter, 1969, Fig. 3).

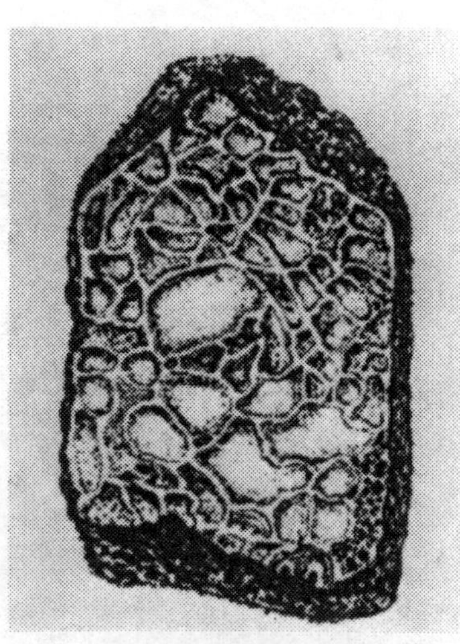

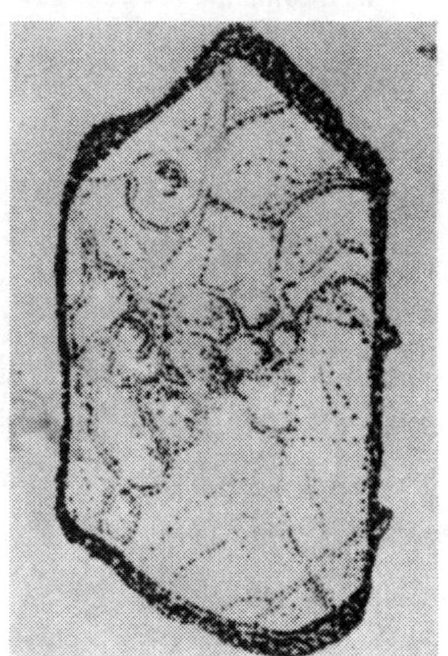

FIGURE 6-5. Bubble-wall texture (Fisher, 1963, Fig. 4). Reproduced by permission of the Journal of Sedimentary Petrology.

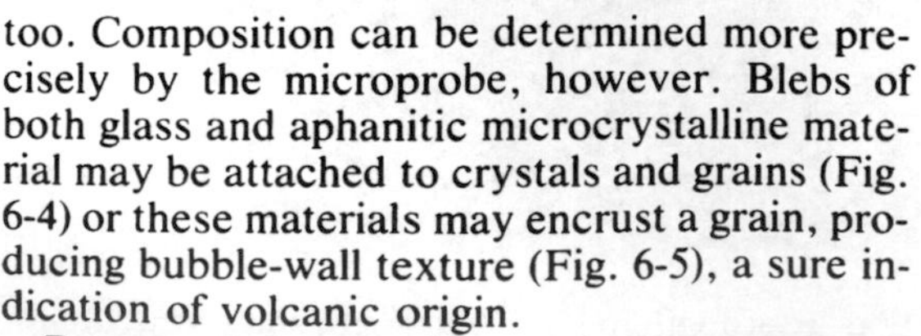

too. Composition can be determined more precisely by the microprobe, however. Blebs of both glass and aphanitic microcrystalline material may be attached to crystals and grains (Fig. 6-4) or these materials may encrust a grain, producing bubble-wall texture (Fig. 6-5), a sure indication of volcanic origin.

Because they crystallize early from a magma, many of the feldspars and mafic minerals are euhedra. In pyroclastic deposits such euhedra are commonly broken. Zoned feldspars (Fig. 6-6) are abundant and reflect either high viscosity or fast cooling in the parent magma, both of

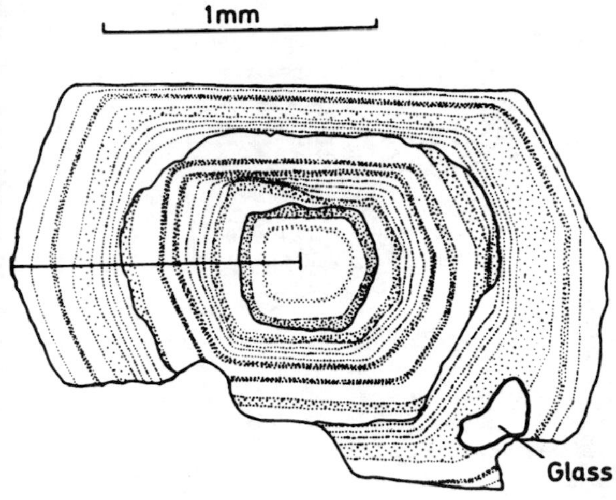

FIGURE 6-6. Zoned plagioclase from a Quaternary ash, New Zealand. Note embayment (Ewart, 1963, Fig. 11). Reproduced by permission of the Journal of Petrology.

which inhibit further reaction of an early formed crystal with the liquid phase. The high-temperature alkali feldspar, sanidine, is characteristic of quickly chilled volcanics and hence is a good indicator of provenance. Plagioclase composition can be determined precisely by microprobe. Such determinations help us map feldspar composition across a basin. Features worthy of note are embayments of quartz and feldspar (Fig. 6-4). These form by resorption by the magma, perhaps shortly before eruption. Another notable feature of the quartz of volcaniclastic sands is that it consists almost entirely of strain-free, monocrystalline grains with unitary extinction. Basaltic or oxy-hornblende—brown in color—in the heavy mineral suite is also another telltale indicator of a volcanic source.

Figures 6-7 and 6-8 show some of the different types of rock fragments in volcaniclastics. Randomly, as well as preferentially, oriented microlites are characteristic of such fragments, which may or may not contain phenocrysts. Glass is also common as interstitial material between the microlites. In a consolidated reworked volcaniclastic sandstone, the combination of fine-grained volcanic rock fragments, detrital glass, and much alteration makes for extremely difficult point counting—where does a rock fragment end and matrix begin? See Vuagnat (1952), Heiken (1974), and Stanley (1977) for helpful guides to the recognition of volcanic rock fragments.

Volcaniclastic sands are more susceptible to diagenesis than any other sand, because of the chemical instability and reactivity of their framework grains: glass, pumice, and fine-grained rock fragments plus feldspathoids and mafic minerals (Chaps. 11 and 12). High porosity and surface area of the fine-grained material also accelerate diagenesis and promote devitrification of volcanic glass. Except where preserved in a "sealed environment," glass is commonly found only in mid-Tertiary and younger rocks. Glass alters to clay minerals, zeolites, and silica. Smectite and halloysite form colloform and vermicular masses which may precipitate in cavity fillings or may simply fill microporosity. And glass can devitrify to a microcrystalline aggregate, which resembles chert; however, microlitic euhedra in such aggregates are a sure sign of their volcanic origin. Nonetheless, distinguishing between normal chert and the devitrification products of volcanic glass is one of the major and difficult problems of thin-section petrology. Complete alteration of an ash may produce bentonite which consists almost entirely of smectite. In outcrop this expandable clay has a very characteristic cracked, flaky weathering surface. Zeolites precipitated in void spaces (Fig. 6-9) and as coatings on framework grains are also a distinguishing earmark of many volcaniclastic sands. Zeolitic minerals may also replace glass. Silica, released from the alteration of volcanic glass, may be present as chalcedony and/or opal as cements or discrete veins, pods, and nodules. Cristobalite is a common secondary feature of many ash flows; it can also be found in coarser-grained rhyolitic rock fragments. Well-crystallized high-cristobalite may form during the cooling of an ash flow whereas opal-CT, a disordered variety of low-cristobalite, forms later in the diagenesis of glass. Much of the alteration of subaerial volcaniclastic deposits may occur early in their history and be related to soil development. See Sand and Mumpton (1978), Fisher and Schmincke (1984, Chap. 12), and Chap. 11 for details of the complex alteration of volcaniclastics after deposition.

Another distinctive feature of volcaniclastic sands, one related to a fluid dynamic rather than a petrographic property, stems from the density contrasts between pumice and scoria, on the one hand, and crystalline material, on the other. Pumice, which may have densities as low as 0.22, may float long distances before becoming waterlogged and thus may be finally deposited far from the more crystalline material with which it was ejected. Because such materials have a low density, coarse-grained pumice may be carried far by turbidity currents and sedi-

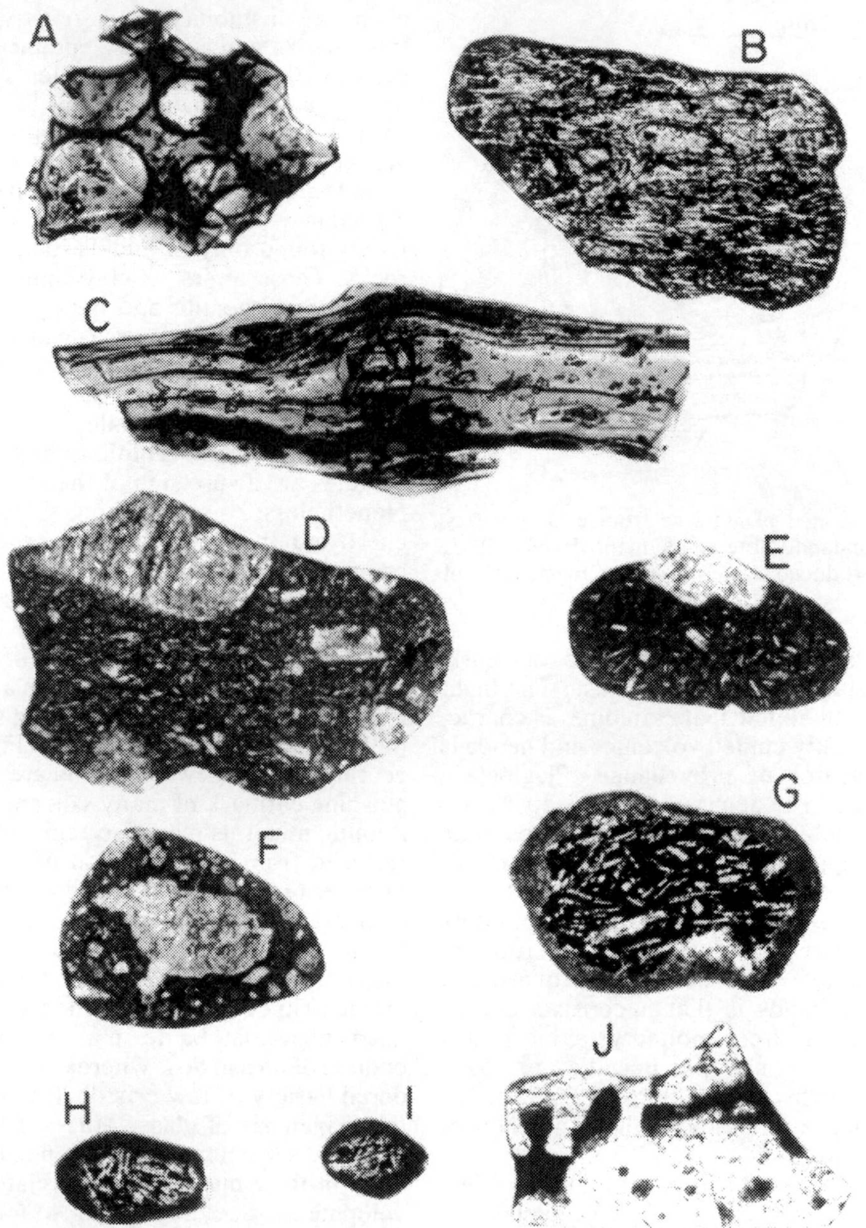

FIGURE 6-7. Photomicrographs of volcanic detritus: A) fragment of glass with ovoid vesicles from modern ash, × 200; B) pumice fragment in Tertiary sandstone, Montana, × 80; C) pumice fragment with fluidal structures in modern ash, × 200; D), E), and F) large phenocrysts in fine-grained groundmass containing euhedral microlites from Judith Fancy Formation (Cretaceous), Virgin Islands, × 40; G) H), and I) microlites of feldspar in black glassy groundmass, fragments coated by carbonate, × 40, Pleistocene caliche, Don Ana Co., New Mexico, U.S.A.; J) large phenocryst embayed by glass (black) from Currabubula Formation (Carboniferous), New South Wales, × 80. Photographs A and C courtesy of R.V. Fisher.

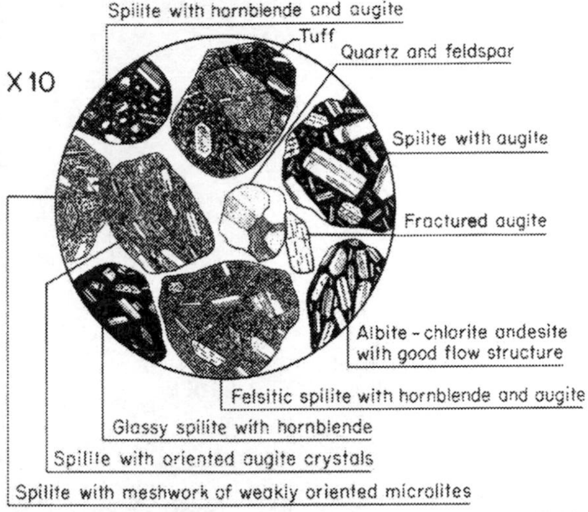

FIGURE 6-8. Volcanic rock fragments from the North Helvetic flysch of Switzerland. (Redrawn from Vuagnat, 1952, Fig. 10).

mented with clay. Subaerial debris flows composed largely of volcanic debris also benefit from the low density of individual particles of ash and pumice. A large river can also carry pumice from its headwaters in a leading margin across a continent far toward a trailing margin as does the Amazon. Should the ash have been deposited in a desert, eolian transport is also an effective segregation mechanism: for a given size fraction the less dense pumice will travel farther and faster than the crystalline material. Within a single ash fall bed, sorting in air can also produce marked changes in grain size and composition from bottom to top of the bed.

Criteria for the field recognition of the major types of volcaniclastic deposits are given later in this chapter (in "Volcaniclastic Deposits").

FIGURE 6-9. Laumontite (a) as a pore filling in a sandstone of the Blairmore Group (Cretaceous), Alberta, × 100.

Petrographic Classification

The classification of a volcaniclastic sand depends on whether it is of pyroclastic or erosional origin.

If redeposited by wind or water, one uses the standard sandstone classification of Fig. 5-3, adding the prefix "tuffaceous." Because rock fragments commonly predominate in these volcanic sands, such sands will be either tuffaceous lithic wackes or arenites. Some workers informally use the term "volcanic-rich" for these sandstones.

If the sand were eroded from a single source, which in turn received pyroclastic debris from a petrologically homogeneous magma, the feldspar composition is an index of rock type so that one can add a specifying adjective such as andesitic, tuffaceous, lithic arenite. Although the feldspar may be identifiable petrographically (Henrich, 1965, pp. 334–70), or by X-ray diffraction, the electron microprobe provides rapid and precise identification of the feldspars. Feldspars are abundant and informative in the study of all volcanic debris. It is usually wise to study a number of samples before attempting to assign a specific compositional name.

Pyroclastic materials are classified by size, composition, and origin (Table 6-1). The size classes used are after Schmid (1981), who also gives a good review of the terminology of volcaniclastic rocks. Rocks composed of particles larger than 2 mm have been generally described as breccia but are subdivided into lapilli (2 to 64 mm) and bombs or blocks (larger than 64 mm). Consult Parson (1969) to learn about the great variety of volcanic breccias and field criteria employed to distinguish them. Ash and tuff both refer to material smaller than 2 mm, the former being unconsolidated, the latter consolidated.

In order to define the composition of the parent magma, one first determines the relative proportions of glass, rock fragments, and crystals in a tuff, and then the composition of the crystal fraction (Fig. 6-10). Having determined how much quartz, how much and what types of feldspar and of other crystalline minerals, the silica content of any glass that may be present, and other compositional data, one uses a standard igneous classification scheme, such as that of Moorhouse (1959, Table 16), Rittmann (1962, pp. 96–103), or Streckeisen (1967, pp. 177–93) to assign a name. But because many pyroclastic rocks are very fine-grained or glassy, a chemical analysis is needed from which one can calculate a probable mode. If none of the above methods are available, a compositional name can be assigned on the basis of only the recognizable mineral content, perhaps only phenocrysts. If so, one prefixes "pheno" to the name. One can describe a rock, for example, as a lithic, phenoandesitic tuff or a vitric, phenodacitic tuff. As is true for the redeposited sandstones, it is usually advisable to determine composition on several samples before assigning a name. Williams *et al.* (1982, Chap. 9) give a good, brief introduction to the petrography of pyroclastics. See Tables 6-2 and 6-3 for selected petrographic and chemical analyses.

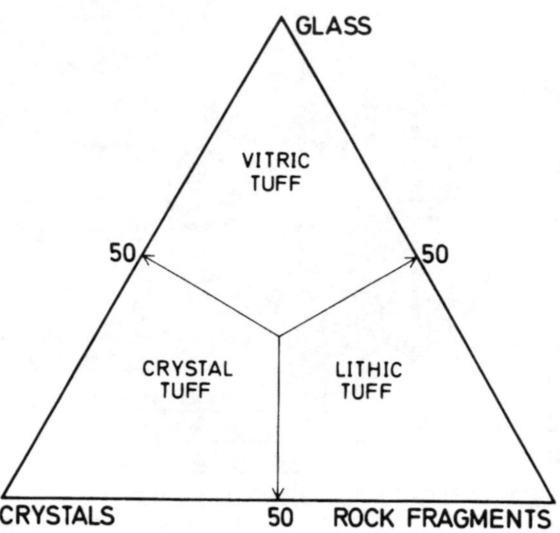

FIGURE 6-10. Triangular diagram and end-members of tuffs.

TABLE 6-1. Classification of pyroclastic deposits.

Size (mm)	Rock name	Origin
Bombs and blocks —64— Lapilli —2— Ash or tuff[a]	Determined by proportion of glass, crystals, and rock fragments plus composition of crystals or via chemical analysis.	Three types, *juvenile* (derived from fresh magma), *accessory* (torn from volcanic neck), *accidental* (fragment of country rock).

[a] Termed ash, if uncompacted, tuff, if compacted.

TABLE 6-2. Some modal analyses of redeposited volcaniclastics.

Unit	Quartz	Feldspar	Rock fragments	Glass	Clay	$CaCO_3$	Silica cement	Remarks	
Modern river sand, Rio Tituaca, Western Chihuahua	19	14	35	17	4	—	—	11% heavy minerals	Drains a mountain region underlain by rhyolithic tuffs
(Webb and Potter, (1969)	13	18	52	12	3	—	—	6% heavy minerals	
Gueydan Formation (Late Tertiary)	47	25	19	—	2	34	—	Caliche cement	Thin fluvial deposit
southeast Texas (McBride *et al.*, 1968)	41	23	35	—	—	—	37	Cement is opal	
Currabubula Formation (Carboniferous)	11	27	17	—	45	—	—	Coarse rhyolitic, vitric, crystal tuff	
New South Wales (Whetten, 1965)	1	52	8	—	32	—	—	Andesitic crystal vitric tuff	
Parry Group (Devonian–Carboniferous)	1	19	65	—	7	5	—	Parry group is 17,000 ft (5520 m) thick	
New South Wales (Crook, 1960)	—	35	47	—	14	—	—	4% heavy minerals in framework	
Umpqua Formation (Eocene) Oregon (Dott, 1966)	33	17	29	—	14	—	—	7% accessories, mostly micas and heavy minerals. Some glass in matrix. Average of 10 analyses.	
Napere and Ouka Sandstones (Miocene)	—	31	7	—	51	—	—	11 and 9% mafic minerals framework	
New Guinea (Edwards, 1947)	—	24	31	—	36	—	—	Thick marine turbidites	

TABLE 6-3. Chemical analyses of volcaniclastics.

	Pyroclastic/deposits			Redeposited volcanic sands	
	Rhyolitic tuff, Peru[1]	Andesitic pumice, Japan[2]	Ignimbrite, Waihi district, New Zealand[3]	Modern rivers, Western Chihuahua, Mexico[4]	Columbia River, USA[5]
SiO_2	70.81	63.53	75.5	73.50	61.69
Al_2O_3	13.12	16.01	13.5	13.3	13.89
Fe_2O_3	1.11	2.50	0.35	1.55	3.82
FeO	0.06	2.81	0.2	0.56	2.20
MgO	0.35	1.44	0.2	0.56	2.20
CaO	1.05	4.84	1.3	1.13	3.10
Na_2O	4.92	4.29	4.35	2.34	2.20
K_2O	4.20	0.84	3.5	4.01	1.88
H_2O^+	2.34	2.49	0.45		
H_2O^-	0.10	0.75	0.45	1.80	1.81
CO_2	0.02	—	—	0.12	—
SO_3	0.31	—	—	—	—
Cl	1.12	—	—	—	—
F	0.08	—	—	—	—
TiO_2	0.19	0.75	0.17	0.34	1.03
P_2O_5	0.00	0.16	0.06	0.02	0.25
MnO	0.07	0.07	0.01	0.04	0.11
BaO	0.09	—	—	—	—
	100.00	100.48		99.3	94.00[6]

[1] Jenks and Goldich (1956, Table 3.1).
[2] Yagi (1962, Table 4.6).
[3] Thompson *et al.* (1965, Table B-L,5).
[4] Webb and Potter (1969, Table 5.6).
[5] Whetten (1966, Table 2, CC 29).
[6] Loss on ignition 5.99.

Tuffs may grade into other volcaniclastics, into nonvolcanic quartzose terrigenous sandstones, or into carbonate sand, or may be interbedded with them. If terrigenous material exceeds volcanic, the mixture is called a tuffaceous sand or sandstone.

Volcaniclastic Deposits

Volcaniclastic deposits form as subaerial pyroclastic flows and air falls, as debris avalanches, and as lahars (debris flows), and they also originate as alluvial and other continental sands by erosion of either recent or ancient volcanic terranes. Volcaniclastics are also deposited in lakes or seas as air falls far from or near their source, by pyroclastic flows and debris flows, and by both normal, shallow marine currents and deep-water density currents (Table 6-4). Comprehensive discussion of all these modes is provided by Fisher and Schmincke (1984). Knowledge of sedimentary structures (Chap. 4), depositional systems (Chap. 10), and volcanic rocks are all needed for effective interpretation. A glossary (p. 243) is always helpful when beginning.

Before considering each of the different volcaniclastic deposits, some salient facts about volcanism are relevant. Most pyroclastic materials are felsic in composition and tend to be rhyolitic, although dacitic and andesitic compositions also occur. The fact that silicic magmas are more viscous and contain more dissolved gas than mafic magmas favors explosive eruptions of the felsic volcanics. In contrast, the volcanic flows that characterize the outpourings of plateau basalts are quieter and do not, as a rule, form stratovolcanoes. As Mount St. Helens has shown, the sequence of explosive eruptions and lava flows may be complex. Initially, country rock fragments may be incorporated into the eruptive material, whereas later the lava completely dominates. Thus in a section deposited over considerable time, a mixture of compositions may be deposited from the same magma.

Pyroclastic Flows

Broad synonyms used for pyroclastic flows include ash flows, pyroclastic surges, *nuée ardentes*, and sand flows. Local names such as sillar (Peru) also abound. The resulting deposit is called an *ignimbrite*. A pyroclastic surge differs from a flow in that a surge is temporary with rapidly decaying kinetic energy whereas a flow has a more continuous energy supply. Ignimbrites represent hot, "catastrophic" events—hot density currents that travel swiftly downhill, some at velocities approaching 161 km/h, commonly following the larger topo-

TABLE 6-4. Volcaniclastic deposits.

Pyroclastic deposits

Flows and surges: Both flows and surges form ignimbrites, which may be subaerial or subaqueous. Flows and surges are formed by hot gas-charged mixtures derived from a volcanic center or fissure. Lateral and vertical zonation is good, clasts are abundant, and welding is common. Sediments are somewhat similar to turbidites and, in a sense, are hot turbidites. Tend to follow valleys and be flat-topped. May extend many kilometers down dip. Also called ash flows, base surges, and nuée ardentes.

Air falls: Ejected into air, consisting of pumice or ash and have strong lateral gradients of thickness and size; some may extend tens to hundreds of kilometers from vent. Cover preexisting topography on land but also are deposited at sea, where they are well preserved, making good chronostratigraphic markers. Well stratified according to size and density.

Debris avalanches

May be hot or cold, consisting of a poorly sorted mixture of debris rapidly moving downslope. Hot avalanche deposits are called glowing avalanches. Both hot and cold avalanche deposits are confined to lows on volcanic slopes. Distinguished from pyroclastic flow by lack of welding (lower temperature) and different sedimentary structures. May be difficult to distinguish from some lahars.

Lahars

Identical to debris flows in form, process, and sedimentary characteristics except composition, which is almost totally volcanic. On land, hot or cold lahars are derived from a volcanic center or a steep volcanic flank, follow valleys, and are elongate, but can be digitate, where they terminate in a wide valley. Similar debris flows occur with turbidites and may be associated with an underwater explosion or with an oversteepened shelf margin.

Redeposited volcaniclastics

Occur in all sedimentary environments with a volcanic source, recent or ancient. Where contemporaneous with volcanism, they occur in all major sandy environments, but alluvial, fan delta, and turbidite deposits are especially common along active continental margins and in oceanic magmatic arcs. With burial, virtually all become wackes because of susceptibility of unstable, volcanic debris to alteration.

FIGURE 6-11. Above: Taupo Ignimbrite (Holocene) fills small channel on North Island, New Zealand (Froggatt, 1981, Fig. 6). Below: Air fall tuffs (white) interbedded with reworked tuffs (red) in the Comondú Formation (Miocene) at San Juan de la Costa, Municipio de la Paz, Baja California Sur, Mexico. Photograph above courtesy of New Zealand Journal of Geology and Geophysics.

graphic lows. The temperatures of this mixture of gas and incandescent particles have been estimated to be between 550° and 950°C. Such deposits may extend as much as 32 to 96 km from their source. Because pyroclastic flows cover the ground surface uniformly, they have flat tops and variable thickness where there is local relief (Fig. 6-11); however, if the caldera is large and the paleolandscape had little relief, as seems to be true in the Sierra Madre Occidental of northern Mexico, the ignimbrites are sheet-like.

In addition, ignimbrites are generally internally zoned, imparting a strongly bedded character to the deposits (Fig. 6-12). Single cooling

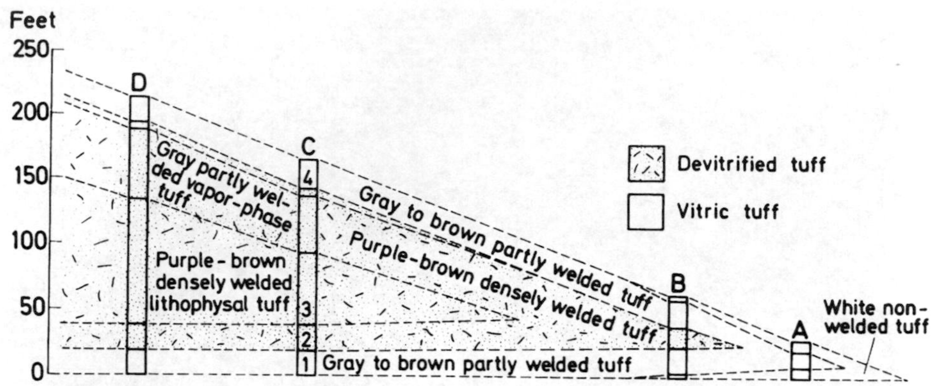

FIGURE 6-12. Internal zoning of a welded tuff. Length of section about 8 km. (Modified from Lipman and Christiansen, 1964, Fig. 2).

units, units representing a single pulse of sedimentation, can be as much as 100 m in thickness, but most vary between 15 and 30 m. Vertical zonation within a unit is defined by color, density, and structures, all of which reflect degrees of welding and crystallization. The internal zoning results from differential cooling between exterior and interior of the unit. One flow may differ from another in number of inclusions, grain size, color, and welding characteristics as well as by mineralogical composition. Change in grain size at contacts between flows is particularly marked. Rapid accumulation of the incandescent tephra prevents rapid dissipation of heat so that the glass shards and pumice fragments are either partly or totally fused or collapsed, forming a welded tuff (Fig. 6-13). The greater the degree of welding, the greater is the density. Definitive field evidence for pyroclastic flows are flattened, stretched pumice fragments which can show tension cracks, pull-aparts, imbrication, and rotation, all of which can be used to infer transport direction. In rare circumstances, movement during or after weld-

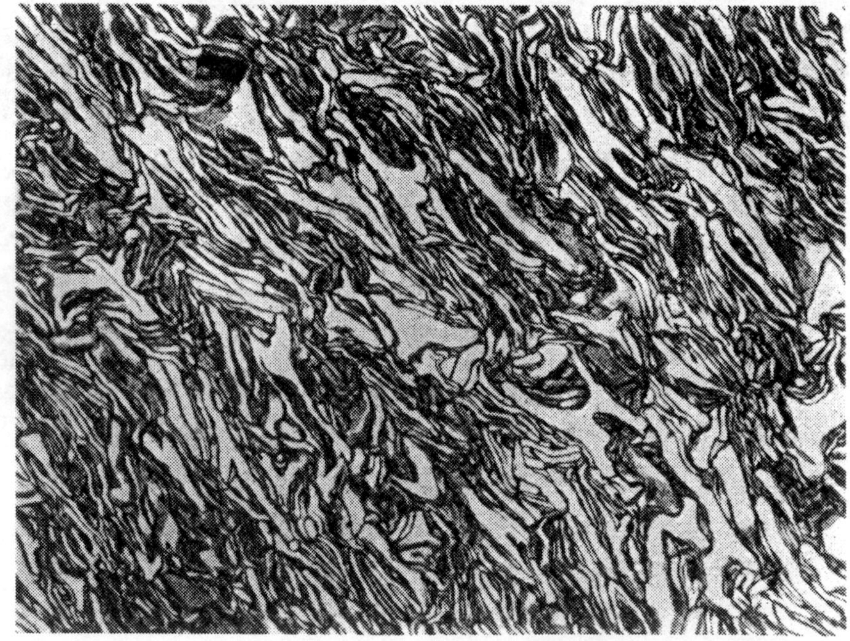

FIGURE 6-13. Welded vitric tuff, × 40. Walcott Tuff (Tertiary), Ferry Hollow, Sec. 6, T.8 S., R.31 E., Power Co., Idaho, U.S.A.

FIGURE 6-14. Thickness and clast size in cooling units 1 and 2 of Picture Gorge Ignimbrite (Oligocene–Miocene) in eastern Oregon. Clast size is from lower 6 meters of deposit. Transport from left to right. (Modified from Fisher, 1966b, Figs. 1 and 4).

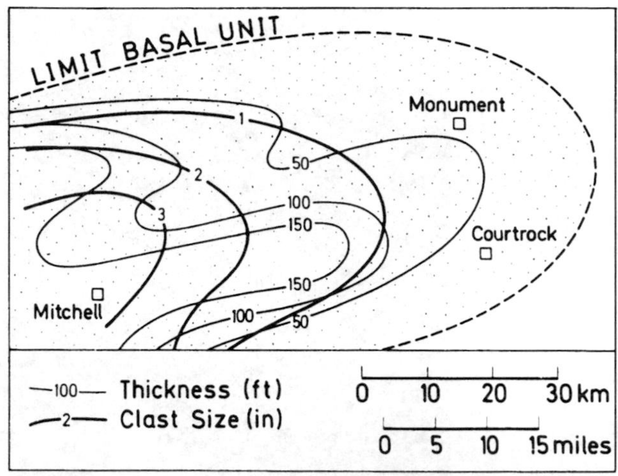

ing produces both deformation and laminar flowage features in the deposits. Features indicative of such movement are illustrated by Schmincke and Swanson (1967). Also useful is the downcurrent decline in flow thickness away from the source (Fig. 6-14), which may be accompanied by decrease in clast size, although marked thickness decrease is not always conspicuous. However, where a flow crossed topography at right angles, clast size and zoning vary markedly (Moore and Sisson, 1981, Fig. 247). Oriented logs are also good paleocurrent indicators (Fig. 6-15).

Sedimentary structures are well described (Fisher and Schmincke, 1984, pp. 107–15; Sheridan and Updike, 1975; Wohletz and Sheridan, 1979). In surface flows these consist of plane, massive, and sand wave beds (Fig. 6-16), whose origin as an incandescent, fluidized density flow is broadly similar to that of cold-water density currents (Chap. 8). As a flow or surge moves downslope, it deflates (defluidizes) and its structures change from proximal to medial to distal—from sand waves to massive to planar beds—as reported by Wohletz and Sheridan (1979), who also note a downcurrent decline in thickness of crossbedding (Fig. 6-17). The types of crossbedding due to antidunes vary widely (Fig. 6-18). The paleocurrent pattern of the crossbedded facies has been mapped with good results.

Fiske and Matsuda (1964) give a good description of submarine pyroclastic flow deposits, as do Wright *et al.* (1980), Fisher (1984), and Yamada (1984). When an explosion occurs in deep water, the pyroclastic flow is quickly chilled and settles to the bottom, where it forms a dense turbidity flow. Such submarine flow deposits are better sorted than their subaerial counterparts, because the greater density and viscosity of water reduce the settling velocity, permitting better segregation by differences of size and density. For example, pumice and

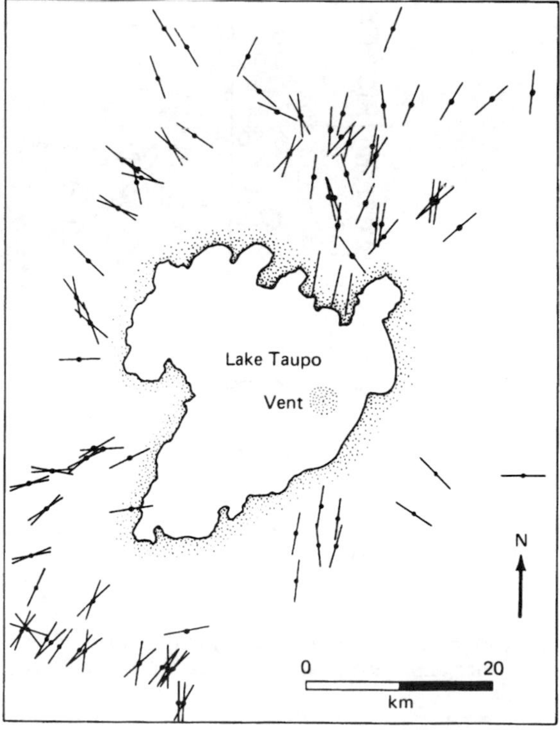

FIGURE 6-15. Oriented logs of the Taupo Ignimbrite on North Island, New Zealand, show flow direction and vent position, as would oriented pumice and other paleocurrent structures. (Redrawn from Froggat *et al.*, 1981, Fig. 2).

FIGURE 6-16. Large sand wave in ash flow deposit of Mount St. Helens, Sec. 27,T.9 N., R.5 E., South Kamina County, Washington, U.S.A. Flow from left to right. Compare with Figure 6-18. Photograph courtesy of Stephen Self.

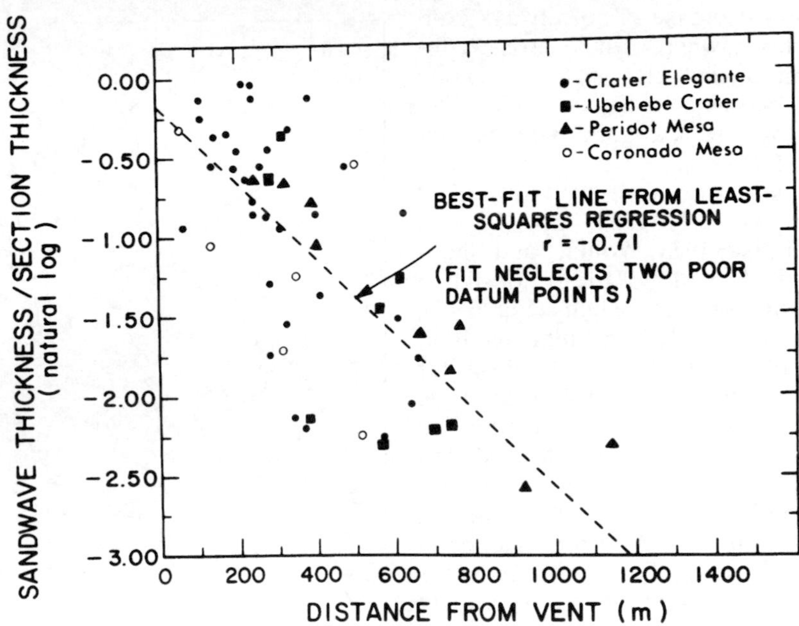

FIGURE 6-17. Plot of average thickness of sand waves at an outcrop versus its distance from vent (Wohletz and Sheridan, 1979, Fig. 5). Published by permission of the authors and the Geological Society of America.

lapilli will float to the surface and be dispersed by surface currents whereas crystals and lithic fragments tend to be deposited first both laterally away from a vent and at the base of a flow unit. Pyroclastic flows also can start on land, follow valleys to a lake or sea, and ultimately end as turbidites, if water depth permits.

Notable studies of modern pyroclastic flows are by Lacroix (1904) of Mount Pelée on the island of Guadaloupe, by Gorshkov and Dubik

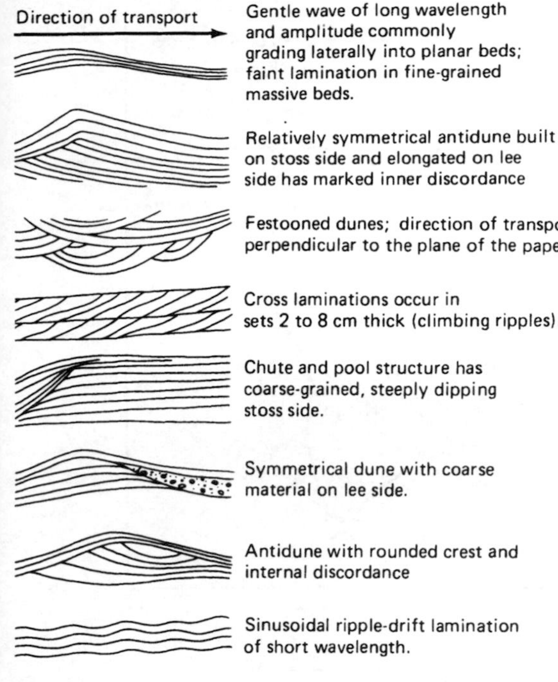

Direction of transport
Gentle wave of long wavelength and amplitude commonly grading laterally into planar beds; faint lamination in fine-grained massive beds.
Relatively symmetrical antidune built on stoss side and elongated on lee side has marked inner discordance
Festooned dunes; direction of transport perpendicular to the plane of the paper.
Cross laminations occur in sets 2 to 8 cm thick (climbing ripples).
Chute and pool structure has coarse-grained, steeply dipping stoss side.
Symmetrical dune with coarse material on lee side.
Antidune with rounded crest and internal discordance
Sinusoidal ripple-drift lamination of short wavelength.

FIGURE 6-18. Types of inclined bedding—antidunes—in pyroclastic flows. (Redrawn from Wohletz and Sheridan, 1979, Fig. 3).

(1970) of the Shiveluck volcano on Kamchatka Peninsula in far eastern Siberia, and those of the 1980 eruption of Mount St. Helens in Washington (Rowley *et al.*, 1981).

Air Falls

Air falls are the products of atmospheric transport from which particles are segregated by their fall velocity as they are transported downwind. Size and density are primary factors that control fall velocity (Chap. 8). Wind direction, velocity, and turbulence, as well as the height to which the particles are ejected, control the resultant fallout pattern. This pattern may be roughly symmetrical or markedly asymmetrical and elongate as in Fig. 6-19. Air falls tend to have an exponential decline of thickness downwind and a corresponding decrease in grain size (Fig. 6-20). In the ideal case, composition, grain size, and thickness are all interrelated and systematically change downwind. Scheidegger and Potter (1968) developed equations relating grain size and thickness to distance from source, but see Suzuki (1983) for more details. Air fall deposits are especially suitable for such study, because they are one of the most simple sedimentation systems in nature—progressive downwind decrease of turbulence in the plume with essentially no complications due to reworking. Air fall deposits are distinguished from flows by their good stratification (Fig. 6-21) and sorting (Fig. 6-20) and, near their source, the presence of volcanic bombs and bomb impact structures, formed in unconsolidated ash and sediment by clasts ejected from a volcano.

As is true of pyroclastic flows, post-depositional alteration of air fall deposits by subaerial weathering is usually extensive and is favored by high porosity and fineness of grain. Hay (1959) described alteration of air fall deposits in the West Indies. As a result of alteration, pre-Cenozoic air falls have been almost always altered to bentonite and are composed primarily of smectite. Silicification is common along the boundaries of bentonites, especially at the base.

Ash falls, stream deposits, lahars, lava, or soil horizons may be found between pyroclastic flows. Crossbedding in associated stream deposits and bent pipe amygdules of lavas (Waters, 1960) can help establish flow direction. Mapping the distribution of types of interbedded lavas also helps establish paleogeography and local paleorelief, as is well shown by Dimroth and Rocheleau (1979, pp. 32–39), who recognized paleochannels and upward shoaling cycles bounded by distal pillow lavas in the Precambrian of Quebec.

Debris Avalanches

The collapse of a crater rim or failure of an oversteepened slope composed of loose, recently deposited volcanic debris on a stratovolcano produces an avalanche deposit, either hot or cold. Topography permitting, such avalanches can travel up to 10 to 20 km or more and consist of angular or crushed, poorly sorted volcanic boulders, cobbles, gravel, and sand. Avalanche deposits are never far from the volcanic center and are deposited in valleys or lows. If hot, they are called glowing avalanche deposits. Davies *et al.* (1978) describe such deposits in Guatemala, where they consisted of two parts—underflows of fine and coarse debris and overlying turbulent gas or dust clouds. These underflows carried blocks as large as 5 m more than 7 km and individual avalanches formed deposits up to 2 m thick. Average speeds of up to 60 km/h were observed. Davies *et al.* (1978) provide complete sedimentologic descriptions and much insight into the fluid mechanics of the glowing avalanches produced by

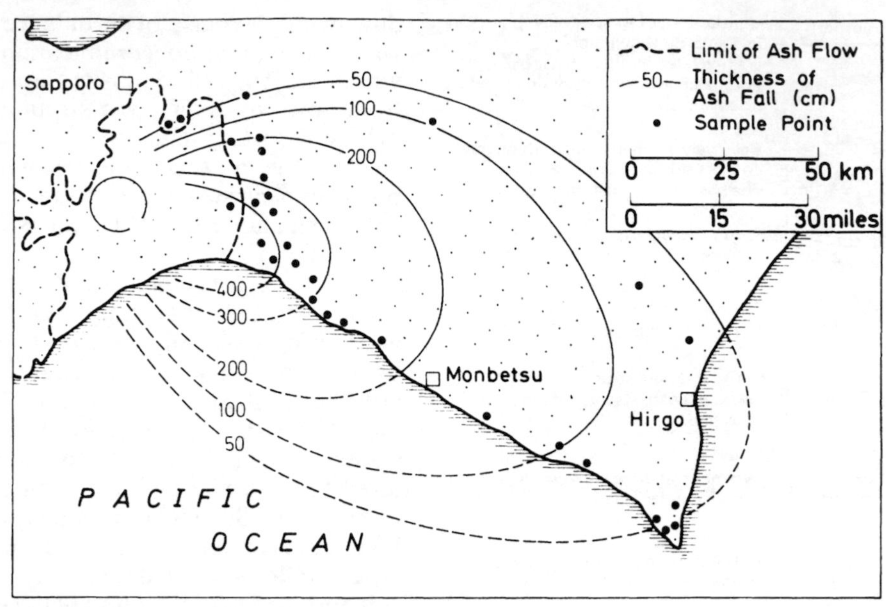

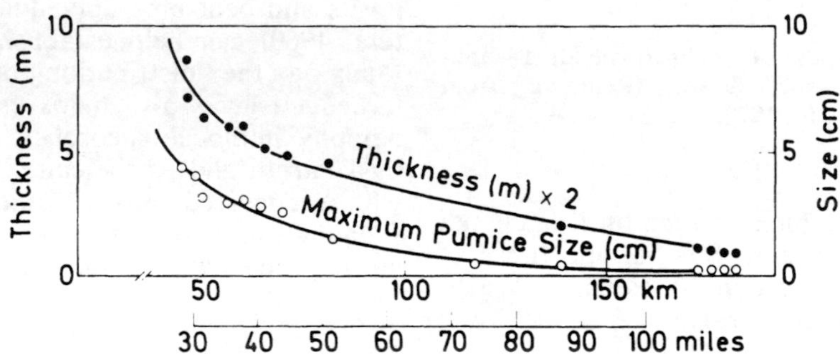

FIGURE 6-19. Downwind thickness and grain size decline of an ash fall. (Modified from Katsui, 1963, Figs. 4 and 5).

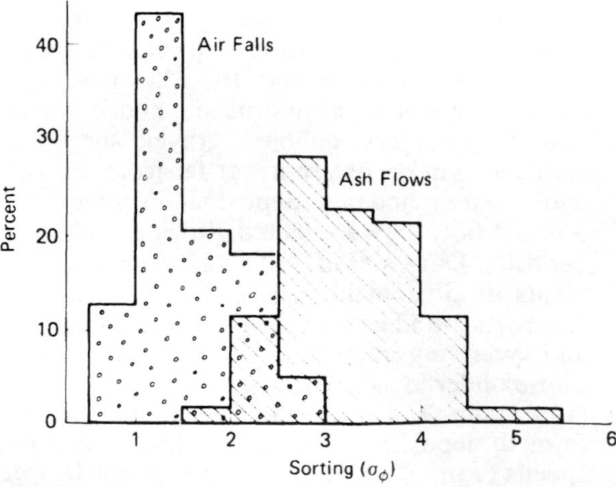

FIGURE 6-20. Sorting contrast between 39 Recent air falls with excellent to fair sorting and 192 ash flows with fair to very poor sorting. Data from Japan (Musai, 1961, Tables 7 and 9).

Fuego in 1974 in Guatemala. Waitt (1981) also describes a debris avalanche at Mount St. Helens. See Hsu's translation (1975) of Albert Heim's classic 1882 description of a nonvolcanic avalanche, a *Sturzstrom*, in Switzerland to appreciate fully the awesome tractive force of debris avalanches. Hsu (1975, pp. 134–38) also gives a full analysis of the mechanics of cold-debris avalanches.

Lahars

Like debris avalanches, lahars are associated with the eruption of a stratovolcano. Volcanic debris of all sizes can be mobilized by water from a crater lake, a heavy rain, or the sudden thawing of ice by an eruption. The Indonesian word *lahar* is used for such deposits. They may be hot or cold: a cold lahar is one not contemporaneous with eruption. Lahars flow downslope

FIGURE 6-21. Quaternary ash fall pumice in the "Taupo" District, New Zealand. Observe perfection of bedding and its parallelism to topography, except below unconformity near base. Photograph courtesy of D.L. Homer and New Zealand Geological Survey.

following valleys for long distances—ten's of kilometers. Thickness ranges from a few meters to perhaps as much as 100 m. Where they leave the mountain front, lahars tend to have a digitate pattern, topography permitting. Lahars have poor sorting typical of mass flow deposits and may contain boulders ranging up to four to five meters or more. Lahars commonly are associated with fluvial deposits at their sides and on top and they may discordantly over- or underlie any of the other volcanic deposits or be intercalated with other lahars.

Descriptions of modern lahars are most informative. Waldron (1967) and Kesel (1973) report on modern lahars in Costa Rica (Fig. 6-22), where velocities ranged from 2.9 to 10 m/s and averaged 5.3 m/s. Janda *et al.* (1981) describe in detail lahars associated with the eruptions of Mount St. Helens in Washington, some of which extended more than 120 km down-valley and had velocities ranging from 1.5 to 40 m/s.

Volcanic debris flows also occur underwater (Wright, 1980) in association with turbidites. The needed requirement is an oversteepened, metastable submarine slope such as an unstable shelf margin or along the sides of a growing underwater volcanic vent. A submarine eruption could also generate turbidity currents.

Lahars can be easily confused with glacial till. Crandell and Mullineaux (1975, Table 2) list features that help distinguish coarse lahars from avalanche and glacial till. Better sorting and stratification plus possible ripple marks and crossbedding help distinguish a coarse fluvial volcanic gravel from a lahar. However, the ratio of load to water is so high in many proximal streams that classic fluvial crossbedding is not always well developed.

Eisbacher and Claque (1984, pp. 29–43 and Table 2) summarize the different types of mass movement deposits associated with volcanoes, their initial temperature, water, and gas contents, lithologic composition, and relationship to eruption. See Innes (1983) for a review of the mechanics of all kinds of debris flows.

Redeposited Volcaniclastic Sandstones

Pyroclastic material and lavas are major primary sources of terrigenous debris along active continental margins and island arcs. Rifts developing on cratons commonly also contain volcanic rocks, although commonly in lesser amount. Redeposited volcanic debris may come to rest on coastal plains, beaches, and shallow marine shelves, or be carried into deep water by turbidity currents. We can trace in some detail this seaward transport of debris, beginning by

FIGURE 6-22. Accelerated channel erosion in headwater stream (above) and resultant debris flow in valley (below) after an eruption and subsequent heavy rainfall in central Costa Rica (Waldron, 1967, Figs. 6 and 10).

considering what happens when lava, ignimbrites, avalanches, lahars, and air falls inundate a landscape, as has happened many times in the geologic past.

Vegetative cover is destroyed, resulting in greatly accelerated sheet, rill, and gully erosion that destroys slope stability. Increased runoff can then greatly enhance channel erosion in the upper reaches of the drainage network (Fig. 6-22, top). Loss of vegetative cover usually is accompanied by landsliding, thus contributing more debris to valley floors. In such a setting a heavy rain can trigger the transport of an enormous quantity of debris. Waldron (1967, pp. 16–29) gives an excellent account of the erosion and transport of volcanic debris following an air fall and heavy rain in central Costa

Rica. Debris flows, possibly containing 35 percent solids, develop in the headwaters and spread quickly downstream, covering lower valleys with a thick, poorly sorted, bouldery deposit (Fig. 6-22, bottom). Sorting averages 3.1 ϕ and ranges from 2.6 to 4.0 ϕ. Huge quantities of sand, silt, and clay are transported farther downstream to form coastal plain deposits and hence to the strandline and the shelf. Air fall and avalanche deposits and lava flows are commonly associated with the redeposited volcanic debris. Where lavas are interbedded, it may be possible to compare paleoslope inferred from both sands and lavas (Sandberg, 1938, pp. 818–30; Schmincke, 1967, Fig. 2; Hammond, 1974, Figs. 4-4, 4-8, 4-9, and 4-13).

Beyond the shelf, volcanic debris is carried into deep water by turbidity currents. Pillow lavas resulting from submarine eruptions may be interbedded with such turbidites, and debris flow deposits are common. Because of the low density of pumice, the Bouma cycle of such turbidites does not fine upward, but has coarse pumice at its top.

Examples

Brief summaries of both modern and ancient volcaniclastic deposits provide insight into their diversity and their role in volcanic stratigraphy and mineral exploration.

The major valleys of mountains along the leading edge of a continent commonly contain volcaniclastics, as is well shown in northern Colombia and on the island of Luzon in the Phillipines. The volcaniclastic sediments of the Magdalena Valley of northern Colombia were derived from the stratovolcanoes of the Cordillera during the late Cenozoic and are preserved for about 450 km along the upper Magdalena Valley, which is up to 50 km wide. These sediments are up to 3000 m thick. This valley fill is well exposed, and has been described by Van Houten (1976). His observations (Table 6-5) are of interest because they enhance our understanding of ancient buried equivalents, of which there are probably many. Down valley, longitudinal transport prevails but marginal fans occur and are related to volcanoes that have been intermittently active since the Middle Miocene. During some periods these fans have forced the Magdalena to the far side of its structurally controlled valley. Lacustrine deposits, although not reported, may also be expected where rapid aggradation by the main stream dammed tributary valleys. These valley fill deposits, except for their volcanic composition and the far-traveled lahars, are broadly similar to those of a glacial outwash valley train. The study of paleocurrents, clast composition, and facies maps would help locate the volcanoes that supplied debris to similar ancient deposits. See Mathiesen and Vondra (1983) for a description of a broadly similar sequence in the Phillipines.

TABLE 6-5. Facies of valley fill of upper Magdalena Valley, Colombia. (Adapted from Van Houten, 1976, p. 494).

Unbedded lahars: Coarse boulders to fine sands, angular, poorly sorted debris, and clasts of lava and pumice. Lighter clasts travel farther than denser ones. Form stratigraphic units traceable for many kilometers.

Interbedded lahars, gravels, and sands: As above, but contain appreciable water-transported sands and gravels and represent greater proportion of normal valley train fill with lesser amounts of unbedded lahars. Medial to distal fans and fills.

Gravels and sands: Broad valley fills deposited by braided streams with minor lahars as well as some overbank, parallel laminated, and crossbedded, pumice-rich sands.

Sedimentologic study of the deposits of the 1974 explosion of Fuego in Guatemala by Vessell *et al.* (1981) shows how the sedimentary facies of volcaniclastic deposits vary with distance of transport from a point volcanic source (Fig. 6-23). Volcanic cycles consist of four phases: an eruptive phase characterized by air fall ash and glowing avalanches, a fan-building phase with lahars, a fluvial-braiding phase that lasts 20 to 40 years after an eruption, and a quiescent phase of 80 to 125 years. Stream incision and delta reworking occur during the quiescent phase. Also distinguished are four volcanic facies—proximal, medial, distal, and offshore deposits (Fig. 6-24). A useful table summarizes the characteristics of these nonmarine deposits (Table 6-6). Davies *et al.* (1978) give a careful sedimentological analysis of transport and abrasion in these deposits. Recognition of these facies combined with studies of clast size and paleocurrents will go far to locate a volcanic vent of economic interest in ancient deposits (Fox, 1977).

Late Quaternary deposits of the Lesser Antilles Arc have been described by Sigurdsson *et al.* (1980) and Sigurdsson and Carey (1981) using both deep-sea cores and outcrops. Here the production of volcaniclastic materials of all kinds is unusually large—527 km^3 in the last 100,000 years—but probably not exceptionally so in a volcanic arc. Over 80 percent is believed to have been deposited in the ocean as ash falls,

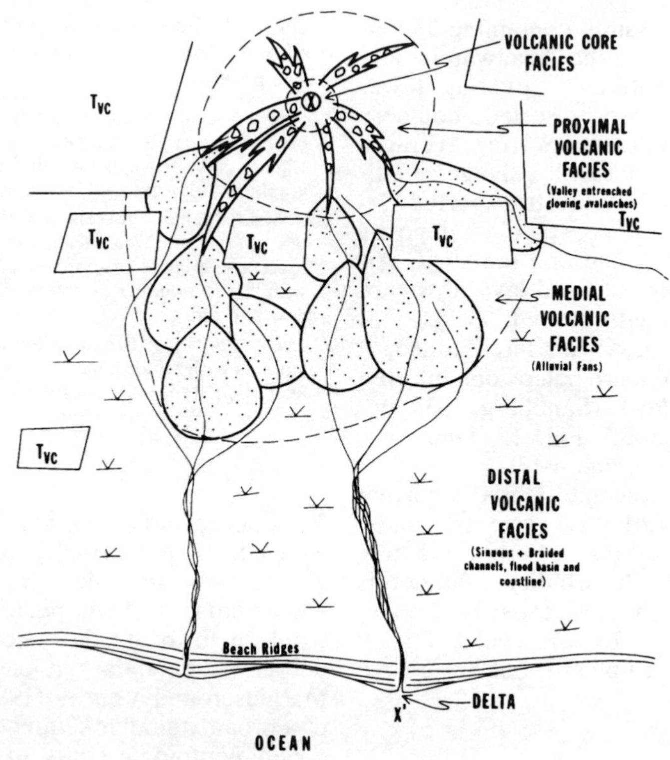

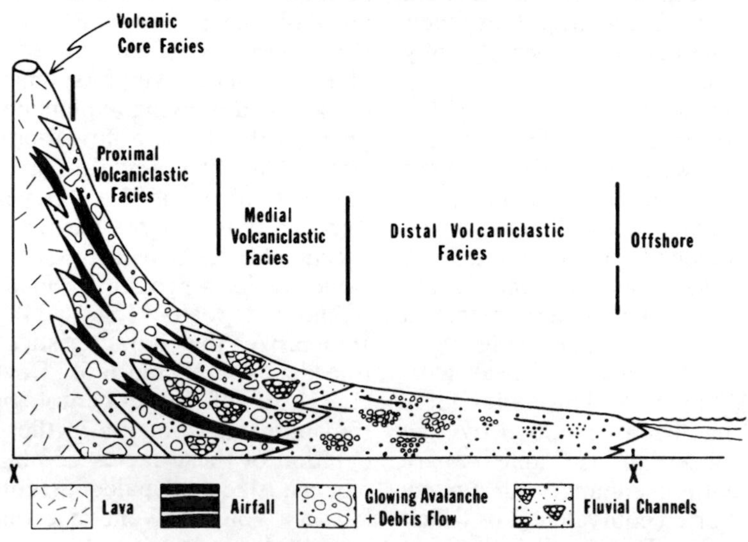

pyroclastic debris flows, and volcanic turbidites. The combination of debris and turbidite flows is estimated to form about 70 percent of the total volume—most of it in the back-arc Grenada Basin (Fig. 6-25). Lahar and flash flood deposits occur on all the islands and locally are the dominant rock type. On Martinique alluvial fans reach 200 m in thickness. Pyroclastic flows also are believed to enter directly into the sea,

FIGURE 6-23. Schematic distribution of nonmarine facies around El Fuego from Vessell and Davies (1981, Figs. 3 and 4): Stratovolcano and its deposits in Guatemala (above); Down-dip cross section of volcanic deposits from vent to sea (below). Published by permission of the authors and the Society of Economic Paleontologists and Mineralogists.

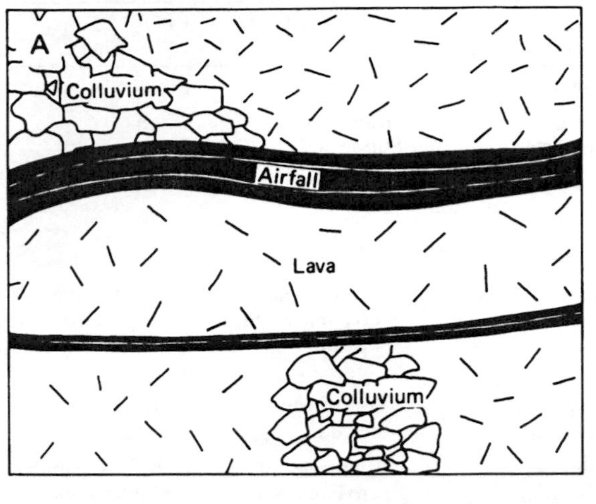

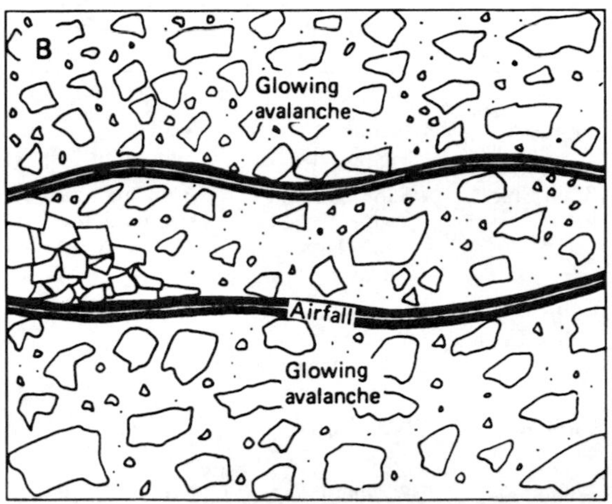

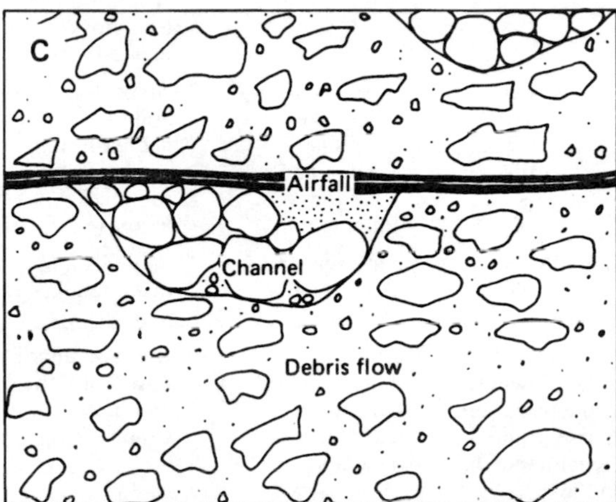

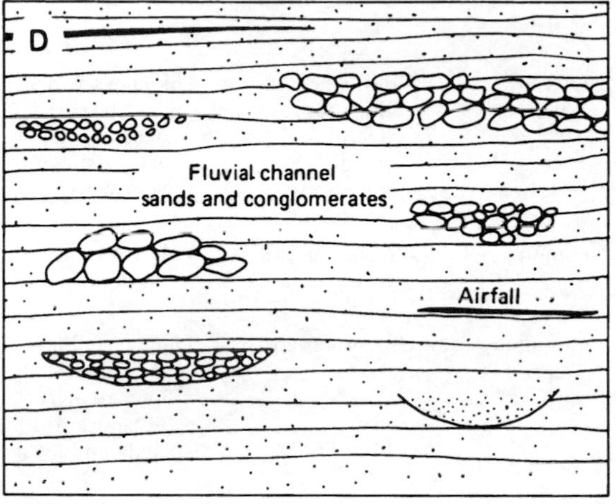

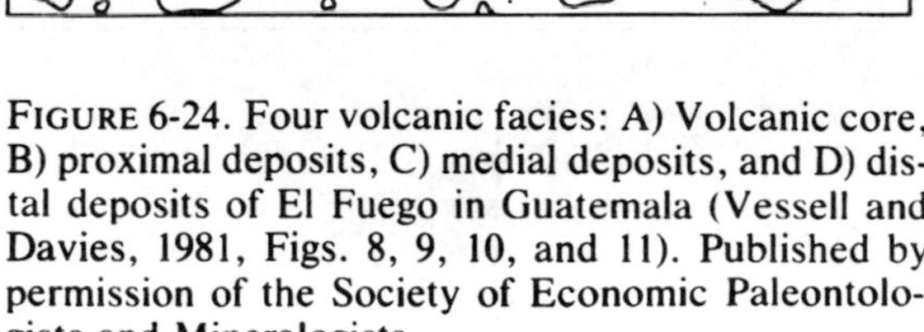

FIGURE 6-24. Four volcanic facies: A) Volcanic core, B) proximal deposits, C) medial deposits, and D) distal deposits of El Fuego in Guatemala (Vessell and Davies, 1981, Figs. 8, 9, 10, and 11). Published by permission of the Society of Economic Paleontologists and Mineralogists.

because most vents are but a few kilometers from it. One of these flows extends 250 km from its source—so widespread that it covered most of the Grenada Basin as a catastrophic submarine event. A schematic cross section is informative (Fig. 6-26) and the role of the wind system in sending most Late Quaternary air falls east of the arc is noteworthy.

The many facets of Cenozoic volcanism in southeastern New Mexico are described in 18 papers which give a good perspective of the great diversity of rocks in a volcanic province on a continent and how sedimentology helps locate volcanic vents (Elston and Northrop, 1976). Thompson *et al.* (1965) describe the central volcanic region of New Zealand whose geologic processes, rocks, and landscapes must have had many equivalents in orogenic zones of the ancient past.

Examples of sedimentologic study of ancient volcaniclastic deposits are becoming fairly common. Niem (1977) describes pyroclastic flow and ash fall deposits in the Mississippian turbidites of the Ouachitas in Oklahoma and Arkansas and shows maps of clast size and thickness. Cas *et al.* (1981) describe a submarine volcanic apron consisting of mass-flow deposits in the Lower Devonian of New South Wales. Dimroth, Rochelau, *et al.* (1985) recognize three stages in the Archean sequence of the Chibougamau area of Quebec: an initial submarine volcanic sea mount, emergence with a volcanic apron, and final uplift and erosion to the level of their subvolcanic plutons. They distinguished different lava types, mapped their distribution, and recognized pyroclastics and reworked volcaniclastics deposited in coastal shelf and deep-water environments. Sedimen-

TABLE 6-6. Sedimentary characteristics of volcaniclastics formed by the 1974 eruption of Fuego in Guatemala (Adapted from Vessell and Davies, 1981, Table 1).

Sedimentary characteristics	Air fall ash and tephra	Glowing avalanche	Debris flow	FLUVIAL DEPOSITS Coarse	Fine
Base	Non-erosive	Non-erosive	Non-erosive	Erosive	Erosive
Fabric	Grain Supported	Matrix Supported	Matrix Supported	Grain Supported	Grain Supported
Grain Size	Sand, granule size particles and large clasts	Boulders to sand size particles	Boulders to sand size	Boulders to sand	Gravel to sand
Vertical Grain Size Change	Commonly graded, may fine or coarsen upward	No change	Coarsest boulders at top	Variable; may coarsen or fine upward	Generally fine upward but variable
Sorting	Well sorted	Very poorly sorted	Very poorly sorted	Very poorly sorted	Poorly to well sorted
Sedimentary Structure	Parallel laminated, may be graded	None	Faint parallel bedding	Structureless and crossbedded	Generally parallel bedded, but also structureless crossbedded
Downflow Changes	Grain size and thickness decrease rapidly. Sorting improves downwind	Thickness variable. No change in grain size, sorting or roundness	Thickness decreases. No change in grain size, sorting or roundness	Thickness and grain size decrease, but sorting and roundness increase	Thickness and grain size decrease, but sorting and roundness increase
Geometry	Sheets conform to topography	Valley fills, with flat tops	Valley fills and lobes on fans; elongate	Valley fills and distal fanlike deposits	Valley fills and distal fanlike deposits
Associated Deposits	Highly variable	Thick airfall ash beds, debris flows, water-transported sands and lava	Thin airfall ash beds, glowing avalanches and coarse grained fluvial deposits and lava	Rare, thin airfalls. Lahars and some glowing avalanches	Airfall and debris flows, or delta and beach deposits

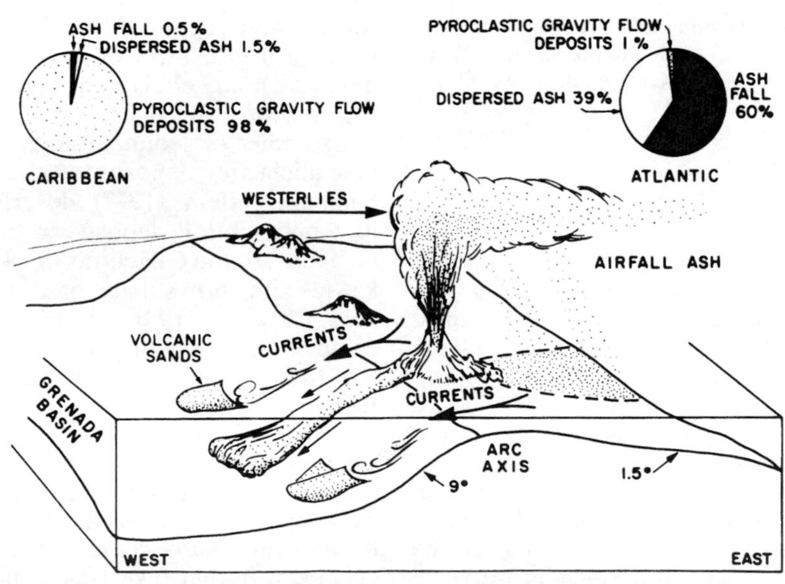

FIGURE 6-25. Idealized block diagram of magmatic arc of the Lesser Antilles (Sigurdsson and others, 1980, Fig. 4). Note asymmetry of turbidity and air fall deposits. Published by permission of the Journal of Geology.

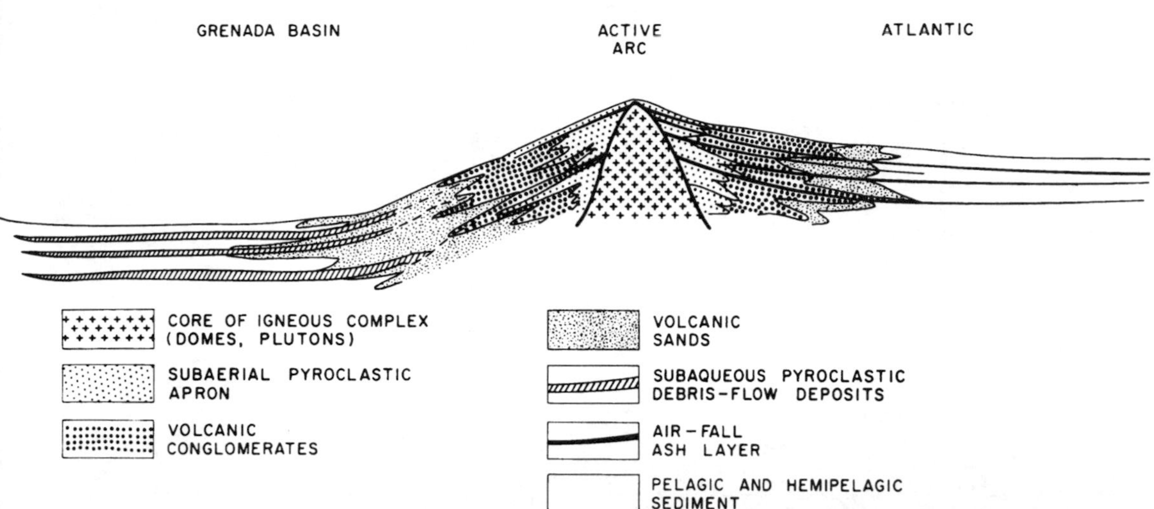

FIGURE 6-26. Idealized cross section of Figure 6-26 (Sigurdsson *et al.*, 1980, Fig. 8). Published by permission of the Journal of Geology.

tary structures and sedimentology are central to their interpretation of this ancient volcanic basin. Juve (1977) describes detrital sulfides deposited by turbidity currents in a fore-arc, volcanic basin of Precambrian age in Sweden and Schermerhorn (1971) documents pyrite emplacement by gravity flow in Spain.

Tectonic Setting

Explanations of the kinds and distribution of volcanism are embedded in plate tectonic theory (Honza, 1983; Shimozuru and Yokoyama, 1983; Uyeda, 1981). The relationships of volcanism to seismicity, plate boundaries, and hot spots are central to the ways in which plates converge and diverge and to plate kinematics (Fig. 6-27A). Basaltic volcanism is an essential part of plate accretion at mid-ocean ridges. Island arc volcanoes are surface expressions for the consumption of oceanic lithosphere at subduction zones. Where oceanic lithosphere is subducted beneath a continental margin, volcanoes such as those of the Andes erupt huge quantities of pyroclastics and lavas. The volcanic "ring of fire" around the Pacific Ocean, containing about 67 percent of the active volcanoes of the world, is formed by the many subduction zones consuming the enormous Pacific plate. Continental volcanism is localized at hot spots and rifts.

At the same time, plate tectonics exerts a strong influence on the kinds and distribution of sedimentary environments, the provenance of detritus, and the diagenesis of sediments buried in sedimentary basins (Chap. 12). Thus the distinctive associations of volcaniclastics and the sediment types that are found with them are the result of the overall tectonics of a region. Below we briefly summarize the kinds of volcanic activity and magmatic compositions that are related to plate tectonic environments and review the sedimentary–volcaniclastic associations of those environments. Johnson *et al.* (1978), Kokelaar and Howells (1984), and Aramaki and Kushiro (1983) discuss aspects of plate tectonics, volcanism, and sedimentary associations in volcanic arcs.

Plate Boundaries and Intraplate Volcanism

Without doubt, the most spectacular volcanoes are those of subduction zones. Eighty percent of the active volcanoes of the world occur in regions of plate convergent such as the island arcs of intra-oceanic subduction zones and in the mountain chain volcanoes of Andean, oceanic–continental convergences. The volcanoes in such regions are known for their violent eruptions—Krakatoa and Mount St. Helens are two well-known examples—and abundant pyroclastics. The horizontal distance from the subduction zone to the volcanic arc and the width of the volcanic belt are variable and depend on the dip of the subducting plate, which is generally steeper under intra-oceanic convergences than at continental margins. This angle controls volcanism at the surface (Marsh, 1979, p. 164).

Subduction zone magmas have a wide range of compositions. The intra-oceanic island arcs

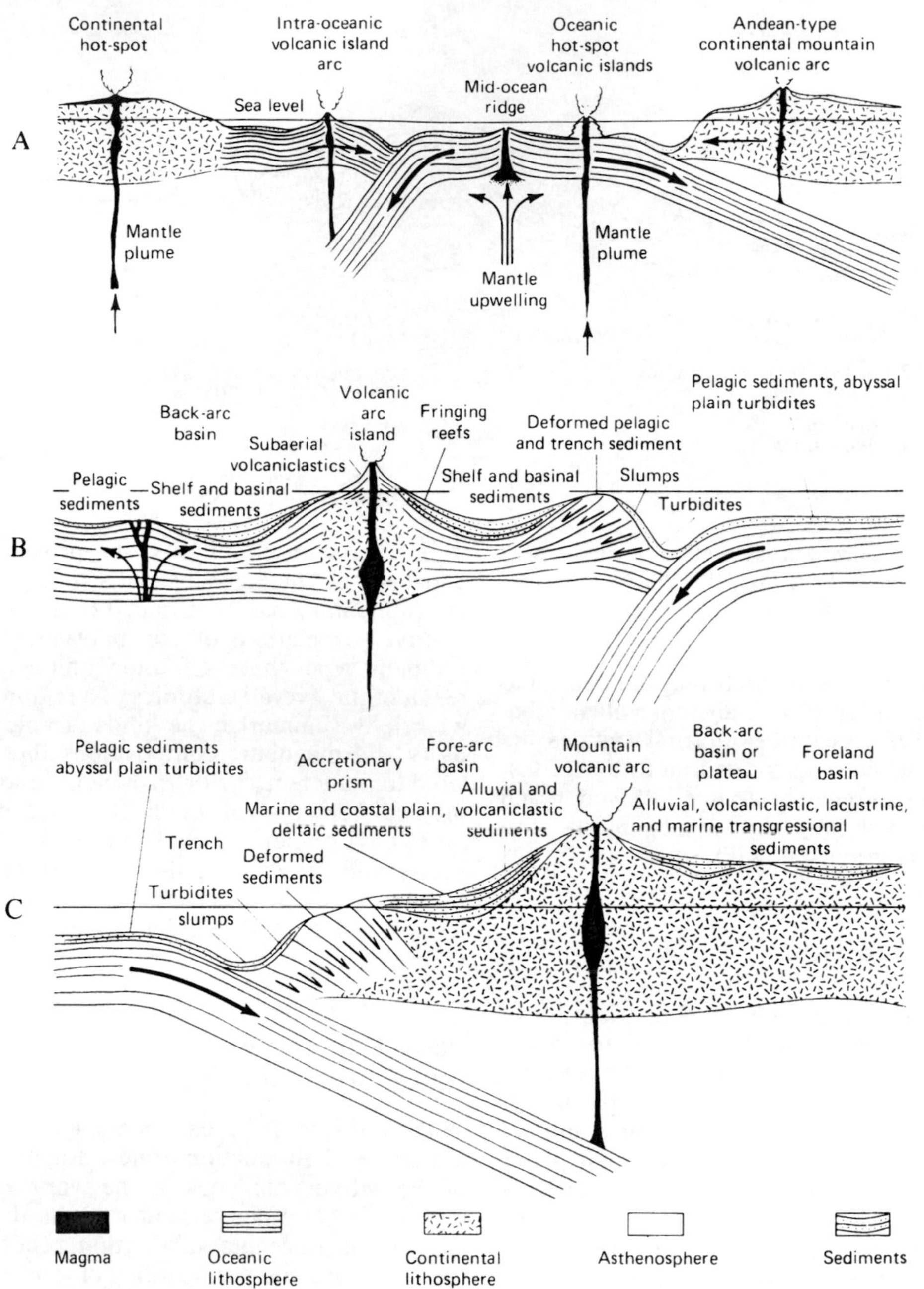

FIGURE 6-27. Volcanism and plate tectonics: Overview (A); tectonics, volcaniclastic deposits, and sedimentary environments of intra-oceanic island arcs, (B); and of Andean-type continental mountain arcs (C).

tend to be dominated by mafic magmas, most commonly basalt to andesite. Andean-type volcanics range from mafic to highly silicic. There is some tendency for magmas farthest from the subduction zone, those derived from a deeper, higher-temperature region of the slab, to be richer in K_2O and SiO_2. The major volcanic suite of subduction zones is the calc-alkaline series, typically andesites, derived from water-rich magmas. The high water content of andesitic magmas, their interaction with downward percolating seawater, and their higher viscosity produce the violent eruptive character of andesitic volcanoes. The andesites tend to be porphyritic with zoned plagioclases and contain small amounts of diopsidic clinopyroxene and hypersthene. Some island arc basalts are tholeiitic, somewhat less mafic and silicic than those of mid-ocean ridges or oceanic islands. They differentiate into Fe-rich andesites to dacites.

The consequence of the violent eruptive activity of andesitic volcanoes is the production of great quantities of volcaniclastic deposits of all kinds. Pyroclastics commonly consist of the andesitic and more silicic differentiated members of the calc-alkaline suite, up to and including rhyodacites and rhyolites. As magmas differentiate and erupt, the arcs become underplated with intrusives. Thus the longer an arc is active, the larger is the volume of differentiated plutonic material in the arc core. With continued evolution, especially as an arc migrates seaward, a mass of essentially continental crust is formed and the geologic structure of an arc massif comes into being.

By far the greatest amount of extrusives, about 3 km^3 per year, are produced at divergent plate boundaries as tholeiitic magmas well up from the mantle and produce new oceanic lithosphere at a rate of approximately 21 km^3 per year (Schmincke, 1982). The extrusives are largely in the form of basalt flows that form pillowed volcanoes (Fig. 6-28) and flow sheet accumulations. The total amount of volcaniclastics is small, including only a little pyroclastic material and some broken fragments of fresh and weathered basalts. Because of the low volatile content of the magma and generally deep water, explosive eruptions are few and the lavas are not vesiculated. For the sandstone petrologist the major interest in rocks of a mid-ocean ridge is probably in some of the varied rocks of the ophiolite suite—pillow basalts, sheeted dikes, and layered gabbros—that originate there and are later emplaced in subduction zones. There they may ultimately be eroded to supply some of the detritus that signals an important and complex history for a sand.

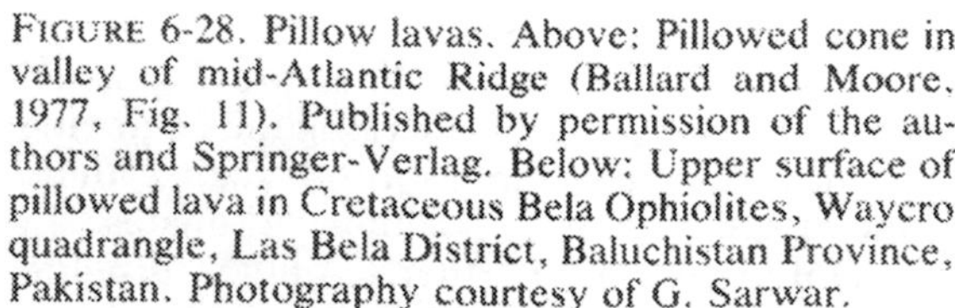

FIGURE 6-28. Pillow lavas. Above: Pillowed cone in valley of mid-Atlantic Ridge (Ballard and Moore, 1977, Fig. 11). Published by permission of the authors and Springer-Verlag. Below: Upper surface of pillowed lava in Cretaceous Bela Ophiolites, Waycro quadrangle, Las Bela District, Baluchistan Province, Pakistan. Photography courtesy of G. Sarwar.

Volcanics of divergent zones include those of the rift valleys on the continents, such as those of East Africa, the Jordan valley, and the Rio Grande rift of the western United States. Pyroclastics may form a significant part of the eruptives, though on a worldwide scale their total volume is small. The magmas of continental rift valleys may be extensively differentiated, alkali-rich, and contain more volatiles than alkali or tholeiitic basalts. Nepheline and leucite are the mineralogical guides to these alkalic volcanics. Rhyolites and trachytes may be abundant and widely distributed as ash falls, flows, and ignimbrites. Flood basalts, only rarely accompanied by pyroclastics, are continental tholeiites typified by those of Cretaceous age in the Paraná Basin of Brazil, the Columbia River basalts of the northwestern United States, and the Deccan basalts of India.

Transform fault plate boundaries, especially where they are a part of a complex transform divergence zone such as the Gulf of California or the Gulf of Aqaba in the northeastern arm of the Red Sea, may include volcanism. The volcanics may be mixtures of basaltic flows along the spreading ridges and more silicic and volatile-rich extrusives along the transform fault portions, where there may be appreciable mixing of magmas with sedimentary material. Although transform basins are not regions of abundant extrusives, they are of interest because of the interaction of pyroclastics with continental margin sediments.

Oceanic intraplate islands, the result of underlying hot spots, are the sites of extensive production of mafic magmas that form sea mounts and shield volcanoes of the Hawaiian type. These ocean island basalts are tholeiitic, somewhat richer in volatiles and alkalis than mid-ocean ridge basalts, and grade into alkali basalts. Some pyroclastics are erupted as the submarine volcanoes build to shallow water depths or break the surface as islands and the more volatile-rich magmas vesiculate. Studies of their detritus are not numerous. See Hammond (1974) for a careful description of the volcanic stratigraphy and sedimentation of Hawaiian volcanoes, however.

On the continents, intraplate volcanism caused by hot spots is an important source of volcaniclastic material. Other than volcanoes related to rifts and transforms, the only volumetrically significant volumes of extrusions are those related to flood basalts. These are tholeiites that tend to be more alkalic than their oceanic counterparts. Pyroclastics are a small part of the total extrusives, occurring as agglutinates at fissures or cinder cones or as brecciated deposits where lavas are exuded under water, in lakes. In some regions, such as Yellowstone Park in the United States, there are extensively differentiated magmas and thus a higher pyroclastic content.

Plate Tectonics and Basin Fill

There are two end-member types of volcanic basins—those filled largely by flows and those mostly filled by water-transported volcaniclastic sandstones. Both such basins may be either below or above sea level so that their dominant fill may be either mostly marine or nonmarine. Between these end-member basins are mixed volcanic basins, whose fill can also be either marine or nonmarine and contain intermediate quantities of lava.

The kinds of volcaniclastic sandstones deposited in or near subduction zones depend strongly on the type of convergence—intra-oceanic or Andean (Fig. 6-27A). Within each type, the nature of the detritus and the sedimentary environment depend on the location with respect to arc and trench, on the general configuration of the volcanic belt, on whether the volcanism occurred above or below sea level, and on the wind directions that disperse airborne tephra.

Intra-oceanic convergences early in their history produce quartz-free, mafic volcanics that constitute the basement sequences of volcanic edifices or accretionary terranes. In these basal sequences pillow basalts are abundant (Fig. 6-28) and pyroclastics rare. As the volcanic archipelago grows to the surface and enlarges, the volcanic islands, fore- and back-arc basins, and accretionary prisms evolve as more silicic differentiates appear. Sandstones only become important in the middle and later part of this sequence.

Fore-arc basins contain the greatest volume of volcaniclastic sandstones and sediments, which may range from ash falls and flows and reworked volcaniclastics from the arc islands to carbonate buildups and reefs in tropical and subtropical regions. Carbonates and volcaniclastic sandstones may be associated in several ways: carbonates may fringe a volcanic cone or submarine sea mount so that fan deltas and shelf sands rich in volcaniclastics pass seaward between reefal barriers or later are deposited over them. Or volcanics, marking the beginning of a magmatic arc, may abruptly cap a thick sedimentary section including carbonates. In the Dolomites of northern Italy, turbidite volcaniclastics infill a deep basin surrounded by a spectacular carbonate platform (Bosellini, 1984). If wind dispersal is toward the back-arc region, the fore-arc may contain relatively few ash falls (see Fig. 6-26). This complicates the recognition of a fore- vs back-arc basin. Towards the trench, the fore-arc region gives way to the prism of pelagic sediments accreting by tectonic erosion of some of the sediments of the subducting slab. Sliced by complex patterns of imbricated faults, the prism may be recognized by its mélange of pelagic cherts and muds, turbidites, and other oceanic sediments. The trench itself is the site of slumps and turbidites. Volcaniclastics commonly form an important part of these sediments. Carey and Sigurdsson (1984) summarize volcanogenic sedimentation in marginal basins.

The fore-arc-trench regime is cannibalistic. Sediments of the accretionary prism, together with shallow-water sediments carried across the fore-arc region, slump and flow down the trench slope and accumulate on the trench floor. There they may be scraped off, underthrust, and reincorporated into the accretionary wedge.

The back-arc region of an intra-oceanic convergence may be filled with volcaniclastics from the arc and a small amount of sediment eroded from the volcanic islands (Fig. 6-27B). The proportion of volcanically derived airborne material is controlled by wind directions. If the back-arc basin is itself spreading from a subsidiary spreading center, sediments and volcaniclastics may be mixed with basalt flows. The major clue to the fore- or back-arc origin of a sequence is the polarity of the arc, the accretionary prism and trench lying on the side towards the subducting plate.

Andean-type convergences are both more complex and more likely to be preserved in the continental rock record (Fig. 6-27C). The fore-arc region tends to be strongly affected by the detritus derived from erosion of the continental-arc mountain chain. The result is a mixture of airborne and waterlaid pyroclastics and eroded older continental marginal sediments and volcaniclastic sandstones. Much of the fore-arc basin may be on land and consist of rapidly subsiding basins filled with alluvial deposits containing both fresh and reworked volcanic debris. Ignimbrites may also be abundant. These fore-arc basins may even trap so much sediment that relatively little reaches the trench floor. The accretionary prism may also include structurally reworked fore-arc and trench debris. Where the fore-arc basin is mainly marine, thick shelf and/or turbidite sequences will accumulate, depending on sediment supply and the proximity of the trench to the shoreline or shelf edge, especially in early stages. With continued progradation of the shoreline or shelf edge, turbidites give way to shelf deposits and the turbidites will be restricted to the trench. The net result of all of these processes may be thick accumulations of turbidite volcanic graywackes overlain by shelf deposits.

The back-arc regions of Andean convergences lie entirely on continental crust. The region may be one of extensive high plateaus such as the antiplano of the central Andes, the Sierra Madre Occidental of Mexico (Fig. 6-29), or broad, lower inland regions, such as the eastern Oregon and Washington plateaus in back of the volcanic chain of the Casacade Mountains. Much depends on the earlier history of this region of the continent. Complex island arc chains that have had long continental crustal histories, such as the Japanese and Phillipine Islands, may include back-arc regions that record a history of many previous volcanic periods and a complex series of terranes. Back-arc basins may lie on continental cratons and develop as foreland basins that subside rapidly in response to loading by detritus eroded from the volcanic-arc mountain chain or basin subsidence may be the direct result of volcanic loading. These basins may contain abundant volcaniclastic-rich alluvial and lacustrine deposits or, when flooded by transgressions, marine shelf sandstones. Or ignimbrites and flows may prevail, as in the Sierra Madre Occidental of northern Mexico where ignimbrites are notably sheetlike (Fig. 6-30). Again, the distribution of airborne volcanic detritus is determined by prevailing winds. The Andes, which extend across several wind belts, illustrate this well. Where the prevailing winds are westerly, ash deposits are sent to the east, to the back-arc region. In the regions of easterly trade winds, the volcanic debris is blown to the west, into the coastal fore-arc basins. Back-arc regions such as the Columbia Plateau in the United States may be further complicated by combinations of arc-derived volcaniclastic and detrital sediments with flood basalts issuing from continental hot spots.

The volcanism associated with continental hot spots such as the Columbia River basalts, as noted above, contains small amounts of pyroclastics, and the accumulations show thick sequences of lava flows. In some places these are interbedded with alluvial deposits with high proportions of reworked basalt grains. Weath-

FIGURE 6-29. Typical dissected landscape of the Sierra Madre Occidental of western Chihuahua, northern Mexico—a vast area of andesitic rhyolitic flows and ignimbrites interbedded with minor alluvial deposits (Webb and Potter, 1969, Fig. 2). Horizontal ignimbrites impart a strong sedimentary aspect to this spectacular region of Tertiary volcanism. The Sierra Madre now supplies fresh, recycled volcanic debris to a wide area on both the Atlantic and Pacific coasts of Mexico.

ered soil zones are common. This picture is dramatically altered, however, in Yellowstone Park in the United States, where pyroclastics of silicic compositions are abundant. Here the interaction of hot-spot volcanism and thick continental crust with an active hydrothermal system of hot springs and geysers has produced a complex of volcanics, alluvial and lacustrine deposits, and siliceous hot-spring precipitates.

Transform fault basins on the continents may include some volcaniclastic sandstones, especially reworked volcanics from earlier convergent plate boundary histories. Where the basin is marine, the volcanic-rich sandy sediments of the near shore may grade into muddy turbidites and silica-rich or carbonate-rich deep-water marine sediments, some of which are cut by dikes and sills. Where the transform basin is continental, such as the transform basins of the San Andreas fault system in California, extensive alluvial fan and plain deposits containing reworked volcanics and some contemporary volcaniclastic sandstones may interfinger with lacustrine sediments. Where the climate is arid and sediment supply low, evaporites can occur in the lacustrine deposits. Where marine and nonmarine sections of a transform system meet, as at the head of the Gulf of California where the delta of the Colorado River is advancing, sedimentation of a volcaniclastic-rich fraction is mixed with continental detritus from a variety of rocks and is deposited in a wide range of environments.

The volcaniclastics of all of these tectonic environments may be telescoped, severely deformed, and metamorphosed by orogenies related to continental collisions and/or accretion of oceanic plateaus. Volcanic-rich melanges may be juxtaposed with fore-arc tuffs, sandstones, and shales. Continental magmatism at the site of the former arc may superimpose continental volcanics on top of former marine fore- and back-arc sequences. Certainly at this stage of our knowledge we cannot define any standard or even typical sequence. Each has to be deciphered on its own in the context of its regional and even subcontinental geologic history.

Much attention has been given to Precambrian volcaniclastics, in part related to continuing discussion of the timing of inception of plate

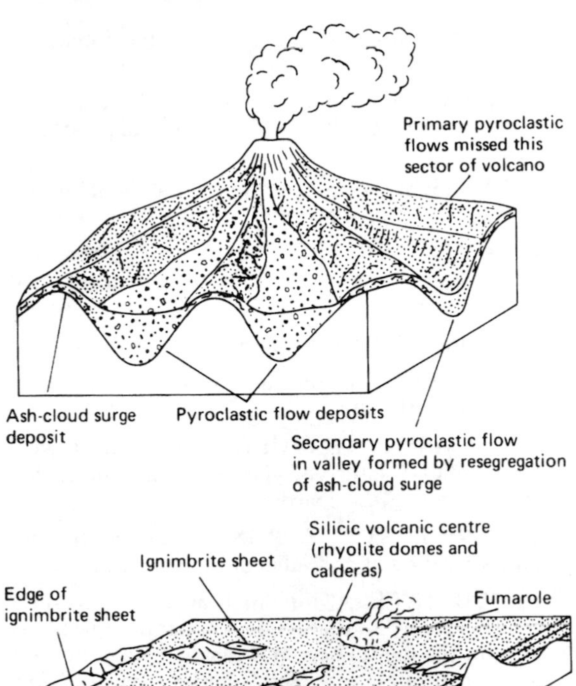

FIGURE 6-30. Contrasting geometry of ignimbrites in an island arc and on continental crust effectively illustrates the role of tectonic setting. (After Wright, 1980, Fig. 2).

tectonics (e.g., Ayres *et al.*, 1985; Dimroth, Rocheleau *et al.*, 1985; Lowe, 1982; Tasse *et al.*, 1978). Archean greenstone belts, some 3 to 4 km thick, and their mineral deposits also provide excellent opportunities for sedimentologic study (Ayres, 1983). Sedimentology has also been applied to reconstruct the proximal and distal facies of mafic flows in the Precambrian which contain shoaling-upward sequences (Dimroth, Imreh, *et al.*, 1985). Many authors believe Archean volcanism was related to a different pre–plate tectonic regime at a time when the total mass of continental crust was small and volcanic centers were widespread. The transition to a plate tectonic regime may have taken place later in the Precambrian (see Chap. 12).

The foregoing discussion relies heavily on inference and deduction. Volcaniclastic sedimentology offers, we believe, many opportunities for future research. Field studies by sedimentologists (Chaps. 4, 9, and 10) closely integrated with regional tectonics and geophysics (Chap. 12) are needed to develop effective insight into volcanic basins and their fill. When combined with more emphasis on petrology—both igneous and sedimentary—a significantly improved earth history and exploration of ancient volcanic basins seems well within reach.

Glossary

Included here are a few of the many terms that are used to describe and interpret the broad spectrum of volcaniclastic deposits.

AGGLOMERATE A deposit composed of bombs (material larger than 64 mm) that largely solidified in flight.

ASH Unconsolidated, pulverized volcanic debris less than 2 mm, which, when fresh, may be acid, glassy, gritty, harsh, and most unpleasant.

AIR FALL Volcanic debris deposited from air on land or in water.

ANDESITE A dark-colored, fine-grained, quartz-free volcanic rock that is the extrusive equivalent of diorite; contains 54 to 62 percent silica.

ASH FLOW A turbulent mixture of gas and pyroclastic materials at high temperature that travels swiftly outward and downslope, having been explosively ejected from a crater or fissure. Same as a pyroclastic flow (after Ross and Smith, 1961, p. 3).

ASH FLOW TUFF Indurated deposit of ash flow.

AXIOLITIC STRUCTURE A fine, linear intergrowth of cristobalite and feldspar that forms from the devitrification of glass shards. Long axes of crystals perpendicular to shard boundary. According to Ross and Smith (1961, p. 37), axiolitic structure has been observed only in ash flows.

BASALT A dark volcanic rock that is generally rich in iron and magnesium, has no free quartz, and contains 45 to 54 percent silica.

BASE SURGE DEPOSIT Occurs at the base of an eruption, travels outward at high speed, and has sand waves, massive beds, and flat bedding (Fisher and Schmincke, 1984, p. 249).

BOMB Molten material ejected from a volcano and solidified in flight; larger than 64 mm.

BOMB IMPACT STRUCTURE Stratification [sedimentary beds and laminae] deformed by impact from clasts ejected during a volcanic eruption.

DEBRIS FLOW OR AVALANCHE A high-density, sediment–water mixture that generally contains a large proportion of coarse fragments.

FLUIDAL TEXTURE Flow structure in an aphanitic rock.

HYALOPILITIC TEXTURE A meshwork of randomly oriented microlites with interstitial glass.

HYDROCLASTIC An adjective used to describe the products of an explosion produced by water-generated steam (Fisher and Schmincke, 1984, p. 231).

HYDROCLASTITE Deposit consisting of shattered, small angular clasts formed when lava is extruded underwater.

IGNIMBRITE A tuffaceous rock of acidic composition deposited by a nuée ardente-type explosion—a "fiery rain cloud" (Marshall, 1935, p. 338).

KRAKATOAN CALDERA A collapsed crater formed by a great and rapid outpouring of pumice-fall and especially pumice-flow deposits.

LAHAR Volcanic debris flow.

LAPILLI Pyroclastic material between 2 and 64 mm.

NUÉE ARDENTE Explosive ejection from a volcano produces overriding dust clouds plus a dense, basal gravity flow of high temperature (Lacroix, 1904, pp. 442–43).

PERLITIC STRUCTURE Concentric, onion-like partings in glass produced by hydration and exfoliation of obsidian (Ross and Smith, 1955, Pl. 1b).

PILLOW LAVA OR BRECCIA Pillow-like structure formed by mafic lavas extruded beneath the water (see Fig. 6-28). Lava surface may be brecciated; pillow lavas and breccias may be interbedded with sediment.

PILOTAXITIC TEXTURE The groundmass of a volcanic rock with a meshwork of weakly oriented microlites. Glass may be present between the crystal meshwork.

PUMICE Highly vesicular, silica-rich volcanic glass. Porosity commonly exceeds 50 percent. Loosely speaking, a froth of volcanic glass.

PUMICE FLOW A pyroclastic flow composed of pumice. Virtually the same as ash flow of some authors.

PYROCLASTIC FLOW Incandescent flow of volcanic debris along a surface; forms a deposit with diverse sedimentary structures depending upon concentration and velocity.

RHYOLITE A fine-grained, glassy, light-colored volcanic rock with phenocrysts of quartz and alkali feldspar; the extrusive equivalent of granite containing 69 percent or more silica.

SCORIA Pumice of intermediate to mafic composition and thus having a very dark brown or black color.

SIDEROMELANE Basaltic glass.

SPILITE An altered basalt with albitized feldspar and secondary chlorite, calcite, epidote, and other low-temperature alteration products; may have pillow structure and is common in the submarine lava flows of orogenic basins.

TEPHRA Volcanic material of any size ejected from a crater through the air (Thorarinsson, 1955, p. 12).

THOLEIITE A silica-saturated basalt with orthopyroxene and/or pigeonite.

TRACHYTIC TEXTURE Flow structure defined by elongate grains of alkali feldspar in a fine-grained groundmass.

TUFF Consolidated equivalent of volcanic ash.

TUFF-BRECCIA A breccia of volcanic origin with abundant tuff matrix.

VITROCLASTIC TEXTURE Pertains to typical fragmental structure of glass-rich volcanic rocks, those rich in glass shards.

VOLCANICLASTIC APRON Volcaniclastic deposits peripheral and adjacent to a volcano.

WELDED TUFF A tuff that accumulates sufficiently rapidly so that its heat is not dissipated; as a result the debris becomes plastic enough to fuse together or melt and exhibit various types of flow structure.

References

Aramaki, S., and Kushiro, I. (Eds.): Arc volcanism. Jour. Volcanology and Geothermal Research, Spec. Issue 18, 631 pp. (1983).

Ayres, L.D.: Bimodal volcanism in Archean greenstone belts exemplified by grawacke composition, Lake Superior Park, Ontario. Canadian Jour. Earth Sci. 20, 1168–1194 (1983).

Ayres, L.D.; Thurston, P.C.; Card, K.D.; and Weber, W. (Eds.): Evolution of Archean supracrustal sequences. Geol. Assoc. Canada Spec. Paper 28, 300 pp. (1985).

Ballard, R.D., and Moore, J.G.: Photographic atlas of the Mid-Atlantic Ridge Rift Valley, 114 pp. New York: Springer-Verlag 1977.

Best, M.G.: Igneous and metamorphic petrology, 630 pp. San Francisco: W.H. Freeman and Co. 1982.

Bosellini, A.: Progradational geometries of carbonate platforms: Examples from the Triassic of the Dolomites, northern Italy. Sedimentology 31, 1–24 (1984).

Botvinkina, L.N.: Geneticheskie tipi otlozhenii oblacte: aktuvnogo vulkanuzma (Genetic types of deposits in the areas of active volcanism), 318 pp. Moscow: Hayka 1974.

Bullard, F.M.: Volcanoes of the earth, Rev. Ed., 579 pp. Austin and London: University of Texas Press 1976.

Carey, S., and Sigurdsson, H.A.: A model of volcanogenic sedimentation in marginal basins. In: Kohelaar, B.P., and Howells, M.F. (Eds.): Marginal basin geology. Geol. Soc. Spec. Pub. 16, 37–58 (1984).

Cas, R.A.F., and Jones, J.G.: Paleozoic interarc basin in eastern Australia and a modern new Zealand analogue. New Zealand Jour. Geol. Geophysics 22, No. 1, 71–85 (1979).

Cas, R.A.; Powell, C.McA.; Fergusson, C.L.; Jones, J.G.; Roots, W.D.; and Fergusson, J., The Lower Devonian Kowmung volcaniclastics; A deep-water succession of mass-flow origin, northeastern Lachlan Fold Belt, N. S. W. Jour. Geol. Soc. Australia 28, 271–288 (1981).

Clapperton, Chalmers M.: Volcanoes in space and time. Prog. Phys. Geog. 1, 375–411 (1977).

Crandell, D.R., and Mullineaux, D.R.: Technique and rational of volcanic-hazard appraisals in the Cascade Range, northwestern United States. Environmental Geol. 1, 23–32 (1975).

Crook, K.A.W.: Petrology of the Parry Group, Upper Devonian–Lower Carboniferous, Tomnorth-Nundle District, New South Wales. Jour. Sed. Petrology 30, 538–552 (1960).

Davies, D.K.; Quearry, M.W.; and Bonis, S.B.: Glowing avalanches from the 1974 eruption of the volcano Fuego, Guatemala. Geol. Sci. America Bull. 89, 369–384 (1978).

Dickinson, W.R.: Sedimentation within and beside ancient and modern volcanic arcs. *In:* Dott, R.H., Jr., and Shaver, R.H. (Eds.): Modern and ancient geosynclinal sedimentation. Soc. Econ. Paleon. Mineral. Spec. Pub. 19, 230–239 (1974).

Dickinson, W.R., and Seeley, D.R.: Structure and stratigraphy of fore-arc regions. Am. Assoc. Petroleum Geologists Bull. 63, 2–31 (1979).

Dimroth, E., and Rocheleau, M.: Volcanology and sedimentology of Rouyn-Noranda area, Quebec, Field Trip A-1. Geol. Assoc. Canada–Mineral Assoc. Canada–Université Laval, 194 pp. (1979).

Dimroth, E.; Rocheleau, M.; Mueller, W.; Archer, P.; Brisson, H.; Fortin, G.; Jutras, M.; Lefebvre, C.; Piché, M.; Pilote, P.; and Simoneau, P.: Paleogeographic and paleotectonic response to magmatic processes: a case history from the Archean sequence in the Chibouqamau area, Quebec. Geol. Rundschau 74, 11–32 (1985).

Dimroth, E.; Imreh, Laszla; Cousineau, P.; Leduc, M.; and Sanschagrin, Y.: Paleogeographic analyses of mafic submarine flows and its use in the exploration of massive sulfide deposits. In: Ayres, L.D., *et al.* (Eds.): Evolution of Archean supracrustal sequences. Geol. Assoc. Canada Spec. Paper 28, 203–222 (1985).

Dott, R.H., Jr.: Eocene deltaic sedimentation at Coos Bay Oregon. Jour. Geology 74, 373–420 (1966).

Edwards, A.B.: The petrology of the Miocene sediments of the Aure Trough Papua. Royal Soc. Victoria Proc. 60, 123–148 (1947).

Eisbacher, G.H., and Claque, J.J.: Destructive mass movements in high mountains: Hazard and management. Geol. Survey Canada Paper 84-16, 230 pp. (1984).

Elston, W.E., and Northrop, S.A. (Eds.): Cenozoic volcanism in southwestern New Mexico. New Mexico Geol. Soc. Spec. Pub. 5, 151 pp. (1976).

Ewart, A.: Petrography and petrogenesis of the Quaternary pumice ash in the Taupo area, New Zealand. Jour. Petrology 4, 392–431 (1963).

Fisher, R.V.: Bubble-wall texture and its significance. Jour. Sed. Petrology 33, 224–227 (1963).

Fisher, R.V.: The geology of an ignimbrite layer, the John Day Formation, eastern Oregon. Univ. Calif. Pubs. Geol. Sci. 67, 58 p. (1966).

Fisher, R.V.: Submarine volcaniclastic rocks. In: Kokelaar, B.P., and Howells, M.F. (Eds.): Marginal basin geology. Geol. Soc. Spec. Pub. 16, 5–28 (1984).

Fisher, R.V., and Schmincke, H.-U. Pyroclastic rocks, 472 pp. Berlin, Heidelberg, New York, Tokyo: Springer-Verlag 1984.

Fiske, R.S., and Matsuda, Tokohiko: Submarine equivalents of ash flows in the Tokiwa Formation, Japan. Am. Jour. Sci. 262, 76–106 (1964).

Fox, J.S.: Rapid pyroclastic mapping in base metal exploration. Canadian Inst. Mining Bull. 70, 173–178 (1977).

Fox, J.S.: Besshi-type volcanogenic sulphide deposits—a review. Canadian Inst. Mining Bull. 77, 57–68 (1984).

French, D.E., and Freeman, K.J.: Tertiary volcanic stratigraphy and reservoir characteristics of the Trip Spring Field, Nyc County, Nevada. In: Newman, G.W., *et al.* (Eds.): Basin and range symposium and Great Basin field conference, pp. 487–502. Rocky Mountain Assoc. Geol.–Utah Geological Assoc., 1979.

Frogatt, P.C.; Wilson, C.J.N.; and Walker, G.P.L.: Orientation of logs of the Taupo Ignimbrite as an indicator of flow direction and vent position. Geology 9, 109–111 (1981).

Frogatt, P.C.: Stratigraphy and nature of Taupo Pumice Formation. New Zealand Jour. Geology and Geophysics 24, 231–248 (1981).

Goodell, P.C., and Waters, A.C. (Eds.): Uranium in volcanic and volcaniclastic rocks. Am. Assoc. Petroleum Geologists, Studies in Geol. Ser., 13, 331 pp. (1981).

Gorshkov, G.S., and Dubik, Y.M.: Gigantic directed blast at Shiveluck volcano (Kamchatka). Bull. Volcanologique 34, 261–288 (1970).

Green, J., and Short, N.M.: Volcanic landforms and surface features, 519 pp. New York–Heidelberg–Berlin: Springer-Verlag 1971.

Hammond, P.E.: Synopsis of volcanic stratigraphy. In: Greeley, Ronald (Ed.): Geologic guide to the Island of Hawaii, A field guide for comparative planetary geology, pp. 38–47. Washington, D.C.: U.S. National Aeronautics and Space Administration, Office of Planetary Programs, 1974.

Hay, R.L.: Origin and weathering of Late Pleistocene ash deposits on St. Vincent B.W.I. Jour. Geology 67, 65–87 (1959).

Heiken, G.: An atlas of volcanic ash. Smithsonian Contributions to the Earth Sciences 12, 101 pp. (1974).

Heim, Albert: Der Bergsturz von Elm. Zeitschr. deutsch geol. Gesell. 34, 74–115 (1882).

Heinrich, E.W.: Microscopic identification of minerals, 414 pp. New York: McGraw-Hill 1965.

Honza, E.: Evolution of arc volcanism related to marginal sea spreading and subduction at trench. In: Shimozuru, D., and Yokoyama, I. (Ed.): Arc volcanism: Physics and tectonics, pp. 177–189. (Proc. IAVCEI, 1981 Symposium). Tokyo: Terra Scientific Pub. Co. 1983.

Hsu, K.J.: Catastrophic debris streams (Sturzstroms) generated by rockfalls. Geol. Soc. America Bull. 86, 129–140 (1975).

Huber, N.K., and Reinhart, C.D.: Some relationships between refractive index of fused glass beads and the petrologic affinity of volcanic rock suites. Geol. Soc. America Bull. 77, 101–110 (1966).

Ingersoll, R.V.: Petrofacies and petrologic evolution of the Late Cretaceous forearc basin, northern and central California. Jour. Geology 86, 335–352 (1978).

Innes, J.L.: Debris flows. Prog. Phys. Geog. 7, 469–501 (1983).

International Association of Volcanology: Catalogue of active volcanoes of the world including solfatera fields. Naples and Rome.

Janda, R.J.; Scott, K.M.; Nolan, K.M.; and Martinson, H.A.: Lahar movement, effects and deposit. In: Lipman, P.W., and Mullineaux, D.R. (Eds.): The 1980 eruptions of Mount St. Helens, Washington. U.S. Geol. Survey Prof. Paper 1250, 461–448 (1981).

Jenks, W.F., and Goldich, S.S.: Rhyolitic tuffs flows in southern Peru. Jour. Geology 64, 156–172 (1956).

Johnson, R.W., MacKenzie, D.E., and Smith, I.E.M.: Volcanic rock associations at convergent plate boundaries: Reappraisal of the concept using case histories from Papua New Guinea. Geol. Soc. America Bull. 89, 96–106 (1978).

Juve, G.: Metal distribution at Stekenjokk: Primary and metamorphic patterns. Geologiska Foreningens i Stockholm Forhandlingar 99, 149–158 (1977).

Katsui, Y.: Evolution and magmatic history of some Krakatoan calderas in Hokkaido, Japan. Jour. Fac. Sci., Hokkaido Univ., Ser. 4, 11, 631–650 (1963).

Kesel, R.H.: Notes on the lahar landforms of Costa Rica. Zeitschr. Geomorph., N.F., Suppl. Bd, 18, 78–91 (1973).

Kokelaar, B.P., and Howells, M.F. (Eds.): Marginal basin geology. Volcanic and associated sedimentary and tectonic processes in modern and ancient marginal basins, 328 pp. Palo Alto, California: Blackwell Scientific Pub. 1984. (Geol. Soc., Spec. Pub. 16)

Lacroix, A.: L'éruption de la montagne Pelée en janvier 1903. C.R. Acad. Sci. (Paris) 136, 442–445 (1904).

Lajoie, S.: Volcaniclastic rocks. In: Walker, R.G. (Ed.), Facies models, 2nd Ed, pp. 39–52. Geoscience Canada Reprint Ser. 1 (1984).

Lipman, P.W., and Christiansen, R.L.: Zonal features of an ash-flow sheet in the Piapi Canyon Formation, southern Nevada. U.S. Geol. Survey Prof. Paper 501B, 74–78 (1964).

Lipman, P.W., and Mullineaux, D.R. (Eds.): The 1980 eruptions of Mount St. Helens, Washington, U.S. Geol. Survey Prof. Paper 1250, 844 pp. (1981).

Lowe, D.R.: Comparative sedimentology of the principal volcanic sequences of Archean greenstone belts in South Africa, Western Australia and Canada: Implications for crustal evolution. Precambrian Research 17, 1–30 (1982).

Lundberg, N.: Evolution of the slope landward of the Middle America trench, Nicoya Peninsula, Costa Rica. In: Leggett, J.K. (Ed.): Trench-forearc geology. Geol. Soc. London Spec. Pub. 10, 131–147 (1982).

Marsh, Bruce D.: Island-arc volcanism. Am. Scientist 67, 161–172 (1979).

Marshall, P.: Acid rocks of Taupo-Rotorua volcanic district. Royal Soc. New Zealand Trans. 64, 323–366, 1935.

Mathiesen, M.E., and Vondra, C.F.: The fluvial and pyroclastic deposits of the Cagayan Basin, Northern Luzon, Phillipines—an example of non-marine volcaniclastic sedimentation in an interarc basin. Sedimentology 30, 369–392 (1983).

Meyer, W.; Stets, J.; and Wurster, P.: Gefüge und Enstehung der Ringdünen in den grauen Tuffen des Laacher Vulkans. Geol. Rundschau 63, 1113–1132 (1974).

Mizutani, S.: Clastic plagioclase in Permian graywacke from the Mugi area, Gifu Prefecture, Central Japan. Jour. Earth Sci., Nagoya Univ. 7, 108–136 (1959).

Moore, J.G., and Sisson, T.W.: Deposits and effects of the May 18 pyroclastic surge. In: Lipman, P.W., and Mullineaux, D.R. (Eds.): The 1980 eruptions of Mount St. Helens, Washington. U.S. Geol. Survey Prof. Paper 1250, 421–438 (1981).

Moorhouse, W.W.: The study of rocks in thin section, 514 pp. New York: Harper 1959.

Musai, I.: A study of the textural characteristic of pyroclastic flow deposits in Japan. Earthquake Research Inst. Bull. 39, 133–284 (1961).

Niem, A.R.: Mississippian pyroclastic flow and ashfall deposits in the deep-marine Ouachita flysch basin, Oklahoma and Arkansas. Geol. Soc. America Bull. 88, 49–63 (1977).

Parson, W.H.: Criteria for the recognition of volcanic breccias, a review. In: Larsen, L.H., Prinz, M.; and Mason, V. (Eds.): Igneous and metamorphic geology. Geol. Soc. America Mem. 115, 263–304 (1969).

Proffett, J.M., Jr.: Cenozoic geology of the Yerington District, Nevada, and implications for the nature of Basin and Range faulting. Geol. Soc. America Bull. 88, 248–266 (1977).

Rittmann, A.: Les volcans et leur activité, 2nd Ed., 461 p. Paris: Masson 1962.

Ronov, A.B.: Common tendencies in the chemical evolution of the earth's crust, ocean and atmosphere. Geochm. Int. V, 713–737 (1964).

Ross, C.S.: Provenance of pyroclastic materials. Geol. Soc. America Bull. 66, 427–434 (1955).

Ross, C.S., and Smith, R.L.: Water and other volatiles in volcanic glass. Am. Mineralogist 40, 1071–1089 (1955).

Ross, C.S., and Smith, R.L.: Ash-flow tuffs—their origin, geologic relations, and identification. U.S. Geol. Survey Prof. Paper 366, 81 pp. (1961).

Rowley, P.D., Kuntz M.A., and Macleod, N.S.: Pyroclastic flow deposits. In: Lipman, P.W., and Mullineaux, D.R. (Eds.): The 1980 eruptions of Mount St. Helens, Washington. U.S. Geol. Survey Prof. Paper 1250, pp. 489–512, 1981

Sand, L.B., and Mumpton, F.A.: Natural zeolites: Occurrence, properties, use, 546 pp. Fairview Park, Elmsford, New York: Pergamon Press 1978.

Sandberg, A.E.: Section across Keweenawan lavas at Duluth, Minnesota. Geol. Soc. America Bull. 49, 795–830 (1938).

Scheidegger, A.E., and Potter, P.E.: Textural studies of graded bedding. Sedimentology 11, 163–170 (1968).

Schermerhorn, L.J.G.: Pyrite emplacement by gravity flow. Instituto Geología Minero, Bol. Geol. Minero 82, 88–92 (1971).

Schmid, R.: Descriptive nomenclature and classification of pyroclastic deposits and fragments: Recommendations of the IUGS Subcommission on the Systematics of Igneous Rocks. Geology 9, 41–43 (1981).

Schmincke, H.-U.: Graded lahars in the type sections of the Ellensburg Formation, south-central Washington. Jour. Sed. Petrology 37, 438–448 (1967).

Schmincke, H.-U.: Flow direction in Colombia River basalt flow and paleocurrents of interbedded sedimentary rocks, south-central Washington. Geol. Rundschau 56, 992–1020 (1967).

Schmincke, H.-U.: Vulkane und ihre Wurzeln. Rhein-Westf. Akadiwiss. Westd., Verlag Opladen, Vorträge N 315, 35–38 (1982).

Schmincke, H.-U., and Swanson, D.A.: Laminar viscous flowage structures in ash-flow tuffs from Gran Canaria, Canary Islands. Jour. Geology 75, 641–664 (1967).

Self, S., and Sparks, R.S.J. (Eds.): Tephra studies, 481 pp. Dordrecht, Holland: D. Reidel Pub. 1981. (Proceedings of the NATO Advanced Study Institute).

Sheets, P.D., and Grayson, D.K. (Eds.): Volcanic activity and human ecology, 644 pp. New York: Academic Press 1979.

Sheridan, M.F., and Updike, R.G.: Sugarloaf Mountain Tephra—a Pleistocene rhyolitic deposit of base-surge origin in northern Arizona. Geol Sci. America Bull. 86, 571–581 (1975).

Shrock, R.R.: Sequence in layered rocks, 507 pp. New York: McGraw-Hill 1948.

Shimozuru, D., and Yokoyama, I. (Eds.): Arc volcanism: Physics and tectonics. 263 pp. Tokyo: Terra Scientific Co. Pub. 1983.

Siebert, L.: Large volcanic debris avalanches: characteristics of source areas, deposits, and associated deposits. Jour. Volcanology and Geothermal Research 22, 163–198 (1984).

Sigurdsson, H.; Sparks, R.S.J.; Carey, S.N.; and Huang, T.C.: Volcanogenic sedimentation in the Lesser Antilles Arc. Jour. Geology 88, 523–540 (1980).

Sigurdsson, H., and Carey, S.N.: Marine tephrachronology and Quaternary explosive volcanism in the Lesser Antilles Arc, pp. 255–280. In: Self, S., and Sparks, R.S.J., (Eds.): Tephra studies: Dordrecht, Holland: D. Reidel Pub. Co., 1981.

Simkin, T., and Fiske, R.S.: Krakatoa, 1883, the volcanic eruption and its effects, 464 pp. Washington, D.C.: Smithsonian Institution, 1983.

Simkin, T.; Siebert, L.; McClelland, L.; Bridge, D.; Newhall, C.; and Latter, J.H.: Volcanoes of the world, 232 pp. Smithsonian Institution (Stroudsburg, Pennsylvania: Hutchinson Ross Pub. Co.) 1981.

Stanley, K.O.: Sandstone petrofacies in the Cenozoic High Plains sequence, eastern Wyoming and Nebraska. Geol. Soc. America Bull. 87, 297–309 (1976).

Strauss, G.K., and Madel, J.: Geology of the massive sulphide deposits in the Spanish-Portuguese pyrite belt. Geol. Rundschau 63, 191–211 (1973).

Streckeisen, A.L.: Classification and nomenclature of igneous rocks (final report of an inquiry). Neues Jahrb. Mineral. Abh. 107, 144–214 (1967).

Suzuki, Takeo: A theoretical model for dispersion of tephra. 95–113. In: Shimozuru, D., and Yokoyama, I (Eds.): Arc volcanism: Physics and tectonics, pp.. 95–113. (Proc. IAUCE, 1981 Symposium). Tokyo: Terra Scientific Pub. Co., 1983.

Tasse, N.; Lajoie, J.; and Dimroth, E.: The anatomy and interpretation of an Archean volcaniclastic sequence, Noranda region, Quebec. Canadian Jour. Earth Sci. 15, 874–888 (1978).

Thompson, B.N.; Kermode, L.O.; and Ewart, A.: New Zealand volcanology, central volcanic region. New Zealand Geol. Survey Handbook, Inf. Ser. 50, 211 pp. (1965).

Thorarinsson, S.: Discussions. Bull. Volcanol. Sect. 216, 11–13 (1955).

Uyeda, S.: Subduction zones and back arc basins—a review. Geol. Rundschau 70, 552–569 (1981).

Van Houten, F.B.: Late Cenozoic volcaniclastic deposits, Andean foredeep, Colombia. Geol. Soc. America Bull. 87, 481–495 (1976).

Vessell, R.K., and Davies, D.K.: Nonmarine sedimentation in an active forearc basin. In: Ethridge, F.G., and Flores, R.M. (Eds.): Recent and ancient nonmarine depositional environments: Models for exploration. Soc. Econ. Paleon. Mineral. Spec. Pub. 31, 31–45 (1981).

Vuagnat, Marc: Pétrographie, répartition et origine

des microbrèches du Flysch nordhelvétique. Beitr. Geolog. Karte Schweiz, N.S., 97, 103 pp. (1952).

Waitt, R.B., Jr.: Devastating pyroclastic density flow and attendent airfall of May 18—Stratigraphy and sedimentology of deposits. In: Lipman, P.W., and Mullineaux, D.R. (Eds.): The 1980 eruptions of Mount St. Helens. Washington, D.C.: U.S. Geol. Survey Prof. Paper 1250, 439–460 (1981).

Waldron, H.H.: Debris flow and erosion control problems caused by the ash eruptions of Irazu Volcano, Costa Rica. U.S. Geol. Survey Bull. 1241-I, 37 pp. (1967).

Waters, A.C.: Determining direction of flow in basalts. Am. Jour. Sci. 258, 350–366 (1960).

Webb, W.M., and Potter, P.E.: Petrology and geochemistry of modern detritus derived from a rhyolitic terrane, Western Chihuahua, Mexico. Bol. Soc. Geol. Mexicana 32, 45–61 (1969).

Whetten, J.T.: Carboniferous glacial rocks from Werrie Basin, New South Wales Australia. Geol. Soc. America Bull. 76, 43–56 (1965).

Whetten, J.T.: Sediments from the lower Colombia River and the origin of graywacke. Science 152, 1057–1058 (1966).

Williams, Howell, and McBirney, A.R.: Volcanology, 400 pp. San Francisco: Freeman, Cooper and Co. 1979.

Williams, H., Turner, F.J., and Gilbert, C.M.: Petrography: an introduction to rocks in thin section, 2nd Ed., 626 pp. San Francisco: W.H. Freeman, 1982

Wilson, C.J.N., and Walker, G.P.L.: Ignimbrite depositional pyroclastic facies: The anatomy of a pyroclastic flow. J. Geol. Soc. London 139, 581–592 (1982).

Wohletz, K.H., and Sheridan, M.F.: A model of pyroclastic surge. In: Chapin, C.E., and Elston, W.E. (Eds.): Ash-flow tuffs. Geol. Soc. America Spec. Paper 180, 211 pp. (1979).

Wright, J.V.: A note on subaqueous ignimbrites and pyroclastic flows (with some reference to Puerto Rico). In: Snow, W.E. (Ed.): 9th Caribbean Geological Conference, Dominican Republic, Santo Domingo, pp. 351–353 (1980).

Wright, J.V., Smith, A.L., and Roobol, M.J.: Ignimbrites from island arcs and continents: A comparison of examples from the Lesser Antilles and Mexico. In: Snow, W.E. (Ed.): 9th Caribbean Geological Conf., Dominican Republic, Santo Domingo, pp. 367–378 (1980).

Yagi, K.: Welded tuffs and related pyroclastic deposits in north eastern Japan. Bull. Volcanol. Ser. 2, 24, 109–128 (1962).

Yamada, E.: Subaqueous pyroclastic flows: Their development, their deposits. In: Kokelaar, B.P., and Howells, M.F. (Eds.): Marginal basin geology. Geol. Soc., Spec. Pub. 16, 29–36 (1984).

Part III. Processes that Form Sand and Sandstone

Having established a fundamental petrographic basis for the study of sand and sandstone, we now examine the subject largely from a process point of view—where does sand come from, what is known of the hydraulics of its transport, deposition and soft sediment deformation, how is it deposited in various environments, and finally how is sand transformed into a lithified sandstone? Unlike Parts I and II, which are mostly geological, the five chapters of Part III draw upon fluid mechanics, engineering hydraulics and solution chemistry and thermodynamics as well as the essentially geological framework of depositional environments. Here too, we have had to make a choice among a great many areas in which new applications of physics and chemistry have illuminated the processes that form sandstone. We seek mainly to show in a general way how, for example, fluid mechanics can be applied to sediment transport and how solution chemistry can be used for understanding diagenetic minerals and textures. In the same way, we need to appreciate the application of the other earth sciences, such as oceanography, to the study of sedimentary environments. Perhaps better than any other portion of the book, Part III shows that the search for fundamental explanations in geology requires knowledge well beyond the domain of geology itself. We hope that Part III will interest sedimentologists to look deeper into neighboring fields in their search for ever more fundamental explanations. We should also be very pleased if Part III did, in fact, attract interested scientists in other fields to some of the problems associated with sand and sandstone as we see them.

CHAPTER 7
Production and Provenance of Sand

Introduction

In this chapter we consider two problems. One has to do with how sand is formed and the other deals with the source or place where sand is produced. In the first, we are concerned with the processes involved in the production of sand, how sand grains come into being. In the second, we wish to know what the source rocks were and where the source area was. It is presumed that a sand body or formation itself contains much of the information needed to answer these questions. We might ask about each and every sand grain—how did it form and where did it come from? This in a nutshell is the theme of this chapter.

The question of provenance is important to us for several reasons. A solution of the problem of provenance will immeasurably increase our understanding of the paleogeography of a region, enabling us to locate and identify possible source lands. We may also be able to trace the movement of materials and thus say something about paleocurrents and paleoslope. The combination of source and paleocurrents leads to inferences of tectonics and climate, the primary controlling influences on sedimentary mineralogy.

Provenance has some practical aspects also. The study of the mineralogy of modern river and beach sands may disclose valuable minerals, such as chromite, tin, and gold, and if provenance is understood, we may know from where the minerals come and so may prospect for placers more intelligently. Sand composition has also been used to determine the nature of the bedrock buried by ice sheets (Kalsbeek *et al.*, 1974). Minerals alien to a particular sand or grains which have been dyed or made radioactive may be added to a sand as tracers in order to follow its movement, especially in the study of shore drift, of engineering importance (Baker, 1956; Kennedy and Kouba, 1970). In such experiments the source is known and the transport pattern determined by systematic sampling. Provenance and porosity of sands also are related through diagenesis: sands derived from granitic sources tend to have a higher porosity than those rich in argillaceous rock fragments, which, when plastically deformed and squeezed into pores, form a barrier to fluid movement and reduce storage capacity; volcanic rock fragments alter and yield pore-filling zeolites, which quickly fill pores.

Though the polarizing microscope remains the primary tool, cathodo-luminescence, the electron microprobe, and the scanning electron microscope enhance our powers of observation so that the future seems even more exciting than the past (Sippel, 1968; Zinkernagel, 1978).

How Sand is Formed

In the broadest sense, sand is any noncoherent, granular material within some generally accepted size limits. Such materials are capable of being transported by currents, sorted and stratified, crossbedded and graded both laterally and vertically. Materials meeting this definition are of diverse origins (Fig. 7-1). Included here are detrital quartz and feldspar, the basaltic dune sands of Moses Lake, Washington, the gypsum dune sands of the Persian Gulf and of White Sands, New Mexico, the greensands rich in glauconite, the calcareous oolitic sands such as those in the Bahamas, calcarenites composed of fossil debris, and even the "clay dunes" of Corpus Christi, Texas. Although all these fit the size definition of sand and are mechanically transported and emplaced in the fabric of the rock as sand grains, they originate in radically different ways and are not all equally abundant in the geologic record.

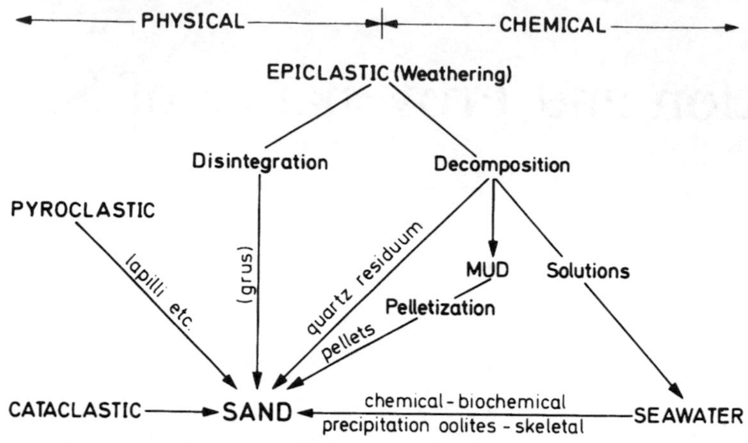

FIGURE 7-1. Processes of sand formation.

There are, perhaps, five basic processes that lead to the formation and release of sand-sized grains. These are (1) weathering, including both disintegration and decomposition, (2) explosive volcanism (pyroclastic), (3) crushing, both by rock movements (cataclastic) and impact, (4) pelletization, and (5) precipitation from solution, both chemical and biochemical. Those which relate to weathering have been termed *epiclastic*.

Weathering

One process is primarily rock disintegration without much, if any, decomposition. Under certain conditions there is a loosening of the grains in a rock, by changes in temperature, frost-action, or hydration so that the rock disintegrates into a loose granular mass. Such destruction of the coarser-grained plutonic rocks produces a residual sand or "grus."

Another process is rock decomposition which may also produce sand—or at least yield a significant sand fraction—as a product of this decomposition. This is probably the origin of the bulk of the quartz sands released by the decomposition of the quartz-bearing plutonic rocks. Most of the rock, mainly feldspar, is converted to fine-grained, clay-sized material from which the inert, undecomposed quartz grains, of sand size, are liberated and concentrated by the movement of fluids.

Most sand-sized quartz is released from the disintegration of quartz-bearing plutonic rocks such as granite, quartz monzonite, and quartz diorite and probably small quantities are derived from quartz-rich volcanics. Such rocks contain an abundance of quartz of the right size to yield a quartz sand. Dake (1921, p. 102) has shown that plutonic rocks seldom yield grains over 1.0 mm. Grains larger than this are apt to be fractured and broken so that only nine percent of the quartz would exceed 1.0 mm. Twenty percent would exceed 0.6 mm.

Disintegration without much decomposition of the associated feldspars seems to be the result of incipient chemical alteration of feldspar and the unstable minerals along cracks and microfractures. This alteration is known even in arid lands. When frost action combines with chemical alteration the process is enhanced. Water penetrates microfractures and on freezing will expand and lead to rock disruption. The physical and biochemical action of plant roots and lichens also contributes to rock disintegration. It has been presumed that heating and cooling would set up stresses within the rock, owing to different rates of expansion of the component minerals. These stresses would lead to loosening of the grains. Both field and experimental studies fail to confirm this concept. Griggs (1936) subjected a granite to wide and rapid changes of temperature, changes greater than those likely to be encountered in nature, without visible results.

Although it is true that the bulk of the epiclastic (siliciclastic) sands are produced by disintegration and/or decomposition of phaneritic rocks, rock fragments of sand size are an important component of many sands—dominant in some. By definition, a sand-sized rock fragment must be that of an aphanitic igneous rock or other equally fine-grained rock such as chert, limestone, or slate. How are such sand-sized particles produced? Not much has been written on this subject. Unlike phanerites, where the cohesion between grains is overcome to yield sand, the problem is not to separate grains but to produce rock rather than mineral particles. There are few quantitative descriptions of grus consisting solely of small-size limestone, slate, or phyllite particles. This is a problem worthy of further study.

Volcanism

The explosive action of volcanoes yields vast quantities of sand-sized debris. According to Garrels and Mackenzie (1971, p. 247) volcanogenic rocks form 26 percent of the total sedimentary mass. Of this a significant part is volcanic sand. Much volcanic material is recycled and mixed with other sands so that its identity is lost. Recognition of volcanogenic materials in sands is one of the most difficult problems facing a petrographer.

The explosion accompanying meteor impact will also shatter rocks. Although the fall-back materials which fill an impact crater are often coarse, a considerable part is sand-sized.

Crushing and Fracturing

Sand-sized materials may be produced by crushing action but not by ordinary abrasion of pebbles. As shown by experimental work, the abrasion of pebbles produces silt and not sand (Marshall, 1927). Moreover, as Kuenen (1959, p. 2) has pointed out, the average pebble is one million times larger than the average sand grain and a freighter load of gravel worn down to sand size would yield only a tumbler full of sand. Crushing or the shattering of grains, however, may in some cases produce a significant volume of sand (Higgins, 1971).

Impact in streams may shatter grains and, if so, it is common enough to be a significant producer of rock fragments of sand size. Davies *et al.* (1978, p. 74–81) suggest that a volcanic rock fragment with phenocrysts in an aphanitic groundmass quickly disintegrates by impact shattering in high-gradient streams into glass, grains composed of small crystals and glass, and large crystals with thin glass crusts. They found volcanic rock fragments to decrease rapidly in the upper reaches of high-gradient streams and suggest that this decrease may correlate with the ratio of fall velocity to turbulent energy of flow, a ratio that governs saltation (see p. 285).

Earth movements may crush rocks and even though the sand yield is small and not likely to form a significant deposit, glacial crushing—a special case of cataclastic action—does produce considerable body of sand-sized material.

Slatt and Eyles (1981) studied sand-sized rock fragments in both glaciers and in their meltwater deposits and concluded that size reduction of a grain occurred along ice-induced fractures by impact during saltation. They also noted that second-cycle sand of glacial origin was richer in feldspar than first-cycle sand. In sum, although both pyroclastic and cataclastic action may be of local importance in the production of sand, such action is dwarfed in importance by the more common but less spectacular action of normal weathering.

Pelletization

Much sand-sized material is also produced by diagenetic action and by chemical and biochemical precipitation. Sands thus formed have been termed *endogenetic* (Grabau, 1904; 1913, p. 271). They are generated *within* the basin of sedimentation and, unlike the epiclastic sands, are not the products of the wastage of a landmass. Some of these intrabasinal sands are built up from finer materials—muds which are both pelitic and micritic. Included here are the sand-sized pellets produced by organisms and the pellets of other origins such as those blown into "clay dunes" (Huffman and Price, 1949). Pryor (1975) showed the importance of fecal pellets in marine sands.

Precipitation

Biochemical and chemical precipitation forms skeletal and oolitic sands that may form a significant deposit. Although intrabasinal carbonate sands, of whatever origin, may form sizable deposits, they are usually included with the limestones rather than the sandstones. For this reason the problems of their generation are omitted. Likewise, the sands of volcanic origin form a special group, the problems of which are quite different from the common sands of epiclastic origin. We recognize, of course, that the epiclastic sands may be mingled in all proportions with these sands of a very different origin. In some cases, the problem becomes one of recognizing the carbonate or volcanic contributions as being intrabasinal or pyroclastic rather than being derived, as an epiclastic sand, by the breakdown of older carbonate or volcanic rocks (Zuffa, 1980).

Summary and Grain-Size Distribution

This brief survey of the ways in which sand is formed should emphasize continental weathering as the first-order producer of sand grains. The major control on the size distribution of detritus of this origin is the size of the source grains formed from the parent rock. Also important is the production of volcaniclastic sand grains, both directly by volcanic ejection and indirectly by weathering and transport of primary volcanic materials. Though there are enormous quantities of volcanogenic sand grains, we have no quantitative estimate of the relative abundances of grains of this origin and those formed by ordinary weathering. Of the other grain-forming processes, only glacial crushing is of quantitative importance and that only at certain times in geologic history. This evaluation justifies a focus on continental weathering and volcaniclastic processes while generally neglecting the minor modes of sand production. We cover the volcaniclastic source materials and sedimentation in Chapter 6.

What is the mechanical composition of the material fed into the sedimentary mill? There are few data which directly answer this question. On the other hand, some studies suggest that all possible sizes are not equally abundant. There seems to be a dearth of certain size grades and higher than normal concentrations of others. This has been interpreted as a reflection of initial abundance and, therefore, a question of provenance. The abundance by weight of grains in the 1 to 8 mm range (especially the 2 to 4 mm grade) seems to be relatively small (Pettijohn, 1940; Hough, 1942, p. 26; Krumbein, 1942, p. 9; Schlee, 1957, p. 1379). This deficiency has also been explained by differences in the competency of the several modes of transport (Wentworth, 1933; Sundborg, 1956, p. 191). Although sorting is effective in separating sizes, it can only result in concentration of particular sizes in different environments. Perhaps mechanical instability in the grains in the 1 to 4 mm range, in which the component mineral grains are large relative to the whole fragment, may be responsible for the destruction of these grades. A considerable quantity of quartz within these grades is produced by rock disintegration (Blatt, 1967a) but such quartz is commonly polycrystalline or is marked by many incipient fractures which likewise may lead to its early destruction during transport.

Russell (1968) rejected these concepts and thought, instead, that the deficiency of very coarse sand and fine gravel in alluvial and in shallow-water deposits was due to a sorting process which concentrated such materials on beaches. He cited many examples of marked concentration of grains in the 1 to 6 mm range on beaches and pointed out that this observation may prove significant in the study of older sandstones and finer conglomerates.

There also appears to be a deficiency at the other end of the sand range. The coarser silt grades seem not to be as abundant as either sand or the finer silts and clays (Udden, 1914, Table p. 741; Pettijohn, 1940; Wolff, 1964). Moss *et al.* (1981, p. 271) support this idea because they found little size reduction below 20 microns. Although this gap has been attributed to the mode of transport, it is more likely that rock decomposition yields primarily two classes of material: (1) sand and (2) clay with little material in the coarse silt or fine sand grades. This concept has been designated the "Sorby Principle" (Folk and Robles, 1964, p. 287) inasmuch as Sorby was perhaps the first to recognize the relation between sediment grain size and the micro- and macrostructure of the source rocks, be they terrigenous or biogenetic.

The most comprehensive investigation of the problem is that of Shea (1974) who considered the evidence for "deficiencies" inadequate owing to sampling and to lack of statistical rigor. Examination of over eleven thousand analyses failed to confirm the supposed gaps.

The Problem of Provenance

Definitions and Concepts

The term *provenance*, derived from the French "provenir," meaning to originate or to come forth, has been used to encompass all the factors relating to the production or "birth" of the sediment. For a given sand, for example, we might ask from what kind of source rock (or rocks) was the sand derived? What was the relief and climate in the source area? How far and in what direction did the source area lie? What was its size? These are pertinent questions which must be answered if we are to achieve a complete paleogeographic reconstruction delineating not only the basin of deposition but also the source lands.

What kind of observations must we make to find an answer to our questions? The nature and character of the source land can be ascertained from a study of the composition of the deposit itself. Sands, more than any other sediment, lend themselves to this approach. The constituent detrital minerals and rock particles are determined by the nature of the source rocks. Some minerals are characteristic of a particular kind or class of source rocks—kyanite, for example, being indicative of a metamorphic source. Others, such as quartz, are ubiquitous but even these display varietal characteristics indicative of a particular source—volcanic quartz, for example. Hence individual mineral species or varieties, or better still, mineral suites, are important sourcerock clues. Rock particles, present in many but not all sands, are invaluable aids in deciphering provenance.

But the composition of the sand is generally not the same as that of the source rocks—the debris has been through a geologic "filter" so that its composition is modified by selective losses and enrichment. Such alterations are related to mineral stability and to selective losses by weathering, abrasion, and sorting during transit and by alteration or solution during diagenesis. To understand these modifications requires knowledge of mineral stability, both mechanical and chemical.

The question of provenance is one of the most difficult the sedimentary petrographer is called on to solve. It is fraught with pitfalls due to the fact that sands are recycled—that is, are derived from preexisting sands which themselves had a provenance as well as a history of transport and rounding. Moreover, the source areas may be multiple, and may have changed with time. The record is further obscured by the complex interplay of relief and climate, the lack of simple relation between the recorded transport direction at the site of deposition and the overall transport path from source to depositional site, and by the effects of diagenesis—particularly intrastratal solution. Blatt (1967b) has written a thoughtful analysis of recycling of sediments and related problems.

It is clear that both the light and heavy minerals of a sand are important in the study of provenance. In fact, it is here, perhaps, that the heavy mineral studies make their principal contribution.

Evidence from Detrital Components

We will first consider the usefulness of individual minerals, especially the varietal characters of the "light" minerals in source rock determination and then discuss the role of mineral suites, especially of heavy minerals. It is with the sands of epiclastic origin—the products of rock disintegration and decomposition—that we are mainly concerned. Such sands are "washed" residues—the transported and sorted products of rock decomposition and disintegration. The character of the source rocks will limit or control what is put into the system. We need to know what kind of materials potential source rocks might be expected to furnish. Volcanic or plutonic quartz? Volume and types of polycrystalline quartz? Heavy mineral suites distinctive of source rock types? Let us review what is known about these and related topics as a guide to provenance.

Detrital Quartz as a Guide to Provenance. Mono- or polycrystalline quartz in sands and sandstones may be derived from plutonic or volcanic igneous rocks, metamorphic rocks, or sedimentary rocks (Table 7-1). By *monocrystalline quartz* is meant a grain consisting of a single crystal; by *polycrystalline* or *composite quartz* is meant a grain consisting of two or more

TABLE 7-1. Genetic classification of detrital quartz.

Genesis	Criteria for recognition
Monocrystalline quartz	
Plutonic rocks	Inclusions, nature of strain
Metamorphic (high rank) rocks	Inclusions, marked strain
Sedimentary rocks (recycled sands)	Worn overgrowths
Volcanic rocks	Extinction sharp, lack of inclusions
	Grain shape (bipyramidal)
Polycrystalline quartz	
Igneous rocks	
Metamorphic rocks	Sutured boundaries,
Low grade	Stretched grains, strong undulose extinction
High grade	Polygonal fabric
Sedimentary rocks	
Quartzites	Relict detrital grains
Cherts (recrystallized opaline and chalcedonic rocks)	Microcrystalline

quartz crystals. Because quartz sand is the principal product of rock disintegration and decomposition and is the dominant constituent of most sands, over the years attempts have been made to utilize quartz as a guide to provenance. Inclusions, grain shape, extinction pattern (undulose or nonundulose), and size and character of internal grains or crystals have all been utilized (Table 7-1).

Among the earliest investigations on the subject were those of Sorby (1880), Mackie (1899), and Dake (1921). One of the most comprehensive analyses of the problem is that of Krynine (1940, 1946a) based on grain shape, inclusions, and extinction. Interest in the subject has grown and there now is a considerable literature (Feniak, 1944; Keller and Littlefield, 1950; Bokman, 1952; Voll, 1960, p. 536; Blatt and Christie, 1963; Conolly, 1965; Moss, 1966; Blatt, 1967a; and most recently Basu *et al.*, 1975; and Young, 1976).

In general, attention has been directed to discriminating between igneous (plutonic) and metamorphic origins of common monocrystalline quartz. It was presumed by Mackie and others that a discrimination could be made based on inclusions, shape, and extinction (undulatory or not). Many workers have found that these criteria are usually difficult to apply (for example, Bokman, 1952), in part because such attributes as shape and extinction show wide variations in the same rock and an assessment of them is very subjective, and in part also because a critical look at the quartz of source rocks shows that the presumed differences in inclusions, shape, and extinction either do not exist or that there is too wide a range of variations and overlap between the quartz of source rock types. Although there may be differences in the statistical average for the several source rock types, it is commonly impossible to assign specific grains in a sandstone to one or another source.

Even though one cannot distinguish with confidence the origins of specific grains of most monocrystalline plutonic quartz, one can frequently identify volcanic quartz. The quartz derived from volcanic rocks—notably the quartz porphyries—may find its way into sediments although only locally is it apt to be important. Volcanic quartz is essentially strain-free, commonly rounded or embayed, the embayments being filled in some cases with the groundmass of the porphyry from which the quartz came. Volcanic quartz also exhibits euhedral form, the grain displaying straight borders of well-developed hexagonal dipyramids without prism faces. Although volcanic quartz is relatively rare, its appearance suggests a volcanic provenance and is likely to be associated with felsic volcanic rock fragments. Modern sands of rhyolitic derivation may contain upwards of ten percent quartz (Webb and Potter, 1971). Volcanic quartz has been recognized in several Eocene formations in the coastal plain of Texas (Todd and Folk, 1957, p. 2550).

Polycrystalline quartz, consisting of several more or less distinct crystalline units seen to have differing extinction positions under the microscope, is mainly derived from igneous and metamorphic rocks. Fine-grained quartzites and recrystallized chert also yield polycrystalline grains. Acid volcanic rocks, lacking phenocrysts, may be mistaken for these very fine-grained particles (Wolf, 1971).

Voll (1960, p. 536) noted that polycrystalline quartz of metamorphic origin is of two types: (1) *polygonized quartz*, in which the component grains form polygonal units, with straight boundaries, which tend to meet at 120 degree angles; and (2) *polycrystalline quartz*, the components of which exhibit sutured boundaries. The former is believed to have formed by "static annealing" of highly strained quartz whereas the latter is a product of "cold working." As Voll points out, these varieties may characterize differing source areas, as in the Scottish Highlands where polygonized quartz characterizes the Western Highlands whereas the quartz of the Moines south of the Moine thrust and much of the eastern Dalradian has been affected by cold rolling and supplies sutured quartz.

Young (1976) further explored the problem of origin of polycrystalline quartz. He noted that quartz deformation is a continuum in which nonundulose quartz will, under increasing pressure and temperature, first exhibit undulatory extinction, then deformation breaks between domains of undulatory extinction, then formation of a polycrystalline aggregate with elongated crystal fabric, and finally a mosaic of polyhedral or polygonized quartz—interpreted as a product of a "static annealing" by Voll. Young showed that one can discriminate in a general way between plutonic, low–medium grade and high-grade metamorphic provenance of detrital quartz.

The ratio of polycrystalline to monocrystalline quartz appears to be a maturity index, because polycrystalline quartz is eliminated by recycling and disintegrates in the zone of weathering, as does strained quartz. Most of the quartz of sand formed in the tropics is

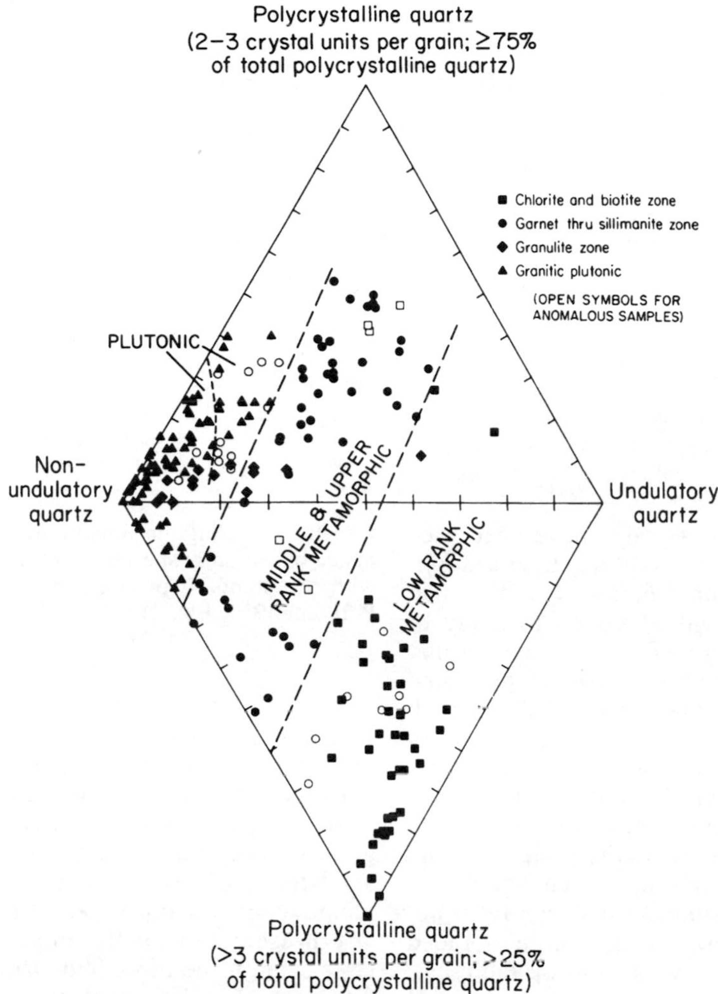

FIGURE 7-2. Four variable plots of nature of quartz population in Holocene sand derived from source rock indicated by symbols. (After Basu *et al.*, 1975, Fig. 6, by permission of the Society of Economic Paleontologists and Mineralogists.)

monocrystalline and strain-free. On the North American craton there is good correlation between the percent of polycrystalline quartz and the mineralogical and textural maturity of the sand (Potter and Pryor, 1961, Fig. 6). Hence, a high ratio of polycrystalline to monocrystalline quartz suggests derivation from a tectonic terrane without significant weathering. Harrell and Blatt's (1978) tumbling barrel studies of quartz types led to the conclusion that the greater abundance of monocrystalline quartz in sandstone was the result of its superior resistance to chemical alteration.

Basu *et al.* (1975) have reexamined the usefulness of quartz as a guide to provenance. They could determine source rocks by undulosity of the quartz even from flat-stage measurements alone. Undulosity of monocrystalline quartz coupled with quantity of polycrystalline quartz and the number of crystal units per grain of the polycrystalline quartz led to discrimination between plutonic, low-rank, and high-rank metamorphic source rocks (Fig. 7-2). Basu *et al.* believe the distinctions would be blurred in the very mature quartz arenites, where monocrystalline, unstrained quartz prevails. Multiple sources will also complicate the problem.

Inferences about the ultimate origin of detrital quartz are clearly subject to considerable uncertainty. But for mapping petrographic provinces and as an index of maturity, quartz varieties are excellent. Moreover, the study of quartz types can permit the petrographic comparison of mature sandstones, perhaps widely separated in both space and time, as was done by Bond and DeVay (1980, Table 2) for quartzose flysch in the western United States.

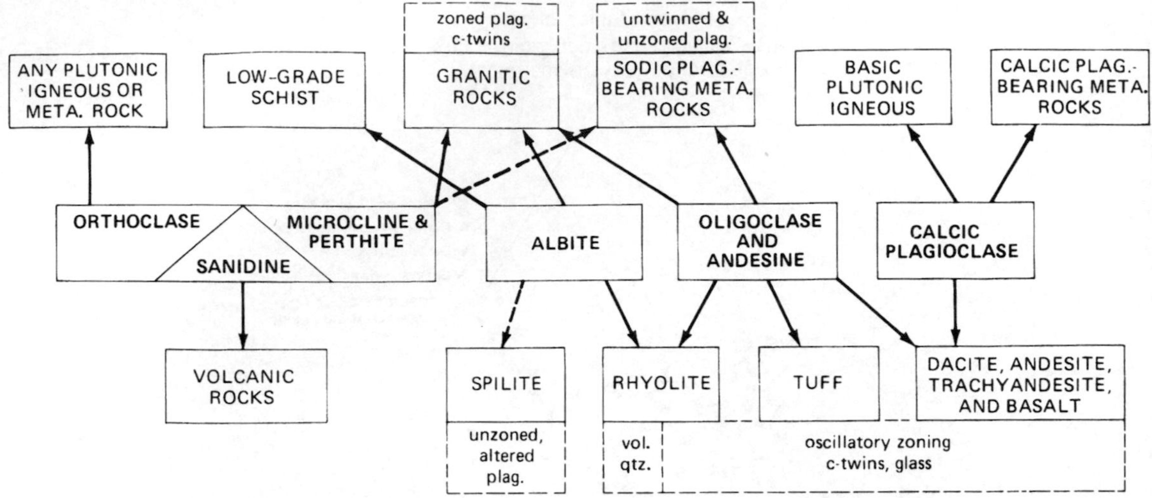

FIGURE 7-3. Diagram illustrating analysis of provenance based on properties of plagioclase determined with the standard petrographic microscope. (After Pittman, 1970, Fig. 5).

The shapes of quartz grains have been used as a guide to provenance—elongate grains presumed to be of metamorphic origin (Krynine, 1946a). A more recent approach to study of grain shape and its value for provenance is that of Ehrlich and Weinberg (1970), Hudson and Ehrlich (1980), and Brown *et al.* (1980). Ehrlich and Weinberg have developed a novel way of measuring shape variations of sand grains. They utilize a Fourier series expansion of the radius about the center of mass utilizing coordinates of peripheral points in the projected grain outline. This approach has been shown by Brown *et al.* to discriminate between relict fluvial sands and those of the more abraded Coastal Plain origin now present on the continental shelf off the Carolinas. Hudson and Ehrlich have extended the technique using Q-mode factor analysis on shape frequency distributions to increase the sensitivity of the method. Application of grain shape analysis to more complex problems of provenance remains to be demonstrated.

For a review of cathodoluminescence of quartz and its bearing on provenance see Matter and Ramseyer (1985). Cathodoluminescence of quartz has still to enjoy wide use for provenance studies, but see Owen and Carozzi (1986) for an application.

In summary, the study of quartz varieties remains an important tool in the study of provenance, the ratio of poly- to monocrystalline quartz being perhaps the most important parameter; following that are the ratio of undulatory to nonundulose grains and grain shape. The quartz varieties are useful in defining petrographic provinces and help in the assessment of source rock types.

Feldspar as a Guide to Provenance. Granular disintegration of acid to intermediate plutonic igneous rocks and the gneisses releases feldspar. In the residual "grus," feldspar may greatly exceed quartz and even in the sand or sandstones of cratons it may equal quartz in abundance. Several types of feldspar are usually present though the more alkali-rich varieties seem to be more abundant than the calcic feldspars. To distinguish the several varieties, especially for point counting, it is imperative that stains be applied particularly for K-feldspar.

In using feldspar as a guide to provenance (Fig. 7-3), one should keep in mind that unlike quartz, feldspar is generally unstable and the several varieties are not all alike in this respect. Feldspar may be lost in the soil profile or in transit or lost by solution; it may, however, form in the buried sediment so it behooves the petrologist to discriminate between detrital and authigenic feldspar. Moreover, the character of the feldspar may be transformed by diagenesis. Conversion of a more calcic plagioclase to nearly pure albite has been reported in buried Eocene sands of the Gulf Coast (Boles, 1982) and in Cambrian sandstones of Quebec (Ogunyomi *et al.*, 1981).

In view of the instability of feldspar in sandstones, it may be prudent to examine the associated mudrocks in which, because of their low permeability, there is a greater chance of preservation of the feldspars (Blatt, 1985).

The provenance of detrital feldspar has been the subject of many recent investigations—in thin section and with the X-ray and microprobe. Rimsaite (1967) studied the varietal features of feldspar in potential source rocks. Nine classes based on zoning, diverse intergrowth habits, twinning, fracturing, and the like could be defined. Gorai (1951) and Slemmons (1962) made careful studies of twinning in plagioclase and its abundance in igneous and metamorphic rocks. Other optical studies have been made by Mizutani (1959), Ermanovics (1967), and Plymate and Suttner (1979).

The feldspars of many sediments display zoning. Oscillatory zoning, progressive zoning, and lack of zoning in plagioclase may be clues to the provenance of the feldspar (Pittman, 1963). The plagioclase in volcanic and hypabyssal rocks is characterized by oscillatory zoning whereas this type of zoning is rare in plutonic igneous and metamorphic rocks. Zoned plagioclase, regardless of type of zoning, is strongly indicative of a volcanic source (Fig. 7-3).

The feldspar of the acid volcanics is likely to be sanidine; that of the acid plutonic rocks is either orthoclase or microcline. Perthitic feldspar is indicative of slow cooling and hence characteristic of the plutonic sources. The feldspars of pyroclastic origin tend to show euhedral forms, commonly broken, and display a thin envelope of glass, whereas the plutonic feldspars are anhedral.

Suttner and Basu (1977) studied disorder in single grains of alkali feldspars using both two and three X-ray peaks. The two-peak method is not as precise as the three-peak method but only requires an hour per grain; Suttner and Basu give full procedural details. Using Holocene sands of known provenance, they found sanidine to be highly disordered, implying rapid cooling, whereas disorder in microcline and orthoclase was less and correlated with a cooling history related to the rate of erosion of the batholith or metamorphic belt.

Trevena and Nash (1979, 1981) used the microprobe to determine the composition of over 2000 feldspars in igneous and metamorphic rocks and plotted them on a standard Ca–Na–K triangular diagram. Much of the plagioclase of volcanic origin fell in a fairly distinct compositional field, though extremely accurate determination is required for discrimination, whereas plagioclase from metamorphic and plutonic rocks did not. Trevena and Nash (1981, pp. 145–147) also thought the Fe and Sr in plagioclase useful for discrimination.

Maynard (1984) has investigated the composition of plagioclase feldspar in modern deep-sea sands. He found that such sands differ from crystalline rocks in that they have a high content of pure albite which he attributes to recycling of authigenic feldspar of sedimentary origin. Maynard also found a correlation between plagioclase composition and tectonic setting and could distinguish five tectonic settings: trailing edge, island arcs, strike-slip, continental margin volcanic arc, and back-arc areas. There was, however, considerable overlap in plagioclase composition between these categories.

What then can be concluded from feldspar and its types in sandstones? Like quartz and its varieties, feldspar is excellent for mapping mineral associations and defining petrographic provinces. Moreover, the abundance of feldspar is a kind of maturity index—much is destroyed by weathering where relief is low and rainfall high (p. 194). Its twinning favors loss by mechanical abrasion (Anwar, 1977; Mack, 1978). Thus, from a survival point of view it is more sensitive than quartz. How much feldspar in a sandstone is recycled—and how we can identify it? We also need to be aware that secondary albitization may be far more extensive than earlier thought, especially in marine sandstones, and thus may falsely color our perception of source area.

Basu (1981) noted that the detrital feldspars of Cambro-Ordovician sandstones are unaltered K-feldspar; plagioclase is absent or severely altered. This Basu attributed to a different soil–water composition in the absence of land plants in the earlier times—a soil–water environment in which only K-feldspar was stable. Thus, in addition to diagenesis, perhaps long-term changes in surface weathering may have affected the abundance and types of feldspar in sandstones.

Rock Particles as Guides to Provenance. Sandstones commonly contain rock particles in addition to the usual quartz and feldspar (see Chap. 2). Rock particle content is strongly size dependent; rock particles are more abundant in the coarser size fractions. Thus it is best to make comparisons with sandstones of comparable grain size. As size decreases, identification becomes more difficult and more subjective (Boggs, 1968). Dickinson (1970, p. 700) suggests use of operational criteria, mainly textural, to classify rock particles. He defines five groups: (1) volcanic rock particles which have the textures of aphanites, (2) clastic rock particles with fragmental textures, (3) tectonite fragments with schistose fabric, (4) micrigranular particles, that have roughly equant, well-sized grains (chert, etc.), and (5) carbonate rock particles. These groups can be further subdivided: the volcanic include felsic, microlitic, lathwork, and vitric; the clastic may be silt-sandy or argillaceous or volcanoclastic; the tectonites include metasedimentary and metavolcanic; the microgranular can be subdivided, often with difficulty, into hypabyssal, hornfelsic, and sedimentary.

The rock particles themselves, perhaps more than any other type of grains, carry their own evidence of provenance. The volcanic assemblage includes glass, crystal fragments, corroded and embayed quartz with envelopes of glass or glass found in the embayed recesses, and volcanic rock particles. It may be difficult to distinguish between the immediate products of explosive volcanism and the reworked materials of this origin and the true epiclastic products of volcanic rock disintegration.

The sedimentary products include carbonate sands derived by the disintegration of the coarser dolomites and marbles, the calcareous debris so characteristic of the sands of the Alpine Molasse (Füchtbauer, 1967, p. 271) and the Miocene Oakville Sandstone of the Texas coastal plain (Folk, 1968, p. 141). More common are the abundant pelitic materials, the shales in particular, which tend to be deformed and molded about the more resistant quartz so as to appear more like pore-filling matrix rather than like a detrital grain (Allen, 1962, p. 669). Chert particles though never dominant are nearly ubiquitous in lithic arenites especially where a carbonate terrane of low to moderate relief is being eroded in a humid climate. In the United States the streams draining the cherty carbonates of the Ozark region of Arkansas and Missouri are particularly rich in chert detritus as are many Cretaceous sands in the northern Rocky Mountains area of the United States and Canada (Suttner, 1969, p. 1401).

Metamorphic materials are primarily schist particles but include also metavolcanic particles such as "greenstones," and not infrequently serpentine fragments which may locally be dominant (Zimmerle, 1968). Krynine (1937, p. 427) described "schist arenites" which contain 35 to 40 percent schist particles.

Rock particles are among the most informative of all detrital components and all efforts to determine provenance should include a close scrutiny of these materials. Good examples of their utility are given by Graham *et al.* (1976), who studied rock fragments to infer common provenance for the Carboniferous sandstones of the Ouachita and Black Warrior basins, two basins separated by over 600 km, and by Makrutzki (1982).

Micas and Provenance. The micas, though conspicuous in some sandstones, are never a major constituent. They are derived from schists and gneisses, from plutonic igneous rocks, and from volcanic sources. Because of stability, muscovite is more common in the sediments than biotite. The latter, if it occurs as pseudohexagonal plates, is probably of volcanic origin. In general, abundant mica suggests a metamorphic provenance for the sand and likely proximity to the source.

Heavy Minerals and Provenance. The heavy minerals—the minor, high-density, accessory detrital minerals of sandstone—have long been used as indices of provenance (Boswell, 1933, p. 47; Milner, 1926, p. 101). That certain species are characteristic of certain source rocks is well known. Sedimentary petrologists have defined certain detrital mineral associations, each indicative of a major class of source rocks (Milner, 1926; Krumbein and Pettijohn, 1938, p. 463; Feo-Codecido, 1956; Baker, 1962, p. 90). These associations are summarized in Table 7-2. For a recent review see Morton (1985a).

Certain species appear in more than one of the major associations as does quartz, for example. Consequently attention has been given to the *varietal* characteristics of these species with the expectation that the several occurrences can be distinguished from one another. Minerals especially scrutinized include tourmaline and zircon. Krynine (1946b) made a thorough study of the varieties of tourmaline and believed these could be related to their provenance. Krynine's approach to the problem was largely inductive whereas the work that has been done on zircons (Vitanage, 1957; Poldervaart, 1956) is based more largely on actual studies of zircon morphology in source rocks. All these efforts have been directed toward establishing generalizations applicable to the several mineral varieties wherever found. The size and predominant characters of zircons are apparently invariant or closely similar throughout a body of magmatic granite. There are marked differences between igneous bodies; the zircons of the extreme alkaline rocks, for example, display bipyramidal habit in contrast to the long

TABLE 7-2. Heavy mineral associations and provenance. (Modified from Feo-Codecido, 1956, p. 997).

Association	Source
Apatite, biotite, brookite, hornblende, monazite, muscovite, rutile, titanite, tourmaline (pink variety), zircon	Acid igneous rocks
Cassiterite, dumortierite, fluorite, garnet, monazite, muscovite, topaz, tourmaline (blue variety), wolframite, xenotime	Granite pegmatites
Augite, chromite, diopside, hypersthene, ilmenite, magnetite, olivine, picotite, pleonaste	Basic igneous rocks
Andalusite, chondrodite, corundum, garnet, phlogopite, staurolite, topaz, vesuvianite, wollastonite, zoisite	Contact metamorphic rocks
Andalusite, chloritoid, epidote, garnet, glaucophane, kyanite, sillimanite, staurolite, titanite, zoisite-clinozoisite	Dynamothermal metamorphic rocks
Barite, iron ores, leucoxene, rutile, tourmaline (rounded grains), zircon (rounded grains)	Reworked sediments

prismatic forms of some granites. As first noted by Mackie (1923), the zircons of very old rocks, especially the Precambrian, tend to be purple (hyacinth). Tomita's work (1954) on colored zircons suggests that such color is produced by prolonged radiation bombardment. A Precambrian source of such zircons is probable (Beveridge, 1960, p. 534). Zimmerle (1972) studied purple zircons in central Europe and found their proportion to increase in older sandstones.

Callender and Folk (1958) believed they could distinguish between volcanic and nonvolcanic zircons in some Tertiary sands of Texas. The zircons of volcanic provenance were more idiomorphic; there was, however, considerable overlap in zircon populations from the various sands. Force (1980) studied the distribution of rutile in source rocks and found this mineral to occur chiefly in metamorphic rather than igneous rocks.

Other studies have been directed toward a study of the minor accessory minerals of specific crystalline bedrock bodies which were or might have been the sources of heavy minerals found in younger sediments. Examples include the classic study of the Dartmoor granites by Brammall (1928) and that of Tyler *et al.* (1940) of the Precambrian granites and gneisses in the Lake Superior region. These and similar studies have made it possible to identify the sources of some of the heavy minerals in the younger sandstones. A notable early example of this type of investigation is that of Mackie (1923) on the purple zircon which is indigenous in the Precambrian Lewisan Gneiss of the north of Scotland but appears in sediments ranging from late Precambrian through Jurassic in age.

Mange-Rajetsky (1981) showed that blue sodic amphibole in Recent sands on the southern coast of Turkey demonstrates the as yet undiscovered presence of a blueschist facies in the hinterland.

In a new application of the microprobe, Morton (1985b) used it to study compositional variations of garnet and defined three mineral associations in Brent Group (Jurassic) of the North Sea.

Mineral Stability, Climate, and Provenance

As noted elsewhere, the composition of a sand is unlike that of the source rocks as it has been through a geologic filter. Many of the minerals in the source rock have been destroyed during weathering, transportation, or following deposition. To read provenance history, therefore, we need to know something of the survival capabilities of the minerals—their *stability*. Although the question of stability was touched on in Chap. 2, we need to reexamine it here as it relates to provenance. There has been a longstanding interest in the subject. Derby (1891) was one of the first to investigate mineral stability. Thoulet (1913) had a clear perception of the problem, discussed the "pathology" of minerals, and even established an order of resistance to destruction—a stability series. Boswell (1933) devoted a chapter to the subject; Milner (1962, p. 434) published an elaborate table in which he made a qualitative assessment of the stability of each mineral species. For more recent reviews of the subject see Pettijohn (1975, p. 489), Nickel (1973) and Luepke (1984).

Stability is both mechanical and chemical. *Mechanical stability* defines the ability of the mineral to withstand abrasion and other size-reduction processes—processes that tend to destroy a grain during transport by wind, streams, and the surf. *Chemical stability* relates to the mineral's ability to resist solution or alteration in a particular chemical environment. Such stability depends both on the nature of the mineral and the chemical character of the surrounding medium, be it air, soil water, groundwater, or seawater.

There are empirical, experimental, and theoretical approaches to the study of mineral stability. For the most part, absolute stability, depending how we define it, is not measured; only the relative stability is established. Goldich (1938), in a classic paper, studied soil profiles and showed that the common rock-forming minerals could be arranged in an order dependent on the degree of weathering. This order was the same as Bowen's reaction series reversed—the minerals formed at the highest temperatures and pressures were found to be the least stable under surface conditions. Another example of this empirical approach is that of Sindowski (1949), who compared the heavy minerals in several Rhine terrace sands. Presumably the older terraces had undergone more leaching and had, therefore, a more restricted heavy mineral suite. Pettijohn (1941), in like manner, compared the mineralogy of sandstones of various geologic ages and, like Sindowski, was able to arrange the heavy minerals in a stability order. The resistance of minerals to intrastratal solution (see Chap. 10) turned out not to be greatly different from their resistance to weathering though there are some important exceptions. Formation waters vary in composition and a particular mineral may be stable in one case and not in another. The stability of the more common heavy mineral species is given in Table 7-3. One should remember, however, that this table, like others published elsewhere, is an overall assessment of stability under "average" conditions and that wide departures from it may be expected.

Of the many experimental studies on relative chemical stability of minerals at low temperatures, few have been focused on common heavy minerals of sandstones. Many leaching experiments, most directed toward other goals, were summarized by Boswell (1933). A more recent review of heavy mineral solubilities is given by Nickel (1973).

There have also been some theoretical approaches to the problem, based largely on thermodynamic properties of minerals, most notably by Gruner (1950), Fersman (cited by Baturin, 1942), Muri (cited by Polynov, 1937), and Curtis (1976), and a summary is given by Stumm and Morgan (1970, pp. 385–92 and 400–402). Experimental and theoretical studies are, thus far, limited by the lack of data on the kinetics of very slow reactions.

As noted, the chemical stability of a mineral will determine its response to weathering and hence its survivability. But weathering, in turn, is influenced by climate. And can we, from the minerals that survive and on the basis of knowledge of their stability, draw any inferences about the climate in the source area? How much, if at all, does climate affect the major components of sand and their proportions? There have been several studies of modern fluvial sands that show a marked contrast in the proportions of quartz, feldspar, and rock fragments under differing climatic conditions (Basu,, 1976; Suttner *et al.*, 1981). These and other studies have shown that sands from similar source rocks are markedly different in humid and arid climates (Fig. 7-4). But as noted long ago by Krynine (1935), the effects of climate may be obscured by relief. Arkose, for example, may be produced even in tropical humid regions if relief is high.

Dickinson's (1982) study of Mesozoic sandstones of the convergent boundaries around circum-Pacific basin shows a remarkable homogeneity that surely is independent of climate. On the other hand, Franzinelli and Potter (1983) found the quartz arenites prevail in the larger rivers of the Brazilian shield—rivers with watersheds in crystalline rocks covered in part by tropical rain forest. Unfortunately, we lack studies of sands in modern watersheds that might help us quantitatively estimate the interaction of climate and relief. We need to examine watersheds with high relief and diverse climates and the opposite, areas of low relief and variable climates. We have indirect evidence of the pattern of chemical weathering in the source areas from geochemical studies of river water, such as those by Stallard and Edmond (1983). From these one can deduce some of the effects of lithology, relief, and climate on the chemical

TABLE 7-3. Stability of some detrital heavy minerals.

Ultrastable	Rutile, zircon, tourmaline, anatase
Stable	Apatite, garnet (iron-poor), staurolite, monazite, biotite, ilmenite, magnetite
Moderately stable	Epidote, kyanite, garnet (iron-rich), sillimanite, sphene, zoisite
Unstable	Hornblende, actinolite, augite, diopside, hypersthene, andalusite
Very unstable	Olivine

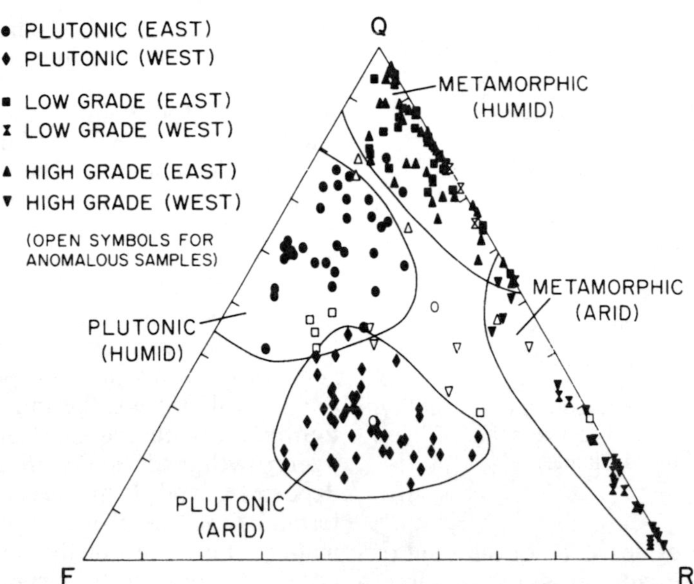

FIGURE 7-4. QFR ternary plot of the composition of the medium-size fraction of Holocene fluvial sand of first-cycle parentage (from Suttner *et al.*, 1981, Fig. 1, by permission of the Society of Economic Paleontologists and Mineralogists).

load and thus by implication on the sand-size suspended load. Extrapolation to the past includes the assumption that the roles of plant cover and partial pressure of carbon dioxide and oxygen were the same then as now (see Chap. 12).

Suttner *et al.* (1981, p. 1244) note that the role of depositional environment also needs consideration and suggest that climatic influence is strongest in first-cycle fluvial sand near the source and is probably minimal in first-cycle marine sands.

As noted by Mann and Cavaroc (1973) in their study of modern sands derived from crystalline rocks of the Piedmont, the mineral composition of the sands is texture dependent. Comparison between sands is valid only if they are of similar grain size; otherwise mineralogical differences related to texture may obscure those related to climate and other factors.

Mechanical stability has also been studied both experimentally and empirically. Early experiments by Friese (1931) and Thiel (1940) determined the relative resistance of various minerals in abrasion mills. A more recent study is that of Dietz (1973). Field data are ambiguous because natural sands are subject to sorting processes based on size, shape, and specific gravity so that the proportions of the several species are altered. Moreover, the proportions are modified by dilution—introduction of new materials. Hence downstream change in the mineralogy is the end result of factors other than selective abrasion (see Chap. 2). The field studies themselves, even when carefully designed, yield contradictory results. Dana Russell's early work on Mississippi River sands (1937) seemed to indicate little or no changes in sand composition due to abrasion. Other studies, such as those of Plumley (1948) and, more recently, Cameron and Blatt (1971), show some selective losses in high-gradient streams. The effects of transportation have been summarized by Suttner *et al.* (1981, pp. 1238–39) who thought that abrasion of feldspar and rock fragments on a marine coastline can be appreciable. Twinning of feldspar seems to be a major factor in its lower abundance in high-gradient streams and coastal deposits. Suttner *et al.* (1981, p. 1240) also recognize that abrasion is more intense in a high-, rather than a low-energy environment. At present we can recognize the role of abrasion but it is difficult to estimate its importance quantitatively when reconstructing the provenance history of an ancient sandstone.

Reading Provenance History

There are many approaches to determination of the provenance of a sand or sandstone—mineralogy, texture, paleocurrent pattern, and facies distribution. We can supplement these with knowledge of paleolatitude (to better estimate paleoclimate) and of regional geology and perhaps even plate tectonics (Fig. 7-5). Our goal is

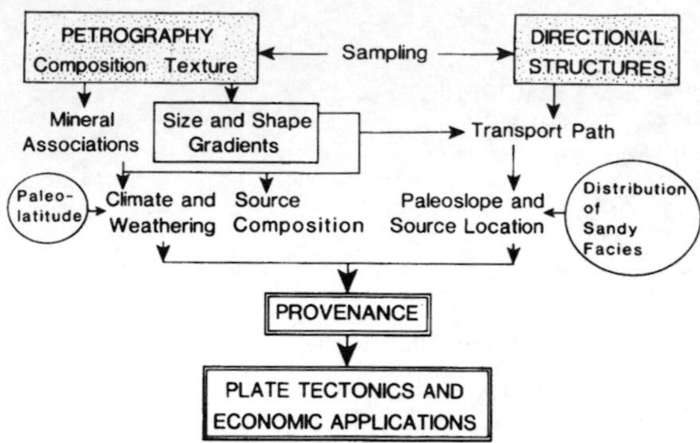

FIGURE 7-5. Provenance methodology and utility. Many studies may only need to use some of these methodologies, but all are available.

to unravel the "line of descent" of the sand or sandstone. Three broad paths exist—direct contributions, recycling, and combinations of the two (Fig. 7-6).

As must now be evident from the preceding sections, reading the provenance history of a sand is a difficult task. Multiple sources and multiple cycles may be both involved (Fig. 7-6). It may be difficult to distinguish between the immediate source (last cycle) and the ultimate source of the sand. There are several approaches to deciphering provenance. Even a single sand grain can shed some light on the problem. A tourmaline grain (Fig. 7-7), for example, may show a secondary overgrowth on a rounded detrital core. The overgrowth may be rounded also. Abraded overgrowths on an abraded core imply (1) formation of a tourmaline crystal in an igneous or metamorphic source rock, (2) weathering and release of this grain followed by transportation and abrasion, (3) deposition in a sand followed by (4) an authigenic or low-rank metamorphic regeneration of the crystal, (5) weathering and release of the tourmaline with transport and abrasion of the overgrowth, and, finally, (6) deposition in a new deposit of sand. It may even be possible to determine from the character of the original tourmaline, the nature of the source rock utilizing criteria formulated by Krynine (1946b).

A more complete analysis of provenance can be made from a sample than from a single grain. What could we infer, for example, from a single sample of Mississippi River sand about the nature and character of the rocks in the Mississippi Basin? A typical sand from the Delta (Russell, 1937, no. 1083-3/4) contains 64 percent quartz, 19 percent rock particles, and 15 per-

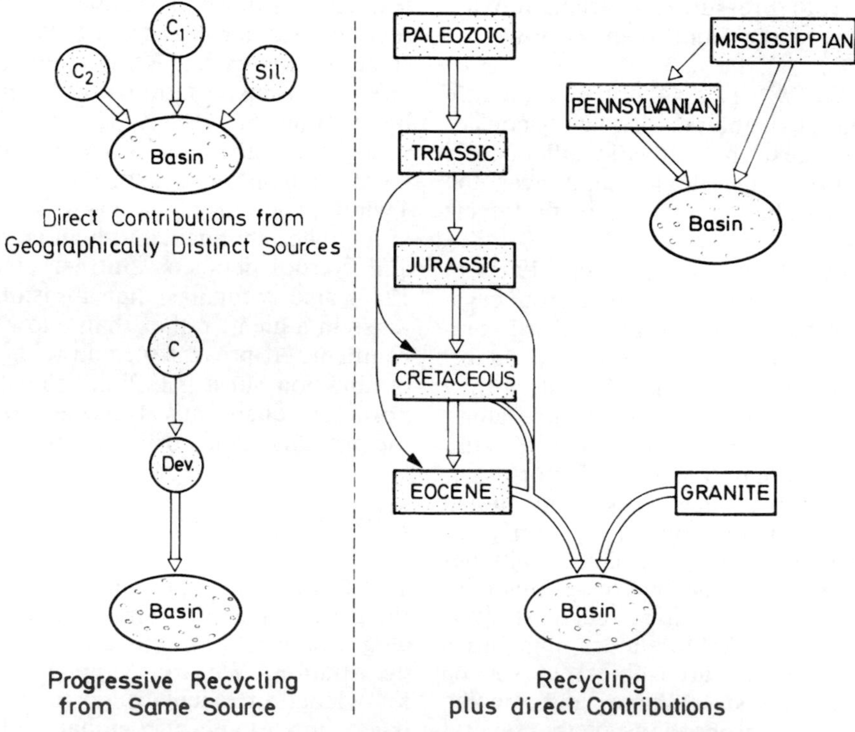

FIGURE 7-6. Some common paths of mineral provenance.

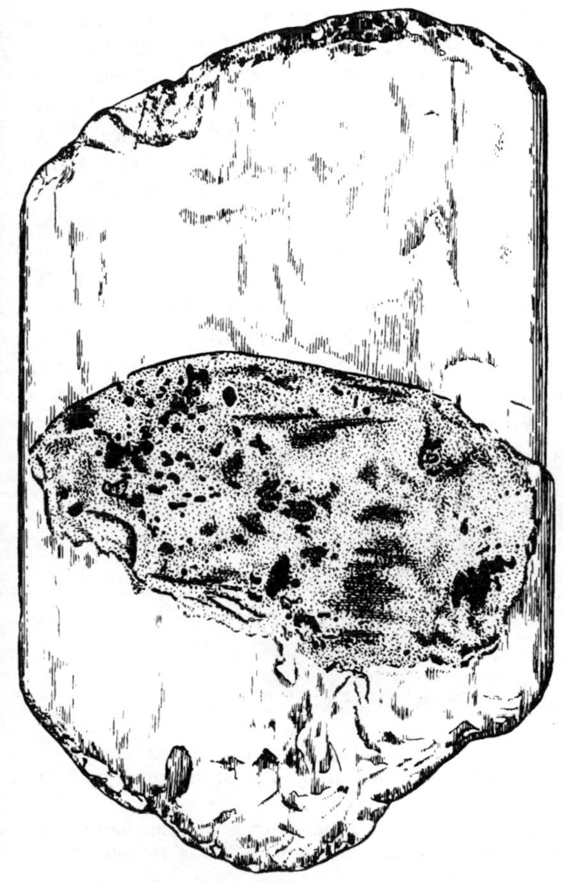

FIGURE 7-7. Abraded tourmaline overgrowth on abraded detrital core, Cretaceous McNairy sand, Henry County, Tennessee, U.S.A. (Redrawn from Potter and Pryor, 1961, Plate 2).

cent feldspar. The quartz is subangular as is the feldspar. The latter is dominantly microcline with a lesser amount of oligoclase and andesine and a little orthoclase and sanidine. Principal rock particles are chert and fine-grained quartzite, but included are several varieties of volcanic rocks, some slates, and schists. The sand contains a little detrital calcite and glauconite. The dominant heavy minerals are ilmenite, pyroxenes, and amphiboles with lesser amounts of garnet, zircon, titanite, rutile, and monazite. From this analysis we might conclude that the dominant source was a granite or granodiorite with minor sources from extrusive volcanics, metamorphic and sedimentary rocks. Clearly such an analysis fails to do justice to the geology of the Mississippi Basin, although it does reflect its glacial source which is characterized by a significant proportion of crystalline rocks. We may identify the source rocks and estimate their relative importance as contributors to the sand but we are unable to assess their areal extent in the drainage basin.

The use of more than one sample can improve our assessment. This is especially true if our samples are properly distributed throughout the basin or, in the case of ancient sandstones, from the whole area of sand accumulation.

The investigator's conclusions are commonly presented as a kind of "flow sheet" or provenance diagram. These interpretative diagrams are of two kinds—the first is based solely on what can be seen in the rock itself and records a judgment about the nature or kind of source rock for each of the detrital components of the sand. It can be as simple as a QFL plot.

The second type of provenance diagram is based not only on a study of the thin section but also upon a knowledge of the regional geology and stratigraphy. It depicts what is possible or, in the judgment of the petrologist, probable, in terms of the source rock contributions. To construct such a diagram, the investigator must piece together various lines of evidence—the distribution and kinds of heavy minerals, paleocurrent information, facies relations, both vertical and lateral, the distribution and kind of rock fragments both in the sands and in the associated conglomerates, knowledge of structure and the larger tectonic elements—in short, any and all manner of geologic data.

Good examples which illustrate this approach and in which there are diagrams summarizing the provenance of the sandstone have been published by Krynine (1940, Fig. 3), Payne (1942, p. 1754), Todd and Folk (1957, p. 2557), Potter and Pryor (1961, Fig. 14), Doty and Hubert (1962, Fig. 11), Hopper (1962, Fig. 8), ten Haaf (1964, p. 135), Suttner (1969, Fig. 15), and Walker and Pettijohn (1971). The diagrams of Krynine, Doty and Hubert, Todd and Folk, and Walker and Pettijohn are diagrams of the first type, largely based on petrology—what can be seen in the sandstone itself—whereas those of Payne, Suttner, Potter and Pryor, and ten Haaf belong to the second category and utilize regional geology, tectonics, and stratigraphy to help tell the story.

In its broadest aspect, the problem of provenance can be considered as a problem of accounting—making an inventory of the different types of grains contributed by different source rocks. To this, one should add the problem of *the same kinds of grains coming from different source rocks* so that, at present, we are generally unable to distinguish the quartz grains derived from a granite from those derived from a gneissic rock. More sophisticated accounting will be possible only when we can better petrographically "tag" grains and better evaluate the number of cycles of erosion in their abrasion history.

Provenance and Plate Tectonics

The relation between sand mineralogy and tectonics has long been and continues to be a central topic in sandstone petrology. The mineral composition is presumed to reflect the tectonic history of both the source area and the site of deposition—an idea pursued especially by Krynine (1943) in America.

The subject has received renewed attention with the advent of plate tectonics. The general notion is that the framework fraction reflects position with respect to plate margins and also the type of plate boundary. Moreover, the composition of the sandstones would not only vary with site but also change through time. The subject or some aspect of it has been explored by a great many, including among others Crook (1974), Schwab (1975, 1981), Potter (1978), Valloni and Maynard (1981), Dickinson (1969, 1982), Dickinson and Suczek (1979), Dickinson and Valloni (1980), Dickinson *et al.* (1982, 1983), and Valloni (1985). We cover it in full in Chap. 12. Here we briefly outline some of the basic ideas in relation to provenance determination.

We start from the idea that the types and proportions of the basic components—quartz, feldspar, and rock fragments—of a sandstone reflect a provenance that varies in an orderly way with plate tectonic setting (Table 7-4).

TABLE 7-4. Provenance and plate tectonics.

Part A. *Tectonic provenances*
(adapted from Dickinson and Suczek, 1979, pp. 2175–78)

Magmatic arc

A. Undissected, volcanogenic highlands associated with active continental and island arc systems. Derived sands are rich in plagioclase feldspar and lithic volcanic fragments, and are deposited in trenches, fore-arc basins, intra-arc basins, and back-arc basins. Little effect of climate.

B. Dissected, mature, continental margin magmatic arc provenance characterized by sands rich in plutonic and volcanic grains deposited in backarc, fore-arc, trench, and abyssal plains basins. Little to moderate effect of climate.

Recycled orogenic

A. Tectonically uplifted subduction complexes with derived sands characterized by abundance of chert exceeding combined quartz and feldspar grains. Sands are redeposited into fore-arc basins or trenches. Little effect of climate.

B. Collision-derived detritus from suture belts. Little effect of climate.

C. Collision orogenic provenances represented by foreland fold–thrust belts, characterized by an abundance of sedimentary–metasedimentary lithic fragments. Little to moderate effect of climate.

D. Uplifted foreland fold–thrust belts with derived sands characterized by variable associations of quartz, chert, and sedimentary lithic fragments deposited in foreland basins. Little to moderate effect of climate.

Cratonic (continental)

A. Quartz arenites many times recycled and minor arkoses above unconformities plus some lithic arenites derived from marginal belts. Potash feldspar more abundant than plagioclase in cratonic derived sands. Cratonic sands then accumulated on passive margins and in bordering oceans. Climate, if tropical, can have significant effect.

B. Fault-bounded basins in continental basement have subarkosic and arkosic sandstones and possibly some minor volcanic debris. Minor influence of climate?

Part B. *Essential petrologic and chemical variables*

QFL: Plots on triangular diagrams essential for all studies and effectively separate quartz-rich, cratonic sandstones from lithic-rich sandstones of collision margins

F_P/F_T: Plagioclase to total feldspar ratio helps distinguish potash-rich cratonic sandstones from plagioclase-rich leading margin sandstones. Microprobe useful supplement to identify provenance of individual feldspars.

Volcanics/total rock fragments: Quick and easy index to leading margins. Microprobe needed to identify exact compositions of plagioclase phenocrysts and give composition of volcanic glass.

$Q_MF_PF_K$: The triad of monocrystalline quartz, plagioclase, and potash feldspar deserves more use and should help identify volcanic terranes.

$Q_PL_VL_S$: Triangular diagram of polycrystalline quartz, volcanic, and sedimentary rock fragments should be routinely plotted as a discriminator of composite grains.

SiO_2/Al_2O_3: Long-time standard index of chemical maturity (also seen as $SiO_2/(Al_2O_3 + SiO_2)$).

K_2O/Na_2O: Simple, approximate chemical index of plagioclase in total feldspar and hence the rough chemical equivalent of F_P/F_T.

Dickinson *et al.* (1982) and Dickinson (1985) define three main classes of provenance—continental blocks, magmatic arcs, and recycled orogens—each of which is divided into subprovinces. The continental block includes the stable platform, uplifts along plate boundaries, and intraplate deformation. The magmatic arc includes the volcanic chain, granite plutons of the arc roots, and the bounding metamorphic envelope. Recycled orogens contain sedimentary strata and subordinate volcanic rocks and their metamorphic derivations which are exposed to erosion by orogenic uplift of fold belts and thrust sheets.

This analysis depends on a suite of samples in which the mineralogy is determined by point counting and plotted on the conventional compositional triangles. The interpretation is dependent on the plotted points forming a well-defined cluster (Fig. 7-8). Individual samples may plot outside the field of points and there are clearly gradational and overlapping field boundaries.

According to Dickinson and Suczek (1979), quartzose sands derived from the cratonic arcas are common within interior basins, on platforms, miogeosynclinal wedges, and open ocean basins bounded by passive margins. Arkosic sands from uplifted basement blocks occur locally in rift grabens and wrench basins related to transform ruptures. Volcaniclastic lithic and more complex volcano-plutonic sands derived from magmatic arcs occur in trenches, forc-arc basins, and marginal seas. Recycled sands, rich in chert and other rock fragments derived from subduction complexes, collision orogenic belts, and foreland uplifts, are found in closing ocean basins, diverse successor basins, and foreland basins.

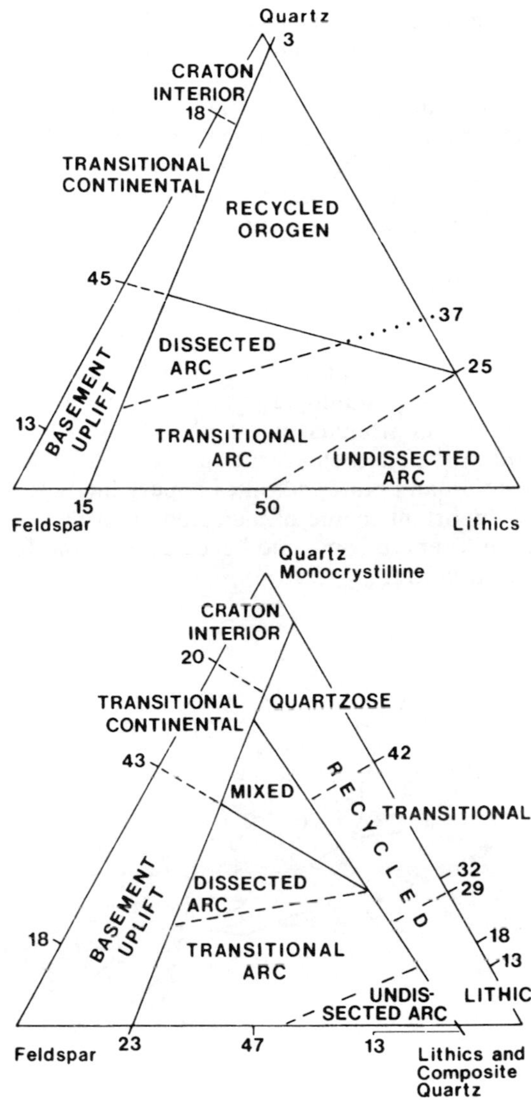

FIGURE 7-8. Relation between framework composition of sandstones and tectonic setting. After Dickinson *et al.* (1983, Fig. 1).

The provenance of exotic terranes accreted to a continent by convergent and transform plate motions is an important factor in their recognition. Terranes such as Wrangellia and others are disjunct from the continental masses to which they are welded in terms of detrital provenance. The exotic nature of these terranes is demonstrated by paleomagnetic evidence of latitudinal motion of the terrane. Thus paleomagnetism is itself a guide to provenance with respect to the single variable, global latitude.

Detailed studies of the west coast of North America have led to the conclusion that significant portions of the Cordillera are exotic domains. The paleomagnetism and the provenance of the sediments within the exotic block can be understood only if the terrane in question originated elsewhere and was transported to its present position and accreted to the continental block. The transport may in some cases have been great—perhaps hundreds of kilometers. Provenance plays a major role in unraveling accretionary tectonics.

And what should be said of cautionary and contrary opinions? Schwab (1981) found sandstone composition in the western Alps to be a good but not infallible index to plate tectonic regime when matched against stratigraphic and structural history. Climate, as assessed by rainfall, also has been considered (Potter, 1978, pp. 445–46; Franzinelli and Potter, 1983, Fig. 7), but probably only has a major effect on composition when relief and tectonism are low, as they are on many cratons and passive margins even though the combination of high relief and high rainfall produces large sediment volumes (Potter, 1978, Fig. 14) (Fig. 7-9). In the distant past of the Proterozoic and earlier, we should also recognize, as have Condie (1982) and others, that plate tectonics may have operated in a different manner, or not at all, and hence inferences based on studies of modern sands and Mesozoic sandstones may not apply. Finally, the anti–plate tectonic school, steadily decreasing in numbers, continues to find some expression (Beloussov *et al.*, 1979). All this said, it is nonetheless true that sandstone composition and plate tectonics appear likely to be an important theme of sandstone petrology for many years to come and hence deserve our fullest attention.

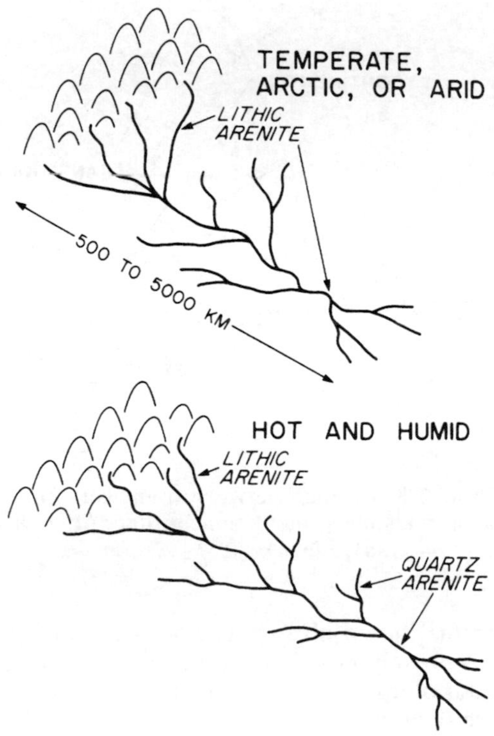

FIGURE 7-9. Possible effect of climate on the composition of sands of a large river draining across a continent from its active to passive margin. Hot humid climates provide greater opportunity for weathering on flood plains and should supply more quartz-rich sand tributaries downstream than does a cold-cool climate in watersheds of comparable relief. (Franzinelli and Potter, 1983, Fig. 7). *The Journal of Geology*, by permission of the University of Chicago Press.

Examples of Provenance Studies

Provenance studies are numerous—some are a small part of a broader study and some are the main theme. Below we have selected forms that give an insight into the diversity that exists. Modern as well as ancient sandstones are included.

Sand from the world's largest river system:

Franzinelli, Elena, and Potter, Paul Edwin: Petrology, chemistry and texture of modern river sands, Amazon River system. Jour. Geology 91, 23–39 (1983).

The Amazon River and its tributaries transport sand to the Atlantic from both the Andes and from the vast reaches of the Brazilian and Guiana Shields. Some rivers of the Amazon system also flow largely in Tertiary molasse, derived from the Andes. Thin-section and chemical study of 95 sand samples of river sand from this basin, a basin with great contrasts in relief and climate, reveals that sands derived from the Andes are lithic arenites, whereas those derived from Precambrian terranes are much richer in quartz and also are coarser. Sands from rivers in Tertiary molasse are the most quartz-rich but are finer-grained like Andean sands.

This study suggests that tropical weathering on low-relief cratons can produce first-cycle quartz arenites and the authors also speculate that climate can significantly alter the composition of sand of a large river draining the back side of a leading-edge mountain range (Fig. 7-9).

Molasse of southern Germany:

Füchtbauer, H.: Sedimentpetrographische Untersuchungen in der älteren Molasse nördlich der Alpen. Eclogae geol. Helvetiae 57, 157–298 (1964).

This monographic work is based on mineralogic analysis of over 1,500 samples collected in an area 450 km long lying north of the Alps, mainly in southern Germany. The Molasse sediments crop out along the tilted southern margin of the basin. They are coarse conglomeratic fans which pass in the subsurface into fine-grained, horizontal deposits within the basin.

The sandstones, rich in feldspars and rock fragments, especially dolomitic rock fragments, have a varied heavy mineral assemblage which can be divided into subassemblages, each correlated with a particular fan, and which reflect a local Alpine source.

These assemblages are identified basinward and their mapping defines the paths of sediment transport. Near the margin, transport is lateral but in the basin, longitudinal transport prevails. There is a small input from the north side of the basin of an alien heavy mineral suite, mainly tourmaline, zircon, and rutile, unlike the dominant Alpine assemblage, which consists principally of garnet, staurolite, apatite, and zircon with an important epidote component in the southeast. The mineralogy of the sands also changes with time; the earlier sands are derived from the Flysch sediments including limestones and dolomites whereas the younger deposits contain significant contributions from the Alpine crystalline rocks.

Füchtbauer's study shows how the petrology of sands, coupled with stratigraphic data, can contribute to the understanding of the filling of a basin and the reconstruction of the paleogeography of the area. It is a nice integration of outcrop and subsurface data and shows what can be done when paleocurrent data are missing.

Great Valley Sequence (Mesozoic) of California:

Dickinson, W.R., and Rich, E.I.: Petrologic intervals and petrofacies in the Great Valley Sequence, Sacramento Valley, California. Geol. Soc. America Bull. 83, 3007–3024 (1972).

This study, unlike that of Füchtbauer just described, relies entirely on the composition of the framework fraction—the "light minerals."

Dickinson and Rich have demonstrated the usefulness of quartz (Q), feldspar (F), and rock particles (L) in deciphering the provenance history of the sandstones of the Cretaceous Great Valley Sequence and the Franciscan sandstones of California. Three general provenance types can be inferred from the composition of the detrital grain population: (1) sandstones derived from andesitic volcanics (L greater than F, Q low, and V/L high, where V/L is the ratio of volcanic rock fragments to total rock fragments), (2) sandstones derived from highlands in which volcanic, sedimentary, and low-grade metamorphic rocks are exposed (L exceeds F, Q is moderate, and V/L is low and C/Q is high where C/Q is the ratio of quartz-chalcedony rock fragments to total quartz), and (3) sandstones derived from deeply eroded areas exposing plutonic igneous rocks and high-grade gneisses (F exceeds L, Q is moderate, and V/L and C/Q are low).

In the Great Valley Sequence the source area was fixed throughout the deposition of some 49,000 ft (15,000 m) of clastic sedimentary strata, mainly interbedded turbidite sands and mudstones. These were deposited concurrently with the erosion of the source area, only the deeper parts of which remain. Hence the only record of the original higher rocks of the source area lies in the lowest members of the sedimentary sequence. The record is inverted, for the detritus first eroded is buried deepest. The record shows an overall progressive change with time, revealing the initial sources to have been volcanic and to have been progressively destroyed to expose the plutonic roots.

There is an irreversible change in the P/F ratio (ratio of plagioclase to total feldspar) from 0.6 to 1.0.

The Franciscan graywackes have framework modes (Q, F, L) that are similar enough to those of the sands of the Great Valley Sequence to establish their derivation from the same source area. Dickinson and Rich believe that there is a close linkage among major magmatic, tectonic, and depositional events in an arc-trench system. The framework composition of a sandstone may indeed be a guide to different plate tectonic settings.

References

Allen, J.R.L.: Petrology, origin and deposition of the highest Lower Old Red Sandstone of Shropshire, England. Jour. Sed. Petrology 32, 657–697 (1962).

Anwar, J.: Destruction of plagioclase twins by stream transport. Jour. Sed. Petrology 47, 933 (1977); reply by E.D. Pittman, *ibid.*, 933–934 (1977).

Baker, George: Sand drift at Portland, Victoria. Roy. Soc. Victoria Proc. 68, 151–197 (1956).

Baker, George: Detrital heavy minerals in natural accumulates. Australian Inst. Mining Metall. Mono. 1, 146 pp. (1962).

Basu, A.: Petrology of Holocene fluvial sand derived from plutonic source rocks: Implications to paleoclimatic interpretations. Jour. Sed. Petrology 46, 694–709 (1976).

Basu, A.: Feldspar dissolution before advent of land plants on earth (Abstract). Am. Assoc. Petroleum Geologists Bull. 65, 896 (1981).

Basu, A.; Young, W.; Suttner, L.J.; James, C.; and Mack, G.M.: Re-evaluation of the use of undulatory extinction and polycrystallinity in detrital quartz for provenance interpretations. Jour. Sed. Petrology 45, 873–883 (1975).

Baturin, V.P.: On stability and formation of minerals of abyssal geospheres in the stratisphere. C.R. Dokl. Akad. Sci. U.S.S.R. 37, 32–34 (1942).

Beloussov, V.V.; Ruditch, E.M.; and Shapiro, M.N.: Intercontinental structural ties and mobilistic reconstructions. Geol. Rundschau 68, 393–427 (1979).

Beveridge, A.J.: Heavy minerals in lower Tertiary

formations in the Santa Cruz Mountains, California. Jour. Sed. Petrology 30, 513–537 (1960).

Blatt, Harvey: Original characters of clastic quartz. Jour. Sed. Petrology 37, 401–424 (1967a).

Blatt, Harvey: Provenance determinations and recycling of sediments. Jour. Sed. Petrology 37, 1031–1044 (1967b).

Blatt, Harvey: Provenance studies and mudrocks. Jour. Sed. Petrology 55, 69–75 (1985).

Blatt, Harvey, and Christie, J.M.: Undulatory extinction in quartz of igneous and metamorphic rocks and its significance in provenance studies of sedimentary rocks. Jour. Sed. Petrology 33, 559–579 (1963).

Boggs, S., Jr.: Experimental study of rock particles. Jour. Sed. Petrology 38, 1326–1339 (1968).

Bokman, John: Clastic quartz particles as indices of provenance. Jour. Sed. Petrology 22, 17–24 (1952).

Boles, J.R.: Active albitization of plagioclase, Gulf Coast Tertiary. Am. J. Sci. 282, 165–180 (1982).

Bond, G.C., and Devay, J.C.: Pre-Upper Devonian quartzose sandstones in the Shoo Fly formation northern California—Petrology, provenance and implications for regional tectonics. Jour. Geology 88, 285–308 (1980).

Boswell, P.G.H.: Mineralogy of sedimentary rocks, 393 pp. London: Murby 1933.

Brammall, A.: Dartmoor detritals. A study in provenance. Proc. Geol. Assoc. 39, 27–48 (1928).

Brown, P.J.; Ehrlich, R.; and Calquhoun, D.J.: Origin of patterns of quartz sand types on the southeastern United States shelf and implications on contemporary shelf sedimentation—Fourier grain shape analysis. Jour. Sed. Petrology 50, 1095–1100 (1980).

Callender, D.L., and Folk, R.L.: Idiomorphic zircon, key to volcanism in the Lower Tertiary sands of central Texas. Am. Jour. Sci. 256, 257–269 (1958).

Cameron, K.L., and Blatt, H.: Durabilities of sand size schist and "volcanic" rock fragments during fluvial transport, Elk Creek, Black Hills, South Dakota. Jour. Sed. Petrology 41, 565–576 (1971).

Condie, K.C.: Plate-tectonic model for Proterozoic continental accretion in the southwestern United States. Geology 101, 37–42 (1982).

Conolly, J.R.: The occurrence and polycrystallinity and undulatory extinction in quartz in sandstones. Jour. Sed. Petrology 35, 116–135 (1965).

Crook, K.A.W.: Lithogenesis and geotectonics: the significance of compositional variation in flysch arenites (graywackes). Soc. Econ. Paleon. Mineral. Spec. Pub. 19, 304–310 (1974).

Curtis, C.D.: Stability of minerals in surface weathering reactions. A general thermochemical approach. Earth Surface Processes 1, 63–70 (1976).

Dake, C.L.: The problem of the St. Peter Sandstone. Univ. Missouri School of Mines and Metall. Bull. Tech. Ser. 6, 158 pp. (1921).

Davies, D.K.; Vessell, R.K.; Miles, R.C., Foley, M.G., and Bonis, S.F.: Fluvial transport and downstream modifications in an active volcanic region, In: Miall, A.D. (Ed.): Fluvial sedimentology. Canadian Soc. Petroleum Geologists Mem. 5, 61–84 (1978).

Derby, C.A.: On the separation and study of the heavy accessories of rocks. Proc. Rochester Acad. Sci. 1, 198–206 (1891).

Dickinson, W.R.: Evolution of calc-alkaline rocks in the geosynclinal system of California and Oregon. Oregon Dept. Geol. Min. Ind. Bull. 65, 151–156 (1969).

Dickinson, W.R.: Interpreting detrital modes of graywacke and arkose. Jour. Sed. Petrology 40, 695–707 (1970).

Dickinson, W.R.: Composition of sandstone in circum-Pacific subduction complexes fore-arc basins. Am. Assoc. Petroleum Geologists Bull. 66, 121–137 (1982).

Dickinson, W.R.: Interpreting provenance relations from detrital modes of sandstones, pp. 333–361. In: Zuffa, G.G. (Ed.): Provenance of arenites. Dordrecht-Boston-Lancaster: D. Reidel Pub. Co. (1985).

Dickinson, W.R., and Rich, E.J.: Petrologic intervals and petrofacies in the Great Valley Sequence, Sacramento Valley, California. Geol. Soc. America Bull. 83, 3007–3024 (1972).

Dickinson, W.R., and Suczek, C.A.: Plate tectonics and sandstone composition. Am. Assoc. Petroleum Geologists Bull. 63, 2164–2182 (1979).

Dickinson, W.R., and Valloni, R.: Plate settings and provenance of sands in modern ocean basins. Geology 8, 82–86 (1980).

Dickinson, W.R.; Ingersoll, R.V.; Cowan, D.S.; Helmhold, K.P.; and Suczek, C.A.: Provenance of Franciscan graywackes in coastal California. Geol. Soc. America Bull. 93, 5–107 (1982).

Dickinson, W.R.; Beard, L.S.; Brakenridge, G.R.; Erjavec, J.L.; Ferguson, R.C.; Inman, K.F.; Knepp, R.A.; Lindberg, F.A.; and Ryberg, P.T.: Provenance of North American Phanerozoic sandstones in relation to tectonic setting. Geol. Soc. America Bull. 94, 222–235 (1983).

Dietz, V.: Experiments on the influence of transport on shape and roundness of heavy minerals. In: Contributions to Sedimentology, No. 1, pp. 99–102. Stuttgart: Schweizerbart'sche Verlag 1973.

Doty, R.W., and Hubert, J.F.: Petrology and paleogeography of the Warrenburg channel sandstone, western Missouri. Sedimentology 1, 7–39 (1962).

Ehrlich, R., and Weinberg, B.: An exact method for characterization of grain shape. Jour. Sed. Petrology 40, 205–212 (1970).

Ermanovics, I.F.: Statistical application of plagioclase extinction in provenance studies. Jour. Sed. Petrology 37, 683–687 (1967).

Feniak, M.W.: Grain sizes and shapes of various minerals in igneous rocks. Am. Mineralogist 29, 415–421 (1944).

Feo-Codecido, Gustavo: Heavy mineral techniques and their application to Venezuelan stratigraphy.

Am. Assoc. Petroleum Geologists Bull. 40, 984–1000 (1956).

Folk, R.L.: Petrology of sedimentary rocks. 170 pp. Austin, Texas: Hemphill's (1968).

Folk, R.L., and Robles, R.: Carbonate sands of Isla Perez, Alacran Reef complex, Yucatan. Jour. Geology 72, 255–292 (1964).

Force, E.R.: The provenance of rutile, Jour. Sed. Petrology 50, 485–488 (1980).

Franzinelli, E., and Potter, P.E.: Petrology, chemistry and texture of modern river sands, Amazon River System. Jour. Geology 91, 23–39 (1983).

Friese, F.W.: Untersuchungen von Mineralen auf Abnutzbarkeit bei Verfrachtung im Wasser. Min. Pet. Mitt 41, n.s., 1–7 (1931).

Füchtbauer, Hans: Sedimenpetrographische Untersuchungen in der älteren Molasse nördlich der Alpen. Eclogae geol. Helvetiae 57, 158–298 (1964).

Füchtbauer, Hans: Die Sandsteine in der Molasse nördlich der Alpen. Geol. Rundschau 56, 266–300 (1967).

Garrels, R.M., and Mackenzie, F.T.: Evolution of sedimentary rocks, 397 pp. New York: W.H. Norton & Co. 1971.

Goldich, S.S.: A study in rock weathering. Jour. Geology 46, 17–58 (1938).

Gorai, M.: Petrological studies on plagioclase twins. Am. Mineralogist 51, 884–901 (1951).

Grabau, A.W.: On the classification of sedimentary rocks. Am. Geol. 33, 228–247 (1904).

Grabau, A.W.: Principles of stratigraphy, 1185 pp. New York: Seiler (1913).

Graham, S.A.; Ingersoll, R.V.; and Dickinson, W.R.: Common provenance for lithic grains in Carboniferous sandstone from the Ouachita Mountains and Black Warrior Basin. Jour. Sed. Petrology 46, 620–632 (1976).

Griggs, D.T.: The factor of fatigue in rock exfoliation. Jour. Geology 44, 783–796 (1963).

Gruner, J.W.: An attempt to arrange silicates in the order of reaction energies at relatively low temperatures. Am. Mineralogist 35, 137–148 (1950).

Haaf, E. ten: Flysch formations of the northern Apennines. In: Bouma, A.H., and Brouwer, A. (Eds.): Turbidites, pp. 127–136. (Developments in sedimentology, Vol. 3). Amsterdam: Elsevier Scientific Pub. Co. 1964.

Harrell, J., and Blatt, H.: Polycrystallinity: Effect on the durability of detrital quartz. Jour. Sed. Petrology 48, 25–30 (1978).

Higgins, M.W.: Cataclastic rocks. U.S. Geol. Survey Prof. Paper 68F, 92 pp. (1971).

Hooper, W.F.: Petrography of Lakota conglomerate. Casper Arch area, Wyoming. In: Symposium on Early Cretaceous rocks of Wyoming and adjacent areas—Wyoming, pp. 131–140. Wyoming Geol. Assoc., 17th Ann. Field Conf., Casper, Wyoming (1962).

Hough, J.L.: Sediments of Cape Cod Bay, Massachusetts. Jour. Sed. Petrology 12, 10–30 (1942).

Hudson, C.B., and Ehrlich, R.: Determination of relative provenance contributions in samples of quartz sand using Q-mode factor analysis of Fourier grain shape data. Jour. Sed. Petrology 50, 1101–1110 (1980).

Huffman, G.G., and Price, W.A.: Clay dune formation near Corpus Christi, Texas. Jour. Sed. Petrology 19, 118–127 (1949).

Kalsbeek, F.; Ghisler, M.; and Thompsen, B.: Sand analysis as a method of estimating bedrock composition in Greenland. Groenlands Geol. Unders. Bull. 111, 32 pp. (1974).

Keller, W.D., and Littlefield, R.F.: Inclusions in the quartz of igneous and metamorphic rocks. Jour. Sed. Petrology 20, 74–84 (1950).

Kennedy, V.C., and Kouba, D.L.: Fluorescent sand as a tracer of fluvial sediment. U.S. Geol. Survey Prof. Paper 562E, E1–E13 (1970).

Krumbein, W.C.: Statistical summary of some alluvial gravels. Rept. Comm. Sedimentation for 1940–1941, Natl. Res. Council, 9–14 (1942).

Krumbein, W.C., and Pettijohn, F.J.: Manual of sedimentary petrography, 549 pp. New York: Appleton-Century 1938.

Krynine, P.D.: Arkose deposits in the humid tropics: A study in sedimentation in southern Mexico. Am. Jour. Sci., Ser. 5, 29, 353–363 (1935).

Krynine, P.D.: Petrography and genesis of the Siwalik Series. Am. Jour. Sci., Ser. 5, 34, 422–446 (1937).

Krynine, P.D.: Petrology and genesis of the Third Bradford Sand. Pennsylvania State College Min. Ind. Expt. Sta., Bull. 27, 134 pp. (1940).

Krynine, P.D.: Diastrophism and the evolution of sedimentary rocks. Pennsylvania Min. Ind. Tech. Paper 84-A, 21 pp. (1943).

Krynine, P.D.: Microscopic morphology of quartz types. Proc. 2nd Pan-Am. Cong. Mining Eng. and Geology 3, 2nd Comm., 35–49 (1946a).

Krynine, P.D.: The tourmaline group in sediments. Jour. Geology 54, 65–87 (1946b).

Kuenen, Ph.H.: Sand—its origin, transportation, abrasion and accumulation. Geol. Soc. South Africa, 62, Annexure, 1–33 (1959).

Luepke, Gretchen: Stability of heavy minerals in sediments, 306 pp. New York: Van Nostrand Reinhold Co. 1984.

Mack, G.H.: The survivability of labile light-mineral grains in fluvial, aeolian, and littoral marine environments: The Permian Cutler and Cedar Mesa formations, Moab, Utah. Sedimentology 25, 587–604 (1978).

Mackie, W.: The sands and sandstones of eastern Moray. Edinburgh Geol. Soc. Trans. 7, 48–172 (1899).

Mackie, W.: The source of the purple zircons in the sedimentary rocks of Scotland. Edinburgh Geol. Soc. Trans. 11, 200–213 (1923).

Makrutzki, J.: Die Gesteinsbruchstücke der Devon-Sandsteine in der Bohrung Schwarzbachtal 1. Senckenbergiana 63, 65–77 (1982).

Mange-Rajetsky, M.A.: Detrital blue sodic amphibole in Recent sediment, southern coast of Turkey. Jour. Geol. Soc. (London) 138, 83–91 (1981).

Mann, W.R., and Cavaroc, V.V.: Composition of sand released from three source areas under humid, low relief weathering in the North Carolina Piedmont. Jour. Sed. Petrology 43, 870–881 (1973).

Marshall, P.E.: The wearing of beach gravels. Trans. Proc. New Zealand Inst. 59, 507–532 (1927).

Matter, A., and Komseyer, Karl: Cathodoluminescence microscopy as a tool for provenance studies of sandstones, pp. 191–211, In: Zuffa, G.G. (Ed.), Provenance of arenites. Dordecht-Boston-Lancaster: D. Reidel Pub. Co. (1985).

Maynard, J.B.: Composition of plagioclase feldspar in modern deep-sea sands: Relationship to tectonic setting. Sedimentology 31, 493–501 (1984).

Milner, H.B.: Supplement to introduction to sedimentary petrography, 157 pp. London: Murby (1926).

Milner, H.B.: Sedimentary petrology, Vol. 2. New York: Macmillan, 715 pp. (1962).

Mizutani, S.: Clastic plagioclase in Permian graywacke from the Mugi area. Gifu Prefecture, central Japan. Jour. Earth Sci., Nagoya Univ. 7, 108–136 (1959).

Morton, A.C.: Heavy minerals in provenance studies. In: Zuffa, G.G. (Ed.): Provenance of arenites, pp. 249–277. Dordrecht–Boston–Lancaster: D. Reidel Pub. Co. 1985a.

Morton,, A.C.: A new approach to provenance studies: Electron microprobe analysis of detrital garnets from Middle Jurassic sandstones of the North Sea. Sedimentology 32, 553–566 (1985b).

Moss, A.J.: Origin, shaping and significance of quartz sand grains. Jour. Geol. Soc. Australia 13, 97–136 (1966).

Moss, A.J.; Green, P.; and Hutka, J.: Static breakage of granitic detritus by ice and water in comparison with flowing water. Sedimentology 28, 261–272 (1981).

Nickel, E.: Experimental dissolution of light and heavy minerals in comparison with weathering and intrastratal solution. In: Contributions to Sedimentology, No. 1, pp. 1–68. Stuttgart: Schweizerbart'sche Verlag 1973.

Ogunyomi, O., Martin, R.F., and Messe, R.: Albite of secondary origin in Charny sandstones, Quebec: A re-evaluation. Jour. Sed. Petrology 51, 597–606 (1981).

Owen, M.R., and Carozzi, A.V.: Southern provenance of upper Jackfork Sandstones, southern Ouachita Mountains: Cathodoluminescence petrology. Geol. Soc. America Bull. 97, 110–115 (1986).

Payne, T.G.: Stratigraphical analysis and environmental reconstruction. Am. Assoc. Petroleum Geologists Bull. 26, 1697–1700 (1942).

Pettijohn, F.J.: Relative abundance of size grades of clastic sediments (Abstract). Program Soc. Econ. Paleon. Mineral. Chicago (1940).

Pettijohn, F.J.: Persistence of heavy minerals and geologic age. Jour. Geology 79, 610–625 (1941).

Pettijohn, F.J.: Sedimentary rocks, 3rd Ed., 628 pp. New York: Harper and Row 1975.

Pittman, E.D.: Use of zoned plagioclase as an indicator of provenance. Jour. Sed. Petrology 33, 380–386 (1963).

Pittman, E.D.: Plagioclase feldspar as an indicator of provenance in sedimentary rocks. Jour. Sed. Petrology 40, 591–598 (1970).

Plumley, W.J.: Black Hills terrace gravels: A study in sediment transport. Jour. Geology 56, 526–577 (1948).

Plymate, T., and Suttner, L.J.: Evaluation of optical and X-ray techniques for detecting source-rock variations in detrital feldspar. Geol. Soc. America, Abstracts with program, 11, 496 (1979).

Poldervaart, Arie: Zircon in rocks, 2 Igneous rocks. Am. Jour. Sci. 254, 521–554 (1956).

Polynov, B.B.: Cycle of weathering (A. Muir, trans.), 220 pp. London: Murby 1937.

Potter, P.E.: Petrology and chemistry of modern big river sands. Jour. Geology 86, 423–449 (1978).

Potter, P.E., and Pryor, W.A.: Dispersal centers at Paleozoic and later clastics of the upper Mississippi valley and adjacent areas. Geol. Soc. America Bull. 72, 1195–1250 (1961).

Pryor, W.A.: Biogenic sedimentation and alteration of argillaceous sediments in shallow marine environments. Geol. Soc. America Bull. 86, 1244–1254 (1975).

Rimsaite, J.: Optical heterogeneity of feldspars observed in diverse Canadian rocks. Schweiz. Min. Pet. Mitt. 47, 61–76 (1967).

Russell, R. Dana: Mineral composition of Mississippi River sands. Geol. Soc. America Bull. 48, 1307–1348 (1937).

Russell, R.J.: Where most grains of very coarse sand and fine gravel are deposited. Sedimentology 11, 31–38 (1968).

Schlee, J.: Uplands gravels of southern Maryland. Geol. Soc. America Bull. 68, 1371–1410 (1957).

Schwab, F.L.: Framework mineralogy and chemical composition of continental margin-type sandstones. Geology 3, 487–490 (1975).

Schwab, F.L.: Evolution of the western continental margin, French-Italian Alps: Sandstone mineralogy as an index of plate tectonic setting. Jour. Geology 89, 349–368 (1981).

Shea, J.H.: Deficiencies of clastic particles of certain sizes. Jour. Sed. Petrology 44, 985–1003 (1974).

Sindowski, F.K.H.: Results and problems of heavy-mineral analysis in Germany; a review of sedimentary-petrographical papers, 1936–1948. Jour. Sed. Petrology 19, 3–25 (1949).

Sippel, R.F.: Sandstone petrology, evidence from luminescence petrography. Jour. Sed. Petrology 38, 530–554 (1968).

Slatt, R.M., and Eyles, N.: Petrology of glacial sands: implications for the origin and durability of lithic fragments. Sedimentology 28, 171–184 (1981).

Slemmons, D.B.: Determination of volcanic and plutonic plagioclase using a three- or four-axis universal stage. Geol. Soc. America Spec. Paper 69, 64 pp. (1962).

Sorby, H.C.: On the structure and origin of non-calcareous stratified rocks. Geol. Soc. London Proc. 36, 46–92 (1880).

Stallard, R.F., and Edmond, J.M.: Geochemistry, the Amazon 2. The influence of geology and weathering environment on the dissolved load. Jour. Geophys. Res. 88, 9671–9688 (1983).

Stumm, W., and Morgan, J.J.: Aquatic chemistry, 583 pp. New York: Wiley-Interscience 1970.

Sundborg, Åke: The River Klaralven—a study of fluvial processes. Geog. Annaler 38, 127–326 (1956).

Suttner, L.J.: Stratigraphic and petrographic analysis of Upper Jurassic–Lower Cretaceous Morrison and Kootenai Formations, southwest Montana. Am. Assoc. Petroleum Geologists Bull. 53, 1391–1410 (1969).

Suttner, L.J., and Basu, A.: Structural state of detrital alkali feldspars. Sedimentology 24, 63–73 (1977).

Suttner, L.J.; Basu, A.; and Mack, G.H.: Climate and origin of quartz arenites. Jour. Sed. Petrology 51, 1235–1246 (1981).

Thiel, G.A.: The relative resistance to abrasion of mineral grains of sand. Jour. Sed. Petrology 10, 103–124 (1940).

Thoulet, J.: Notes sur lithologie sous-marine. Ann. Inst. Oceanogr. 5, 1–14 (1913).

Todd, T.W., and Folk, R.L.: Basal Claiborne of Texas, record of Appalachian tectonism during Eocene. Am. Assoc. Petroleum Geologists Bull. 41, 2545–2566 (1957).

Tomita, T.: Geologic significance of the color of granite zircon and the discovery of the Pre-Cambrian in Japan. Kyushu Univ. Mem. Fac. Sci., Ser. D., Geology 4, 135–161 (1954).

Trevena, A.S., and Nash, W.P.: Chemistry and provenance of detrital plagioclase. Geology 7, 475–478 (1979).

Trevena, A.S., and Nash, W.P.: An electron microscope study of detrital feldspar. Jour. Sed. Petrology 51, 137–149 (1981).

Tyler, S.A.; Marsden, R.W.; Grout, F.F.; and Thiel, G.A.: Studies of the Lake Superior Pre-Cambrian by accessory-mineral methods. Geol. Soc. America Bull. 51, 1429–1538 (1940).

Udden, J.A.: Mechanical composition of clastic sediments. Geol. Soc. America Bull. 25, 655–744 (1914).

Valloni, Renzo: Reading provenance from modern marine sands. In: Zuffa, G.G. (Ed.): Provenance of arenites, pp. 309–332. Dordrecht–Boston–Lancaster: D. Reidel Pub. Co. 1985.

Valloni, R., and Maynard, J.B.: Detrital modes of recent deep-sea sands and their relation to tectonic setting: A first approximation. Sedimentology 28, 75–83 (1981).

Vitanage, P.W.: Studies of zircon types in Ceylon Pre-Cambrian complex. Jour. Geology 65, 117–138 (1957).

Voll, G.: New work on petrofabrics. Liverpool and Manchester Geol. Jour. 3, 503–567 (1960).

Walker, R., and Pettijohn, R.J.: Archean geosynclinal basin: Análysis of the Minnitaki Basin, northwestern Ontario. Geol. Soc. America Bull. 82, 2099–2129 (1971).

Weaver, C.E.: Geologic interpretation of argillaceous sediments. Am. Assoc. Petroleum Geologists Bull. 42, 254–271, 272–309 (1958).

Webb, W.M., and Potter, P.E.: Petrología y geoquímica de detritos derivados de un terreno riolítico de la región occidental de Chihuahua Mexico. Bol. Soc. Geol. Mexicana 32, 45–61 (1971).

Wentworth, C.K.: Fundamental limits to the sizes of clastic grains. Science 77, 633–634 (1933).

Wolf, Karl H.: Textural and compositional transitional stages between various lithic grain types (with a comment on "Interpreting detrital modes of graywacke and arkose"). Jour. Sed. Petrology 41, 328–332 (1971).

Wolff, R.G.: The dearth of certain size of materials in sediments. Jour. Sed. Petrology 34, 320–327 (1964).

Young, S.W.: Petrographic textures of detrital polygonal quartz as an aid to interpreting crystalline source rocks. Jour. Sed. Petrology 46, 595–603 (1976).

Zimmerle, Winfried: Serpentine graywackes from the North Coast basin, Colombia, and their geotectonic significance. Neues Jahrb. Mineral. Abh. 109, 156–182 (1968).

Zimmerle, Winfried: Sind detritische Zirkone rötlicher Farbe auch in Mitteleuropa Indikatoren für präkambrische Liefergebiete? Geol. Rundschau 61, 116–139 (1972).

Zinkernagel, U.: Cathodoluminescence of quartz and its application to sandstone petrology. Contr. Sedimentology 8, 49 pp. (1978).

Zuffa, G.G.: Hybrid arenites: Their composition and classification. Jour. Sed. Petrology 50, 21–29 (1980).

CHAPTER 8
Transportation and Deposition of Sand

Introduction

Gravity acting alone moves sand down the slip slope of a dune, of an oversteepened submarine slope, and powers a debris flow. Continental and alpine glaciers also move sand and probably a significant number of the world's sand grains have, at one time or another in their history, been transported by ice. But the dominant agents are water and wind. Although one can but surmise, probably every sand grain on earth, from loose sand on a modern Chilean beach to sand in an early Precambrian quartzite in Canada, has been transported millions of times by water and thousands of times by wind, either on a continental desert, in coastal dunes, or on a dry floodplain. Thus streams and rivers, tidal and longshore currents in shallow water, waves in shallow water, contour currents in deep water along continental margins, density currents in oceans and deep lakes, wind, and ash flows and falls from volcanic explosions are the dominant agents that transport and sort sand and segregate it from mud, silt, and coarse particles.

These diverse processes can be viewed and studied in many different ways. Hydraulic engineers have, for example, considered the mechanics of transport of individual grains and have also developed semi-empirical formulas for sand transport along a beach that yield estimates of tons per unit area per year. Others have investigated the mechanics of turbidity currents, and many engineers and sedimentologists have made experimental studies in flumes. A few sedimentologists have also used computers to model larger features such as deltas, subsea fans, spits, and beaches. But sand is also transported and deposited in laminae, beds, bedforms, and sedimentary structures, many of which occur in associations that tend, to some degree, to be characteristic of specific processes and depositional environments. The objective of this chapter is to explore these processes, especially those most relevant to the formation of sedimentary structures. Thus in this chapter we emphasize the hydraulic origin of sedimentary structures because they, along with grain size, are the best keys to paleohydraulics, paleogeography, and paleoceanography available to the sedimentologist.

The first part of this chapter, "Fluid Flow and Entrainment," provides the minimal background needed for an insight into the transport, emplacement, and deformation of sand. Here we begin by first considering, because it is relatively simple, fluid flow without particles, and subsequently the interaction of fluid and grains. To supplement "Fluid Flow and Entrainment" a reader who is not familiar with elementary fluid mechanics may benefit from consulting the annotated references, which contain a brief introduction to much of the hydraulic engineering and fluid dynamic literature, and the glossary, which contains commonly recurring fluid dynamic and related terms. A few of the many hydraulic texts include Bogardi (1974), Chapman and Wright (1981), Graf (1971), Morris and Wiggert (1972), Raudkivi (1976), Scheidegger (1970), Streeter (1971), Vennard (1961), Yalin (1977), and Zanke (1982). Vennard's book is especially recommended for initial reading as is that by Giles (1962). Also useful is Embleton *et al.* (1979, pp. 39–72), and monographic treatment is provided by Allen (1982a and b).

In our selection and treatment of topics we have been strongly influenced by the fact that as geologists our prime interest is an understanding of the past. For example, to make a careful study of the fluid dynamics of a cross section of a modern river reach, one measures discharge or average velocity, slope, determines bottom profile and median diameter of the sediment, and fluid properties such as viscosity and density. In a beach study one would include offshore bottom profile, breaker angle, period, wavelength and amplitude of the waves, median diameter of the sand, and fluid properties. Many more and complex fluid dynamic parameters are in use today but almost all can be derived from these few. In both examples one specifies the process variables to understand the resultant phenomena—the size distribution, bedforms, and perhaps the morphology of the sand accumulation. In contrast, the geologist looking at an ancient river or beach deposit has only the resultant to work with, for the fluid phase has long since vanished and his job is to infer at least some of its characteristics. Thus when working with ancient sands and sandstones, the geologist sees formative processes through a "geologic filter," one that strongly conditions his view of transportation and deposition. Essentially only the size distribution, sedimentary structures, bedding, the vertical sequence, and information to be gleaned from the shape of the deposit are available. In other words, if the application of sediment transport mechanics to ancient sandstones is to have maximum value, it must be based on variables that leave a tangible record.

Fluid Flow and Entrainment

Aspects of Fluid Flow

There are two types of flow, *laminar* and *turbulent*, based on the kinematic and dynamic properties of fluids. In the former, *streamlines* (imaginary curves connecting the tangents to the directions of motions of a fluid particle) are parallel and separate from one another, the flow moves in laminae parallel to the boundary, and no mass transfer between layers takes place. Flow is relatively slow and velocity components other than those in the principal flow direction are negligible. In turbulent flow, on the other hand, the streamlines are complex and intertwined and continually changing with time in an unpredictable way so that rather than fluid moving in well-defined laminae or sheets there are instead complex eddies which superimpose an irregular, random motion upon the larger general movement of the flow. In turbulent flow, masses of water move up, down, and laterally with respect to general flow direction, transferring mass and momentum. Although the irregular motions of such masses or lumps of fluid have velocities that deviate only a few percent from average velocity, they nonetheless exert a decisive effect upon the flow for it is turbulence which keeps particles in suspension, either constantly, as are the clays and silts of rivers and the sand of turbidity currents, or intermittently, as are most sand grains in rivers, beaches and dunes.

Turbulence entrains particles in a combination of two ways: by added fluid force and by a related reduction of the local pressure as the turbulent eddy passes overhead. Both promote entrainment of sand along the bottom. Almost all the flows that transport sand in nature are fully turbulent ones. Turbulence is principally generated in rivers by shear along the boundary as water flows, and is increased significantly by bottom roughness; along shorelines and at sea it is produced by waves, surface wind stress, and shear between currents. In air it is turbulence that carries debris from a volcanic explosion as it is transported downwind. The magnitude of turbulent motion varies from micro to macro, the latter being easily seen in rivers as large complex eddies or boils that impinge against the surface, persist for some seconds, and are then replaced by others. Matthes (1947) gives a useful classification of macroturbulence.

Whether laminar or turbulent, the effect that a flowing fluid exerts on its boundary at the sediment interface depends on fluid properties: *density*, ρ, and *dynamic* or *absolute viscosity*, μ; some measure of fluid dynamic properties of *acceleration* or *velocity*; and on the geometry of the sediment–water boundary. From density and dynamic viscosity one obtains two other widely used properties: *specific weight*, $\lambda = \rho g$, where g is the acceleration of gravity, and *kinematic viscosity*, $\nu = \mu/\rho$. Commonly recurring symbols along with their dimensions are given in Table 8-1. Knowledge of the dimensions of these and other hydraulic parameters provides helpful insight into the role they play in hydraulics.

Every real fluid such as water or air has an internal frictional resistance to flow called viscosity. The viscosity of a fluid is a measure of its resistance to shear—the extent to which a slower-moving fluid mass retards a faster-moving one so that a *shearing stress*, τ, is generated at the boundary of the two. Shearing stress is measured in force per unit area and always acts parallel to direction of movement (Fig. 8-1). The concept of a shear stress acting either between two fluid masses or between a fluid and a solid, such as the shear stress exerted by a stream on its boundary or by waves breaking on a beach, is fundamental to sediment transport

TABLE 8-1. Commonly recurring symbols and their dimensions.

d	grain diameter, L	γ	specific weight, $ML^{-2}T^{-2}$
D	depth of flow, L	μ	dynamic viscosity, $ML^{-1}T^{-1}$
F	Froude number	ν	kinematic viscosity, L^2T^{-1}
g	gravity, LT^{-2}	ρ	density, ML^{-3}
h	water depth, L	ρ_s	solid density, ML^{-3}
H	wave height, L	ρ_f	fluid density, ML^{-3}
L	length, L	σ_s	shear strength of a nonfluid substance, $ML^{-1}T^{-2}$
M	mass, M		
P	pressure, $ML^{-1}T^{-2}$	τ	shear stress, $ML^{-1}T^{-2}$
p	power, $ML^{-2}T^{-3}$	τ_0	shear stress between a fluid and its boundary, $ML^{-1}T^{-2}$
R	Reynolds number		
R_*	boundary Reynolds number	$\tau_0 U$	stream power per unit area, $ML^{-2}T^{-3}$
T	time, T	τ_i	shear stress between a turbidity current and its upper, *fluid* boundary, $ML^{-1}T^{-2}$
U	velocity, LT^{-1}		
U_*	friction or shear velocity, LT^{-1}	τ_b	shear stress between two unconsolidated beds, $ML^{-1}T^{-2}$
α	angle of inclination		
w	grain settling velocity, LT^{-1}	τ_t	threshold shear stress, $ML^{-1}T^{-2}$

mechanics. In laminar flow the shear stress or drag force per unit area between two such sliding masses is given by

$$\tau = \mu(dU/dy), \tag{8-1}$$

where dU/dy is the velocity gradient normal to the boundary, y is the distance from the bottom, and U is the velocity (Fig. 8-2).

Dynamic viscosity is a fluid property that does not vary with the state of motion of fluid, but is very temperature dependent, decreasing with increase in temperature. The temperature dependence of viscosity is the result of the fundamental molecular forces of a fluid that are described by viscosity terms. This temperature effect is important to the sedimentologist, because viscosity plays a significant role in the settling or fall velocity of a particle; the colder and more viscous the water, the greater its resistance to deformation and consequently the slower the fall velocity. Change in fall velocity can thus significantly alter the capacity of a stream to carry sand, and influences the microrelief of the sand-water interface. Another factor affecting dynamic viscosity is concentration of suspended fine clay. The viscosity of the clay-water mixture is called *apparent viscosity*; it is larger than that of clear water alone and has an important effect in retarding the fall velocity of suspended particles. This, together with the ability of a muddy stream to exert a greater drag force, for a given discharge and temperature, accounts for the capability of a turbid stream to transport more sand than a comparable clear-water stream. It is vital to an understanding of the mechanics of turbidity currents.

In turbulent flow one must consider not only the viscosity generated by the frictional forces as described above but also that generated by the turbulent eddies as well so that the drag force is now defined by

$$\tau = (\mu + \eta)dU/dy \tag{8-2}$$

where η is called the *eddy viscosity* and is the rate of exchange of fluid mass between adjacent

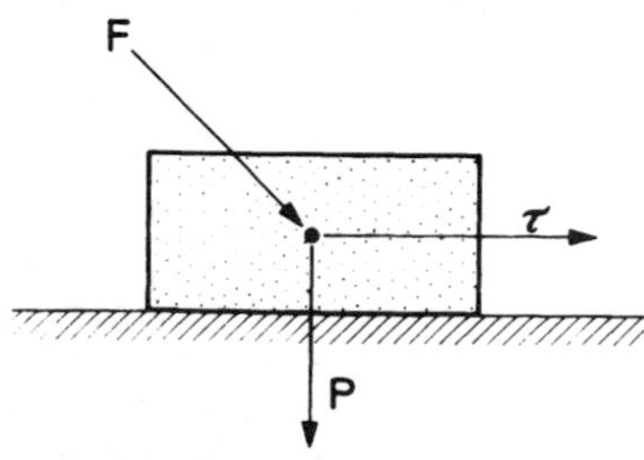

FIGURE 8-1. Applied force, F, resolved into pressure, P, and shear stress, τ, both of which have dimensions of pressure ($ML^{-1}T^{-2}$).

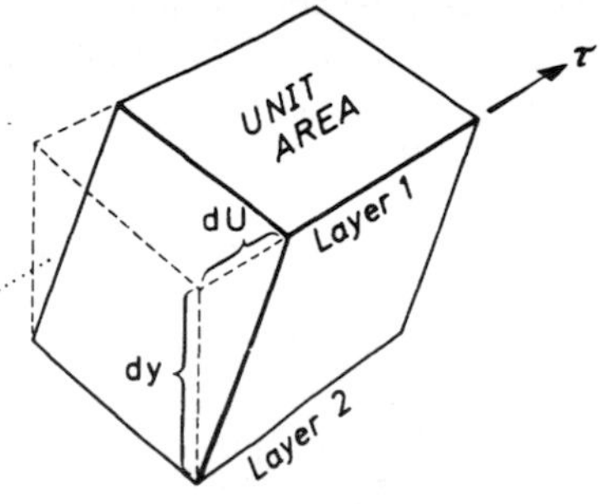

FIGURE 8-2. A geometrical representation of laminar shear stress between two sliding masses of fluid. Layer 1 slides over layer 2 and τ, the shear stress, may be thought of as the force that produces a change in velocity, du, relative to height, dy.

water masses, that is, the viscosity resulting from turbulent momentum transfer. Because of the complex random motion imparted to the fluid by the turbulent eddies, one uses the time-average velocity, $\bar{U}$, at any point to determine the gradient $d\bar{U}/dy$, rather than dU/dy as with laminar flow. The random interchange of material between masses not only transfers momentum, thus slowing faster particles and accelerating the slower ones, but also transfers sediment and diffuses salt concentrations. Eddy viscosity depends on the state of motion of the fluid and is much less temperature dependent than is dynamic viscosity. Numerically, eddy viscosity greatly exceeds dynamic viscosity in fully turbulent flows except very near the boundary.

To distinguish between turbulent and laminar flow the *Reynolds number*, R, is used and is defined by the equation

$$R = \frac{\rho UL}{\mu}, \tag{8-3}$$

where L is some measure of length, sometimes called a *hydraulic radius* (frequently taken as depth for rivers). Reynolds number is the ratio of the inertial force of a fluid to its viscous force and as such it represents the ratio between a driving and a retarding force. Sir Osborne Reynolds, an English physicist, pointed out in 1883 the relationship between inertial and viscous forces that led to this dimensionless number. For a given flow geometry a large Reynolds number, certainly one in excess of 2000, indicates that the flow is turbulent and that inertial forces greatly exceed viscous forces. Thus the larger the Reynolds number, the less important the influence of dynamic viscosity upon flow pattern. Conversely, if R is small, much less than 500, viscous forces are dominant and the flow is laminar. For any given boundary, liquid, and temperature there will be a range of values that define a zone of transitional flow between the two types of flow, laminar and turbulent. The transition zone also depends on geometry of the flow and surface roughness of the boundary.

Another dimensionless number that plays an important role is the *Froude number*, F, the ratio of inertial force to gravity force. Froude number is defined as

$$F = \frac{U}{\sqrt{hg}} \tag{8-4}$$

where U is velocity, h is depth, and g is acceleration of gravity. Because it is the ratio of inertial to gravity force, Froude number is used in fluid flow problems when the flow has an unconfined or free surface as in a stream or in a shallow tidal estuary. The concept of free surface flow has been extended by some to apply to submarine turbidity currents, whose "free" surface is considered to be the interface between the denser turbid current below and the clear water above. In free surface flow gravity plays a significant role unlike, for example, pressure flow in a closed pipe where pressure and not gravity is the driving force. The Froude number also distinguishes between two types of flow, *shooting* or supercritical ($F > 1$) and *tranquil*, sometimes called streaming, or subcritical ($F < 1$). The transition between the two is called *critical flow* ($F = 1$). In shooting flow a surface wave will be carried downstream by the current whereas in tranquil flow the wave front will move upstream against the current. Shooting flow occurs in rapids, constrictions, some floods, and where breakers rush up a beach. The type of flow—whether it is tranquil or shooting—plays an important role in molding the microrelief, the bedforms that produce many sedimentary structures on sandy bottoms. Two flows are dynamically similar if both Reynolds and Froude numbers are the same. Dynamic similarity is an important concept for model studies, for it allows the scaling down of natural large systems so that they can be studied in the laboratory.

A fluid flowing over a boundary exerts a shearing force on it and, conversely, the boundary exerts a retarding force on the flowing fluid. The zone where the fluid is appreciably retarded by the frictional drag of the boundary is called the *boundary layer* (Fig. 8-3). Depending upon the relief or roughness of the boundary and the velocity and viscosity of the flow, the thickness of the boundary layer may vary from a small fraction of a millimeter to several millimeters, or it can, under certain conditions, such as a shallow flow over a scour trough or a rough gravel bed, greatly expand to become an appreciable fraction or all of the total depth of the stream. Flow within the boundary layer may be either turbulent or laminar but, in any case, it is in the boundary layer that the flowing fluid loses much of its kinetic energy. If the flow is laminar in the boundary layer, the term *laminar sublayer* is used. A laminar sublayer in an otherwise turbulent boundary markedly reduces frictional resistance and thus may protect many small grains from entrainment by a turbulent flow. Turbulent boundary layers prevail, however, in most natural flows that carry sand.

The idea of the boundary layer allowed the conceptual separation of the fluid into two

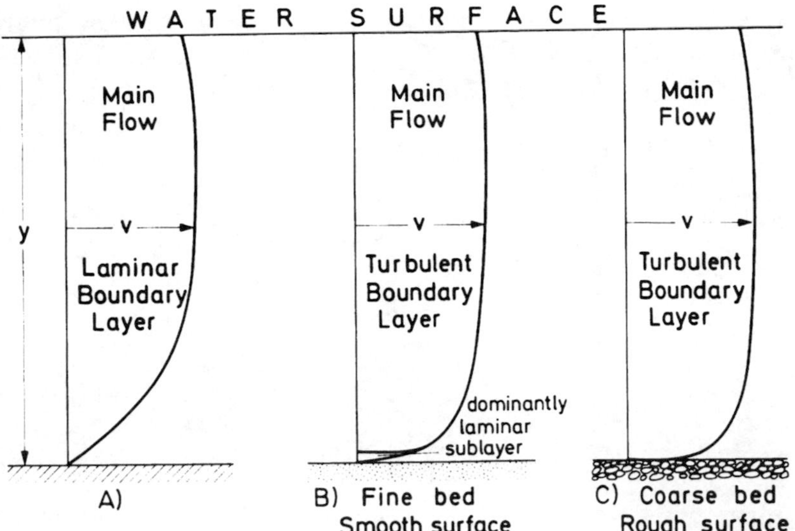

FIGURE 8-3. Laminar boundary layer (A), turbulent boundary layer with dominantly laminar sublayer (B), and fully turbulent layer (C).

parts—one that is strongly affected by the boundary and a main flow that is largely isolated from it. The idea of separation of the total flow into these two parts that was proposed by Ludwig Prandtl in 1904 was a major achievement, for it permits one to apply the classical laws of fluid dynamics to the main body of flowing fluid and then combine this solution with a separate analysis of the flow in the boundary layer itself. For sedimentologists the boundary layer has a threefold significance. It is in the boundary layer that the maximum exchange between the loose grains of the interface and those temporarily in suspension takes place, obviously a key zone for transportation and deposition. Secondly, the characteristics of the boundary layer have a marked influence on the shear stress exerted by the moving fluid on the bottom. Finally, the boundary layer may become expanded downstream from a sand wave or behind a boulder, where the main flow separates from the boundary. This phenomenon is called *flow separation* and is characteristic of the lee sides of obstacles such as sand waves or pebbles and cobbles that project above the bottom (Fig. 8-4).

For fully turbulent flow over a rough surface such as a sand bed, the vertical velocity profile is given by an equation developed by von Kármán

$$\frac{\overline{U}}{U_*} = 8.5 + \frac{2.3}{k} \log \frac{y}{y_0}, \qquad (8\text{-}5)$$

where $\overline{U}$ is the average velocity at a distance y above the bed, U_* is the shear or friction velocity, k is von Kármán's dimensionless constant, determined by experiment to be 0.4 for most fluids but may be as low as 0.2, and y_0 is a roughness length which depends upon the magnitude of the irregularities or roughness of the bottom. *Friction velocity* is related to shear stress by $U_* = \sqrt{\tau_0/\rho}$ and has the dimensions of velocity. One should note that τ_0 is the shear stress between the fluid and the boundary and not the shear stress within the fluid. The above velocity law applies to both air and water.

The roughness or relief of the boundary is some fraction of grain diameter for a flat bed of sand but is much larger should the bed be rippled or duned. Roughness has a pronounced effect on the velocity profile; the greater the roughness, the greater the decrease in the velocity gradient or profile in the boundary layer because, as roughness increases, the increased frictional resistance of the boundary retards the flow near it (Fig. 8-5). Resistance resulting from the deformation of the boundary is called *form resistance* or *form drag*. Ripples and dunes are features that give rise to form drag.

The von Kármán equation is important to sedimentologists because it shows that the average velocity at any point in a flow depends on the height of that point above the bottom, on the boundary roughness, and on friction velocity. The laminar sublayer plays a rọle in the von Kármán equation for it can alter $\overline{U}$ via boundary roughness and friction velocity. Shear stress, an impelling force acting at the boundary, is more commonly used than velocity when studying the entrainment of sand and the origin of bedforms, because velocity varies logarithmically as the height.

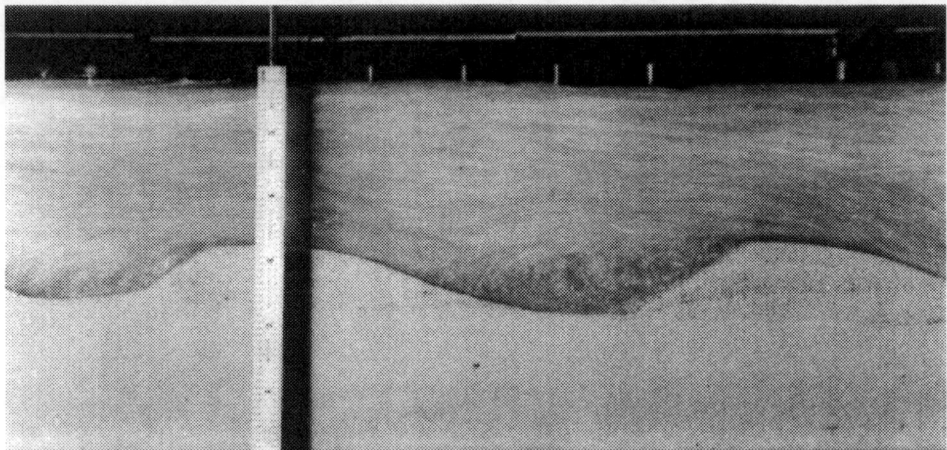

FIGURE 8-4. Side view of asymmetrical ripples in a flume (current from left to right). Aluminum powder defines main flow, flow separation, and its turbulent eddies. (Photo by A. V. Jopling).

Settling Velocity

Four physical parameters dominate sediment transport mechanics: *settling velocity*, the *state of fluid flow* described by its turbulent structure and boundary shear stress, *gravitational sliding*, and *porewater pressure*. Irrespective of depositional environment or process, these four commonly play an important role. In the sedimentological literature, settling velocity has received the most attention.

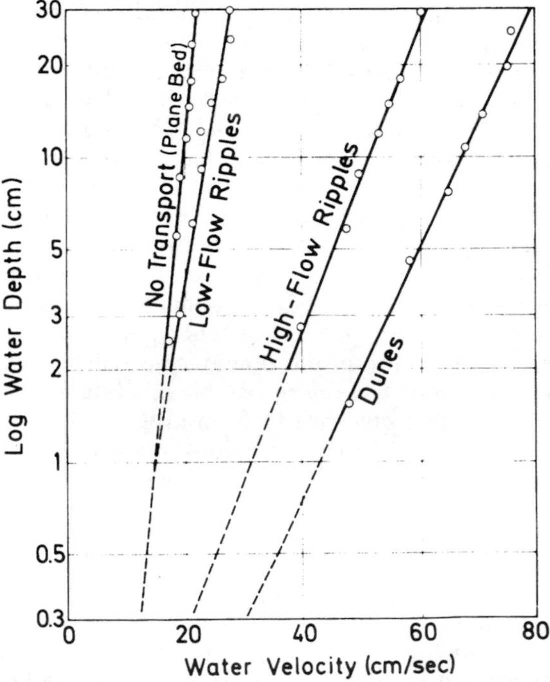

FIGURE 8-5. Velocity gradients and forms of bed roughness observed in a flume. Note change in vertical gradients as bedform roughness increases. (Modified from Harms, 1969, Fig. 10).

The properties of grains that are significant in relation to fluid flow are size, density, and shape. The combined effect of these variables as well as the density and viscosity of the fluid is summed up by settling or fall velocity, which is the terminal or steady state velocity of a particle falling in the fluid. Settling velocity is an essential feature of practically all theories of sedimentation, whether it is for bedload (sand and gravel moving along the bottom), saltation (intermittent suspension transport), or complete suspension. Settling velocity is also the primary variable that determines interaction between bed material and its transporting fluid. Certainly settling velocity is a convenient and meaningful way to sum up the interaction of a fluid and a single grain.

The equation for settling velocity, w, is given by the governing equation of motion: the algebraic sum of the downward force, the weight of the particle, and the upward or resisting forces of buoyancy and frictional resistance or drag are equated to $M \frac{dw}{dt}$ to obtain

$$M \frac{dw}{dt} = F_W - F_B - F_D \tag{8-6}$$

where M is particle mass, F_W is the weight of the particle, F_B is its buoyancy, and F_D is the drag force of the liquid on the falling particle. At terminal velocity there is no acceleration so $dw/dt = 0$ and thus $F_D = F_W = F_B$. Assuming that the drag force exerted by a fluid on a falling grain is proportional to the density of the fluid, cross-sectional area of the grain, and velocity, one obtains for spheres (Scheidegger, 1970, pp. 176–79)

$$F_D = C_D \pi \frac{d^2}{4} \frac{\rho_f w^2}{2}, \tag{8-7}$$

where C_D is the drag coefficient, which depends on Reynolds number and the shape of the particle, d is grain diameter, ρ_f is the fluid density, and w is the settling velocity. C_D can be thought of as a measure of the relative resistance to the flow of differently shaped bodies with the same cross-sectional area or length under identical flow conditions. F_W and F_B are given by

$$F_W = \frac{d^3}{6} \pi g \rho_s \text{ and } F_B = \frac{d^3}{6} \pi g \rho_f, \tag{8-8}$$

where ρ_f and ρ_s are fluid and solid densities. Substituting Eqs. (8-7) and (8-8) into (8-6), for $\frac{dw}{dt} = 0$, we have:

$$w = \left(4/3 \frac{dg}{C_D} \frac{\Delta\rho}{\rho_f}\right)^{1/2} \tag{8-9}$$

where $\Delta\rho = (\rho_s - \rho_f)$. In the common sedimentologic situation of quartz grains falling in water, for a given drag coefficient, Eq. (8-9) reduces to w varying as a constant times the square root of the diameter, where the constant term includes the drag coefficient, the densities, and g. A similar formula can also be obtained for nonspherical particles, but empirically determined shape factors must be used.

The drag coefficient, C_D, because of its dependence on R, varies with the type of flow. If the flow is fully turbulent (inertial), C_D is approximately one-half for spheres. If it is slow laminar (viscous), C_D is found to be $24/R$ from which we obtain Stokes' law by substituting Eq. (8-3) into (8-9),

$$w = \frac{1}{18} \frac{d^2 g}{\mu} \Delta\rho = \frac{2}{9} r^2 g \frac{\Delta\rho}{\mu} \tag{8-10}$$

and substituting for Reynolds number, where the length term of R is particle diameter, and converting diameter to radius, r. For Stokes' law (formulated by G. G. Stokes in 1845) to apply, the fluid must be isothermal, free of boundary effects, and completely nonturbulent. It works well for particles up to 0.18 mm that have viscous (laminar) settling but not for larger ones that have inertial (turbulent) settling.

For larger particles—and different approaches to settling velocities—articles by Rubey (1933), Stringham *et al.* (1969), Graf and Acaroglu (1966), Lerman (1979, pp. 263–67), and Baba and Komar (1981) are instructive. Drawing upon these sources, the following picture emerges. Somewhere in the range from 0.1 to 0.5 mm measured settling velocities become progressively less than those predicted by Stokes law. Discrepancies are caused by a change from inertial to turbulent settling and by irregular and nonspherical shape.

To obtain the settling velocity of a natural quartz grain whose diameter is specified by intermediate radius (measured in centimeters by binocular or by midpoint of a sieve class), the following procedure of Baba and Komar is used. First use the equation of Gibbs *et al.* (1971, p. 10) to compute the fall velocity of an equivalent quartz sphere

$$w_s = \frac{-3\mu + \sqrt{9\mu^2 + gr^2\rho(\Delta\rho)(0.015476 + 0.19841r)}}{\rho(0.011607 + 0.14881r)} \tag{8-11}$$

where all the variables are defined as above and then use either

$$w_{m_b} = 0.977 w_s^{0.913} \tag{8-12}$$

when intermediate diameters are determined by binocular micrometers (Fig. 8-6) or

$$w_{m_s} = 1.473 w_s^{0.842} \tag{8-13}$$

when they are determined by midpoint of a sieve class. Equation 8-11 was developed experimentally by Gibbs *et al.* using glass spheres, which ranged in diameter from 50 to 5000 microns, and hence gives settling velocity over a very wide range of sizes. Baba and Komar found settling velocity to be largely independent of roundness, but to vary significantly with sphericity—less spherical grains settle more slowly than spherical ones because C_D is greater (Eq. 8-9).

All of the above equations are sensitive to changes in water temperature and viscosity—as water cools, its viscosity increases and settling velocity decreases so that "cold water" can support larger particles in suspension than warm water for the same degree of turbulence. Similarly, settling velocity is lower in highly turbid water.

Hallermeier (1981) provides a dimensionless graph that incorporates settling velocity and compiles much of the published empirical data on which it is based.

In recent years another approach has developed—why not directly analyze the observed

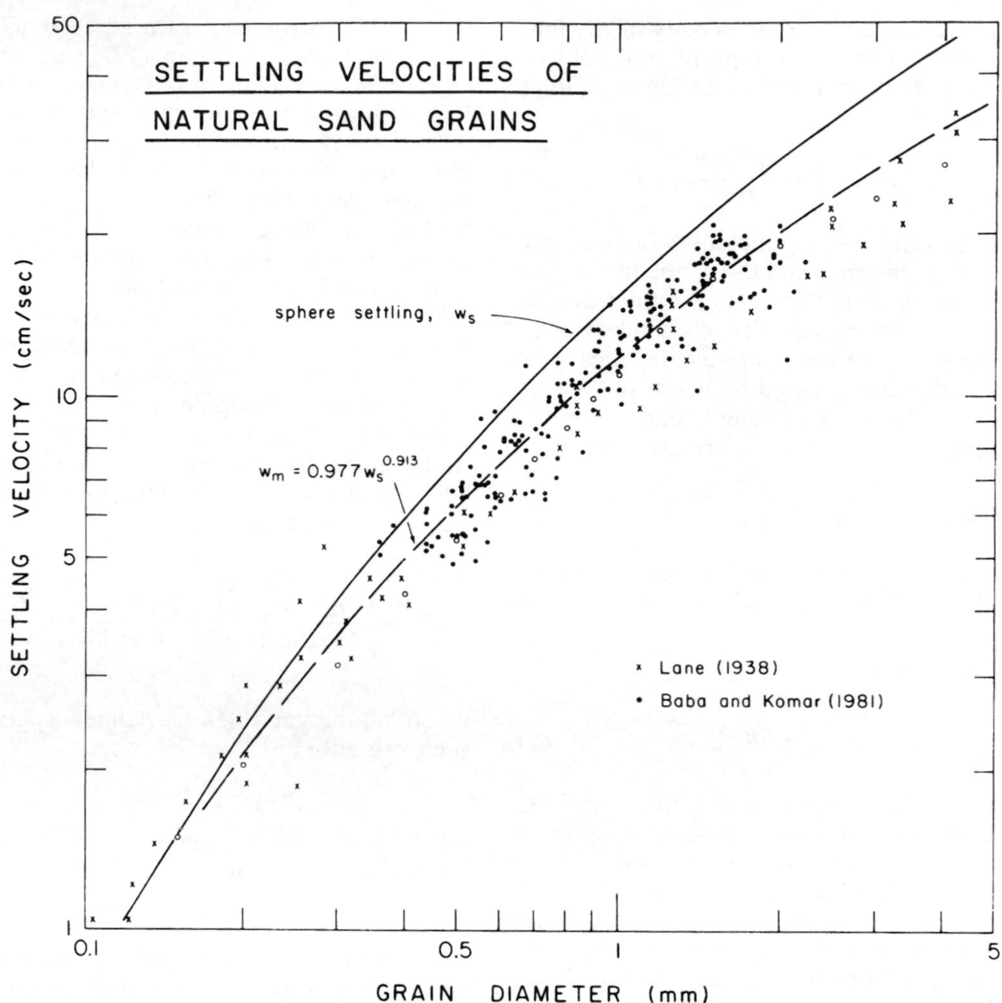

FIGURE 8-6. Settling velocities of glass spheres (solid line) and of natural quartz grains with an equivalent intermediate diameter (dotted line). (After Baba and Komar, 1981, Fig. 3).

distribution of settling velocities rather than the distribution of equivalent spheres (Middleton, 1967, p. 484; Reed *et al.*, 1975, p. 1323–25)? Underlying this viewpoint is the idea that it is *settling velocity that reacts with flowing water and not grain diameter*. Hence let us examine settling velocity in more detail.

Settling velocity not only affects entrainment of sand, but is also the chief control on the vertical concentration gradient of suspended solid in a fluid as can be seen from data collected from the Mississippi River, in which finer sizes with their smaller fall velocities have the flatter (smaller) concentration gradients (Figs. 8-7 and 8-8). The governing equation is

$$-\frac{Qdc}{dy} = cw \qquad (8\text{-}14)$$

which can be integrated to obtain the relative concentrations of c_y and c_a

$$\frac{c_y}{c_a} = \left(\frac{(9D - y)a}{(D - a)y}\right)^Z \qquad (8\text{-}15)$$

where $Z = w/RU_*$, D is depth of flow, and y and a are depths, a being the depth having concentration c_a. Scheidegger (1970, pp. 1881–92) gives complete details.

Settling velocity also plays a role in the *hydraulic equivalence* of grains. Two different mineral grains are said to be hydraulically equivalent if they have the same fall velocities (Rittenhouse, 1943, p. 1741). For the light minerals, size and shape are the principal factors affecting fall velocity. But with the heavy minerals density is all important. Because of their greater density, heavy minerals are finer than their associated light minerals (Fig. 8-9). Briggs *et al.* (1962) found that shape has importance for the settling of heavy minerals also.

Although the sedimentological origin of

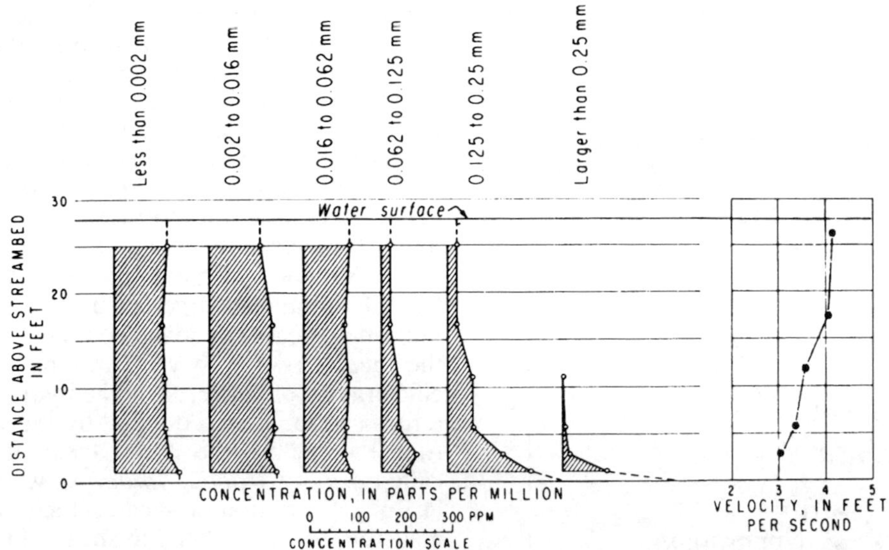

FIGURE 8-7. Vertical distribution of suspended grains and velocity profile of the Mississippi River near St. Louis, Missouri. Note the steeper concentration gradients of the coarser grains (Colby, 1964, Fig. 7).

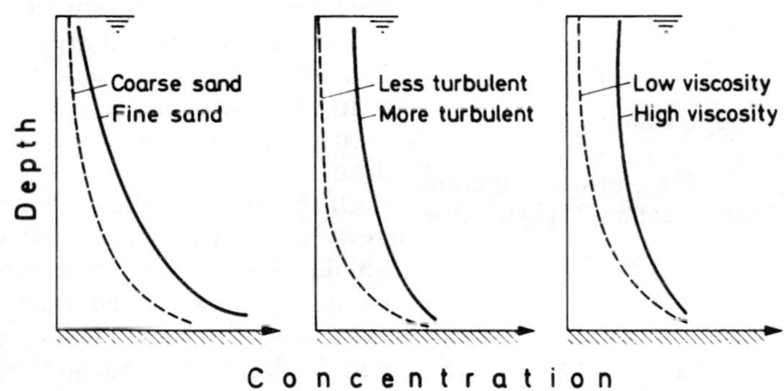

FIGURE 8-8. Vertical distribution of suspended sediment as a function of grain size, turbulence, and viscosity in streams (Fig. 7-12 of Handbook of Applied Hydrology by V.T. Chow, Ed. Copyright, 1964. Used with permission of McGraw Hill Book Co.).

placer concentrates and sands is poorly known, settling velocity and hydraulic equivalence no doubt play critical roles. Also needed for a placer is an abundant supply of heavy minerals in the source area. Immature, easily erodable, unconsolidated sands probably provide the best source. Repeated reworking under broadly similar hydraulic conditions also seems to be necessary. The origin of placers deserves much more attention than it has received.

Entrainment of Single Grains

One can obtain insight into the transport process and some of its problems by considering streams, where the process has been most thoroughly observed.

In streams the finest material, that less than 20 or 30 microns, is transported entirely in suspension, is supported by the upward components of turbulence, and moves approximately at the velocity of the water. The term "wash load" is sometimes used for suspended load. Supply, ultimately from the weathering of the source area, controls concentration of suspended material, for even at low discharge the fall velocity of 20 to 30 micron particles is sufficiently small so that they can be supported by available upward components of the turbulence.

Sand is transported by a different process. It moves mostly along the bottom as "bed load" going but briefly into suspension. Such short jumps are referred to as *saltation*. Instead of moving continuously, sand grains are likely to rest at the interface most of the time, the probability of entrainment being proportional to the

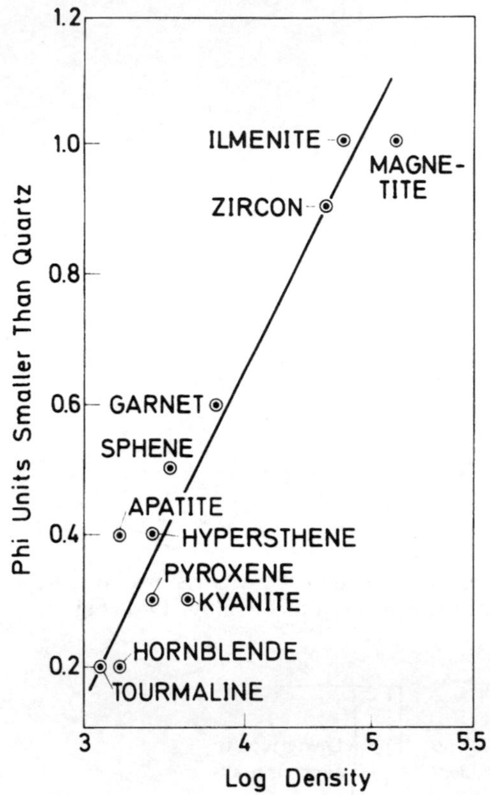

FIGURE 8-9. Hydraulic equivalance of some different heavy minerals. Data from Rittenhouse (1943, Table 7).

ratio of fall velocity to turbulent uplift. The acting forces are those of gravity (weight, downward, and buoyancy, upward), turbulent uplift (perpendicular to the current), and drag (parallel to it). Lift commonly does not appear explicitly in many analyses, but is represented in empirically determined constants.

Close observation of flumes clearly reveals the statistical, random nature of entrainment of grains. At the velocity at which a grain of given size and density will move, the critical velocity, is approached, a few of the smallest and lightest grains in different parts of the bed will move at irregular intervals. This irregular movement may result from random impingement of turbulent eddies on the bottom (Sutherland, 1967). The instantaneous velocity fluctuations associated with these eddies entrain the grains, starting with the smallest and lightest. As flow intensity increases, frequency and magnitude of the eddies increase and movement becomes more general; finally all of the grains, including the largest, are in motion everywhere on the bed and grain-to-grain collisions are the rule. General movement is accompanied by the progressive formation of ripples and dunes.

Different but interdependent measures of flow intensity in open channels have been used by different experimenters. Initially, velocity was used, but because it varies as the logarithm of the distance from the bottom, it is not commonly the most convenient measure, unless one can be satisfied with average velocity obtained from dividing discharge per unit time by cross-sectional area. The most common measure is the *shear stress*, $\tau_0 = \gamma R_h S$, where γ is the specific gravity of water, R_h is the hydraulic radius (cross-sectional area divided by the wetted perimeter), and S is the slope. Shear stress is also called *drag* or *tractive force*. As we have seen in Eq. 8-5, friction or shear velocity is another measure, for it is related to shear stress and has the dimensions of velocity. It describes flow intensity at the fluid-solid boundary and can be used together with shear stress as measures related to the mean velocity of the flow. Another measure of flow intensity is *flow* or *stream power* per unit of bed area, which is defined by Bagnold (1968, p. 46) as $\overline{U}\tau_0$ where $\overline{U}$ is average velocity. Flow power per unit area has the dimensions of power (work per unit time per unit area).

Of the many studies of the beginnings of grain movement in flumes, those based on work by Shields (1935), an American working in Germany, are among the most widely cited. He used fully turbulent flows and began each run with a flat bed. Repeated runs with different grain sizes and different water temperatures permitted him to establish the threshold of movement for a variety of conditions. Implicit in his approach was the idea that the beginning of motion is determined by τ_0 (shear stress), $\Delta\gamma$ (the difference of specific weight between fluid and grain), d, ρ, and μ. He combined these variables into two dimensionless quantities, τ_* and R_*, frequently called Shields' parameters (Fig. 8-10). τ_* is a dimensionless shear stress and R_* the boundary Reynolds number, a measure of turbulence at the boundary. Unlike ordinary Reynolds number, the velocity term of R_* is the shear velocity, U_*, and is a measure of turbulent eddying; the length term of R_* is particle diameter, d. One notes that both τ_* and R_* describe the hydraulic conditions at the boundary, which is the important site of initiation of grain movement.

The curves of Figure 8-10 separate the graphs into two fields: if a point lies above the curve, the grains on the bed are fully in motion

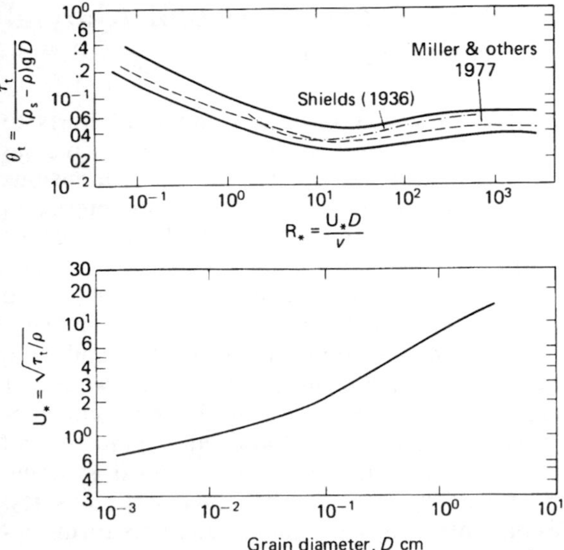

FIGURE 8-10. Modifications of Shields' diagram by Miller *et al.* (1977, Figs. 2 and 3).

whereas a point below it indicates no movement. The curve itself indicates threshold or critical conditions, when movement is just beginning. From the standpoint of fluid flow, Shields obtained more generality by plotting τ_* against R_* than against d, as one might at first be tempted to do. By using R_* one can, for example, find the value of τ_*, the critical dimensionless shear stress, for different values of $\Delta\gamma$ or of $\nu = \mu/\rho$, the latter a temperature-dependent quantity. Thus, for example, Shields' curve can be used for wind as well as water. Hence a plot of τ_* against R_* conveys maximum information in a single graph. *When and only when* ρ, $\Delta\gamma$, and ν *are either constant or assumed to be so*, as, for example, in freshwater streams in a given season during average flow, can one plot U versus d as shown in Fig. 8-10. Figure 8-10 also shows, in striking manner, how τ_* varies with temperature because of the temperature dependence of ν, especially for sand between 0.1 and 0.5 mm.

Shields' curve is concave upward with a poorly defined minimum between values of R_* of 8 and 15. For water at standard conditions this minimum corresponds to a size range of about 0.2 mm to 0.5 mm: larger grains require stronger currents for entrainment, as do smaller grains. The paradox of smaller particles requiring stronger dimensionless drag force for entrainment is explainable in terms of their shear strength according to Sundborg (1956, pp. 177–180). If the clays and silts are unconsolidated they will require only a small shear velocity for entrainment, but if they are consolidated or semiconsolidated, a greater drag force is needed for entrainment. Hiding of small particles in the laminar sublayer may also be a factor. Cohesion of clay beds and thus shear resistance may be importantly affected by chemical variables, such as clay mineralogy and composition of the fluid, as well as the conditions under which the clay bed settled. However, the entrainment of silt and clays is not as yet fully understood and Sundborg's studies and those of the Committee on Sedimentation (1968) should be consulted for fuller details.

Slope of the bed and groundwater conditions in natural streams also affect the entrainment of grains. A smaller drag force is needed to entrain a grain of given size on a surface sloping downcurrent than upcurrent, gravity making the difference. In natural streams one should also keep in mind that the drag force needed to entrain grains will vary depending on the inflow or outflow of water perpendicular to the bed. Inflow to the stream decreases the drag force needed whereas outflow from it increases it. The porewater pressure of the bed is the determining factor.

In its simplicity of variables, its generality, and its experimental reliability, the significance of Shield's work has not been greatly altered even though more recent experimental results are available.

Suspension and Discontinuities in Grain-Size Curves

Now that we better understand fall velocity and have discussed turbulence, let us consider a criterion for suspension versus bedload transport of sand that has been proposed by Middleton (1976, pp. 409–10) to explain the inferred bedload versus suspension segments of a cumulative grain-size distribution (Fig. 3-2). To do so, we need to look more closely at turbulence.

The instantaneous velocity, U', of a mass of turbulent flowing water has three components of turbulence,

$$U' = \sqrt{(U'_x)^2 + (U'_y)^2 + (U'_z)^2},$$

one in the x direction (vertical), another in the y direction (parallel to flow), and the third in the z direction (transverse to the main flow). Only the magnitude of the vertical components of turbulent flow, specified by its standard deviation (also called the root mean square), is important for suspension, and it seems reasonable to believe that when it exceeds the settling velocity of a grain, suspension occurs; in other words, when

$$w < [(U'_x - \overline{U}'_x)^2/n]^{0.5} \quad (8\text{-}16)$$

where n is the number of observations. Figure 8-11 shows diagrammatically the different trajectories of salting and suspended grains.

Middleton (1976), building on earlier work by Lane and Kalinske (1939) and Inman (1949, p. 59), related the standard deviation of U'_x to the shear velocity, U_*, at the boundary to explain the breaks in the grain size curve that are commonly observed (Fig. 3-2). As we have seen, shear velocity, like average velocity or stream power, is a measure of flow intensity. Middleton argues that when U_* exceeds w, grains on the bed go into suspension or at least begin to saltate along the bed. Middleton (1976, pp. 410–12) shows how to calculate the size breaks for various shear velocities. Because most sand is transported at high stream discharge, the corresponding shear velocities should be used. Later, Blatt *et al.* (1980, pp. 100–101) related U_* to variables that can be calculated in an open channel: U_* can be obtained from the von Kármán velocity profile or from

$$U_* = \sqrt{ghS} \quad (8\text{-}17)$$

where h is depth and S is slope, and steady, uniform flow is assumed. See Bridge (1981) for another approach to the hydraulic interpretation of grain size curves.

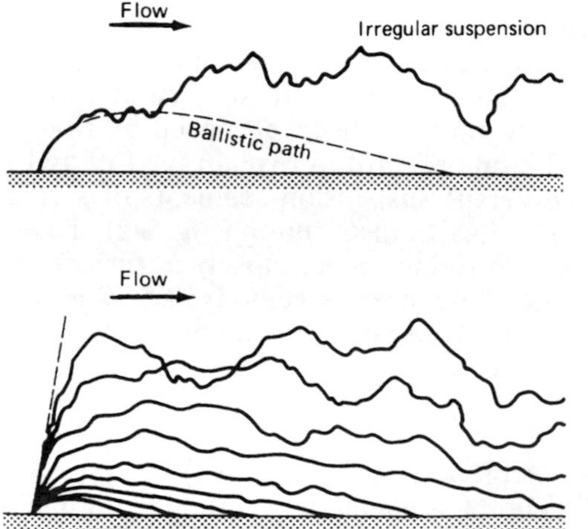

FIGURE 8-11. Ballistic versus irregular suspension paths; two grains (above) and a family of grains (below) with different fall velocities, w. (Redrawn from Yalin, 1977, Fig. 1.12).

Bedforms in Flumes and Alluvial and Tidal Channels

As flow intensity increases and sand starts to move, there is a systematic and progressive development of bedforms, whose dimensions range from millimeters to many kilometers and whose names are many. Such bedforms are best known from streams, tidal channels, in deserts, and in flumes, but they also have been studied on marine shelves, continental slopes, and submarine canyons and in the deep sea—and have been well documented on Mars—wherever sand is transported by a fluid. Here we consider mostly modern alluvial and tidal channels and evidence from flumes. Later we will examine bedforms on marine shelves and in deserts. Sedimentologists are interested in bedforms because, as they migrate, they produce an internal structure which is preserved in the geologic record, whereas the form commonly is not. Alluvial bedforms belong to a broad hierarchy (Table 8-2).

The question has been asked, "Why isn't the interface between moving sand and air or water flat?" Although much has been written (see for example, Yalin, 1977, Chap. 7), we can still only say that either some instability of the flow—almost certainly related to its turbulent structure—starts the process on an initially flat bed of sand or that a bedform develops around some initial irregularity on the interface itself—a stray pebble, a shell, perhaps even an irregularity such as a fecal pellet or trace fossil. But accepting, as our point of departure, that periodic bedforms are an inherent part of the transport of sand by air and water, we still face questions such as "Is there a universal and progressive sequence of bedforms related to increasing flow intensity and to grain size? Is depth of flow a factor? How do bedforms differ in different environments?" Today we can answer some, but not all, of these questions. Some landmark studies of bedforms and their significance include those by Sorby (1851) and Cornish (1901), who made early field observations, those by Bagnold (1941) and Wilson (1972) on eolians dunes, and the experimental work of Gilbert (1914), Simons and Richardson (1961), and Guy *et al.* (1966), and the many studies by J.R.L. Allen, mostly with unidirectional currents of water. Virtually every modern engineering text in hydraulics contains a section of bedforms in alluvial channels. Thus the study of bedforms began with field observation and progressed into experiments. Today,

TABLE 8-2. Bedforms and the alluvial hierarchy.

Meander and braided belts

Formed by lateral migration of river channel.
Point, side, and various midstream bars.
Thickness up to 40 m or more and widths up to tens of kilometers.
Alluvial meander belts tend to have serrate margins and meander belts have more mud than braided stream deposits.
Paleocurrent dispersion moderate in braided belts and moderate to great in meander belts.
Commonly heterogeneous with many internal disconformities and cutouts.

Channels

Straight and curved reaches.
Thicknesses comparable to belt sandstone bodies.
Chiefly point and mid-channel bars.
Depth of scour during high discharge or tidal range controls thickness of sands.
Good correlation between paleocurrents and channel trend.

Point, side, and midstream bars

Wide range of thicknesses which are largely controlled by depth of stream scour. Point bars of meandering streams have well-defined vertical sequence with broad range of grain sizes and structures.

Sinuous, 3D megaripples

Occur in systems and have sinuous and lunate crests; variable heights from 5 to perhaps 100 cm.
Lee scour pits well developed; trough crossbedding.
Slip slope tends to be tangential.
Moderate to large variance of foreset dip direction.

Straight, 2D megaripples

Occur in systems in thalweg and on bars and form in lower flow regime.
Crests straight to slightly sinuous; heights from 5 to 100 cm or more.
Lee slip slope at angle of repose.
Lee scour pits absent; tabular or planar crossbedding.
Small variance of foreset dip direction.

Ripples

Smallest asymmetric, transverse bedforms occur in systems with many diverse morphologies.
Heights less than 5 cm and wavelengths less than 30 cm.
Occur in all sandy alluvial environments and commonly superimposed on all larger bedforms.
Orientation highly variable.

Parallel lamination

Common, especially in fine sand and silt, and a key element for reconstruction of flow sequence; occurs on both upper and lower flow regime.

however, because of the recognized scale limitations of most flumes and because of improved techniques of current measurements in the field as well as field studies by side-scan sonar, bedforms are best studied in rivers and tidal estuaries, where their diverse morphology is most easily related to hydraulics. Of the two, the tidal environment is probably the better, especially where tidal range is more than 5 m, because bedforms are well exposed twice a day. Flume experiments, on the other hand, provide good insight into the origin of bedforms, because the relative importance of different variables can be better assessed, at least qualitatively. In one such study (Southard *et al.*, 1980), by using water at 85°C viscosity was decreased—and hence fall velocity increased (p. 281)—permitting an "up-scaling" of the bedforms generated in the flume so that experimental conditions more closely approximated natural ones. Here we focus on those features of bedforms that are well preserved in ancient sandstones; e.g., planar vs. trough crossbedding, thickness of crossbeds, tangential vs. angular foresets, variation of ripple morphology, parting lineation, etc. The processes that produce most of these features and structures also apply to the marine realm.

When water of increasing flow intensity (velocity, discharge, stream power, etc.) flows over a loose bed of sand, general movement begins and the sand bed is characteristically molded by the flow into a distinctive sequence of bedforms, each of which is recognizable in ancient sandstones. This movable boundary is both a response to flow and a conditioner of it, for as bedforms and microrelief change they in turn influence—through their changed roughness—the resistance of the boundary to the flow (Fig. 8-5). Middleton (1965) did much to make sedimentologists aware of flow regime and its significance for sedimentary structures.

The flow in an alluvial channel has been divided by Simons and Richardson (1961) into lower, transition, and upper regimes based on bed morphology, sediment concentration, mode of transport, type of roughness, and phase relation between bed and water surface (Table 8-3). As a first approximation in flume studies, F is used to distinguish the three regimes: $F < 1$ (tranquil) for the lower regime, $F = F_{critical} = 1$ for the transitional regime, and $F > 1$ (shooting) for the upper regime. As discharge or stream power increase, the following generalized sequence of bedforms develops in sand less than 0.6 mm, as demonstrated by both flumes and field studies: initial flat bed → asymmetrical ripples → asymmetrical, larger transverse bedforms (large ripples or megaripples) → their washed-off equivalents → plane bed → antidunes → chutes and pools. Scale of transverse bedforms ranges over several orders of magnitude in alluvial and tidal environments (Fig. 8-12) and is best specified by wavelength and amplitude.

Simons *et al.* (1965, p. 46) early suggested that under both constant discharge and sediment supply, bedform type is a function of the following factors:

Bedform = f[water depth, slope (flow intensity), grain size, concentration of sand–water mixture at the boundary, and fall velocity]

These factors have largely been confirmed by Vanoni (1974). Today, flow intensity, depth of flow, and grain size are universally considered to be the most important factors (Fig. 8-13). Depth of flow is a factor, because high flow intensity in shallow water inhibits formation by shearing off the tops of bedforms. The joint response of sedimentary structures and grain size was early recognized in ancient sandstones (Fig. 8-14).

Terminology is not standardized but transverse bedforms are separated into two broad groups: those less than 3 cm in height are called *ripples* (small ripples), and those greater are called *megaripples* (large ripples). Megaripples are divided into those with straight crests, called two-dimensional or 2D megaripples, and those with sinuous crests called three-dimensional or 3D megaripples (Harms *et al.*, 1982, pp. 2–18).

How do crest spacing, height, and average grain size of transverse sandy bedforms in alluvial and tidal channels depend on fluid dynamic conditions? Although complete answers remain in the future, we do know that ripples and megaripples behave differently. Yalin (1964) and Raudkivi (1976, pp. 57–66) concluded that current ripples have a spacing proportional to particle size and are independent of depth of flow. Certainly observation of both modern and ancient sands and many experiments support the idea that ripple height and spacing are largely independent of water depth. Observation also shows that ripples can grade from one pattern to another on the same bedding plane in relatively short distances. Allen (1968, Fig. 4-61) has discussed how ripple form responds to variation in grain size, flow and depth in shallow water and suggested that relatively straight and continuous patterns change to curved and discontinuous patterns as depth and velocity increase in the shallow waters of streams and the strand as have Harms *et al.*, (1982, pp. 2–18). Sediment transport over the front of a transverse bedform has been studied experimentally (Fig. 8-15). Allen (1968, Fig. 4–61) has discussed how ripple form responds to variation in grain size, flow intensity, and water depth in shallow water. He noted, for example, that patterns are more regular at low rather than high velocities and that finer grain size permits more

TABLE 8-3. Flow regime and bedforms.

Flow regime	Bedform	Remarks
Lower	Ripples Ripples on megaripples	$F < 1$. Grains move in discrete steps and form roughness predominates. Water and bed surface are out of phase. Bedforms move downcurrent.
Transition	Washed-out megaripples	$F = 1$. Both grain and form roughness.
Upper	Plane bed Antidunes Chutes and pools	$F > 1$. Grains move continuously and grain roughness predominates. Water and bed surface are in phase. Antidunes move upcurrent, remain stationary, or go downcurrent.

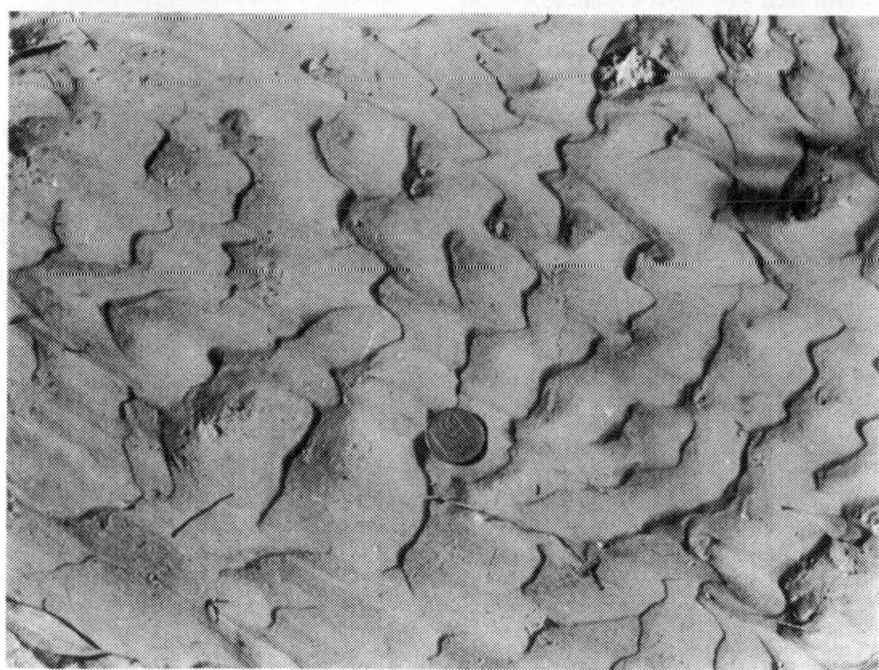

FIGURE 8-12. Asymmetrical alluvial bedforms of contrasting scale. Above: A large sand wave field in the Luusnan River (river flows from right to left) northeast to Järvsö, Sweden (Lundquist, 1963, Fig. 64). Below: Small asymmetrical, sharp-crested ripples (current from left to right) in irrigation ditch, Mason Verte, Hassi Messaoud, Perfecture de Oasis, Algeria.

suspension transport, which dramatically alters morphology and internal lamination leading to climbing ripples.

The contrast between planar (tabular) and trough crossbedding has long been recognized in ancient sandstones, and modern sediment study shows that it depends upon the presence or absence of an erosional scour at the base of the slip slope of transverse bedforms. Two-di-

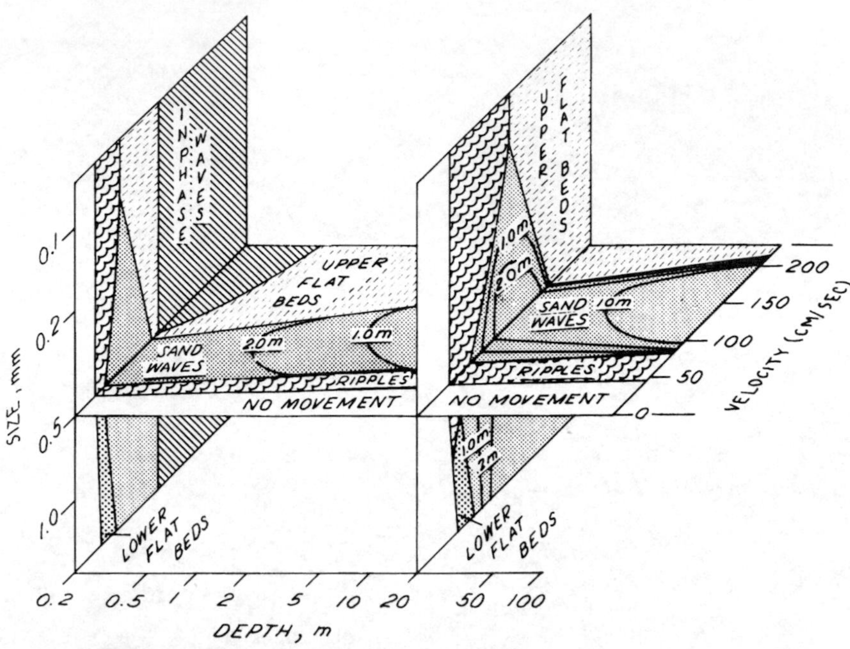

FIGURE 8-13. Stability fields of bedforms in San Francisco Bay (redrawn from Rubin and McCulloch, 1980, Fig. 11) harmonize well with those of flumes.

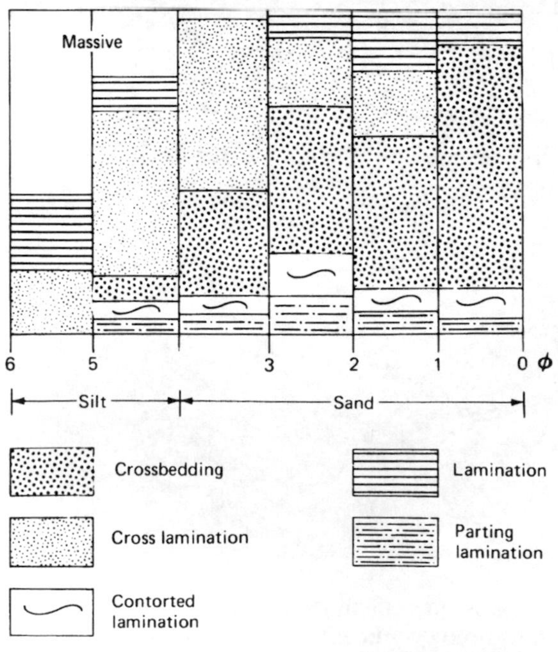

FIGURE 8-14. Correlation between grain size and structures (grain size) in the Old Red Sandstone (Devonian) of Spitsbergen (redrawn from Friend, 1965, Fig. 4) is a good example based on measurements of what exists in all sandstones. See also Scheidegger and Potter (1967, Fig. 1) for the relation between crossbed thickness and grain size.

mensional megaripples with straight crests of uniform height, lack erosional scours at their slip slopes, and produce planar crossbedding whereas three-dimensional megaripples with sinuous crests of variable height have scour troughs at their base and produce trough crossbedding. As a rule, the variance of planar crossbedding is smaller than that of trough crossbedding (Meckel, 1967). Careful description of sandy sequences commonly shows that planar and trough crossbedding tend to be segregated into distinct zones. Higher flow intensity is associated with three-dimensional bedforms.

Whether foresets are tangential or angular depends upon flow intensity and water depth, both of which affect flow separation downcurrent of the crest (Figs. 8-16 and 8-17). For a given water depth, great flow intensity produces more suspension transport beyond the crest rather than avalanching as does shallower depth for a given flow intensity. Another feature—reactivation surfaces—are produced by erosion of the crest and are also related to flow separation. Allen (1980) has effectively illustrated the increase in complexity of crossbedding and how flow separation controls the relative importance of avalanching versus suspension transport, reactivation surfaces,

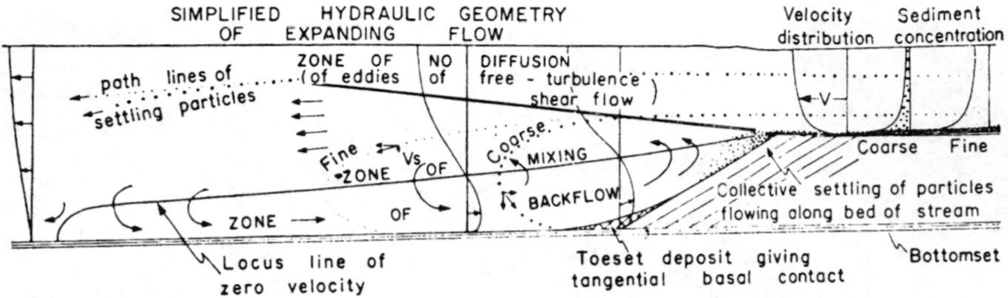

FIGURE 8-15. Hydraulic conditions at the lip of a small delta produced in a flume. The flow structure developed over the slip slope is the same as that produced over the slip slope of a dune or ripple (Jopling, 1963, Fig. 2).

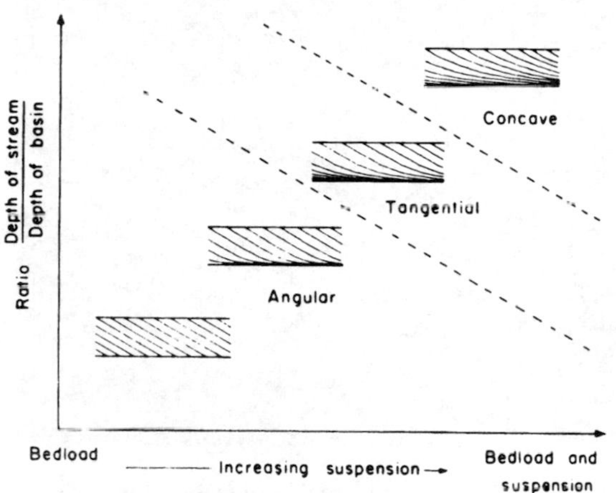

FIGURE 8-16. Dependence of foreset geometry on ratio of bed to suspended load and ratio of depth of stream to depth of basin (Allen, 1968, Fig. 16.5). Reproduced by permission of North Holland Pub. Co.

and the presence or absence of ripples and perhaps even plane beds (Fig. 8-18).

Many sedimentologists equate the flat bed of the transitional flow regime of flumes with parting lineation, but some field evidence suggests that not all was so formed. For example, sand beds that are transitional to shale commonly are very fine-grained and current-lineated and suggest deposition from low-intensity flows. At issue here is the uniqueness of structure and flow condition. Is every sedimentary structure the product of a unique fluid dynamic process with fixed hydraulic and grain size parameters? Flume study failed to find lamination in sands of the plane bed of the upper flow regime (McBride *et al.*, 1975). Instead, they found that lamination developed from low-relief bedforms. Bridge (1978, Table II) has summarized the many inferred processes involved to produce lamination and also suggested a new one, *bursting*, a name applied to flow variation in the turbulent boundary layer. Lamination in sands occurs in all the major environments and possibly has several origins whose details are obscure. Lamination is also produced by avalanching (grain flow) down the slipface of a transverse bedform, a process easily seen in the eolian and alluvial environments (Smith, 1972, Fig. 4).

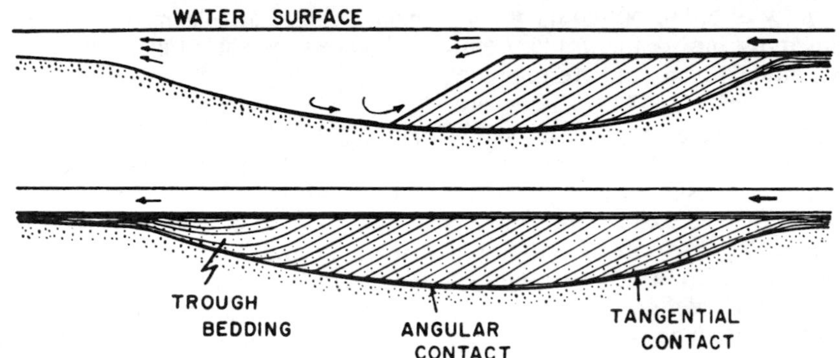

FIGURE 8-17. Changing geometry of foresets with change in water depth. Trough axis parallel to page. Depth ratio of basin downcurrent from crest to affects flow separation which controls geometry (Jopling, 1965, Fig. 9). Published by permission of the Society of Economic Paleontologists and Mineralogists.

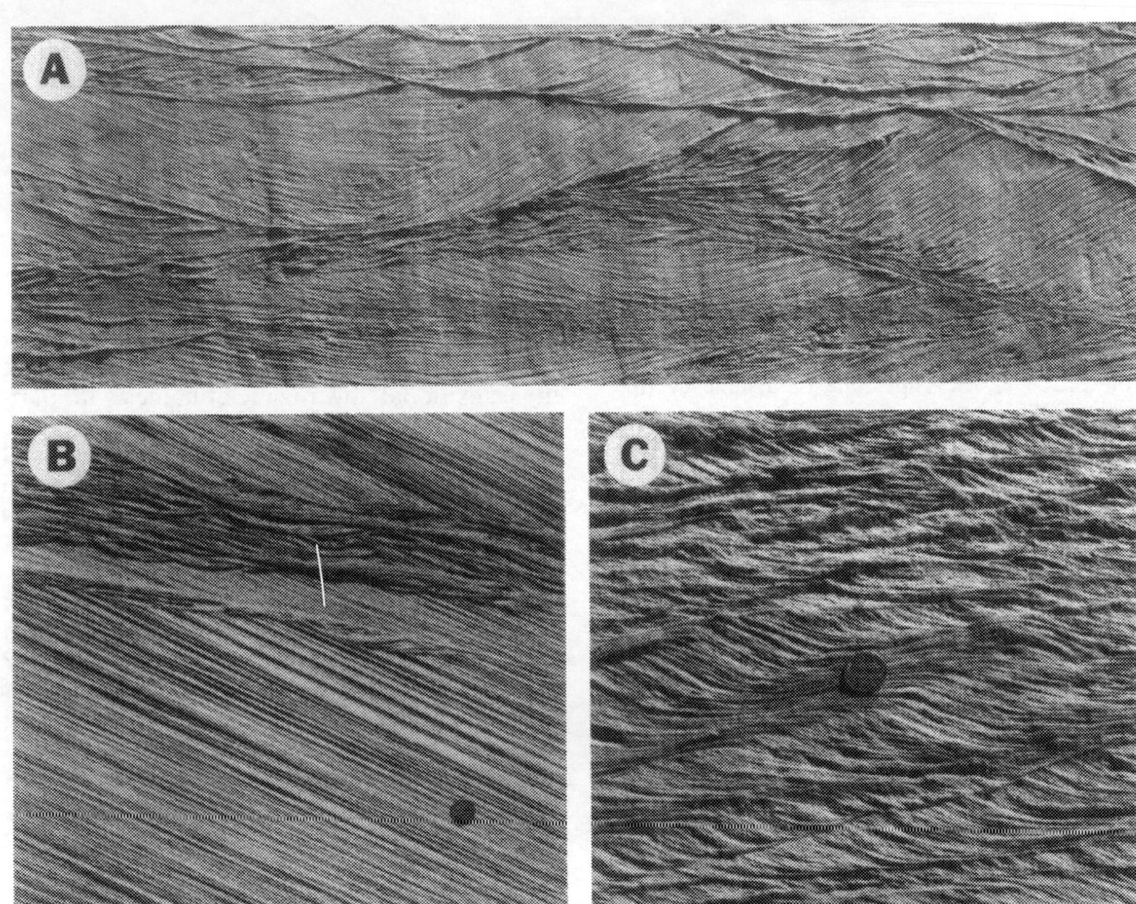

FIGURE 8-18. Variants of inclined bedding in point bar of Red River north of Gainesville, Cooke County, Texas: A) section perpendicular to current has obliquely filled troughs and cross-lamination, some of which shows transport up foreset slope; B) section parallel to flow has gentle reactivation surface and cross-lamination; and C) climbing ripples. Current from left to right in B and C. Epoxy films courtesy of R. Moiola and the Mobil Field Research Laboratory. See also Figure 10-19.

However it forms, the sand grains of parting lineation (Fig. 4-20) are commonly well oriented and originate from bedload washing along the bottom in ridges a few grains thick. Grain orientation closely parallels ridge orientation. Allen (1982a, Fig. 6-14) suggested that the distinctive ridges of parting lineation owe their origin to a transverse instability of a thin boundary layer.

Antidunes can move upstream, remain stationary, or move downstream, and are in phase with the water surface (Fig. 8-19). They are rarely preserved in water-deposited sandstones, although they occur in base surge deposits of debris-laden density flows related to shallow, phreatic, volcanic explosions (Fisher and Waters, 1970). Preservation of antidunes requires an abrupt cessation of flow. Antidunes commonly have a poorly defined internal bedding, although they produce what has been called "backset bedding" (Power, 1961; Skipper and Bhattacharjee, 1978).

Good descriptions of alluvial bedforms and structures are provided by Blodgett and Stanley (1980), Cant (1978), Coleman (1969), Collinson (1970), McGowen and Garner (1970), Levey (1978), and Smith (1972); for bedforms and structures in tidal estuaries see Allen (1973), Allen and Klingebiel (1974), Bajard (1966), Boersma and Terwindt (1980), Boothroyd and Hubbard (1975), Dalrymple *et al.* (1978), Evans (1965), Hayes (1980), Reineck (1963), and Rubin and McCulloch (1980). The more recent of these studies commonly relate bedform to hydraulic measurements. Although little more than speculation, it seems that the bedforms of tidal deposits have greater morphologic variability than those of rivers. Nonetheless, based

FIGURE 8-19. Above: Standing waves and antidunes in phase with water surface in a small flume with flow and antidune motion from left to right (Committee on Sedimentation, 1966, Fig. 8c). Below: Standing waves in the Rio Puerco during a flood close to Bernardo, Socorro County, New Mexico. (Photo by J.P. Beverage, United States Geological Survey).

only on kinds of sandy sedimentary structures, it is difficult to distinguish the two—except where tidal deposits have bipolar, reversing paleocurrents (p. 393).

Paleohydraulics and Open Channel Flow

What is paleohydraulics and why do sedimentologists study it?

Paleohydraulics is, in its fullest sense, the estimation of discharge, stream velocity, and water depth from an ancient fluvial deposit with perhaps lesser attention to variables such as drainage basin area, sinuosity, and meander wavelength. Prime attention has been focused on fluvial deposits, but paleovelocities and paleodischarge of turbidites have also received some attention. Paleohydraulics is based on the fundamental hydraulic equations of Table 8-4 with knowledge of channel cross section, grain size, and type of sedimentary structures. Hence it logically follows discussion of alluvial bedforms.

Why is paleohydraulics of interest? There are at least three answers, possibly more. On a regional scale, paleohydraulics can help verify paleogeography. For example, a petrologist finds metamorphic rock fragments in a fluvial sandstone which suggest a source area about 300 km distant. Such an inference would be stronger when supported by an independent estimate of size of the drainage basin. Or suppose a pebbly sand represented the distal part of a catastrophic flood—how large was maximum discharge? Or consider a channel on a modern subsea fan—what was the velocity and discharge of the last episodic, turbidite flow in the main channel just downcurrent from canyon mouth? In a broad sense, the numerical estimates of paleohydraulics simply represent the continuing and expanding process of quantification that

TABLE 8-4. Fundamental equations of open channel flow.

Equation	Description
$\overline{U} = Q/A$	Average velocity equals discharge divided by cross-sectional area perpendicular to the flow. Average velocity can also be estimated from the von Karman equation which also permits an estimate of τ_0.
$\tau_0 = \gamma RS$	Derived deductively for steady, uniform flow, this formula corresponds to an eraser sliding down an inclined plane at constant velocity: τ_0 is the average shear stress of the fluid on the boundary, γ is the specific weight of the fluid, and R is the hydraulic radius (cross-sectional area divided by walled perimeter). In nature, S, the slope, is hard to measure even in modern streams (and far more in ancient ones!), but still this equation is very fundamental because it is used as a stepping stone to other relations such as $\tau_0 + f\rho U^2/8$, whose terms are defined below.
$h = f \cdot \frac{L}{D} \cdot \frac{U^2}{2g}$	A semi-empirical formula developed for flow in pipes where h is the head loss in a pipe (difference in water level), f is a friction factor called the Darcy–Weisbach f (f is a function of Reynolds number and surface roughness), D is pipe diameter, L is pipe length, g is gravity, and U velocity. This formula, developed in the early 19th century, for open pipes, is critical to a better understanding of flow in rivers. Thus theory of flow in open pipes was a prerequisite to a better understanding of flow in rivers.
$U = \sqrt{\frac{8g}{f}}\sqrt{RS} = C\sqrt{RS}$	The famous empirical Chézy equation developed in 1775 for flow in channels, where C is the Chézy coefficient. Books give values of C which depend on both the roughness of the boundary and upon its shape. The formula has also been applied to channelized turbidity currents, the upper interface being considered, as a first approximation, as "open."
$Q = (1.49/n)AR^{2/3}S^{1/2}$	The Manning equation estimates discharge, Q, and is an empirical equation, where n is known as Manning's n after the Victorian English engineer who developed it. Books give pictures and appropriate values of n. If the SI system is used, the constant 1.49 is eliminated.
$U = \frac{R^{2/3}S^{1/2}}{n}$	Utilizing the Chézy equation and Manning's n, this equation provides another way to estimate average velocity.

we see everywhere in the study of sand and sandstone and the progressively greater use of engineering methods.

Paleohydraulics has been applied to a single exposure using only bedforms, grain size, and bedding thickness but is more effective when channel width, channel depth, and map pattern are also used. Jopling's (1965) study of paleoflow in an esker illustrates the possibilities and problems associated with a single outcrop, whereas the study of Padgett and Ehrlich (1976) of an Upper Carboniferous fluvial system in Morocco is based on exhumed meander scrolls, thickness, and lithology, and provides greater insight.

A classic paper is that of Schumm (1972) and good general sources are Starkel and Thornes (1981) and Gregory (1983). In all this work fluvial geomorphology provides fundamental background (Leopold *et al.*, 1964; Schumm, 1977).

Paleohydraulics is based on the fundamental hydraulic equations of Table 8-4 because they relate flow intensity in open channels to discharge, slope, and boundary roughness. The sedimentologist must also specify grain size, sedimentary structures, vertical sequence, and channel cross section in an ancient deposit if estimates of paleoflow are to be made. The derivation of the fundamental equations of Table 8-4 is given in all hydraulic texts and by Blatt *et al.* (1980, pp. 98–101). Today, it is hard to read much of the sedimentologic literature without some familiarity with the equations of Table 8-4.

Two studies illustrate many basic ideas.

Gardner (1983) studied a paleochannel and point bar of the Pennsylvanian Harold Sandstone in eastern Kentucky, where exposure is complete. He used crossbedding orientation to obtain true channel width from oblique outcrop sections (Fig. 8-20); bankfull width was estimated by thickness of the entire fining-upward cycle and confirmed by thickness of the lateral accretion surfaces of point bar fill. The continuity equation, $Q = VA$, provided estimates of paleodischarge based on a trapezoidal cross section, and paleovelocity was obtained from breaks in the grain size curve following Middleton (1976) and Bridge (1981). The velocities thus obtained, ranging from about 23 to 45 cm/s, agree fairly well with velocities based on the entrainment curve, bedforms, and bedding (Figures 8-10 and 8-13). Gardner also obtained good agreement between his estimate of paleodischarge from the continuity equation and velocities provided by an entirely different method used by Schumm (1972, pp. 100–102).

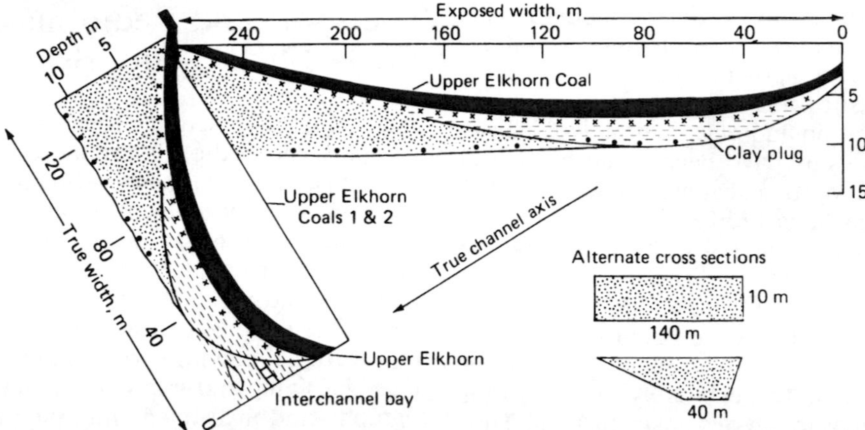

FIGURE 8-20. Abandoned meander in Harold Sandstone (Pennsylvanian) in eastern Kentucky. (Redrawn from Gardner, 1983, Fig. 7).

Gardner's Table 2 summarizes all the different paleostream parameters he estimated and shows their great diversity, although the key ones are discharge, velocity, slope stream length, and drainage basin area. His calculations suggest that a river about 10 m deep and only 140 m wide may have had an upstream drainage basin area of about 30,000 km and a stream length of about 200 to 300 km. Gardner's study is recommended as a source from which to learn needed methods and is desirable to emulate where good exposures exist. A general methology for fluvial deposits is summarized in Figure 8-21.

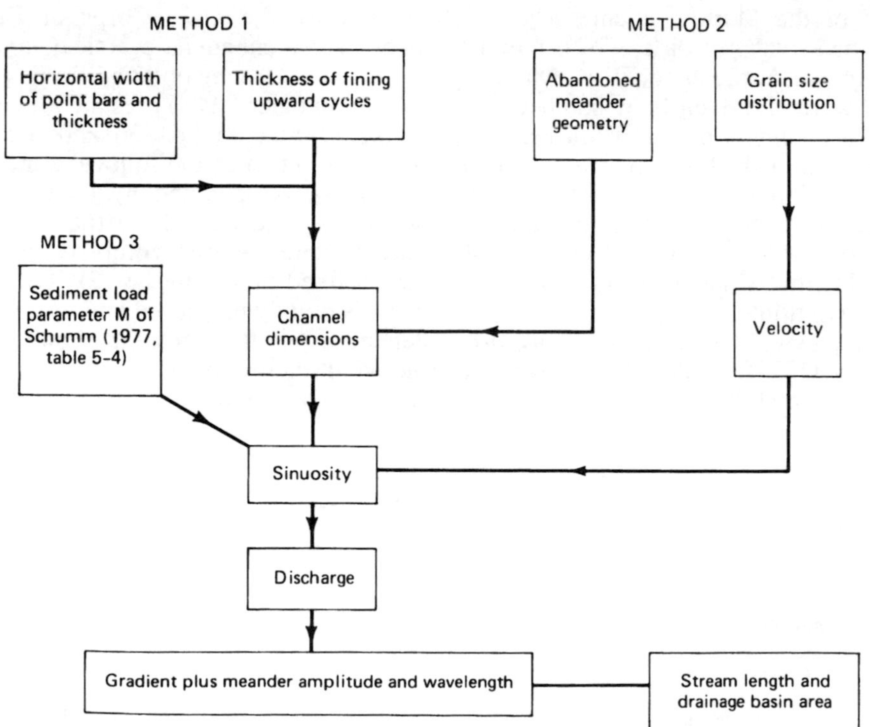

FIGURE 8-21. Three methods for estimating paleohydraulic parameters. (Adapted from Gardner, 1983, Fig. 8).

Compare rivers on Earth, the modern channels of deep-sea fans, and channels on Mars? Not many years ago this would have been a totally fantastic dream but today, thanks to space exploration and paleohydraulics, informative comparisons have been made by Komar (1979). Large Martian channels range up to 1000 km long, many have widths of 20 to 100 km, and consist of a large channel with perhaps two or three minor tributaries. These channels are fairly straight and a few are braided (Table 8-5). Komar's analysis focuses on their formation by water and, this assumption made, he believes Martian channels to be broadly similar to the deeper channels found on subsea turbidite fans. The low effective value of dilute turbidite flows on Earth simulates the gravity field of Mars and is the prime reason he believes that deep channels on Earth seem to offer good analogs to Martian channels. His hydraulic analysis starts with

$$\tau_0 + \tau_i = \Delta\rho g D S$$

where τ_0 and τ_i are bottom and interface shear stress, $\Delta\rho$ is the density contrast between the turbidity current and its overlying water mass, g is gravity, D is flow depth, and S is slope. His analysis uses modifications of the Chezy and Manning equations and is based on an estimate of the slope of the Martian channel; he also assumes a "bed load" of blocks of 0.1 to 1.5 meters in diameter in the channel! Komar concludes that if water depth on Mars had been 100 to 150 m, the channels probably formed in a short time whereas a shallower flow would have required a much longer time. Komar estimates the transport rate of Mangala Channel and, dividing this rate into its estimated eroded volume, concludes that it might have been cut in two years to a century.

Other examples of paleohydraulics are provided by Baker (1973), Miall (1979), and Saunderson and Jopling (1980).

Transport and Bedforms on Sandy Shelves and Beaches

The 1950s saw sedimentologists study sand transport and bedforms produced by unidirectional flow, study that has greatly advanced our understanding of ancient sandstones. Later, sedimentologists began to study how waves and oscillatory currents transport sand and develop bedforms and apply this knowledge to ancient marine shelf and lacustrine sandstones. As with unidirectional flow, theory and experiments are needed. Diving and improved underwater photography and instrumentation plus experimental studies provide the basis for the present understanding of oscillatory wave motion, threshold of sand movement, and sequence of bedforms. We begin with wave theory and conclude with a brief review of shoreline processes.

Initial background reading is provided by Bascom (1980), Barber (1969), Shepard (1973, Chap. 3), and Tricker (1974), and more advanced reading by King (1972, Part II), Morris and Wiggert (1972, Chap. 14), Komar (1976, Chaps. 3 and 4), and Wiegel (1964, Chaps. 2, 7, and 14). Inman's (1963) account is still worth careful attention and *The Shore Protection Manual* by the U.S. Corps of Engineers is highly recommended, practical, and broad in scope, as is Greenwood and Davis (1984). Clifton and Dingler (1984) present a lucid summary of theory and application and are the prime source of much of the following methodology.

A wave is specified by its wavelength, L, height, H, and period, T, the latter being the time needed for one complete wavelength to pass a fixed point (Fig. 8-22). The period of a wave is readily measured and, in a general way, depends upon the size of the body of water and the wind energy imparted to it: wave period is short in lakes and bays, commonly two to four

TABLE 8-5. Flow parameters estimated for Mangala Channel of Mars compared to Earth analogs. (Adapted from Komar, 1977, Table VI).

	Slope	Depth (m)	Stream power[a] ($\tau_0 u$)	Discharge[b]
Mangala Channel of Mars	0.003	10	5×10^5	7×10^6
		50	6×10^6	8×10^6
		100	2×10^7	2×10^7
		150	3×10^7	2×10^7
Mississippi River	0.0005	12	$1\text{–}2 \times 10^4$	$2\text{–}3 \times 10^4$
Deep-sea channels	0.003	100	5×10^5	1×10^7
	0.002	200	1×10^7	3×10^7

[a] ergs/cm^2/s [b] m^3/s

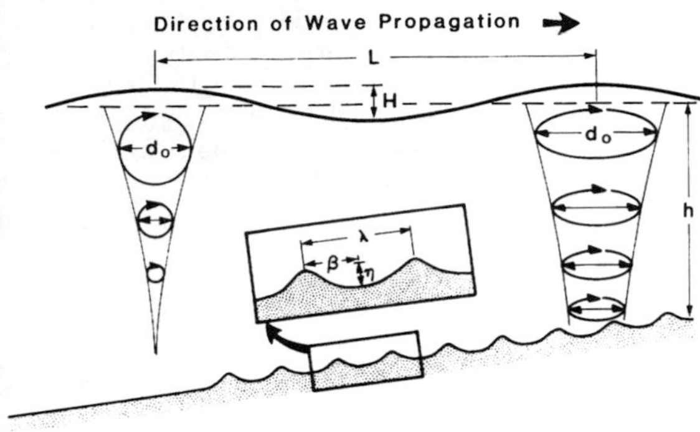

FIGURE 8-22. Definition diagram of a shoaling wave (Clifton and Dingler, 1984, Fig. 1). (Published by permission of the authors and Elsevier Scientific Publishers).

seconds, whereas intermediate wave periods of five to eight seconds are characteristic of the margins of enclosed and semi-enclosed seas. In contrast, waves in large oceans may have periods of ten to fifteen seconds. The great navigators and explorers of the fifteenth and sixteenth centuries used the period of ocean swell to guess at the size of an as yet unexplored sea or ocean. Wave period is all important because, unlike unidirectional flow, the entrainment velocity of a given grain size varies with wave period.

Water waves are of several kinds and include surface waves as well as internal waves between two water layers of contrasting density. Surface waves form when wind blows over water and produces a tangential stress. Wind transmits energy to the water and produces surface waves and current of many types (Table 8-6). Once formed, surface waves can travel long distances, and the longer their period, the faster they travel. There are a variety of supporting mathematical theories that describe the behavior of shoaling waves from deep-water, sinusoidal waves to breakers; wave height, water depth, and period determine which theory is to be applied (Fig. 8-23).

In a shoaling wave, water particles move in elliptical orbits which wave theory relates to wave height, wavelength, period, and water depth. Orbital diameter, d_0, is the diameter of the orbit followed by a water particle beneath the wave's crest. Orbital diameter decreases downward beneath the crest and, near the bottom, the ellipsoid becomes nearly flat. The maximum orbital velocity, U_m, is the peak speed attained by the water particles as they follow this orbit, and ΔU_m is the maximum orbital *asymmetry* or difference in velocity beneath the crest and trough of a wave of a water particle following this ellipsoidal path. Orbital asymmetry of flow occurs when a unidirectional current—perhaps storm surge, a tidal current, or even an intruding ocean current—is superimposed on a wave-generated current. Orbital asymmetry produces *combined flow ripples* (Harms, 1969, p. 366)—those whose external asymmetry and internal lamination indicate a net movement of sand in one direction (Fig. 8-24A). In shoal water Newton (1968) reports landward symmetry of ripple marks—an observation that could be of greater value in paleogeographic studies, although more documentation is desirable. Clifton (1976, p. 132) suggests that asymmetrical ripples develop when ΔU_m is greater than 5 cm/s. Ripple types vary with orbital diameter (Fig. 8-24B). Both Harms *et al.* (1982, pp. 2-25 and 2-41) and Clifton and Dingler (1984) provide full details of ripple types and the procedure and problems of inferring wave condition from them. Both also stress that there remains much to learn.

Entrainment of sand in lakes, seas, and oceans is commonly related to maximum orbital velocity, U_m, rather than τ_0, because U_m can be calculated knowing wave height, wave period, and water depth, whereas τ_0 depends on additional factors. An important additional factor,

TABLE 8-6. Waves and currents near shore and on shelves.

Type	Occurrence	Transport
Oscillatory (wave-generated)		
Symmetric	Inner and outer shelf	None
Asymmetric	Inner shelf and near shore	Shoreward
Unidirectional		
Wave-driven longshore and rip	Near-shore	Parallel to shore
Wind stress	Inner and outer shelf	Seaward
Tidal	Inlets, bays, and restricted shelf	Variable
		Variable, bipolar
Storm surge return	Inner and outer shelf	Seaward
Geostrophic	Outer shelf	Parallel to contours of shelf

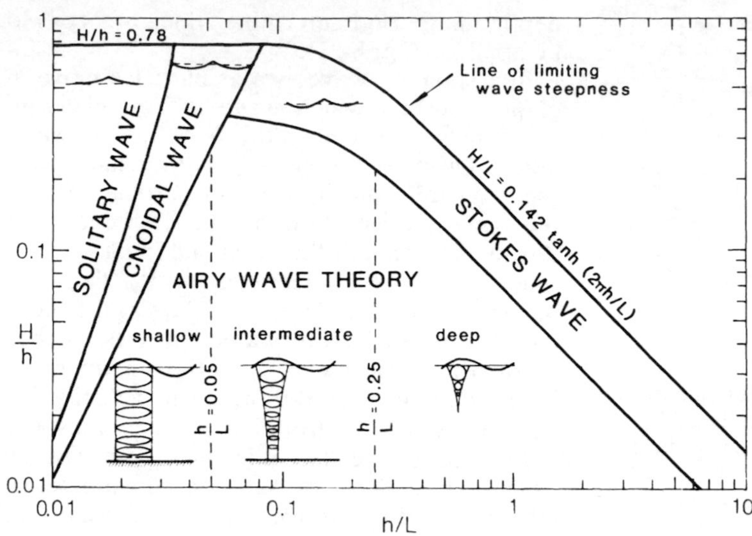

FIGURE 8-23. Applicability of wave theory to variation of water depth and wavelength, H (Clifton and Dingler, 1984, Fig. 9). Published by permission of Elsevier Scientific Publishers.

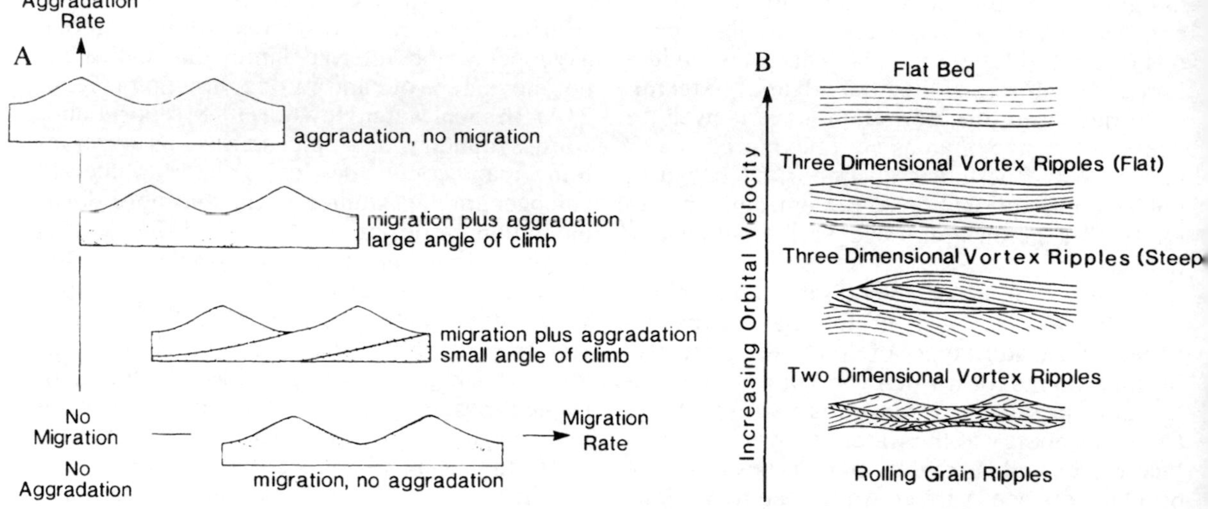

FIGURE 8-24. Wave-formed ripples: A) combined flow ripples and B) variation of ripple types with increasing orbital velocity, U. (Redrawn from Harms *et al.*, 1982, Figs. 3-13 and 3-16).

one not present in unidirectional flow, is wave period, T. In oscillating flow, entrainment depends not only on grain size, d, and U_m, but also on wave period. According to Fig. 8-25, an orbital velocity of about 55 cm/s will ripple coarse sand, but produce flat bedding in very fine sand. Thus bedforms are direction functions of the following:

$$\text{Bedform} = f(d, U_m, \Delta U_m, T)$$

and the general sequence is initial oscillatory laminar boundary layer → development of turbulence → initial motion of some grains → formation of various symmetrical wave ripples → sheet flow (parallel lamination) (Fig. 8-25). This bedform sequence is fundamental to understanding hummocky crossbedding.

Hummocky crossbedding (Fig. 4-9) appears to be widespread in fine-grained to very fine-grained marine and lacustrine sands, especially in the lower shoreface. Although variants exist (Fig. 8-26), the principal type of hummocky crossbedding forms a cycle with a gentle erosional scour at its base and is composed of weakly inclined, poorly oriented laminae capped by ripples and burrows, which may be followed by mud. Sand flutes may be present on this scour surface. Studies of both ancient shelf sandstones (Bourgeois, 1980; Wright and Walker, 1981; Mount, 1982) and modern shelfs

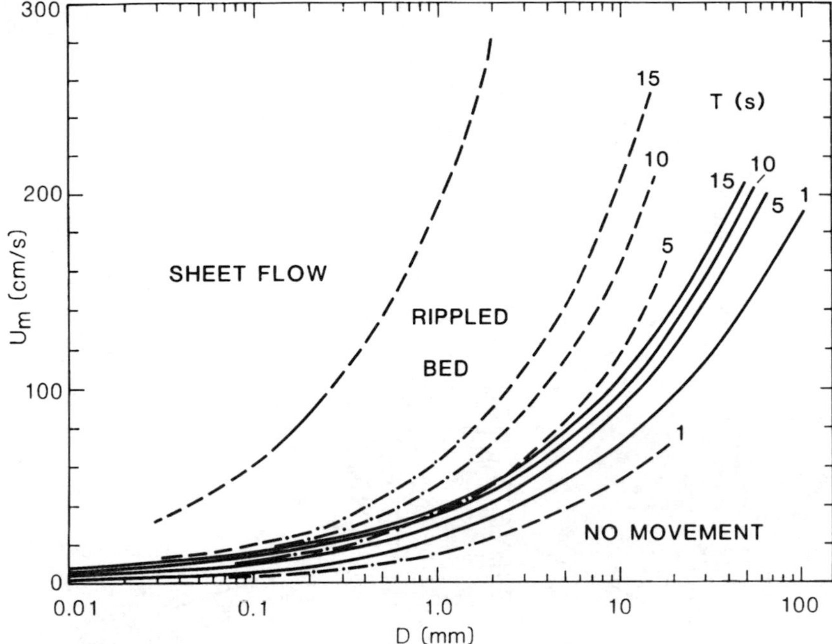

FIGURE 8-25. Entrainment of sand by oscillatory waves depends on wave period as well as orbital velocity (Clifton and Dingler, 1984, Fig. 4). Compare with Fig. 8-13. Published by permission of Elsevier Scientific Publishers.

(Swift *et al.*, 1983) support a storm or storm-related origin for hummocky crossbedding by storm surge and ebb currents: abrupt increase in U_m initially scours the bottom and is followed first by sheet flow (parallel, weakly inclined lamination) and subsequently by ripples as storm-ebb currents decrease with passage of time. Burrowing at the top of the cycle indicates low sedimentation and a return to fair weather conditions. Paleocurrent measurements should always be separated by fair versus storm bedding in such deposits. Care should also be taken to distinguish cyclic sandy bedding of storm surge origin from that of turbidite origin (Valecka, 1984).

Estimating the combinations of wave height (H) and water depth (h) from sedimentary structures and their grain size is still in its infancy but important progress has been made and is well summarized by Clifton and Dingler (1984). Four steps are needed:

1. estimate orbital diameter or U_m from sedimentary structures and grain size using formulas given by Clifton and Dingler (1984),
2. select the correct wave theory (Fig. 8-23),
3. construct water depth–wave depth diagram (Fig. 8-27) with a range of values, and
4. impose restraints from paleogeography and the breaking limits of waves to restrict further the range of h–H values.

An h–H diagram requires several steps and commonly several trial solutions. Some or all of these steps are illustrated by Dott (1974), Clifton (1981), Allen (1981, 1984), and Clifton and Dingler (1984); these and other studies should be reviewed carefully to understand all needed steps before an h–H diagram is attempted.

Two examples illustrate the conclusions from such studies. Studying a Miocene shoreline sandstone from California, Clifton (1981, pp. 176–77) inferred sheet flow deposition in about 8 m of water with an open ocean wave period of 8 to 12 seconds. In Devonian lacustrine sandstones of the Shetland Islands, Allen (1981) estimated water depths to be mostly about 5 meters and, from a short wave period of about 3 to 4 seconds, inferred a lake about 20 km wide. Recognition of vertical sequence defined by grain size, primary structures (parallel lamination, ripple types, hummocky bedding, and crossbedding) plus trace fossils is essential for all such studies.

Large-scale sand waves and ribbons—some more than hundreds of meters long—also are present on most shallow sandy shelves and in many sandy estuaries and are described by a vast, scattered literature. Controlling factors are not fully known but, in a general way, bed-

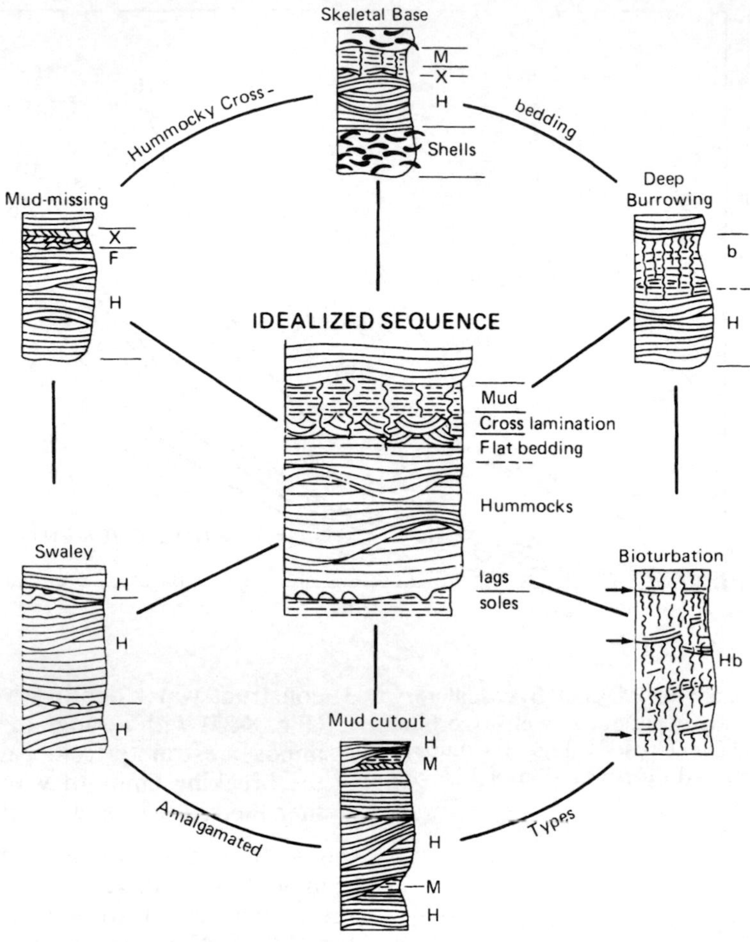

FIGURE 8-26. Varieties of hummocky bedding. (Redrawn from Dott and Bourgeois, 1983, Fig. 1).

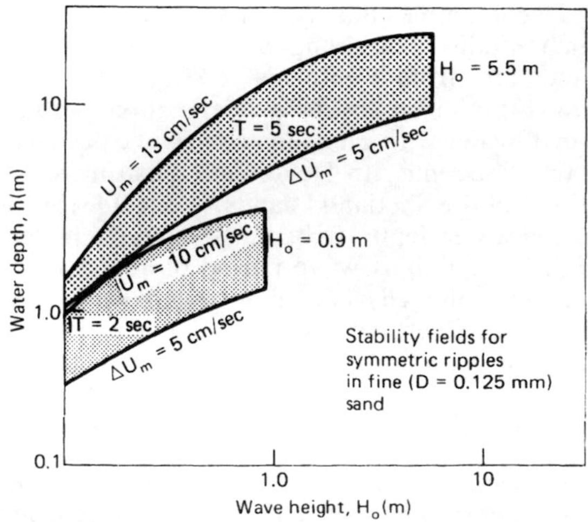

FIGURE 8-27. Probable stability fields for symmetric ripples in 0.125-mm sand for waves with periods, T, of 2 and 5 seconds. (Modified from Clifton, 1976, Fig. 8). Note how fields vary with wave period.

form varies with the velocity of the tidal current (Fig. 8-28). Tidal currents tend to be particularly strong in narrow inlets and straits, but judging from the bimodal crossbedding orientation of some cratonic sandstones and limestone, bipolar reversing currents existed far onto cratons. Notable early studies of modern sand waves include those of Cloet (1954), Reineck (1963), and Stride (1963). A few of many later studies are those by Belderson *et al.* (1972), Boggs (1974), So *et al.* (1974), Werner and Newton (1975), Swift (1975), Swift *et al.* (1979), and Boersma and Terwindt (1980). There is also a very complete summary with excellent illustrations provided by Vanney (1977, p. 178–227). Both longitudinal and transverse bedforms are present and their properties are summarized in Tables III and IV of Vanney. Helical motion produces long sand ribbons and longitudinal sand ridges (Vanney, 1977, Fig. 116).

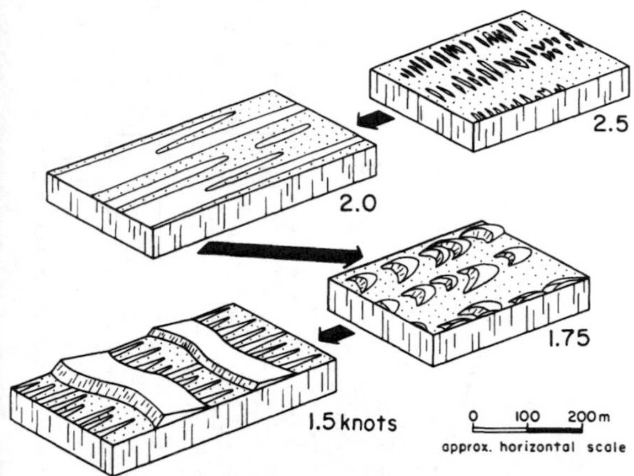

FIGURE 8-28. Bedforms and currents in European tidal seas (Redrawn from Kenyon, 1970, Fig. 3).

The sand waves described above produce prolifically crossbedded tidalites and marine shelf sandstones, almost all of which are well-winnowed and well-sorted. Some crossbeds can be very thick and exceed 10 meters. Many shelf sandstones are supermature, especially on ancient cratons. Bimodal transport prevails where ebb and flood times are of equal magnitude, but where one exceeds the other, unimodal transport prevails. Whether bi- or unimodal, the modes nearly always indicate depositional dip, either transport across the shelf or up or down a sandy estuary, or up and down the axis of a sandy tidal inlet. On the other hand, paleocurrents parallel to depositional strike can exist in a sandy bay whose long axis parallels shoreline. Hence attention should probably always be directed to the strongest currents—those of tidal inlets and flood deltas. Generally, however, there appears to be net transport toward the shelf edge (Fig. 8-29).

The term *tidalite* is used to describe marine and shoreline sands that respond in part to tidal influences, and thus such sands can belong to both subtidal and intertidal realms. Klein (1972) estimated tidal range from the thickness of a tidal sequence, and Allen (1981a,b) has estimated tidal velocities based on spacing of mud drapes on foresets in tidally formed crossbeds. Allen suggests that the mud drapes represent deposition from slack water in the tidal cycle. Tidal range based on thickness of a tidal cycle has proven controversial and tidal velocities are still speculative. Tidalites, however, are now widely recognized in ancient sandstones.

Swift (1970), writing about the development of a shelf, has built on the classic work of Johnson (1919, pp. 199–271). As summarized by Swift (1970, p. 6), an equilibrium shelf profile has the shape of a concave-upward, exponential curve whose steepest portion is at the shore. Seaward both slope and grain size become less. The profile shows a good correlation between grain size and slope just as does a graded stream. This is the classic idea largely inferred from studies of ancient sediments. If normal wind-induced waves are to accomplish grading, water depth should probably not exceed 20 m, the depth at which wave refraction patterns are first commonly observed. For deeper shelves, Swift (1970, p. 12) favors the idea that storm-generated wave drift is a major factor in sand dispersal and that it acts in an essentially random manner, moving sediment relatively short distances, first one way and then another, as it passes. Certainly crossbedding in many ancient marine sandstones is more variable in direction than most fluvial and deltaic crossbedding (Chap. 4) and may in fact partially be the response of storm-induced bedforms. Current directions produced by hummocky crossbedding, for example, are diverse. One should note, however, that in many marine sandstones the net crossbedding vector is commonly downslope, as inferred from paleogeographic evidence. Given a linear sediment source at the shoreline, a largely random movement of sand

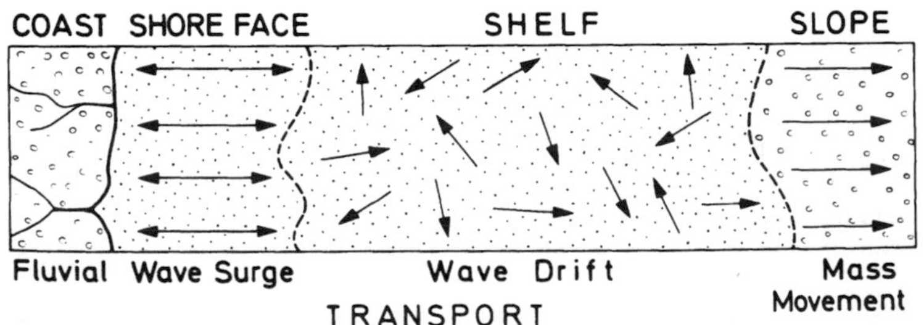

FIGURE 8-29. Diagrammatic model of the different transport processes on shelves and their relation to the shoreline. (Modified from Swift, 1970, Fig. 5).

on the shelf, and a linear sink at the shelf break, as shown in Fig. 8-30, a net transport of sand across the shelf by some type of weakly anisotropic diffusion process seems necessary (Swift, 1970, pp. 12–13), although return surges from storms may be the major factor. While details are not fully clear, some type of progressive sorting mechanism whereby smaller size fractions are eventually concentrated downslope in the direction of the shelf break seems also needed. Donahue *et al.* (1966, Fig. 2) found some evidence to support the idea of an incipient equilibrium shelf off the New Jersey coast, where a thin layer of sand, probably in part reworked from older relict sediments, becomes finer seaward. Keulegan and Krumbein (1949) wrote an early theoretical paper on the equilibrium profile of a graded shelf. McCave (1985) provides a comprehensive review of both the deposits and processes of Recent shelf sands.

Beaches are also of great interest, because here waves are converted to surf, and sorting and abrasion are maximal (Fig. 8-31). Petrologists have long discussed how feldspar and weak rock fragments may be differentially eliminated by this abrasion (Mack, 1978). Sand movement on beaches has been much studied and is reviewed by Komar (1976, Chaps. 3 and 4), and a good overview is that of Inman and Brush (1973). Field and laboratory studies have shown that in addition to the oscillatory movement of grains, storms tend to remove sand from the beaches whereas long swells tend to add sand to it. Sand shuttles back and forth on the beach as a result of both continual wave action and variations in weather and, depending upon the magnitude of the longshore current, will move laterally along it. Repeated shuttling promotes good sorting and intense abrasion, the

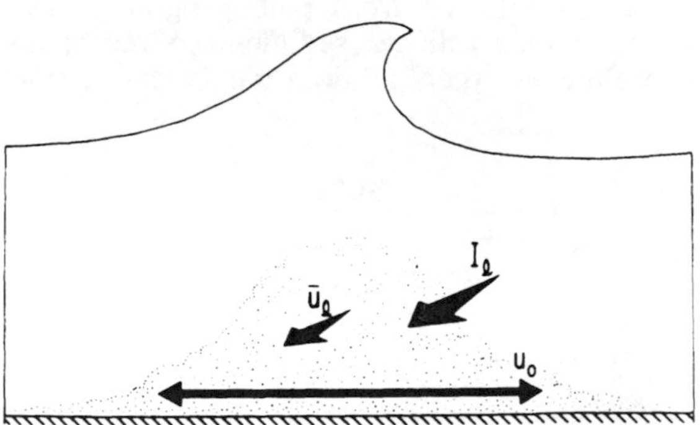

FIGURE 8-30. Sand transport in the surf zone is a function of both wave turbulence and the long shore current. (From Komar, 1976, Fig. 8-3). Published by permission of John Wiley and Sons.

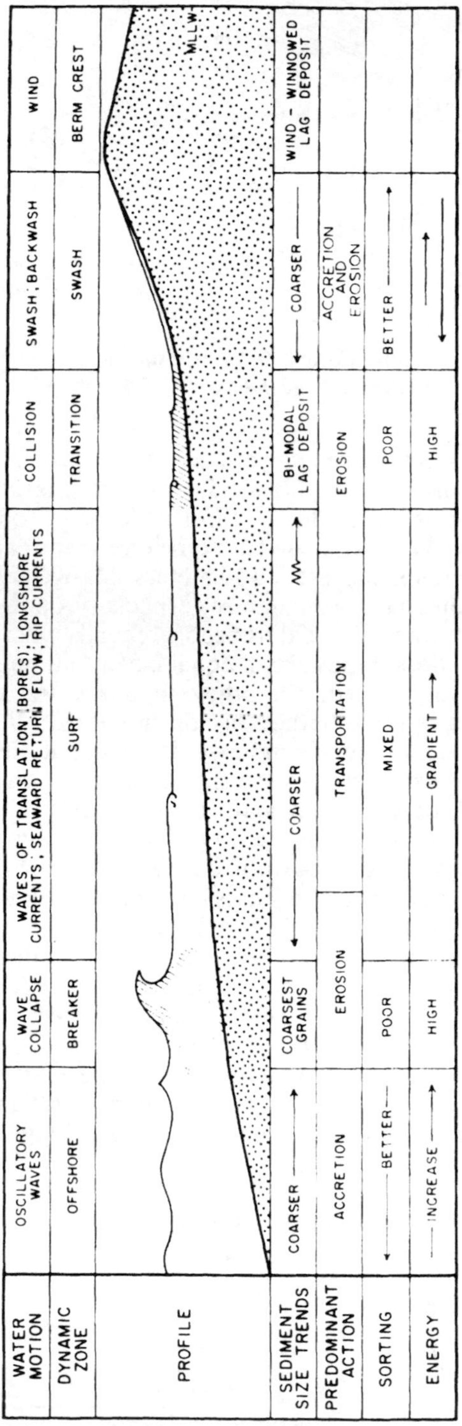

WATER MOTION	OSCILLATORY WAVES	WAVE COLLAPSE	WAVES OF TRANSLATION (BORES); LONGSHORE CURRENTS; SEAWARD RETURN FLOW; RIP CURRENTS	COLLISION	SWASH; BACKWASH	WIND
DYNAMIC ZONE	OFFSHORE	BREAKER	SURF	TRANSITION	SWASH	BERM CREST
PROFILE						MLLW
SEDIMENT SIZE TRENDS	COARSER →	COARSEST GRAINS	← COARSER ⟿→	BI-MODAL LAG DEPOSIT	← COARSER	WIND - WINNOWED LAG DEPOSIT
PREDOMINANT ACTION	ACCRETION	EROSION	TRANSPORTATION	EROSION	ACCRETION AND EROSION	
SORTING	← BETTER —	POOR	MIXED	POOR	BETTER →	
ENERGY	— INCREASE →	HIGH	— GRADIENT →	HIGH	↕	

FIGURE 8-31. Summary of geometry, water motion, dynamic zones, and sedimentation characteristics of a beach. Hachured areas are zones of high suspended grain sedimentation. MLLW = mean lower low water (Ingle, 1966, Fig. 116). Reproduced by permission of Elsevier Scientific Publishers.

latter perhaps eliminating mechanically weak particles such as argillaceous rock fragments. Slope of the beach face varies with grain size and wave energy (Wiegel, 1964, Fig. 14-17).

Waves play an important role in sediment transport along the beach, because they put sand into suspension and thus permit transport by a longshore or tidal current which by itself may be too weak to entrain sand. In other words, wave action supplies the energy to put sand in motion and a net current moves it (Fig. 8-30). The greater the shear stress imparted by the wave, the more sand that goes into suspension and the greater the transport. An additional result of waves impinging on a shore at an angle is the longshore drift that is a consequence of grains being thrown up by swash at an angle and returning back with backwash normal to the shore.

Just as with streams, engineers have attempted to formulate theories of bedload transport for beaches. Examples of this approach are provided by Inman and Bagnold (1963) and Abou-Seida (1965), the latter using a modification of Einstein's (1950) bedload theory first proposed for streams. For some California beaches Bowen and Inman (1966, p. 8) obtained a semi-empirical formula for longshore transport rate, S_t (measured in cu. ft/s),

$$S_t = 1.13 \times 10^{-4} p \qquad (8\text{-}18)$$

where p is the longshore component of wave power and depends on some nine variables including density of the water, gravity, wave height and period, angle the breaking waves make with the shore, etc. Bowen and Inman's approach not only brings out the very large number of variables involved, but emphasizes the added difficulty of predicting transport rate rather than simply analyzing forces on single grains. Changes in beach topography and the growth of shore features such as spits and bars are directly related to longshore currents and their intensity. Relevant theory is scant, although some work has been done. Longshore transport is one of the most important processes by which sand may be carried long distances parallel to, rather than down, a paleoslope.

Seaward dipping, well-defined lamination formed by the swash zone of beaches, has a characteristic inverse grading which grades upward from fine, heavy-mineral-rich to coarse, heavy-mineral-poor sand (Clifton, 1969), an internal texture that helps distinguish beach lamination from some other types. Other sedimentary structures are numerous and are summarized by Komar (1976, pp. 363–84) and Reineck and Singh (1980, pp. 360–68) and include trace fossils. Paleocurrents are complex and, as in other environments, should always be related to subfacies.

Subaqueous Gravity Flows

Density differences caused by fine suspended sediment, temperature, or salinity produce currents down submarine slopes which, once started, can travel far. A *density current* is defined as the movement under gravity of a stream of fluids under, through, or over another fluid, the density of which differs by a small amount from that of the primary current (Committee on Sedimentation, 1963, p. 78). When suspended sediment, usually fine clay, produces the enhanced density, the term *turbidity current* is used. Turbidity currents are strong enough to carry sand and gravel, and the sand may be terrigenous, skeletal debris, or even detrital pyrite! Hot ash flows of volcanic origin produce ignimbrites and are another variety of turbidites, the suspended ash producing the enhanced density. Turbidity currents also existed in some dense magmas, judging by their primary structures (Parsons and Butterfield, 1981).

Deposits of turbidity currents, called *turbidites*, are widely recognized and occur in lakes, seas, and oceans but are most important along the margins of deep marine basins. The essential requirement is an influx of muddy water moving downslope below wave base. The flooding river of a delta, storm waves impinging on a coast line, high swell, or perhaps the oversteepening of a shelf margin may initiate turbidity currents. One important factor to remember is that it is unlikely that turbidity currents can get started and continue on wide, shallow shelves, where wave action would destroy separate, far-traveling turbid layers.

Today, turbidity currents are recognized as but one end-member of a wide spectrum of subaqueous gravity flows, which differ from one another in fluid behavior and sediment concentration (Fig. 8-32). To a large degree, sedimentary structures reflect these different processes. Turbidity currents that transport sand, fluidized sediment flows, grain flows, and debris flows are all *sediment surges*—processes that act intermittently, when equilibrium upslope is disrupted. Slower turbidity currents that transport mostly fine clays may flow more continuously, however.

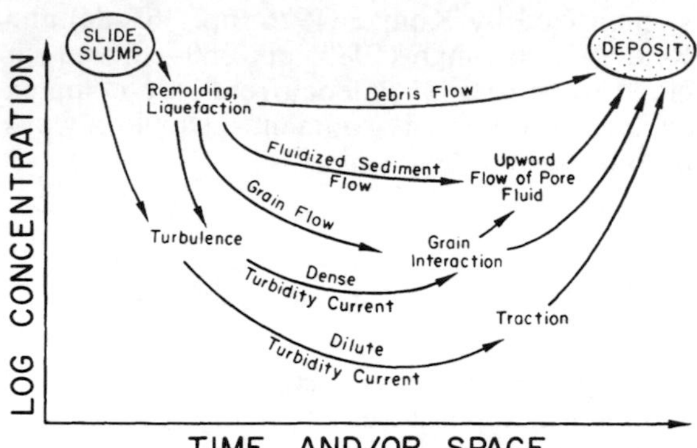

FIGURE 8-32. Classification of subaqueous gravity flows that transport sand. (Redrawn from Middleton and Hampton, 1973, Fig. 1).

Turbidity currents were first recognized in Switzerland in Lake Constance (Forel, 1885) and subsequently in many other reservoirs, and an early, comprehensive study was made by Gould (1951) on Lake Mead. Field observations, in modern lakes and oceans and in many ancient sandy basins, experiments (Kuenen, 1973; Middleton, 1966a, 1966b, 1967; Kuenen, 1970), and analytical and computer studies (Komar, 1977) have all contributed to our present understanding as have studies by engineers, oceanographers and sedimentologists, and micropaleontologists (identification and mapping of displaced faunas). Simpson (1982) provides a detailed review.

Figure 8-33 shows diagrammatically how a turbidity current forms in fresh water lakes and reservoirs, the simplest case. The river current is sufficiently slowed in velocity as it enters the lake so that it does not mix completely with the surrounding lake water. Insofar as it does not mix, it is maintained as a separate layer of water with a significantly higher density than the clear lake water. Under the force of gravity the turbid layer then moves downslope along the floor of the lake until it loses momentum, either by being stopped at the dam forming the reservoir, as in the case of Lake Mead, or by dissipation by frictional losses after a flat bottom is reached and the current spreads out in all directions. As the current loses velocity, its suspended material gradually settles out. The nature of the fluid flow is complex because it depends on the density of the suspension, the slope and roughness of the bottom, the amount of turbulence and mixing with the overlying layer, and the size distribution of the suspended material, the coarser grains of which settle out of the flow as its velocity decreases.

We can describe the conditions under which a two-dimensional turbidity current will flow as

$$\tau_0 + \tau_i = \Delta\rho g t \alpha \tag{8-19}$$

where τ_0 is the shear stress at the bottom, τ_i is the shear stress between the turbidity current and the overlying layer, $\Delta\rho$ is the density difference between the two layers, called the *effective density*, t is the thicknessof the flow, and α the slope of the bottom. The current will flow when the impelling gravitational force caused by the density difference operating over a slope exceeds the shear stresses on the fixed boundary below and the moving interface with the layer above. The depth of the turbidity current is assumed to be small with respect to the depth of the layer above. The interface of the two layers is assumed to be sharp and smooth without significant mixing.

Most workers who have developed theory for turbidity current flow have concentrated on the characteristics of a steady, uniform flow, though Middleton (1966a) has worked on the velocity of the head of the flow. For the steady uniform flow case, using a Chézy type of equation developed for flow in open channels (Leopold *et al.*, 1964, pp. 156–59), Middleton (1966b, p. 628) gives

$$U = C'\sqrt{\frac{\Delta\rho}{\rho} t\alpha} \tag{8-20}$$

where u is the velocity of uniform flow of the turbidity current, C' is a modified Chézy coefficient (a term combined from specific weight and flow resistance that relates mean velocity to hydraulic radius and slope), $\Delta\rho$ is the effective density, ρ is the density of the flow, t is flow thickness, and α is the bottom slope. Middleton discusses ways of estimating C' from bed roughness, Reynolds number, and the flow resistance offered by the fluid interface above the current. Estimation of the resistance of the fluid interface is complicated because the interface goes through stages from sharp and smooth, to wavy, to breaking waves, with turbulent mixing at the interface increasing through all these stages as the flow intensity increases. Keulegan (1949) proposed flow conditions for the formation of waves and mixing at the interface to be dependent on the *densimetric Froude number*, F_ρ, where

$$F_\rho = \frac{U}{\sqrt{\frac{\Delta\rho}{\rho} gt}} \tag{8-21}$$

and refers to gravitationally induced waves at the interface in an analogous way to the defini-

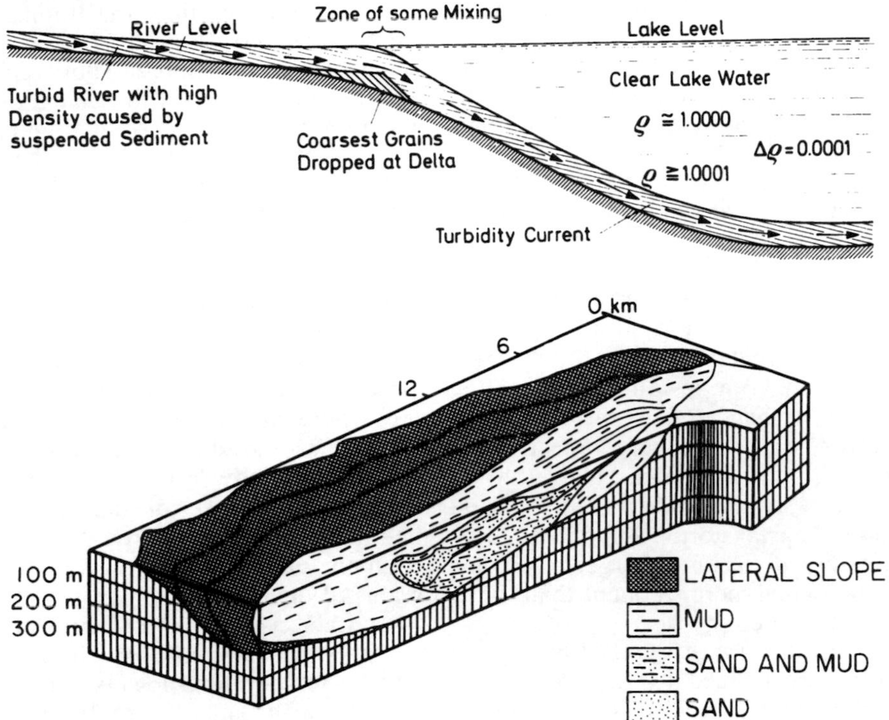

FIGURE 8-33. Above: Thc formation of a turbidity current by a stream with a high concentration of suspended sediment entering a lake. Below: Turbidity flow in Lake Constance in Switzerland. (Redrawn from Houbolt and Jonker, 1968, Fig. 1).

tion of F for free surface flow of water under air. The stability of the interface, θ, was measured by Keulegan in terms of F_ρ and Reynolds number

$$\theta = \frac{1}{(F_\rho{}^2 R)^{1/3}} \tag{8-22}$$

Keulegan experimentally determined an average critical value for θ of 0.18 for the turbulent range of flows. If θ exceeds this value, no mixing should occur.

It is now possible using Eqs. (8-19) and (8-22) to see how a turbidity current might behave as the effective density and slope are varied. The greater the suspended load and thus the greater the effective density, the faster the current moves. As the slope increases, the densimetric Froude number, F_ρ, increases and so controls the stability of the interface and, through its effect on flow resistance of the interface, the Chézy coefficient, which affects the velocity. The faster and more turbulent the flow, the more mud, silt, and sand can be kept in suspension. But at the same time, the greater the density and viscosity of the suspension, the more turbulence is damped, so that there is an optimal Reynolds number for maximum current competence and capacity. As the velocity increases, so does F_ρ and mixing at the upper surface of the flow; as the critical value of θ is exceeded, waves form at the interface, eventually breaking and mixing more turbulently with the overlying clear water. The mixing decreases the effective density, offers more flow resistance, and so acts to dissipate the current. Thus a turbidity current traveling down a slope is poised between a lower bound of velocity, turbulence, and F_ρ too low to keep much coarse material in suspension and an upper bound of the same parameters that would destroy the current by extensive mixing with the overlying water. When a turbidity current meets the flat floor of a lake or an abyssal plain of the ocean, it starts to decay as its inertial momentum is lowered by internal and boundary resistance, turbulence lessens, and suspended material gradually sediments as the laws of settling velocity become dominant. More complicated formulas are reviewed by Komar (1977) and Lüthi (1980). Komar (1985) applied one criterion of suspension, $w/U_* < 1$ proposed by Middleton (1976), to vertical variation of grain size and structures in a single turbidite bed without complete success. Thus much still remains to be learned.

Most surge-type currents are based on the concept of a turbidity current having a head, neck, and body (Fig. 8-34). Computer simulation shows how velocity and thickness of the head and body probably vary down a long channel (Fig. 8-34).

As shown by the vast sediment accumulations at the bases of many submarine canyons, much sediment, especially clay, is carried from the shelf to the mouths of the canyons, where it accumulates in subsea fans. Current meters in such canyons (Shepard and Marshall, 1979) show that most currents rarely exceed 50 cm/s, and that they flow both up- and down-canyon and commonly transport fine silt and mud. These currents seem to largely reflect tides. Faster currents, up to 200 cm/s, believed to be turbidity currents, transport down-canyon sand and respond to coastal storms and flooding rivers. Currents in canyons near larger rivers are commonly faster and more frequent than those of sediment-starved coasts.

Turbidity currents are recognized by Bouma cycles, which are defined by an upward sequence of sedimentary structures and companion decrease in grain size (Fig. 4-13). In the

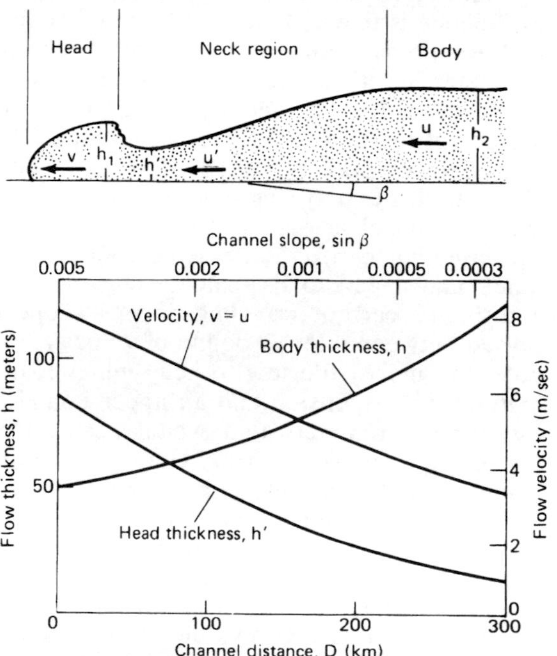

FIGURE 8-34. Above: Head, neck and body of a turbidity current on slope β. (Redrawn from Komar, 1977, Fig. 1). Below: Downslope changes in velocity and thickness of head and body inferred from computer simulation. (Redrawn from Komar, 1977, Fig. 8).

downcurrent direction, the Bouma sequence is base truncated (Fig. 8-35). The Bouma sequence has been related to flow regime and time (Fig. 8-36). Allen (1982b, Chap. 10) provides a recent summary of current thinking about the origin of Bouma sequences but, because experimentation on a scale to produce vertical sequences of structures is virtually impossible, interpretation remains speculative. Flute marks, however, have been studied experimentally (Dzulynski and Walton, 1963).

Two types of upward size decline are recognized—*distribution grading*, where the entire size fraction fines upward, and *coarse-tail grading*, where only the coarsest grains fine upward. *Reverse size* grading, coarsening upward, also exists, especially in volcaniclastic and skeletal turbidites, because of density segregation—light and hydraulically rough glass shards, for example, have a smaller fall velocity than either grains of quartz or feldspar (p. 219).

Another type of bottom current are *contour currents*, which result from thermohaline circulation—temperature-density differences—on a rotating earth. These currents develop along the western sides of deep oceanic basins near the continental rise. Contour currents are believed capable of transporting fine sand and forming well-sorted, cross-laminated sands. Hollister and Heezen (1972) initially set forth the characteristics of contour currents and compared them with turbidites. A more recent review is that of Stow and Lovell (1979), whose summary tables show that the recognition of contourites in the ancient is not easy. Ideally the transport direction of contourites will be at ninety degrees to that of the episodic turbidity currents flowing downslope.

Middleton and Hampton (1976) provided an overview of three additional gravity-related processes that transport sand—fluidization, grain flow, and debris flow, although none are as important as turbidity currents. Other terminology used is reviewed by Carter (1975) and Lowe (1982), who both provide insight into processes.

Fluidization of sand occurs when water flowing vertically in a bed of sand is strong enough to transport upward at least some grains, and produces narrow channels, pillars, and even sand volcanoes. Dish structure (Fig. 4-24) is another water escape structure (see Lowe, 1975, for an extended discussion). Fluidization probably is not a major factor in sediment transport, because excess porewater pressures are soon dissipated, but it probably does contribute to convolute lamination, ball-and-pillow struc-

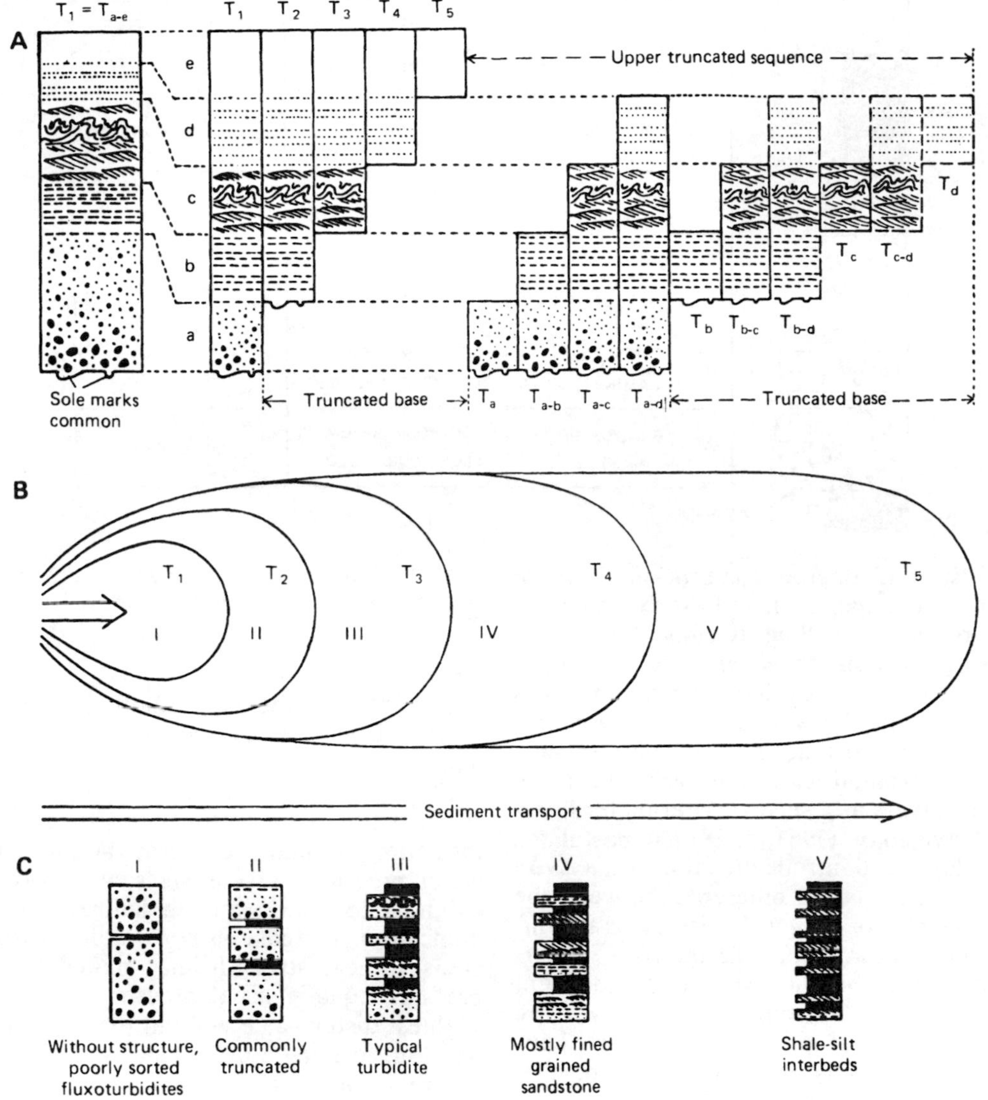

FIGURE 8-35. Early schematic representation of base truncated Bouma sequence and changing proportions of conglomerate sandstone and shale in a turbidite (Einsele, 1963, Fig. 6).

tures, overturned crossbedding, and slumps. Allen (1982b, Chap. 8) discusses fluidization at length. Liquification is also caused by upward flow of water and is the same process except it represents an unsteady state.

Grain flow is believed to result from the dispersive pressure between grains, wet or dry, an idea early advocated by Bagnold (1954). Grain flow is mass emplacement, is common on the slip slopes of dunes, and seems to be a local feature. Characteristic sedimentary structures are not yet well established. Middleton and Hampton (1976, pp. 204–207) provide a good summary.

Debris flows are both subaerial and subaqueous and are episodic, sluggish, downslope movements of poorly sorted debris, somewhat similar to wet concrete (Middleton and Hampton, 1976, p. 209). Large boulders can be floated or rafted by the matrix even on low slopes of one to two degrees. Debris flows are common on both alluvial and subsea fans. Johnson (1984, Chap. 8) provides an analytical model.

Specific weight and shear between layers are two other factors that provide insight into soft sediment deformation. Two adjacent layers are unstable if the upper has a larger specific weight than the lower ($\gamma_u > \gamma_l$), where γ is the weight per unit volume. Differences in specific weights may result from (1) original differences in pack-

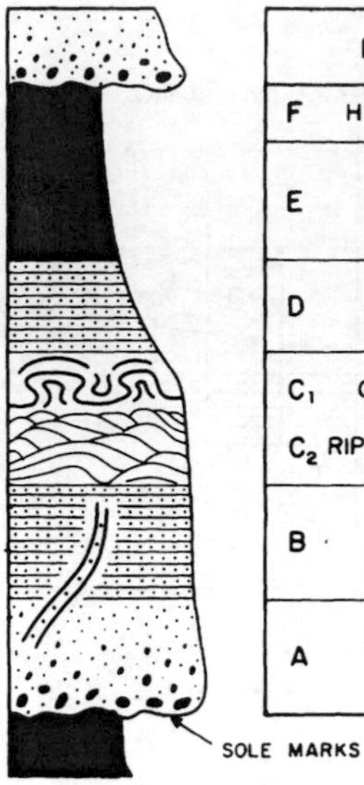

DIVISIONS	INTERPRETATION	POSSIBLE TIME
F HEMI - PELAGIC	INTERTURBIDITE	1 TO 1000 YEARS
E PELITIC	LOW DENSITY TAIL OF TURBIDITY CURRENT	DAYS TO MONTHS
D UPPER LAMINATED	LOWER FLOW REGIME, PLANE BED	HOURS
C_1 CONVOLUTED C_2 RIPPLE - LAMINATED	LOWER FLOW REGIME	
B LOWER LAMINATED	UPPER FLOW REGIME, PLANE BED	
A GRADED OR MASSIVE	NON EQUILIBRIUM FLOW, QUICK BED	MINUTES

SOLE MARKS

FIGURE 8-36. Bouma sequence, flow regime, and possible time span. (Modified from Hesse, 1975, Fig. 1).

ing (porosity), (2) degree of saturation (the more saturated, the smaller γ), and (3) expansion of clays, producing swelling. Because the contact surface of two adjacent layers rarely defines a perfect plane, there are always random deviations in elevation between high and low points at the interface so that one can relate specific weight, differential elevation, and shear between the layers as a second criterion for instability. Artyushkov (1963, p. 28) has postulated that one layer will intrude the other when $\Delta\gamma\Delta e > \tau_b$, where $\Delta\gamma$ is the difference between the specific weights of the two layers, Δe is the difference of elevation along the interface, and τ_b is the shear *between* the layers, that is, tangent to their interface. The dimensions of $\Delta\gamma\Delta e$ are $ML^{-1}T^{-2}$, those of pressure or shear. Thus when $\Delta\gamma\Delta e$ exceeds τ_b, deformation results. This analysis assumes that the shear strengths of both beds, σ_u and σ_l, are much greater than $\Delta\gamma\Delta e$, the latter generally being small.

Shear strength, σ_s, of a sediment is given by

$$\sigma_s = C + (P_n - P_w) \tan \phi, \quad (8\text{-}23)$$

where C is cohesion resulting from physicochemical bonding between particles and is closely related to composition, P_n is the component of pressure normal to the interface, P_w the porewater pressure (given by ρgh, h being the height of water that would be measured if a manometer were inserted into the bed), and ϕ the angle of internal friction, which is a measure of the mechanical resistance to sliding and overturning of one particle with another. The internal friction angle depends on particle shape, sorting, and relative bulk density of the material; the greater the density, the greater ϕ. Wetting has little effect on ϕ in sands but may be important in some clays, notably the "quick clays." Cohesion will generally increase with diagenesis.

The *effective pressure*, P, is defined by $P_n - P_w$. It is the major factor in most soft rock deformation through short-term variation in porewater pressure and, as we saw previously, is essential to understanding fluidization and liquification. Increase in porewater pressure decreases shear strength and, should porewater pressure equal normal pressure, $\sigma_s = C$, the slightest disturbance will put the sand into suspension, forming a quicksand. Alternatively, if failure occurs, perhaps because of undercutting due to river bank erosion or tidal currents, there is an increase of porewater pressure associated with strain beyond failure which thus causes liquifaction (Andresen and Bjerrum, 1967, p. 226). If the sand goes partially into suspension, shear strength is reduced and pressure from the overlying bed will exceed $\sigma_{s_{max}}$, promoting additional flowage. Should complete suspension occur on a slope, a viscous, high-density flow, such as a debris flow or a turbidity current, will form.

Fine grained sand and siltstones are very susceptible to soft sediment deformation (Fig. 8-37) primarily because of their low permeability, that is, they cannot adjust quickly to exterior pore water pressure changes due perhaps to variation of water depth in the basin, to sudden change in groundwater flow or to sudden deposition of a layer of sand on top of an uncompacted bed. Via specific weight, gravity is the

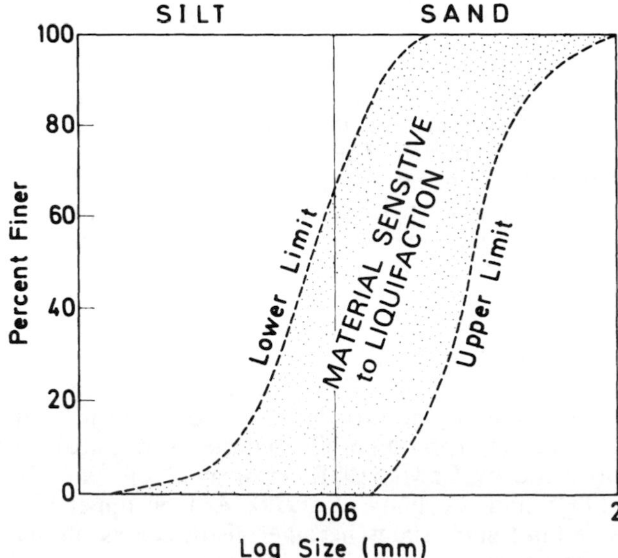

FIGURE 8-37. Size limits of silt and sand most commonly found in slides. (Modified from Andresen and Bjerrum, 1967, Fig. 8).

driving force wherever the upper layer subsides into the underlying one, as in load cast formation. Pore water overpressure may well have been responsible for reduction of shear strength, however. Soft sediment deformation is favored where sedimentation is rapid (normal compaction does not have time to squeeze out pore water), where sands and muds are interstratified (contrasts in specific weight), where the sands have porosities in excess of 44 percent (Andresen and Bjerrum, 1967, p. 226), and where initial slopes are high.

Overturned crossbedding, crossbedding whose foresets are overturned in the downcurrent direction prior to deposition of the overlying unit, also results from local upward groundwater seepage reducing τ_c, the critical drag force needed to entrain grains, so that the drag force of the current overturns the foresets downcurrent. Underseepage has also been offered as one explanation of different bedforms in stream reaches that otherwise appear to be fluid dynamically homogeneous: change in porewater pressure changes τ_0 and thus may alter bedforms.

Slides in subaqueous slopes of loose sand and silt can occur on slopes as low as 1° and 2°, although the steeper the slope the more likely the failure. Morgenstern (1967, Table 1) cites some earthquake-induced slumps with slopes of 3.5 to 10°. Andresen and Bjerrum (1967, Fig. 1) recognized two types of slides: retrogressive flow slides and spontaneous liquifaction (Fig. 8-38). The former starts in the lower part of a

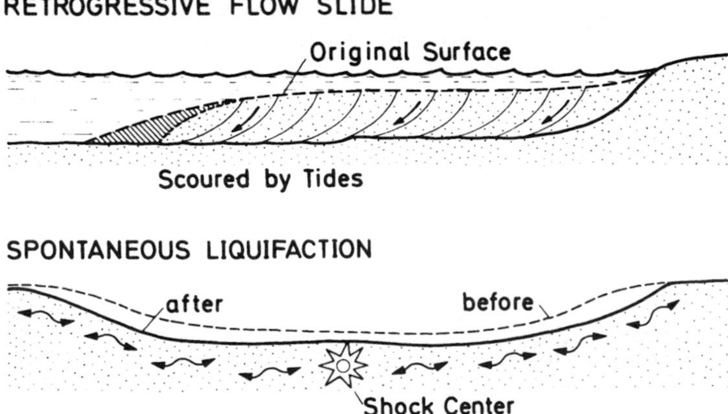

FIGURE 8-38. Retrogressive flow and spontaneous liquifaction slides. (Modified from Andresen and Bjerrum, 1968, Fig. 1).

slope and expands, slip by slip, updip because of liquifaction associated with strain beyond failure. In this kind of liquifaction, the sand flows away, leaving an unsupported face, and the process continues until more resistant material is encountered. The potential of a sand for liquifaction *after* the initial slip is critical to the formation of a retrogressive flow slide. Spontaneous liquifaction slides, on the other hand, are caused by an initial shock so that liquifaction is believed to start at one point and spread in all directions uphill as well as down. Resedimentation follows the initial collapse.

Submarine slides are logical precursors for large, catastrophic turbidity currents. Description and interpretation of submarine slides are provided by Lewis (1971), Prior and Coleman (1979), and Nardin *et al.* (1979). At their upper size limit such slides become olistostromes of giant proportions that are easily recognized on seismic cross sections.

Wind

Wind is a geologically important agent in sedimentation, for it transports vast quantities of sand in continental deserts, carries inland the terrigenous or carbonate sands of beaches that are supplied by longshore currents, and carries silt and clay as dust storms far from their source and supplies much of the pelagic sediment of the deep oceans. Hence wind transport is a major part of sedimentology, even though it perhaps has received less attention than it deserves. General reviews are given by Raudkivi (1976, pp. 46–53), Scheidegger (1970, pp. 383–401), the Committee on Sedimentation (1965), and Warren (1979). Sundborg's (1956) article is also very helpful. Much of the following is based on their work and thus indirectly on the classic study of Bagnold (1941). The exploration of Mars has, with the discovery of its dune fields, reawakened a general interest in eolian processes (Sagan and Bagnold, 1975; Iversen and White, 1982) and in the recognition of ancient eolianites on Earth.

Wind transports sand markedly differently than it does silt and clay. Sand travels close to the ground and mostly by successive jumps whereas silt and clay travel long distances in suspension (Table 8-7).

When wind reaches a critical velocity over loose, dry sand, grains begin to roll and accelerate and after a few centimeters may jump into the air, travel many times their diameter, and finally return to the surface in a long parabolic path. When the grain moves upward into the faster-flowing air, it acquires additional energy and moves as a projectile at about the velocity of the wind. The impact angle, α, is commonly between 10° and 16° and is given by $\tan \alpha = U_g/U$, where U_g is the terminal velocity of the grain and U is wind velocity. Successive jumps of a grain are called saltation. The saltating sand of sand storms, in contrast to the silt and clay, has a clearly defined upper limit, usually about a meter. This upper limit will vary depending upon the surface; the harder the surface, the higher the grains bounce, and the softer, the lower. As a rough approximation 13 mph (537 cm/s), has been suggested as the minimum velocity needed to entrain dry sand. When a grain returns to the surface it may bounce back up, its impact may entrain another grain, or the grain may simply roll forward. Surface creep refers to the bedload transport of sand and has been estimated to account for about 25 percent of total sand transport. It is surface creep that permits grains too large to move by saltation to move downwind. Surface creep and saltation, like many other sedimentary processes, are gradational to one another.

Although saltation and surface creep occur in both air and water transport, saltation is much more pronounced in air. Kalinske (1943, p. 47) estimated the saltation jump in water to be only about 1/800 of that in air. It is the density contrast between air and water that is chiefly responsible for the greater role of saltation in air, which in turn alters some aspects of eolian sand

TABLE 8-7. Calculated flight time and height of particles[a] (Committee on Sedimentation, 1965, Table 2-1.3).

Diameter mm	Fall velocity cm/s	Flight time	Distance	Maximum height
0.001	0.00824	9–90 yr.	4–40 × 10^6 km	6.1–61 km
0.01	0.824	8–80 yr.	4–40 × 10^2 km	61–610 m
0.1	82.4	0.3–3 s	46–460 m	0.61–6.1 m

[a] For a wind of 1500 cm/s.

wave morphology, for at 18°C the density of water is 869 times the density of dry air.

The drag force obtained from Eq. (8-7) permits us to obtain an estimate of the differing momentum and kinetic energies imparted to grains by air and water transport. For entrainment of the same grain in the two media, their drag forces must be equal so that

$$C_{D_a}(\rho_a U_a^2/2) = C_{D_w}(\rho_w U_2^2/2). \quad (8\text{-}24)$$

But C_{D_a} and C_{D_w} are roughly the same because D_D is a function of Reynolds number and changes little for spheres for R values of 10^{-2} to 10^{-5} (Raudkivi, 1975, Fig. 2-1), so we can write

$$\rho_a U_a^2/2 = \rho_w U_w^2/2 \quad (8\text{-}25)$$

so that the velocities must be, as a first approximation, in the ratio of

$$U_a/U_w = \sqrt{\rho_w/\rho_a} = 29.3 \quad (8\text{-}26)$$

to obtain the same drag force. Thus the relative momentum of an airborne grain of a sand grain with mass M is 29.3 M times greater in air than in water. The corresponding kinetic energy is $(29.3)^2\ M/2$ or approximately 430 times greater in air than in water. This greater kinetic energy of windblown sand explains the stronger abrasion by wind than by water transport that Kuenen (1960) and others have reported. On the basis of experiments, Kuenen (1960, p. 442) estimated sand to lose weight 100 to 1000 times faster by eolian rather than water transport. He also suggested that the much smaller absolute viscosity of air—only 1.73×10^{-6} of that of water—means that cushioning in air prior to impact is also minimal. Presumably, the greater rounding that has been attributed by some to eolian sands also stems from the higher kinetic energy of saltation transport by wind. Studies with the SEM also suggest a distinctive pattern of surface marks (Krinsley *et al.*, 1976, p. 130). The greater kinetic energy of wind-driven sand is also responsible for the ventifacts cut by it. When saltating grains of high kinetic energy hit others on the ground, the added momentum may be enough to cause entrainment. This has been called the *impact threshold velocity*.

Von Engelhardt (1940, Figs. 3 and 4) suggested a difference in hydraulic equivalence between quartz and magnetite and quartz and garnet for wind-versus water-deposited sands. Hand (1967) has modified this idea and tested it with some success using New Jersey beach and dune sands, as has Bart (1977, pp. 171–74).

An entrainment function for sand has been investigated in most detail by Bagnold (1941, Fig. 29), Chepil (1945), Zingg (1953) and Iversen and White (1982). This function is concave upward and has a minimum value for grains having diameters of about 0.075 mm. The increase in shear velocity for smaller particles is the result of their greater cohesion.

Dune and dust storms have also been recognized on Mars and estimates of wind speeds for entrainment in the Martian atmosphere have been made. Imagery of Mars has shown that its dune fields are similar to those on Earth.

Study of eolian transport in modern dunes has distinguished three different processes—grain fall, grain flow, and traction (Table 8-8), which Hunter (1977, 1981) describes in detail as do Kocurek and Dott (1981). They report that the structure that appears to be most diagnostic of eolian transport is a type of climbing ripples called *subcritical climbing ripples* (Hunter, 1981, Fig. 2). These differ from their subaqueous counterparts in three ways: they are thinner and more uniform, are only weakly cross-laminated with few visible foresets, and are inversely graded (Fig. 8-39). Small-scale bedding structures such as subcritical climbing ripples are attractive as possible discriminators of eolian transport, because they can be used bed by bed in the field, as was done by Glennie (1970, Fig. 36), or when studying a core as can *adhesion ripples* (Fig. 8-40), which form when windblown sand comes to rest on a wet surface.

As with water transport, the genesis of eolian bedforms is complex and is far from understood

TABLE 8-8. Processes that form eolian stratification. (Simplified from Hunter, 1977, Table 1).

Grain fall

Settling from air caused by flow separation at crest of transverse bedforms produces low, inclined laminae or beds of both large and small scale. Conformable base, tabular. Dip angle typically 20–30°, but ranges from 0 to 40°.

Grain flow

Gravity transport on slip slope produces tongue-shaped bodies 2 to 5 cm thick with discordant bases in dry sand and microslides and slumps in wet sand. Dry grain flows may be crudely cross-stratified. Inverse grading in dry sand flow except at toe. Dip angle typically 28–34°. Open packing.

Traction

Saltation and creep produce smooth (plane) bed lamination and ripples of several types of which climbing ripples are common. Angle of climb depends on ratio of ripple migration to rate of sedimentation. Typical thicknesses range from 1 to 15 mm and inverse sizing is common. Dip angle is variable but typically 5 to 20°. Close packing.

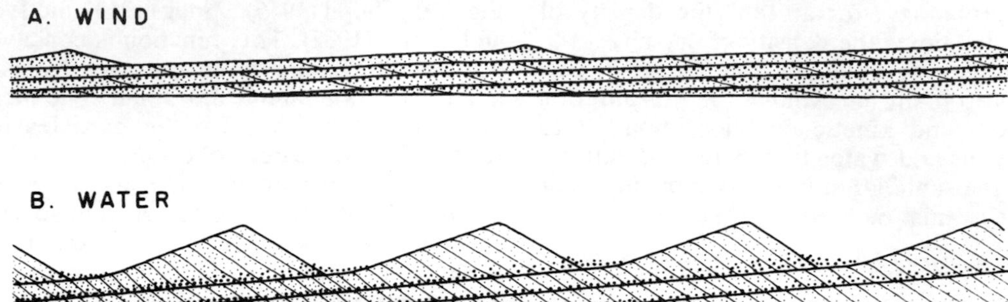

FIGURE 8-39. Contrasting climbing ripples—wind versus water according to Kocurek and Dott (1981, Fig. 3). Published by permission of the Society of Economic Paleontologists and Mineralogists.

but, because depth of flow is greater, large heights are not uncommon—some up to 200 m! Like their aqueous counterparts, eolian bedforms occur in hierarchies and there is a relationship between grain size and wavelength (Wilson, 1972, Fig. 2). Bedform morphology can also be thought of as a function of wind strength and stability, surface materials, and topography. McKee (1979) gives much new data and examines all aspects of eolian transport, bedforms, and structures.

Eolian bedforms have proven difficult to classify and McKee (1979, Table 1) recognizes 11 major types, although linguoid, crescentic, and meandering bedforms, as well as networks and linear forms, are probably the most common.

FIGURE 8-40. Adhesion ripples on bedding plane of the eolian Cedar Mesa Sandstone (Permian) in southeastern Utah (Loope, 1984, Fig. 12B). Published by permission of the Society of Economic Paleontologists and Mineralogists.

Fabric

When sand is deposited, be it by surface creep, saltation, or suspension, it acquires a preferred orientation which, unless disturbed by animal activity or slumping, is preserved in consolidated sandstones. Rees (1968) and Hamilton *et al.* (1968) have considered fabric from the viewpoint of Bagnold's theory (1956). Although more details of its application to fabric are needed, Bagnold's theory does seem particularly appropriate, because it deals with forces on aggregates of grains rather than on single ones. Below we briefly restate the salient points of Rees and Hamilton.

The fluid force impelling a shearing mass of sand can be resolved into a shear stress τ in the plane of shear (parallel to depositional surface) and a dispersive pressure P normal to it. The dispersive pressure depends on τ and the ratio of τ/P depends on the size and density of the grains, the viscosity of the intragranular fluid, and the intensity of shearing of the grains. Observationally, we know that the longest axis of a grain lies in the plane of τ and its shortest is parallel to P. As a result, the long axis that defines the imbrication angle of a sand grain can be considered as a function of the ratio of τ and P, its angle with the plane of shear being $\tan^{-1} \tau/P$. Bagnold (1956, p. 243) found $\tan^{-1} \tau/P =$ 0.32 for grain suspensions in a fully turbulent flow and $\tan^{-1} \tau/P = 0.75$ for laminar flow, both values being for low-viscosity media such as water or air. The former corresponds to a value of 18° and the latter to 37°. An imbrication of 18° from the depositional surface closely approximates that observed in the grains of most sandstones, a striking agreement between theory and observation, and thus lends support to

Bagnold's theory of dispersive pressure generated by the flow of granular materials. Allen (1982a, pp. 175–236) gives a complete description.

No one has yet formulated a quantitative theory on the complete orientation—in contrast to the imbrication—of aggregates of grains in unidirectional flow, but observational information is available on grain shape orientation in relation to current flow (Potter and Pettijohn, 1977, pp. 51–55, 70–72).

Glossary

ANTIDUNE Symmetrical sand and water surface waves that occur in trains and which are in phase and may move upstream, downstream or remain stationary (Simons *et al.*, 1965, pp. 40–41). Characteristic of upper flow regime.

APPARENT VISCOSITY The viscosity of a clay-water mixture that is greater than that of the clear water alone.

BOUNDARY LAYER Zone where the fluid is appreciably retarded by the frictional resistance of the boundary. May be turbulent throughout or may have a laminar sublayer next to boundary.

BURSTING Name for poorly understood flow variation in turbulent boundary layer and believed to be responsible for at least some lamination.

COHESION Roughly speaking, the bonding forces between particles. Appreciable for clays but virtually absent for sand and most silt.

COMBINED FLOW RIPPLES Ripples produced by a combination of currents and waves (Harms, 1969, p. 387).

CONTOUR CURRENT Proposed by Hollister and Heezen (1972, p. 60) for currents that flow parallel to the strike of the continental slope and transport fine sand and silt. Contour currents produced by thermohaline ocean circulation.

CREEP Forward movement of sand grains caused by impact of a saltating grain upon one lying on the bed.

CRITICAL TRACTIVE FORCE, τ_c Shear stress exerted by a fluid on particles of a given size necessary to start movement. Same as critical shear stress.

CRITICAL VELOCITY That velocity at some height above the bed necessary to initiate grain movement.

DENSIMETRIC FROUDE NUMBER, F_ρ Used for a flow with an interface with a fluid of slightly lower density, such as a turbidity flow, and defined by $F_\rho = \dfrac{U}{\sqrt{\dfrac{\Delta\rho}{\rho} gd}}$.

DEPTH OF FLOW, D Roughly the same as average depth, but is commonly restricted to be the depth to the top of a dune or ripple field. Depending on the point of view, one may have as many "depths of flow" in a stream as "velocities."

DISPERSIVE PRESSURE, P Proposed by Bagnold as an important factor in transport of bedload material. Acts normal and upward to bed and is generated by shearing of moving, colliding grains.

DRAG FORCE The force of frictional resistance offered by a fluid to a body moving through it. Also the frictional force exerted by a fluid moving over a fixed boundary.

DUNE A large sand wave formed by a fluid flow over a granular bed. Has gentle slope upcurrent and steep slip slope downcurrent and thus a roughly asymmetric triangular cross section. Related to but larger than ripples.

DYNAMIC VISCOSITY, μ Molecular internal resistance to flow, a kind of internal friction characteristic of the particular fluid. Defined by shear stress divided by the rate of shear or deformation.

EDDY VISCOSITY, η Internal resistance of a fluid to motion due to turbulence. Defined by shear stress divided by the average rate of turbulent shear and so a function of turbulence rather than the kind of fluid.

EFFECTIVE DENSITY, $\Delta\rho$ The difference between the density of a turbidity flow and that of the static body of fluid under which it flows.

EFFECTIVE PRESSURE, P' Defined as the difference between normal overburden pressure and porewater pressure. The principal variable determining shear strength in saturated sands.

FALL VELOCITY, w Same as settling velocity. Rate at which a particle falls through a fluid. Depends on density, diameter, and shape of particle plus density and viscosity of water.

FLAT BED See plane bed.

FLOW INTENSITY Measured in different ways—discharge, average velocity, friction velocity, etc.

FLOW SEPARATION General flow separates from boundary as a result of abrupt change in geometry of boundary. In area of separation there are many turbulent eddies which make no net contribution to forward movement. Common downcurrent from bottom obstacles such as dunes.

FLOW REGIME Concept developed by engineers for streams and defined by a range of flows having similar bedforms, resistance to flow, and mode of transport.

FLUIDIZATION Water flows vertically in a sand bed and is strong enough to produce channels, pillars, and sand volcanoes.

FORM DRAG The resistance to a flow resulting from the deformation of the boundary layer, typically by ripples and dunes in streams. Also called form resistance.

FORM ROUGHNESS Roughness of boundary generated by microrelief such as small ripples. Form roughness may be included as an important part of form drag.

FRICTION VELOCITY, U_* Defined as $\sqrt{\frac{\tau_0}{\rho}}$, it appears in many fluid dynamic equations. Same as shear velocity. Applies to effects on boundary only.

FROUDE NUMBER, F Dimensionless ratio of inertial to gravitational force, defined as $\frac{U}{\sqrt{Dg}}$ where U is velocity, D is commonly water depth, and g is the force of gravity. Also see densimetric Froude number.

GRAIN FLOW A pseudolaminar structural flow of a sheared mass of water with a high concentration of dispersed grains. A variety of bedload movement inferred from ancient deposits (Stauffer, 1967, p. 500).

HELICAL FLOW Rotating flow, parallel to average transport direction, that occurs in both air and water and produces linear grain fabrics and sandy ridges parallel to average flow direction.

HYDRAULIC EQUIVALENCE Two grains are said to be hydraulically equivalent if they have the same fall velocity. Commonly given with respect to a quartz sphere as a standard.

HYDRAULIC RADIUS, R_h Cross-sectional area divided by wetted perimeter. Commonly taken as depth in rivers.

INTERNAL ANGLE OF FRICTION, ϕ Depends on shape, sorting, and density of material; the denser the packing, the greater ϕ. Wetting to saturation apparently has little effect on ϕ.

KINEMATIC VISCOSITY, v Dynamic viscosity divided by density.

LAMINAR FLOW Fluid moves in parallel sheets or laminae with no mixing between layers.

LAMINAR SUBLAYER A basal zone of laminar flow in the boundary layer.

LIQUIFACTION Sand loses all strength because porewater pressure in the beds equals or exceeds gravity forces between grains.

PLANE BED A smooth, firm sand bottom lacking ripples or dunes and thus having a low resistance to flow. Maximum grain size determines roughness.

POREWATER PRESSURE Defined as ρgh, where h is hydraulic head. Has a decisive bearing on shear strength. If porewater pressure equals or exceeds normal pressure, sand goes into suspension.

RESISTANCE Force exerted by the boundary on the fluid flow.

REYNOLDS NUMBER, R Dimensionless ratio of inertial force to internal viscous retarding force. Depending on geometry and roughness, in most natural flows if R is large (>2000), flow is turbulent; if it is small (<500), flow is laminar.

RIPPLE Small sand waves, transverse to flow, that may be symmetrical or asymmetrical depending on whether formed by currents or waves.

ROUGHNESS Loosely speaking, the microtopographic "relief" on a fluid's boundary due either to grains, sand waves, or debris.

SALTATION The jumping behavior of grains that are temporarily entrained in a flow and then fall back to the bed.

SAND WAVE A constructional ridge on the bottom formed by current movement of silt, sand, or gravel; ridge is commonly normal to flow. Includes ripples and dunes.

SCOUR Net removal of sediment from interface.

SHEAR STRENGTH, σ_s Component of force tangent to a boundary and defined as $\sigma_s = C + P' \tan \phi$ where C is cohesion, P' the effective pressure, and ϕ internal angle of friction. Shear strength is the critical value, which when exceeded, results in deformation.

SHEAR STRESS, τ Acts parallel to bed to transport particles and is defined as the product of dynamic viscosity and rate of shear or deformation. It is also equal to γSD where γ is the specific weight of the fluid, S the slope of uniform flow, and D the depth of flow. The force per unit area that acts parallel to the bottom.

SHEET FLOW Refers to the parallel lamination produced by waves at higher values of orbital velocity, U_m, but threshold is grain size dependent (Fig. 8.24).

SHOOTING FLOW A type of fluid flow in which gravity waves are swept downcurrent, and characterized by Froude numbers greater than 1.

SPECIFIC WEIGHT, γ A ratio of weight to volume defined as product of density and gravity.

STREAMLINE Defined by tangent to path of moving fluid particle.

SURFACE CREEP The generalized bedload movement of grains, many of which are too heavy to saltate.

TRACTION CARPET Coined by Dzulynski and Sanders (1962, p. 88) to describe the dense zone of saltation at the base of a turbulent flow.

TRACTIVE FORCE, τ_0 A shear stress exerted on the boundary of a particle by a moving fluid.

TRANQUIL FLOW A type of fluid flow in which Froude numbers are less than 1. Gravity waves will be propagated upcurrent. Also called streaming flow.

TURBIDITY CURRENT A density current flowing in consequence of the load of suspended sediment it is carrying, the load giving it the excess density.

TURBULENT FLOW Characterized by irregular flow streamlines with much lateral and vertical mixing.

UNIFORM FLOW Velocity (or mean velocity) does not vary with time in a cross section.

VISCOSITY Internal resistance of a fluid to flow. See dynamic, eddy, and kinematic viscosity.

WASH LOAD Material carried in a stream by suspension, commonly clay and fine silt.

References

Abou-Seida, M.M.: Bedload function due to wave action. Univ. California Hydraulic Eng. Lab. Tech. Rept. Hel-2-11, 58 pp. (1965).

Allen, G.P.: Etude des processes sédimentaires dans l'estuaire de la Gironde. L'Institut de Géologie du Bassin d'Aquitaine, Mémoire V, 310 pp. (1973).

Allen, G.P., and Klingebiel, A.: La sédimentation estuarienne: exemple de la Gironde. Bull. Centre Rech. Pau-SNPA 8, 263–293 (1974).

Allen, J.R.L.: Current ripples, 433 pp. Amsterdam: North Holland 1968.

Allen, J.R.L.: Sand waves: A model of origin and internal structure. Sedimentary Geology 26, 281–328 (1980).

Allen, J.R.L.: Lower Cretaceous tides revealed by cross-bedding with mud drapes. Nature 289, 579–581 (1981a).

Allen, J.R.L.: Paleotidal speeds and ranges estimated from cross-bedding sets with mud drapes. Nature 293, 394–396 (1981b).

Allen, J.R.L.: Sedimentary structures, Vol. 1, 593 pp. Amsterdam–Oxford–New York: Elsevier Scientific Pub. Co. 1982a.

Allen, J.R.L.: Sedimentary structures, Vol. 2, 663 pp. Amsterdam–Oxford–New York: Elsevier Scientific Pub. Co. 1982b.

Allen, P.A.: Wave-generated structures in the Devonian lacustrine sediments of southeast Shetland and ancient wave conditions. Sedimentology 28, 369–379 (1981).

Allen, P.A.: Miocene waves and tides in the Swiss Molasse. In: Greenwood, B., and Davis, R.A., Jr. (Eds.): Hydrodynamics and sedimentation in wave-dominated coastal environments. Marine Geology 60, 455–473 (1984).

Andresen, A., and Bjerrum, L.: Slides in subaqueous slopes in loose sand and silt. In: Richards, A.F. (Ed.): Marine geotechnique, pp. 221–229. Urbana, Illinois: Univ. Illinois Press 1967.

Artyushkov, Ye. V.: O vozmozhnosti vozniknoveyiya I obshchikh zakonomernostyakh razvitiya knvektivnoy neustoychivosti v osadochnykh porodakh (Possibility of convective instability in sedimentary rocks and the general laws of its development) Dokl. Akad. Nauk SSSR 153, 162–165 (1963).

Baba, Jumpei, and Komar, P.D.: Measurements and analysis of settling velocities of natural quartz sand grains. Jour. Sed. Petrology 51, 631–640 (1981).

Bagnold, R.A.: The physics of blown sand and desert dunes, 265 pp. London: Methuen 1941.

Bagnold, R.A.: Experiments on a gravity-free dispersion of large solid spheres in a Newtonian fluid under shear. Roy. Soc. London Phil. Trans., Ser. A, 225, 49–63 (1954).

Bagnold, R.A.: Flow of cohesionless grains in fluids. Royal Soc. London Phil. Trans., Ser. A, 249, 235–297 (1956).

Bagnold, R.A.: Deposition in the process of hydraulic transport. Sedimentology 10, 45–56 (1968).

Bajard, Jacques: Figures et structures sédimentaires dans la zone intertidale de la partie orientale de la Baie du Mont-Saint-Michel. Rev. Géog. Phys. Géol. Dynamique 8, 39–111 (1966).

Baker, Victor R.: Paleohydrology and sedimentology of Lake Missoula flooding in eastern Washington. Geol. Soc. America Spec. Paper 144, 79 pp. (1973).

Bart, H.A.: Sedimentology of cross-stratified sandstone in the Arikaree Group, Miocene, southeastern Wyoming. Sedimentary Geology 19, 165–184 (1977).

Barber, N.F.: Water waves, 142 pp. London: Wykeham Pub. 1969.

Bascom, Willard: Waves and beaches (revised and updated), 366 pp. Garden City, New York: Anchor Press/Doubleday 1980.

Belderson, R.H.; Kenyon, N.H.; Stride, A.H.; and Stubbs, A.R.: Sonographs of the sea floor. A picture atlas, 185 pp. Amsterdam: Elsevier 1972.

Bird, E.C.F.: Coastal landforms: An introduction to coastal geomorphology with Australian examples, 193 pp. Canberra: Australian National University Press 1964.

Blatt, Harvey; Middleton, Gerard; and Murray, Raymond: Origin of sedimentary rocks, 2nd Ed., 782 pp. Englewood Cliffs, New Jersey: Prentice-Hall 1980.

Blodgett, R.H., and Stanley, K.D.: Stratification, bedforms and discharge relations of the Platte braided river system, Nebraska. Jour. Sed. Petrology 50, 139–148 (1980).

Boersma, J.R., and Terwindt, J.H.J.: Neap-spring tide sequences of intertidal shoal deposits in a mesotidal estuary. Sedimentology 28, 151–170 (1980).

Bogardi, Janos: Sediment transport in alluvial streams, 826 pp. Budapest: Akademiai Kiado 1974.

Boggs, S., Jr.: Sand-wave fields in Taiwan Strait. Geology 2, 251–253 (1974).

Boothroyd, J.C. and Hubbard, D.K.: Genesis of bedforms in mesotidal estuaries. In: Cronin, L.E. (Ed.): Estuarine research, Vol. II, Geology and engineering, pp. 217–234. New York: Academic Press 1975.

Bourgeois, Joanne: A transgressive shelf sequence exhibiting hummocky stratification: The Cape Sebastian Sandstone (Upper Cretaceous), southwestern Oregon. Jour. Sed. Petrology 50, 681–702 (1980).

Bowen, A.J., and Inman, A.D.: Budget of littoral sands in the vicinity of Point Arquello, California. U.S. Army Coastal Eng. Research Center, Tech. Memo, 19, 41 pp. (1966).

Bridge, J.S.: Origin of horizontal lamination under turbulent boundary layers. Sedimentary Geology 20, 1–16 (1978).

Bridge, J.S.: Hydraulic interpretation of grain size distribution using a physical model for bedload transport. Jour. Sed. Petrology 51, 1109–1124 (1981).

Briggs, L.J.; McCullock, D.S.; and Moser, F.: The hydraulic shape of sand particles. Jour. Sed. Petrology 32, 645–656 (1962).

Cant, Douglas J.: Bedforms and bar types in the South Saskatchewan River. Jour. Sed. Petrology 48, 1321–1330 (1978).

Carter, R.M.: A discussion and classification of subaqueous mass-transport with particular application to grain-flow, slurry-flow, and fluxoturbidites. Earth Sci. Rev. 11, 145–177 (1975).

Chapman, R.E., and Wright, E.P.: Geology and water: An introduction to fluid mechanics, 228 pp. Boston: Kluwer 1981.

Chepil, W.S.: Dynamics of wind erosion, I. Nature of movement of soil by wind; II. Initiation of soil movement; III. The transport capacity of wind. Soil Sci. 60, 305–370, 397–471, 475–480 (1945).

Chow, V.T. (Ed.): Handbook of applied hydrology, various paging. New York: McGraw-Hill Book Co. 1964.

Clifton, H. Edward: Beach lamination: nature and origin: Marine Geology 7, 553–559 (1969).

Clifton, H. Edward: Wave-formed sedimentary structures—a conceptual model. In: Davis, R.A., Jr., and Ethington, R.L. (Eds.): Beach and nearshore sedimentation. Soc. Econ. Paleon. Mineral. Spec. Pub. 24, 126–148, 1976.

Clifton, H.E.: Progradational sequences in Miocene shoreline deposits, southeastern Caliente Range, California. Jour. Sed. Petrology 51, 165–184 (1981).

Clifton, H.E., and Dingler, J.R.: Wave-formed structures and paleoenvironmental reconstruction. In: Greenwood, B., and Davis, R.A., Jr. (Eds.): Hydrodynamics and sedimentation in wave-dominated coastal environments. Marine Geology 60, 165–198, 1984.

Cloet, R.L.: Sand waves in the southern North Sea and in the Persian Gulf. Jour. Inst. Navigation 7, 272–279 (1954).

Colby, B.R.: Discharge of sands and mean-velocity relationships in sand-bed streams. U.S. Geol. Survey Prof. Paper 462-A, 1–47 (1964).

Coleman, James M.: Brahmaputra River: Channel processes and sedimentation. Sedimentology (Spec. Issue) 3, 129–239 (1969).

Collinson, John D.: Bedforms of the Tana River, Norway. Geog. Annaler 52A, 31–56 (1970).

Committee on Sedimentation: Sediment transport mechanics: density currents. Am. Soc. Civil Eng. Proc., Jour. Hydraulics Div. 89, 77–87 (1963).

Committee on Sedimentation: Wind erosion and transportation. Am. Soc. Civil Eng. Proc., Jour. Hydraulics Div. 91 Hy 2, Paper 4261, 267–287 (1965).

Committee on Sedimentation: Nomenclature for bed forms in alluvial channels. Am. Soc. Civil Eng. Proc., Jour. Hydraulics Div. 92, Paper 4823, 51–64 (1966).

Committee on Sedimentation: Erosion of cohesive sediments. Am. Soc. Civil Eng. Proc., Jour. Hydraulics Div. 94, Paper 6044, 1017–1049 (1968).

Cornish, Vaughn: On sand-waves in tidal currents. Geog. Jour. 18, 170–202 (1901).

Dalrymple, R.W.; Knight, R.J.; and Lambaise, J.J.: Bedforms and their hydraulic stability relationships in a tidal environment, Bay of Fundy, Canada. Nature 275, 100–104 (1978).

Donahue, J.G.; Allen, R.C.; and Heezen, B.C.: Sediment size distribution profile on the continental shelf off New Jersey. Sedimentology 7, 155–159 (1966).

Dott, R.H., Jr.: Cambrian tropical storm waves in Wisconsin. Geology (Boulder) 2, 243–246 (1973).

Dott, R.H., Jr., and Bourgeois, Joan: Hummocky stratification: Significance of its variable bedding sequences: Reply. Geol. Soc. America Bull. 94, 1249–1251 (1983).

Dzulynski, S., and Sanders, John: Current marks on firm mud bottoms. Connecticut Acad. Arts Sci. Trans. 42, 57–96 (1962).

Dzulynski, S., and Walton, E.K.: Experimental production of sole markings. Trans. Edinburgh Geol. Soc. 19, 279–305 (1963).

Einsele, Gerhard: Über Art und Richtung der Sedimentation im klastischen rheinischen Oberdevon (Famenne). Abh. hess. L.-Amt. Bodenforsch. 43, 1–60 (1963).

Einstein, H.A.: The bedload function of sediment transport in open channel flows. U.S. Dept. Agriculture Tech. Bull. 102, 71 pp. (1950).

Embleton, C.; Thornes, J.B.; and Warren, A.: The nature of fluid motion. In: Embleton, C., and Thornes, Johns (Eds.): Process in geomorphology, pp. 39–72. New York: John Wiley and Sons 1979.

Engelhardt, Wolf von: Die Unterscheidung wasser- und windsortierter Sande auf Grund der Korngrössenverteilung ihrer leichten und schweren Gemengteile. Chemie Erde 12, 445–465 (1940).

Evans, Graham: Intertidal flat sediments and their environments of deposition in the Wash. Geol. Soc. London Quart. Jour. 121, 209–245 (1965).

Fisher, Richard V., and Waters, Aaron C.: Base surge bedforms in maar volcanoes. Am. Jour. Sci. 268, 157–180 (1970).

Forel, F.: Les ravins souslacustres des fleuves glaciaires. C. R. Acad. Sci. Paris 101, 725–728 (1885).

Friend, P.F.: Fluviatile sedimentary structures in the

Wood Bay Series (Devonian) of Spitsbergen. Sedimentology 5, 39–68 (1965).

Gardner, T.W.: Paleohydrology and paleomorphology of a Carboniferous meandering fluvial sandstone. Jour. Sed. Petrology 53, 991–1006 (1983).

Gibbs, R.J.; Matthews, M.D.; and Link, D.A.: The relationship between sphere size and settling velocity. Jour. Sed. Petrology 41, 7–18 (1971).

Gilbert, G.K.: The transportation of debris by running water. U.S. Geol. Survey Prof. Paper 86, 263 pp. (1914).

Giles, R.V.: Fluid mechanics and hydraulics, 2nd Ed., 274 pp. New York: Schaum Publ. Co. 1962.

Glennie, K.W.: Desert sedimentary environments, 222 pp. (Developments in sedimentology, Vol. 14). Amsterdam: Elsevier Scientific Pub. Co. 1970.

Gould, H.R.: Some quantitative aspects of Lake Mead turbidity currents. In: Turbidity currents and transportation of coarse sediments to deep water—a symposium. Soc. Econ. Paleon. Mineral. Spec. Pub. 2, 34–52 (1951).

Graf, Walter, H.: Hydraulics of sediment transport, 513 pp. New York: McGraw-Hill 1971.

Graf, W.H., and Acaroglu, E.R.: Settling velocity of natural grains. Intern. Assoc. Sci. Hydraulics 6, 27–43 (1966).

Greenwood, B., and Davis, R.A., Jr.: Hydrodynamics and sedimentation in wave-dominated coastal environments. Marine Geology 60 (Spec. Issue), 473 pp. (1984).

Gregory, K.J. (Ed.): Background to paleohydrology, 350 pp. New York: John Wiley and Sons 1983.

Guy, H.P.; Simons, D.B.; and Richardson, E.V.: Summary of alluvial channel data from flume experiments, 1956–1961. U.S. Geol. Survey Prof. Paper 462-I, 96 pp. (1966).

Hallermeier, R.J.: Terminal settling velocity of commonly occurring sand grains. Sedimentology 28, 859–865 (1981).

Hamilton, N.; Owens, W.H.; and Rees, A.J.: Laboratory experiments on the production of grain orientation in shearing sand. Jour. Geology 76, 465–472 (1968).

Hand, Bryce, M.: Differentiation of beach and dune sands, using settling velocities of light and heavy minerals. Jour. Sed. Petrology 37, 514–520 (1967).

Harms, J.C.: Hydraulic significance of some sand ripples. Geol. Soc. America Bull. 80, 363–396 (1969).

Harms, J.C.; Southard, J.B.; and Walker, R.G.: Structures and sequences in clastic rocks. S.E.P.M. Short Course No. 9, various paging (1982).

Hayes, M.O.: General morphology and sediment patterns in tidal inlets. In: Bouma, A.H., *et al.* (Eds.): Shallow marine processes and products. Sedimentary Geology, Spec. Issue 26, 139–156 (1980).

Hesse, R.: Turbidite and non-turbidite mudstones of Cretaceous flysch sections of the East Alps and other basins. Sedimentology 22, 387–416 (1975).

Hollister, C.D., and Heezen, B.C.: Geologic effects of ocean bottom currents, western North Atlantic. In: Gordon, A.L. (Ed.): Studies in physical oceanography, pp. 37–66. New York: Gordon and Breach 1972.

Houbolt, J.J.H.C., and Jonker, J.B.M.: Recent sediments in the eastern part of Lake Geneva (Lac Leman). Geol. Mijnbouw 47, 131–148 (1968).

Hunter, Ralph E.: Basic types of stratification in small eolian dunes. Sedimentology 24, 361–387 (1977).

Hunter, Ralph E.: Stratification styles in eolian sandstones: some Pennsylvanian and Jurassic examples from the western Interior U.S.A. *In:* Ethridge, F.G., and Flores, Romero M. (Eds.): Recent and ancient nonmarine depositional environments: Models for exploration. Soc. Econ. Paleon. Mineral. Spec. Pub. 31, 315–330 (1981).

Ingle, J.C., Jr.: The movement of beach sand, 232 pp. New York: Elsevier 1966.

Inman, Douglas L.: Sorting of sediments in the light of fluid mechanics. Jour. Sed. Petrology 19, 51–70 (1949).

Inman, Douglas L.: Ocean waves and associated currents. In: Shepard, F.P., Submarine geology, 2nd Ed., pp. 48–81. New York: Harper and Row 1963.

Inman, D.L., and Bagnold, R.A.: Littoral processes. Part II. In: Hill, N.M. (Ed.): The sea, Vol. 3, pp. 329–543. New York: Interscience 1963.

Inman, D.L., and Brush, B.M.: The coastal challenge. Science 181, 20–32 (1973).

Iversen, J.D., and White, B.R.: Saltation threshold on Earth, Mars and Venus. Sedimentology 29, 111–119 (1982).

Johnson, A.M.: Debris flow. In: Brundsen, D., and Prior, D.B. (Eds.): Slope instability, pp. 257–361. New York: John Wiley 1984.

Johnson, D.W.: Shore processes and shoreline development, 584 pp. New York: John Wiley 1919.

Jopling, A.V.: Hydraulic studies on the origin of bedding. Sedimentology 2, 115–121 (1963).

Jopling, A.V.: Hydraulic factors controlling the shape of laminae in laboratory deltas. Jour. Sed. Petrology 35, 777–791 (1965).

Jopling, A.V.: Some principles and techniques used in reconstructing the hydraulic parameters of a paleo-flow regime. Jour. Sed. Petrology 36, 5–49 (1966).

Kalinske, A.A.: Turbulence and the transport of sand and silt by wind. N.Y. Acad. Sci. Ann. 44, 41–54 (1943).

Kenyon, N.H.: Sand ribbons of European tidal seas. Marine Geology 9, 367–369 (1970).

Keulegan, G.H.: Interfacial stability and mixing of stratified flows. Natl. Bur. Standards Jour. Res. 43, 487–500 (1949).

Keulegan, G.H., and Krumbein, W.C.: Stable configuration of bottom slope in a shallow sea and its bearing on geological processes. Trans. Am. Geophys. Union 30, 855–861 (1949).

King, Cuchlaine A.M.: Beaches and coasts, 2nd Ed., 510 pp. London: Edward Arnold 1972.

Klein, George deVries: Determination of paleotidal range in clastic sedimentary rocks. In: McLauren, D.J., and Middleton, G.V. (Conveners): Stratigraphy and sedimentology. 24th Intern. Geol. Cong. 6, 397–405 (1972).

Kocurek, Gary, and Dott, R.H., Jr.: Distinction and uses of stratification types in the interpretation of eolian sand. Jour. Sed. Petrology 51, 579–596 (1981).

Komar, Paul D.: Beach processes and sedimentation, 429 pp. Englewood Cliffs, New Jersey: Prentice-Hall 1976.

Komar, Paul D.: 15. Computer simulation of turbidity current flow and the study of deep-sea channels and fan sedimentation. In: Goldberg, E.D., *et al.* (Eds.): The sea, Vol. 6, pp. 603–621. New York: John Wiley and Sons 1977.

Komar, P.D.: Comparisons of the hydraulics of water flows in Martian outflow channels with flows of similar scale on Earth. Icarus 37, 156–181 (1979).

Komar, P.D.: The hydraulic interpretation of turbidites from their grain sizes and sedimentary structures. Sedimentology 32, 395–407 (1985).

Krinsley, D.H.; Friend, P.F.; and Klimentidis, R.: Eolian transport textures on the surfaces of sand grains of Early Triassic age. Geol. Soc. America Bull. 87, 130–132 (1976).

Kuenen, Ph.H.: Experiments in connection with Daly's hypothesis on the formation of submarine canyons. Leidse Geol. Mededeel. 8, 327–351 (1937).

Kuenen, Ph.H.: Experimental abrasion 4. Eolian action. Jour. Geology 68, 427–449 (1960).

Kuenen, Ph.H.: Experimental marine suspension currents, competency and capacity. Geol. Mijnbouw 49, 89–118 (1970).

Lane, E.W., and Kalinske, A.A.: The relation of suspended to bed material in rivers. Trans. Am. Geophys. Union 20, 637–641 (1939).

Leeder, M.R.: Folk's bedform theory. Sedimentology 24, 863–874 (1977).

Leopold, L.B.; Wolman, M.G.; and Miller, J.P.: Fluvial processes in geomorphology, 522 pp. San Francisco: W.H. Freeman 1964.

Lerman, A.: Geochemical processes, water and sediment environments, 481 pp. New York: John Wiley and Sons 1979.

Levey, R.A.: Bedform distribution and internal stratification of coarse-grained point bars Upper Congaree River, South Carolina. In: Miall, A.D. (Ed.): Fluvial sedimentology. Canadian Soc. Petroleum Geologists Mem. 5, 105–127 (1978).

Lewis, K.B.: Slumping on a continental slope inclined at 1°–4°. Sedimentology 16, 97–110 (1971).

Loope, D.B.: Eolian origin of upper Paleozoic sandstones, southwestern Utah. Jour. Sed. Petrology 54, 563–580 (1984).

Lowe, Donald R.: Water escape structures in coarse-grained sediments. Sedimentology 22, 157–204 (1975).

Lowe, Donald R.: Sediment gravity flows: II. Depositional models with special references to the deposits of high density turbidity currents. Jour. Sed. Petrology 52, 279–297 (1982).

Lundquist, G.: Beskrivning till Jourdartskarta over Galveborgs lan. Sveriges Geologiska Undersokning, Ser. Ca, 42, 191 pp. (1963).

Lüthi, S.: Some new aspects of two-dimension turbidity currents. Sedimentology 28, 97–105 (1981).

Mack, G.H.: The survivability of labile light mineral grains in fluvial, eolian, and littoral marine environments: The Permian Cutter and Cedar Mesh Formations, Moab, Utah. Sedimentology 25, 587–606 (1978).

McBride, E.F.; Shepherd, R.G.; and Crawley, R.A.: Origin of parallel, near-horizontal laminae by migration of bedforms in a small flume. Jour. Sed. Petrology 45, 132–139 (1975).

McCave, I.N.: Recent shelf clastic sediments. In: Brenchley, P.J., and Williams, B.P.J. (Eds.): Sedimentology, recent developments and applied aspects. Geol. Soc. London, Spec. Paper 18, 49–65 (1985).

McGowen, J.H., and Garner, L.E.: Physiographic features and stratification types of coarse grained point bars: Modern and ancient examples. Sedimentology 14, 77–111 (1970).

McKee, E.D. (Ed.): A study of global sand seas. U.S. Geological Survey Prof. Paper 1052, 429 pp. (1979).

Matthes, G.: Macroturbulence in natural stream flow. Amer. Geophys. Union Trans. 28, 255–265 (1947).

Meckel, L.D.: Tabular and trough cross-bedding: Comparison of dip azimuth variability. Jour. Sed. Petrology 37, 80–86 (1967).

Miall, A.D.: Mesozoic and Tertiary geology of Banks Island, Arctic Canada. Geol. Survey Canada Mem. 387, 235 pp. (1979).

Middleton, Gerard V. (Ed.): Primary sedimentary structures and their hydrodynamic interpretation. Soc. Econ. Paleon Mineral. Spec. Pub. 12, 265 pp. (1965).

Middleton, G.V.: Experiments on density and turbidity currents, I. Motion of the head. Canadian Jour. Earth Sci. 3, 523–546 (1966a).

Middleton, G.V.: Experiments on density and turbidity currents, II. Uniform flow of density currents. Canadian Jour. Earth Sci. 3, 627–637 (1966b).

Middleton, G.V.: Experiments on density and turbidity currents, III. Deposition of sediment. Canadian Jour. Earth Science 4, 475–505 (1967).

Middleton, G.V.: Hydraulic interpretation of sand size distributions. Jour. Geology 84, 405–426 (1976).

Middleton, G.V., and Hampton, M.A.: Subaqueous sediment transport and deposition by sediment gravity flows. In: Stanley, D.J., and Swift, D.J.P. (Eds.): Marine sediment transport and environmental management, pp. 197–218. New York: John Wiley and Sons 1976.

Miller, M.C.; McCave, I.N.; and Komar, P.D.:

Threshold of sediment motion under unidirectional currents. Sedimentology 24, 507–527 (1977).
Morgenstern, N.R.: Submarine slumping and the initiation of turbidity currents. In: Richards, A.F. (Ed.): Marine geotechnique, pp. 189–220. Urbana, Illinois: Univ. Illinois Press 1967.
Morris, Henry M., and Wiggert, James M.: Applied hydraulics in engineering, 2nd Ed., 629 pp. New York: Ronald Press 1972.
Mount, Jeffery F.: Storm-surge-ebb origin of hummocky cross-stratified units of the Andrew Mountain Member, Campito Formation (Lower Cambrian), White-Inyo Mountains, eastern California. Jour. Sed. Petrology 52, 941–958 (1982).
Nardin, T.R.; Hein, F.J.; Gorsline, D.S.; and Edwards, B.D.: A review of mass movement processes, sediment and acoustic characteristics, and contrasts in slope and base-of-slope systems versus canyon-fan-basin floor systems. In: Doyle, J., and Pilkey, O.D. (Eds.): Geology of continental margins. Soc. Econ. Paleon. Mineral. Spec. Pub. 27, 61–74 (1979).
Newton, R.S.: Internal structure of wave-formed ripple marks in the near-shore zone. Sedimentology 11, 275–292 (1968).
Padgett, G.V., and Ehrlich, R.: Paleohydrologic analysis of a late Carboniferous fluvial system, southern Morocco. Geol. Soc. America Bull. 87, 1101–1104 (1976).
Parsons, I., and Butterfield, A.W.: Sedimentary features of the Nunarssuit and Klokken syenites, S. Greenland. Jour. Geol. Soc. (London) 13, 289–306 (1981).
Potter, P.E., and Pettijohn, F.J.: Paleocurrents and basin analysis, 2nd Ed., 425 pp. New York: Springer-Verlag, 1977.
Power, W.R., Jr.: Backset beds in the Coro Formation, Inyo County, California. Jour. Sed. Petrology 31, 603–607 (1961).
Prior, D.B., and Coleman, J.M.: Submarine landslides—geometry and nomenclature. Zeitschr. Geomorph., N.F. 23, 415–426 (1979).
Raudkivi, A.J.: Loose boundary hydraulics, 2nd Ed., 397 pp. New York: Pergamon Press 1975.
Reed, W.E.; LeFever, R.; and Moier, G.J.: Depositional environmental interpretation from settling-velocity (psi) distributions. Geol. Soc. America Bull. 86, 1321–1328 (1975).
Rees, A.I.: The production of preferred orientation in a concentrated dispersion of elongated and flattened grains. Jour. Geology 76, 457–465 (1968).
Reineck, H.-E.: Sedimentgefuge im Bereich der sudlichen Nordsee. Abh. Senckenberg naturf. Gesell. Nr. 505, 1–138 (1963).
Reineck, H.-E., and Singh, I.B.: Depositional sedimentary environments, 2nd rev. updated ed., 549 pp. Berlin · Heidelberg · New York: Springer-Verlag 1980.
Rittenhouse, Gordon: The transportation and deposition of heavy minerals. Geol. Soc. America Bull. 54, 1725–1780 (1943).
Rubey, W.W.: Settling velocity of gravel, sand, and silt particles. Am. Jour. Sci., 5th Ser., 25, 325–338 (1933).
Rubin, D.M., and McCulloch, D.S.: Single and superimposed bedforms: A synthesis of San Francisco Bay and flume observations. Sedimentary Geology 26, 207–231 (1980).
Sagan, Carl, and Bagnold, R.A.: Fluid transport on Earth and eolian transport on Mars. Icarus 26, 209–218 (1975).
Saunderson, H.C., and Jopling, A.V.: Palaeohydraulics of a tabular, cross-stratified sand in the Brampton Esker, Ontario. Sedimentary Geology 25, 169–188 (1980).
Scheidegger, A.E.: Theoretical geomorphology, 2nd Ed., 435 pp. Berlin–Heidelberg–New York: Springer-Verlag 1970.
Scheidegger, A.E., and Potter, P.E.: Bed thickness and grain size. Sedimentology 8, 39–44 (1967).
Schumm, S.A.: Fluvial paleochannels. In: Rigby, J.K., and Hamblin, W.K. (Eds.): Recognition of ancient sedimentary environments. Soc. Econ. Paleon. Mineral. Spec. Pub. 16, 98–107 (1972).
Schumm, S.A.: The fluvial system, 338 pp. New York: John Wiley and Sons 1977.
Shepard, Francis P.: Submarine geology, 3rd Ed., 517 pp. New York: Harper and Row 1973.
Shepard, F.P., and Marshall, N.F.: Currents in submarine canyons and other sea valleys. In: Stanley, D.J., and Kelling, G. (Eds.): Sedimentation in submarine canyons, fans, and trenches, pp. 3–14. Stroudsburg, Pennsylvania: Dowden, Hutchinson, and Ross 1979.
Shields, A.: Anwendung der Ähnlichkeitsmechanik und der Turbulenzforschung auf die Geschiebebewegung. Mitt. Preuss. Versuchsanstalt für Wasserbau und Schiffbau 26, 26 pp. (1935).
Simons, D.B., and Richardson, E.V: Forms of bed roughness in alluvial channels. Am. Soc. Civil Eng. Proc., Jour. Hydraulics Div. 87 (HY-3), 87–105 (1961).
Simons, D.B.; Richardson, E.V.; and Nordin, C.F., Jr.: Forms generated by flow in alluvial channels. In: Middleton, G.V. (Ed.): Primary sedimentary structures and their hydrodynamic interpretation. Soc. Econ. Paleon. Mineral. Spec. Pub. 12, 34–52 (1965).
Simpson, J.E.: Gravity currents in the laboratory, atmosphere and ocean. Ann. Rev. Fluid Mech. 14, 213–234 (1982).
Skipper, Keith, and Bhattacharjee, S.B.: Backset bedding in turbidites: A further example from the Cloridorme Formation (Middle Ordovician), Gaspe, Quebec. Jour. Sed. Petrology 48, 193–202 (1978).
Smith, N.D.: Some sedimentological aspects of planar cross-stratification in a sandy, braided river. Jour. Sed. Petrology 42, 624–634 (1972).
Smith, N.D.: Roman hydraulic technology. Scientific American 238, 154–158 (1978).
So, C.L.; Pierce, J.W.; and Siegel, F.R.: Sand waves

in the Gulf of San Matias, Argentina. Geog. Annaler 56, 227–235 (1974).

Sorby, H.C.: On the oscillation of the currents drifting the sandstone beds of the southeast of Northumberland, and on their general direction in the coal field in the neighborhood of Edinburgh. Proc. West Yorkshire Geol. Soc. 3, 232–240 (1851).

Southard, John B.; Boguchwal, Lawrence A.; and Romea, Richard D.: Test of scale modeling of sediment transport in steady unidirectional flow. Earth Surface Processes 5, 17–23 (1980).

Starkel, L., and Thornes, J.B.: Paleohydrology of river basins. Brit. Geomorphol. Res. Group Tech. Bull. 28, 107 (1981).

Stauffer, P.H.: Grain-flow deposits and their implications, Santa Ynez Mountains, California. Jour. Sed. Petrology 37, 487–508 (1967).

Stow, D.A.V., and Lovell, J.P.B.: Contourites: their recognition in modern and ancient sediments. Earth Sci. Rev. 14, 251–291 (1979).

Streeter, Victor L.: Fluid mechanics, 5th Ed., 755 pp. New York: McGraw-Hill 1971.

Stride, A.H.: Current-swept sea floors near the southern half of Great Britain (with discussion). Geol. Soc. London Quart. Jour. 119, 175–199 (1963).

Stringham, G.E.; Simons, D.B.; and Guy, H.P.: The behavior of large particles falling in quiescent liquids. U.S. Geol. Survey Prof. Paper 562-C, 36 pp. (1969).

Sundborg, Åke: The River Klaralven, a study in fluvial processes. Geog. Annaler, Ser. A., No. 11, 127–316 (1956).

Sutherland, A.J.: Proposed mechanism for sediment entrainment by turbulent flows. Jour. Geophys. Research 72, 6183–6194 (1967).

Swift, D.J.P.: Quaternary shelves and the return to grade. Marine Geology 8, 5–30 (1970).

Swift, D.J.P.: Tidal sand ridges and shoal-retreat massifs. Marine Geology 18, 105–133 (1975).

Swift, D.J.P.; Freeland, G.L.; and Young, R.A.: Time and space distribution of megaripples and associated bedforms, Middle Atlantic Bight, North American Bight, North American Atlantic Shelf. Sedimentology 26, 389–406 (1979).

Swift, D.J.P.; Fiqueiredo, A.G., Jr.; Freeland, G.L.; and Oertel, G.F.: Hummocky cross stratification and megaripples: A geological 1983 double standard? Jour. Sed. Petrology 53, 1295–1317 (1983).

Tricker, R.A.T.: Water waves. Encyclopedia Britannica (Macropedia) 19, 654–660 (1974).

U.S. Corps of Engineers: Shore protection manual, 4th Ed., Vols. I and II, various paging. U.S. Army Coastal Engineering Research Center, 1984. (Superintendent of Documents, U.S. Government Printing Office, Washington, D.C.).

Valecka, Jaroslav: Storm surge versus turbidite origin of the Coniacian to Santonian sediments in the eastern part of the Bohemian Cretaceous Basin: Geol. Rundschau 73, 651–682 (1984).

Vanney, Jean Rene: Géomorphologie des plateformes continentals, 300 pp. Paris: Doin Editeurs, 1977.

Vanoni, V.A.: Factors determining bedforms of alluvial streams. Am. Soc. Civil Eng. Proc., Jour. Hydraulics Div. 100, 363–377 (1974).

Vennard, John K.: Elementary fluid mechanics, 4th Ed., 570 pp. New York: John Wiley and Sons 1961.

Warren, Andrew: Aeolian processes. In: Embleton, C., and Thornes, J. (Eds.): Process in geomorphology, pp. 325–351. New York: John Wiley 1979.

Werner, F., and Newton, R.S.: The patterns of large scale bedforms in the Langeland Belt (Baltic Sea). Marine Geology 19, 29–59 (1975).

Wiegel, R.L.: Oceanographical engineering, 532 pp. Englewood Cliffs, New Jersey: Prentice-Hall 1964.

Wilson, Ian G.: Aeolian bedforms—their development and origins. Sedimentology 19, 173–210 (1972).

Wright, M.E., and Walker, R.G.: Cardium Formation (U. Cretaceous) at Seebe, Alberta—storm-transported sandstones and conglomerates in shallow marine depositional environments below fair-weather wave base. Can. J. Earth Sci. 18, 795–809 (1981).

Yalin, M.S.: Geometrical properties of sand waves. Amer. Soc. Civil Engineers Proc. Jour. Hydraulics Div. 90, Paper 4055, 105–119 (1949).

Yalin, M.S.: Mechanics of sediment transport, 2nd Ed., 298 pp. Oxford–New York: Pergamon Press 1977.

Zanke, Ulrich: Grundlagen der Sedimentbewegung, 402 pp. Berlin–Heidelberg–New York: Springer-Verlag, 1982.

Zingg, A.W.: Wind tunnel studies of movement of sedimentary material. 5th Ann. Hydraulics Conf. Proc., Iowa City, Iowa, 111–135 (1953).

CHAPTER 9
Paleocurrents and Dispersal

Introduction

Between the production of sand and its deposition and diagenesis is its dispersal. Questions we can ask about dispersal concern possible modifications imprinted on the sand by the transport process (textural and compositional changes due to abrasion and sorting), the agent of transport, and the distance and direction of transport.

What geological observations can we make to answer these questions? A part of the answer belongs to "paleocurrents"—a topic which has attained prominence particularly since the 1950s. We have already discussed in some detail the mechanics of transport and its relation to textures and primary structures of sand.

Paleocurrents are those currents of wind, water, or ice that have moved or transported the materials deposited in some earlier period. We are concerned here with the directional and other primary properties of sand which enable us to reconstruct these ancient current systems. As noted over 100 years ago by Sorby (1857, p. 285), "The examination of modern seas, estuaries, and rivers, shows that there is a distinct relation between their physical geography and the currents present in them; currents so impress themselves on the deposit found under their influence that their characters can be ascertained from those found in ancient periods. Therefore their physical geography can be inferred within certain limits." This concept was clearly understood and pursued by Brinkman (1933) and Hans Cloos (1938). The state of the art was summarized by Pettijohn (1962) and the subject was dealt with in depth by Potter and Pettijohn (1977).

What benefits result from paleocurrent measurement? There are many. Knowledge of current direction can (1) predict direction of elongation of a sandstone body, (2) outline the paleocurrent system of a basin and thus contribute to better understanding of the arrangement of its sedimentary fill, (3) help to determine if a structural feature in a basin was active or not during deposition, (4) help locate source regions that lie beyond a basin margin, (5) help identify the allochthonous crustal fragments accreted to a continental block, (6) aid paleoecology by establishing the direction of supply of nutrient-carrying currents, (7) contribute to some types of stratigraphic correlation problems, and (8) can yield information on bulk geophysical properties such as electrical and thermal conductivity, because current direction can exercise the dominant control, through grain fabric, on these properties. For effort expended, measurement of directional structures is probably more rewarding to the field geologist than anything else. Crimes (1970) gives a good example of the role of sedimentary structures, particularly directional structures, in his facies analysis of the Cambrian sediments in the Caledonian trough in Wales. Other references include the volumes by Shawa (1974, 1975), where sedimentary structures in western Canada are chiefly used to help identify the agent of deposition; the report by Johnson (1975), who distinguished between tide- and wave-dominated coastal processes in the Precambrian of Norway; and the paper by Baker (1973), who used sedimentary structures and grain size to help reconstruct the paleohydraulics of the gigantic Spokane flood and channeled Scablands in Washington (see also Bretz, 1969).

Paleocurrent analysis seeks to derive inferences about current direction, sediment dispersal, and depositional slope for specific sedimentary deposits. As now practiced, such analysis is based very largely upon empirical studies of ancient strata, although, as noted below, there is much interest in sediment dispersal patterns of modern seas.

There are many assumptions underlying paleocurrent analyses, some explicit, others implied but unstated. Allen (1966, 1967) has explored the subject largely from a theoretical point of view. Klein (1967) has made a study of modern marine dispersal patterns and Selley (1968) has made an analysis of sediment dispersion based chiefly on the stratigraphic record. In this book, the subject of paleocurrent analysis is limited to sand and sandstone.

The emphasis on sand is fully justified, for most of the criteria of paleoflow reside in sandstones. Shales, except for those containing oriented organic structures such as elongate plant fragments and fossils, yield but little paleocurrent information. Likewise limestones contain few evidences of current action except those which are in fact sands—carbonate, skeletal, or oolitic sands which exhibit all the usual structures of sands. It is to the sands, therefore, that we look for much of the criteria of paleoflow (Table 9-1).

Most useful and important are the primary structures such as crossbedding and ripple marks and, in the case of the turbidite sands, the various sole marks. The latter, though produced on mud surfaces, are preserved only as "casts" on the base of the superjacent sandstones. Of course, a sedimentary structure only tells us in which direction the current flowed at a particular place at some instant in time. Only by making many observations and integrating them over a larger area can we discern the regional paleocurrent pattern. In addition to the primary sedimentary structures, we can learn something of the dispersal system from the texture and composition of the sands—especially the latter. Here the areal distribution of a particular suite of minerals reflects the transport system as well as the nature and location of the sediment source.

Dispersal Patterns Defined by Composition and Texture

The reconstruction of paleocurent patterns in modern sands differs markedly from that in ancient sands. In the Recent underwater sands, there is minimal opportunity to use primary current structures and hence the mineralogy of the modern sand is the principal guide to its dispersal patterns; only exceptionally do ripple marks or other bedforms provide a clue to direction of sand drift. In ancient sandstones, now exposed and dissected by erosion, the direction of paleoflow is obtained with much less effort from crossbedding and other current structures.

The study of a modern environment is well illustrated by the sands of the shore and nearshore areas of the Gulf of Mexico (Goldstein, 1942; Davies and Moore, 1970). Here there are definable heavy mineral provinces which are, from east to west, the Eastern Gulf Province, the Mississippi River Province, the central Texas Province, the Rio Grande Province, and the Mexican Province (Fig. 9-1). Each of these is defined by a heavy mineral assemblage. The Eastern Gulf Province is characterized by a high content of staurolite and kyanite. The Mississippi River Province is primarily an amphibole–epidote–pyroxene province, essentially containing the heavy minerals carried by the river today. The mineral content of the Western Gulf Province is similar to that of the Missis-

TABLE 9-1. Criteria for paleoflow.

Facies patterns

Some facies patterns or facies relationships—for example, the transition of proximal to medial to distal turbidites, orientation of delta fingers in low-energy deltas, and some fluvial–marine transitions as well as marine shallow shelf to deep basin transitions—are so well defined that when proper correlations exist between them, the direction of facies change can be used to infer a paleocurrent direction. Seismic sections interpreted stratigraphically also provide paleocurrent information.

Directional structures

Chiefly crossbedding (shallow-water sands) and sole marks (deep-water sands). For crossbedding some environmental significance can be obtained from variability in orientation. Commonly studied in outcrop, but more and more studies are using oriented cores and/or dipmeter logs. Ripple mark, fossil, orientation, and other criteria can be used, but in most sandstones they are not abundant enough to be consistently mappable. In turbidites, sole marks (flutes and grooves) are almost universally mapped.

Areal variation of scalar properties

Chiefly the mapping of mineral associations (light and heavy minerals, fossil debris, and clay minerals) and to a lesser degree grain size (average modal grain size of the sand fraction or "maximum" size of associated pebbles, the latter being more widely used).

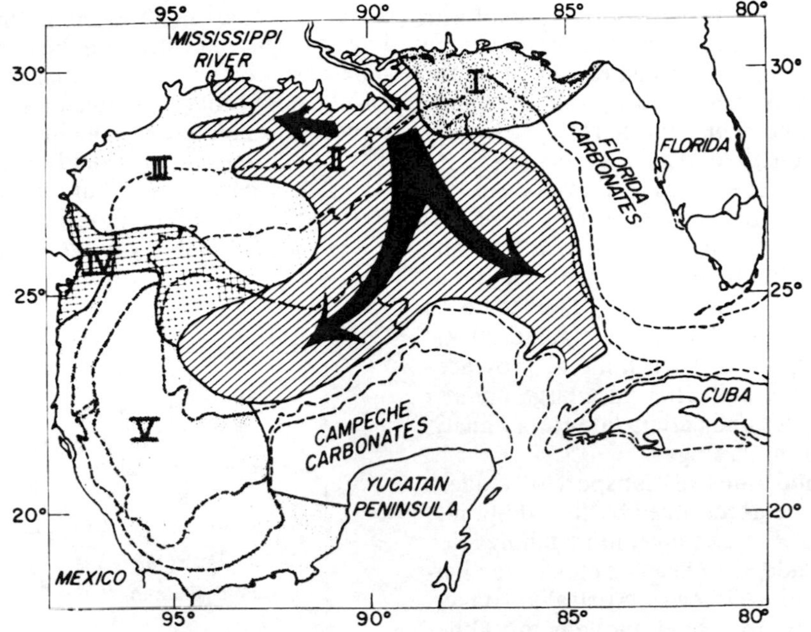

FIGURE 9-1. Major heavy mineral provinces, inferred areal distribution of Mississippi River sediment, and principal sediment dispersal directions in the Gulf of Mexico. I—Eastern Gulf Province, II—Mississippi Province, III—Central Texas Province, IV—Rio Grande Province, V—Mexican Province. Data for map compiled from various sources by Davies and Moore (1970, Fig. 1).

sippi and may have been derived mainly from the Mississippi sources at an earlier time. The Rio Grande Province is characterized by a basaltic hornblende–pyroxene assemblage. The distribution of these provinces is best explained by a slow drift carrying the Mississippi mineralogy westward but not eastward. In the western Gulf this mineralogy is modified and diluted by materials brought to the shore by the smaller rivers of Texas (Brazos, Trinity, Colorado). The Rio Grande contributes a larger volume of sand and thus defines a province of distinctive character of its own. We don't know, of course, what directional structures these sands would display. It is possible that these might be discordant with the net transport due to longshore drift.

Similar studies have been made for other off-

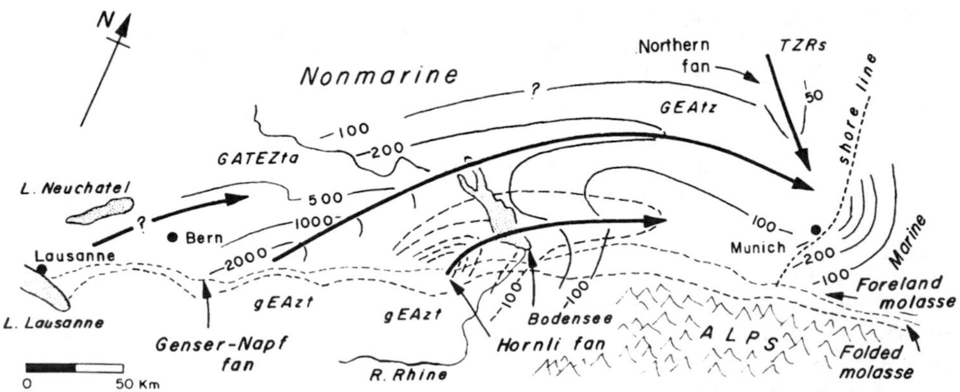

FIGURE 9-2. Paleocurrent systems and mineral provinces deduced from heavy minerals in Lower Freshwater Molasse (Aquitanian) north of Alps. E: epidote; G: garnet; A: apatite; T: tourmaline; Z: zircon; S: staurolite. Capital letters denote greater abundance; lower case letters denote lesser abundance. (After Füchtbauer, 1958, Fig. 14a).

shore areas such as those of the Cascadia Basin and Blanco Fracture Zone off the west coast of the United States (Duncan and Kulm, 1970), the North Sea (Baak, 1936), the Gulf of Maine (Ross, 1970), and along the northwest part of the Adriatic coastline (Gazzi *et al.*, 1973) to name but a few. The sands of Chesapeake Bay also provide a good example (Firek *et al.*, 1977), and on a far smaller scale, Sturm and Matter (1972) effectively used heavy minerals to map mineral provinces in Thunersee in Switzerland, a glacial lake only some 20 km long. The mapping of such mineral provinces depends upon the collection of a large number of bottom samples and careful laboratory analysis of the sand mineralogy.

The paths and rates of transport of modern sands can also be ascertained by the addition of tracer minerals. For example, mine tailings rich in pyrite were added to beach sands of the harbor at Portland, Victoria, Australia (Baker, 1956) to follow the sand movement. Other workers have used dyes or irradiated quartz to "tag" sand for identification, monitoring the tracer with a scintillometer (Duane, 1970; Yasso, 1978).

A similar approach to paleocurrent analysis can be used to study ancient sandstones though in general a paleocurrent reconstruction based on crossbedding or other primary structures yields better results with less effort. Some sandstones, however, are known in the subsurface where generally mineralogy alone is used although more and more oriented cores or dipmeters are becoming available. An example of the mineralogical approach is that of Füchtbauer (1958) whose work on the Molasse of southern Germany is outstanding (Fig. 9-2). His study is based on mineralogic analysis of over 1,500 samples collected in an area 450 km long lying north of the Alps, mainly in southern Germany. The molasse sediments crop out along the tilted southern margin of the basin. They are coarse conglomerate fans which pass, in the subsurface, into the fine-grained, horizontal deposits within the basin. Most of the samples were from wells in the basin.

The sandstones rich in feldspar and rock fragments have a varied heavy mineral assemblage which can be divided into subassemblages, each correlated with a particular fan and derived from a local Alpine source. These assemblages are identified basinward and their mapping defines the paths of sediment transport (Fig. 9-2). Near the margin, transport is lateral but in the basin, longitudinal transport prevails. There is a small input from the north side of the basin of an alien heavy mineral suite, mainly tourmaline, zircon, and rutile, unlike the dominant Alpine assemblage which consists principally of garnet, staurolite, apatite, and zircon with an important epidote component in the southwest. The mineralogy of the sand also changes with time; the earlier sands are derived from the flysch sediments, including limestones and dolomites, whereas the younger deposits contain

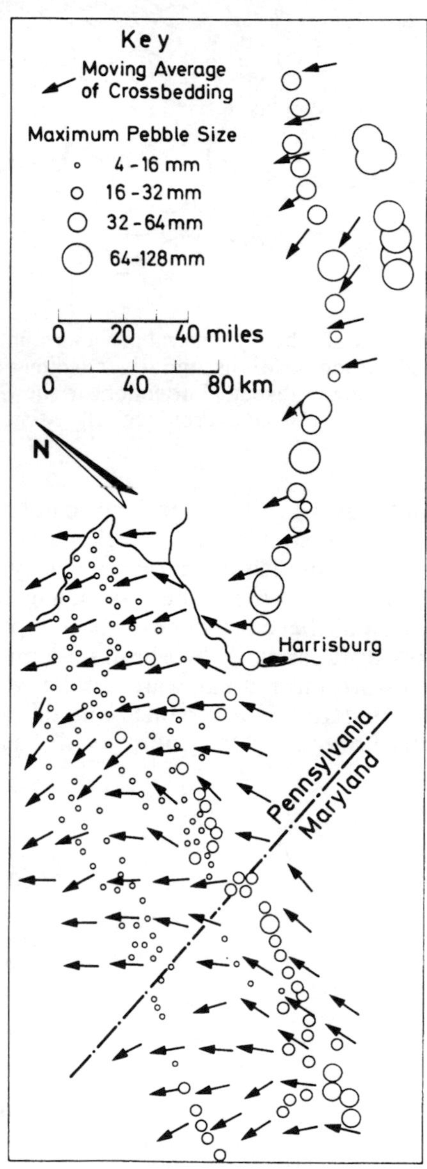

FIGURE 9-3. Downcurrent decline of maximum pebble size and moving average of crossbedding in the fluvial Tuscarora quartzite (Silurian) in the Appalachian Basin. (Redrawn from Yeakel, 1962, Figs. 8 and 14).

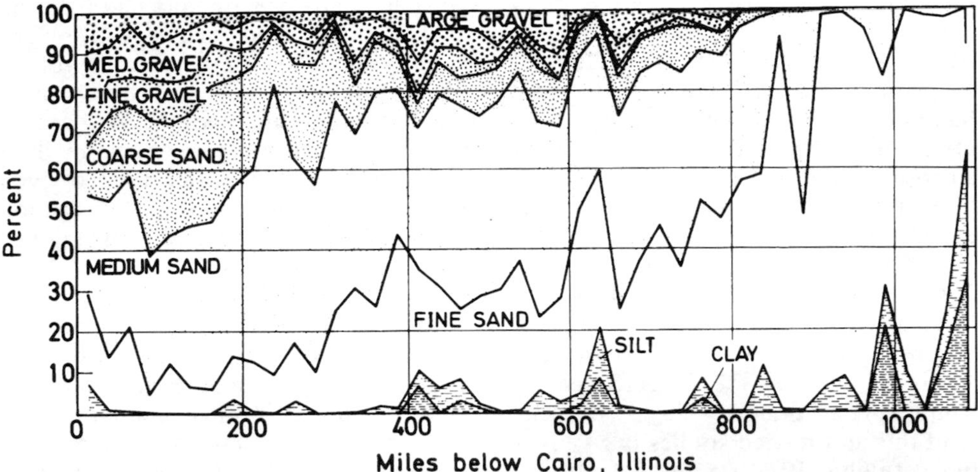

FIGURE 9-4. Grain size of Mississippi River sands, Cairo, Illinois, to Gulf of Mexico. (Modified from U.S. Corps of Engineers, 1935, pl. 58).

significant contributions from the Alpine crystalline rocks.

Füchtbauer's study shows how the petrology of the sands, coupled with stratigraphic data, can contribute to the understanding of the filling of a basin and the reconstruction of the paleogeography of the area. It is a nice integration of outcrop and subsurface data and shows what can be done when the usual criteria for paleoflow are missing.

Textural attributes, primarily grain size and orientation, have also been used to determine the transport direction of both modern and ancient sands. In general, the results have not been commensurate with the efforts expended. A measure of central tendency can be mapped or in some cases, especially in gravels, the maximum observed size can be readily measured in the field. The latter has the advantage of requiring no sample collecting or time-consuming laboratory analyses. It is, however, subject to larger sampling fluctuations than the central tendency. Size trends have been successfully identified in gravels to delineate transport patterns and flow direction. A striking example is the Silurian Tuscarora quartzite of the central Appalachians (Fig. 9-3) as well as the Pennsylvanian Pottsville of the same basin (Meckel, 1970, Fig. 6).

But are there size trends in the typically fine- to medium-grained sands deposited on broad, low-gradient coastal plains and on shallow marine shelves? If the lower portion of the Mississippi River, between Cairo, Illinois, and the Gulf of Mexico, is accepted as a reasonable model for a large alluvial river (Fig. 9-4), such a size gradient may be present in alluvial sandstones deposited on coastal plains. Unfortunately, the search for size trends has not been very successful in most ancient medium- to fine-grained sandstones, perhaps because it can be difficult to secure an adequate sample for a good estimate of size. Pelletier (1965, Figs. 4 and 6), however, did demonstrate a systematic textural gradient in the sands of the Triassic Grey Beds of British Columbia. In addition, there is a marked downcurrent decline in sand size in the fluvial Buntsandstein of Germany over a distance of 400 km (Fig. 9-5). Curray (1961) achieved some success in tracing sand movement on the present marine shelf of the western Gulf of Mexico by dissecting the size distributions of some 750 samples of these sands into subpopulations, each with its own mode. In similar manner, van Andel (1973) used texture to trace sand and silt dispersal in the modern sands of the Panama Basin.

The grain orientation fabric is known to correlate with current flow and theoretically, if mapped, should yield a paleocurrent pattern, although there are pitfalls to this approach. Not only is the collection of oriented samples diffi-

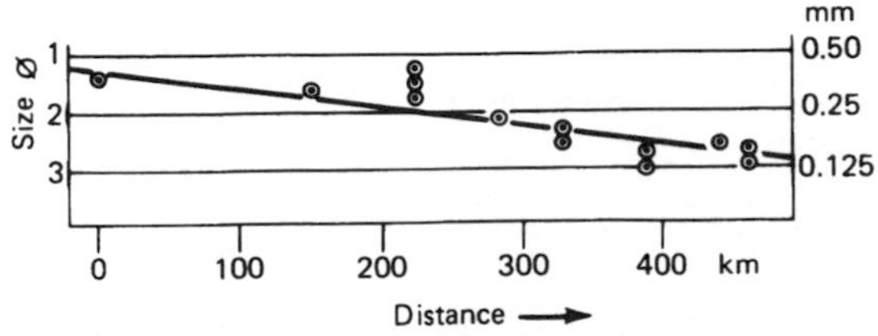

FIGURE 9-5. Decline of grain size in Triassic Buntsandstein of western Germany based on 2633 analyses (Redrawn from Leggewie *et al.*, 1977, Fig. 6).

cult and the thin-section analysis of grain orientation tedious, but the results may be ambiguous in the case of deformed rocks. In deformed rocks the fabric observed may be tectonic rather than primary and sedimentary.

Dielectric anisotropy, which is correlated with grain orientation and hence with current flow, can also be used to reduce the tedium of measurements but here again, oriented samples are required (Nanz, 1960). However, these are now becoming more common, and Shelton *et al.* (1974) clearly demonstrated the correlation between primary structures in a modern sand bar and grain fabric using the dielectric anisotropy of impregnated oriented plugs. A recent summary of this and related studies is given by Potter and Pettijohn (1977, pp. 70–71). An application to oriented cores has been made by Hsu (1977, Figs. 8, 9, and 10).

Dispersal Patterns Inferred from Primary Structures

Most paleocurrent analysis depends on the mapping of primary sedimentary structures, which are current generated. For a directional structure to be useful, it must be easy to measure and widespread. It should also correlate with the direction of the principal current. The vast majority of maps have been made with crossbedding in either fluvial-deltaic, eolian, and shelf sands or with sole marks in the deep-water turbidite sands (Table 9-2). Overall, ripple mark has played a very minor role. Exceptional structures include orientation of long axes of concretions, with the long axes parallel either to current direction shown by other sedimentary structures (Colton, 1967) or even to the axis of the sandstone body itself (Jacob, 1973). Backset bedding associated with antidunes is so rare as not to constitute a problem—even though it has been observed in a number of environments, all of which have in common brief periods of high-intensity flow which produce antidunes (Skipper and Bhattacharjee, 1978). Backset bedding is most common in volcaniclastic base surge deposits, where it is easily recognized so that its paleocurrent significance can be correctly interpreted and mapped (Crowe and Fisher, 1973). Such maps help identify the source of base surge sands and thus may be of interest in sulfide exploration (Fox, 1977). In a sequence of lavas interbedded with volcaniclastic sands it is useful to measure the directional structures of both (Schmincke, 1967).

Directional structures are usually measured "as they come" in an outcrop as one moves either vertically up-section or horizontally along it as in a road cut or along a creek bed. Usually a half dozen or so measurements of crossbedding per outcrop will suffice; but if orientation is very variable, as in some tidally influenced sands, 20 or more readings may be

TABLE 9-2. Common directional structures: measurement and occurrence.

Structure	Measurement	Occurrence
Crossbedding	Trough axis and maximum dip direction of foreset are direction of current flow. One reading per bed. The most widely measured directional structure.	Present in almost all traction-transported sands. Beds thicker than 30 cm present in all but turbidite sands. Rarely solitary.
Flutes and grooves	Long dimension is parallel to current; blunt end of flute points upcurrent. Measure trend on each bed. Second most commonly measured structure.	Only abundant in turbidite sands but present in all except eolianites. Rarely solitary.
Ripple mark	Direction of steep slope is direction of current in asymmetrical ripples; in symmetrical ripples strike of crest is strike of wave.	Found everywhere but perhaps most abundant and most varied types occur near the strandline.
Parting lineation	Trend of parting parallels flow; measure trend of each set.	In all environments but rarely systematically measured because of infrequent occurrence.

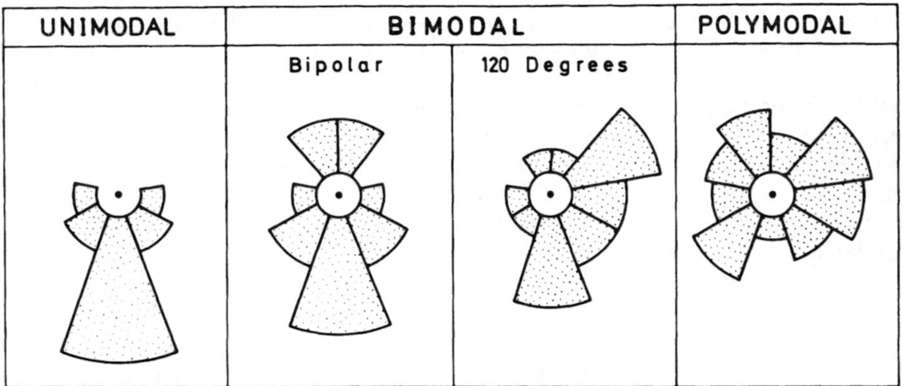

FIGURE 9-6. Types of current roses. (Modified from Selley, 1968, Fig. 2).

necessary to establish a pattern of bimodality. Vertical profiles of current direction can be informative in assessing current stability; pronounced back and forth alternation of ripple mark and crossbedding suggests a strong tidal influence. In order to specify the convergence, divergence, or uniformity of paleocurrents most effectively, it is generally best to have more outcrops with fewer readings than a smaller number of outcrops each with numerous readings. It is best to compute and plot an average for each outcrop. The vector mean is a satisfactory average (Steinmetz, 1962; Jones, 1968) for all except markedly bimodal distributions. For the latter, either midpoints of the two modes should be plotted or special computations are needed (Jones and James, 1969).

Basically then, a paleocurrent analysis of a sedimentary formation involves collection of field data at outcrops (such as sole mark orientation) that determine the direction of current flow, construction of a regional paleocurrent map from such data, and deduction of depositional or paleoslope in conjunction with regional facies relations. Several questions arise during such an analysis. First, what is the effect of deformation on current structures? Second, what is the relation between the structure and the direction of current flow? Finally, what is the relation of flow direction to paleoslope? We discuss each briefly.

The measurement of directional structures in deformed rocks presents special problems. The data collected must be corrected for the tilt of the bed—a correction easily made for gently tilted strata. The correction becomes more difficult and less reliable in strongly folded beds, especially in steeply plunging folds. The structures themselves, such as crossbedding, become markedly deformed but some success has been achieved even in severely deformed Archean terranes (Henderson, 1972, p. 898).

Deducing Current Direction from Sedimentary Structures

A given structure, such as a crossbed, flute, or lineation, tells us in what direction the current went at a particular instant in time at a particular place. If we collect many measurements we find a spread or scatter of the current azimuths. In general, the data from a given locality yield one of several distribution patterns, namely, unimodal, bimodal, or polymodal (Fig. 9-6).

Unimodal patterns generally characterize fluviatile deposits, but they are found also in eolian and turbidite sands. It is generally assumed that the unimodal crossbedding azimuth corresponds to the mean direction of the depositing current and in general this is probably true. But there are possible exceptions. Some rare types of crossbedding are of antidunal origin and dip upcurrent (rare backset bedding in some flood deposits and turbidites, and in base surge deposits, where it is fairly common). Lateral accretion bedding of bars is at right angles to the prevailing current. Turbidite deposits present a special case. There may be two intersecting sets of sole marks or a difference in orientation between the sole marks and ripple-bedding in the overlying sand. Contour currents, currents that parallel depositional strike, may be present in a few deep basins preserved on continents or their margins and, if so, may indicate paleoflow at right angles to that of the turbidity current (Stow and Lovell, 1979). The general consistency of many paleocurrent patterns leads to the conclusion that these exceptions are rare and of very minor importance.

Bimodal patterns, with mutually opposed modes, have been described from presumed littoral or shallow marine deposits, where it is assumed that the bipolar axis coincides with the axis of alternating tidal currents. This is in general correct but bimodal patterns can also arise from other causes. When dune and antidunes are both present in streams, a bipolar pattern may result (Power, 1961). And as McKee and Tibbitts (1964) have noted, recent seif dunes display bipolar patterns with the two opposed modes essentially perpendicular to the wind direction. Bipolar azimuthal patterns with perpendicular modes have been noted, especially in turbidite deposits (Kelling, 1964, p. 76; Hsu, 1964; Stow and Lovell, 1979). These suggest gravity controlled currents alternating with contour currents.

As rightly noted by Allen (1967, p. 80), sedimentary structures differ in scale by several orders of magnitude. A shoestring sand is a geometrical form generated in response to a flow system just as much as a crossbed set or a grain lineation. But these differ not only in scale but also in reliability as a guide to paleocurrent flow. Some are better estimators than others of the prevailing flow direction. The axes of trough crossbeds have a smaller variance than do the individual crossbedding measurements themselves. It is possible to arrange structures, or the bedforms responsible for them, in a hierarchical order; for example, in a fluvial system, ripples → crossbeds → point bars → channels → meander belts. In a general way, the higher the position in this order, the lower the azimuthal variance of the structure. The orientation of a large sand wave, or the crossbedding it produces, is a better guide to prevailing flow than small-scale ripples; a channel is a better estimator than a large-scale ripple. Allen (1966) also raised the question of bias introduced by the preservation potential of the structure. Experience indicates, however, that such a bias is not a factor in regionally consistent paleocurrent patterns.

Paleoslope and Current Directions

The second question—the relation of current flow, established by many measurements, and paleoslope—is more difficult. The paleoslope is the initial sedimentary dip or slope at the time of deposition. One might expect that sand-transporting currents would flow downslope and this is perhaps generally true, but there are important exceptions where the currents bear no relation to the slope of the depositional surface and they may, in fact, flow parallel to the sedimentary strike—contour currents—or even move upslope.

On the land, surface water runs downhill and hence in the fluviatile system the mean direction of flow is downslope. The paleoslope of alluvial deposits is generally unambiguous although, owing to the meandering of streams, a single observation is not sufficient to establish the slope direction.

So also with turbidity currents. A density current behaves as a heavy liquid and underflows clear water and moves downslope. But there are complications. The current motion is, in some sandstones (Potter and Pettijohn, 1977, p. 215), at right angles to the movement of a slumped bed. Presumably the latter moved downslope. Also there can be two intersecting sets of sole marks on the base of a turbidite bed. If one indicates downslope, does the other represent a diagonal motion on the same slope? This latter problem was discussed by Potter and Pettijohn (1977, p. 173), who thought that the two sets were the normal result of an expanding lobe of turbid waters on a near-flat surface, a development analogous to an expanding glacial lobe which also leaves two intersecting sets of striations.

The relation of currents and slope in the littoral zone and shallow marine shelf is much more complicated (Klein, 1967; Selley, 1968). There may be no discernible relation between the two. Movement is not always downslope. Selley (1968, pp. 104–105) cites examples from the geologic record of transport parallel to the sedimentary strike and even of upslope transport. A good example of the latter is the Loyalhanna Limestone (Adams, 1970, p. 16) in which abundant crossbedding indicates a landward movement. On the other hand, in a great many marine deposits the net transport is seaward; this transport system is reflected, in some basins, by the mineral dispersal fans and in others by crossbedding or other structures indicating downslope movement (Pelletier, 1965; Sedimentation Seminar, 1966).

Hubert (1966) and Hubert *et al.* (1978, Fig. 46), following the method of Hansen (1965), have inferred paleoslope from the orientation of the fold axes of slump structures that can be present on delta front deposits of either epicontinental seas or lakes. Because slumps are gravity induced, their orientation, if it can be accurately ascertained, should be a better measure of paleoslope than directional structures, some of which may be oblique or even parallel to the old shoreline and its slope into deeper water. Woodcock (1976) provides a very complete anal-

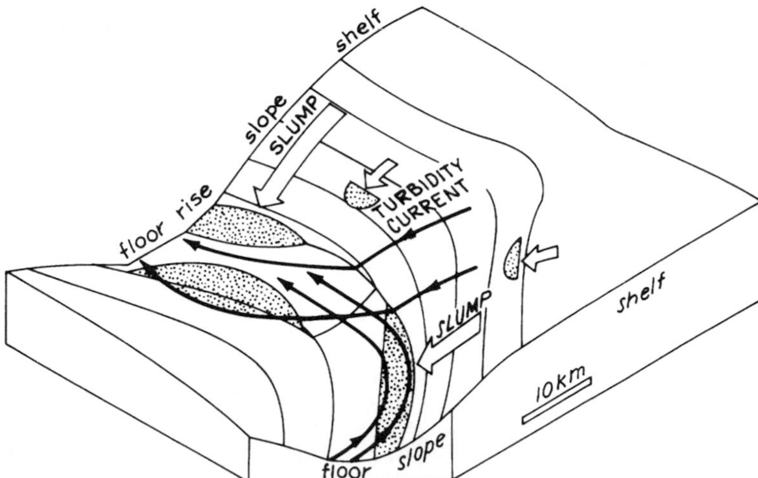

FIGURE 9-7. Direction of movement of slumps and paleocurrents in Silurian turbidites of Montgomery Trough in Wales (Modified from Woodcock, 1976, Fig. 9).

ysis of this problem, based on his studies of the Silurian of Wales (Fig. 9-7).

The relations of the orientation of ripple marks and the paleoslope are a special problem. Ripple marks are of two kinds, one wave-generated and the other produced by unidirectional flow. It is in the littoral and shallow marine areas that the interpretive problems arise. Owing to the complicated shoreline configurations, both waves and currents are variable and hence it would be presumed that the orientation of ripples, of whatever origin, would be quite random and not amenable to interpretation. Such an *a priori* conclusion, however, is not entirely justified. The Berea Sandstone (Mississippian) of Ohio, studied long ago by Hyde (1911), displays a remarkably consistent orientation of oscillation ripples over a distance of 185 km (115 mi). This and other similar examples (see Pelletier, 1965, p. 238) suggest that randomness is not necessarily the rule. The Berea ripples were believed to be parallel to the shore and hence parallel to the sedimentary strike. Wave crests are refracted and tend to become parallel to the shore regardless of the wind direction and their original orientation (Kindle, 1917, p. 53; Scott, 1930, p. 53). This conclusion has been borne out by study of present-day ripples off the Florida coast (Vause, 1959) and elsewhere in both modern (Davis, 1965) and ancient sandstones (Picard and High, 1968), although some exceptions have been noted (Stricklin and Amsbury, 1974, Fig. 3). Moore (1982) showed how much information could be obtained from ripple marks in a fine-grained, epicontinental deposit of Cambrian age in Australia.

In eolian deposits there is no causative connection between wind direction and the depositional slope. Coastal dunes, however, tend to migrate landward, and hence upslope, but the large, more important, inland dune fields show no such relation to slope.

Paleocurrent Models

Review of the literature suggests that there are a number of regional paleocurrent patterns which occur time and time again and which are related to the depositional environments in which the sediments accumulated (Fig. 4-38). Each pattern can be defined in terms of the environment and depositional process, the azimuthal paleocurrent pattern, and the relation of paleocurrents to paleoslope. These eight patterns chiefly occur in alluvial, shallow marine, turbidite or deep marine, eolian, and volcaniclastic sands. Within each are subclasses. And when the paleocurrent map is regional in scope, basin architecture also influences the pattern.

Alluvial

The alluvial environment is divided into fan, alluvial plain, and delta. Sand is a component of all three but it may be subordinate to coarser debris in the fan and to finer silts and clays in the delta.

Fans are characterized by a divergent radial pattern of transport (Fig. 9-8). The pattern will be reflected in crossbedding orientation, pebble imbrication, and a marked size decline of the gravel component.

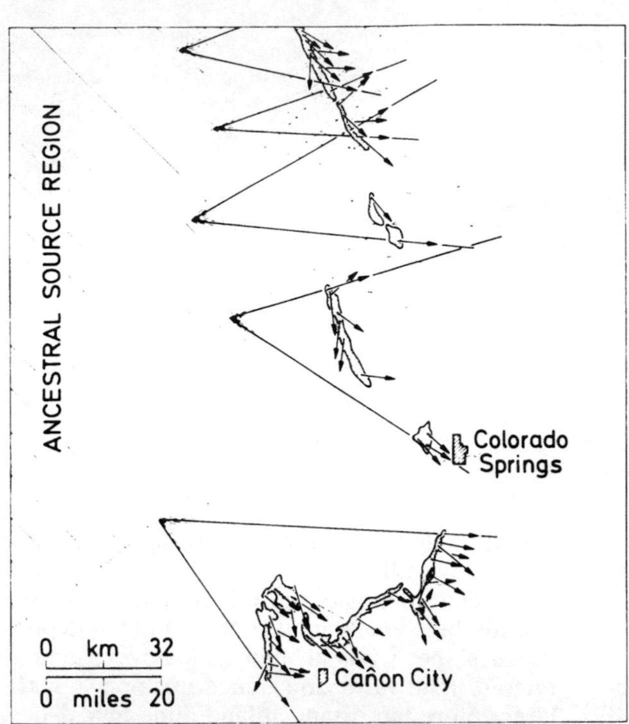

FIGURE 9-8. Imaginative interpretation of crossbedding orientation suggests deposition from point sources of alluvial sands at base of mountain front. Fountain Sandstone (Pennsylvanian) of central Colorado. (Redrawn from Howard, 1966, Fig. 3).

Where the basin is bordered by highlands and the margin abrupt, there may be a series of coalescing fans which pass into an extensive alluvial fill. The latter is traversed by braided and/or meandering streams. These generate crossbedding with a unimodal distribution and a mean downslope orientation. However, wide deviations from the mean are probable. It has been suggested that the variance increases downstream as the slope diminishes and meandering becomes dominant (Brinkman, 1933, p. 5–9; Pelletier, 1958, p. 1046). To date, it is not well established whether braided stream deposits have a lower variance of crossbedding orientation than meandering streams, although it seems logical that they should. However, crossbedding from marine shelf sandstones is more variable, as a rule, than in fluvial environments (Potter and Pettijohn, 1977, Table 4-2; Long and Young, 1978, Fig. 1).

Fluvial patterns can be converging, diverging, or parallel. They are converging where several streams drained toward the longitudinal axis of either a weak or well-defined elongate structural depression (elongate, broad intracratonic basins or intermontaine basins bordered by high relief). A weak centripetal pattern is shown by the Precambrian Athabasca Formation of Saskatchewan (Fahrig, 1961, Fig. 2). Potter *et al.* (1958, Figs. 15 and 17), mapped a similar pattern for the Mississippian Chesterian sandstones of the Illinois basin. Both examples are intracratonic basins with weakly defined longitudinal axes. Diverging patterns prevail where drainage was radial away from a highland of either low or high relief. Another source of diverging patterns is the delta distributaries of low-energy deltas, where crossbedding orientation is approximately parallel to the orientation of individual delta fingers (Hrabar and Potter, 1969, Figs. 3 and 9). Parallel patterns exist where there was a long linear source bordering a depositional basin with wide uniform slope. Under these conditions fluvial deposits can form a transgressive blanket above a uniform, wide paleoslope. An example of this pattern is shown by the Silurian of the Appalachian Basin (Yeakel, 1962; Whisonant, 1977).

Shoreline and Shallow Marine Shelf

Klein (1967) has summarized the dispersal patterns of Recent marine sediments. As he notes, the patterns are likely to be complex because the direction of sediment transport is a function of basin margin currents from deltas, wind-generated drift currents, tidal currents, and oceanic currents. (See Table 8-6). It is difficult to sort out the relative importance of these several current systems. Hence let us consider what sedimentologists have actually found by mapping directional structures in sandstones of ancient coastal and shelf environments.

In coastal barriers, which may be composites of beach deposits capped by eolian dunes, crossbedding orientation is commonly complex and the greatest care should be taken to segregate the eolian environment from that of the beach itself (cf. Chap. 11). Crossbedding in coastal eolianites is variably oriented but most commonly mean direction is landward, although not necessarily at right angles to the coastline. Regional studies of coastal dunes along 5000 km of the coastline of Brazil by Bigarella and his coworkers (1970/71) show a general oblique landward transport. In the beach environment itself, crossbedding—as opposed to the well-laminated, gently seaward dipping beds of the shoreface which may be interrupted by small channels—has variable orientation (Carter, 1978, Figs. 3 and 4; Clifton *et al.*, 1971, Fig. 26).

Estuarine sands also have a wide variety of bedforms, but their average crossbedding tends to roughly parallel that of the axis of the estuary; it may, however, dip landward as well as seaward because of tidal currents so that unimodal as well as bipolar currents prevail (Klein, 1970, Figs. 16 and 17). Sandy coastlines and estuaries, when seen at low tide, provide fine lessons of why paleocurrent observations should be collected at as many outcrops as possible rather than only at a few.

Beyond the shoreline there are two general cases, one in which fluviatile influence is present and the other in which fluviatile currents are absent (Fig. 9-9). Each of these can be subdivided according to whether marine transport is shoreward, longshore, or seaward. A common situation seen along modern coastlines is where the sediment, introduced by a

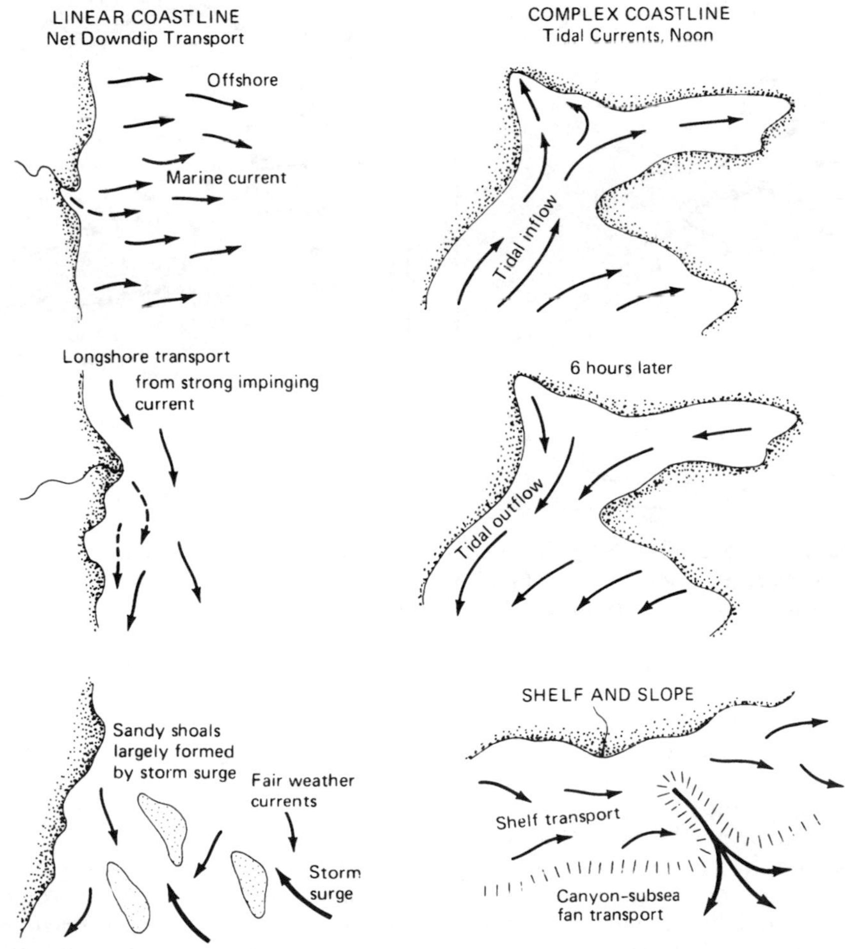

FIGURE 9-9. Shoreline paleocurrent models. Arrows show surface currents.

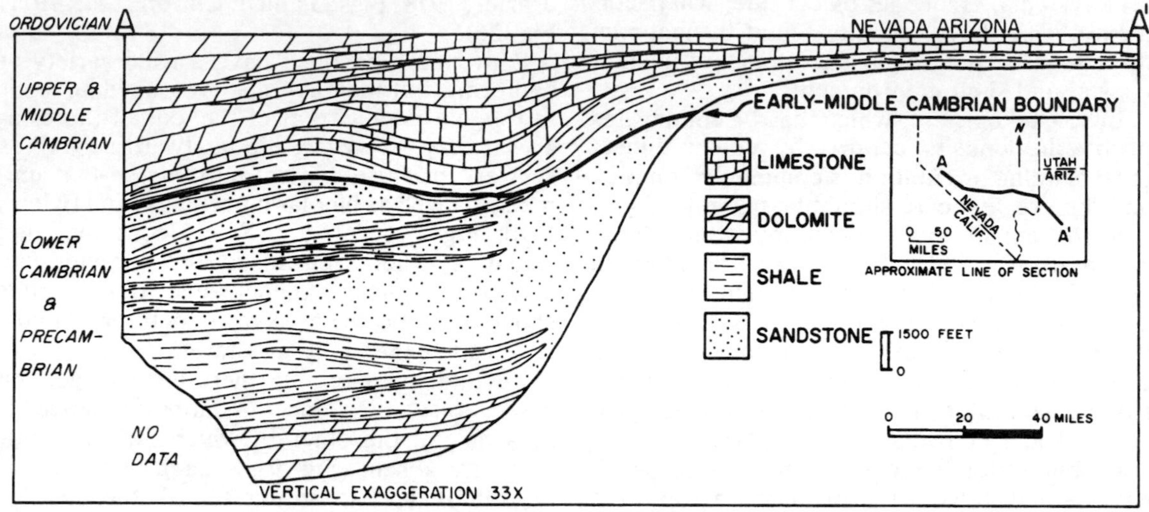

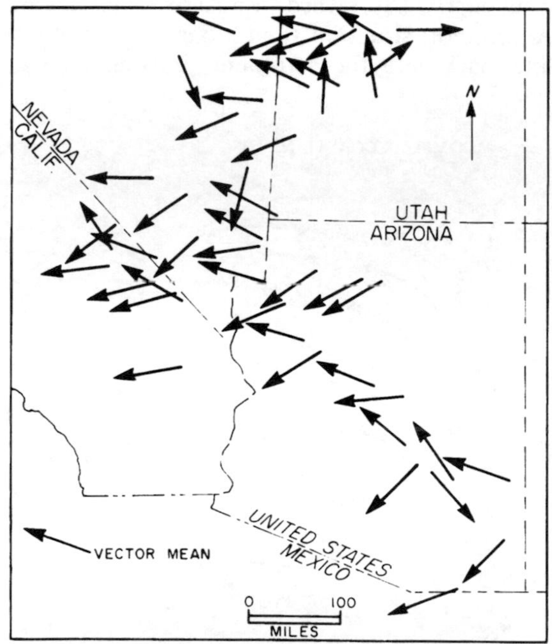

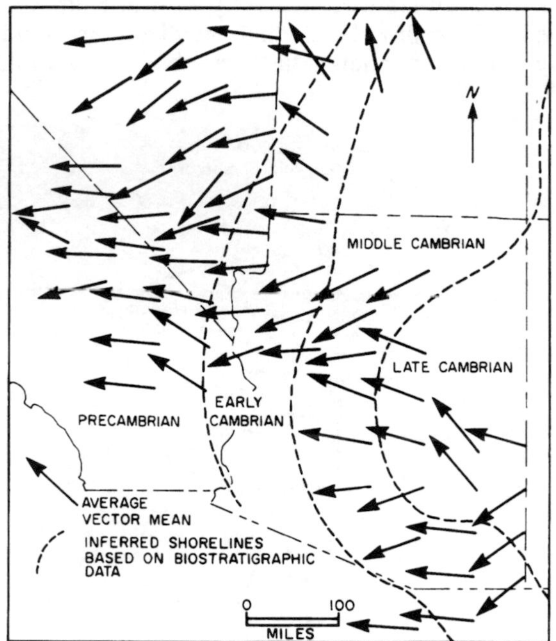

FIGURE 9-10. Marine paleocurrents (Precambrian to Cambrian) on shelf to basin transition in southwestern United States show persistence of paleoslope as sedimentation overlapped craton on the east (Seeland, 1969, Figs. 2 and 3).

river, is deflected by prevailing longshore currents. This is illustrated today in the Gulf of Mexico where the Mississippi River sands are carried westward along the coast. A fossil example of the same is described by Tanner (1955).

In most ancient sandstones of shelf origin net transport appears to be downslope or at some small angle to it, but with greater variability than in fluvial environments (Long and Young, 1978, Fig. 1); variance greater than 4000 is suggestive of a marine shelf environment. Bimodal current patterns are common, especially in sandy limestones and limestones (Young and Jefferson, 1975, Fig. 5) but are much rarer in sands and sandstones, although they do occur.

Owing to the complexities introduced by shoreline configurations, bottom topography, and bottom current systems of diverse origin, not all marine paleocurrent patterns can be fitted into these six cases. Polymodal crossbedding patterns have been recorded which defy analysis (Wermund, 1965). Perhaps these complexities are greatest proximal to the shoreline, in deposits which have the least preservation

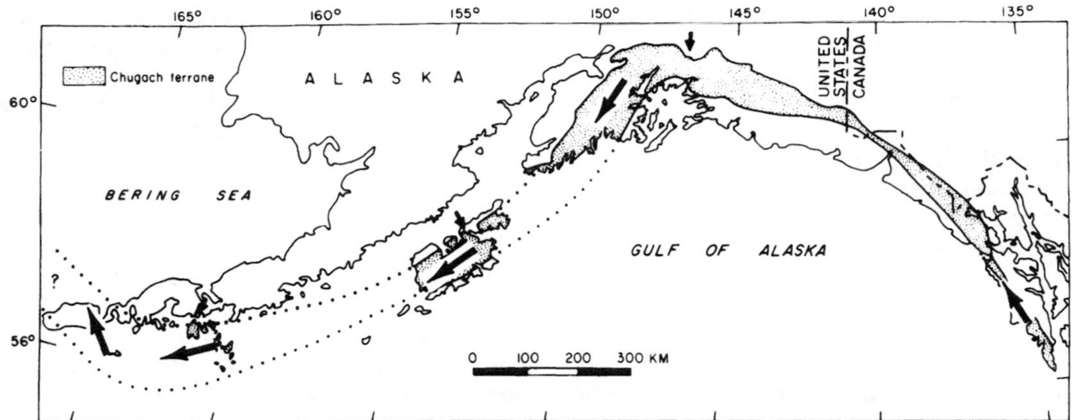

FIGURE 9-11. Paleoflow in turbidite sandstones of the Chugach Terrane, a Cretaceous trench-fill deposit, southern Alaska (Neilson and Zuffa, 1982).

potential, and more commonly the deeper marine shelf patterns are less influenced by the vagaries of tidal and other currents. Certainly the pattern of many marine sandstones shows a regional uniformity of seaward transport (Figs. 8-29 and 9-10).

The recognition of fair weather versus storm events—roughly speaking ripple marks and bioturbation versus hummocky bedding and tempestites (Fig. 8-26)—and tidal currents offers promise of resolving the total directional variability of a marine shelf sequence into genetically different parts that can be fitted into an overall model of shelf circulation. Fundamental to this approach is the separation of the different paleocurrent indicators in a marine shelf bedding sequence according to their origin—fair weather, normal shelf transport, storm events, and tides. See Duke (1985) and Boyles and Scott (1982) for two examples with full details and Chapter 8.

Basinal Turbidite Models

Paleocurrent systems generated by turbidity flows are, in many ways, analogous to fluvial systems. When a turbidity flow emerges from a submarine canyon it may build a submarine fan (Sullwold, 1960). The distal portions of such a fan merge into a basin plain, perhaps an abyssal plain. The fan has a radial paleocurrent pattern, and indeed such fans have been identified in the geologic record. In other basins turbidite deposits display a regional pattern astonishingly uniform over wide areas, as is the case with the Chugach Terrane (Fig. 9-11). Both lateral and longitudinal flow are present. Variance is quite small. Turbidity currents are unimodal and, except for the anomaly of intersecting sets of sole marks on the same bedding plane, are a record of the direction of flow. Where such uniformity is lacking and variance is greater, possibly the patterns are related to several supply points with their overlapping distribution systems and even the possibility of contour currents should be recognized.

The principal dilemma is posed by deposits where the direction of movement of slumped beds, as determined from orientation of fold axes, is about at right angles to the prevailing currents (Anketell and Lovell, 1976; Woodcock, 1976). Where slump movement was downslope, as certainly it *must have been*, the sand-depositing currents seem to have moved obliquely or parallel to the slope. Currents which behave in this manner have been called "contour currents," at least where the water was considered deep (Heezen *et al.*, 1966). Some workers have considered these to be normal oceanic currents—not turbidity currents. As the deposits resemble turbidites in most respects but apparently behaved in a manner quite unlike that postulated for such currents, some workers have concluded that the role of turbidity currents has been exaggerated (Hubert, 1966; Scott, 1966; Klein, 1967, p. 377). Others have considered the deposits to be turbidites, formed from currents moving longitudinally in the basin on a flat or near-flat floor, and the slump deposits to be derived from lateral slopes which provide enough initial velocity and momentum to carry the slumped materials out into the basin into the realm of turbidite deposition. Attempts have been made to draw distinctions between the deposits of turbidity currents and the presumed contour currents. The similarities seem to be greater than the differences. And although contour currents capable of moving sand seem to exist today in some modern ocean bottoms, they do not seem to provide a convincing mechanism for the production of the rhythmical alternation of graded sand and shale so characteristic of the deposits of the geologic record (Kuenen, 1967). It can only be said here that longitudinal, unimodal paleocurrent patterns characterize many flysch troughs (Kuenen, 1957; Hesse, 1973, Fig. 13), and that whether they are due to contour currents or turbidity currents is still a topic of contention. An excellent review is that of Stow and Lovell (1979).

While a basic model of laterally and axially filled marine troughs is now clearly established, the processes operating are still not fully understood.

Eolian Paleocurrent Patterns

Eolian crossbedding has been used as a guide to paleowind direction. The best review is that of Bigarella (1972), which is very comprehensive. Continental eolian sand sheets are among the most laterally extensive of the terrigenous sandstones—the Botucatú paleodesert of the Paraná Basin in Brazil is over 2500 km wide—and all are profusely crossbedded, many on a giant scale. Regionally consistent patterns of sand movement, both in the Recent and deduced from crossbedding in ancient eolianites, exist (Reiche, 1938; Shotten, 1937; Opdyke and Runcorn, 1960; Yaalon and Laronne, 1971), including some of Precambrian age (Ross, 1983). Such large-scale, consistent patterns have been related to high-pressure cells around which the winds circulate, clockwise in the northern hemisphere and the reverse in the southern. Eolian transport related to high-pressure systems in ancient dunes has been described by Shotten (1956), Bigarella and Salamuni (1964) and Poole (1964).

Volcanic

The paleoflow pattern of ash flow sands from a volcanic cone is the only common one that has a radial pattern—at least on a small scale—and

TABLE 9-3. Depositional environments, characteristic directional structures, and dispersal patterns.

Environment	Structure	Dispersal Pattern
Eolian	Crossbedding	Independent of paleoslope, crossbedding orientation if mapped over large areas, may indicate global atmospheric circulation especially when combined with paleolatitude based on paleomagnetics.
Alluvial	Crossbedding	Unimodal and downslope. Virtually always yields unambiguous results.
Beach	Crossbedding	Bimodal to variable orientation with respect to strandline although gently inclined lamination of berm dips seaward.
Low-energy delta	Crossbedding	Parallel to delta fingers except in splay deposits which are at right angles to distributary channels.
High-energy delta (estuary)	Crossbedding	Unimodal and bimodal with modes parallel to axis of estuary and hence perpendicular to regional shoreline. Tidal influence strong.
Shallow shelf	Crossbedding	Most variable with both bimodal and unimodal and rare random patterns. Net transport commonly needed for confirmation. Strong tidal influence. Consider storm events and hummocky crossbedding also.
Turbidite	Sole and ripple marks and slide structures	Unimodal, but complications possible from lateral as well as longitudinal flow, and some rare contour currents on ancient continental rises. Fold axes of slumps may yield best evidence of paleoslope.
Volcanic	Antidune bedding; crossbedding in epiclastic sands; sole marks in turbidites	Radial pattern in base surge deposits, fluvial pattern in epiclastic sands, and turbidite pattern in submarine volcanics.

only then when the entire deposit is preserved and/or the flows were symmetrically developed (cf. Crowe and Fisher, 1973, Fig. 4). Interbedded lavas will also have radial patterns as well as mudflows. On the distal flanks of such a cone some fluvial sands may also have crossbedding with radial pattern. Epiclastic volcaniclastic sands, those derived from the erosion of a volcanic terrane, may also have any of the other basic flow patterns.

On a much larger scale, one might expect radial paleoflow away from the center of an old, very large, stable craton that did not have bounding mobile belts. Continent-wide glacial-fluvial deposits can also have radial patterns.

Summary

The relations between the environment, directional structures, and paleocurrent patterns are summarized in Table 9-3. By combining paleocurrent data with facies patterns, one nearly always obtains an unambiguous interpretation, but a more refined analysis requires recognition of subenvironments and their corresponding subfacies within each of the major environments of Table 9-3. Insight into the origin of each of these facies is obtained by noting its paleocurrent direction and comparing it to that of its neighbors, as was done by Cant and Walker (1976, Table 5). By so doing, downslope channel orientation may be distinguished from overlying lateral accretion deposits, etc. In shallow water marine and lacustrine sands one should always distinguish paleocurrents produced in fair weather—the "normal" currents—from those produced by storm events.

Paleocurrents and Time

Stability of paleocurrent systems has been noted by several writers. Crossbedding in the Moine Series of Scotland displays a uniformity of orientation through a 12,000-ft (3660 m) section. This fact led Sir Edward Bailey to describe this uniformity as ". . . the most surprising single phenomenon" displayed by these strata (Wilson *et al.*, 1953, p. 386). The Lorrain quartzite, a Precambrian formation, shows no significant variation in crossbedding orientation through nearly 8000 ft (2440 m) of section. Current directions, determined largely from crossbedding, are remarkably persistent for most of the sandstones of Paleozoic age in the central and southern Appalachians; marked changes or reversals are few and such as did occur relate to major geologic events (Fig. 9-12). Potter and Pryor (1961, Fig. 8) demonstrated an equally stable and persistent transport pattern recorded in sandstones of the upper Mississippi Valley ranging in age from Early Paleozoic to the Tertiary so that the grand mean of all formations differs only slightly from that of present-day streams. Another example of a stable paleocurrent system was described by Barrett and Kohn (1975), who identified but one major change from Devonian through the Triassic in Antarctica.

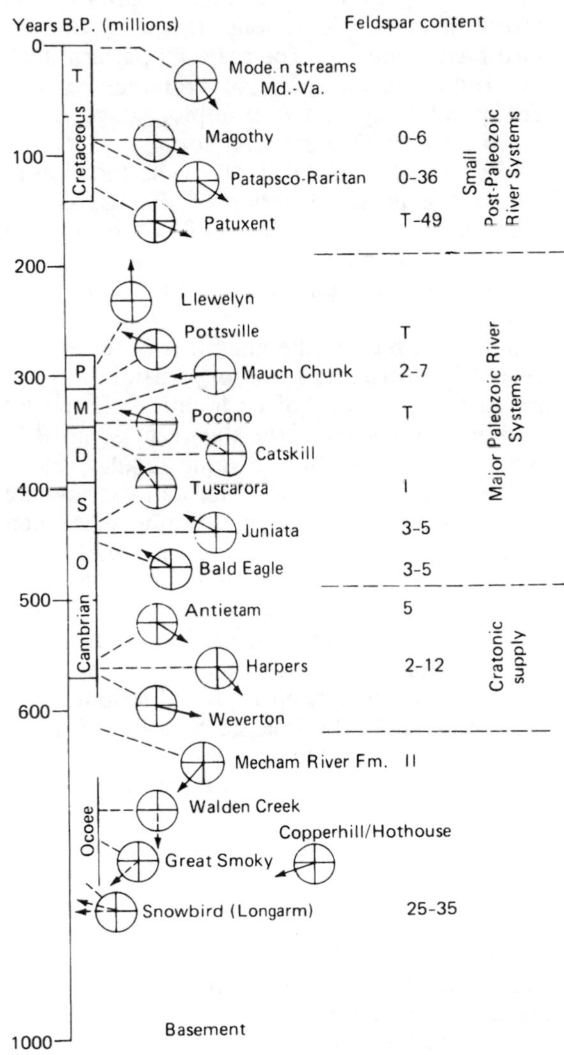

FIGURE 9-12. Paleocurrent data from sandstones, the central and southern Appalachians. (Data from J. Glaser, L. Yeakel, L. Meckel, B. Pelletier, J. Mellen, J. Whitaker, and F. Schwab).

Great variability occurs, however, in the turbidites of some mobile belts which have had a complex tectonic history of deep furrows and linear highs. The best mapped examples occur in the Carpathians (Contescu, 1969, Figs. 3 and 4; Marschalko, 1978).

Paleocurrents and Basin Analysis

As foreseen by Sorby, mapping of paleocurrents can assist in paleogeographic reconstructions. Such mapping of itself is no panacea but, used in conjunction with facies patterns and stratigraphic considerations, it can go far toward basin analysis. The regional pattern itself may reflect basin shape or architecture—the centripetal pattern, for example, suggesting a closed basin with peripheral supply.

A complete analysis involves the location of basin margins, the location of the sources of sediment supply, delineation of the shoreline if the basin is in part marine, and elucidation of the relation between the structural and bathymetric axes.

In ancient basins, the margin is generally not present—deformation and/or erosion has removed the evidence of its location. If the fill contains gravels and if the clast size diminishes downcurrent according to some regular rule—such as Sternberg's law—one can extrapolate in the upcurrent direction to some presumed reasonable initial size. The source ledges could have been no closer to the basin than the estimated distance. Presumably this is at or near the margin of the basin itself. In general, conglomeratic facies are most common in the proximal areas as illustrated by the distribution of conglomerates in the Triassic Newark basins of the eastern United States.

The paleocurrents help establish the polarity of a basin. The proximal margin is often marked by alluvial wedges which, in fact, form the bulk of the clastic fill. The distal margin is commonly indeterminate. The central Appalachian basin in mid-Paleozoic is an example (Meckel, 1970). The clastic fill is alluvial, thickest at the eastern edge, which fines westward and passes into marine shales and limestones. The paleocurrent pattern is consistent with this concept and indeed is part of the evidence for it.

In deep basins, largely marine, the sand input may be from the sides, though Kuenen (1957) believed many were end-fed by large rivers. The paleocurrent pattern of the turbidite sands in such basins is largely longitudinal.

The shoreline itself is the zero contour on the paleoslope. If the latter is defined by paleocurrent analysis, the strike of the shoreline can be determined. If the shoreline can be located by paleontologic or other criteria, at any one point, it can be extrapolated by drawing it normal to the dip of the paleoslope.

The nature of the basin, its tectonics and general architecture, is the subject of a later chapter where the questions of basin classification, relation to crustal plates, and larger topics are explored. Sand is an important component of almost all basin fills—except those of the deep ocean far removed from land—and the paleocurrent patterns disclosed by study of these sands bear some rational relation to the basin architecture and source area. The petrology of the sands must also reflect the source area, both its lithology and geomorphology.

Paleocurrents and Plate Tectonics

Planetary winds and continental glaciers are on a large enough scale to produce a paleocurrent pattern of continent-wide scope. Hence rifting and separation of continental blocks would disrupt and displace these systems. Mapping the paleocurrent systems thus disrupted should be consistent with the presumed separations and provide confirmation of such fragmentation. When an individual cratonic basin lies astride a zone of rifting, paleocurrent analysis should also assist in identifying the separated parts of such a basin. It is also possible that some old cratons that do not have bordering fold belts may have a continent-wide radial pattern preserved in their sandstones. A partial demonstration of such a pattern for the Devonian of the North Atlantic is shown in Fig. 9-13.

Bigarella (1973) has discussed the relation of paleocurrents and the problem of continental drift. In addition, attempts have been made to model and/or infer oceanic circulation patterns for continental configuration that may have prevailed in the past (Luyendyk *et al.*, 1972, Fig. 5). Although most of these patterns concern the open ocean, they do have a bearing on the study of paleocurrent patterns in sandstones of continental margins and conceivably might even have some effect on the interior seas of cratonic basins.

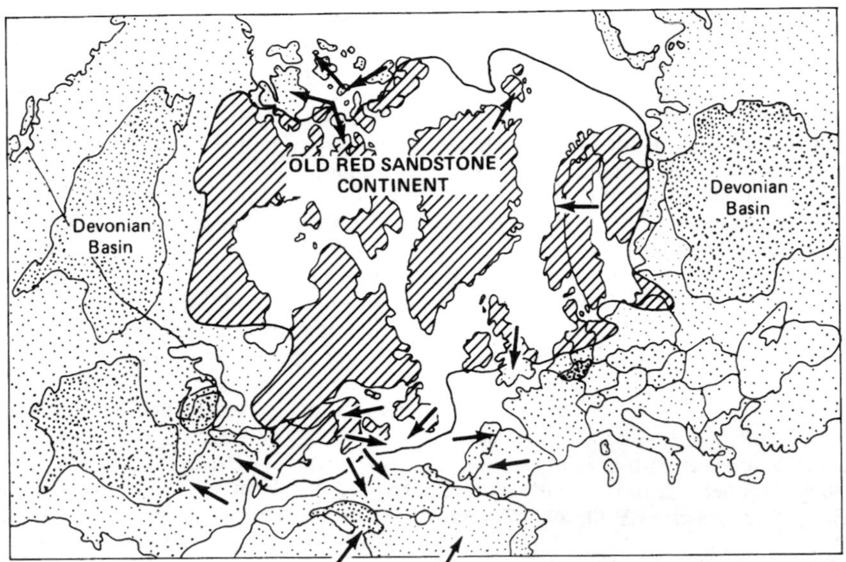

FIGURE 9-13. Paleocurrents and inferred reconstruction of the Old Red Sandstone Continent. (Redrawn from House, 1969). Arrows show generalized paleoflow (chiefly from Greiner, 1978).

References

Adams, R.W.: Loyalhanna Limestone—crossbedding and provenance. In: Fisher, G.W.; Pettijohn, F.J.; Reed, J.C.; and Weaver, K.N. (Eds.): Studies of Appalachian geology—central and southern, pp. 83–100. New York: Interscience 1970.

Allen, J.R.L.: On bed forms and paleocurrents. Sedimentology 6, 153–190 (1966).

Allen, J.R.L.: Notes on some fundamentals of palaeocurrent analysis, with reference to preservation potential and sources of variance. Sedimentology 9. 75–88 (1967).

Andel, Tj. H. van: Texture and dispersal of sediments in the Panama Basin. Jour. Geology 81, 434–457 (1978).

Anketell, J.M., and Lovell, P.H.: Upper Llandoverian Grogal Sandstones and Aberystwyth Grits in the New Quay area, central Wales: A possible upward transition from contourites to turbidites. Geol. Jour. 11, 101–108 (1976).

Baak, J.A.: Regional petrology of the southern North Sea, 127 pp. Wageningen, 1936.

Baker, G.: Sand drift at Portland, Victoria. Proc. Roy. Soc. Victoria 68, 151–197 (1956).

Baker, V.R.: Paleohydrology and sedimentology of Lake Missoula flooding in eastern Washington. Geol. Soc. America Spec. Paper 144, 79 pp. (1973).

Barrett, P.J., and Kohn, B.P.: Changing sediment transport directions from Devonian to Triassic in the Beacon Super-Group of South Victoria Land, Antarctica. In: Campbell, K.S.W. (Ed.): Gondwana geology, pp. 15–35. Canberra: Australia Natl. Univ. Press (1975).

Bigarella, J.J.: Wind pattern deduced from dune morphology and internal structures. Bol. Paranaense de Geosciencias 28/29, 73–113 (1970/71).

Bigarella, J.J.: Eolian environments—their characteristics, recognition and importance. In: Rigby, J.K., and Hamblin, W.K. (Eds.): Recognition of ancient sedimentary environments. Soc. Econ. Paleon. Mineral. Spec. Pub. 16, 12–62 (1972).

Bigarella, J.J.: Paleocurrents and the problems of continental drift. Geol. Rundschau 62, 447–477 (1973).

Bigarella, J.J., and Salamuni, R.: Paleowind patterns in the Botucatú Sandstone (Triassic–Jurassic) of Brazil and Uruguay. In: Nairn, A.E.M. (Ed.): Problems of palaeoclimatology, pp. 406–409. New York: Interscience 1969.

Boyles, J.M., and Scott, A.J.: A model for migrating shelf-bar sandstones in Upper Mancos Shale (Campanian), northwestern Colorado. Amer. Assoc. Petroleum Geologists Bull. 66, 491–508 (1982).

Bretz, J Harlen: The Lake Missoula floods and the channeled Scablands. Jour. Geology 77, 503–543 (1969).

Brinkman, R.: Über Kreuzschichtung im deutschen Buntsandsteinbecken. Nacht. Ges. Wiss. Göttingen, Math.-Physik. Kl. Bachgruppe IV 32, 1–12 (1933).

Cant, D.J., and Walker, R.G.: Development of a braided-fluvial facies model for the Devonian Battery Point Sandstone, Quebec. Canadian Jour. Earth Sci. 13, 102–119 (1976).

Carter, C.H.: A regressive barrier and barrier protected deposit: Depositional environments and geographic setting of the Late Tertiary Cohansey Sand. Jour. Sed. Petrology 48, 933–950 (1978).

Clifton, H.E.; Hunter, R.E.; and Phillips, H.: Depositional structures and processes in the non-barred high energy nearshore. Jour. Sed. Petrology 41, 651–670 (1971).

Cloos, H.: Primäre Richtungen in Sedimenten der rheinischen Geosynkline. Geol. Rundschau 29, 357–367 (1938).

Colton, G.W.: Orientation of carbonate concretions in the Upper Devonian of New York. U.S. Geol. Survey Prof. Paper 575B, 857–859 (1967).

Contescu, L.R.: Variation transversal des textures internes dans les flyschs albiens du bassin de la Bistrita, et son importance pour les models paléogéographiques. Rev. Roum. Geve. Geophys. Geog., Ser. Geo. 13, 81–96 (1969).

Crimes, T.P.: Facies analysis of the Cambrian of Wales. Palaeogeog. Palaeoclimatol. Palaeoecol. 7, 113–170 (1970).

Crowe, B.M., and Fisher, R.V.: Sedimentary structures in base-surge deposits with special reference to cross-bedding, Ubehehe Craters, Death Valley, California. Geol. Soc. America Bull. 84, 857–866 (1973).

Curray, J.K.: Tracing sediment masses by grain size modes. Intern. Geol. Cong. 21st, Copenhagen, 1960, Proc. Pt. 23, 119–130 (1961).

Davies, D.K., and Moore, W.R.: Dispersal of Mississippi sediment in Gulf of Mexico. Jour. Sed. Petrology 40, 339–353 (1970).

Davis, J.R., Jr.: Underwater study of ripples, southeastern Lake Michigan. Jour. Sed. Petrology 35, 857–866 (1965).

Duane, D.B.: Synoptic observation of sand movement. Proc. 12th Conf. Coastal Eng., Washington, D.C., 799–813 (1970).

Duke, W.L.: Hummocky cross-stratification, tropical hurricanes, and intense winter storms. Sedimentology 32, 167–194.

Duncan, J.R., and Kulm, L.D.: Mineralogy, provenance and dispersal history of Late Quaternary deep-sea sands in Cascadia Basin and Blanco Fracture Zone off Oregon. Jour. Sed. Petrology 40, 874–887 (1970).

Fahrig, W.F.: The geology of the Athabasca Formation. Geol. Survey Canada Bull. 68, 41 pp. (1961).

Firek, F.; Shidler, G.L.; and Fleischer, P.: Heavy-mineral variability in bottom sediments of the Lower Chesapeake Bay, Virginia. Marine Geology 23, 217–235 (1977).

Fox, J.S.: Rapid pyroclastic mapping in base metal exploration. Canadian Inst. Mining Bull. 70, 173–178 (1977).

Füchtbauer, H.: Die Schüttungen im Chatt and Aquitan der Alpenvorlandsmolasse. Eclogae géol. Helvetiae 51, 928–941 (1958).

Gazzi, S.P.; Zuffa, G.G.; Gandolfi, G.; and Paganelli, L.: Provenienza a dispersione litoranea delle sabbie delle spiagge adriatiche fra le faci dell' isonzo e del faglia: Inquadramento regionale. Mem. Soc. Italia 12, 1–37 (1973).

Glaser, J.D.: Nonmarine Cretaceous sedimentation in the middle Atlantic coastal plain. Ph.D. dissertation, Johns Hopkins Univ., Baltimore, 359 pp. (1967).

Goldstein, August, Jr.: Sedimentary petrologic provinces of the northern Gulf of Mexico. Jour. Sed. Petrology 12, 77–84 (1942).

Greiner, H.: Late Devonian facies interrelationships in bordering areas of the North Atlantic and their paleogeographic implications. Palaeogeog. Palaeoclimatol. Palaeoecol. 25, 241–263 (1978).

Hansen, E.: Methods of deducing slip-line orientation from the geometry of folds. Annual Rept. Director Geophys. Lab., 1964–1965, Carnegie Institution of Washington, D.C., pp. 386–405 (1965).

Heezen, B.C.; Hollister, C.; and Ruddiman, W.F. Shaping of the continental rise by deep geostrophic contour currents. Science 152, 502–508 (1966).

Henderson, J.B.: Sedimentology of Archean turbidites at Yellowknife, Northwest Territory. Canadian Jour. Earth Sci. 9, 882–902 (1972).

Hesse, R.: Flysch Gault und Falknis-Tasna-Gault (Unterkreide): Kontinuierlicher Übergang von der distalen zur proximalen Flyschfazies auf einer pennischen Trogebene der Alpen. Geologica et Palaeontologica SB2, 90 pp. (1973).

House, M.R.: Continental drift and the Devonian system, 24 p. Hull: Univ. Hull. Inaugural Lecture 1968.

Howard, J.D.: Patterns of sediment dispersal in Fountain Formation of Colorado. Mountain Geologist 3, 147–153 (1966).

Hrabar, S.V., and Potter, P.E.: Lower West Baden (Mississippian) sandstone body of Owen and Greene Counties, Indiana, Am. Assoc. Petroleum Geologists Bull. 53, 2150–2160 (1969).

Hsu, K.J.: Cross-lamination in graded bed sequences. Jour. Sed. Petrology 34, 379–388 (1964).

Hsu, K.J.: Studies of Ventura Field, California. Facies, geometry and genesis of Lower Pliocene turbidites. Am. Assoc. Petroleum Geologists Bull. 61, 137–168 (1977).

Hubert, J.F.: Sedimentary history of Upper Ordovician geosynclinal rocks, Girvan, Scotland. Jour. Sed. Petrology 36, 677–699 (1966).

Hubert, J.F.; Butera, J.G.; and Rice, R.F.: Sedimentology of Upper Cretaceous Cody-Parkman delta, southeastern Powder River Basin, Wyoming. Geol. Soc. Am. Bull. 83, 1649–1670 (1977).

Hubert, J.F.; Reed, A.A.; Dowdall, W.L.; and Gilchrist, J.M.: Guide to the Mesozoic redbeds of central Connecticut. State Geological and Natural History Survey of Connecticut, Dept. Environmental Protection, Guidebook 4, 124 pp. (1978).

Hyde, J.E.: The ripples of the Bedford and Berea Formations of central Ohio, with notes on palaeogeography of that epoch. Jour. Geology 19, 237–269 (1911).

Jacob, A.F.: Elongate concretions as paleocurrent indicators. Tongue River Formation (Paleocene), North Dakota. Geol. Soc. America Bull. 84, 2127–2132 (1973).

Johnson, H.D.: Tide- and wave-dominated inshore and shoreline sequences from the late Precambrian, Finnmark, North Norway. Sedimentology 22, 45–74 (1975).

Jones, T.A.: Statistical analysis of orientation data. Jour. Sed. Petrology 38, 61–67 (1968).

Jones, T.A., and James, W.R.: Analysis of bimodal orientation data. Mathematical Geology 1, 129–135 (1969).

Kelling, G.: The turbidite concept in Britain. In: Bouma, A.H., and Brouwer, H.A. (Eds.): Turbidites, pp. 79–91. Amsterdam: Elsevier Publ. Co. (1964).

Kindle, E.M.: Recent and fossil ripple marks. Bull. Mus. Geol. Survey Canada 25, 1–56 (1917).

Klein, G. deVries: Paleocurrent analysis in relation to modern marine sediment dispersal patterns. Am. Assoc. Petroleum Geologists Bull. 51, 366–381 (1967).

Klein, G. deVries: Depositional and dispersal dynamics of intertidal sand bars. Jour. Sed. Petrology 40, 1095–1127 (1970).

Kuenen, Ph.H.: Longitudinal filling of oblong sedimentary basins. Verhandl. Koninkl. Nederlandse Geol. Mijn. Genootschap. Geol. Ser. 18, 189–195 (1957).

Leggewie, R.; Füchtbauer, H.; and El-Najjar, R.: Zur Bilanz des Buntsandsteinbeckens (Korngrössenverteilung und Gesteinsbruchstücke). Geol. Rundschau 66, 551–577 (1977).

Long, D.G.F., and Young, G.M.: Dispersion of cross-stratification as a potential tool in the interpretation of Proterozoic arenites. Jour. Sed. Petrology 48, 857–862 (1978).

Luyendyk, B.P.; Forsyth, D.; and Phillips, J.D.: Experimental approach to the paleocirculation of the oceanic surface waters. Geol. Soc. America Bull. 83, 2649–2664 (1972).

Marschalko, R.: Vývoj sedimentárnych bazenov A paleotektonické rekonstrukcie Západnych Karpát (Evolution of sedimentary basins and paleotectonic reconstructions of the West Carpathians). In: Vozar, J. (Ed.): Paleograficky vývoj Zapadnych Karpát. Geol. Ustav Dionyza Stura, Bratislava, Czechoslovakia, pp. 49–80 (1978).

McKee, E.D., and Tibbitts, G.C.: Primary structure of a seif dune and associated deposits in Libya. Jour. Sed. Petrology 34, 5–17 (1964).

Meckel, L.D.: Palaeozoic alluvial deposition in the central Appalachians: A summary. In: Fisher, G.W.; Pettijohn, F.J.; Reed, J.C., Jr.; and Weaver, K.N. (Eds.): Studies of Appalachian geology—central and southern, pp. 49–68. New York: Wiley Interscience 1970.

Mellen, J.: Pre-Cambrian sedimentation in the northeast part of Cohutta Mountain Quadrangle, Georgia. Georgia Mineral Newsletter, Georgia Geol. Survey, 9, 46–61 (1956).

Moore, P.S.: Ripple-mark analysis of a fine-grained epeiric-sea deposit (Cambrian, South Australia). Jour. Geol. Soc. Australia 29, 71–81 (1982).

Nanz, R.H., Jr.: Exploration of earth formations associated with petroleum deposits. U.S. Patent 2,963,641 (1960).

Neilson, T.H., and Zuffa, G.G.: The Chugach Terrane, a Cretaceous trenchfill deposit, southern Alaska. In: Leggett, J.K. (Ed.): Trench-forearc geology. Geol. Soc. London Spec. Pub. 10, 213–228 (1982).

Opdyke, N.D., and Runcorn, S.K.: Wind direction in the western United States in the Late Paleozoic. Geol. Soc. America Bull. 71, 959–972 (1960).

Pelletier, B.R.: Pocono paleocurrents in Pennsylvania and Maryland. Geol. Soc. America Bull. 69, 1033–1064 (1958).

Pelletier, B.R.: Paleocurrents in the Triassic of northeastern British Columbia. In: Middleton, G.V. (Ed.): Primary sedimentary structures and their hydrodynamic interpretation. Soc. Econ. Paleon. Mineral. Spec. Publ. 12, 233–245 (1965).

Perkins, B.F. (Ed.): Aspects of Trinity Division Geology. Geoscience and Man 8, 53–66 (1974).

Pettijohn, F.J.: Paleocurrents and paleogeography. Am. Assoc. Petroleum Geologists Bull. 46, 1468–1493 (1962).

Picard, M.D., and High, L.R., Jr.: Shallow marine currents on the early(?) Triassic Wyoming shelf. Jour. Sed. Petrology 38, 411–423 (1968).

Poole, F.G.: Paleowinds in western U.S.A. In: Nairn, A.E.M. (Ed.): Problems in paleoclimatology, pp. 390–405. New York: Interscience 1964.

Potter, P.E.: Sedimentology: Past, present and future. Die Naturwissenschaften Jahrgang 61(11), 461–512 (1974).

Potter, P.E., and Pettijohn, F.J.: Paleocurrents and basin analysis, 2nd Ed., 425 pp. New York: Springer-Verlag 1977.

Potter, P.E., and Pryor, W.A.: Dispersal centers of Paleozoic and later clastics of the Upper Mississippi Valley and adjacent areas. Geol. Soc. America Bull. 72, 1195–1250 (1961).

Potter, P.E.; Nosow, E.; Smith, N.M.; Swann, D.H.; and Walker, F.H.: Chester crossbedding and sandstone trends in Illinois Basin. Am. Assoc. Petroleum Geologists Bull. 42, 1013–1046 (1958).

Power, W.R.: Back set beds in the Coco Formation, Inyo County, California. Jour. Sed. Petrology 31, 603–607 (1961).

Reiche, P.: An analysis of cross-lamination; the Coconino Sandstone. Jour. Geology 46, 905–932 (1938).

Ross, D.A.: Source and dispersion of surface sediments in the Gulf of Maine—Georges Bank Area. Jour. Sed. Petrology 40, 906–920 (1970).

Ross, G.M.: Aeolian quartz arenites from the Hornby Bay Group, Northwest Territories, Canada: Implications for Precambrian aeolian processes. Precambrian Research 20, 149–160 (1983).

Schmincke. H.-U.: Flow direction in Columbia River basalt flow and paleocurrents of interbedded sedimentary rocks, southcentral Washington. Geol. Rundschau 56, 992–1020 (1967).

Schwab, F.L.: Mechum River Formation: Late Precambrian(?) alluvium in the Blue Ridge Province of Virginia. Jour. Sed. Petrology, 44, 862–871 (1974).

Scott, G.: Ripple marks of large size in the Fredericksburg rocks west of Fort Worth, Texas. In: Contributions to geology 1930. Univ. Texas Bull. 3001, 53–56 (1930).

Scott, K.M.: Sedimentology and dispersal patterns of a Cretaceous flysch sequence, Patagonian Andes, southern Chile. Am. Assoc. Petroleum Geologists Bull. 50, 72–107 (1966).

Sedimentation Seminar: Cross-bedding in the Salem Limestone of central Indiana. Sedimentology 6, 95–114 (1966).

Seeland, D.A.: Marine current directions in Upper Precambrian and Cambrian rocks of the southwestern United States. In: Geology and natural history of the Grand Canyon Region. Four Corners Geol Soc. Guidebook, 4th Field Conference Powell River Expedition (Durango, Colorado), pp. 123–126 (1969).

Selley, R.C.: A classification of paleocurrent models. Jour. Geology 76, 99–110 (1968).

Shawa, M.S. (Ed.): Use of sedimentary structures for recognition of clastic environments, 66 p. Calgary: Canadian Soc. Petroleum Geologists 1974.

Shawa, M.S. (Ed.): Guidebook to selected sedimentary environments in southern Alberta, Canada. Canadian Soc. Petroleum Geologists Field Conf., 1975, 56 pp. (1975).

Shelton, J.W.; Burman, H.R.; and Noble, R.L.: Directional features in braided meandering stream deposits, Cimarron River, northcentral Oklahoma. Jour. Sed. Petrology 44, 1114–1117 (1974).

Shotten, F.W.: The lower Bunter Sandstone of north Worcestershire and east Shropshire. Geol. Mag. 74, 534–553 (1937).

Shotten, F.W.: Some aspects of the New Red Desert in Britain. Liverpool and Manchester Geol. Jour. 1, 450–465 (1956).

Skipper, K.., and Bhattacharjee, S.B.: Backset bedding in turbidites: A further example from the Cloridorme Formation (Middle Ordovician) Gaspé, Québec. Jour. Sed. Petrology 48, 193–202 (1978).

Soares, Paulo C.: Divisão estratigráfica do Mesozóico no Estado de São Paulo. Rev. Brasileira Geosciências 5, 229–251 (1975).

Sorby, H.C.: On the physical geography of the Tertiary estuary of the Isle of Wright. Edinburgh New Philosophical Jour. 5, 275–298 (1857).

Steinmetz, R.: Analysis of vectorial data. Jour. Sed. Petrology 32, 801–812 (1962).

Stow, D.V.A., and Lovell, J.P.B.: Contourites: Their recognition in modern and ancient sediments. Earth Sci. Rev. 14, 251–291 (1979).

Stricklin, F.L., Jr., and Amsbury, D.L.: Depositional environments on a low-relief carbonate shelf, middle Glen Rose Limestone, central Texas. In: B. F. Perkins (Ed.): Aspects of Trinity Division Geology. Geoscience and Man 8, 53–66 (1974).

Sturm, M., and Matter, A.: Sedimente und Sedimentationsvorgange in Thunersee. Eclogae geol. Helvetiae 65, 563–590 (1972).

Sullwold, H.H., Jr.: Tarzana fan, deep submarine fan of late Miocene age, Los Angeles County, California. Am. Assoc. Petroleum Geologists Bull. 44, 433–457 (1960).

Tanner, W.F.: Paleogeographic reconstruction from cross-bedding studies. Am. Assoc. Petroleum Geologists Bull. 39, 2471–2483 (1955).

U.S. Corps of Engineers: Studies of river bed materials and their movement, with special reference to the Lower Mississippi River. Vicksburg, U.S. Waterways Expt. Sta. Paper 17, 161 pp. (1935).

Vause, J.E.: Underwater geology and analysis of Recent sediments of the northwest Florida coast. Jour. Sed. Petrology 29, 555–563 (1959).

Wermund, E.G.: Cross-bedding in the Meridian Sand. Sedimentology 5, 69–80 (1965).

Whisonant, R.C.: Lower Silurian Tuscarora (Clinch) dispersal pattern in western Virginia. Geol. Soc. America Bull. 88, 215–220 (1977).

Whitaker, J.C.: Direction of current flow in some Lower Cambrian clastics of Maryland. Geol. Soc. America Bull. 66, 763–766 (1955).

Wilson, G.; Watson, J.; and Sutton, J.: Current-bedding in the Moine Series of northwestern Scotland. Geol. Mag. 90, 377–389 (1953).

Woodcock, N.H.: Ludlow Series slumps and turbidites and the form of the Montgomery Trough, Powys, Wales. Proc. Geol. Assoc. 87, 169–182 (1976).

Yaalon, D.H., and Laronne, J.: Internal structures in eolianites and paleowinds, Mediterranean Coast, Israel. Jour. Sed. Petrology 41, 1059–1064 (1971).

Yasso, W.E.: Tracer techniques in sediment transport. In: Fairbridge, R.W., and Bourgeois, J. (Eds.): The encyclopedia of sedimentology, pp. 813–816. Stroudsburg, Pennsylvania: Dowden, Hutchinson, and Ross 1978.

Yeakel, L.S., Jr.: Tuscarora, Juniata, and Bald Eagle paleocurrents and paleogeography in central Appalachian. Geol. Soc. America Bull. 73, 1513–1540 (1962).

Young, G.M., and Jefferson, C.W.: Late Precambrian shallow water deposits, Banks and Victoria Islands, Arctic Archipelago. Canadian Jour. Earth Sci. 12, 1734–1748 (1975).

CHAPTER 10
Sandy Depositional Systems

Introduction

Why is the study of sandy depositional systems central to the understanding of sand and sandstone? From earliest times geologists have wanted to know where and under what conditions a sandstone was deposited—to better envision its areal extent, its internal properties, and its relation to associated sediments. Sandy depositional systems also play a key role in helping us assign a basin to its proper type and in relating it to a plate tectonic or other genetic framework. With the exception of some carbonates, sandstones are commonly the most permeable of all sedimentary rocks—a key element in their economic significance for oil, gas, groundwater, and later mineralization by uranium, copper, zinc, and lead. In addition, placer deposits such as gold, rutile, and ilmenite are directly related to sandy facies. Identifying the correct depositional model and using it to map and predict the areal extent of a depositional system or one of its parts is the essential key to understanding the origin of known economic sandy deposits and finding new ones. Identification and mapping of sandy deposits in and across ancient basins also play a key role in classifying basins, help us better understand them, and help us place ancient basins in their proper paleoplate setting. Thus, from applied and theoretical viewpoints, we need to be able to identify and interpret ancient sandy deposits. In the 1970s and early 1980s studies on the depositional systems became a focal point for many sedimentologists with the result that today we routinely use both outcrops and subsurface data—wireline logs and cores—to distinguish better the diverse environments of sandy deposits.

Sand, generated principally by the erosion of continents and their margins, occurs throughout the geologic column in eight major environments—alluvial fans; meandering, anastomosing and braided streams; deltas; fan deltas; coastal barriers; marine shelves; deep marine basins; and eolian dune fields. Subenvironments that contain sand are numerous and form a natural hierarchy (Fig. 10-1). Studies of both Holocene and ancient sandy environments have contributed to today's understanding, and each major environment has a *depositional model*—an idealization of the key features of the environment best displayed by block diagrams and cross sections and commonly summarized in a table. A depositional model, as defined by Harms and Tackenberg (1972, p. 46), is

> . . . a simplified summary of the properties of strata deposited in a general environment. Within any environment, only certain processes of erosion, transport, and deposition of sediment operate. These processes impose restrictions on the thickness, continuity, and spacing of beds, the nature of contacts, and the type and rate of vertical or lateral change.

This definition stresses both formative processes and resulting products, applies to all environments, and is especially useful in seismic stratigraphy. As pointed out by Walker (1984, p. 5), a depositional model of a sedimentary environment serves as a *norm* for comparison, is a *framework* and *guide* for future observations, and is a *predictor* in exploration.

Fisher and McGowen (1967, Fig. 9) introduced the concept of a *depositional system*, a genetically related group of depositional environments (Table 10-1). The seven depositional systems are largely independent of scale so that in different basins the same depositional system can have different dimensions and thickness. To appreciate this, consider, for example, deltaic sandstones on a vast, stable craton versus those of a small, discrete, rapidly subsiding, intermontane basin along a major fault; the deltaic sandstones of the craton will be thin and

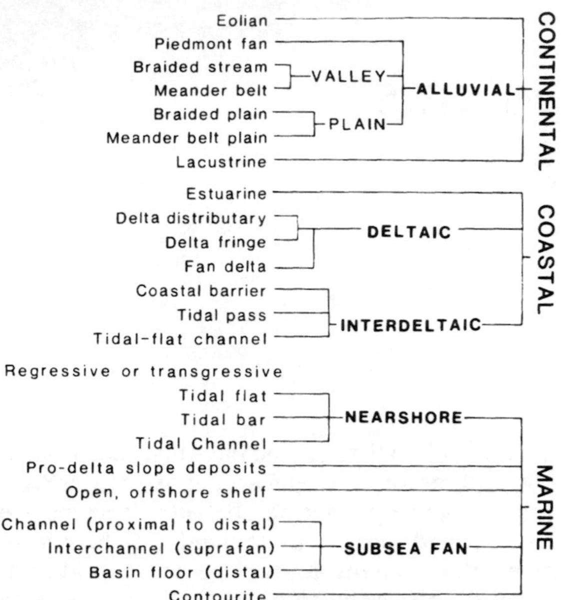

FIGURE 10-1. Natural hierarchy of environments that contain sand.

TABLE 10-1. Sandy depositional systems and their common components.

Part A: Major sandy facies
Alluvial fans
Channel and interchannel gravels and sands of braided and ephemeral streams, debris flows, minor muds and, in arid climates, dunes plus associated playas and saline lakes and bajadas.
Fluvial Systems
Sands and sand gravels of braided and meandering streams, levee and floodplain deposits including clay plugs and dunes plus lakes.
Deltas
River-, tidal- and wave-dominated deltas plus possible interdistributary swamps, lakes, and bays, floodplains, crevasse fills, delta front deposits, coastal barriers, and dunes—much variety depending upon the type of delta and its climate. Carbonates possible in abandoned delta lobes
Fan deltas
Braided and minor meandering stream deposits, debris flows, coastal and possible turbidite sands; may be transgressed by shelf carbonates.
Sandy coastlines
Barrier beaches, coastal dunes, washover fans, lagoons, inshore marine sands and muds; tidal and fluvial sands and muds; carbonates.
Marine shelves
Submarine dune fields, sand sheets, tidal sands, deep transverse channel fills; carbonates and marine muds.
Deep basin and slope
Subsea fans and base of slope deposits, debris flows and slides; deep marine muds.
Part B: Specialized facies
Desert
Large and small dune fields of inland and coastal deserts, wadis, coastal barriers, sabkhas, and ephemeral lakes and lagoons; minor carbonates.
Glacial
Sandy to clay-rich tills and associated outwash plus lake deposits; minor dunes and peats.
Lacustrine
Coastal barriers, sheet sands, deltas and marginal dunes.

widespread whereas those of the intermontaine basin will be thick and stacked. Moreover, the proportions of the different environments in the same depositional system will also vary with factors such as climate, river discharge, inshore wave power, and ratio of terrigenous supply to basin subsidence. Thus the concept of a depositional system has not only universality but flexibility as well, and has wide use in seismic stratigraphy, in the study of sedimentary basins, and in single reservoirs.

The elements of depositional models have been recognized for some time (Table 10-2) and their vertical variation is an important key to the recognition of ancient environments. Vertical variation is defined by grain size, type and abundance of sedimentary structures, bed thickness, and proportions of interbedded rocks. We study both modern and ancient sandy deposits, in outcrop and in the subsurface, by both direct observation of outcrops and cores and by indirect observation using downhole geophysical logs and seismic cross sections. Indirect methods are used more and more but all rely on the fundamental depositional models to transform fragmentary and incomplete data to maps or cross sections and predict thickness, geometry, porosity and permeability, or mineral content. Today, many sandy deposits can be rather easily assigned to a depositional model, especially if many properties can be studied. Nonetheless, we have much to learn and perhaps 20 to 40 percent of all sandstones still pose challenging problems of environmental recognition. We should also recognize that, in the past, conditions at the earth's surface, such as plant cover, climate, and perhaps even the scale or intensity of major depositional processes could have been different, so we should not be surprised that some ancient sandy deposits are hard to fit into Holocene models (Schumm, 1968, p. 1585).

There are many excellent references on sandy deposits including Medeiros *et al.* (1971),

TABLE 10-2. Elements of stratigraphic models

GEOMETRY	INTERNAL CHARACTERISTICS
Dimensions: length, width and thickness.	Sedimentary structures: kinds, abundance, and orientation. Biogenetic structures.
Orientation: parallel or perpendicular to depositional strike; random.	Bedding: type, thickness, and homogeneity.
Position in basin: relation to major paleophysiographic features such as shelf edge, shoreline, etc.	Constituents: framework grains, fossils, and cements.
	Texture: mean, sorting, and skewness.
BOUNDING LITHOLOGIES	ORGANIZATION
BASAL CONTACT	Putting the parts together, the essential key to it all. Based on modern and ancient analogues.

LeBlanc (1972), Rigby and Hamblin (1972), Shelton (1973), Reading (1986), Walker (1984), Klein (1980), Reineck and Singh (1980), Scholle and Spearing (1982), Galloway and Hobday (1983), and Davis (1983) plus a host of more specialized books and articles. A good summary in Spanish is by Spalletti (1980). Much can also be learned from the literature of regional summaries of sand facies; e.g., Hyne (1979), Campbell (1981), and Galloway *et al.* (1983), all three of which are exceptionally informative.

Concepts and Methods

The methods used to study sand deposits are diverse. Outcrops in an Andean mountain chain or Precambrian shield demand one set of techniques, study of a densely drilled basin uses others, while the exploration of a virtually undrilled offshore continental margin requires a still different approach. To cope with this diversity, a wide range of disciplines are needed (Fig. 10-2). Ample outcrops or abundant subsurface information are equally useful.

Depositional Strike, Paleoslope, and Walther's Law

Depositional strike plays an important role in the study of sandstones, because it helps us correctly orient the depositional model or system and relate it to the shape of the basin. Paleoslope, at right angles to depositional strike, can be used equally as well.

In a modern basin, depositional strike is defined by the orientation of the shoreline, by the trend of bathymetric contours offshore, and by the transport paths of the basin's alluvial deposits. In an ancient basin, size gradients, paleocurrent system, and facies distribution are used to infer depositional strike. A coastal barrier is a good example of a single sedimentary environment that defines depositional strike and separates marine from nonmarine deposits. Depositional strike may be uniform, curvilinear, or ovate (Fig. 10-3). Knowledge of depositional strike in an ancient basin helps predict its facies distribution, orientation of its sandstone bodies and is essential for the construction of cross sections, which are most informative when they are parallel or perpendicular to depositional strike.

To predict the facies patterns of an interval, we combine knowledge of depositional strike with Walther's Law (Walther, 1894, p. 979), which states that lithologic transitions in vertical section reflect laterally migrating environments. Shaw (1964, p. 50) and Visher (1965, pp.

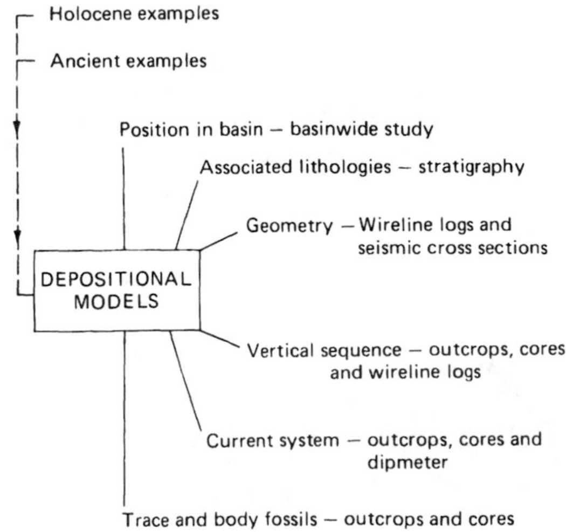

FIGURE 10-2. Principal sources and disciplines needed for the study of sandy environments.

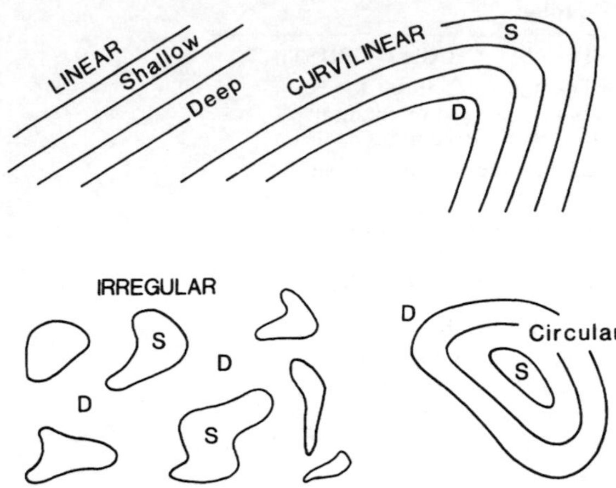

FIGURE 10-3. Contrasting patterns of depositional strike in shallow and deep water.

41–42) reemphasized Walther's law as an effective means of predicting lateral occurrence from a vertical sequence. This procedure assumes that there are no erosional breaks in the section and that a vertical dependence or memory does in fact exist in the sequence.

Whether a vertical section displays a memory (see p. 346) and whether memory is short or long (does the presence or absence of a lithology or bed depend only on the immediately preceding or on several preceding lithologies?) is primarily a consequence of the spatial distribution of environments at the time of deposition (Fig. 10-4). Suppose, for example, that a series of depositional environments is related to water depth which increases systematically offshore and that each environment produces a distinctive sediment type. Ideally, slow transgression will result in a stratigraphic sequence wherein all or nearly all of the environments will be represented in the same order as they occur offshore. As the different environments transgress shoreward, they leave behind a strongly linked sequence, one with a memory extending back one or more units or steps. Suppose, on the other hand, that the different depositional environments occur more or less at random in a "crazy quilt" or haphazard pattern rather than having systematic relations to one another or to the strandline (Fig. 10-4). Shoreward transgression will now produce a stratigraphic section without strong vertical dependence—essentially a "memoryless" section.

These two hypothetical and rather extreme examples, assuming stratigraphic sections free of depositional breaks, suggest the following: sections with strong vertical memory accumulated in basins which had well-defined, strong systematic environmental zonation and, conversely, sections with little or no memory accumulated in basins with little or no systematic zonation of environments.

Effective prediction of a sand body or any other lithology with Walther's Law in a new borehole requires (1) that memory does in fact exist in the sequence, (2) knowledge of depositional strike (paleoslope), (3) keeping in the same cycle, and (4) that map pattern remains the same during facies migration. Successful prediction is enhanced if boundaries between environments are fairly straight and if units are thick rather than thin.

Geometry and cross section of sandstone bodies vary greatly among different sandy envi-

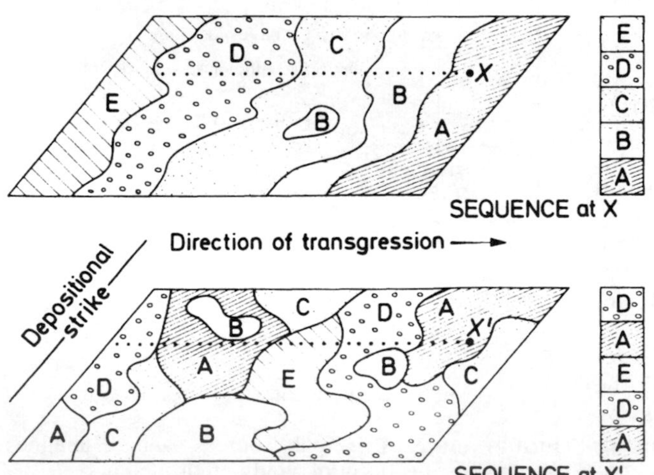

FIGURE 10-4. Map pattern of depositional environments and resultant vertical column—regular pattern yields regular, vertical section but "crazy quilt" pattern does not.

ronments, and terminology, like that of bedforms, is not universally agreed upon, although the terms sheet, wedge, ribbon, pod, belt, and distributary are useful (Fig. 10-5).

Sheets are many times wider than thick, may be up to 100 km or more wider, and tend to have uniform thickness and cross section. Sheets may be thick or thin and commonly occur as a transgressive or regressive deposit or be of eolian origin. Wide sheets tend to be laterally composite—when carefully studied they tend to be found to consist of several environments. Wedge geometries are typical of alluvial and subsea fans and fan deltas. Ribbons are long, elongate narrow sand bodies characteristic of either coastal barriers, stream channels, or isolated marine shelf sandstone bodies; cross section varies with origin. Pods are small, isolated sandstone bodies that vary from ovate to slightly elongate to irregular and commonly occur in the marine shelf environment, where they may be marginal to a marine shelf sandstone. Belts are wide, elongate composite sandstones formed by braided or meandering stream- or density-cut channels with boundaries that can either be fairly straight or meandering. Tributaries or distributaries may be present. Cross section is usually concave downward and in wide belts may be quite irregular. The branching or distributary geometry is a common one and nearly always has a concave-downward cross section.

The terms *multistory* and *multilateral* are used to describe vertical and lateral composite sandstone bodies, which are common.

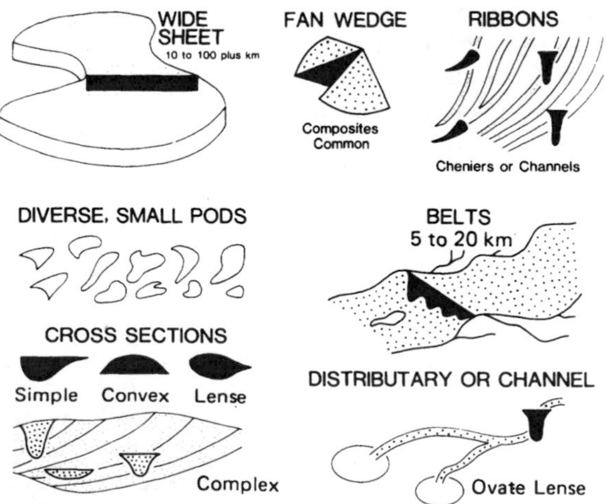

FIGURE 10-5. Idealized geometry of sandstone bodies.

Associated Lithologies, Vertical Sequence, and Memory

Sandy deposits are virtually always interbedded with other sedimentary lithologies, grade into them, or may rest unconformably on them. Such associated beds commonly contain important clues to the origin of the sandstones in question. Consider, for example, two crossbedded sandstones—one associated with many shales and some coal beds and the other with tillites. Both indeed are probably fluvial, but the first was probably deposited on a muddy coastal plain by a meandering stream and the other probably on a braided, glacial outwash apron. Thus the nature and abundance of associated shales and carbonate—indeed, all other rocks—always need full attention.

Knowledge of the vertical sequence of individual sandstone deposits is also needed and forms the essential foundation of environmental recognition. Vertical sequence is defined by grain size, bed thickness, and sedimentary structures, and is recognizable on wireline logs and perhaps even on a seismic trace. This idea was developed in the 1950s by a Shell research group in Houston. The underlying concept is that of a *cycle*, a recurring sequence of rock types. If established, the vertical sequence in a cycle helps us identify the correct depositional model and thus better contour and extrapolate total thickness and distribution of the included sandstones. Originally the study of vertical sequences started with outcrops and cores, but today wireline logs and seismic cross sections are also used. We refer here to "autocyclic" deposits that have cycles imposed on them by the depositional process—not cycles externally imposed due to climate or tectonics.

Sandstone bodies may fine upward so that thin, fine-grained beds overlie coarse, thick ones, they may coarsen upward, or they may be uniform and lack well-defined vertical trends; individual sandstone beds may be thin and irregularly interbedded with shale. The average grain size of the framework fraction may systematically change or remain much the same. Perhaps only the proportion of interstitial clay may change. As a rule, kinds and proportions of primary sedimentary structures, bed thickness, and initial permeability and porosity all systematically change. These interrelationships are all controlled by the original transport agent (Visher, 1969): a powerful current or strong wave entrains coarse grains and produces a distinctive set of sedimentary structures and bed thicknesses, whereas a weak current or wave

A)

	S_1	S_2	S_3	
S_1	n_{11}	n_{12}	n_{13}	n_1
S_2	n_{21}	n_{22}	n_{23}	n_2
S_3	n_{31}	n_{32}	n_{33}	n_3

B)

$$\begin{matrix} n_{11}/n_1 & n_{12}/n_1 & n_{12}/n_1 \\ n_{21}/n_2 & n_{22}/n_2 & n_{23}/n_2 \\ n_{31}/n_3 & n_{32}/n_3 & n_{33}/n_3 \end{matrix}$$

C)

$$\begin{bmatrix} p_{11} & p_{12} & p_{13} \\ p_{21} & p_{22} & p_{23} \\ p_{31} & p_{32} & p_{33} \end{bmatrix}$$

D)

$$\begin{bmatrix} 0.3 & 0.2 & 0.5 \\ 0.4 & 0.4 & 0.2 \\ 0.3 & 0.6 & 0.1 \end{bmatrix}$$

E)

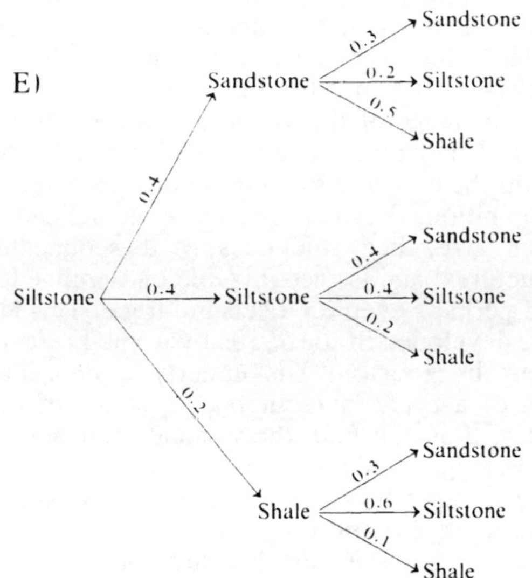

FIGURE 10-6. Above: Three-component sedimentation system of sandstone (S_1), siltstone (S_2), and shale (S_3). Steps in calculating a transition matrix: A) matrix of numbers of transitions, n_{ij}. B) calculations of the p_{ij}. C) 3 × 3 transition probability matrix, and D) as in C) but with specific values of p_{ij}'s. Below: Tree diagram E shows graphic pattern. From such a diagram one can see and calculate probabilities for different paths. For example, the probability of the sequence siltstone → siltstone → sandstone is obtained by $p_{22} \times p_{21} = 0.4 \times 0.4 = 0.16$, whereas the probability of the sequence siltstone → shale → sandstone is $p_{23} \times p_{13} = 0.2 \times 0.3 = 0.06$.

entrains only fine grains and produces a different set of structures that commonly occur only in thin beds (Figs. 8-10 and 8-13).

Another important consideration is bed contacts and transitions—are they gradational or abrupt? If abrupt, is the contact even and conformable or is it erosional and disconformable? Erosional and disconformable contacts may form barriers to fluid flow between beds.

Sequential relations of one bed to another—vertical *lithologic transitions*—are today widely used to better understand the depositional process. By lithologic transition is meant the order to deposition—what lithology or bed follows what. Lithologic transitions may be used on the different rock types of the sandy body or they may be used on different styles of bedding within the sand body itself. Instead of being counted at each lithologic or bedding change, transitions can also be counted at arbitrary, regularly spaced intervals. In either case, one can determine the *probability* of one type of sediment succeeding another. These probabilities are summarized in a *transition matrix* (Fig. 10-6), wherein each element p_{ij} of the matrix gives the probability of lithology i being followed by lithology j. In a three-component system of sandstone (S_1), siltstone (S_2), and shale (S_3), one simply determines the number of times sandstone follows itself (n_{11}), the number of times it follows siltstone (n_{21}), and the number of times it follows shale (n_{31}), and similarly for other rock types, where n_{ij} is the pair in question. To obtain probabilities, one divides each n_{ij} by the total number of beds of a particular kind n_i so that $n_{11}/n_1 = p_{11}$, $n_{21}/n_2 = p_{21}$, $n_{32}/n_3 = p_{32}$, etc. Each row of the matrix sums to one, because *something* follows each lithology. A transition matrix is constructed either by recording the rock type at fixed intervals or by recording the frequency of the different types of transitions at bed contacts (Krumbein, 1967, pp. 3–5; Potter and Blakely, 1968, pp. 155–56).

The final step involves examination of the

empirically established relations to see whether or not there is a recurring sequence. A recurring sequence implies a cycle and a "memory"—a dependence of one lithology on what was deposited previously. Memory may be long and extend back through a series of older beds, lithologies, or environments. Conversely, it may depend only on the preceding bed, or there may be no memory at all. The latter means that a particular lithology occurs independently of previous sedimentary events. No memory appears to be most unusual, however, for the vast majority of sedimentary sections without breaks are either the result of a slow, orderly, lateral migration of related environments or the result of perhaps a rapid, episodic pulse, rather than the progression of haphazard, unrelated events. Within the general framework of slow, progressive change with time, however, many smaller random events may occur—events which we may not be able to assign specific causes. Vertical variation of crossbedding thickness in a fining-upward, fluvial sandstone, one in which bed thickness as well as grain size gradually decreases upward, is such an example. Here we have a good grasp of the overall process, but cannot give a full deterministic explanation of successive bed-to-bed variation of crossbedding thickness. It is in such a context that knowledge or recognition of the probability of a transition from one sedimentary event to another provides additional insight into the actual mechanism of transport and deposition of sand by a waning current. If memory is short and extends back but one step, the sedimentation process is called Markov-1; if it extends back two steps, Markov-2; and so on. Informative references include Selley (1969) and Miall (1973); two recent mathematical discussions are provided by Carr (1982) and Powers and Easterling (1982). From a slow beginning, applications are now fairly routine, especially for the study of cyclic sequences in coal measures and for studies of sequence of sedimentary structures and stratification type. Regardless of length of memory, however, analysis is made clearer by constructing a tree diagram (Fig. 10-6E) from a transition matrix or by formulating an ideal cycle in which one uses the transition matrix to estimate probabilities between the different steps of the cycle.

The vertical profile and type of cycle helps us formulate a model, which in turn is based on a specific sedimentary process. Such a model, if well constructed, makes possible a more adequate environmental reconstruction than would be possible without it. Visher (1969) early recognized six sand models and stressed their recognition through vertical profiles.

Geophysical Logs

Wireline or geophysical logs are of many types and today their use and understanding are absolutely indispensible for the study of buried sandstones—for both environmental recognition and for identification of reservoir type and quality. Good references include LeRoy and LeRoy (1977) and Allen (1975) plus the excellent manuals provided by logging companies. Here we concentrate on the utility of geophysical logs for environmental recognition based on self-potential and resistivity logs and on the gamma-ray neutron log, all of which identify the rock types and provide a rough measure of grain size.

Log Signature. Log forms, motifs, or signatures are of three types—blocky, "fining" upward, and "coarsening" upward—all of which may be serrated or smooth (Fig. 10-7). An additional variable is whether the nature of the contact is abrupt or gradational. The log signature, because it reflects vertical sequence, helps identify the environment of deposition. We emphasize "help" because signatures are not unique to each depositional environment—one needs to know position in the basin, "basinal setting," to distinguish, for example, a coarsening-upward turbidite from a coarsening-upward coastal barrier. Hence knowledge of regional geology, perhaps gleamed from a seismic section that shows a large clinoform, as well as sedimentologic study of a core or evidence from micropaleontology are needed to improve discrimination. We make ample use of log signatures in the discussion of individual environments. For best results, log signatures should be combined with a map of sandstone thickness (Fig. 10-8) in all basins except unstable ones, where a thickness map reflects local differential subsidence along an active fault or shale diapir rather than the initial geometry of the sandy depositional environment. Ruoff (1976) and Magara (1978) utilize computer methods to identify depositional environment from wireline logs. Zamora (1977) relates, diagrammatically, log form to geometry of the sandstone body.

Dipmeter. The dipmeter is a wireline logging device that measures both sedimentary and structural dip downhole and, although it was first developed in the early 1930s, is still not as widely used for the study of sandstones as it

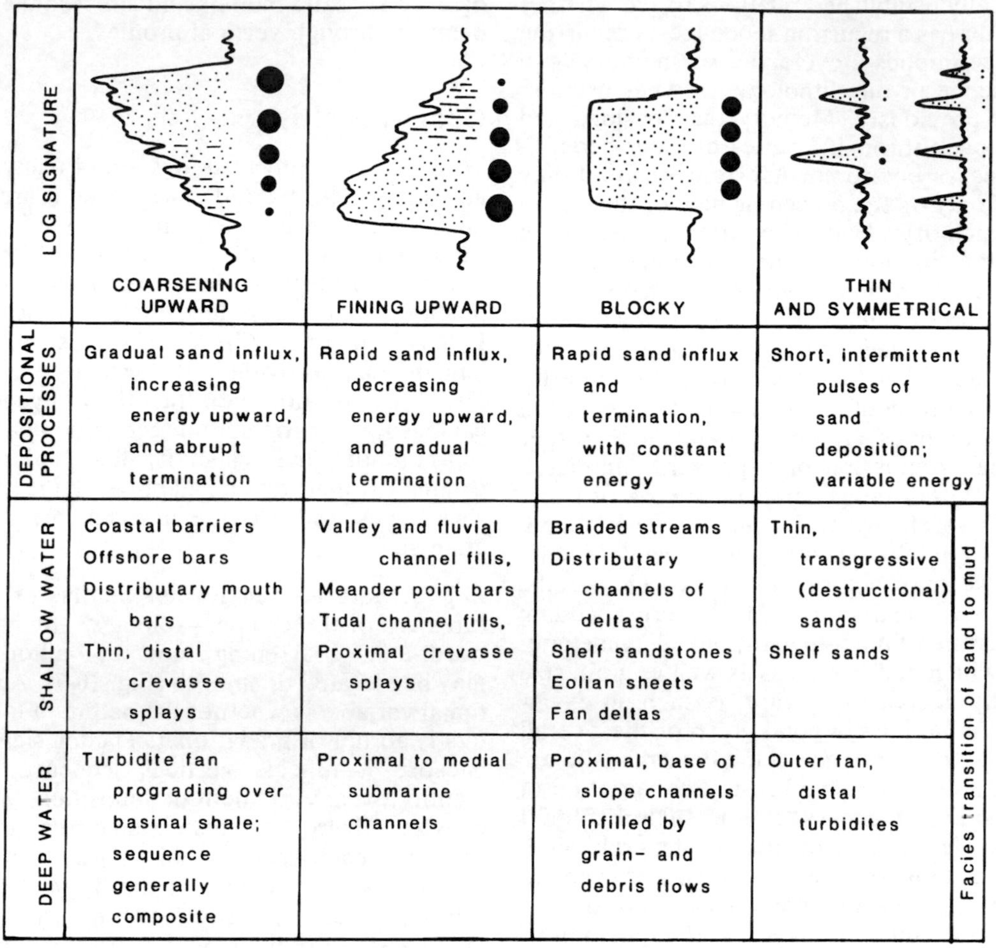

LOG SIGNATURE	COARSENING UPWARD	FINING UPWARD	BLOCKY	THIN AND SYMMETRICAL	
DEPOSITIONAL PROCESSES	Gradual sand influx, increasing energy upward, and abrupt termination	Rapid sand influx, decreasing energy upward, and gradual termination	Rapid sand influx and termination, with constant energy	Short, intermittent pulses of sand deposition; variable energy	
SHALLOW WATER	Coastal barriers Offshore bars Distributary mouth bars Thin, distal crevasse splays	Valley and fluvial channel fills, Meander point bars Tidal channel fills, Proximal crevasse splays	Braided streams Distributary channels of deltas Shelf sandstones Eolian sheets Fan deltas	Thin, transgressive (destructional) sands Shelf sands	Facies transition of sand to mud
DEEP WATER	Turbidite fan prograding over basinal shale; sequence generally composite	Proximal to medial submarine channels	Proximal, base of slope channels infilled by grain- and debris flows	Outer fan, distal turbidites	

FIGURE 10-7. Basic log signatures and depositional processes in shallow and deep water.

should be. Sedimentary or initial dip measurable by the dipmeter is produced by primary sedimentary structures, chiefly crossbedding, and by deposition on or over inclined surfaces (Table 10-3). Dips as low as zero can be measured and current roses automatically plotted. Good sources on this topic are Gilreath *et al.* (1985) and Holt (1982).

Effective use of the dipmeter requires the ability to discriminate between fluvial, deltaic, coastal, and marine deposits and to recognize structural dip produced by faults and unconformities. The dipmeter seems to work best where sandy sequences are not too thin.

There are four basic patterns—the structural baseline, a sequence of increasing dips, a sequence of decreasing dips, and random dips (Fig. 10-9). The structural baseline is the lowest dip of the section and will be zero in flat-lying strata and 10 degrees where the strata dip to 10 degrees. The dipmeter is easiest to use, from a sedimentologist's viewpoint, where faulting is largely absent and sandstone bodies are thick rather than thin. As is true for environmental recognition based on electric or gamma-ray logs, the sequence of increasing or decreasing dip does not have a unique interpretation, but when a gamma-ray or electric log and a dipmeter are used together, much can be achieved.

Lenticular Sandstone Bodies. Let us first see how linear, tidal sandstone bodies or channel fills illustrate basic principles. In both examples shown in Fig. 10-10, subsequent sedimentation tends to level or smooth out initial topography—whatever its origin. Crossbedding within a lenticular sandstone body can be recognized (Fig. 10-11), especially if the crossbeds are thicker than 50 cm, as well as the dip of the beds that fill and drape into a channel.

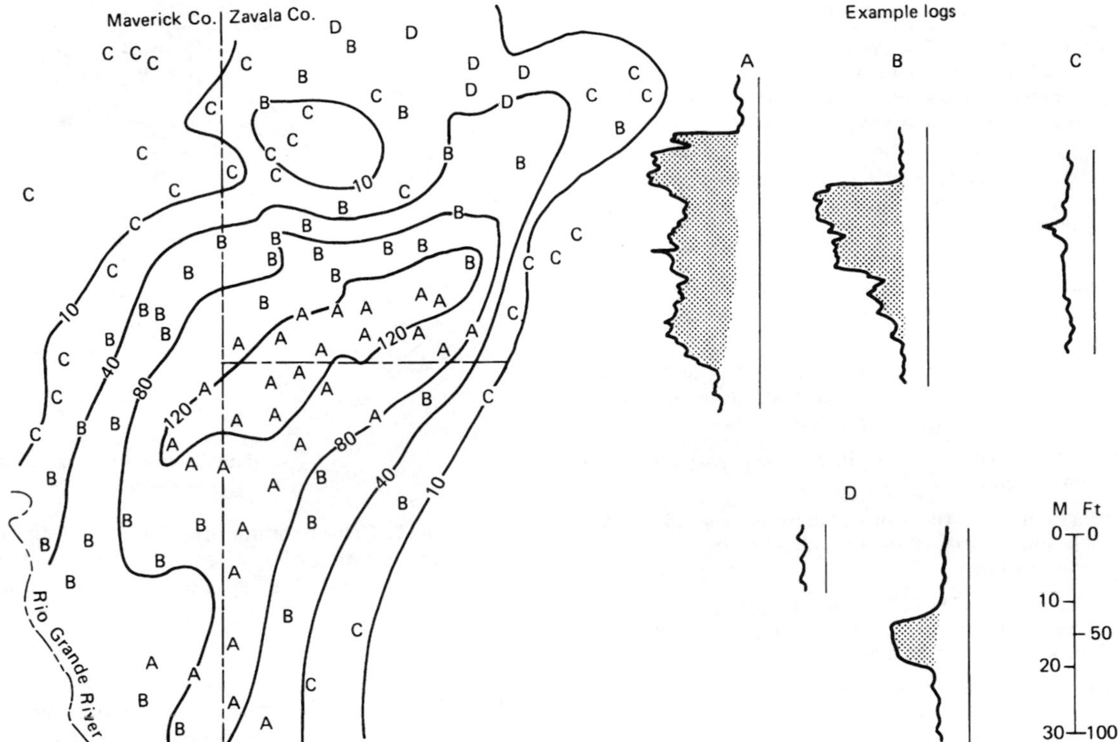

FIGURE 10-8. Distribution of self-potential patterns in Cretaceous San Miguel Formation, unit G (after Weise, 1980, Fig. 15). Published by permission of the author and the Texas Bureau of Economic Geology.

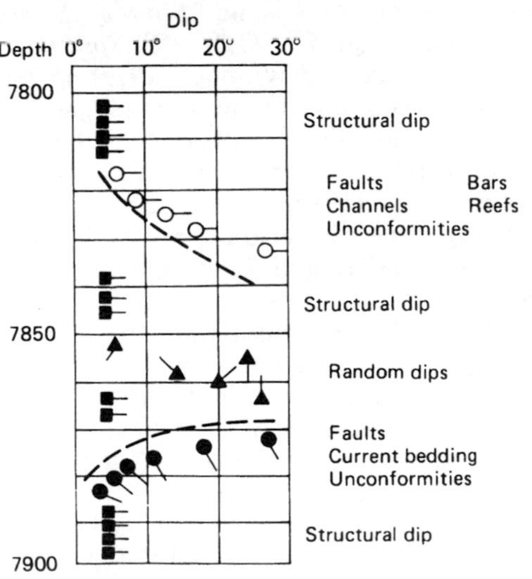

FIGURE 10-9. Dip patterns and their origins (Schlumberger, 1981, Fig. 7). "Tadpoles" show direction and angle of dip: squares on tadpoles indicate structural dip or parallel dipping foresets; open circles indicate "blue" pattern of upward increasing dip; and solid circles indicate "red" pattern of downward increasing dip. Published by permission.

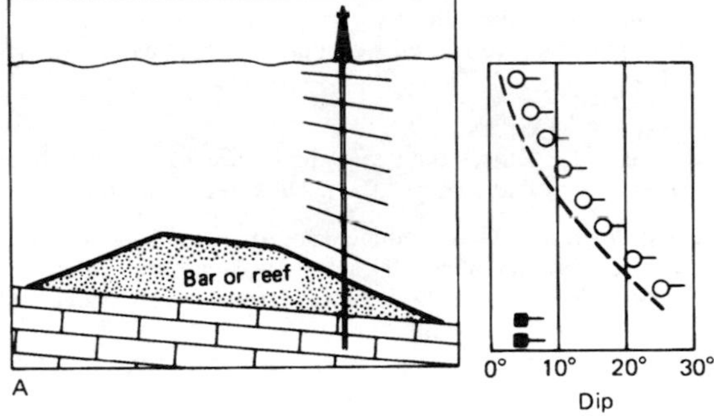

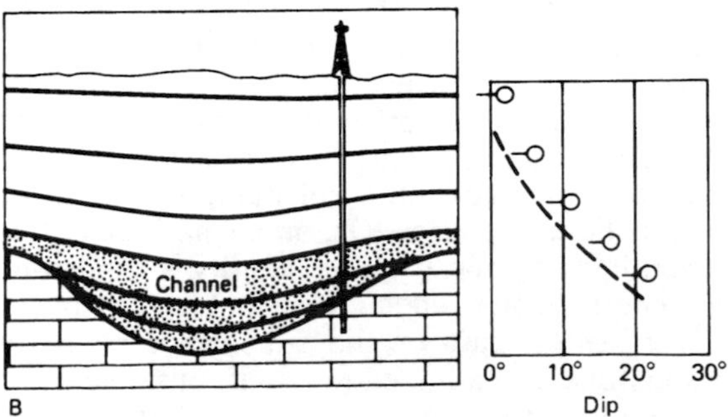

Figure 10-10. Lenticular sandstones and their dipmeter patterns (Schlumberger, 1981, Fig. 8). Published by permission.

TABLE 10-3. Sedimentary dip.

PRIMARY SEDIMENTARY STRUCTURES

Crossbedding: Chiefly beds thicker than 30 cm and may have minor bimodal dips. When beds are thick, see if they have a dip pattern of large, eolian dunes.

Slumps and slides: Dip direction is erratic (many random events) to poorly oriented down slope but dip angle can be very high. Thick slides or slumps may be confused with complex faulting.

INCLINED DEPOSITIONAL SURFACES

Channels: Initial fill closely follows original cut surface of channel; dip systematically decreases upward within the channel.

Point bars: Inclination produced by the slope of the bar into the channel—the epsilon crossbedding of Allen (1963, Fig. 3D)—and easily seen by clay drapes.

Fans: Subaerial, subsea, or both; sedimentary dip is away from a well-defined point source.

Delta front slope: Dip is produced by low-angle, large-scale deposition in front of river mouth—the so-called "Gilbert delta" bedding.

Coastal barriers: Dips are chiefly seaward in foreshore but some are landward into swamp and lagoon and all are steepest near surf line.

Marine bar: Dips are away from top of bar and systematically decrease above it.

Clinoforms: Inclined surface of shelf-to-basin transition, the largest geomorphic feature that commonly produces sedimentary dip. Dip systematically increases upward, but may be interrupted by slumps and slides and debris flow.

MAJOR EROSION

Buried hills: Dips, some very high, are away from high point; reefs have same effect. Dips decrease upward.

Unconformities: Can produce discordance in structural dip and commonly have characteristic signature. Dips into closed depressions above buried karst, into channels, and away from buried hills.

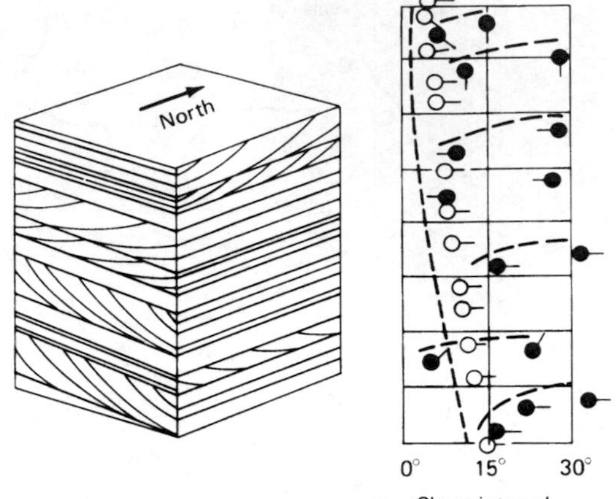

FIGURE 10-11. Crossbedding orientation and dip inferred from a dipmeter log when correlation is short (Schlumberger, 1981, Fig. 9). Published by permission.

Seismic Stratigraphy

Reflection seismology plays an ever-increasing role in exploration of sandy deposits, both regionally and locally. For example, it is widely used where subsurface data are scarce or nonexistent, as in the offshore or when exploring the deep part of a basin. Here distinctive patterns permit recognition of many of the major depositional systems. Or, on a small scale, this technique can be used to identify a sandstone pinchout. Thus reflection seismology is today an essential part of the subsurface study of sandstone and all need to use it, but, like most methods, special terminology and theory are needed. The methodology of reflection seismology evolved over many years and was initially concerned with structure but, beginning in the early 1970s, the subdiscipline of *seismic stratigraphy* developed. The pioneer paper of Harms and Tackenberg (1971) sets forth the main ideas and is still a good starting point. Another historic landmark reference is that by Payton (1977) and excellent references are Anstey's (1980) *Seismic Exploration for Sandstone Reservoirs*, which includes much on sandy depositional environments, his *Simple Seismics* (1982), and Sheriff's (1980) *Seismic Stratigraphy*. Sangree and Widmier (1979) have prepared a detailed table on how to recognize clastic facies and their sandstone bodies from seismic cross sections. Dobrin (1976) and McQuillin *et al.* (1979) provide needed geophysical background. See also the review by Neidell and Beard (1984) and Berg and Woolverton (1985).

Contrasts in seismic behavior between a sandstone–shale contact or a sandstone–carbonate contact produce reflections, and the magnitude and polarity of the reflection depend upon the porosity, cementation, density, and fluid content—gas, oil, or water—of the sandstone in a fairly complex way as well as on the lithology of the adjacent bed. Moreover, energy is absorbed and reflected with depth so that resolution decreases with increasing depth. Nonetheless, within limits, it is possible to recognize fining- and coarsening-upward sandy sequences by their seismic signature and to relate the signature to a depositional model—just as with wireline logs.

Understanding of three terms is essential—acoustic impedance, the reflection coefficient, and the polarity of a reflection.

Acoustic impedance, ρv, is the product of density and velocity, and can be thought of as a kind of "acoustical hardness." At any point along the contact of a bed, there is a characteristic acoustic impedance, which may remain constant over a wide area or may change laterally as cementation, porosity, or fluid content changes. The acoustic impedance of adjacent beds defines the *reflection coefficient*, *R*, whose strength and polarity produce the seismic signature of a contact. The reflection coefficient is defined as $(\rho_2 v_2 - \rho_1 v_1)/(\rho_2 v_2 + \rho_1 v_1)$ where the subscript 1 indicates the upper bed and 2 the lower. The reflection coefficient ranges from positive through zero to negative; when zero, the contact is seismically invisible, when positive, the reflection is *positively polarized* and, when negative, *negatively polarized* (Fig. 10-12). Reflection coefficients near one are termed "strong" and those near zero "weak." The polarity of a reflection and its strength *R* are used to discriminate between porous gas- and liquid-filled sandstones and those that lack porosity. *Thus the strength or amplitude of the reflection and its polarity depend on the acoustical impedance of adjacent beds.* Interpretation is easiest where the sandstone is sufficiently thick that reflections from its top and bottom are distinct. Finally, we note that because density and velocity change with depth—and at different rates—the seismic signature of a sandstone body at shallow depth is different from one at deeper depths.

Seismic cross sections aid the study of sandy deposits in at least three ways—regional sections help us identify where in the basin major sandstones occur and give us clues to their environment of deposition, *and* local sections can help us identify and map individual geometries and pinchouts. If the sandstone body is thick, if wavelets have high frequency and top and bottom can be resolved, it may be possible to distinguish sharp from gradational contacts (Fig. 10-13). Because fluid content and density and velocity contrasts of adjacent beds can have a dramatic effect on the seismic signature of individual sandstone bodies, experience and caution are needed.

Compaction

Differential compaction of mud around a sand body distorts its original shape, thus complicating an interpretation of its origin. This is most apparent when cross sections of a sandy channel or valley fill, one known to truncate underlying marker beds, are made. Because of compaction, choice of the first level line above the body may impart a synclinal appearance to lower marker beds. On the other hand, representing the top of the body as level assumes that the upper surface of sand deposition was in fact so. As shown in Fig. 10-14, an underlying marker is probably the best, for in this example it shows several underlying coals, 5A and 6, to be approximately horizontal, as in fact they are, and nicely displays the compactional bump or high on top of the body. An underlying marker has the great disadvantage, however, that it is

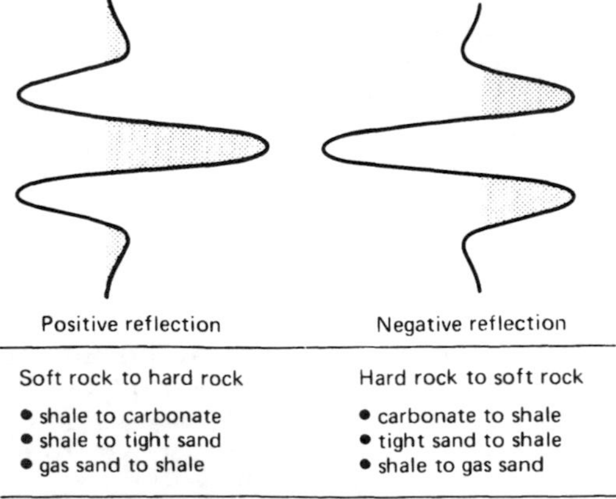

FIGURE 10-12. Acoustic wavelet positively polarized (left) and negatively polarized (right). The reflection coefficient, *R*, determines both polarity and amplitude of the reflection.

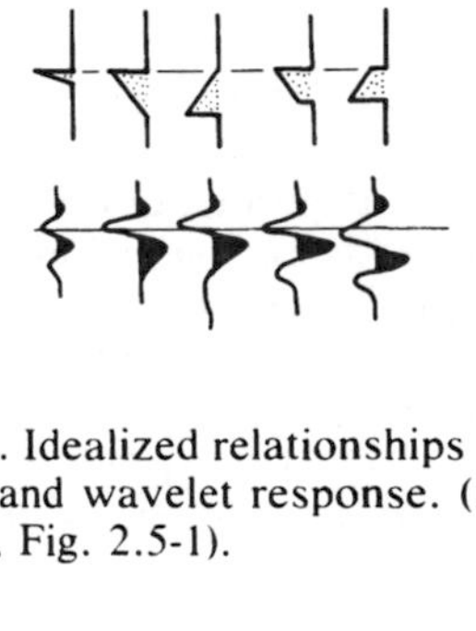

FIGURE 10-13. Idealized relationships between sandstone bodies and wavelet response. (Redrawn from Anstey, 1980, Fig. 2.5-1).

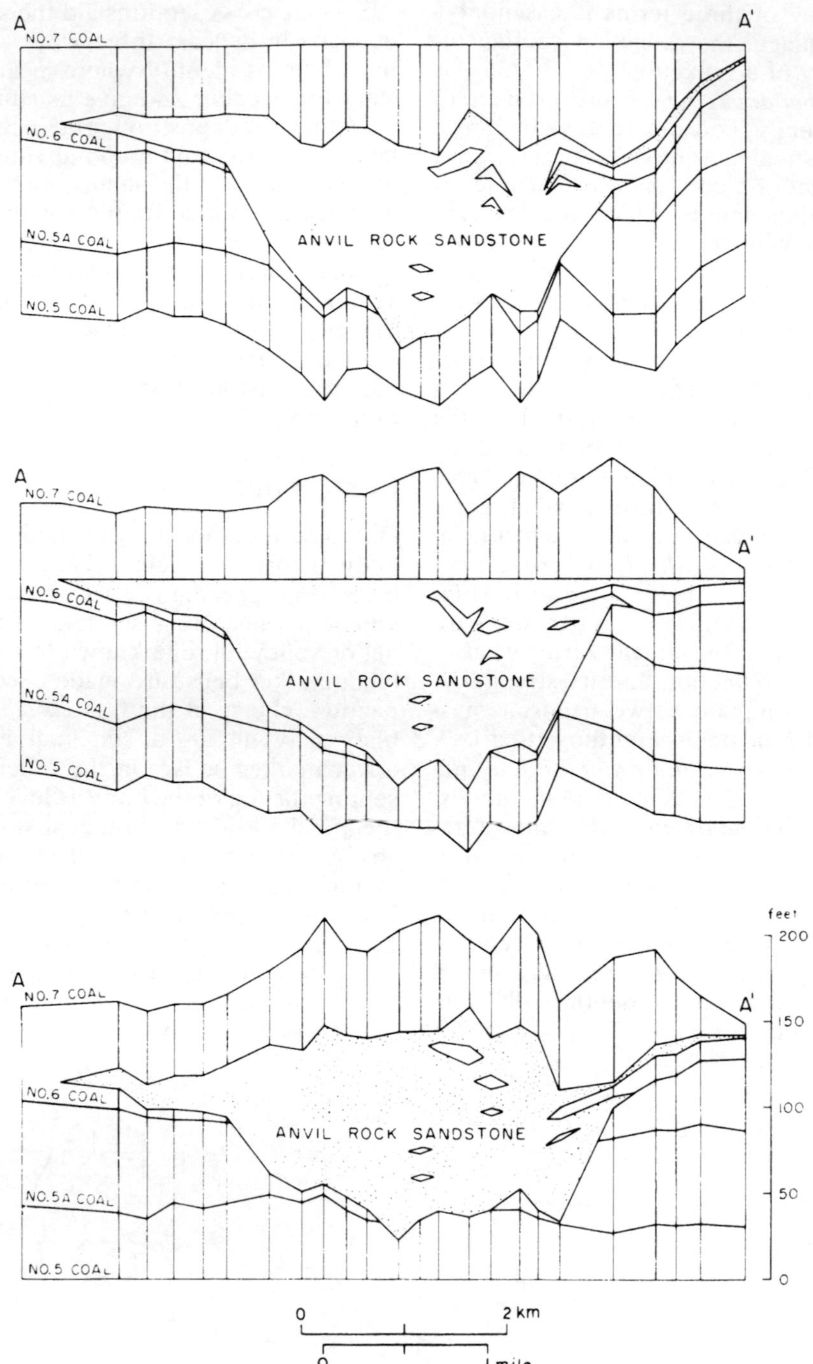

FIGURE 10-14. Cross sections of Anvil Rock Sandstone (Pennsylvanian); Danville (No. 7) Coal Member used as level line (top), top of sandstone as level line (middle), and Harrisburg (No. 5) Coal Member (bottom); from Potter (1963, Fig. 37B).

under the target of interest so that fewer wells reach it.

The effect of differential compaction also depends on type and consolidation of associated sediments. Unfortunately, our recent knowledge of compaction and its effect on sandbody shape is far from adequate. Should a sand be deposited on consolidated mud or shale, subsidence of the sand into the mud will be minimal. If the underlying mud is but little consolidated, however, differential compaction under the sand can appreciably modify its base and can produce unusually steep dips in the associated muds. Such dips can be detected by dipmeter and thus help locate the sandstone body. The key to correct interpretation of cross-sectional shape is associated marker beds—do they intersect or bend around the body? If they intersect the body, the sand is a valley or channel fill deposit, but if they deform or bend beneath it, they indicate deformation and subsidence of the body so that present cross-sectional shape is not the original one. Rittenhouse (1961, pp. 7–8) discusses some of the interrelations between compaction within a sandstone, the topographic relief of the depositional surface, and compaction of the enclosing shale. Brown (1975) provides a general review, and Magara (1978) has published graphs for estimating compaction of associated clays and claystones and used them in decompaction calculations of stratigraphic sections that contain sandstones. Shelton (1962) approached this problem directly by studying small, contorted sandstone dikes in a Cretaceous shale of the western United States, where he estimated the shale to have compacted approximately 2.6 times more than sandstone. Because compaction changes shale density, velocity and acoustic impedance change and thus directly affect interpretation of the seismic cross sections.

References

Allen, J.R.L.: The classification of cross-stratified units with notes on their origin. Sedimentology 2, pp. 93–114 (1963).

Allen, D.R.: Identification of sediments—their depositional environment and degree of compaction—from well logs. In: Chilingarian, G.V., and Wolf, K.H. (Eds.): Compaction of coarse-grained sediments, I, pp. 349–401. Amsterdam–Oxford–New York, Elsevier Scientific Pub. Co. 1975.

Anstey, N.A.: Seismic exploration for sandstone reservoirs, 138 pp. Boston: IHRDC Publications 1980.

Anstey, N.A.: Simple seismics, 168 pp. Boston: IHRDC Publications 1982.

Berg, O.R., and Woolverton, D.G. (Eds.): Seismic stratigraphy II. Amer. Assoc. Petroleum Geologists Mem. 39, 276 pp. (1985).

Brown, L.F., Jr.: Role of sediment compaction in determining geometry and distribution of fluvial and deltaic sandstones. In: Chilingarian, G.V., and Wolf, K.H. (Eds.): Compaction of coarse-grained sediments, I, pp. 247–292. Amsterdam–Oxford–New York: Elsevier Scientific Pub. Co. 1975.

Carr, T.R.: Log-linear models, Markov chains and cyclic sedimentation. Jour. Sed. Petrology 52, 905–912 (1982).

Davis, R.A., Jr.: Depositional systems, a genetic approach to sedimentary geology, 672 pp. Englewood Cliffs, New Jersey: Prentice Hall, 1983.

Dobrin, M.B.: Introduction to geophysical prospecting, 3rd Ed., 630 pp. New York: McGraw-Hill 1976.

Fisher, W.L., and McGowen, J.H.: Depositional systems in the Wilcox Group of Texas and their relationship to occurrence of oil and gas. Trans. Gulf Coast Assoc. Geol. Societies 17, 105–125 (1967).

Galloway, W.E., and Hobday, D.K.: Terrigeneous clastic depositional systems, 416 pp. New York: Springer-Verlag 1983.

Galloway, W.E.; Ewing, T.E.; Garret, C.M.; and Tyler, N.: Atlas of Texas oil reservoirs, 139 pp. Texas Bur. Econ. Geol., Austin (1983).

Gilreath, J.A., Cox, J.W., Fett, T.H., and L.M. Grace: Practical dipmeter interpretation, various paging. Houston: Schlumberger Educational Services 1985.

Harms, J.C., and Tackenberg, P.: Seismic signatures of sedimentation models. Geophysics 37, 45–58 (1972).

Holt, Olin R.: Relating displays to practical geology, 69 pp. Houston, Texas: Dresser Atlas, Dresser Industries Inc. 1982.

Hyne, N.J. (Ed.): Pennsylvanian sandstones of the Mid-Continent. Tulsa Geol. Soc. Spec. Pub. 1, 360 pp. (1979).

Klein, George de Vries: Sandstone depositional models for exploration for fossil fuels, 2nd Ed., 149 pp. Minneapolis, Minnesota: DEPCO Division, Burgess Pub. Co. 1980.

Krumbein, W.C.: Fortran IV computer programs for Markov chain experiments in geology. Kansas Geol. Survey, Computer Contr. 13, 38 pp. (1967).

LeBlanc, R.J.: Geometry of sandstone bodies. In: T.D. Cook (Ed.): Underground waste management and environmental implications. Am. Assoc. Petroleum Geologists Mem. 18, 133–190 (1972).

LeRoy, L.W., and LeRoy D.O. (Eds.): Subsurface geology, 4th Ed., 941 pp. Golden, Colorado: Colorado School of Mines 1977.

Magara, Kinji: Compaction and fluid migration, 319 pp. (Developments in Petroleum Science, Vol. 9), Amsterdam–Oxford–New York: Elsevier Scientific Pub. Co. 1978.

McQuillin, R.; Bacon, M.; and Barclay, W.: An introduction to seismic interpretation, 199 pp. Houston: Gulf Pub. Co. 1979.

Medeiros, R.A.; Schaller, H.; and Friedman, G.M.: Facies sedimentares. Petrobras, Centro de Pesquisas e Desenvolvimento, Pub. 5, 123 pp. (1971).

Miall, A.D.: Markov chain analysis applied to an ancient alluvial plain succession. Sedimentology 20, 347–364 (1973).

Neidell, N.S., and Beard, J.H.: Progress in stratigraphic seismic exploration and definition of reservoirs. Jour. Petrol. Tech. 36, 709–726 (1984).

Payton, C.E. (Ed.): Seismic stratigraphy—applications to hydrocarbon exploration. Am. Assoc. Petroleum Geologists Mem. 26, 516 pp. (1977).

Potter, P.E.: Late Paleozoic sandstones of the Illinois Basin. Illinois Geol. Survey, Rept. Invs. 217, 92 pp. (1963).

Potter, P.E., and Blakely, R.F.: Random processes and lithologic transitions. Jour. Geology 76, 14–170 (1968).

Potter, P.E.; DeReamer, John; Jackson, Dana; and Maynard, J.B.: Lithologic and environmental atlas of the Berea sandstone (Mississippian) in the Appalachian Basin. Appalachian Geol. Soc. Spec. Pub. No. 1, 157 pp. (1984).

Powers, D.W., and Easterling, R.G.: Improved methodology for using embedded Markov chain to describe cyclical sediments. Jour. Sed. Petrology, 52, 913–923 (1982).

Reading, H.G. (Ed.): Sedimentary environments and facies, 2nd ed. 615 pp. New York: Elsevier Scientific Pub. Co. 1986.

Reineck, H.-E., and Singh, I.B.: Depositional sedimentary environments, 2nd ed., 549 pp. Berlin–Heidelberg–New York: Springer-Verlag 1980.

Rigby, J.K., and Hamblin, W.K. (Eds.): Recognition of ancient sedimentary environments. Soc. Econ. Paleon. Mineral. Spec. Pub. 16, 340 pp. (1972).

Rittenhouse, Gordon: Problems and principles of sandstone-body classification. In: Peterson, J.A., and Osmond, J.C. (Eds.): Geometry of sandstone bodies, pp. 3–12. Tulsa, Oklahoma: Am. Assoc. Petroleum Geologists (1961).

Ruoff, W.A.: A technique for interpreting depositional environments of sandstones for the SP log utilizing the computer. Log. Analyst 17, 3–10 (1976).

Rust, B.R.: Depositional models for braided alluvium. In: Miall, A.D. (Ed.): Fluvial sedimentology. Canadian Soc. Petroleum Geologists Mem. 5, 605–621 (1978).

Sangree, J.B., and Widmier, J.M.: Interpretation of depositional facies from seismic data. Geophysics 44, 131–160 (1979).

Scholle, Peter A., and Spearing, D. (Eds.): Sandstone depositional environments, 410 pp. Tulsa, Oklahoma: Am. Assoc. Petroleum Geologists 1982.

Schumm, S.A.: Speculations concerning paleohydrologic controls of terrestrial sedimentation. Geol. Soc. America Bull. 79, 1573–1588 (1968).

Selley, R.C.: Studies of sequence in sediments using a simple mathematical device. Geol. Soc. London Quart. Jour. 125, 557–581 (1969).

Shelton, J.W.: Shale compaction in a section of Cretaceous Dakota Sandstone, northwestern Black Hills. Jour. Sed. Petrology 32, 847–877 (1962).

Shaw, A.B.: Time in stratigraphy, 365 pp. New York: McGraw-Hill 1964.

Shelton, John W.: Models of sand and sandstone deposits: A methodology for determining sand genesis and trend. Oklahoma Geol. Survey Bull. 118, 122 pp. (1973).

Sheriff, R.E.: Seismic stratigraphy, 227 pp. Boston: IHRDC Pub. Co. 1980.

Spalletti, Luis A.: Paleoambientes sedimentarios (en siliclastic sequences). Assoc. Geol. Argentina Ser. B, Didactic Complementaria 8, 175 pp. (1980).

Walther, Johannes: Einleitung in die Geologie als historische Wissenschaft. III Lithogenesis der Gegenwart. Beobachtung über die Bildung der Gesteine an der heutigen Erdoberfläche, pp. 536–1055. Jena: G. Fischer 1893/1894.

Visher, G.S.: Use of vertical profile in environmental reconstruction. Am. Assoc. Petroleum Geologists Bull. 49, 41–61 1965.

Visher, G.S.: Grain size distributions and depositional processes. Jour. Sed. Petrology 39, 1074–1106 (1969).

Walker, R.G. (Ed.): Facies models. Geological Association of Canada Publications (Business and Economic Services, Ltd., Toronto, Ontario, Canada M5V 2H1) 1984.

Weise, B.R.: Wave-dominated deltaic systems of the Upper Cretaceous San Miguel Formation, Maverick Basin, South Texas. Texas Bur. Econ. Geol., Rept. Invs. 107, 39 pp. (1980).

Zamora, L.G.: Uso de perfiles en la identificación de ambientes sedimentarios del Eocene del Lago de Marcaibo. V. Congreso Geológico Venezolano Caracas, Memoria 5. 4, tema 5, 1359–1376 (1977).

Alluvial

Alluvial sandstones form a significant part of many ancient basins and knowledge of their geometry and thickness is essential for study of petroleum, groundwater, and coal resources. They may be hosts for secondary uranium and copper, and may contain detrital gold and diamonds. In addition, thick alluvial deposits are direct indicators of tectonic activity. Miall (1978), Pitty (1979), Ethridge and Romero (1981), and Collinson and Lewin (1983) provide three excellent sedimentologic summaries, as does Schumm (1977).

Alluvial sandstones occur as alluvial fans and as valley fills of braided, anastomosing, and meandering streams. The scale of these deposits varies from a few square kilometers for a small fan to thousands of square kilometers for the molasse in front of a mountain chain. Still other examples include the deposits of a large river following the axis of a regional syncline or of a megafracture spanning much of a continent. The strata associated with alluvial sandstones are mostly shales, conglomerates, and some debris flow deposits, but minor carbonates as well as coals, lignites, and evaporites can also occur. Proximity to source, intensity of discharge, ratio of basin subsidence to input, and climate all determine the nature of the associated beds.

Because of easy accessibility, more is known about modern alluvial processes than other sandy deposits, though our knowledge of fluvial processes in very large rivers and in tropical climates is still far from complete.

Variability is an earmark of alluvial deposits. Grain size varies from the coarse conglomerates of alluvial fans to the fine sands and silts characteristic of many of the deltas of the world's large rivers that drain much of the continent. And, instead of but one vertical profile, there are at least three—fining upward (meandering and anastomosing), no vertical trend (braided), and coarsening upward (uncommon but observed in some glacial outwash). Other distinguishing characteristics of alluvial sandstones include abrupt transitions to shale, multistory sandstone bodies, and limited lateral continuity. Mineral composition varies widely from quartz arenites to arkoses to lithic arenites and with burial and diagenesis some become wackes. Sorting and rounding varies widely but river sands with long transport can be well sorted. Sands and sandstones of alluvial deposits are probably the most immature both texturally and mineralogically.

But within the alluvial basin itself, what is the relation between channel morphology, channel stability, and sediment load? We ask this question, because it is these factors which produce much of the great stratigraphic variability of alluvial deposits which formed meandering, braided and anastomosing streams (Table 10-4). Drawing on Schumm's (1981) perceptive summary (Fig. 10-15) based on many field and experimental studies, the following generalizations emerge: transitions from one channel type to another are common and usually are abrupt and result from changes in gradient, stream power ($\tau_o \overline{U}$), sediment load, and even vegetative cover along the banks. It is also clear from Fig. 10-15 that there is good correlation between channel pattern and sediment load so that the sandstone–shale *ratio* of an ancient alluvial section gives a fairly clear view of stream type.

Meandering Streams

The alluvial valley with many meander loops (Fig. 10-16) has many subenvironments–sandy and muddy channel fills, sandy lateral accretion deposits such as point and midstream bars, side bars along straight reaches, overbank deposits such as levees and floodplain deposits, lakes, crevasse splays, and, climate permitting,

TABLE 10-4. Characteristics of braided, anastomosing and meandering stream deposits.

Meandering	Braided	Anastomosing
Cyclicity		
Well-defined classic point bars; multistory; bell-shaped log forms.	Weak to poor for both grain size and structures; blocky log forms; multistory	Well-defined but interruptible fining-upward to blocky cycles; multistory.
Lithology		
Sand, silt and mud plus peat, climate permitting	Gravel, sandy gravel, and sand with negligible mud; some debris flows	Sand, silt, and mud plus peat, climate permitting
Map pattern		
Narrow to broad belts with sharply scalloped boundaries resulting from low channel stability. Many splays.	Narrow to broad linear belts plus fans, all with fairly straight boundaries; many multiple, unstable interconnected channels	Narrow to broad belts with fairly straight channels and boundaries resulting from high channel stability; some multiple interconnected channels
Size gradients down dip		
Weak for sand fraction because of low gradients	Commonly strong trends, because of high gradients	Weak for sand fraction because of low gradients
Paleocurrents		
High variance; epsilon crossbedding common	Low variance	Low variance
Fossils		
Never abundant, but trace fossils, logs, and freshwater invertebrates possible	Rare, but some vertebrate debris possible as channel lags	Never abundant, but trace fossils, logs, and freshwater invertebrates possible

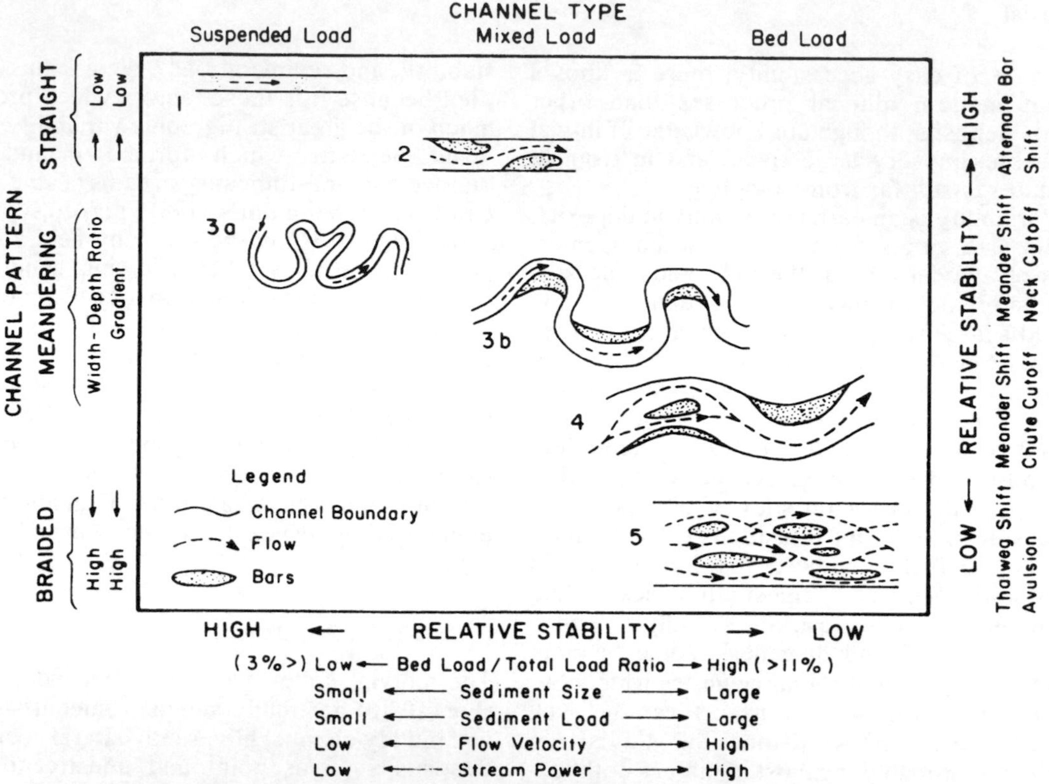

FIGURE 10-15. Relationships between channel pattern and stream stability (Schumm, 1981, Fig. 4). Published by permission of the author and the Society of Economic Paleontologists and Mineralogists.

FIGURE 10-16. Above: Meanders in lowlands east of Andes, Peru. (Photo courtesy of Servicio Aerfotografico Nacional, Lima Peru). Opposite page: Spectacular radar image of Rio Solimões, Amazona, Brazil (Projecto Radan SA-20-YA). Rio Solimões is up to 3 km wide. Note meander scrolls, oxbow lakes, and midstream and point bars.

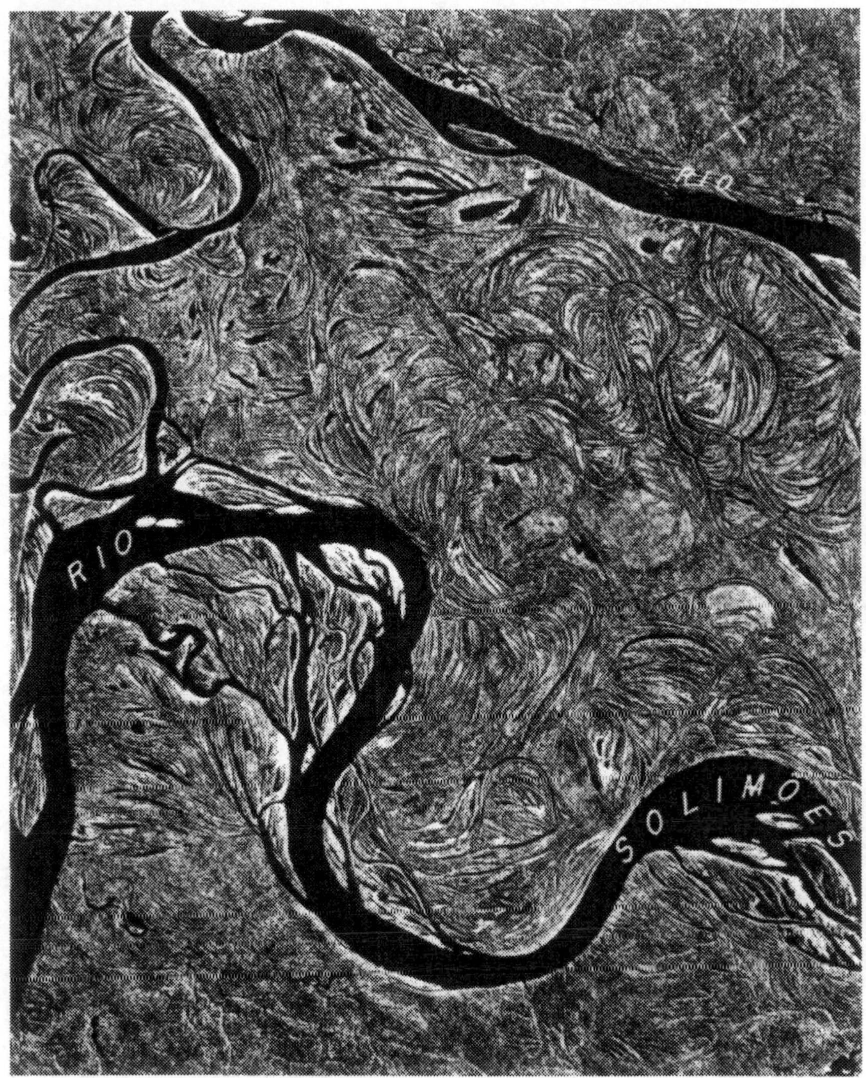

FIG. 10-16 *continued*

dunes. Small alluvial fans may also be preserved along valley margins. The vertical, fining-upward sequence of the sands deposited by meandering streams (Fig. 10-17) is easily recognized on wireline logs, if not truncated by erosion. Basal contact is sharp, lag gravels and pebbles are commonly concentrated along it, and sedimentary sequences range from coarse-grained, massive and crossbedded (planar and trough) basal beds passing upward to finer-grained, horizontally laminated and rippled beds. Soft sediment deformation is also present as are abundant clay clasts. The latter are derived from clay drapes deposited during high water. Epsilon crossbedding is fairly easy to recognize in outcrops transverse to flow (Fig. 10-17) and dips channelward at right angles to the smaller planar and trough crossbedding formed by dune migration downstream in the channel. Variance of crossbedding produced by migration of sand waves is directly related to channel sinuosity and can be large. Superimposed, or multistory, sandstone bodies are common and complexity can be great even in single outcrops. Length of single point bars varies from a few hundred to several thousand or more meters in a large river. Thickness of a point bar sand is a direct function of channel scour—a single point bar cycle cannot exceed depth of scour at high water. Single cycles thicker than 30 to 40 meters are rare and most are much less. As a rule, exploration for isolated point bar deposits is difficult because they are small and curved. A dipmeter is useful be-

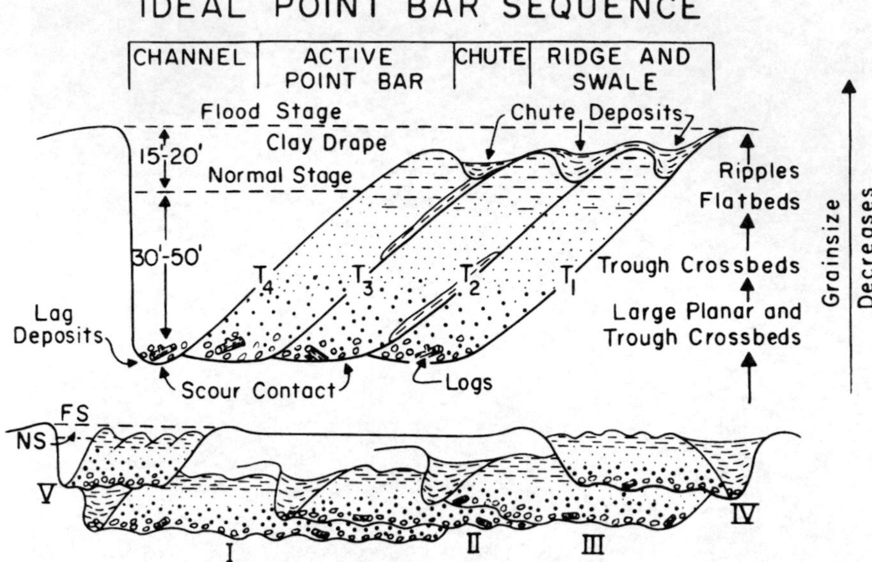

FIGURE 10-17. Ideal point bar sequence (Sedimentation Seminar, 1978, Fig. 11).

cause it can distinguish epsilon from sand wave crossbedding, thus permitting one to better recreate paleocurrent pattern and paleogeomorphology. *Multistory* sandstone bodies are stacked, one on top of the other, and *multilateral* sandstone bodies form sheets. Sheets are lateral compositions (Fig. 10-17) formed as the river meanders back across its former floodplain so that the final result may be a well-defined, sandy meander belt identifiable by its scalloped margins, complex pattern of paleoflow, and many internal discontinuities (Fig. 10-18). In contrast, point bar sandstones in some alluvial deposits may be largely isolated in overbank shale. *Sandy splays*, the deposits formed by a crevasse, have abrupt basal contacts, and form lobes at right angles to levee and channel orientation. Some splays coarsen upward and others fine upward. Whether the splay does or not depends on what part is observed. Because of the high velocity of the discharge when the levee broke, the splay may be separated from the river. Rootlets, plant debris, peats, and even tree trunks are abundant in humid climate deposits and some bioturbation may be present, plus possible vertebrate remains.

The descriptions by Fisk (1944) of modern alluvial valleys and alluvium of the lower Mississippi Valley are classic as is Shantser's (1951) study of the alluvium of Russian rivers, the study by Attia (1954) of Nile alluvium, and

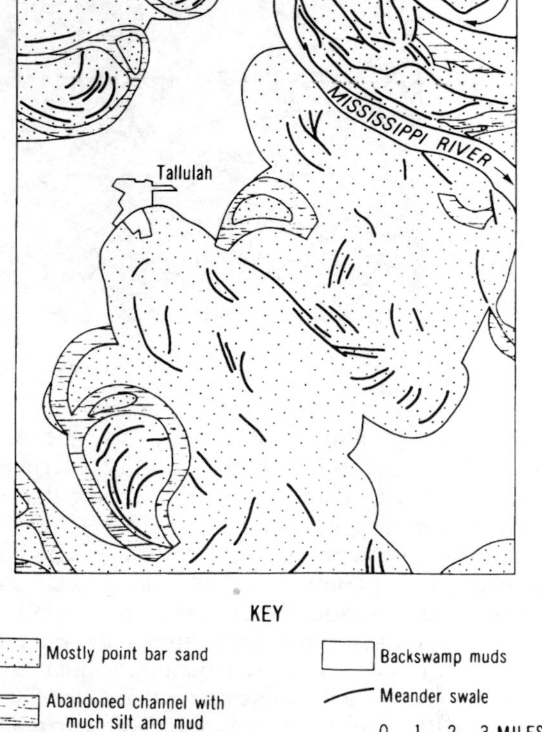

FIGURE 10-18. Belt sandbody of former course of Mississippi River (Hrabar and Potter, 1969, Fig. 11). Note serrated edge and compare with Fig. 10-20. Reproduced by permission of the authors and the American Association of Petroleum Geologists.

FIGURE 10-19. Sand bars and their sedimentary structures of the Mississippi River in Fulton and Hickman Counties, far western Kentucky, U.S.A.: A) midstream and side bar at miles 928 to 930, B) side- and midstream bar at New Madrid Bend, miles 886 to 888, C) front of large advancing sand wave on side bar at mile 809, D) crossbedding in sand wave of point bar at French Point, mile 913, E and F) small megaripples and their internal structure on inner side of long side bar at mile 909, and G) clay drape with mudcracks below front of large sand wave at mile 909.

the study of the Niger by NEDECO (1959). The many reports on the Mississippi River of the U.S. Corps of Engineers Waterways Experiment Station contain a wealth of data. See also the maps of Holocene alluvial valley fill in Poland by Szumanski (1983) to fully appreciate the complexity that may exist in ancient equivalents. All of these, however, are studies of rivers in temperate climates and more are needed of tropical rivers, as emphasized by Baker (1978).

Descriptions of modern point bar deposits are given by Sundborg (1956), McGowen and Garner (1970), Jackson (1976), and Morton and McGowen (1980, pp. 10–54). Jackson (1978) summarizes many of the variations of lithology and stratification that exist in modern meandering streams. See also Potter *et al.* (1987) for description of large point bars at low water along the Mississippi River in western Kentucky (Fig. 10-19).

Descriptions of ancient point bar deposits are fairly numerous. J.R.L. Allen (1970) early described and illustrated point bar deposits and emphasized the processes that form them. Molina (1979) described a point bar in the Tertiary of Spain, Hulse (1979) identified, with logs and core, meander deposits in the Bartlesville Sandstone (Pennsylvanian) of Kansas, and Swanson (1979, pp. 120–21, 132–134) reviewed criteria for the identification of point bar deposits and provides examples in the Upper Morrow (Pennsylvanian) reservoirs of the Anadarko Basin of the southern Mid-Continent in the United States. Visser and Johnson (1978) identified fining-upward point bar sequences in the Late Pliocene Upper Siwalik molasse in the Jhelum Re-Entrant of Pakistan. Galloway (1977, pp. 5–23), using both the subsurface and outcrops, related uranium mineralizations to permeability and bedding of point bar deposits. Excellent detail of the lateral accretion surfaces of an ancient Pennsylvanian point bar is provided by Beutner *et al.* (1967, Pl. 1). Meandering stream deposits in shaly basins have distinctive map patterns (Fig. 10-20).

Fining-upward sequence and epsilon crossbedding similar to the fluvial point bar also occur in estuarine and tidal sandstones so that one always needs to investigate sand body ge-

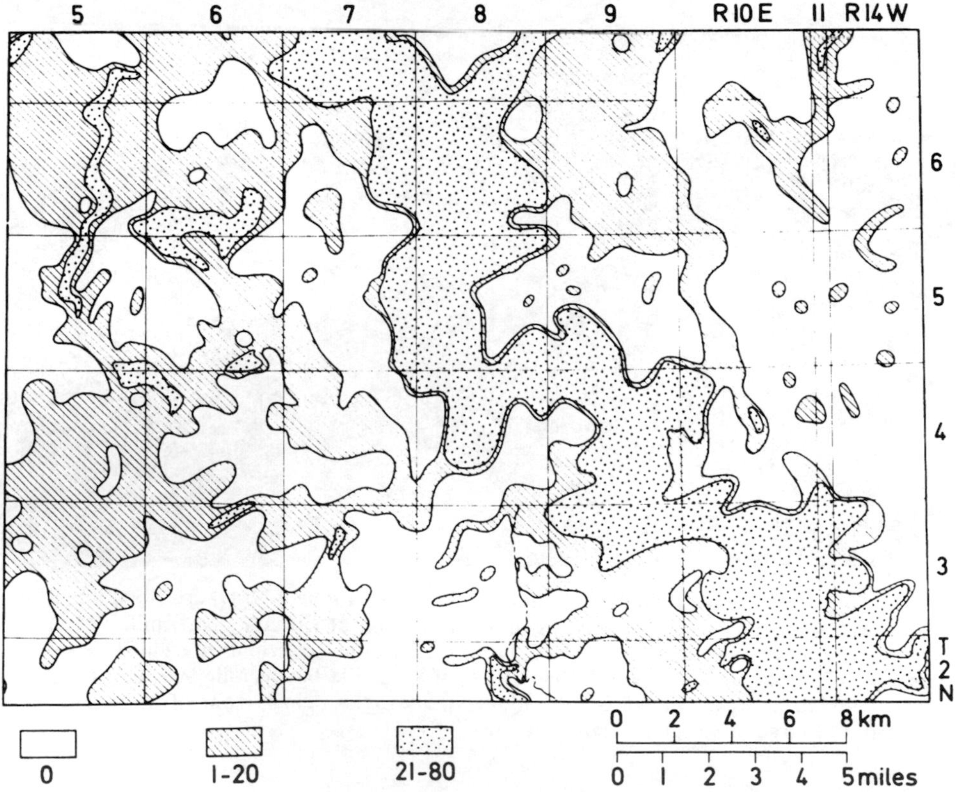

FIGURE 10-20. Pennsylvanian sandstone in Illinois Basin with pronounced meandering pattern (Potter, 1963, Fig. 51).

ometry and identify all the associated environments of the depositional system rather than making a decision based on vertical profile alone.

Braided Streams and Alluvial Fans

A braided stream is one with a network of interconnected mobile channels (Figs. 10-15 and 10-21). Braided stream deposits occur in small to large rivers and almost without exception on alluvial fans and fan deltas.

Subenvironments of braided streams are fewer than those of meandering streams, and include chiefly channels, some lakes, and minor overbank deposits. A clear vertical cycle is lacking because overbank deposits are not common and because of great channel instability. Hence log signatures tend to be blocky. The dominant structure is crossbedding in sandstones and pebbly sandstones. Variance of crossbedding is low and fossil debris of all kinds is rare.

Sandy sedimentation above widespread unconformities commonly shows a sequence of braided stream deposits associated with rapid sedimentation, followed by meandering streams or tidal deposits, where there is a transition to marine deposition. Thin eolian sands may also be present in such braided deposits. Selley (1972) carefully describes the transition from braided to marine sandstones in the Cambrian of Jordan.

The streams of alluvial fans are virtually all braided. Poorly sorted gravel and even boulders are typical, and sand may be locally subordinate, especially near the source (Fig. 10-22). These materials are very immature and their

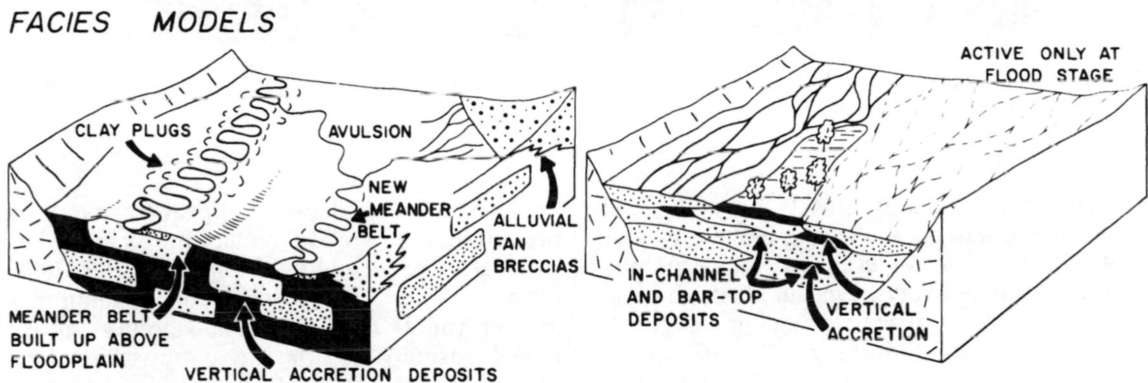

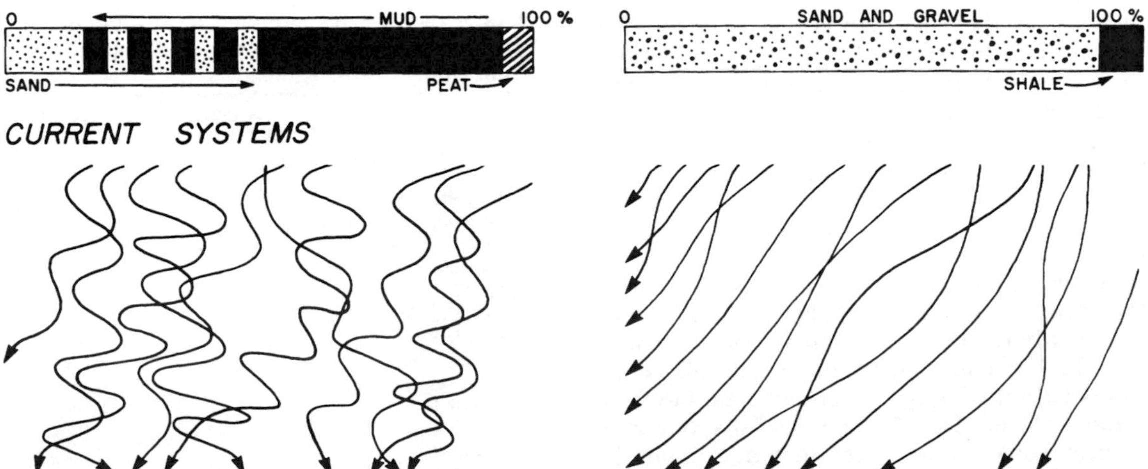

FIGURE 10-21. Comparison of meandering and braided stream deposits. (Redrawn from Walker, 1976, Fig. 8).

FIGURE 10-22. Surface of alluvial fan debouching from coastal mountains east of Highway 1, south of Tocopilla, Antofagasto, Chile.

composition is usually closely linked to bedrock. Size gradients are strong in the downcurrent direction. As determined from crossbedding dip, this is away from fan head. In many fans, size decline of pebbles and gravel can be approximated by Sternberg's law (Chap. 3). Lenticular bedding, cut-and-fill, and intrasandbody channels are common. Textural variation is marked. In terms of geometry, alluvial fan deposits wedge out down dip and commonly are composites—one will be stacked on or overlap another. By lateral coalescence, they may form aprons many kilometers long parallel to a line source such as a rising fault block. Thickness of the resulting wedge, which may extend into the thousands of meters, is controlled by the subsidence of the down-thrown block.

Paleocurrent pattern radiates away from a linear source. Such fans may develop along a major mountain front. Alluvial fans are characteristic of arid basins but also occur in humid basins where they have a permanent through stream. Associated playa, dune, and caliche deposits help distinguish an arid climate fan from a humid one. Debris flow deposits occur in both, but may be more common in arid fan deposits. Braided deposits may also be markedly linear when they fill a narrow paleovalley.

Braided stream deposits appear to have been predominant in pre-Devonian time as judged by a literature review of fluvial style (Cotter, 1978) (Table 10-5). The absence of vegetation—greater runoff and low bank stability—is believed responsible for the contrasting abundance of braided and meandering deposits before and after the Devonian Period (Schumm, 1968). More studies such as Cotter's are

TABLE 10-5. Changing proportions of braided and meandering stream deposits. (Simplified from Cotter, 1978, Table 2).

	Number	Percent
Tertiary		
Braided	8	44
Meandering	10	56
Mesozoic		
Braided	17	49
Meandering	18	51
Paleozoic		
Silurian to Permian		
Braided	22	44
Meandering	28	66
Cambro-Ordovician		
Braided	5	100
Meandering	0	0
Precambrian		
Braided	12	86
Meandering	2	14

needed, and one wonders if other sandy environments show long-term changes in depositional style.

Literature is abundant and Miall's (1977) review is a good place to start. Studies of modern braided streams have been made by Coleman (1969) for the Brahmaputra River, by Smith (1971) for the Platte, and by Cant and Walker (1978) for the Saskatchewan River, and Schwartz (1978) has documented the transition from braided to meandering deposits in the Red River along the Texas–Oklahoma boundary. All of these studies emphasize bedforms, paleocurrents, and, to some degree, vertical sequence. Paleocurrents and facies have also been studied in glacial outwash in the Canadian Arctic (Church, 1972), in southern Ontario (Eynon and Walker, 1974), and in Iceland (Bluck, 1972), where braided aprons are called *sandurs*. Boothroyd and Ashley (1975) describe in detail how structures and facies change downstream in braided outwash fans bordering the Gulf of Alaska. Nadji (1974) has mapped the modern alluvial basins of Iran which are widespread and have thicknesses in excess of 500 m; here the sandy channels of these fans are important aquifers called "water veins."

A comparison and contrast between braided and meandering deposits of Tertiary age in Cook Inlet of Alask has been made by Hayes *et al.* (1976). Paleovalleys, especially narrow ones that act as a sluice, are commonly filled by braided stream deposits (Sedimentation Seminar, 1978, Fig. 18). Trevena (1977) used stratification style in Triassic sandstones to demonstrate a downstream transition from braided to meandering. Excellent insight into a Cambrian alluvial fan in West Texas is given by McGowen and Groat (1971), with many informative illustrations. Alluvial fans and perhaps fan deltas are the site of gold placers in the Witswatersrand of South Africa (Minter, 1978). Teisseyre (1975) reports on alluvial fans of Kulm age in a small intermontaine basin in southern Poland, where a 7000-m section is believed to have about a dozen fans. His conclusions stem from outcrop study of texture, structures, and petrology. Fan deposits up to 25 km thick in a small synorgenic Devonian basin in Norway are described by Steel *et al.* (1977), where they distinguished marginal fans from alluvial plain deposits; here the fluvial cycles of conglomerates and sandstones coarsen upward because of periodic subsidence of the basin. In arid regions, dune sands may be interbedded with or perhaps completely cover sands and gravels of alluvial fans. In the Triassic Buntsandstein of the Eifel region of Germany, Mader (1983) recognized eight sequences of braided stream deposits; the associated environments include local alluvial fans, debris flows, lakes, and dunes. His environmental analysis is exceptionally complete and convincing.

Anastomosing Streams

Low-sinuosity, bifurcating but stable sandy stream channels in a mud-rich section (Fig. 10-23) are called *anastomosing* and have been described by sedimentologists in both the modern (Smith and Smith, 1980) and ancient (Putnam and Oliver, 1980; Smith and Putnam, 1980). Garner (1974, pp. 435–41) has also called attention to anastomosing streams, especially in tropical regions. Deposits of anastomosing streams have only been recognized recently and hence studies are few.

The sandstones of this environment have sharp boundaries except where bordered by a crevasse, form fairly straight elongate bodies, and their structures include massive beds and crossbeds, horizontal lamination, and ripples. Sandstones formed by anastomosing streams differ from those of braided streams in two major ways—there is much shale in the section and they have a fining-upward grain size and log signature (Putnam and Oliver, 190, p. 497). Off-channel deposits include thin-sheet sandstones, perhaps largely formed by splays, levee deposits, and floodplain shales. Plant debris and rootlets are common in deposits of humid climates and bioturbation may be abundant. Deposits of anastomosing streams are probably much more common than has been thought.

The dipmeter and gamma-ray or self-potential logs go far to help discriminate among the alluvial sandstones (Figs. 10-7 and 10-24).

Coal Measures and Other Economic Deposits

In the vast literature of alluvial deposits, studies of coal measures are foremost and commonly include down-dip deltaic deposits as well. Coal cutouts or washouts (Fig. 10-25) are the direct result of erosion by paleostream channels; coal thickness and quality are related to flooding and the introduction of mud, silt, and sand along the borders of the coal swamp, as are mine roof quality and groundwater transmissivity (Vereda, 1970; Moebs, 1977; McCabe and Pascoe, 1978; Donaldson, 1979). A general reference to coal measure sedimentation is that by Horne *et al.* (1978). Seismic cross sections are

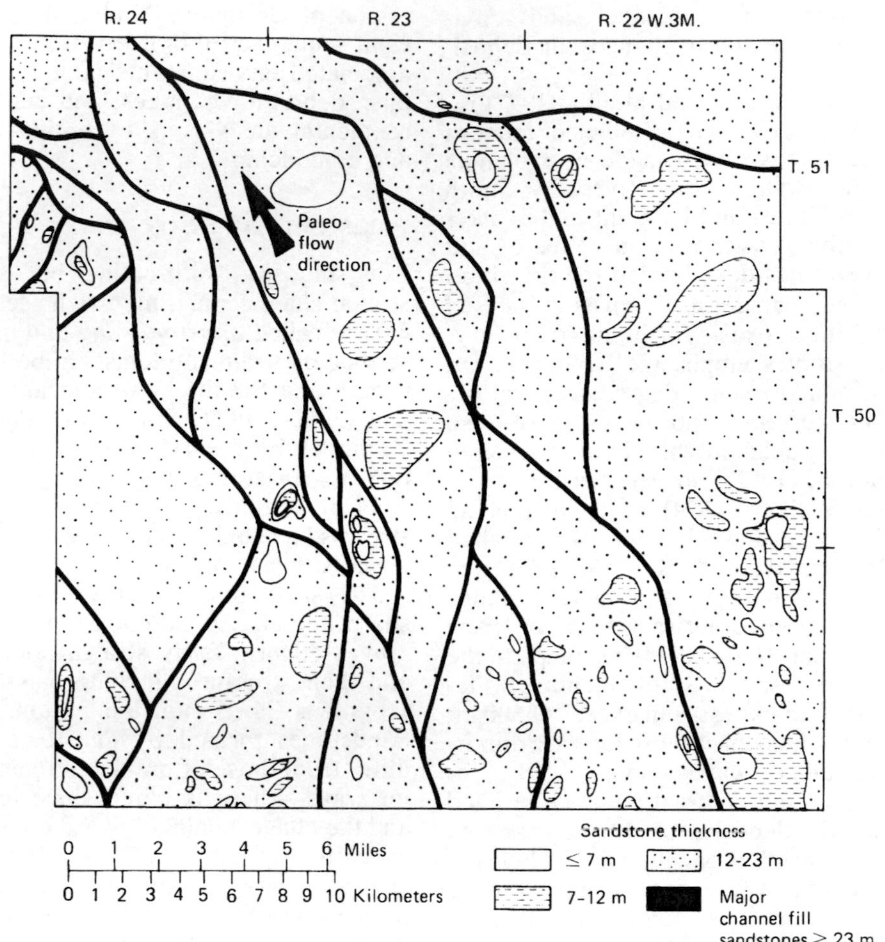

FIGURE 10-23. Anastomosing map pattern of lower part of Cretaceous Mannville Group in Alberta (Putman, 1982, Fig. 14). Published by permission of the author and the American Association of Petroleum Geologists.

increasingly used to map sandstone channel systems in coal measures (Fig. 10-26).

Sandstone-type uranium deposits are largely alluvial and the transfer of knowledge about the geometry and origin of sandstones in coal measures to uranium exploration is direct. A good summary paper is by Qudwai and Jensen (1979); specific studies are those of Galloway (1977) and Cole (1980).

In a report on the Witwatersrand basin of South Africa, Smith and Minter (1980) summarize the sedimentological aspects of gold and uranium in a giant alluvial paleoplacer, as does Pretorius (1975). Redbed copper deposits are commonly associated with alluvial sandstones. See Galloway and Hobday (1983) for full discussion of these and other mineral resources in alluvial deposits.

Summary

Alluvial sandstones are important in many sedimentary basins as fan and valley fill deposits, and also occur in deltas and fan deltas. Variability is characteristic and pronounced—alluvial sandstones range from small, isolated bodies to thick, multiple stacked bodies, to continuous composite sheets with many internal discontinuities. Grain size varies widely—from the well-sorted sands of a large river to poorly sorted sands on the proximal part of an alluvial fan. Crossbedding and ripple marks are the chief sedimentary structures followed by channels and channel scours. Environments include braiding, anastomosing, and meandering streams. Associated deposits include debris flows on fans, peat and coal, a wide range of shales, and evaporites where climate was dry. Eolian deposits occur as well. The latter may be but a few centimeters thick and occur more

FIGURE 10-24. Self-potential (SP) or gamma-ray log (GR) and dipmeter log for major continental sandstones. See Fig. 10-9 for explanation. Published by permission of J. A. Gilreath and Schlumberger Offshore Services, New Orleans.

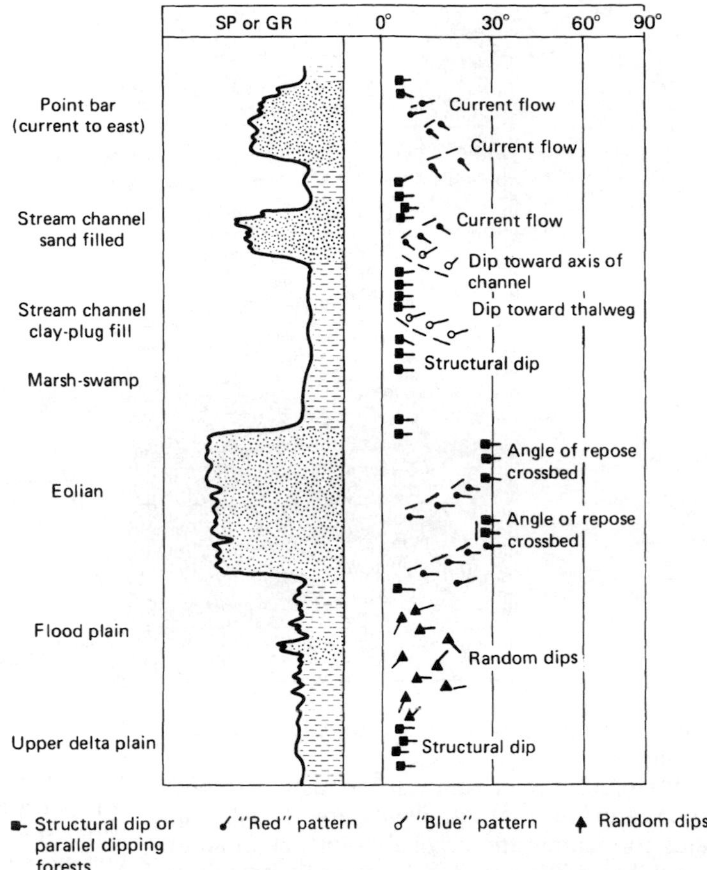

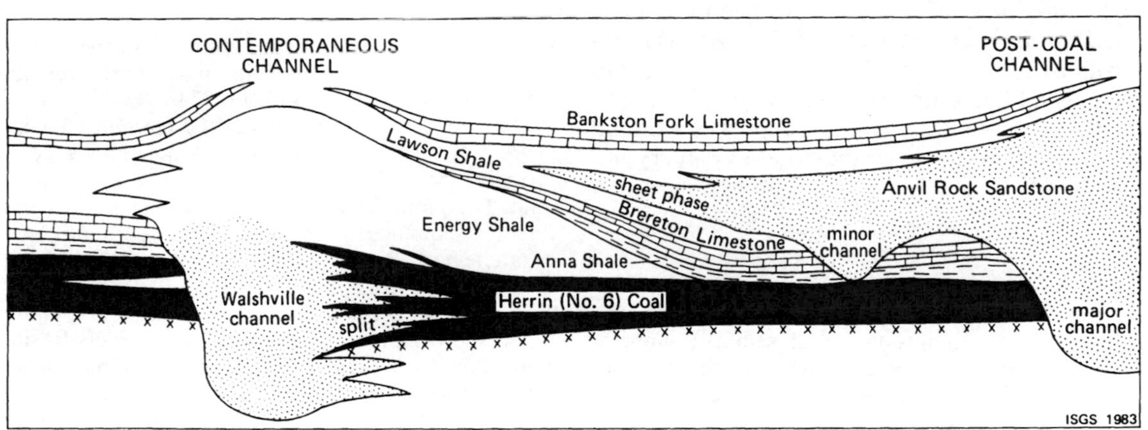

FIGURE 10-25. Pennsylvanian channel replaces Herrin (No. 6) Coal Member and forms well-defined, linear compactional high (Nelson, 1983, Fig. 4).

or less randomly throughout the deposit, or they may be tens of meters thick. Lacustrine deposits may also be present.

Identification of ancient river type is largely based on sandstone–shale ratio, vertical profile, stratification style, grain size, and paleocurrents, and as these variables are systematically mapped in an alluvial sandstone across a basin (or upward through a sandy sequence), they can be expected to change as the river evolved from braiding to meandering or perhaps to anatomosing. Complete fluvial cycles are also valuable

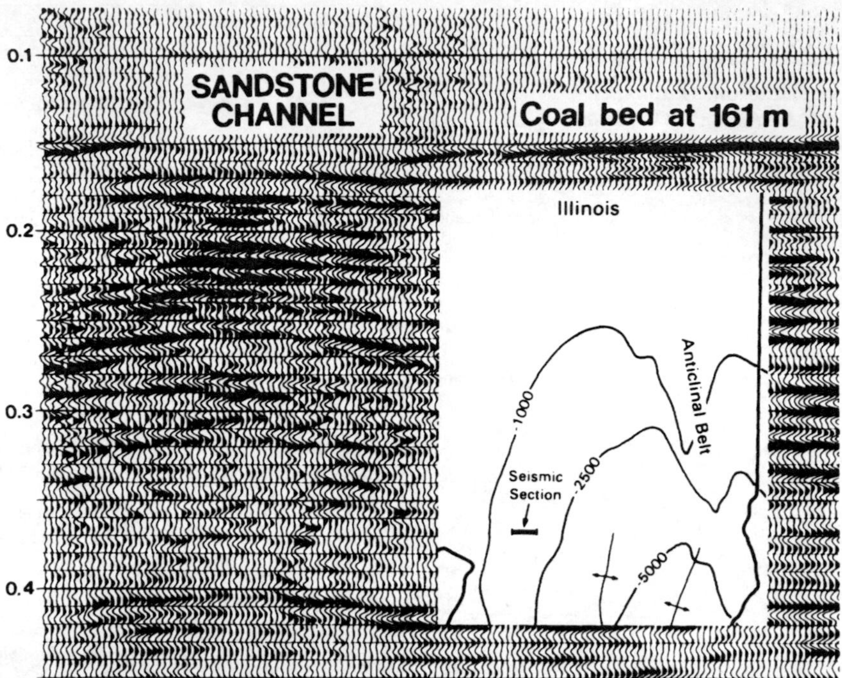

FIGURE 10-26. High-frequency virbroseis time section shows sandstone channel replacing coal bed at 161 m in Pennsylvanian sediments of Illinois Basin (Chapman and Shafers, 1983, Fig. 1). Published by permission of the authors and the American Association of Petroleum Geologists.

for the insight they provide into paleohydraulics (Chap. 8).

Recognition of alluvial sandstones, their type, and their size gradients are all important aids to locating the original boundary of an ancient basin, and thus help identify its source. Though easy to do, only rarely have all the fluvial sandstones of a large basin been compiled (cf. Dennison and Wheeler, 1975). Alluvial deposits occur in every plate tectonic setting, but are especially common along major rifts, continental sutures, and active margins. They should also be looked for and studied above every unconformity.

References

Allen, J.R.L.: Studies in fluvial sedimentation. A comparison of fining upwards cyclothems, with special reference to coarse member composition and interpretation. Jour. Sed. Petrology 40, 298–323 (1970).

Attia, M.I.: Deposits in the Nile Valley and delta, 356 pp. Cairo: Geological Survey of Egypt 1954.

Baker, V.C.: Adjustment of fluvial systems to climate and source terrain in tropical and subtropical environments. In: Miall, A.D. (Ed.): Fluvial sedimentology. Canadian Soc. Petroleum Geologists Mem. 5, 211–230 (1978).

Beutner, E.C.; Fleuckinger, L.A.; and Gard, T.M.: Bedding geometry in a Pennsylvanian channel sandstone. Geol. Soc. America Bull. 78, 911–916 (1967).

Bluck, B.J.: Structure and directional properties of some valley sandur deposits in southern Iceland. Sedimentology 21, 533–554 (1974).

Bryhni, Inge: Flood deposits in the Hornelen Basin, West Norway (Old Red Sandstone). Norsk geol. tidalkr. 58, 273–306 (1978).

Cant, D.J., and Walker, R.G.: Fluvial processes and facies sequences in the sandy, braided South Saskatchewan River, Canada. Sedimentology 25, 625–648 (1978).

Chapman, W.L., and Shafers, C.J.: Shallow channel, sand channel, coal exploration, Illinois Basin. In: Bally, A.W. (Ed.): Seismic expression of structural styles, a picture and work atlas, Vol. 1. Am. Assoc. Petroleum Geologists, Studies in Geology 15.

Church, M.: Baffin Island sandurs: A study of Arctic fluvial processes. Geol. Survey Canada Bull. 216, 208 pp. (1972).

Cole, D.I.: Aspects of the sedimentology of some uranium-bearing sandstones in the Beaufort West area, Cape Province. Trans. Geol. Soc. South Africa 83, 375–390 (1980).

Coleman, James M.: Brahmaputra River: Channel processes and sedimentation. Sedimentary Geology 3, 129–239 (1969).

Collinson, J.D., and Lewin, J. (Eds.): Modern and

ancient fluvial system. Intern. Assoc. Sedimentologists Spec. Pub. 6, 587 pp. (1983).

Cotter, Edward: The evaluation of fluvial style, with special reference to the central Appalachian Paleozoic. In: Miall, A.D. (Ed.): Fluvial sedimentology. Canadian Soc. Petroleum Geologists Mem. 5, 361–383 (1978).

Dennison, J.M., and Wheeler, W.H.: Stratigraphy of Precambrian through Cretaceous strata of probable fluvial origin in southeastern United States and their potential as uranium host rocks. Southeastern Geol. Spec. Pub. 5, 210 pp. (1975).

Donaldson, A.C.: Origin of coal seam discontinuities. In: Donaldson, A.S.; Presley, M.W.: and Renton, J.J. (Eds.): Carboniferous coal. West Virginia Geol. Econ. Survey Bull. B-37-1, 102–132 (1979).

Ethridge, Frank G., and Romero, Flores M. (Eds.): Recent and ancient nonmarine depositional environments: Models for exploration. Soc. Econ. Paleon. Mineral. Spec. Pub. 31, 349 pp. (1981).

Eynon, G., and Walker, R.G.: Facies relationships in Pleistocene outwash gravels, southern Ontario: A model for bar growth in braided rivers. Sedimentology 21, 43–70 (1974).

Fisk, H.N.: Geological investigations of the alluvial valley of the lower Mississippi River, 82 pp. Vicksburg, Mississippi: Mississippi River Comm. 1944.

Gagliano, Sherwood, and von Beck, J.L.: Geologic and geomorphic aspects of deltaic processes, Mississippi Delta system. In: Hyrologic and geologic studies of Coastal Louisiana. Louisiana State Univ., Center for Wetland Resources, Rept. 1, 140 pp. (1970).

Galloway, W.E.: Catahoula Formation of the Texas Coastal Plain: Depositional systems, composition, structural development, groundwater flow history and uranium distribution. Texas Bur. Econ. Geol., Rept. Invs. 87, 59 pp. (1977).

Galloway, W.E., and Hobday, D.K.: Terrigeneous clastic depositional systems, 423 pp. New York–Berlin–Heidelberg–Tokyo: Springer-Verlag 1983.

Garner, H.F.: The origin of landscapes, 734 pp. New York: Oxford Univ. Press 1974.

Hayes, J.B.; Harms, J.C.; and Wilson, T.: Contrasts between braided and meandering stream deposits, Beluga and Sterling Formations (Tertiary), Cook Inlet, Alaska. In: Miller, T.P. (Ed.): Recent and ancient sedimentary environments in Alaska, pp. J1–J27. Anchorage: Alaska Geol. Soc. 1976.

Horne, J.C.; Ferm, J.C.; Caruccio, F.T.; and Baganz, B.P.: Depositional models in coal exploration and mine planning in Appalachian region. Am. Assoc. Petroleum Geologists Bull. 62, 2379–2411 (1978).

Hrabar, S.V., and Potter, P.E.: Lower West Baden (Mississippian) sandstone body of Owen and Greene Counties, Indiana. Am. Assoc. Petroleum Geologists Bull. 53, 2150–2160 (1969).

Hulse, W.J.: Depositional environment of the Bartlesville Sandstone in the Sallyards Field, Greenwood County, Kansas. In: Hyne, N.J. (Ed.): Pennsylvanian sandstones of the Mid-Continent. Tulsa Geol. Soc. Spec. Pub. 1, 327–336 (1979).

Jackson, Roscoe G., II: Depositional model of point bars in the lower Wabash River. Jour. Sed. Petrology 46, 579–594 (1976).

Jackson, Roscoe G., II: Preliminary evaluation of lithofacies models for meandering alluvial streams. In: Miall, A.D. (Ed.): Fluvial sedimentology. Canadian Soc. Petroleum Geologists Mem. 5, 543–586 (1978).

Mader, Detlef: Evolution of fluvial sedimentation in the Buntsandstein (Lower Triassic) of the Eifel (Germany). Sedimentary Geology 37, 1–84 (1983).

McCabe, K.W., and Pascoe, W.: Sandstone channels: Their influence on roof control in coal mines. U.S. Dept. Labor, Mine Safety and Health Administration, Informational Rept. 1096, 24 pp. (1978).

McGowen, J.H., and Garner, L.E.: Physiographic features and stratification types of coarse-grained point bars: Modern and ancient examples. Sedimentology 14, 77–112 (1970).

McGowen, J.H., and Groat, C.G.: Van Horne Sandstone, West Texas: An alluvial fan model for mineral exploration. Texas Bur. Econ. Geol., Rept. Invs. 72, 57 pp. (1971).

Miall, A.D.: A review of the braided-river depositional environment. Earth Sci. Rev. 13 1–62 (1977).

Miall, A.D. (Ed.): Fluvial sedimentology. Canadian Soc. Petroleum Geologists Mem. 5, 859 pp. (1978).

Miall, A.D.: Cyclicity and the facies model concept. Canadian Soc. Petroleum Geologists Bull. 28, 59–80 (1980).

Miall, A.D.: Alluvial sedimentary basins: Tectonic setting and basin architecture. In: Miall, A.D. (Ed.): Sedimentation and tectonics in alluvial basins. Geol. Assoc. Canada Spec. Paper 23, 1–33 (1981).

Moebs, N.N.: Roof rock structures and related roof support problems in the Pittsburgh Coal bed of southwestern Pennsylvania. U.S. Dept. Interior, Bur. Mines, Rept. Invs. 8230, 30 pp. (1977).

Molina, Diaz Margarita: Características sedimentológicas de los paleocanales de la unidad detrítica superior al N. de Hueta (Cuenca). Estudios Geol. 35, 241–251 (1979).

Morton, R.A., and McGowen, J.H.: Modern depositional environments of the Texas Coast. Texas Bur. Econ. Geol. Guidebook 20, 167 pp. (1980).

NEDECO: River studies and recommendations on improvement of the Niger and Benue, 1000 pp. Amsterdam: North-Holland 1959.

Nadji, Mehdi: Terrestrische Sedimente der Intermontanbecken im Iran. Geol. Rundschau 63, 897–904 (1974).

Nelson, John: Geologic disturbances in Illinois coal seams. Illinois Geol. Survey Circ. 530, 47 pp. (1983).

Pitty, A.F. (Ed.): Geographical approaches to fluvial processes. Geo-Abstracts, Univ. East Anglia, Norwich, 280 pp. (1979).

Potter, P.E.: Late Paleozoic sandstones of Illinois Basin. Illinois Geol. Survey Rept. Inv. 217, 92 p. (1963).

Potter, P.E.; Pryor, W.A.; Smith, Lawson M; and Rich, David: Alluvial processes and sedimentation of the Mississippi River, Fulton County, Kentucky and Lake County, Tennessee. Ann. Field Conf., Kentucky Geol. Soc., Sept. 27, 28, 29, 1984. Ky. Geol. Soc. 38 pp. (1987).

Pretorius, D.A.: Depositional environment of the Witwatersrand goldfields: A chronological review of speculations and observations. Minerals Sci. Eng. 7, 18–47 (1975).

Putman, P.E.: Fluvial channel sandstones within Upper Mannville (Albian) of Lloydmister Area, Canada—geometry, petrography and paleographic implications. Am. Assoc. Petroleum Geologists Bull. 66, 436–459 (1982).

Putman, P.E., and Oliver, T.A.: Stratigraphic traps in channel fills within the Upper Mannville subgroup (Albian), east-central Alberta. Canadian Soc. Petroleum Geologists Bull. 28, 489–508 (1980).

Qudwai, H.A., and Jensen, M.L.: Methodology and exploration for sandstone-type uranium deposits. Mineralium Deposita 14, 137–152 (1979).

Schumm, A.S.: Speculations concerning paleohydrologic controls of terrestrial sedimentation. Geol. Soc. America Bull. 79, 1573–1588 (1968).

Schumm, S.A.: The fluvial system, 338 pp. New York: Wiley Interscience 1977.

Schumm, S.A.: Evolution and response of the fluvial system, sedimentological implications. In: Ethridge, F.G., and Flores, R.M. (Eds.): Recent and ancient nonmarine depositional environments: Models for exploration. Soc. Econ. Paleon. Mineral. Spec. Pub. 31, 19–29 (1981).

Schwartz, D.E.: Hydrology and current orientation analysis of a braided-to-meandering transition; the Red River in Oklahoma and Texas, U.S.A. In: Miall, A.D. (Ed.) Fluvial sedimentology. Can. Soc. Petroleum Geologists Mem. 5, 231–255 (1978).

Sedimentation Seminar: Sedimentology of the Kyrock Sandstone (Pennsylvanian) in the Brownsville paleovalley, Edmonson and Hart Counties, Kentucky. Kentucky Geol. Survey Ser. X, Rept. Invs. 21, 24 p. (1978).

Selley, R.C.: Diagenesis of marine and non-marine environments from the Cambro-Ordovician sandstones of Jordan. Jour. Geol. Soc. (London) 128, 135–150 (1972).

Shantser, Ye. V.: Alliuvii ravninnykh rek umerennogo poiasa i ego anachenie dlia poznaniia zakonomernostei stroeniia i formirovaniia alliuvial'nyk svit (Alluvium of the river plains of the temperate zone and its significance for knowledge of regularities of structure and formation of alluvial deposits). Akad. Nauk SSSR, Inst. Geol. Sci. 135, Geol. Ser. 55, 273 pp. (1951).

Smith, Gerald G., and Putman, P.E.: Anastomosed river deposits: Modern and ancient examples in Alberta, Canada. Canadian Jour. Earth Sci. 17, 1396–1406 (1980).

Smith, D.G., and Smith, N.D.: Sedimentation in anastomosed river systems. Examples from alluvial valleys near Banff. Jour. Sed. Petrology 49, 157–164 (1980).

Smith, N.D.: Transverse bars and braiding in the lower Platte River, Nebraska. Geol. Soc. America Bull. 81, 3407–3420 (1971).

Smith, N.D., and Minter, W.E.L.: Sedimentological control of gold and uranium in two Witwatersrand paleoplacers. Econ. Geology 75, 1–14 (1980).

Steel, J.J.; Maehle, S.; Nilsen, H.; Roe, S.L.; and Spinnanger, A.: Coarsening-upward cycles in the alluvium of the Hornelen Basin (Devonian) Norway: Sedimentary response to tectonic events. Geol. Soc. America Bull. 88, 1124–1134 (1977).

Sundborg, Ake: The river Klaralven, a study of fluvial processes. Geog. Annaler 38, 127–136 (1956).

Swanson, Donald C.: Deltaic deposits in the Pennsylvanian Upper Morrow Formation of the Anadarko Basin. In: Hyne, N.E. (Ed.): Pennsylvania sandstones of the Mid-Continent. Tulsa Geol. Soc. Spec. Pub. 1, 115–168 (1979).

Szumański, A.: Paleochannels of large meanders in the river valleys of the Polish Lowland. In: Palaeohydrology of the temperate zone, Poznań, Poland '81, Polish Acad. Sci. Quaternary Studies in Poland, 4, 207–216 (1983).

Teisseyre, A.K.: Sedymentologia i paleogeografia Kulmu starszego w zachodniej czesci niecki srodsudeckiej (Sedimentology and paleogeography of the Kulm alluvial fans). Geol. Sudetica 9, 1–135 (1975).

Trevena, A.S.: Determining fluvial channel patterns from stratification style. A Triassic example from the Colorado Plateau. Contrib. Geology, Univ. Wyoming 16, 45–63 (1977).

Vereda, V.S.: Physical and mechanical properties of the sandstones of the Donets Basin: Lithology and Mineral Resources (pp. 607–610) (Lithogiya i Bleznye Iskopaeniye, pp. 115–119).

Visser, C.F., and Johnson, G.D.: Tectonic control of Late Pliocene molasse sedimentation in a portion of the Jhelum Re-Entrant, Pakistan. Geol. Rundschau 67, 15–37 (1978).

Walker, R.G.: Facies model 3. Sandy fluvial systems. In: Walker, R.G. (Ed.): Facies models. Geoscience Canada Reprint Ser. 1, 23–31 (1976).

Deltas

Deltas and their varieties have been and remain the chief focus of most research on terrigenous deposits, primarily because of their economic resources. Deltas, recognized since Herodotus first named and described them in the fifth century B.C., have an exceptionally large literature whose geological papers are ably reviewed by LeBlanc (1975). This large literature reflects several factors. First, delta's abound on the world's present coastline and much of it is delta influenced. Second, ancient deltas are a major source of both petroleum and coal and thus are attractive exploration targets. Sandstones are always present, although their kind and proportion vary greatly with the type of delta. Like their up-dip alluvial equivalents, the scale of deltas varies by several orders of magnitude—from a few square kilometers for a small, freshwater delta deposited in an ephermal glacial lake to thousands of square kilometers for the delta of a large river. The largest deltas develop on passive continental margins or along the borders of small oceans, and are more mineralogically mature than deltas of active margins.

In examining the sandy facies of deltas we first recognize their many subenvironments, then consider major formative processes, and finally turn our attention to deltaic deposits in ancient basins. If one could study but one major terrigenous environment, it might well be deltas, because of their great diversity and importance in the geologic record.

Fluvial channel fills and levees, splays, freshwater lakes, floodplains, marshes, and delta-front deposits are subenvironments of deltas. Delta-front deposits include distributary mouth sand bars and coastal barriers as well as the salt marshes and shallow marine bays and offshore marine, delta-front slope deposits of marine sand and mud. Also present are thin, transgressive marine sandstones with glauconite and shell lags, thin marine shales, and perhaps carbonates. These develop with the destructional phase of a delta when supply is reduced, most commonly by switching of a major distributary. Because abandonment of one distributary is always accompanied by development of another with attendant seaward progradation, a river-dominated delta always features simultaneous transgression and regression of the coastline. In a shallow epeiric sea, tidal sandstones of an adjacent shelf may also represent the destructional phase of a delta. Where sand supply is large and climate semi-arid to arid, eolian dunes may also be present. Commonly it is not easy to draw a precise boundary between alluvial and down-dip deltaic deposits in an ancient basin. That boundary migrates up and down dip as supply, subsidence rate, and sea level vary.

Wright (1977), building on the earlier study of Albertson *et al.* (1950) and Bates (1953), describes sediment transport at river mouths and relates river mouth morphology to both map pattern and cross section of the distributary mouth bar. Central to this development is the idea of a free jet or plume entering a lake or ocean (Fig. 10-27). Considering only quiet water, there are three types of river mouths: (1) inertia-dominated mouths, high outflow velocities ($F \gg 1$ and τ_0 moderate to large), which occur in deep water and have narrow, jetlike plumes; (2) mouths that are dominated by bed friction with waters that are fully turbulent and have short, flaring plumes in shallow water ($F \gg 1$ and τ_0 very large); and (3) those with high outflow buoyancy ($F \cong 1$ and τ_0 low) which are density stratified producing narrow, river-mouth bars with few bifurcations. (See Chap. 8 for definition of symbols.) The first of these corresponds to the Gilbert-type delta, the second is characteristic of crevasse splays, and the last corresponds to "birdfoot" deltas of the Mississippi type.

Wright and Coleman (1973, 1974), in landmark studies, were the first to show how river discharge, tidal flux (as estimated from tidal range), and inshore wave power explain the great diversity of delta morphology. They showed how morphology is related to the sand–mud proportion of a delta and to kind and orientation of its sandstone bodies, which form its essential framework (Table 10-6A, Fig. 10-28). Inshore wave power is related not only to fetch and wind strength but also to offshore subaqueous profile. A long, shallow offshore profile dissipates wave energy so that branching distributaries develop whereas a short, steep, offshore profile results in longshore transport with many coastal barriers and limited seaward progradation. Fluvial-dominated deltas followed by wave-dominated deltas are the best known and their literature is voluminous.

Modeling by computer of the different types of deltas, modeling that involves estimates of inshore wave power, tidal range, and eustatic sea level changes, can be expected to grow in the years ahead. Examples are provided by Komar (1973) and Horowitz (1976).

Before we examine these three types of deltas, let us consider two down-dip sections—first, an idealized facies section, and second, some simplified seismic sections.

The idealized section of Fig. 10-29 and Table 10-6B show the great number of major terrigenous environments that may be present in a large delta debouching into deep water and how they relate to one another. Distal deep-water fans border delta-front turbidites, followed by shallow-water, delta-front marine sands and muds, and distributary mouth bars and coastal barriers. All are finally capped by delta plain deposits of river channels, levees, splay, and floodplain muds and peats. Such ideal, complete progradational sequences are commonly interrupted either by abandonment of deltaic

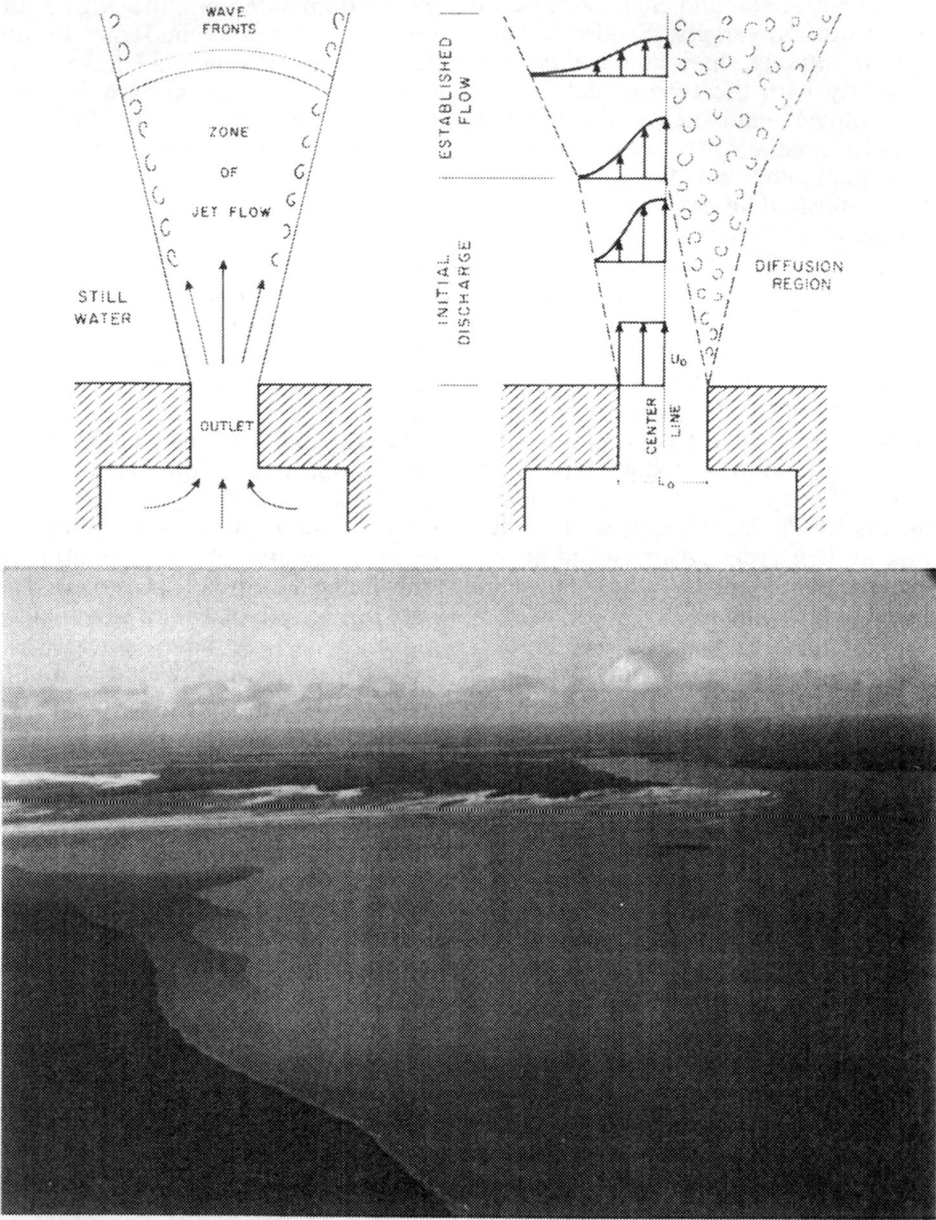

FIGURE 10-27. Hydraulics of free jet entering standing water (above) and actual plume of Rio Parnaíba, Parnaíba delta, Piauí, Brazil (below).

lobes leading to transgressive shelf sands, by decrease in supply, or by change in sea level, any of which can interrupt the cycle. Thus most thick deltaic sequences consist of many cycles, most of which have been interrupted even though the overall trend is a regional down-dip, seaward progradation of the shoreline. Should a delta develop in shallow water, turbidites will be absent. Later, the delta may advance to deep water, perhaps the edge of a continental shelf, and at that stage, turbidites will develop. Climate can play a role in two ways: if it is arid, the subaerial parts of the delta will have many dunes and possible salinas and, if the delta lies in low latitudes, interbedded carbonates may be expected. Or, if the climate is humid, peats may develop. As in other environments, climate has a strong influence on supply and on associated

TABLE 10-6. Characteristics of deltaic depositional systems. (Modified from Galloway, 1975, Table 2).

Part A. Major features

	Fluvial-dominated	Wave-dominated	Tide-dominated
Geometry	Elongate to lobate	Arcuate	Estuarine to irregular
Channel type	Straight to sinuous distributaries	Meandering distributaries	Flaring straight to sinuous distributaries
Lithology	Muddy to sand and mud	Sandy	Variable
Framework facies	Distributary mouth bar and channel fill sands, delta margin sand sheet; minor coastal barriers	Coastal barrier and beach ridge sands; dunes commonly abundant	Estuarine fill and tidal sand ridge system
Framework orientation	Parallels deposition	Parallels depositional strike	Parallels depositional slope

Part B. Environments
(Adapted from McGowen *et al.*, 1979, Table 1)

Fluvial environments	Deltaic environments	Lacustrine	Valley Fill Soil Beach
Meanderbelt	Distributary channel	Lacustrine	
Point bar	Abandoned distributary	Lacustrine–mudflat	
Channel lag	Channel-mouth bar	Mudflat	
Levee	Delta front	Lacustrine–deltaic	
Crevasse splay	Frontal splay		
Floodplain	Interdistributary		
Abandoned channel fill	Lacustrine–interdistributary		
Braided stream	Interdeltaic		
	Delta platform		
	Delta forests		
	Crevasse splay—splay delta		

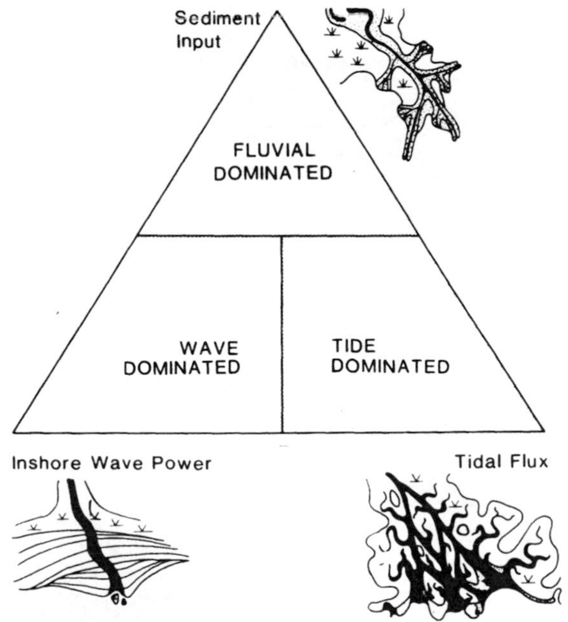

FIGURE 10-28. Fluvial-, wave-, and tide-dominated deltas. (Redrawn from Galloway, 1975, Fig. 3).

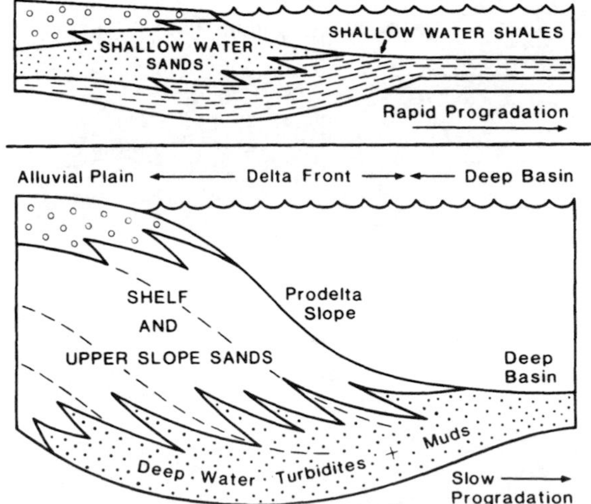

FIGURE 10-29. Idealized down-dip cross sections of a river-dominated delta debouching into a shallow-water basin (above) and into a deep-water basin (below). See Table 10-6B for complete listing of the many depositional environments that can occur in a delta.

lithologies in the delta and on sand composition—warm, wet climates are hostile to feldspar and rock fragments.

The simplified, seismic sections of Fig. 10-30 show commonly occurring longitudinal variations of both overall geometry and internal clinoform structure, as defined by major seismic reflectors. A clinoform is a sloping depositional surface formed when sediments prograde into deep water. Recognition of clinoforms and discrimination between wave- and river-dominated deltas is more difficult where a basin has markedly different differential subsidence within it.

In many ancient sections one type of delta may locally predominate over another, but when traced up and down paleoslope, differences in deltaic facies can be anticipated. Examples of down-dip changes in type of deltas are given by Klein *et al.* (1972) in Brazil, by Homewood and Allen (1981) in the Miocene Molasse of western Switzerland, and by Kingsley (1981) in the Permian of South Africa. Thus in every delta one needs to be aware of all possible variants, including variable amounts of marine sandstone, shale, and carbonates.

Large deltas and their rivers have tectonic significance and occur in intracratonic basins, along rifted and passive continental margins, in small seas or oceans (marginal basins), and at the boundary between cratons and bordering fold belts (Audley-Charles *et al.*, 1977).

River-Dominated Deltas

Much mud, many branching distributaries, marshes, lakes, levees, and splays, as well as bays and some minor barriers are the earmarks of the river-dominated delta (Fig. 10-31). Sands are deposited as fairly straight channel fills that fine upward and cut deeply into the sequence as the distributary progrades seaward. Distributary mouth or lunate sand bars are also present and grade downward into marine mud. Inshore, sand may be deposited in splays formed during floods, and minor coastal barriers may be present. Their orientation is down dip and at right angles to coastal barriers. Distributary channels cut into delta sands (Fig. 10-32) and a dipmeter log can help distinguish the distributary mouth bar from a superimposed fluvial channel (Fig. 10-25), Gilreath *et al.*, 1985). Prodelta sand and mud have pronounced clinoform structure where water is sufficiently deep. Such sand and mud contain marine fossils, abundant bioturbation, ripple marks of wave origin, and perhaps hummocky crossbedding formed during storms; where water depth exceeds wave base, some turbidites may be present. Where sedimentation is rapid, mud lumps and shale diapirs will develop, because of incomplete dewatering.

A large, river-dominated delta contains several lobes, only one of which is active at any one time. A lobe is abandoned when a major

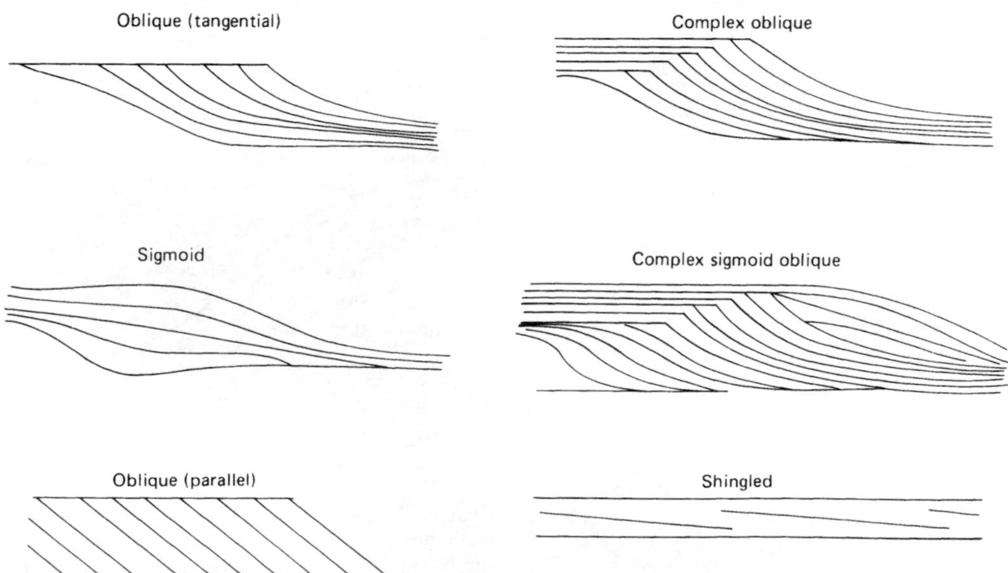

FIGURE 10-30. Idealized variations of down-dip cross sections of deltas simplified from seismic cross sections. (Redrawn from Mitchum *et al.*, 1977, Fig. 6).

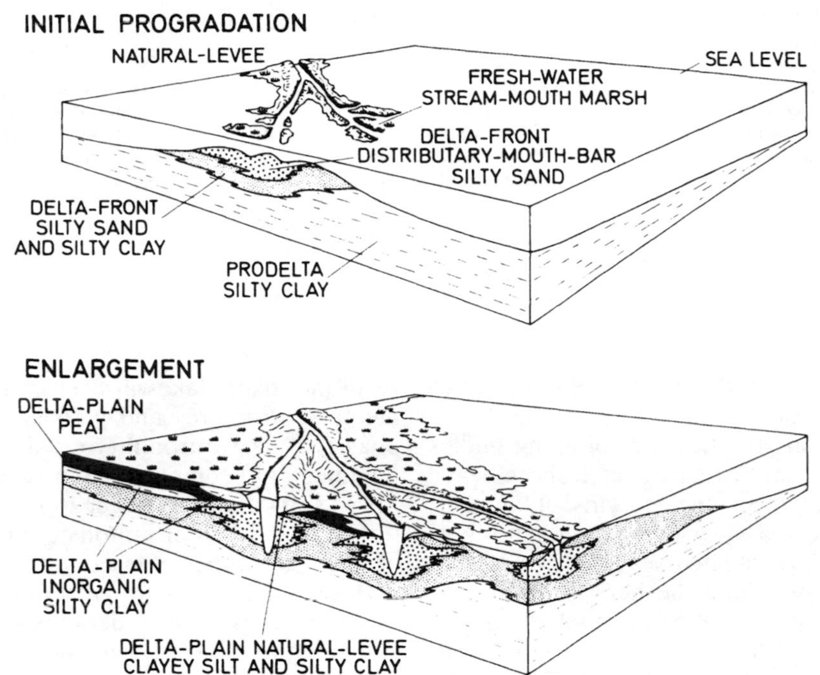

FIGURE 10-31. River-dominated delta with well-developed distributary mouth bar channel sand and much mud. (Modified from Frazier, 1967, Fig. 2).

SOUTHEAST

SHELL KUMAK J-06

GULF MOBIL TOAPOLOK O-54

GULF MOBIL TOAPOLOK H-24

GULF MOBIL YA YA P-53

3800 M 4000 4200 4400 2

2800 3000 3200 3400 2

2600 2800 3000 3200 ? 2 ?

GM ? 4600 4800 5000 5200

4 Km. 5 Km. 5.2 Km.

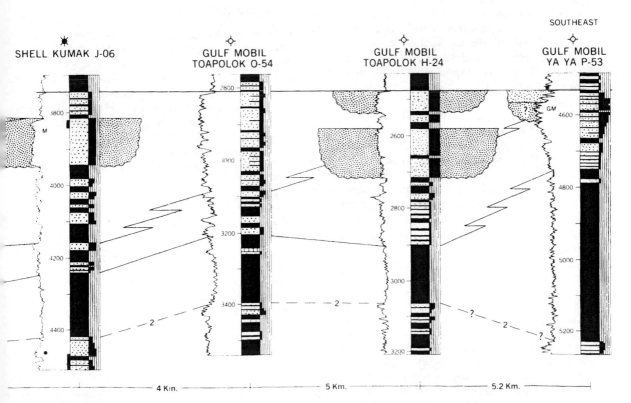

FIGURE 10-32. Surface cross section of river-dominated Eocene Taglu delta, Beaufort-MacKenzie Basin, Northwest Territories, Canada (Dixon, 1981, Fig. 2). Published by permission of the author and the Geological Survey of Canada.

TABLE 10-7. Characteristics of micro-, meso-, and macrotidal deltas. (Hayes, 1975)

Microtidal	Mesotidal	Macrotidal
Tidal range less than 2 m, deltas are small, are formed largely by storms, and are subordinate to wave-formed coastal barriers, beaches, and river deltas. Dominant transport processes are wind and waves rather than tides.	Tidal range is 2 to 4 m, principal sandbodies are large tidal deltas that are either *flood deltas* deposited inward of a coastal barrier or *ebb deltas* deposited seaward of the tidal inlets. Barrier islands short and stubby. Meandering, tidal channels with point bars behind barriers. Dominant transport processes are tides.	Tides exceed 4 m and are the overwhelming transport process. Linea sandbodies, perpendicular to depo tional strike parallel tidal currents broad, funnel-shaped estuary, who shore may have wide, muddy tidal flats.

crevasse occurs as the river seeks a shorter course to the sea.

River-dominated deltas prevail along bodies of water with small to modest inshore wave power or tidal flux—the margins of ancient, broad cratonic seas—or along continents bordered by small, protected seas. Where subsidence is great or where the delta extends into deep water beyond the continental margin, it can be thousands of meters thick and forward, down-dip growth is small. Here growth faults are abundant. But where subsidence is minimal, as on cratons, the delta sequence will be thin and progradation down paleoslope can be hundreds of kilometers. Thus the tectonics of the basin and its type strongly influence the overall geometry of the deltaic sequence.

References to modern river-dominated deltas include the classic papers on the Mississippi by Fisk (1961) and Frazier (1967), the study of the Catumbo delta of Lago Maracaibo (Hyne and Dickey, 1979), and the report by Young (1978) on the Mackenzie delta of northwestern Canada. The summary table of Swanson (1975) provides a very detailed guide to the facies of a river-dominated delta. Gilbert-type deltas marginal to glacial lakes are products of rapid sedimentation, have overlapping lobes (which may be steeply dipping), and density flows of turbid meltwater commonly are important (Gustavson, 1975).

References to ancient, river-dominated deltas are numerous and include a study by Nelson and Glaister (1978) of the oil sands of the McMurray Formation based on 19 cores and many wireline logs, a study which utilized sedimentary structures, grain size, and log signature to define reservoir quality (net oil-sand thickness, porosity, and oil saturation). Stanley and Surdam (1978), in a study based on outcrops, describe a delta deposited in an Eocene lake, where foresets dip up to 20 degrees and form units 10 to 25 meters thick. The entire sequence shows a pronounced coarsening-upward trend resulting from lakeward progradation. Sedimentary structures and paleocurrents are emphasized. The Carboniferous deltas of eastern Kentucky are compared to those of the modern Mississippi delta by Baganz *et al.* (1975).

The thin deltas of cratons commonly consist of deposits superimposed on or intercalated with shelf deposits—the result of small changes in sea level, as is well demonstrated by Potter (1963) in the Illinois Basin and by Brown (1979) in the Pennsylvanian sequence of the Mid-Continent of the United States (Fig. 10-33).

Tide-Dominated Deltas

Tidal range along a coastline is a major factor affecting delta morphology and is divided into three types: micro- (<2 m), meso- (2–4 m), and macrotidal (>4 m) (Table 10-7). Davies (1964, 1980) provides a worldwide summary of coastal tidal and wave regime. High tidal ranges can be expected in deeply indented gulfs and in long straits, although exceptions occur, such as at the mouth of the Amazon. Tidal deltas are of two major types: those that protrude mildly seaward like a wave-dominated delta but have prominent tidal channel distributaries and those whose tidal currents are so strong that the familiar form of a delta changes to an estuary with the result that the coastline is indented rather than protruded seaward (Fig. 10-28). Typically estuaries have a mouth with bell-shaped, flaring map pattern, one with abundant shifting sandy shoals. Tidal currents may be as high as 3 m/s and thus exceed river currents except during floods. As a result, sediment transport and delta

▷

FIGURE 10-33. Schematic representation of fluvial deltaic, coastal, and shelf sandstones of Pennsylvanian age in the Mid-Continent (Brown, 1979, Fig. 12).

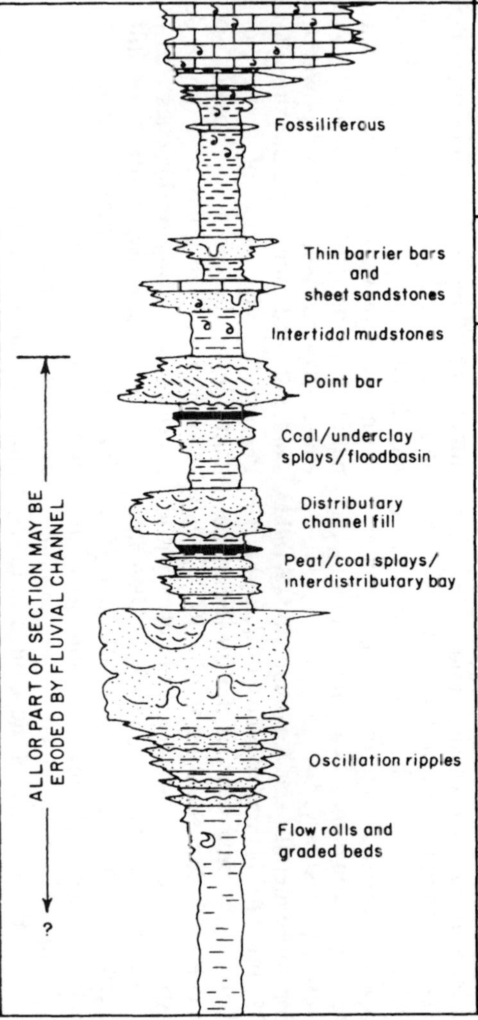

ENVIRONMENTS/FACIES				IDEALIZED LOG PATTERN AND LITHOLOGY	DEPOSITIONAL PHASES		DESCRIPTION
SHELF SYSTEM	SUBMARINE	SHALLOW MARINE	OPEN-MARINE LIMESTONE		MARINE TRANSGRESSION	SUBMARINE AGGRADATION	Commonly mixed biomicrites, fusulines near base, grades upward into algal limestone, well bedded, very fossiliferous, persistent, grades downdip into shelf-wide limestones, grades updip into brackish shales and littoral sandstones.
			TRANSGRESSIVE SHALE	Fossiliferous			Shale becomes more calcareous and fossiliferous upward, assemblage becomes less restricted, highly burrowed. In northern and eastern Mid-Continent, phosphatic black shale common at base.
		SHOALS: WAVES AND TIDES	BARRIER BAR, STORM BERMS, SHEET SAND	Thin barrier bars and sheet sandstones Intertidal mudstones	DELTA DESTRUCTION		Local barrier-bar sandstone: thin, coarsening upward, commonly fringe abandoned delta. Sheet sandstone: widespread, coarsening upward, burrowed, oscillation ripples on top. Storm berm: local, shelly bars composed of broken shells. Intertidal mudstone: laminated, red/olive.
DELTA SYSTEM	SUBAERIAL	UPPER DELTA PLAIN MID- AND LOWER DELTA PLAIN	POINT BAR; DISTRIBUTARY CHANNEL-FILL; CREVASSE SPLAYS; FLOODBASIN/ INTERDISTRIBUTARY BAY; MARSH/ SWAMP PEAT	Point bar Coal/underclay splays/floodbasin Distributary channel fill Peat/coal splays/ interdistributary bay ALL OR PART OF SECTION MAY BE ERODED BY FLUVIAL CHANNEL ?	DELTA CONSTRUCTION	SUBAERIAL AGGRADATION	Point-bar sandstone: fining upward from conglomerate lag to silty levees, upward change from large trough-filled crossbeds to tabular crossbeds and uppermost ripple crossbeds. Distributary channel-fill sandstone: fine- to medium-grained, trough-filled crossbeds, local clay, clast conglomerate, abundant fossil wood. Crevasse splay sandstone: coarsening upward, trough and ripple crossbeds, commonly burrowed at top. Floodbasin/interdistributary mudstone: burrowed, marine fossils, grade updip to non-marine, silty near splays. Coal/peat: rooted, overlie underclay (soil).
	SUBMARINE	DELTA FRONT	BAR CREST			PROGRADATION	Well-sorted, fine- to medium-grained sandstone, plane beds (high flow regime) common, channel erosion increases updip, distal channel fill plane-bedded, some contemporaneous tensional faults.
			CHANNEL-MOUTH BAR				Fine- to medium-grained sandstone, trough-filled crossbeds common, commonly contorted bedding, local shale or sand diapirs in elongate deltas.
			DELTA FRINGE	Oscillation ripples			Fine-grained sandstone and interbedded siltstone and shale, well-bedded, transport ripples, oscillation ripples at top of beds, growth faults in lobate deltas, some sole marks and contorted beds at base.
		PRODELTA	PROXIMAL	Flow rolls and graded beds			Silty shale and sandstone, graded beds, flow rolls, slump structures common, concentrated plant debris.
			DISTAL				Laminated shale and siltstone, plant debris, ferruginous nodules, generally unfossiliferous near channel mouth, grades downdip into marine shale/limestone, grades along strike into embayment mudstones.

morphology radically differ from those of fluvial- and wave-dominated deltas. High velocity, reversing tidal currents imply strong abrasion of the sand fraction in tidal deltas or confined inlets as confirmed by the data of Ovenshine et. al. 1976 in Alaska.

Sediment transport is bidirectional, although flood transport is commonly stronger than ebb transport, so that crossbedding oriented inshore is common, and the thickest crossbeds formed by the strongest currents can be expected to dip shoreward. Marine faunas transported inshore are common. Seaward of the mouth, bidirectional transport follows different paths and produces linear sand ridges generally perpendicular to shoreline—a morphology completely different from the distributary mouth bars of the fluvially dominated delta. Depending upon depth of water, these ridges may be 10 to 20 meters high, and are composed of crossbedding of different styles and scales. Landward from the mouth, river channels in the tidal zone are fairly straight, but above the tidal zone, meanders can occur. Crevassing and splays are present here and form at the highest tides. Low, flat islands contain marshes, lakes, and possible overbank deposits and, depending upon climate, eolian dunes may be prominent along the shores of the estuary.

For more details see Coleman and Wright (1975, pp. 110–11) and Wright (1977, pp. 864–66); an outstanding study is that of the Mahakam delta in Indonesia (Allen *et al.*, 1979). Tidal transport and deposition are well described in the Gironde estuary of Aquitaine (Allen and Klingebiel, 1974). Hayes (1975) effectively relates tidal range to changing morphology of tidally dominated deltas, chiefly using examples from the Atlantic coast of the United States.

The Niger River has a large tide-dominated delta (Allen, 1965) that protrudes seaward with smooth, ovate outline and tidal channel distributaries. Descriptions of its sand deposits and how they relate to the numerous growth faults of the delta are reported by Weber (1971), who used geometry and vertical sequence seen on gamma-ray logs to distinguish tidal- and point-bar sandstone bodies from coastal barriers. Oomkens (1974) gives additional details. Distributary tidal channels, marine tidal bar sands, tidal flats, and coastal barriers of the Colorado delta at the head of the Gulf of California were distinguished by Meckel (1975) using texture, structures, fauna, grain size, and paleocurrents. Earlier, classic studies of tidal sand bodies were made by Dutch investigators (Terwindt *et al.*, 1963). The Fraser delta, which is strongly influenced by tides, is well described by Claque *et al.* (1983).

Ancient estuarine sandstones have been described by Kasino and Davies (1979), whose Table 3 summarizes characteristics of the estuarine complex found in the Pennsylvanian Morrow sandstones of Oklahoma. Here other associated sandy environments include tidal flats along the coast and widespread coastal marshes and, where the sequence is a regressive one, fluvial deposits. Along an arid coastline, coastal dunes will be present. Thus many different environments could be present in the depositional system. Ludwick (1974) has studied the relation between tidal currents and zigzag sandy shoals in a wide estuary of the eastern United States.

Wave-Dominated Deltas

The characteristics of the wave-dominated delta are sand deposited chiefly as laterally coalesced coastal barriers, some marshes and lakes (saline and fresh), and, depending upon aridity, dunes. This sequence is interrupted by fairly straight, fluvial channels, but distributary mouth bars are poorly developed because of strong inshore wave power. The wave-dominated delta protrudes seaward but little, because strong, inshore wave power transports sand, silt, and mud along the shore away from the mouth. The sands of a wave-dominated delta should be, as a rule, better sorted and more mature than those of a fluvial delta, because of reworking on beaches and in dunes, although quantitative studies are still lacking. Shaly zones in the coastal barriers of a wave-dominated delta have good continuity along depositional strike and become thicker and more abundant seaward of each beach sequence. The wave-dominated delta has a seismic signature that is fairly characteristic (Fig. 10-30).

Wave-dominated deltas occur wherever inshore wave energy is high, as along a narrow shelf with strong onshore winds. They are common in trade wind belts where the continental shelf is narrow and where tidal range is small to moderate, and are correspondingly rare along the margins of most ancient, epeiric seas.

Excellent, well-described examples of modern wave-dominated deltas occur along the Brazilian coast from Rio de Janeiro to Recife, where they are spectacular (Lamego, 1940; Dominguez *et al.*, 1981; Martin *et al.*, 1980). Wave-dominated deltas prevail here because the shelf is narrow and perpendicular to strong, southeast trade winds. Troll and Schmidt-

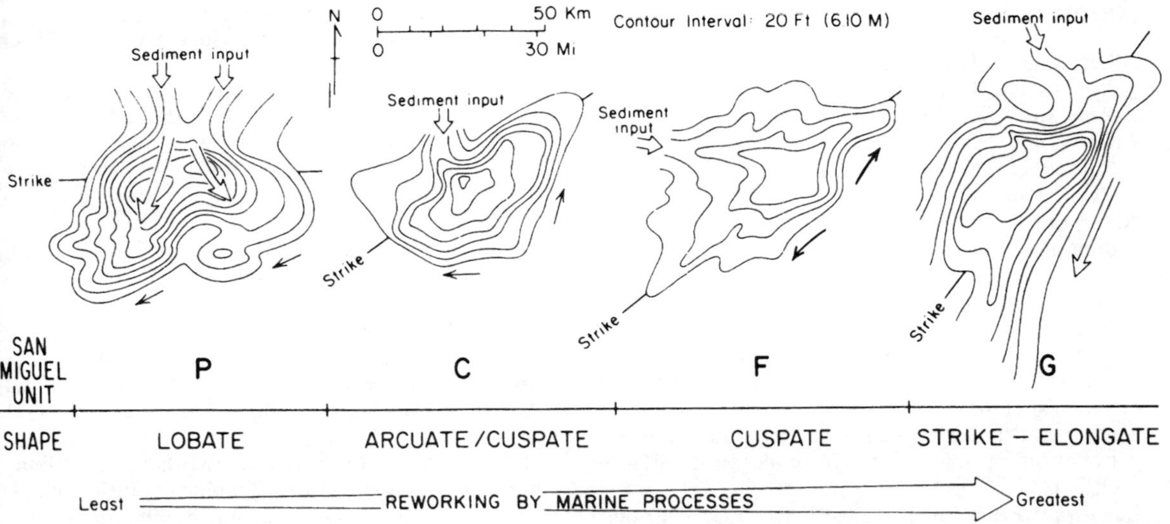

FIGURE 10-34. Spectrum of wave-dominated deltas on a stable platform; Cretaceous San Miguel delta, Maverick Basin, south Texas, U.S.A. (Weise, 1980, Fig. 28). Published by permission of the author and the Texas Bureau of Economic Geology.

Kraepelin (1965) show the historic evolution of a small, lobate, wave-dominated delta along the Caribbean coast of Colombia, where a small river breached a high-energy coastline. A modern, wave-dominated delta in arid western Australia, having red-brown muddy flood deposits, is described in Johnson (1982).

A very complete, in-depth study of a related group of wave-dominated deltas of Cretaceous age in Texas is that of Weise (1980), a study which contains examples of the variants needed for effective exploration (Fig. 10-34). A thick sequence of over 3000 meters in the Devonian of South Africa has five regression sequences, each forming an arcuate deltaic lobe that coarsens upward from prodelta shelf to tidal inlets (Tankard and Barwis, 1982).

Summary

Deltas and their varieties have been and remain the chief focus of most research on terrigenous deposits, primarily because of their coal and petroleum.

Deltas and delta-influenced coastlines significantly alter much of the world's present coastline and are of three types: river-dominated, wave-dominated, and tide-dominated, the latter forming an estuary. River-dominated deltas protrude seaward with bifurcating channels and develop where both inshore wave power or tidal flushing is small in relation to river discharge. Wave-dominated deltas protrude seaward but little and commonly have but one major river channel cutting through a complex of coastal barriers. Strong inshore wave power is needed for a wave-dominated delta and conditions are most favorable where there are strong onshore winds and a narrow shelf. Wave-dominated deltas consist mostly of sandstone whereas river-dominated deltas are shale-rich. Where tides exceed 4 meters, a funnel-shaped estuary with tidal sandbodies, some islands, and mudflats develops. Rainfall influences deltas in several ways: arid climates favor dune development and sabkhas whereas wet climates favor vegetation and peat development. Many different subenvironments are present in deltas and, when traced up and down paleoslope, differences in facies are commonly pronounced. With cessation of supply, carbonates develop in low-latitude deltas and thus can be expected in some ancient deltas as minor interbeds. Of the three types of deltas, tide-dominated deltas have been the least studied.

Ancient deltas form significant parts of many sedimentary basins and are found in all tectonic settings, although the largest develop along passive continental margins or where a large river entered a small, protected sea. Deltas are also characteristic of basins that rim the margins of cratons over which some have prograded great distances. Ancient river-dominated deltas are rich in coal and petroleum. Petroleum is common because source bed shales are abundant and are interbedded with and down-dip from good sandstone reservoirs, which tend to be stacked.

References

Albertson, M.L.; Dai, Y.B.; Jensen, R.A.; and Rouse, H.: Diffusion of submerged jets. Trans. Am. Soc. Civil Eng. 115, 639–697 (1950).

Allen, G.P., and Kingebiel, A.: La sédimentation estuarienne. Exemple de la Gironde. Bull. Centre Rech. Pau—SNPA 8, 263–293 (1974).

Allen, G.; Laurier D.; and Thouvenin, J.: Etude sédimentologique du delta de la Mahakam. Compagnie Française des Pétroles (Paris), Notes et Mémoires 15, 156 p. (1979).

Allen, J.R.L.: Late Quarternary Niger delta, and adjacent areas. Sedimentary environments and lithofacies. Am. Assoc. Petroleum Geologists Bull. 49, 547–600 (1965).

Audley-Charles, M.G.; Curray, J.R.; and Evans, G.: Location of major deltas. Geology 5, 341–344 (1977).

Baganz, B.P.; Horn, J.C.; and Ferm, J.C.: Carboniferous and Recent Mississippi lower delta plains: A comparison. Gulf Coast Assoc. Geol. Soc. Trans. 25, 183–191 (1975).

Bates, C.C.: Rational theory of delta formation: Amer. Assoc. Petrol. Geologists Bull. 37, 2119–2162 (1953).

Berg, O.R.: Seismic detection and evolution of delta and turbidite sequences. Their application to exploration for the subtle trap. Am. Assoc. Petroleum Geologists Bull. 66, 1271–1288 (1982).

Boothroyd, J.C., and Ashley, G.M.: Processes, bar morphology, and sedimentary structures on braided outwash fans. In: Jopling, A.V., and McDonald, B.F., (Eds.): Glaciofluvial and glaciolacustrine sedimentation. Soc. Econ. Paleon. Mineral. Sp. Pub. 23, 193–222 (1975).

Brown, L.F., Jr.: Deltaic sandstones of midcontinent. In: Hyne, N.J. (Ed.): Pennsylvanian sandstones of the Mid-Continent. Tulsa Geol. Soc. Spec. Pub. 1, 35–63 (1979).

Campbell, C.V.: Model for beach shoreline in Gallup Sandstone (Upper Cretaceous) of northwestern New Mexico. New Mexico Bur. Mines and Mineral Resources Circ. 164, 39 pp. (1979).

Claque, J.J.; Luternauer, J.L.; and Hebda, R.J.: Sedimentary environments and postglacial history of the Fraser delta and lower Fraser valley, British Columbia. Canadian Jour. Earth Sci. 20, 1314–1326 (1983).

Coleman, J.M., and Wright, L.D.: Modern river deltas: Variability of processes and sand bodies. In: Broussard, M.L. (Ed.): Delta models for exploration. pp. 99–149. Houston: Houston Geol. Soc. 1975.

Davies, J.L.: A morphogenic approach to world shorelines. Zeitschr. Geomorph. 8 (Sonderheft), 127–142 (1964).

Davis, J.L.: Geographical variations in coastal development, 2nd Ed., 212 pp. London: Longman 1980.

Dixon, James: Sedimentology of the Eocene Taglu delta, Beaufort-Mackenzie Basin: Example of a river-dominated delta. Geol. Survey Canada Paper 80-11, 11 pp. (1981).

Dominguez, J.M.L.; Bittencourt, A.C.S.P.; and Martin, L.: Esquema evolutivo da sedimentacao Quaternária nas feicoes deltaicas dos Rios São Francisco (SE/AL), Jequitinhonha (Ba), Doce (ES), e Paraiba do Sul (RJ). Rev. Brasileira Geociências 11, 227–237 (1981).

Fisk, H.N.: Bar-finger sands of Mississippi delta. In: Peterson, J.A., and Osmond, J.C. (Eds.): Geometry of sandstone bodies, pp. 29–52. Tulsa, Oklahoma: Am. Assoc. Petroleum Geologists 1961.

Frazier, D.E.: Deltaic deposits of the Mississippi River: Their development and chronology. Gulf Coast Assoc. Geol. Soc. Trans. 17, 287–311 (1967).

Galloway, W.E.: Process framework for describing the morphologic and stratigraphic evolution of deltaic depositional systems, pp. 87–98. In: Broussard, M.L. (Ed.) Deltas, models for exploration. Houston, Tx: Houston Geol. Soc. 1975

Gilreath, J.A.; Cox, J.W.; Fett, T.H.; and Grace, L.M.: Practical dipmeter interpretation, various paging. Houston: Schlumberger Educational Services 1985

Gustavson, T.C.: Sedimentation and physical limnology in proglacial Malaspina Lake, southeastern Alaska. In: Jopling, A.V., and McDonald, B.C. (Eds.): Glacial fluvial and glaciolacustrine sedimentation. Soc. Econ. Paleon. Mineral. Spec. Pub. 23, 249–263 (1975).

Hayes, M.D.: Morphology of sand accumulation in estuaries. An introduction to the symposium. In: Cronin, L.E. (Ed.): Estuarine research, Vol. 2, Geology and engineering, pp. 3–22. New York: Academic Press 1975.

Homewood, Peter, and Allen, Phillips: Wave-, tide-, and current-controlled sand bodies of the Miocene Molasse, western Switzerland. Bull. Am. Assoc. Petroleum Geologists 65, 2534–2545 (1981).

Horowitz, D.H.: Mathematical modeling of sediment accumulations in prograding deltaic systems. In: Merriam, D.F. (Ed.): Quantitative techniques for the analysis of sediments (Proceedings of an International Symposium held at the IX International Sedimentological Congress, Nice, France on 8 July, 1975), pp. 105–120. Oxford: Pergamon Press 1976.

Hyne, N.J., and Dickey, P.A.: El delta contemporaneo del Rio Catumbo, Lago Maracaibo, un modelo para explicar antigos deltas intermontanas. V. Congreso Geológico Venezolano, Caracas, Memoria, Tomo 1, Tem 1, 327–337 (1977).

Johnson, D.P.: Sedimentary facies of an arid zone delta. Gascoyne delta, western Australia. Jour. Sed. Petrology 52, 547–563 (1982).

Kasino, R.E., and Davies, D.K.: Environments and diagenesis, Morrow Sands, Cimmaron County (Oklahoma), and significance to regional exploration, production and well completion practices. In:

Hyne, N.E. (Ed.): Pennsylvanian sandstones of the Mid-Continent. Tulsa Geol. Soc. Spec. Pub. 1, 169–194 (1979).

Kingsley, C.S.: A composite submarine fan-delta-fluvial model for the EECA and Lower Beaufort Groups of Permian Age in the Eastern Cape Province, South Africa. Trans. Geol. Soc. South Africa 84, 27–40 (1981).

Klein, G. deVries; DeMelo, U.; and Della Favera, J.C.: Subaqueous gravity processes on the front of Cretaceous deltas, Reconcavo Basin, Brasil. Geol. Soc. America Bull. 83, 1469–1492 (1972).

Komar, Paul D.: Computer models of delta growth due to sediment input from rivers and longshore transport. Geol. Soc. America Bull. 84, 2217–2226 (1973).

Lamego, Alberto Ribeiro: Restingas na costa do Brazil. Div. Geol. Minerl. Bol. 96, 63 pp. (1940).

LeBlanc, R.J.: Significant studies of modern and ancient deltaic sediments. In: Broussard, M.L. (Ed.): Deltas, models for exploration, pp. 13–85. Houston: Houston Geol. Soc. 1975.

Ludwick, John C.: Tidal currents and zig-zag sand shoals in a wide estuary entrance. Geol. Soc. America Bull. 85, 717–726 (1974).

Martin, Louis; Bittencourt, A.C.S.P.; Boas, G.S.V.; and Flexor, J.M.: Mapa geológico do Quaternário costeiro de Estado da Bahia, Escala 1 : 250,000, Texto explicativo. Secretaria das Minas e Energia, CPM, 57 pp. (1980).

McGowen, J.H.; Granata, G.E.; and Serri, S.J.: Depositional framework of the Lower Dockum Group (Triassic), Texas Panhandle. Texas Bur. Econ. Geol., Rept. Invs. 97, 60 pp. (1979).

Meckel, L.L.: Holocene sand bodies in the Colorado delta area, northern end of California. In: Broussard, M.L. (Ed.): Deltas, models for exploration. pp. 239–265. Houston: Houston Geol. Soc. 1975.

Mitchum, R.M., Jr.; Vail, P.R.; and Sangree, J.B.: Seismic stratigraphy and global changes in sea level, Part 6: Stratigraphic interpretation of seismic reflection patterns in depositional sequences. In: Payton, C.E. (Ed.): Seismic stratigraphy—application to hydrocarbon exploration. Am. Assoc. Petroleum Geologists Memo. 26, 117–133 (1977).

Nelson, H.W., and Glaister, R.P.: Subsurface environmental facies and reservoir relationships of the McMurray Oil Sands, northeastern Alberta. Canadian Assoc. Petroleum Geologists Bull. 26, 177–207 (1978).

Oomkens, E.: Lithofacies relations in the Late Quaternary Niger Delta complex. Sedimentology 21, 195–222 (1974).

Ovenshine, A.T.; Bartsch-Winkler, S.R.; O'Brian N.R., and Lawson, D.E., Intertidal sedimentation in the Upper Turnagain Arm, Alaska, p. M1–M26. In: Miller, T.P.(Ed.): Recent and ancient sedimentary environments in Alaska. Alaska Geol. Soc. Symposium Proceedings, Anchorge, 1976.

Potter, Paul Edwin: Late Paleozoic sandstones of the Illinois Basin. Illinois Geol. Survey, Rept. Inv. 217, 92 pp. (1963).

Schlumberger: Dipmeter interpretation, Vol. 1—Fundamentals, 61 pp. Schlumberger Ltd., 277 Park Ave., NY 10017, 1981.

Stanley, K.O., and Surdam, R.C.: Sedimentation on the front of Gilbert-type deltas, Washakie Basin, Wyoming. Jour. Sed. Petrology 48, 557–573 (1978).

Swanson, D.C.: Delta workshop exercise notebook, 110 pp. A.A.P.G.–S.E.P.M. Convention, April 7–9. Dallas, Texas: A.A.P.G. Marine Geology Committee Workshop on Finding and Exploring Ancient Deltas in the Subsurface 1975.

Tankard, A.J., and Barwis, J.H.: Wave-dominated deltaic sedimentation in the Devonian Bokkeveld Basin of South Africa. Jour. Sed. Petrology 52, 959–974 (1982).

Terwindt, J.H.J.; deJong, J.D.; and van der Wilk, E.: Sediment movement and sediment properties in the tidal area of the Lower Rhine (Rotterdam Waterway). Koninkl. Nederlandse Geol. Mijnb. Genoot Verh., Geol. Ser. 21-2, 242–258 (1963).

Troll, Carl, and Schmidt-Kraepelin, Ernst: Das neue Delta des Rio Sinu an der karibischen Küste Kolumbiens. Erdkunde Arkiv Wissenschaftliche Geographie 19, 14–23 (1965).

Weber, K.J.: Sedimentological aspects of oil fields in the Niger delta. Geol. Mijnbouw 50, 559–576 (1971).

Weise, Bonnie R.: Wave-dominated delta systems of the Upper Cretaceous San Miguel Formation, Maverick Basin, south Texas. Texas Bur. Econ. Geol., Rept. Inv. 107, 39 pp. (1980).

Wright, L.D., and Coleman, J.M.: Variations in morphology of major river deltas as functions of ocean wave and river discharge regimes. Am. Assoc. Petroleum Geologists Bull. 57, 370–398 (1973).

Wright, L.D., and Coleman, J.M.: Mississippi River mouth processes. Effluent dynamics and morphologic development. Jour. Geology 82, 751–778 (1974).

Wright, L.D.: Sediment transport and deposition at river mouths, a synthesis. Geol. Soc. America Bull. 88, 857–868 (1977).

Young, F.G., Ed.: Geological and geographical guide to the MacKenzie delta area, 159 pp. Calgary, Alberta: Canadian Soc. Petroleum Geologists 1978.

Fan Deltas

Fan deltas are alluvial fans that debouche into a lake, sea, or ocean, commonly from an adjacent highland (Fig. 10-35). They were first defined by Holmes (1965, pp. 553–54) and an early paper was that by McGowen (1970). Since the early 1970s fan deltas have been widely recognized in

FIGURE 10-35. Small fan delta from a fore-arc embayment, Lower Cook Inlet, Alaska (Hayes and Michel, 1982, Fig. 4C). Published by permission of the authors and the Society of Economic Paleontologists and Mineralogists.

many basins with high relief directly bordering fresh or marine water. Judging by the occurrence of modern fan deltas, most occur where the coastline is wave-dominated and have 100 to 300 cm of rainfall. Morphology varies, depending largely upon wave energy and sediment supply, and five types have been recognized (Fig. 10-36). Because they are favored by a high-relief coastline, their very presence is a type of tectonic indicator and, as a rule, they should be expected along the margins of rifts and fault scarps, along rejuvenated, sharply defined regional uplifts, along the borders of an intermontaine basin, or along the rugged coastline of the leading edge of a continent or an island arc. The sands and sandstones of fan deltas are texturally and mineralogically immature as a rule, especially small ones. To be unaware of fan deltas would be to miss much in the geologic record. An excellent review is by Wescott and Ethridge (1980), who report on many modern and ancient examples. Ricci-Lucchi *et al.* (1981) also provide much general information, especially in their Table I. Other good descriptions of ancient fan deltas include those by Dutton (1979) and Hanford (1980) in the Pennsylvanian and Permian of north Texas and by Link *et al.* (1979) in the Eocene of the San Diego area (Fig. 10-37), and it seems likely that the gold of the Precambrian Witswatersrand of South Africa was deposited in fan deltas bordering a large lake (Pretorius, 1975, Fig. 5).

The fan delta produces the juxtaposition of braided stream deposits and marine sand on a steeply sloping, narrow shelf, deposits which tend to pass into deep water so that the distal

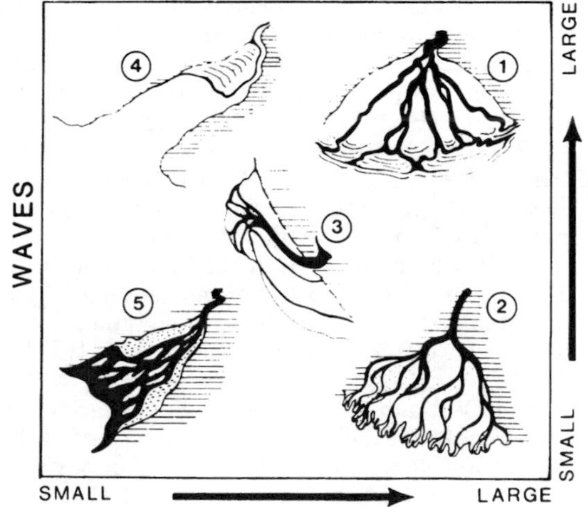

FIGURE 10-36. Morphologic variants of fan deltas (Hayes and Michel, 1982, Fig. 13). Published by permission of the authors and the Society of Economic Paleontologists and Mineralogists.

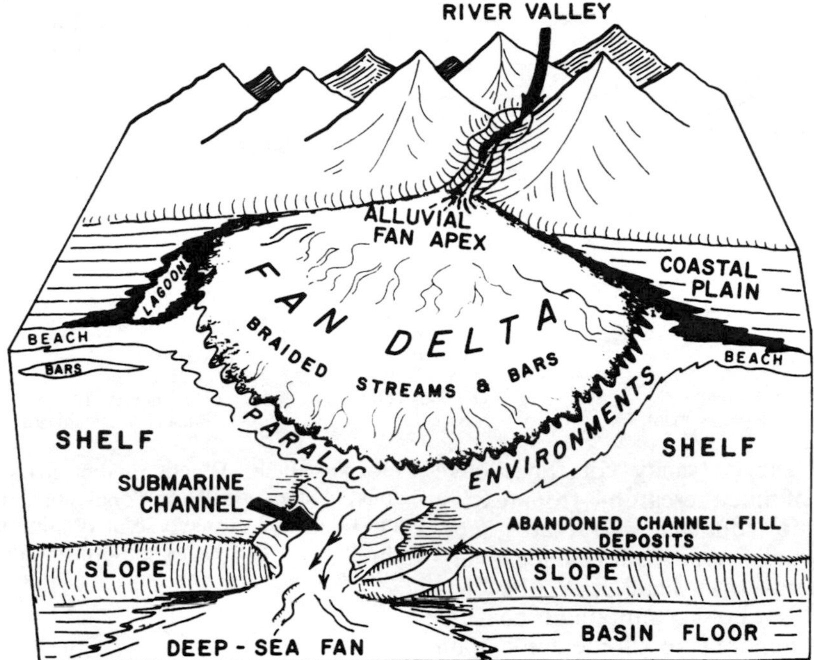

FIGURE 10-37. Fan delta inferred in Eocene sandstones near San Diego, California (Link *et al.*, 1979, Fig. 4). Published by permission of the authors and the Society of Economic Paleontologists and Mineralogists, Pacific Section.

facies of the fan delta may be represented by turbidites. Minor coastal barriers may also be present. Fan deltas show strong radial size gradients, have chiefly semi-radial paleocurrent patterns, and debris flows may be present near their head. Carbonates may overlap and be interbedded with the terrigenous deposits of the fan delta when its river is inactive. Depending upon the persistence of the highland and lake or sea, and a given adequate rainfall, the thickness of a fan delta can be measured in hundreds of meters.

References

Dutton, Shirley P.: Pennsylvanian fan delta sandstones of the Palo Duro Basin Texas. In: Hyne, N.H. (Ed.): Pennsylvanian sandstones of the Mid-Continent. Tulsa Geol. Soc. Spec. Pub. 1, 235–246 (1979).

Hanford, C.R.: Lower Permian facies of the Duro Basin, Texas: Depositional systems, shelf-margin, evolution, paleogeography and petroleum potential. Texas Bur. Econ. Geol., Rept. Invs. **10**, 31 pp. (1980).

Hayes, M.O., and Michel, J.: Shoreline sedimentation within a forearc embayment, Lower Cook Inlet, Alaska. Jour. Sed. Petrology 52, 251–263 (1982).

Holmes, Arthur: Principles of physical geology, 2nd Ed., 1288 pp. New York: The Ronald Press 1965.

Link, M.H. Jr.; Peterson, G.L.; and Abbott, P.L.: Eocene depositional systems, San Diego, California. In: Abbott, P.L. (Ed.): Eocene depositional systems, San Diego, California, pp. 1–8. Los Angeles: Soc. Econ. Paleon. Mineral., Pacific Section 1979.

McGowen, J.H.: Gum Hollow fan delta, Nueces Bay, Texas. Texas Bur. Econ. Geol., Rept. Invs. 72, 57 pp. (1970).

Pretorius, D.A.: Depositional environment of the Witwatersrand gold fields: A chronological review of speculations and observations. Minerals Sci. Eng. 7, 18–47 (1975).

Ricci-Lucchi, F.; Corletta, A.; Ori, G.G.; and Ogliana, F.: Pliocene fan deltas of the Intra-Appennic Basin, Bologna. Intern. Assoc. Sedimentologists 2nd European Regional Meeting, Excursion Guidebook 4, 81–162 (1981).

Wescott, W.A., and Ethridge, F.C., Fan-delta sedimentology and tectonic setting—Yallahs fan delta. Am. Assoc. Petroleum Geologists Bull. 64, 374–399 (1980).

Sandy Coastlines and Shelves

Sandy Coastlines

The sandy coastlines and shelves of a regressive shoreline include coastal, near-shore, and offshore zones (Fig. 10-38, Table 10-8).

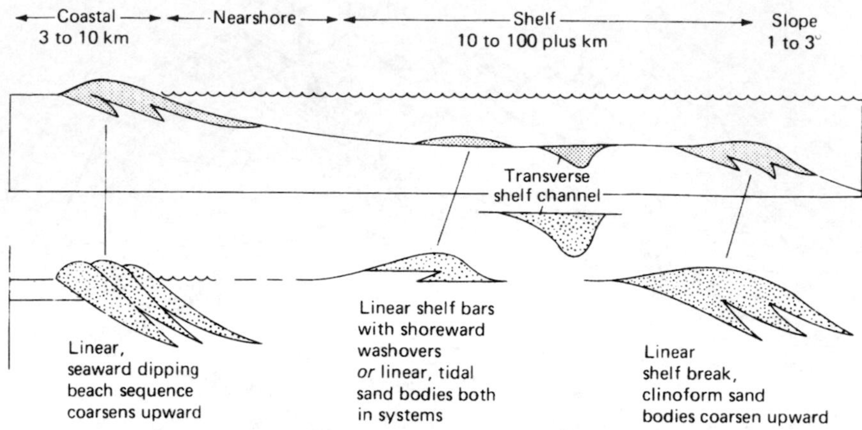

FIGURE 10-38. Diagrammatic cross section from coast to shelf edge shows coastal barrier, linear shelf and tidal bars, transverse or oblique shelf channels, and linear, shelf-break clinoform sandstone bodies. Intervening sediments may be either shales or carbonates.

Modern and ancient sandy coastlines have been the focus of much research—from texture and mineralogy to sedimentary structures, morphology, and beach stability. Morphologic variation is enormous—from small pocket beaches in the protected bays of a mountainous coast to coastal barriers hundreds of kilometers long on the coastal plains of marginal seas and trailing continental margins. Coastal barriers are conspicuous along some, but not all, sandy coastlines, whereas others are marked by wide tidal flats. Tidal range, inshore wave power, and sand supply are the three factors that determine whether or not a sandy coastline exists and its morphologic and textural characteristics.

The beach is the key feature of a sandy coastline for it is a linear sand generator and segregates sand from mud and silt to yield well-sorted framework grains. The beach is also worthy of study, because most sand grains, other than those in fluvial deposits derived directly from igneous rocks and those of some volcaniclastics, may have passed through the beach environment numerous times as they have been progressively recycled throughout geologic history. Abrasion on the beach is also believed to eliminate many sand-sized rock fragments and enhance mineralogic maturity and grain roundness, although both have been difficult to fully document. As a rule, however, beach sands are well washed and sorted, but mineralogy varies widely—from pure quartz sands along passive margins with a wet climate to lithic arenites along active margins to rich, heavy mineral placers of olivine on volcanic islands. Fossil debris and even glauconite are not uncommon.

TABLE 10-8. Definition, environments, and characteristics of coastal, near-shore, and offshore zones of regressive shoreline. (Adapted from Campbell, 1979, pp. 27–29).

Coastal	Near-shore	Offshore
Landward from the usual limit of storm tides to where stream deposits prevail; lower part may be flooded by hurricane and storm tides. Environments include eolian dunes, swamps, salinas (depending on climate), fluvial and tidal channels, levees, and lakes. Climate affects salinity of lagoonal waters, dune development, and cementation.	From upper limit of storm tides to seaward limit of sand deposition, which for many modern beaches is about 10 m. Environments include land-attached beaches, tidal flats and deltas, washover fans, estuaries, bays and sounds, plus marine islands during destructional phase of delta. Complex paleocurrent patterns from diverse sedimentary structures. Morphology determined by inshore wave power and tidal range. Climate affects carbonate production.	Seaward from near-shore sand limit to edge of continental shelf or shelf break with widths from 10s to 100s of kilometers and depths to 100 m. Isolated, elongate sandstone bodies oriented either down dip or subparallel to shoreline in siltstone or shale, or carbonate. Paleocurrents complex and fair-weather currents may have different orientation and structures than those produced by storms. Impinging oceanic currents and tides may be major factors along continental margins. Trans shelf channels possible.

Important references include Komar (1976) on beach processes, Davis and Ethington (1976), Davis (1985), on coastal sedimentary environments and landforms, Snead (1981), Tanner (1977), and Schwartz (1982). Greenwood and Davis (1984) emphasize coastal hydrodynamics and sedimentation. Three other sources, all of which emphasize sandy deposits, are Ginsburg (1975, pp. 2–197), Klein (1977a), the review of wave-dominated clastic shoreline deposits by Heward (1981), and a monograph on siliclastic shelf sediments (Tillman and Siemers, 1984). Useful source books include Komar (1983) and Bird and Schwartz (1985).

Coastal Barriers. Coastal barrier systems are of interest because they define depositional strike, separate marine from nonmarine deposits, make excellent reservoirs for petroleum and groundwater, are important sources of placer minerals, and are highly rated as recreational resources. The different environments of a coastal barrier system include the foreshore or beach, which commonly has dunes, washover fans, the shoreface, tidal inlets, and marshes and lagoonal deposits behind the barriers, which together can form a complex sequence. Both fluvial and shelf deposits may also be closely associated. Coastal barriers that are now stranded on and form part of a modern coastal plain are also called *cheniers* (Fr.) and *restingas* (Port., Span.).

The original worldwide maps of modern barriers, lagoons, and tidal ranges were made by Gierloft-Emden (1961). Other early studies were made by Dutch investigators and show well the complex evolution of coastal barriers (Fig. 10-39). Key papers are those of Hayes (1979, 1980), from which much of the following is taken.

Coastal barriers are most abundant on the coastal plains of the trailing edges of continents and on marginal seas, especially where tidal range is less than 4 m on micro- and mesotidal coasts (Table 10-7) and where unconsolidated sands are abundant (Glaeser, 1978). Modern coastal barriers are usually 10 to 15 m or less thick, a thickness that seems less than that of many ancient barriers, which probably are composites. Widths vary from a few hundred meters to several kilometers or more but lengths can be many kilometers. In Rio Grande do Sul of southern Brazil, a coastal barrier is over 500 km long, width varies from 10 to 25 km, and there is but one tidal inlet along this microtidal coast. Modern barriers nearly always are capped by sand dunes.

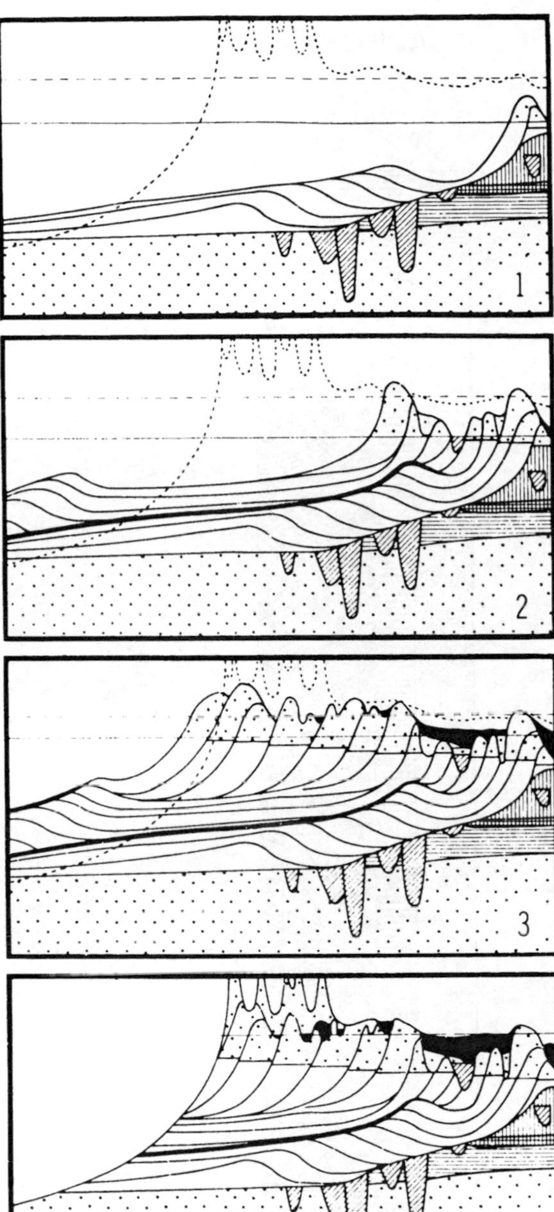

FIGURE 10-39. Complex sequential development of coastal dunes in a barrier island along the North Sea (Van Straaten, 1965, Fig. 26).

Coastal barriers and beaches have a distinctive coarsening-upward sequence passing upward from offshore silts and muds to fine and then to coarser, well-sorted sand at their top. They also have a distinctive sequence of sedimentary structures from bioturbation at their base passing upward to wave and current ripples and finally to crossbedding and lamination,

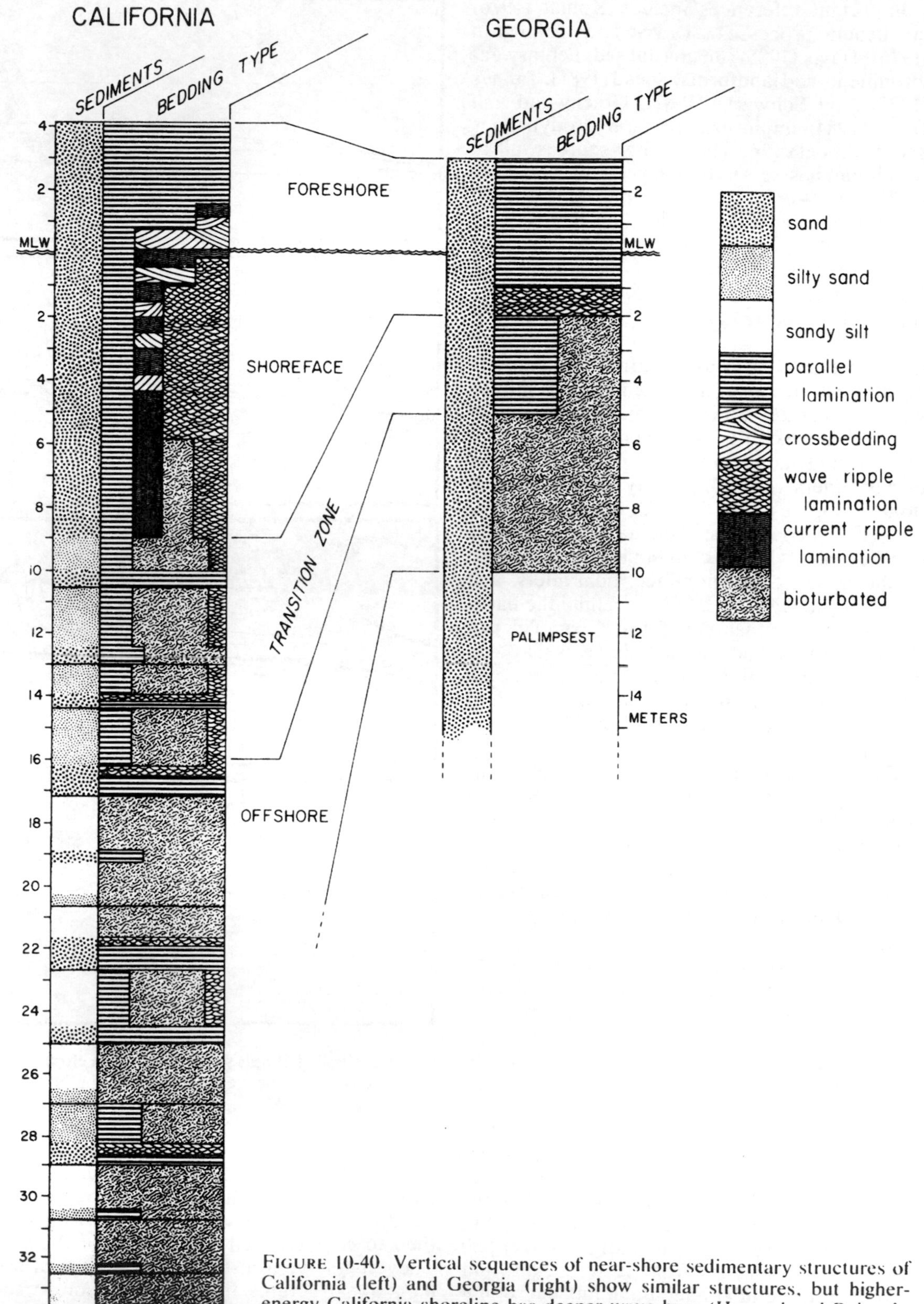

FIGURE 10-40. Vertical sequences of near-shore sedimentary structures of California (left) and Georgia (right) show similar structures, but higher-energy California shoreline has deeper wave base (Howard and Reineck, 1981, Fig. 19). Published by permission of the authors and the American Association of Petroleum Geologists.

FIGURE 10-41. Heavy mineral concentrates and even, well-defined lamination in Lake Erie beach sand at Cedar Point, Erie County, Ohio, U.S.A. Such lamination dips gently lake- or seaward at an angle that varies with grain size.

a sequence that has been very well described (Howard *et al.*, 1972; Chowduri and Reineck, 1978; Howard and Reineck, 1981). The above sequence has been used to define facies limits on both Georgia's low-energy coast and on California's high-energy coast; in California the boundary of the near-shore facies is deeper, because wave base is deeper (Fig. 10-40). Delicate to pronounced heavy mineral laminae are commonly well developed in the swash zone (Fig. 10-41), and should be looked for in ancient deposits. A key dynamic boundary is always the transition from abundant wave ripples generally, striking parallel or subparallel to the shoreline and developed in fair weather, to bioturbation which prevails below wave base. Kumar and Sanders (1976) recognize this transition and show how storm deposits may interrupt it. Because the upper part of the coastal barrier may be eroded prior to final burial, only the lower foreshore may be preserved and the distinction from a near-shore or open shelf sandstone becomes difficult.

Washover or spillover lobes of sand, formed chiefly by storms that break the barrier (Fig. 10-42), may extend into the lagoon behind it. Later these breaks may become enlarged tidal inlets.

Trace fossils—grazing trails and much bioturbation—distinguish the deeper foreshore from the surf zone, where vertical burrows predominate and a few, such as *Callianassa*, are widely used as near-shore trace fossils (De-Windt, 1974). Trace fossil zonation is a key feature of most Mesozoic and modern beach deposits, but may be less diagnostic in the lower Paleozoic. Paleocurrents of a barrier sequence are commonly complicated and care must be taken to relate them to each subenvironment—to recognize possible dune deposits, washover fans (Fig. 10-42), the swash zone, and the complex patterns offshore, and, if possible, distinguish fair-weather- from storm-deposited sand. Carey's (1978) study of the Permian Snapper Point Formation in Australia is a good example of the integrated study of trace fossils and paleocurrents in a beach sequence.

References dealing with Holocene coastal barriers and restingas include Lamego (1940), Bigarella (1965), Horn (1965), Kraft and John (1979), Martini (1976), and Roy *et al.* (1980). Deery and Howard (1977) report on washover fans along the Georgia coast.

Studies of ancient coastal barriers are numerous. Weber (1971) reports on the barriers of the Niger delta (Fig. 10-43), Hobday and Jackson (1979) in the Pleistocene of South Africa, Rautman (1977) in the Jurassic of the Black Hills, Shelton (1965) describes a barrier in the Cretaceous Eagle Sandstone of Montana, Ambrose and Flint (1981) describe a regressive barrier

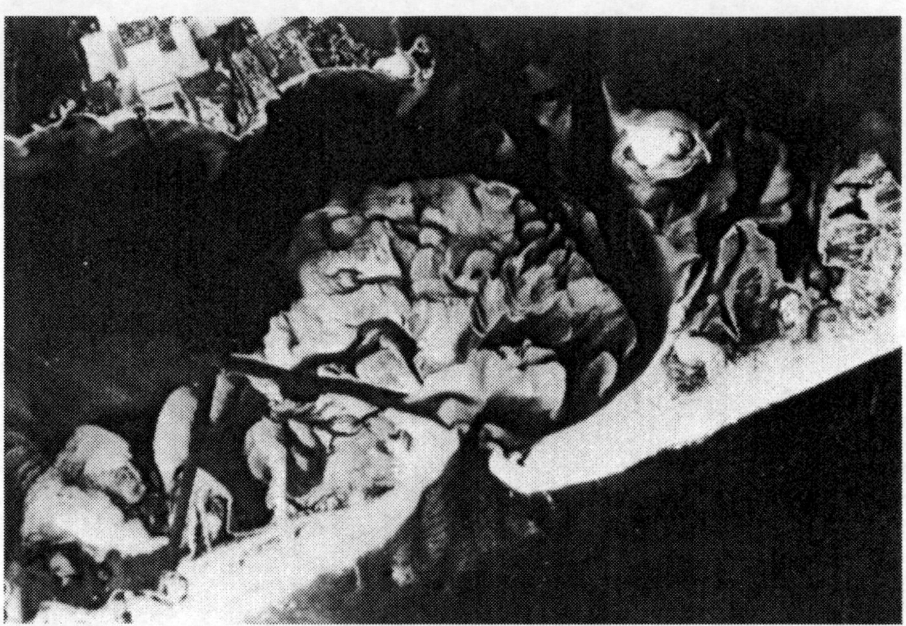

FIGURE 10-42. Moriches Inlet, Long Island, New York in September, 1947. Tidal or washover delta developed after hurricane breached barrier island. Photograph by U.S. Dept. of Agriculture, courtesy M. M. Nichols.

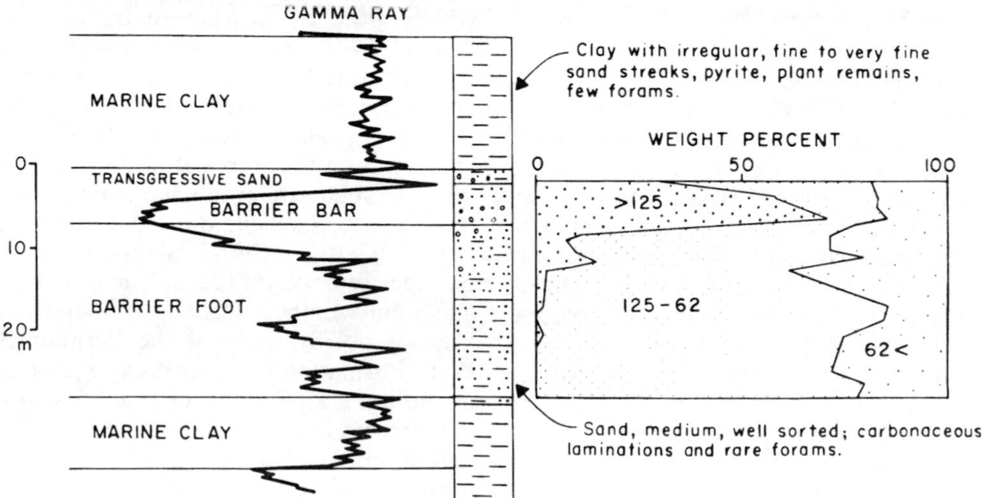

FIGURE 10-43. Vertical, coarsening-upward sequence of a coastal barrier of the Niger delta. (Redrawn from Weber, 1971, Fig. 12).

system in a Miocene lake of Australia, and an outstanding but little known early study is that of a Devonian barrier in eastern Pennsylvania by Kelley (1967), whose maps and cross sections still provide an excellent model for exploration. Campbell's (1979) study of the Cretaceous Gallup Sandstone in New Mexico combines both outcrop and subsurface data and is rich in general applicability.

Beaches and barriers have economic importance not only as petroleum and groundwater reservoirs, but also as sites of heavy mineral placers (Fig. 10-41). May (1973) suggested that on beaches of moderate wave energy, more heavy minerals than light minerals are transported landward and heavy minerals are thus concentrated on beaches. Literature is volumi-

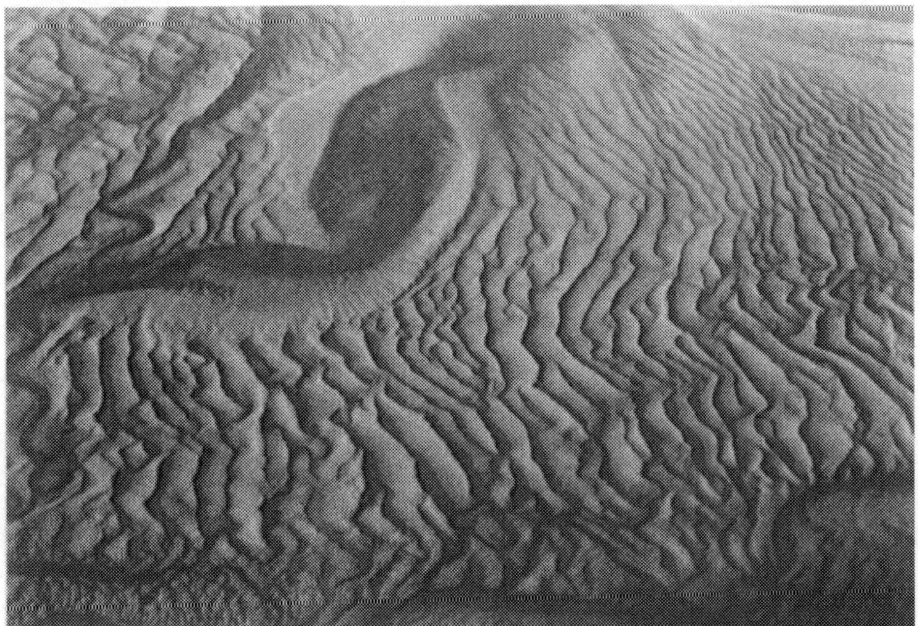

FIGURE 10-44. Complex-megaripples of Selmak tidal bar formed by flood tide (Knight, 1980, Fig. 10.7A). Published by permission of the author and the Geological Survey of Canada.

nous and here only two of many references are cited. Behrn (1965, pp. 19–30) interprets the heavy mineral placers of the Ecca Stage of the Karoo System in South Africa as regressive beach deposits and summarizes most of the relevant English-language literature of modern and ancient beach placers up to the mid-1960s; Houston and Murphy (1977) interpret the "black sands" of Cretaceous sandstones of the western United States as beach deposits. Zimmerle (1973, Table 2) reviewed fossil heavy mineral concentrations, found them most abundant in littoral zone, and notes that ancient placers are less well known than modern ones.

Sandy Coasts without Barriers. Sandy coasts also exist without conspicuous coastal barriers and have several different modes. One major code occurs as wide tidal flats on stable platforms and has been inferred in lower Paleozoic and Proterozoic cratonic sequences. *Tidal Sedimentation* by Hobday and Eriksson (1977) reports on both modern and ancient tidal deposits of this type and is a rich source for sedimentary structure and paleocurrent data as is *Shallow Marine Processes and Products* by Bouma *et al.* (1980).

Where tidal range is moderate to high, coastal sands and gravels are daily reworked by strong tidal currents and molded into large fields of spectacular bedforms with great diversity of type and scale (Fig. 10-44), paleocurrents are complex, and bipolarity is conspicuous (Table 10-8). Well-defined tidal channels remove water from inundated areas and have fining-upward sequences virtually identical to those of fluvial deposits except for the presence of bipolar paleocurrents and marine fossils. Such marine fossils are mostly transported inshore and are common as fossil lags; bioturbation is abundant to spectacular. Hayes (1980) provides a good summary. Because wide areas of seabottom are exposed daily, tidal flats have been much studied—for example, along the North Sea by Reineck (1963, 1972) and McCave and Geiser (1979), and on the Bay of Fundy of Maritime Canada by Klein (1970) and Knight (1980).

Wide sandstone sheets with crossbedding, ripple marks, and shallow channels are found mostly on ancient lower Paleozoic and Proterozoic platforms. Both unidirectional and bimodal crossbedding patterns exist. Loer Paleozoic deposits have bioturbation, whereas Proterozoic ones do not. Such widespread, thin deposits offer many challenging problems, chiefly because modern analogues of comparable scale do not exist (see Dott and Byers, 1982, pp. 341). More study of sedimentary structures and paleocurrent systems would be helpful for improved knowledge of ancient tidal ranges (Klein, 1977a,b) and would significantly contribute to improved understanding of coastal paleogeography.

FIGURE 10-45. Tidal channel at low tide on sandy coastline of eastern Marajó Island, Pará, Brazil. The tidal channel probably has fairly simple bimodal current system whereas current system outside of channel is more complex.

Low-lying muddy or sandy shorelines may be interrupted by channels flushed by tidal currents (Fig. 10-45). Onshore these channels are sinuous and may have point bars virtually identical to those of rivers except for possible bipolar currents. Another occurrence is a thin, perhaps isolated deposit between underlying fluvial and overlying marine shelf sandstones—where a major transgression covers an erosion surface of low to moderate relief.

Sandy Shelves

Sandstones on broad ancient platforms and cratons, either as wide sheets or as thin, isolated bodies, are fairly common in the geologic record. Because many are closely associated with carbonates and marine shales, most are believed to have been deposited in shallow, ancient seas or shelves. Such sandstones contain good reservoirs for petroleum and groundwater and some serve as glass sands. Mineralogically, shelf sands and sandstones tend to be mature to submature, may be well or poorly sorted, and in the Phanerozoic may have fossils and glauconite. In North Ameria, sandstones of this type were the first to be studied by sedimentologists early in the century (Dake, 1921).

Environmental recognition of such sandstones has been difficult because our knowledge of the character and processes of modern shelves is far from complete and possibly because of another more fundamental problem—perhaps modern sandy shelves provide but poor analogues for their presumed ancient equivalents. The series of statements below summarize much of what is known about the sands and sandstones of shallow shelves and are followed by several case histories. Useful general references which provide background for the statements below include Vanney's (1977) book on the geomorphology of continental platforms, the informative summary of an SEPM research conference (Dott and Byers, 1982), comparison of modern shelves and ancient epicontinental seaways by Bouma *et al.* (1980), *Sedimentary Dynamics of Continental Shelves* by Nittrouer (1981), and the philosophical overview by Brenner (1980). The concept of a depositional system is especially valuable for the study of sandy shelves.

1. Shallow, marine sandy facies appear to have three major modes:
a) The classic quartz–arenite–carbonate association, which is notably poor in shale, is most extensive in the early Paleozoic and middle to late Precambrian. Associated sandy environments include tidal flats, eolian, and fluvial deposits.
b) Both sandstone sheets and isolated sandstone bodies are common in the deposits of Mesozoic epicontinental seas and commonly there is much shale with both.

c) Intensely current-washed sandstones deposited in narrow straits and restricted seas by strong tidal currents have been chiefly recognized on modern shelves.
2. Sedimentary structures are everywhere dominated by crossbedding and ripple marks in the early Paleozoic and the Precambrian, although bioturbation becomes significant in the Mesozoic. On modern shelves sonographs show both transverse (amplitudes generally less than 20 meters and wavelengths of hundreds of meters) and longitudinal bedforms (sand ridges, ribbons and furrows with amplitudes in meters and wavelengths in meters to kilometers).
3. Paleocurrents
a) Crossbedding generally has large variance and orientation ranges from unimodal down paleoslope to bipolar (up and down paleoslope) and includes, less commonly, longshore trends.
b) Ripple marks are of both wave and current origin and the strike of wave ripple marks closely parallels shoreline or shoals in shallow water.
c) Sedimentary structures and associated ichnofossils and bioturbation are increasingly used to distinguish fair-weather- from storm-produced deposits and even to provide rough estimates of water depth and size of the ancient sea or ocean (p. 296).
4. Shelf channels are steadily becoming more recognized and are important both as reservoirs and as keys to paleogeography (Lewis, 1982).
5. Mineralogically mature sandstones are typical of many of the shelf sandstones of the lower Paleozoic and later Precambrian whereas submature to mature sandstones are more common in the Mesozoic and later.

Quartz Arenite–Carbonate Association. Texturally mature, quartz-rich sandstones commonly associated with shelf carbonates occur on cratons and their margins and are best known in the early Paleozoic and middle to late Precambrian. This association does not match modern shelves well for at least three reasons—sands and carbonates are rarely deposited together on modern shelves, few modern shelves are as wide as many of these ancient ones were, and modern shelves chiefly border deep oceans whereas many ancient shelves did not. Moreover, most of the sand of modern shelves is relict sand, that is, only slightly reworked fluvial sand deposited in the low-water stage of the last glacial cycle.

The best way to identify the depositional environment of the sandstones of the quartz arenite–carbonate association as marine in the Paleozoic is with fossils, where they exist, and with trace fossils, which are more common. In the Precambrian, prime reliance is upon sedimentary structures, sorting, and association with carbonates and stromatolites. Selley (1972) used trace fossils and sedimentary structures plus lateral continuity of minor interbedded shales and channel dimensions to distinguish between pebbly, coarse, sandy, braided stream deposits and overlying finer sands believed to have been deposited as marine sands in bars and tidal channels. This couplet forms a vast sand blanket that extends across much of North Africa from the Atlantic into the Middle East, an accumulation notably lacking in shale, and also one that invites more study. The distinction between its fluvial and marine phases has implications for their reservoir characteristics—the marine sandstones lack large channels, are better sorted than the underlying fluvial ones, and have thin shales with greater lateral continuity than those associated with the underlying fluvial sandstones. Similar relations probably exist for many of the sandy transgressive units above major unconformities in other sandy basins.

In Wisconsin, Driese *et al.* (1981) used sedimentary structures (crossbedding, ripple marks, clay drapes and clasts, scour channels, grain size, and trace fossils) to distinguish braided stream deposits from tidal channels and tidal flats from marine shore deposits. In sandstones of Paleozoic and later age, trace fossil associations not only help differentiate freshwater sandstones (*Scogenia*) from marine sandstones but distinguish also between shallow, turbulent waters (*Skolithos* and *Arenicolites*) and shallow, less turbulent waters (*Cruziana, Daldalus, Phycodes,* and *Rusophycus*). See Crimes (1981) for discussion of trace fossils in the Lower Paleozoic sandstones of Africa and their use in environmental discrimination and correlation. Other informative references dealing with environmental studies of the marine sandstones of the quartz arenite association include Fraser (1976) and Visser (1974).

Mesozoic Epicontinental and Marginal Seas. The Cretaceous seaway across North America is perhaps the best studied ancient epicontinental sea, because of its petroleum resources and consequent abundant subsurface data. West of this seaway was a subduction zone and associated rising foreland, which supplied sediment to eastward progarding deltas bordering an elongate, muddy epeiric sea that extended from the Arctic to the Gulf of Mexico.

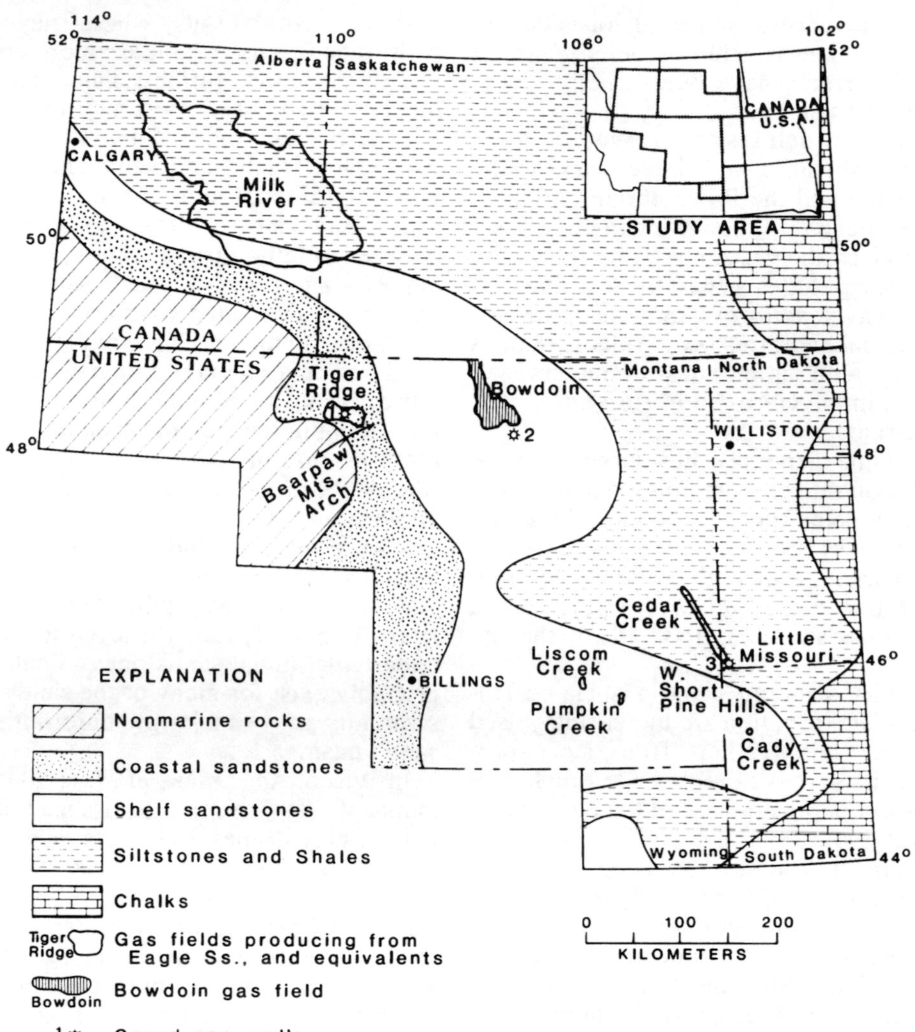

FIGURE 10-46. Facies distribution of part of Cretaceous seaway in northern United States and adjacent Alberta and Saskatchewan (Gautier and Rice, 1982, Fig. 1). Published by permission of the authors and the Society of Petroleum Engineers.

Many studies have contributed to the establishment of its broad facies patterns (Fig. 10-46) from nonmarine through littoral to an offshore muddy domain that contains isolated, sandstone bodies. Boyles and Scott (1982) have summarized the properties of many of these sandstone bodies in the Upper Mancos Shale of Colorado (Table 10-9). Storm-created turbulence superimposed on a general circulation seems to be all-important in concentrating sand and forming isolated, elongate sandbodies which are believed to have been at least 10 km from shore—at least in the Cretaceous Upper Mancos Shale. These sandstone bodies are shoals as much as 40 or more meters above the sea bottom. Basal contact tends to be gradational and the sandstone body coarsens upward. Storms produced channel scours, eroded clay drapes to form clay clasts, and produced unidirectional crossbedding. During fair weather, bioturbation and wave ripples developed. In many shelf sandstones, hummocky crossbedding should be looked for (Figs. 10-47, 4-9, and 8-26). See Swift and Rice (1984) for additional discussion of how these isolated sandstone bodies might have formed.

Other examples include the sandstone bodies of the Cretaceous Colorado Group in Saskatchewan, described by Simpson (1982), where grain size, sedimentary structures, trace fossils, and geometry are related to reservoir properties and depositional environment; the sandstone bodies of the Cretaceous Milk River and Lea

TABLE 10-9. Characteristics of isolated, marine shelf sandstone bodies.

Shoal	Tidal
Elongate bodies may be oblique or subparallel to shoreline and commonly coarsen upward but may interfinger with shelf muds, or carbonates. Length up to 8 to 16 km. Bioturbation common plus glauconite and marine fossils. Sedimentary structures include crossbedding, diverse ripple marks, hummocky bedding, and shallow, erosional scour channels. Spillover lobes developed by storms.	Well-washed, crossbedded and ripple-marked, isolated bodies with abrupt basal contacts and complex internal structure. May be parallel to shoreline in narrow straits or be perpendicular to it on broad shelves. Vertical burrows, glauconite and transported fossils. Formed by mutually evasive tidal currents plus storm surge. Widths up to 5 to 6 km.

FIGURE 10-47. Hummocky crossbedding of Pimenteiras Formation of Lower Devonian age along Transamazonica Highway near Picos, Parnaíba Basin, Brazil. Photograph courtesy of J.C. della Favara, PETROBRAS.

Park Formations of Alberta studied by Meijer Dress and Mhyr (1981); and the Mississippian Knifley Sandstone located on the crest of the Cincinnati Arch in south-central Kentucky (Sedimentation Seminar, 1972), all of which coarsen upward (Fig. 10-48). Other example are provided by Swann (1951) for the Chesterian sandstones of the Illinois Basin.

Shelf sandstone bars, where isolated in marine shales, may be localized along active, but weak tectonic highs or may be more widespread and reworked from underlying sheet sandstones. More likely, it may be difficult to find an explanation for their location. In any case, sand transport occurs when storms combine with tidal or shelf currents to transport sand. In shallow water, storm surge produces crossbedding and channels; in deeper parts of the shelf, however, a Bouma-type cycle capped by wave ripples results and is called a *tempestite*. These are best developed below average wave base but gradually decrease in abundance as the shelf becomes deeper. Tempestites—an example of "event stratification"—are becoming more widely recognized and are a supplement to normal, more continuous, fair-weather sedimentation (Einsele and Seilacher, 1982).

Paleocurrent measurements should be related to fair versus storm facies of all shelf and shoreline sandstones, for only in this way can different major processes of shelf sedimentation be segregated and properly assessed.

Sandstone bodies on marine shelves may be oriented parallel or subparallel to shoreline, where tidal currents are not of major importance. Le Fournier (1980) describes offshore bars subparallel to the French coast bordering the English Channel. The geometry of modern heavy mineral sands on the continental shelf of New Zealand is described by Carter (1980). Swift *et al.* (1978) compare shoreface connected

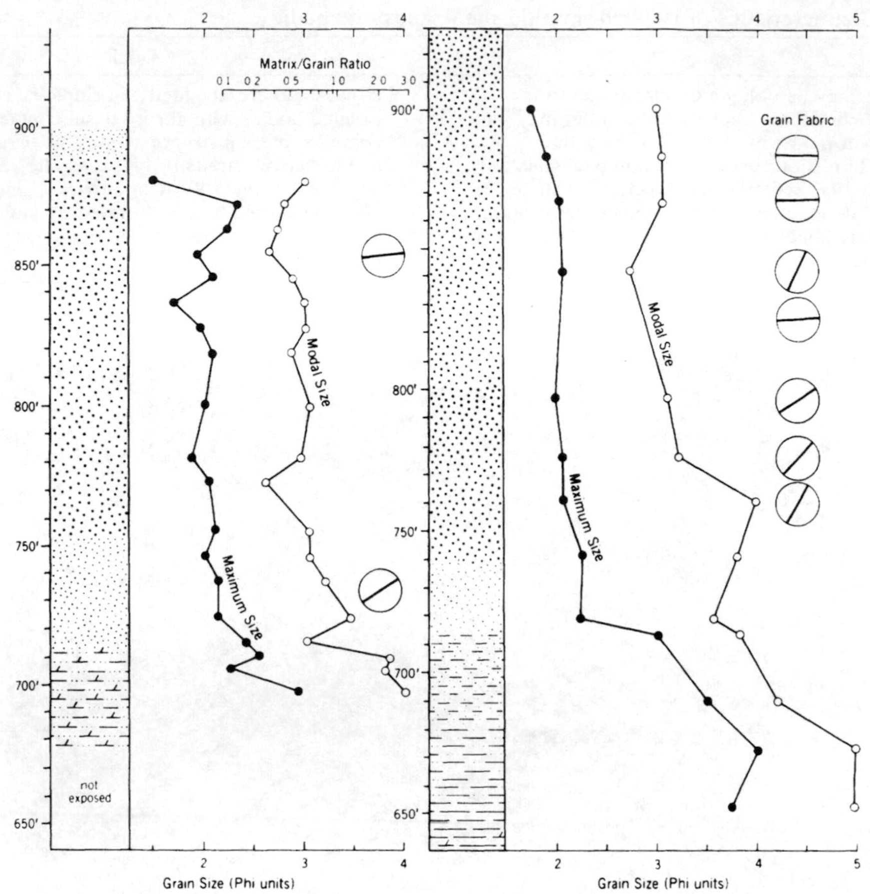

FIGURE 10-48. Coarsening-upward shelf break, Knifely Sandstone of Mississippian age in south-central Kentucky (Sedimentation Seminar, 1978, Fig. 11). Grain fabric is oblique to strike of shelf but is remarkably uniform in this sandstone with abundant bioturbation.

sandbodies on European and American continental shelves, and Hutnance (1982) attempts a mathematical analysis of their origin and spacing.

Thin, transgressive sandy deposits from one to a few meters thick may occur in many basins. They may be continuous or discontinuous, and commonly consist of well-sorted, stable minerals that have glauconite and shells and clasts at their base. Grain size may decrease weakly upward into shale or be fairly uniform, and the gamma-ray curve tends to be blocky. The chief structures are ripple mark, bioturbation, small-scale crossbedding, and possibly hummocky bedding. Such deposits, which represent the destructional phase of a delta, formed after a delta distributary was abandoned and where the ratio of inshore wave power to fluvial input was temporarily high.

Probably less common are thick transgressive sandy deposits, which have strong upward size decline; structures change from planar beds at the base to crossbedding upward and then into finer-grained sandstones with more bioturbation and marine fossils, hummocky bedding, and ripple marks. The gamma-ray curve is upward fining or blocky. Wave ripples near the base parallel the shoreline. Where strong coastal currents existed, crossbedding may predominate and be subparallel to shoreline, but downslope paleocurrent patterns also exist. The sandstone body may be 50 to 100 m or more thick and commonly has a sharp base. See Bourgeois (1980) and Hunter and Clifton (1982) for description and interpretation. In rapidly subsiding tectonic basins such sandstones may have limited lateral extent but should be much more widespread on passive margins and cratons. Care should be taken to look for thin interstratified beaches or perhaps even fluvial facies in this type of marine sandstone (cf. Evans, 1970; Selley, 1972).

The Tidal Shelf. Tidally swept seas are best known from modern shelves, chiefly from the seas around the British Isles, where strong tidal currents exist far from shore and transport, winnow, and deposit sand into well-defined elongate bodies up to 15 m thick, 2 km wide, and 10 to 15 km long, bodies that commonly occur in groups (Stride, 1982). Off (1963) was the first to emphasize isolated tidal bars. Orientation parallels or subparallels the shoreline in narrow seas and straits, but is perpendicular to the shoreline on wide shelves and basins. Sand waves and ripples are superimposed on these tidal bodies, whose internal structures include inclined bedding; crossbedding orientation may be uni- and bimodal, but ripples are likely to be bimodal. Locally, gravels may occur, but the sand is typically medium- to coarse-grained. Vertical size gradients are weak so that the gamma-ray curve tends to be blocky. Basal contact is commonly sharp. Tidal bars such as these appear to be formed by mutually evasive tidal currents—the ebb and flood currents follow different paths. Excellent ancient examples include the Cretaceous sandstones (Fig. 10-49) of the Lloydminster Sparky Pool (Burnett and Adams, 1977) and the tidally influenced shallow-water sandstone bodies of late Precambrian age in Norway (Johnson, 1977). Bipolar current orientation, both in single outcrops as well as in mutually evasive areas, also occurs in widespread sandstone sheets (Thompson and Potter, 1981) and is a key to inferring deposition from ancient tidal currents. Thick deposits of sandstones deposited on tidally influenced shelves have also been recognized and may be characteristic of deposition along an unstable continental margin bordering a large ocean (Anderton, 1976; Levell, 1980).

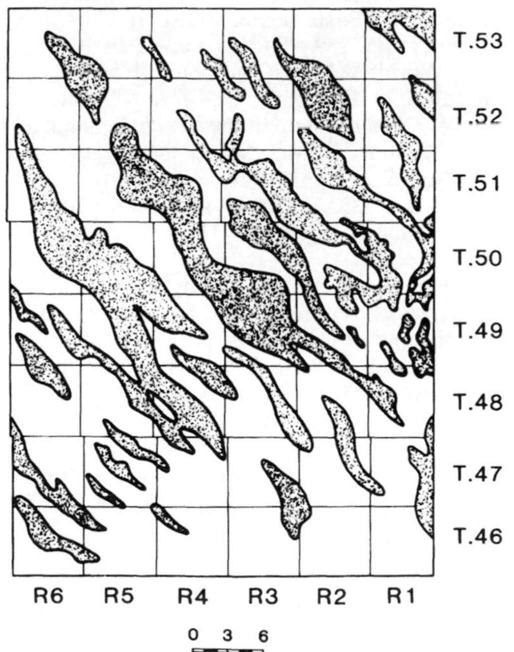

FIGURE 10-49. Map pattern of Cretaceous sandstone bodies in Lloydminster Sparky Pool, Alberta, Canada, is believed to have been formed by tidal currents. (Redrawn from Burnett and Adams, 1977, Fig. 30).

How far landward tides existed on ancient cratons is uncertain, although bipolar crossbedding, especially in ancient oolites, commonly supports the view that tides extended far onto ancient shelves (Carr, 1973). Klein (1982) provides a deductive analysis for the probable sequential arrangement of facies on cratons based on the assumption of tidal seas. His analysis emphasizes tidal processes both on the shelf and near shore to explain how marine shelf deposits are intercalated with those of other environments.

Summary

Ancient sandy coastlines and shelves are key elements of many sedimentary basins

Coastal barriers are one of the best understood of all the sedimentary models and, because they define depositional strike, provide a key element in the reconstruction of a basin. Coastal barriers coarsen upward, may be capped by dunes at their top, and provide good reservoirs for petroleum and can be the sites of rich placers. On the other hand, some basins are notably lacking in coastal barriers for reasons that we do not fully understand. Has basin margin been everywhere eroded or did fan deltas and rivers somehow pass directly into shelf sandstones because inshore wave power and tidal range were too small?

Offshore, sandy shelves are far less well understood, because of the diversity of their sand deposits and the lack of well-defined facies models, especially Holocene ones. Here thin or thick widespread sheet sandstones may be present or isolated and ovate to elongate sandstone bodies may prevail. Vertical profile may be uniform or coarsen or shoal upward. Paleocurrent system varies from well-oriented to variable to bimodal, the latter the result of reversing tidal currents. Reconstruction of water depth and estimates of ocean or sea size are now being attempted based on grain size and type and sequence of sedimentary structures (see Chap. 8).

A recurring problem is how, in an ancient deposit, to separate coastal sandstones from those of the open shelf and, on the shelf itself, to distinguish between wave- and current-swept shelves. More studies of modern near-shore sands such as those by Hunter and Clifton (1982) and Aigner and Reineck (1982) are needed, and possibly computer models (George and Hand, 1977) may be useful.

References

Aigner, T., and Reineck, Hans-Erick: Proximality trends in modern North Sea storm sand from Helgoland Bight (North Sea) and their implications for basin analysis. Senckenbergiana Marit. 14, 183–215 (1982).

Ambrose, G.J., and Flint, R.B.: A regressive Miocene lake system and silicified strandlines in northern South Australia: Implications for regional stratigraphy and silicrete genesis. Jour. Geol. Soc. Australia, 28, 81–94 (1981).

Anderton, R.: Tidal-shelf sedimentation: An example from the Scottish Dalradian. Sedimentology 23, 429–458 (1976).

Behrn, S.H.: Heavy-mineral beach deposits in the Karoo System. Geol. Survey South Africa, Mem. 56, 116 pp. (1965).

Beuf, S.; Biju-Duval, B.; deChapal, D.; Rognon, P.; Gariel, O.; and Bennacef, O.: Les grès du Paléozoïque inférieur au Sahara, 464 pp. Paris: Editions Technip, 1971.

Bigarella, J.J.: Sand ridge structures from Paraná coastal plain. Marine Geology 3, 269–278 (1965).

Bird, E.C.F., and Schwartz, M.L. (Eds.): The World's coastline, 1071 pp. New York: Van Nostrand Rheinhold, 1985.

Bouma, A.H.; Gorsline, D.S.; Monty, C.; and Allen, G.P. (Eds.): Shallow marine processes and products. Sedimentary Geology, Spec. Issue 26, 1–328 (1980).

Bourgeois, Joanne: A transgressional shelf sequence exhibiting hummocky stratification: The Cape Sebastian Sandstone (Upper Cretaceous), southwestern Oregon. Jour. Sed. Petrology 50, 681–682 (1980).

Boyles, J.M., and Scott, A.J.: A model for migrating shelf-bar sandstones in Upper Mancos Shale (Campanian), northwestern Colorado. Am. Assoc. Petroleum Geologists 66, 491–508 (1982).

Brenner, R.L.: Construction of process-response models for ancient epicontinental seaway depositional systems using partial analogs. Am. Assoc. Petroleum Geologists Bull. 64, 1223–1244 (1980).

Burnett, A.I., and Adams, K.C.: A geological, engineering, and economic study of a portion of the Lloydminster Sparky Pool, Lloydminster, Alberta. Bull. Canadian Petroleum Geology 25, 342–366 (1977).

Campbell, C.V.: Model for beach shoreline in Gallup Sandstone (Upper Cretaceous) of northwestern New Mexico. New Mexico Bur. Mines and Mineral Resources Circ. 164, 39 pp. (1979).

Carey, J.: Sedimentary environments and trace fossils of the Permian Snapper Point formation, Southern Sydney Basin. Jour. Geol. Soc. Australia 25, 433–458 (1978).

Carr, Donald D.: Geometry and origin of oolite bodies in the Ste. Genevieve Limestone (Mississippian) in the Illinois Basin. Indiana Geol. Survey Bull. 48, 81 pp. (1973).

Carter, L.: Ironsand in continental shelf sediments off western New Zealand—a synopsis. New Zealand Jour. Geol. Geophysics 23, 455–468 (1980).

Chowduri, K.R., and Reineck, H.-E.: Primary sedimentary structures and their sequence in the shoreface of barrier island Wangerooze (North Sea). Senkenbergiana Marit. 10, 15–29 (1978).

Cotter, Edward: Shelf, paralic, and fluvial environments and eustatic sealevel fluctuations in the origin of the Tuscarora Formation (Lower Silurian) of central Pennsylvania. Jour. Sed. Petrology 53, 25–49 (1983).

Cram, J.M.: The influence of continental shelf width on tidal range: Paleooceanographic implications. Jour. Geology 87, 442–447 (1979).

Crimes, T.P.: Lower Paleozoic trace fossils of Africa. In: Holland, C.H. (Ed.): Lower Paleozoic of Middle East, Eastern and Southern Africa, and Antarctica, pp. 189–198. (Lower Paleozoic Rocks of the World, vol. 3). Chichester, England: John Wiley & Sons 1981.

Dake, C.L.: The problem of the St. Peter Sandstone. Univ. Missouri School Mines and Metall. Bull. Tech. Ser. 6, 158 pp. (1921).

Davis, R.A., Jr. (Ed.): Coastal sedimentary environments, 2nd ed., 716 pp. New York: Springer Verlag 1985.

Davis, Richard, A. Jr., and Ethington, R.L.: Beach and nearshore sedimentation. Soc. Econ. Paleon. Mineral. Spec. Pub. 24, 187 pp. (1976).

Deery, J.R., and Howard, J.D.: Origin and character of washover fans on the Georgia coast, USA. Gulf Coast Assoc. Geol. Soc. Trans. 27, 259–271 (1977).

DeWindt, T.J.: *Callianassia* burrows as indicators of subsurface beach trends. Jour. Sed. Petrology 44, 1136–1139 (1974).

Dott, R.H., Jr., and Byers, C.W. (Conveners): SEPM research conference on modern shelf and ancient cratonic sedimentation—the orthoquartzite–carbonate suite revisited, Aug. 11–16, 1980. Jour. Sed. Petrology 51. 330–347 (1982).

Driese, S.G.; Byers, C.W.; and Dott, R.H., Jr.: Tidal deposition in the basal Upper Cambrian Mt. Simon Formation in Wisconsin. Jour. Sed. Petrology 51, 367–381 (1981).

Einsele, G., and Seilacher, A. (Eds.): Cyclic and

event stratification, 536 pp. Berlin–Heidelberg–New York: Springer-Verlag 1982.

Evans, W.E.: Imbricate linear sandstone bodies of Viking formation in Dodsland-Hoosier area of southwestern Saskatchewan, Canada. Am. Assoc. Petrol. Geologists Bull. 54, 469–486 (1970).

Field, M.E.; Nelson, C.H.; Cacchione, D.A.; and Drake, D.E.: Sand waves on an epicontinental shelf: Northern Bering Sea. Marine Geology 42, 233–258 (1981).

Fraser, G.S.: Sedimentology of a Middle Ordovician quartz arenite–carbonate transition in the Upper Mississippi Valley: Geol. Soc. America Bull. 87, 833–845 (1976).

Gautier, D.L., and Rice, D.D.: Conventional and low permeability reservoirs of shallow gas in the northern Great Plains. Jour. Petrol. Tech. 34, 1600–1608 (1982).

George, D.J., and Hand, D.M.: Computer simulation of barrier island migration. Computers and Geoscience 3, 469–473 (1977).

Gierloff-Emden, H.G.: Nehrungen und Lagunen: Gesetzmässigkeiten ihrer Formenbildung und Verbreitung. Petermanns Georg. Mitt. 105, 81–92 and 161–176 (1961).

Ginsburg, R.N. (Ed.): Tidal deposits, 428 pp. New York: Springer Verlag 1975

Glaeser, J. Douglas: Global distribution of barrier islands in terms of tectonic setting. Jour. Geology 86, 283–297 (1978).

Greenwood, B., and Davis, R.A., Jr. (Eds.): Hydrodynamics and sedimentation in wave-dominated coastline environments, 473 pp. Amsterdam New York; Elsevier Sci. Pub. 1984.

Hayes, M.O.: Barrier island morphology as a function of tidal and wave regime. In: Leatherman, S.B. (Ed.): Barrier islands, pp. 1–27. New York: Academic Press 1979.

Hayes, M.O.: General morphology and sediment patterns in tidal inlets. Sedimentary Geology 26, 139–156 (1980).

Heward, A.P.: A review of wave-dominated clastic shoreline deposits. Earth Sci. Rev. 17, 223–276 (1981).

Hobday, D.K., and Eriksson, K.A. (Eds.): Tidal sedimentation. Sedimentary Geology 18, 1–287 (1977).

Hobday, D.K., and Jackson, M.P.A.: Transgressive shore zone sedimentation and syndepositional deformation in the Pleistocene of Zululand, South Africa. Jour. Sed. Petrology 49, 145–158 (1979).

Horn, Dietrich: Zur geologischen Entwicklung der sudlichen Schleimundung im Holozan. Meyniana 15, 42–58 (1965).

Houston, R.S., and Murphy, J.F.: Depositional environment of Upper Cretaceous block sandstones of the Western Interior. U.S. Geol. Survey Prof. Paper 994-A, 29 pp. (1977).

Howard, J.D.; Frey, R.W.; and Reineck, H.-E.: Georgia coastal region, Sapelo Island, U.S.A.: Sedimentology and biology. Senckenbergiana Marit. 4, 223 pp. (1972).

Howard, J.D., and Reineck, H-E.: Depositional facies of high-energy beach-to-offshore sequence: Comparison with low-energy sequence. Am. Assoc. Petroleum Geologists Bull. 65, 807–829 (1981).

Hunter, R.E., and Clifton, H.E.: Cyclic storm deposits and hummocky cross-stratification of probable storm origin in Upper Cretaceous of the Cape Sebastian area, southwestern Oregon. Jour. Sed. Petrology, 52, 127–143 (1982).

Hutnance, J.M.: On one mechanism forming linear sand banks. Estuarine, Coastal Shelf Sci. 14, 79–99 (1982).

Johnson, H.D.: Shallow marine sand bar sequences: An example from the Late Precambrian of Norway. Sedimentology 24, 245–270 (1977).

Kelley, D.R.: Geology of the Red Valley Sandstone in Forest and Venango Counties, Pennsylvania. Penn. Geol. Survey, 4th Ser. Bull. M57, 49 pp. (1967).

Klein, G. deVries: Depositional and dispersal dynamics of intertidal sand bars. Jour. Sed. Petrology 40, 1095–1127 (1970).

Klein, G. deVries: Clastic tidal facies, 149 pp. Urbana, Illinois: Continuing Education Pub. Co. 1977a.

Klein, G. deVries: Epilogue. Sedimentary Geology, Spec. Issue 18, 283–287 (1977b).

Klein, G. deVries: Probable sequential arrangement of depositional systems on cratons. Geology 10, 17–22 (1982).

Knight, R.J.: Linear sand bar development and tidal current flow in Cobequid Bay, Bay of Fundy, Nova Scotia. In: McCann, S.B. (Ed.): The coastline of Canada. Geol. Survey Canada Paper 80-10, 153–180 (1980).

Komar, P.D.: Beach processes and sedimentation, 429 pp. Englewood Cliffs: Prentice-Hall 1976.

Komar, Paul D. (Ed.): CRC handbook of coastal processes and erosion, 305 pp. Boca Raton, Fla.: CRC Press 1983.

Kraft, J.C., and John, C.J.: Lateral and vertical facies relations of transgressive barriers. Am. Assoc. Petroleum Geologists Bull. 63, 2145–2163 (1979).

Kumar, N., and Sanders, J.E.: Characteristics of shoreface storm deposits: Modern and ancient examples: Jour. Sed. Petrology 46, 145–162 (1976).

Lamego, A.R.: Restingas na costa do Brazil. Dir. Geologia e Mineralogia, Bol. 96, 63 pp. (1940).

LeFournier, J.: Modern analogue of transgressive sand bodies off eastern English Channel. Bull. Centres Rech. Explor.-Prod. Elf-Aquitaine 4, 99–118 (1980).

Levell, B.K.: A late Precambrian tidal shelf deposit, the Lower Sandfjord Formation, Finnmark, north Norway. Sedimentology 27, 539–557 (1980).

Lewis, D.S.: Channels across continental shelves: Corequisites of canyon-fan systems and potential petroleum conduits. New Zealand Jour. Geol. Geophysics 25, 209–225 (1982).

Martini, I.: Sedimentology of a lacustrine barrier systems at Wasaga Beach, Ontario, Canada. In: Segundo Congreso Latino Americano de Geología, Caracas, Venezuela, 11 al 16 de Noviembre de 1973, República de Venezuela Ministerio de Minas e Hidrocarbons, Geología Bol. Geologia, Pub. Especial 7, 1227–1241 (1976).

May, J.P.: Selective transport of heavy minerals by shoaling waves. Sedimentology 20, 203–211 (1973).

McCave, I.N., and Geiser, A.C.: Megaripples, ridges and runnels on intertidal flat of the Wash, England. Sedimentology 26, 353–369 (1979).

Meijer Dress, N.C., and Mhyr, D.W.: The upper Cretaceous Milk River and Lea Park Formations in southeastern Alberta. Bull. Canadian Petroleum Geology 29, 42–74 (1981).

Nittrouer, C.A. (Ed.): Sedimentary dynamics of continental shelves. Marine Geol. Spec. Issue 42, 1–449 (1981).

Off, Theodore: Rhythmic linear sand bodies caused by tidal currents. Am. Assoc. Petroleum Geologists Bull. 47, 324–341 (1963).

Rautman, C.A.: Sedimentology of Late Jurassic barrier-island complex—Lower Sundance Formation of Black Hills. Am. Assoc. Petroleum Geologists Bull. 62, 21275–22289 (1977).

Reineck, H.-E.: Sedimentgefüge im Bereich der südlichen Nordsee. Seckenbergische Naturforschende Gesell. Abh. 505, 138 pp. (1963).

Reineck, H.-E.: Tidal flats. In: Rigby, J.K., and Hamblin, Wm. Kenneth (Eds.): Recognition of ancient sedimentary environments. Soc. Econ. Paleon. Mineral. Spec. Pub. 16, 146–159 (1972).

Roy, P.S.; Thom, B.G.; and Wright, L.D.: Holocene sequences on an embayed high-energy coast: An evolutionary model. Sedimentary Geology 26, 1–19 (1980).

Schwartz, M.L. (Ed.): The encyclopedia of beaches and coastal environments, 960 pp. (Encyclopedia of Earth Sciences, vol. 15). New York: Van Nostrand Reinhold Co. 1982.

Sedimentation Seminar: Sedimentology of the Mississippian Knifley Sandstone and Cane Valley Limestone in south-central Kentucky. Kentucky Geol. Survey Ser. X, Rept. Invs. 13, 30 pp. (1972).

Selley, R.C.: Diagnosis of marine and nonmarine environments from the Cambro-Ordovician Sandstones of Jordan. Jour. Geol. Soc. (London) 128, 135–150 (1972).

Shelton, J.W.: Trend and genesis of lowermost sandstone unit of Eagle Sandstone at Billings, Montana. Am. Assoc. Petroleum Geologists Bull. 49, 1385–1397 (1965).

Simpson, Frank: Sedimentology, palaeoecology and economic geology of Lower Colorado (Cretaceous) strata of west-central Saskatchewan. Saskatchewan Geol. Survey Rept. 15, 183 pp. (1982).

Snead, R.E.: Coastal landforms and surface features, 272 pp. New York: Academic Press 1981.

Stride, A.H. (Ed.): Offshore tidal sands, 216 pp. Andover Hants, England: Chapman and Hall 1982.

Swann, D.H.: Waltersburg Sandstone oil pools of the Lower Wabash area, Illinois. Am. Assoc. Petroleum Geologists Bull. 35, 2561–2581 (1951).

Swift, D.J.P., Parker, G., Lanfreidi, N.W., Perillo, G., and Figge, K.: Shoreface-connected sand ridges on American and European shelves: A comparison. Estuarine and Coastal Marine Science 7, 257–273 (1978).

Swift, D.J.P.; Young, R.D.; Clarke, T.L.; Vincent, C.E.; Niedoroda, A.; and Lesht, B.: Sediment transport in the Middle Atlantic Bight of North America: Synopsis of recent observations. In: Nio, S.C.; Schuttenhelm, R.T.E.; and Van Weering, T.C.E. (Eds.): Holocene sedimentation in the North Sea Basin. Intern. Assoc. Sedimentologists Spec. Pub. 5, 361–383 (1981).

Swift, D.J.P. and Rice, D.D.: Sand bodies on muddy shelves: A model for sedimentation in the Cretaceous Western Interior Seaway, North America. In: Tillman, R.W. and Siemers C.T., (Eds.): Silisclastic shelf sediments. Soc. Econ. Paleo. Mineral. Sp. Pub. 34, pp. 43–62 (1984).

Tanner, W.F. (Ed.): Coastal sedimentology, 313 pp. Tallahassee, Florida: Florida State Univ. Press 1977.

Thompson, S., III, and Potter, P.E.: Paleocurrents of Bliss Sandstone (Cambrian-Ordovician), southwestern New Mexico and western Texas, pp. 36–51. In: Kottlowski, F.K., and others: Annual Report, 1 July 1979 to 30 June 1980, New Mexico Bur. Mines and Mineral Resources, 1981.

Tillman, R.W., and Siemers, C.T. (Eds.): Siliclastic shelf sediments. Jour. Sed. Petrology Spec.Pub., 268 pp. (1984).

Vanney, Jean-René. Géomorphologie des plates-formes continentales, 300 pp. Paris: Doin 1977.

van Straaten, L.M.J.U.: Coastal barrier deposits in south- and north Holland in particular in the areas around Scheveningen and Ignviden. Medelel. Geol. Stichtung 17, 412–474 (1965).

Visser, J.N.J.: The Table Mountain Group: A study in the deposition of quartz arenites on a stable shelf. Trans. Geol. Soc. South Africa 77, 124–143 (1974).

Weber, K.J.: Sedimentological aspects of oil fields in the Niger delta. Geol. Mijnbouw 50, 559–576 (1971).

Zimmerle, W.: Fossil heavy mineral concentrations Geol. Rundschau 62, 536–548 (1973).

Slope and Deep Basin

The study of turbidite deposits is central to sedimentology, and turbidite models are successfully used to solve many diverse problems of petroleum and mineral exploration and geologic history. Sandstones of turbidite origin range widely in composition from those that are quartz-rich on passive margins to lithic- and feldspathic-rich turbidites on active continental margins in deep rifts or in oceanic magmatic arcs. Carbonate turbidites also occur. All have the same basic spectrum of stratification (Table 10-10). These are fairly well-established log and dipmeter patterns (Fig. 10-50).

Turbidite models are grounded partially in the study of modern deposits, mostly along continental margins and in a few deep lakes, and in the many studies of ancient deposits, both on- and offshore. Offshore seismic sections, both shallow and deep, have contributed much to our understanding of facies geometry, and piston and box cores have helped us explore the upper parts of modern subsea fans. Onshore, turbidites have been identified throughout the geologic record both in outcrop and subsurface. To some degree, our perception of the "turbidite model" depends on our experience—marine geology versus the study of ancient turbidite basins.

The literature of turbidites is vast and perhaps second only to that of deltas—of which many turbidites are the most distal facies. The now classic paper of Kuenen and Migliorini (1950) initiated the study of turbidites. Good recent summaries include those of Parker (1977), Walker (1978; 1984), Normark (1978), Stanley and Kelling (1978), Siemers *et al.* (1981), and Bouma *et al.* (1985). Italian sedimentologists have made notable contributions beginning with Migliorini and followed by many others, but especially Mutti (1972) and Ricci Lucchi (1975).

Turbidite Model

Let us first review how and where turbidity currents form, examine the turbidite model, and then comment on selected examples from the vast literature.

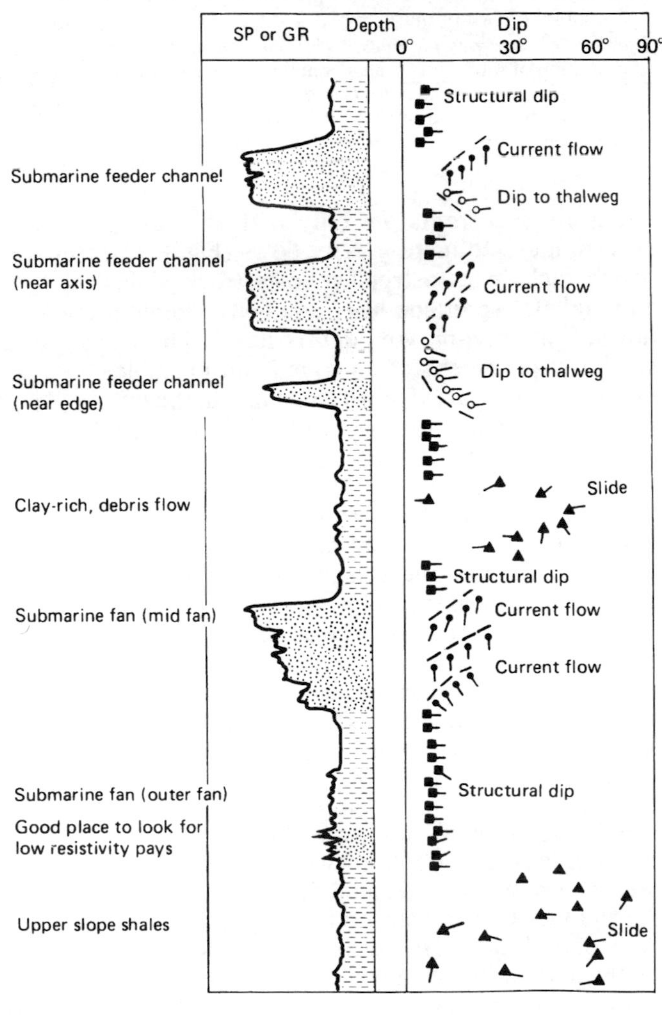

FIGURE 10-50. Idealized log and dipmeter patterns of deep-water sandstones and their associated sediments (courtesy of J. A. Gilreath and Schlumberger Offshore Services). See Fig. 10-9 for definition of dip symbols used in dipmeter analysis.

TABLE 10-10. Principal turbidite facies. (Adapted from Mutti and Ricci-Lucchi, 1972, Table 1).

Conglomerates and sandy conglomerates: Massive, graded and reverse graded, densely packed conglomerates—the so-called "deep-water" conglomerates—and pebbly sandstones. Lenticular, individual beds up to 10 m. Commonly occur where section is very sandy in channels or canyons. Sliding, slumping, and debris flows. Chaotic blocks may be present.

Sandstone: Mostly fine to medium sandstone in thick beds, abundant lamination, erosional channels, and soft sediment deformation. Few shale interbeds. Deposition by "grain flow" and strong turbidity currents.

Sandstone and shale: Classical turbidites with sandstone beds up to one meter, good Bouma cycles and good lateral continuity and sole marks, but only minor channels. Much interbedded shale. Deposition by turbidity currents of moderate strength.

Shaly sandstone: More shale than sandstone, which is chiefly rippled and laminated, few beds exceed 0.5 m thickness, but have excellent lateral continuity. Deposition by weak turbidity currents.

Chaotic beds: Blocks many meters thick transported by sliding and bedding plane slips plus minor to complete disruption of two or more beds.

Hemipelagic and pelagic beds: "Normal", deep-sea sedimentation of mud, which may contain carbonate; microfauna, silt content, and color help distinguish normal mud from that transported by turbidity currents.

Turbidity currents are only part of a larger spectrum of sediment gravity flows (Fig. 8-32), which include catastrophic flows (rock slides and falls), spontaneous, liquified sediment flows, and slow-moving debris flows. The essential requirement of all four is an unstable, up-dip sedimentary deposit that fails. In the marine realm, sands and muds may slowly accumulate on a shelf or in a canyon for many years until metastable equilibrium is reached and perhaps a large storm or rare seismic tremor causes failure. The bluff of a terrace bordering a deep lake or fjord, the oversteepened front of a fan delta (Fig. 10-35), the delta of a large river, the sides of a submarine canyon, the edge of a shelf or the boundary of a large carbonate platform or reef all provide examples where the potential energy of the sedimentary deposit is catastrophically converted into kinetic energy and the resedimented mass moves downslope underwater. See Slaczka and Thompson (1981) for details based on field studies in Poland. The end result of failure is a bottom-seeking density current which, where unconfined, spreads out and decelerates to form an underwater fan at the base of a slope. Some turbidites are deposited on the lower part of slopes and others travel far onto abyssal plains before deposition. The surfaces of modern subsea fans dip from 5° or more near their head to less than 0.25° at their margins.

The base of slope of passive continental margins is a common site of subsea fans. Fan development is favored by abundant sediment supply from a large river or by a submarine canyon, which intersects lateral transport on the shelf. Such sources can form fans with dimensions of tens of thousands of square kilometers, especially where the river is large (Amazon, Bengal, Congo, and Indus). Other major occurrences are trenches marginal to continents or magmatic arcs, where marginal fans contribute to axial flow (Fig. 10-51). Or, where basin tectonics are complex, a series of small elongate basins and channels exist such that there are ponded turbidites and channels as well as elongate fans or lobes. Deeply flooded rifts also contain turbidites, and longitudinal transport fed by lateral supply prevails. Turbidites have been but rarely reported from cratons but can overlap them on their borders.

The underwater fan of turbidite deposition may be represented by several alternative models, although Fig. 10-52 seems fairly typical. Variants occur because size, shape, channel system, grain size, and lithic proportions differ so much. For example, some subsea fans can be markedly elongate rather than fanlike, channels may be fairly straight, meandering, or

SUBMARINE FAN-CONTINENTAL RISE

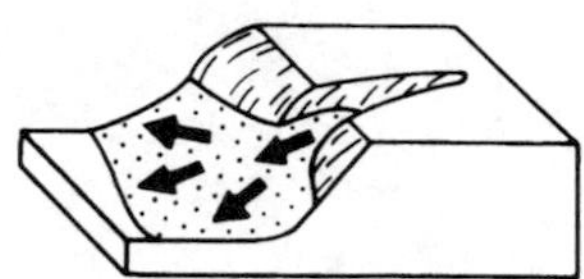

SLOPE OR FOREARC BASIN

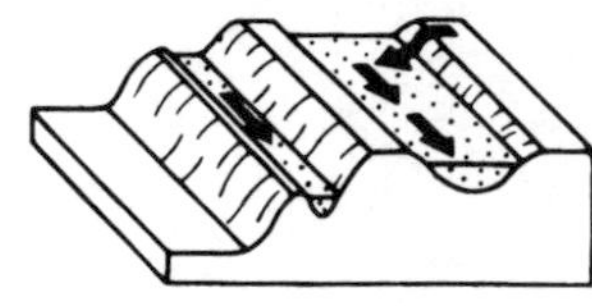

TRENCH

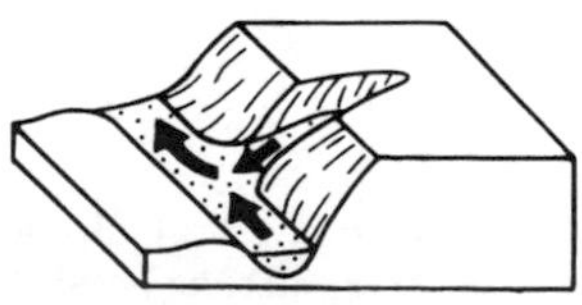

FIGURE 10-51. Simplified occurrence of turbidite deposits along continental rise, in fore-arc basins and in trenches. (Redrawn from Moore, 1973, Fig. 12).

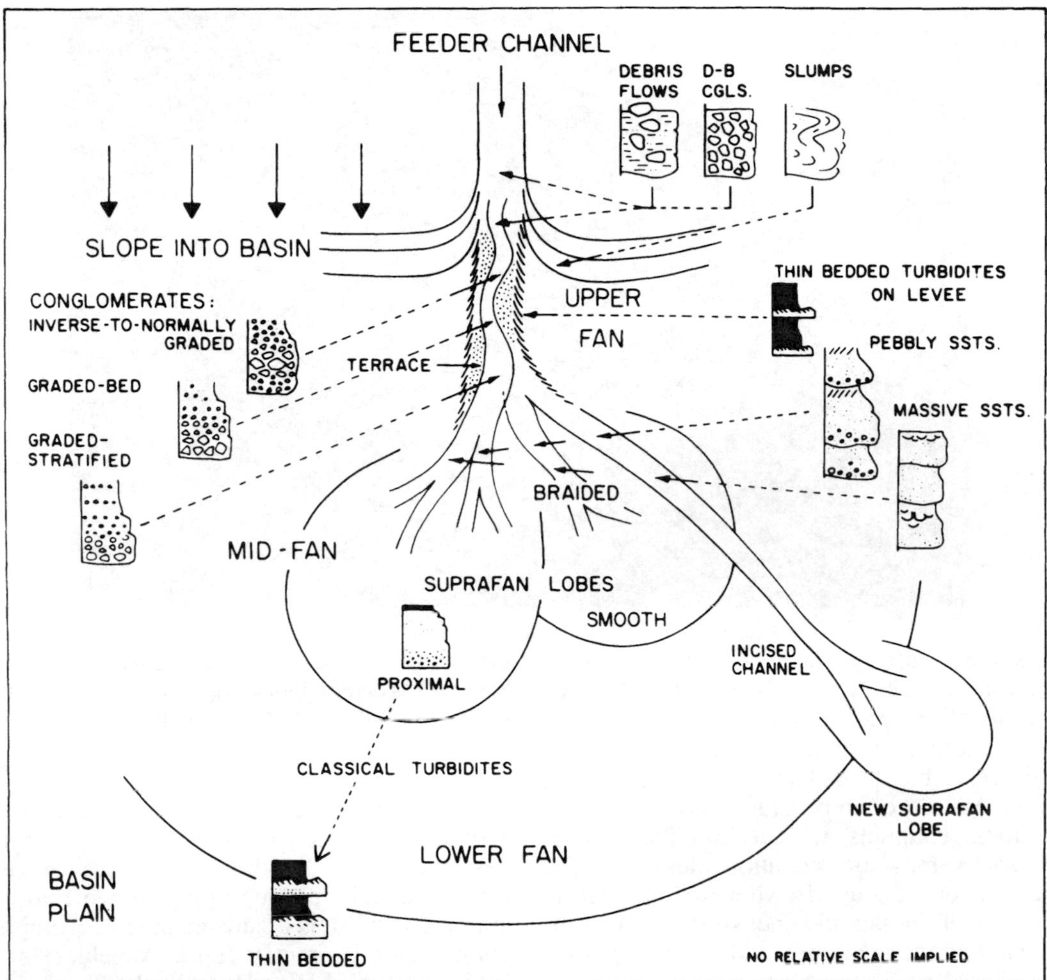

FIGURE 10-52. Morphology of typical subsea fan and terminology according to Walker (1978, Fig. 13). Published by permission of the author and the American Association of Petroleum Geologists.

braided, and, in addition, some fans are rich in sand and gravel whereas others are dominantly silt and mud. Sediment supply, water depth, and basin tectonics all contribute to these variations, although just how is not yet clear. Discussion of fan models is provided by Nilsen (1980, pp. 1101–12) and Stow (1981, pp. 389–91). Incomplete knowledge of both modern and ancient subsea fans is still a major problem and, as our knowledge increases, several models of subsea fans may emerge—some for abundant sand in water of modest depth, some for lesser amounts in very deep water, etc.

Channel deposits include thick, structureless, massive beds, conglomerates, and chaotic facies; these may occur both in the "feeder canyon" as well as in the inner and middle parts of the fan. But should a channel be abruptly abandoned, its fill may be mostly hemipelagic. Channels range from small ones that can be seen in outcrop (Fig. 10-53) to large ones that have been mapped in the subsurface or on the sea floor. In the Laurentian Fan of the Canadian Maritimes, channel depth is 800 m, and over 500 m of sediment has been eroded (Stow, 1981, p. 378).

As transport distance from fan head increases, there is an overall decline of bed thickness and grain size, change in proportions and kind of structures, and more mud until finally hemipelagic, basin plain deposits prevail (Table 10-10). Slaczka and Unrug's (1976) study of the Oligocene Cergowa Sandstone in the Carpathians beautifully illustrates these relations. Classical turbidites with sole marks and complete Bouma cycles, the sandstone–shale facies of Table 10-10, predominate near much of the outer part of the fan and the interbedded muds

FIGURE 10-53. Small turbidite channel in Cretaceous Rosario Formation along Highway 1, km 92, north of Ensenada, Baja California, Mexico.

of this facies always have greater continuity than its sandstones. However, the same "proximal to distal" sequence also occurs at right angles to an active distributary so that facies is not always relatable to actual distance from fan head. Overall, the fan progrades down dip and, as it does, channels are cut into its surface. Contourites may be common near the fan fringe, but can occur elsewhere on it. During the history of the fan, channel shifts occur, producing abandoned channels and lobes which are then covered by a mud blanket.

Mapping the facies shown in Table 10-10 and relating each to paleocurrents measured in outcrops, cores, or with the dipmeter helps reconstruct fan morphology (Fig. 10-52) and relate fan origin to paleoslope (Fig. 10-54). Some nine paleocurrent models have been recognized. Mutti and Ricci Lucchi (1972, Table 1) show how facies are related to fan morphology as does Walker (1978, p. 945–47). Thickening- or thinning-upward sequences exist and should be noted; for example, most channels have a blocky to fining-upward trend whereas outer fan deposits coarsen and thicken upward (Figs. 10-50 and 10-55).

Ancient Turbidites

Turbidites were first studied in outcrop (Fig. 10-56) and such studies still are the most numerous. Classic areas include the Apennines, the Carpathians, the northern Alps, Wales, New Zealand, California, the southern Pyrenees, and Quebec.

The many field studies of Cretaceous and Tertiary turbidites in Italy are rich in careful descriptions of bedding sequences, have informative field photographs, and include diagrammatic models of subsea fans with emphasis upon facies geometry (e.g., Angelucci *et al.*, 1967; Mutti and Ricci Lucchi, 1972). Here turbidites may reach thicknesses of several thousand meters and some beds have been traced more than 100 km. Paleocurrent studies show how paleocurrents can reverse in the elongate troughs of fore-arc basins. Other informative studies include that of Harms (1974) on the Permian Brush Canyon Formation in west Texas, that of Nilsen and Abbott (1981) on the Cretaceous Rosario Formation of southern California and Baja California, and Hicks' (1981) study of Late Triassic metasediments of the Torlesse zone of New Zealand. Hicks recognized several facies, most of which correspond to those of Table 10-10. The turbidite fill of deeply cut canyons has also been studied, in the Precambrian of Australia by Borch *et al.* (1982) and in the Tertiary Annot Sandstone of France by Stanley *et al.* (1978). Distinguishing deep from shallow water conglomerates is discussed by Walker (1975).

There are a few outcrop studies of single beds traceable 10 to 20 km or more wherein thickness, paleocurrents, grain size, and bedding

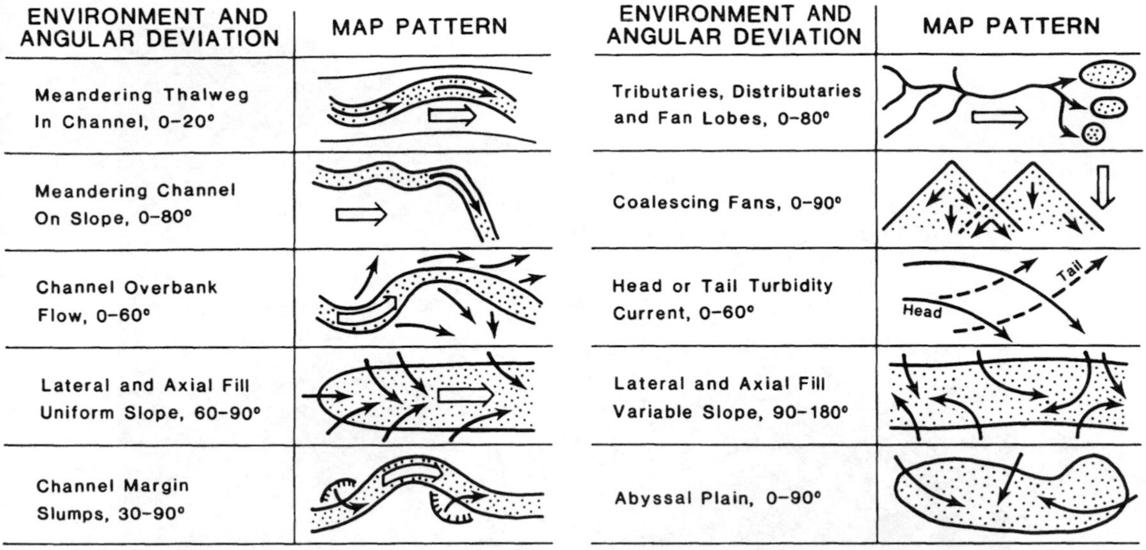

ENVIRONMENT AND ANGULAR DEVIATION
MAP PATTERN
Meandering Thalweg In Channel, 0–20°
Meandering Channel On Slope, 0–80°
Channel Overbank Flow, 0–60°
Lateral and Axial Fill Uniform Slope, 60–90°
Channel Margin Slumps, 30–90°
ENVIRONMENT AND ANGULAR DEVIATION
MAP PATTERN
Tributaries, Distributaries and Fan Lobes, 0–80°
Coalescing Fans, 0–90°
Head or Tail Turbidity Current, 0–60°
Head
Tail
Lateral and Axial Fill Variable Slope, 90–180°
Abyssal Plain, 0–90°

FIGURE 10-54. Paleocurrents, patterns of turbidites, and contourites and their relation to paleoslope (open arrows). (Redrawn from Lovell and Stow, 1981, Fig. 2).

SL
DF
CGL
CGL
PS
(CGL)
MS
PS
CPT
MS
CPT
CDT
CDT
MUD
UPPER FAN
CHANNEL FILL
CHANNELLED
SUPRAFAN
LOBES
CHANNELLED
SMOOTH
SMOOTH
SUPRAFAN
LOBES
MID FAN
LOWER FAN
BASIN PLAIN

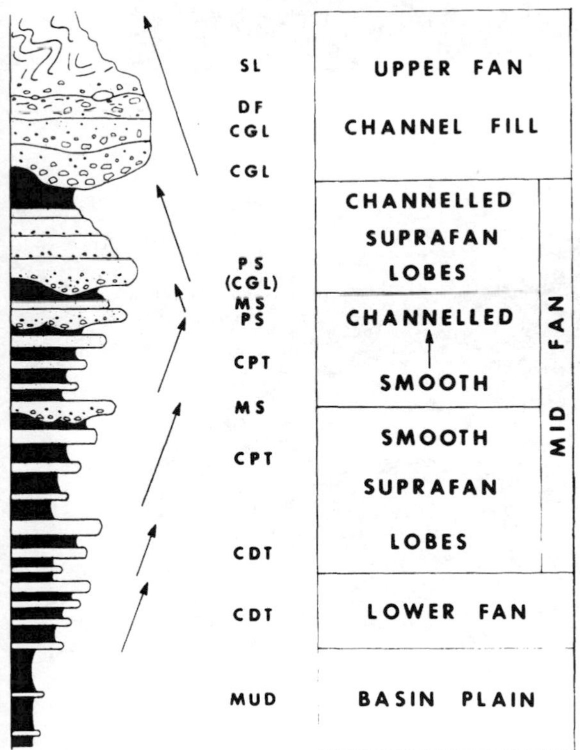

FIGURE 10-55. Idealized vertical sequence of prograding fan and its turbidite sandstones. (Modified from Walker, 1978, Fig. 14). Arrows indicate thickening- or coarsening- or fining-upward sequences; CDT, classic distal turbidites; CPT, classic proximal turbidites; MS, massive sandstones; PS, pebbly sandstones; CGL, conglomerate; DF, debris flow; and SL, slumps.

types have been mapped (Hirayama *et al.*, 1969; Hirayama and Nakajima, 1977; Sturm and Matter, 1978; Maynard and Lauffenburger, 1978, Fig. 4). Hesse (1973, Tafel 2) and Ricci Lucchi (1975, pp. 64–65) report on outcrops where beds have been traced over 100 km. More such studies are needed, especially in the subsurface.

Subsurface studies include those of the Pliocene turbidites of the Ventura Field in California (Hsu, 1977), of turbidites of Pennsylvanian slope sediments in north Texas (Galloway and Brown, 1972), of the Great Valley of California (Garcia, 1981), and of turbidites in the North Sea graben (Hill and Wood, 1980). Seismic sections are also used to recognize and map ancient turbidites (Fig. 10-57) and full details are given by Mitchum (1985).

Cratonic turbidites are exceptional, but have been reported by Lineback (1968) in the subsurface of the Illinois Basin and by Kepferle (1977) in its outcrop. Cratonic turbidites along the western margin of the Appalachian Basin are described by Bryan (1983).

Global changes in sea level are now recognized to have dramatic effect on shelf, slope, and base of slope sedimentation. Shanmugam and Moiola (1982) found 10 oil-bearing submarine fans of Tertiary age to have all developed at low stands of world sea level and attribute this to minimal sediment trapping on shelves. This correlation needs testing elsewhere in older pre-Tertiary turbidites and may not exist in basins with active local tectonism.

FIGURE 10-56. Above: Turbidites seen in outcrop. Medial turbidites with long, even sandstone beds interbedded with shale. Marnoso-arenacea (Miocene), valley of Santerno River, near Castel del Rio, Emilia Romagna, Italy. Below: Distal turbidites of the Guaricio Formation (Eocene) along Highway 9, km 212, Penalver, Anzolatequir, Venezuela.

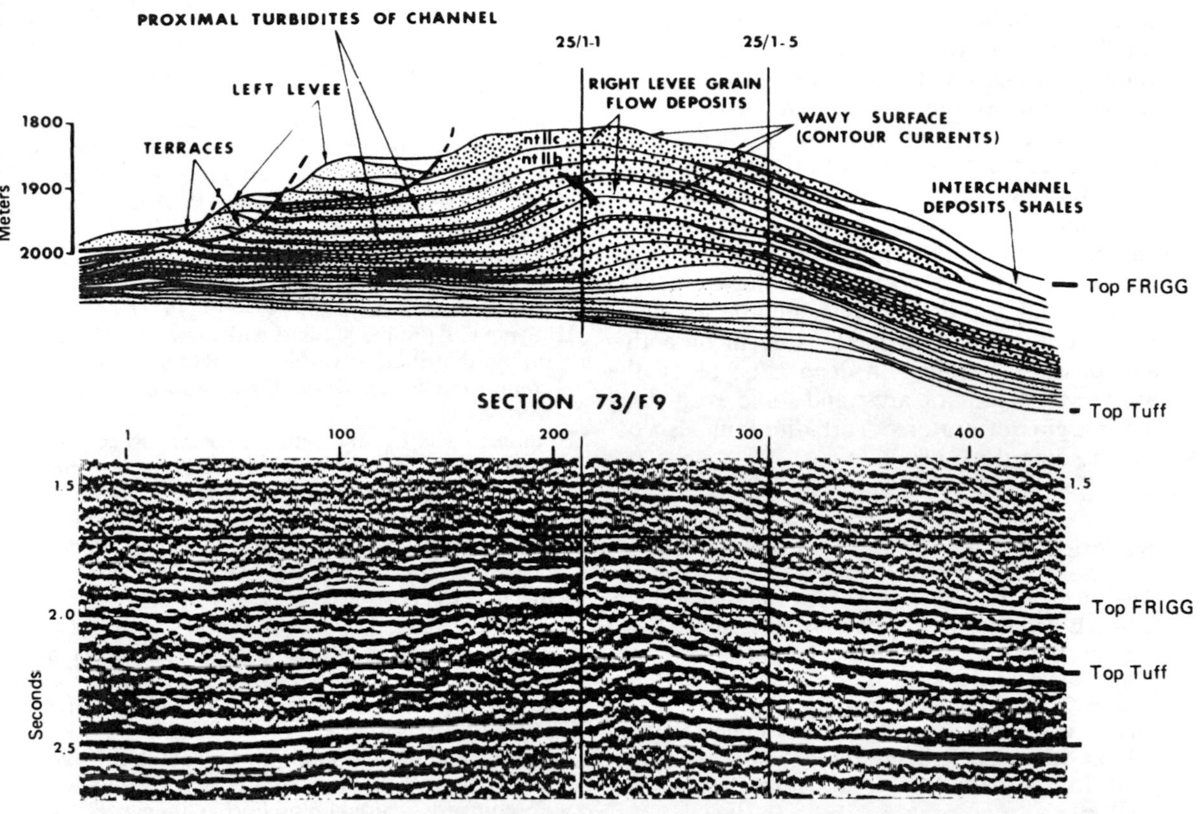

FIGURE 10-57. Large submarine fan of Frigg Field in North Sea: interpretatioin and seismic section from Hertier *et al.* (1980, Fig. 13). Published by permission of the authors and the American Association of Petroleum Geologists.

There are a few studies of turbidites in modern deep lakes, whose facies correspond well to those of the deep sea (Houbolt and Jonker, 1968, Fig. 1; Normark and Dickson, 1976; Hsu *et al.*, 1980, Fig. 8). Turbidites have even been reported in tributaries of rivers that have had catastrophic floods (Baker, 1973, Fig. 31).

Contourites, fine-grained, ripple-bedded sandstones interbedded with shale, may be common along the outer parts of some subsea fans (Fig. 10-54); a good review is provided by Stow and Lovell (1979). Such beds are thin, tend to have paleocurrents perpendicular to paleoscope, tend to be bioturbated, have both down- and along-slope provenance and concentrates of heavy minerals (Lovell and Stow, 1981), and are believed to develop best at low stands of world sea level (Shanmugam and Moiola, 1982, Table 2).

Emplacement of detrital pyrite by gravity flow in volcanigenic basins is reported by Schermerhorn (1971) and Juve (1977).

Summary

The discovery and evolution of the turbidite model and its several variants is one of the most outstanding achievements of physical sedimentology since Kuenen and Migliorini related graded bedding to turbidity currents over 35 years ago. The turbidite model seems to fit and explain many ancient deposits—terrigenous and carbonate alike—and thus has very wide applicability, and deposits range from the earliest Archean to the recent. Turbidite models are all the more notable because they are based on partial and incomplete evidence from both ancient and Holocene deposits.

Underwater slope failure along the front of an oversteepened delta or fan delta, along a shelf edge, along the flanks of a volcanic arc where volcaniclastics rapidly accumulated, where pyroclastic flows entered the sea, or along the margins of a steep carbonate bank or reef can generate density underflows, where water was sufficiently deep. Low rather than high stands of world sea level seem to favor marine turbidites. As turbidity currents flow basinward, they finally decelerate and deposit progressively thinner, fine-grained beds whose classic characteristics are Bouma cycles and diverse sole marks. Such successive flows, perhaps a few every hundred or thousand years, produce subsea fans that range from small to hundreds of thousands of square kilometers. Such fans have channels and well-defined proximal, medial, and distal facies. Associated lithologies include pelagic shales and limestones and debris flows near fan head. Submarine fans occur on active and passive margins, in deep rifts, along the borders of magmatic arcs, and some even along the margins of cratons. Turbidite fans also occur in some deep lakes.

References

Angelucci, A.; DeRosa, E.; Fierro, G.; Gnaccolini, M.; LaMonica, G.B.; Martinis, B.; Parea, G.C.; Pescatore, T.; Rizzini, A.; and Wezel, F.C.: Sedimentological characteristics of some Italian turbidites Geol. Romana 6, 345–420 (1967).

Baker, V.R.: Paleohydrology and sedimentology of Lake Missoula flooding in eastern Washington. Geol. Soc. America Spec. Paper 144, 79 pp. (1973).

Borch, C.C. von der; Smit, R.; and Gradym, A.E.: Late Proterozoic submarine canyons of Adelaide Geosyncline, South Australia. Am. Assoc. Petroleum Geologists Bull. 66, 332–347 (1982).

Bouma, A.H.; Normark, W.R.; and Barnes, N.E. (Eds.): Submarine fans and related turbidite systems, 368 pp. (Frontiers in Sedimentary Geology, Vol. 1). New York: Springer-Verlag 1985.

Bryan, Timothy M.: Depositional history provides clues. Northeast Oil Reporter 3, 43–46 (1983).

Galloway, W.E., and Brown, L.F., Jr.: Depositional systems and shelf-slope relationships in Upper Pennsylvanian rocks, North-central Texas. Texas Bur. Econ. Geol., Rept. Invs. 75, 62 pp. (1972).

Garcia, R.: Depositional systems and their relation to gas accumulation in Sacramento Valley, California. Am. Assoc. Petroleum Geologists Bull. 65, 653–673 (1981).

Harms, J.C.: Brush Canyon Formation, Texas: A deep-water density current deposit. Geol. Soc. America Bull. 85, 1763–1788 (1974).

Hertier, F.E.; Lossel, P.; and Wathne, E.: Frigg Field—large submarine-fan trap in Lower Eocene rocks of the Viking Graben, North Sea. In: Halbouty, M.T. (Ed.): Giant oil and gas fields of the decade 1968–1978. Am. Assoc. Petroleum Geologists Mem. 30, 59–79 (1980).

Hesse, Reinhard: Flysch-Gault und Falknis-Tasna-Gault (Unterkreide): Kontinuierlicher Übergang von der distalen zur proximalen Flyschfazies auf einer pennischen Trogebene der Alpen. Geologica et Palaeontologica SB, 90 pp. (1973).

Hicks, D.M.: Deep-sea fan sediments in the Torlesse zone, Lake Ohan, South Canterbury, New Zealand. New Zealand Jour. Geol. Geophysics 24, 209–230 (1981).

Hill, P.J., and Wood, G.V.: Geology of the Forties Field, U.K. continental shelf, North Sea. In: Halbouty, M.L. (Ed.): Giant oil and gas fields of the decade 1961–1971. Am. Assoc. Petroleum Geologists Mem. 30, 81–93 (1980).

Hirayama, J., and Nakajima, T.: Analytical study of turbidites, Otadai Formation, Baso Peninsula, Japan. Sedimentology 24, 747–779 (1977).

Hirayama, J.; Fujii, K.; and Nakajima, T.: Granulometric distribution within a single sandstone bed composing flysch. Bull. Geol. Survey Japan 20, 669–684 (1969).

Houbolt, J.J.H.C., and Jonker, J.B.M.: Recent sediments in the eastern part of the Lake of Geneva (Lac Leman). Geol. Mijnbouw 47, 131–148 (1968).

Hsu, H.J.: Studies of Ventura Field, California, 1: Facies geometry and genesis of Lower Pliocene turbidites. Am. Assoc. Petroleum Geologists Bull. 61, 137–168 (1977).

Hsu, K.J.; Kelts, K.; and Valentine, J.W.: Resedimented facies in Ventura basin California and model of longitudinal transport of turbidity currents. Am. Assoc. Petroleum Geologists Bull. 64, 1034–1051 (1980).

Juve, G.: Metal distribution at Stekenjokk: Primary and metamorphic patterns. Geologiska Foreningens i Stockholm Forhandlingar 99, 149–158 (1977).

Kepferle, R.C.: Stratigraphy, petrology, and depositional environment of the Kenwood Siltstone Member, Borden Formation (Mississippian), Kentucky and Indiana. U.S. Geol. Survey Prof. Paper 100, 49 pp. (1977).

Kuenen, Ph.H., and Migliorini, C.I.: Turbidity currents as a cause of graded bedding. Jour. Geology 58, 91–117 (1950).

Leggette, J.K. (Ed.): Trench-forearc geology: Sedimentation and tectonics on modern and ancient active plate margins. Geol. Soc. London Spec. Paper 10, 576 pp. (1982).

Lineback, Jerry A.: Turbidites and other sandstone bodies in the Borden Siltstone (Mississippian) in Illinois. Illinois Geol. Survey Circ. 42, 29 pp. (1968).

Lovell, J.P.B., and Stow, D.A.V.: Identification of ancient sandy contourites. Geology 9, 347–349 (1981).

Maynard, J.B., and Lauffenburger, S.K.: A marcasite layer in prodelta turbidites of the Borden Formation (Mississippian) in eastern Kentucky. Southeastern Geol. 20, 47–58 (1978).

Mitchum, R.M. Jr.: Seismic expression of submarine fans in: Berg O.R., and Woolverton, D.G. (Eds.) Seismic stratigraphy II. Amer. Assoc. Petroleum Geologists Mem. 39, 117–136 (1985).

Moore, J.C.: Cretaceous continental margin sedimentation. Geol. Soc. Amer. Bull. 84, 595–614 (1973).

Mutti, Emiliano, and Ricci Lucchi, F.: Le torbiditi

dell' Appennino settrionale: introduzione all'analisi di facies. Soc. Geol. Italiana Mem. 11, 161–199 (1972). See Intern. Geol. Rev. 20, 125–166 (1978) for English translation.

Nilsen, T.H.: Modern and ancient submarine fans: Discussion of papers by R.G. Walker and W.R. Normark. Am. Assoc. Petroleum Geologists Bull. 64, 1094–1112 (1980).

Nilsen, T.H., and Abbott, P.L.: Paleogeography and sedimentology of Upper Cretaceous turbidites, San Diego, California. Am. Assoc. Petroleum Geologists Bull. 65, 725–1284 (1981).

Normark, W.R.: Fan valleys, channels and depositional lakes on modern submarine fans: Characters for recognition of sandy turbidite environments. Am. Assoc. Petroleum Geologists Bull. 62, 912–931 (1978).

Normark, W.R., and Dickson, F.H.: Sublacustrine fan morphology in Lake Superior. Am. Assoc. Petroleum Geologists Bull. 60, 1021–1036 (1976).

Parker, J.R.: Deep-sea sands. In: Hobson, G.D. (Ed.): Developments in petroleum geology—1, pp. 225–242. London: Applied Science Publishers Ltd. 1977.

Ricci Lucchi, F.: Miocene paleogeography and basin analysis in the Periadriatic Apennines. In: Squyres, Coy (Ed.): The Periadriatic Apennines, pp. 1–111. Tripoli: Petroleum Explor. Soc. Libya 1975.

Schermerhorn, L.J.G.: Pyrite emplacement by gravity flow. Instituto Geología Minero, Bol. Geol. Minero 82, 88–92 (1971).

Shanmugam, G., and Miola, R.J.: Eustatic control of turbidites and winnowed turbidites. Geology 10, 231–235 (1982).

Siemers, C.T.; Tillman, R.W.; and Williamson, C.R. (Eds.): Deep-water clastic sediments, a core workshop, 416 pp. Soc. Econ. Paleon. Mineral. Core Workshop No. 2, San Francisco, May 30 and 31, 1981.

Slaczka, A., and Thompson, S. III: A revision of the fluxoturbidite concept based on type examples in the Polish Carpathian flysch. Ann. Soc. Geol. Pologne 51, 3–44 (1981).

Slaczka, A., and Unrug, R.: Trends of textural and structural variation in turbidite sandstones: The Cergowa Sandstone (Oligocene, outer Carpathians). Ann. Soc. Geol. Pologne 46, 55–75 (1976).

Stanley, D.J., and Kelling, G. (Eds.): Sedimentation in submarine canyons, fans, and trenches, 395 pp. Stroudsburg, Pennsylvania: Dowden Hutchinson and Ross, 1978.

Stanley, D.J.; Palmer, H.D.; and Dill, R.F.: Coarse sediment transport by mass flow and turbidity current processes and downslope transformation in Annot Sandstone Canyon—fan valley systems. In: Stanley, D.J., and Kelling, G. (Eds.): Sedimentation in submarine canyons, fans, and trenches, pp. 85–115. Stroudsburg, Pennsylvania: Dowden Hutchinson and Ross 1978.

Stow, D.A.V.: Laurentian Fan: Morphology, sediments, processes, and growth pattern. Am. Assoc. Petroleum Geologists Bull. 65, 375–393 (1981).

Stow, D.A.V., and Lovell, J.P.B.: Contourites: Their recognition in modern and ancient sediments. Earth Sci. Rev. 14, 251–291 (1979).

Sturm, M., and Matter, A.: Turbidites and varves in Lake Brienz (Switzerland): Deposition of clastic detritus density currents. In: Matter, A., and Tucker, M.E. (Eds.): Modern and ancient lake sediments. Intern. Assoc. Sedimentologists Spec. Pub. 2, 147–168 (1978).

Tokuhashi, S.: Three dimensional analysis of a large sandy-flysch body, Mio-Pliocene Kiyosumi Formation, Baso Peninsula, Japan. Kyoto Univ. Mem. Fac. Sci., Geol. Mineral., 46, 60 pp. (1979).

Unrug, R.: Ancient contourites in the Menilite Beds (Ogliocene) of the Carpathian flysch, Poland. Ann. Soc. Geol. Pologne 50, 175–182 (1980).

Walker, R.G.: Generalized facies models for resedimented conglomerates of turbidite association: Geol. Soc. Amer. Bull. 86, 737–748 (1975).

Walker, R.G.: Deep-water sandstone facies and ancient submarine fans: Models for exploration for stratigraphic traps. Am. Assoc. Petroleum Geologists Bull. 62, 932–966 (1978).

Walker, R.G.: Turbidites and associated coarse clastic deposits, pp. 171–188. IN: Walker, R.G., Ed.: Facies models, 2nd ed. Geoscience Canada Reprint Ser, 1984 1 (Bus. and Econ. Ser. Ltd, 111 Peter St. Suite 509, Toronto, Ont. MSV. 2H1), 1984.

Eolian

Sand transported by wind may dominate basin fill where multiple, thick formations are of eolian origin as in the southwestern United States (Poole, 1964), although a single eolian formation in a basin is more likely. What seems to be most common by far, however, is that only small parts of a sandy deposit may be eolian. For example, one need only walk along a desert wadi or a large point bar to see how windbown and water-transported sand can be intercalated on the scale of centimeters. Another example is a subaqueous dune exposed at low water with a scree of windblown sand on its slip slope.

Eolian deposits, both ancient and modern, have received much attention in recent years and today there is a modest literature. Shotten (1937, 1956) and McKee (1934) made pioneering studies of ancient eolianites. Another early pioneer was Knight (1929). Major sources from which much of the following is taken include Glennie (1970, 1972), Bigarella (1972), McKee (1979), and Brookfield and Ahlbrandt (1983). Important background is also provided by atlases of desert geomorphology (McGinnies *et al.* 1968; Cooke and Warren, 1973). Solle (1966) published an informative comparison of modern

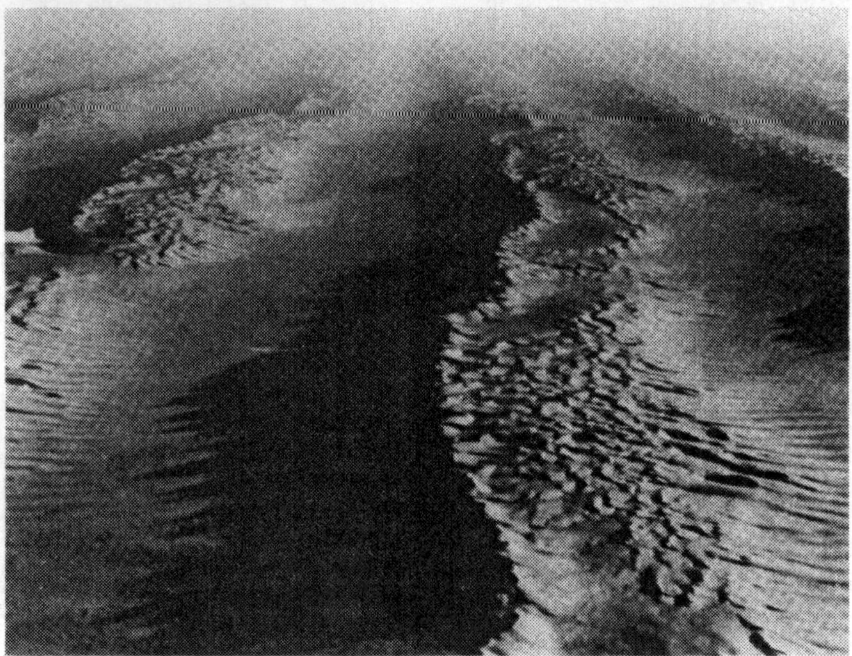

FIGURE 10-58. Coastal dune system flooded by Paraniba River near Paraniba, Piaui, Brazil. South Atlantic Ocean in far distance (*above*). Longitudinal dunes of the Rub'al Khali in Arabian Peninsula as seen from airplane (photo courtesy Arabian American Oil Co.). These longitudinal dunes are about 1 to 2 km wide, 100 meters high, and up to 200 km long (*below*).

FIGURE 10-59. Compound barchan in coastal desert of northwestern Peru. Note small transverse dunes at lower left (*above*). Foto Servicio Aerofotográfico Nacional de Peru. "Sand mountains" in the Rub'al Khali of the Arabian Peninsula (photograph courtesy of the Arabian American Oil Co.) (*below*). Sand mountains may reach heights of as much as 250 m. Large areas of desert are swept bar of sand to produce concentrations as shown in Figs. 10-58 and 10-59 so that there are also areas of exposed bedrock and gravel plains in most large continental deserts.

and ancient eolian deposits and Bigarella did much to enhance our knowledge both of ancient deposits, such as the widespread Triassic Botucatú Sandstone in southern Brazil (Bigarella and Salamuni, 1961), and of coastal Holocene dunes (Bigarella *et al.*, 1969; Bigarella, 1975). A mixed eolian-fluvial redbed sequence of Cretaceous age in Mongolia, the Barun Goyot Formation, is described in detail by Gradzinski and Jerzkiewicz (1974), who give careful descriptions of five bedding facies of general interest, some of which are fluvial.

Dune forms are diverse (Figs. 10-58 to 10-60) and exceptionally they can reach heights of over 100 m and longitudinal dunes reach lengths of 100 km or more (Figs. 10-58 and 10-59). Dune forms have also been found on Mars (Fig. 10-60). Large dune fields require a major source of sand—appreciable longshore transport for a coastal dune field or a large, topographic basin with centripetal drainage from surrounding cuestas or highlands for a large inland sand sea. Associated deposits of a continental desert include alluvial fans with sands and gravels and wadis that transport sand, silt, and mud long distances until deposition finally occurs in a temporary lake or inland sabkha. On sabkhas sand, silt, mud, and detritus become encrusted with crystals of gypsum or halite or intercalated with nodules or thin beds thereof. Sparse versus abundant evaporites may help distinguish between an inland versus coastal sabkha in ancient deposits as would the absence or presence of marine carbonates which are common along low-relief coastlines. Mud curls and chips are common in associated fluvial and lacustrine deposits as are caliches. Adhesion ripples (Chap. 8) may also be present where the surface was wet as in interdune deposits when the water table was temporarily high or in ephemeral lakes. Debris flows occur in the alluvial fan and wadi deposits. Coastal dunes may be intercalated with coastal sabkha and lagoonal, lacustrine, and carbonate deposits, as dunes migrate along the coast or inland over other environments. Along arid coasts with strong winds and ample sand, migrating coastal dunes disrupt drainage and create many lakes. Coastal dune sand can also be blow far inland and, where winds are strong, climb far up the slopes of arid mountains. Lag deposits of gravel and pebbles are common, and after initial deflation of sand, silt, and mud, form an armored surface on which ventifacts develop.

FIGURE 10-60. Dunes, believed to be transverse, found on Mars (Carr, 1981, Fig. 11.3). Published by permission of the author and Yale University Press.

Crossbedding in dunes is generally of medium to large scale and mostly of tabular and wedge types. Dip of foresets is commonly steep, say greater than 20°, and maximum angle is about 30 to 34° but complexity can be great, as shown by the summaries of McKee (1979) and Bigarella (1972) and the study of Tsoar (1982, Figs. 9 and 13) of crossbedding in longitudinal dunes. Minor structures of dunes include ripple marks, rain pits, and deformational structures. Eolian ripples have a distinctive pattern with ripple indices greater than 15. Hunter (1977, pp. 372–73) coined the phrase "subcritical, climbing translatent stratification" for a small-scale structure he believed to be uniquely characteristic of eolian transport (Fig. 8-39). Deformational structures include folds, faults, and breccias, most of which form by avalanching; wet crusts play an important role in their formation. Animal tracks also occur and are best preserved on slip slopes.

Sand also occurs in interdune areas where horizontal lamination seems characteristic and

FIGURE 10-61. Eolian crossbedding of Triassic Piramboia Formation along Rodavia Castello Branco, km 167, São Paulo State, Brazil. Man stands at base of Mesozoic Erathem (Soares, 1975, Fig. 3). Published by permission of the author and the Sociadade Brasileira de Geologia.

adhesion ripples are present. Interdune deposits should also be good sites to look for plant and animal fossils. Interdune deposits, because of possible differences in size, cements, and bedding, can be recognized on wireline logs and on the dipmeter through their lack of crossbedding.

Sandstones believed to be eolian are, according to McKee (1979, pp. 191–94), mostly light-colored, abundantly crossbedded, fine- to medium-grained (0.125 to 0.25 mm), well sorted, and occur as large sheets representing continental deserts or as smaller deposits possibly along ancient coastlines. Boundaries are mostly abrupt rather than gradational, because dunes override fluvial, lacustrine, or marine environments rather than grade into them. Most ancient deposits believed to be eolian are generally mineralogically mature and many are quartz arenites; some crossbeds can be as thick as 30 m (Fig. 10-61) and commonly dip at 20° to 30°. One may question whether maturity results from the elimination of physically unstable grains such as rock fragments or from recycling from older sandstones. Variance of paleocurrents is moderate so that most paleocurrent maps from ancient eolianites are fairly homogeneous; some transport paths of eolian deposits have been related to paleolatitudes and paleowinds. High ripple indices are common on the dunes and the presence of adhesion ripples in interdune deposits provides confirming evidence of eolian origin. Reptilian tracks and rain-

TABLE 10-11. Criteria for eolian vs. water deposition usable in cores. (Chiefly from Glennie, 1970, Table 1).

Part A. Wind-deposited sandstones

1. Thickness from centimeters to several hundred meters with horizontal to dipping laminae up to 34°.
2. Wedge crossbedding common in medium to thick sets which may be isolated in horizontal lamination or vertically stacked; crossbeds commonly have well-defined orientation.
3. Mostly planar lamination but some ripples on foresets.
4. Good sorting in individual laminae, differences in maximum grain size common between laminae, and some bimodal textures. Applicable to cuttings.
5. Larger grains well rounded. Applicable to cuttings.
6. Clay and silt generally less than 5 percent authigenic. Applicable to cuttings.
7. Clay drapes rare.
8. Shallowly buried quartz sands friable and lightly cemented with hematite.
9. Adhesion ripples associated with more clay and gypsum and anhydrite; wind ripples on bedding planes.
10. Deflation gravels, some ventifacts.

Part B. Water-deposited sandstones

1. Crossbedding of small to large scale, and mostly trough and planar with good to poor orientation.
2. Mostly current ripples.
3. Cements of calcite, gypsum, and anhydrite. Applicable to cuttings.
4. Hematite coatings. Applicable to cuttings.
5. Conglomerates common and deflation may produce a sandfill matrix at top.
6. Debris flow conglomerates.
7. Sharp upward increase in grain size from abrupt change in discharge.
8. Clay clasts and curls common.
9. Mud cracks with sandy infill.
10. Sandstone dikes, some also in eolian sandstones.

drop impressions should be looked for. A summary description of ancient eolianites is provided by McKee (1979, pp. 187–240) and by Brookfield and Ahlbrandt (1983, p. 407–660). Sanderson's (1974) description of the Navajo Sandstone (Early Jurassic?) in San Rafael Swell in Utah is very complete as in Klemmensen's (1985) study of the Bunter Sandstone in the German Basin. Eolian deposits also occur on Mars (Mutch *et al.*, 1976, Chap. 7).

Glennie (1970) has compiled a list of characteristics to identify eolian sandstones in cores (Table 10-11) as does Mader (1983, p. 589–90). To these should be added evidence from the dipmeter, which appears to be most convincing—thick sandstone intervals free of shale breaks (Figs. 10-7 and 10-62).

Summary

Eolian sands occur today in large fields in the continental deserts of North Africa and Australia, on the Arabian Peninsula, along arid coastlines such as those of Peru and Chile, and along semi-arid and temperate coastlines as well. Other sites of modern eolian sand occur on the alluvial fans of arid and semi-arid climates and along the borders of braided streams. On a small scale, a walk at low water along a sandy point bar will show many wind ripples and patches of eolian sand that are water stratified with its water-transported sand. Coastal dunes not only cap barriers but may transgress landward damming rivers and thus are juxtaposed with fluvial and lagoonal deposits.

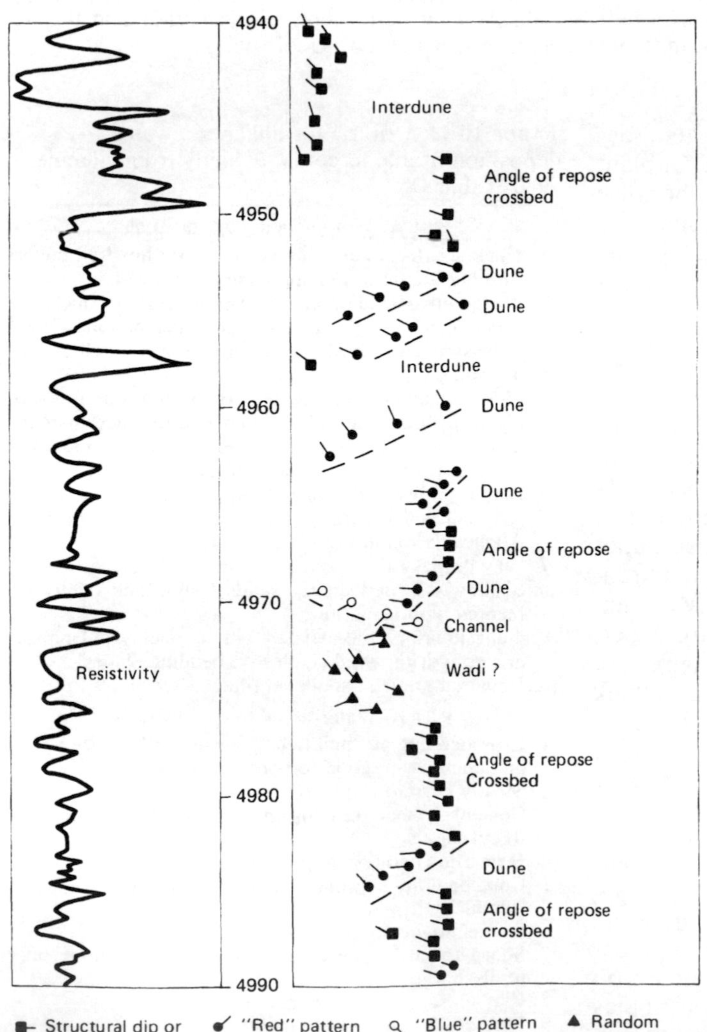

FIGURE 10-62. Rotliegendes Sandstone (Permian) of eolian origin in Well 49/26-2 of Leman Gas Field, North Sea. (Redrawn from van Veen, 1975, Fig. 8). Compare dipmeter pattern with Figs. 10-9 and 10-24.

Ancient equivalents of all of these are present in the geologic record and are chiefly recognized by their lithologic associations and sedimentary structures. Ancient eolianites tend to be mineralogically and texturally mature. Ancient continental eolianites tend to develop in areas of centripal drainage and have interbedded alluvial and lacustrine or sabkha deposits. Profuse crossbedding, some in thick sets and some in thin truncated sets, adhesion ripples, and good sorting are characteristic of all such sandstones. Recognition of ancient eolianites requires unusually careful observation and detail, and possibly many more exist in the geologic record, especially on a small scale, than are reported in the current literature.

References

Bigarella, J.J.: Eolian environments: Their characteristics, recognition, and importance. In: Rigby, J.K., and Hamblin, W.K. (Eds.): Recognition of ancient sedimentary environments. Soc. Econ. Paleon. Mineral. Spec. Pub. 16, 12–62 (1972).

Bigarella, J.J.: Structures developed by dissipation of dune and beach ridge deposits. Cantena 2, 107–152 (1975).

Bigarella, J.J., and Salamuni, R.: Early Mesozoic wind patterns as suggested by dune bedding in the Botucatu Sandstone of Brazil and Uruguay. Geol. Soc. America Bull. 72, 1089–1106 (1961).

Bigarella, J.J.; Becker, R.D.; and Duarte, G.M.: Coastal dune structures from Paraná (Brazil). Marine Geology 7, 5–55 (1969).

Brookfield, M.E., and Ahlbrandt, T.S. (Eds.): Eolian sediments and processes. (Developments in Sedimentology, Vol. 38). Amsterdam: Elsevier Scientific Pub. Co. 1983.

Carr, M.H.: The surface of Mars, 232 p. New Haven: Yale University Press, 1981.

Cooke, R.U., and Warren, A.: Geomorphology of deserts, 374 pp. Berkeley and Los Angeles: Univ. California Press 1973.

Glennie, K.W.: Desert sedimentary environments, 222 pp. (Developments in Sedimentology, Vol. 14). Amsterdam–London–New York: Elsevier Scientific Pub. Co. 1970.

Glennie, K.W.: Permian Rotliegendes of Northwest Europe interpreted in the light of modern desert sedimentation studies. Amer. Assoc. Petrol. Geologists, Bull. 56, 1048–1071 (1972).

Gradziński, R., and Jerzkiewicz, T.: Results of the Polish–Mongolian palaeontological expeditions—Part V. Sedimentation of the Barun Goyot Formation. Palaeontologia Polonica No. 30, 111–146 (1974).

Hunter, R.E.: Basic types of stratification in small eolian dunes. Sedimentology 24, 361–387 (1977).

Knight, S.H.: The Fountain and Casper Formations of the Laramie Basin—A study of the genesis of sediments. Wyoming Univ. Spec. Pub. Geol. 1, 82 pp. (1929).

Klemmensen Lars B.: Desert sand plain and sabkha deposits from the Bunter Sandstone Formation (L. Triassic) at the nothern margin of the German Basin. Geol. Rundschau 74, 519–536 (1985).

Mader, D.: Aeolian sands terminating an evolution of fluvial depositional environment in Middle Buntersandstein (Lower Triassic) of the Eifel, Federal Republic of Germany. In: Brookfield, M.E., and Ahlbrandt, T.S. (Eds.): Eolian sediments and processes, pp. 583–612. (Developments in Sedimentology, Vol. 38). Amsterdam: Elsevier Scientific Pub. Co. 1983.

McGinnies, W.G.; Goldman, B.J.; and Payne, P.: Deserts of the world, 788 pp. Tucson: Univ. Arizona Press 1968.

McKee, E.D.: The Coconino Sandstone—Its history and origin. Carnegie Inst. Washington Pub. 440, 77–115 (1934).

McKee, E.D. (Ed.): A study of global sand seas. U.S. Geol. Survey Prof. Paper 1052, 429 pp. (1979).

Mutch, T.A.; Arvidson, R.E.; Head, J.W., III; Jones, K.L.; and Saunders, R.S.: Geology of Mars, 400 pp. Princeton, New Jersey: Princeton Univ. Press 1976.

Poole, F.G.: Paleowinds in the western United States. In: Narin, E.M. (Ed.): Problems in paleoclimatology, pp. 394–405. NATO Palaeoclimates Conf., Univ. Newcastle-upon-Tyne and Durham, England, 1963, Proc. New York: Interscience 1964.

Sanderson, I.D.: Sedimentary structures and their environmental significance in the Navajo Sandstone, San Rafael Swell: Utah: Brigham Young University Studies 21, 215–246 (1974).

Shotten, F.W.: The Lower Bunter Sandstones of north Worcestershire and east Shropshire (England). Geol. Mag. 74, 534–553 (1937).

Shotten, F.W.: Some aspects of the New Red desert in Britain. Liverpool and Manchester Geol. Jour. 1, 450–465 (1956).

Soares, P.C.: Divisão estratigráfica do Mesozóico no Estado de São Paulo. Rev. Brasileira Geosciências 5, 229–251 (1975).

Solle, G.: Rezente und fossile Wüste. Notizbl. Hess. Landesamtes Bodenforsch 94, 54–121 (1966).

Tsoar, Haim: Internal structure and surface geometry of longitudinal (sief) dunes. Jour. Sed. Petrology 52, 823–831 (1982).

van Veen, F.R.: Geology of the Leman Gas-Field. In: Woodland, A.W. (Ed.): Petroleum and the continental shelf of northwest Europe, Vol. 10. pp. 223–231. New York: John Wiley and Sons 1975.

Sandy Reservoirs

One of the most important economic aspects of sandbodies is their capacity to store and transmit water, oil, gas, and mineralizing solutions that deposit copper, lead, zinc, uranium, and

Specific Permeability, K.(Darcys)

10^5 10^4 10^3 10^2 10 1 10^{-1} 10^{-2} 10^{-3} 10^{-4} 10^{-5}

Material	Clean gravel	Clean sands; mixtures of clean sands and gravels	Very fine sands; silts; mixtures of sand, silt, and clay; glacial till; stratified clays; etc.	Unweathered clays
Flow characteristics	Good aquifers		Poor aquifers	Impervious

FIGURE 10-63. Magnitude of specific permeability, k, of gravel, sand, silt, and clay. (Modified from Todd, 1960, Fig. 34).

other metallics. Good reviews of sandstone reservoirs are provided by LeBlanc (1977), Taylor (1977), and Galloway and Hobday (1983). A classic early study of a sandstone reservoir is that of Hewitt and Morgan (1965).

The amount and distribution of porosity and permeability are key controls in both fluid-filled reservoirs and mineralized sandstones. Variation in permeability is enormous and can range from hundreds of darcys in well-washed, unconsolidated terrace and river sands to virtually zero in a sandstone fully cemented by quartz, the quartz-cemented sandstone having a permeability comparable to that of some silts and clays (Fig. 10-63). Thus, the range of variation

TABLE 10-12. Rock properties and flow response.

Rock Property	Effects on Permeability and Porosity
TEXTURE	
Grain Size	Permeability decreases with grain size; porosity unchanged.
Sorting	Permeability and porosity decrease as sorting becomes poorer.
Packing	Tighter packing favors both lesser permeability and porosity.
Fabric	Permeability is maximum parallel to mean shape fabric.
Cement	The more cement, the less permeability and porosity.
SEDIMENTARY STRUCTURES	
Parting lineation	Maximum permeability generally parallels fabric in plane of bedding.
Crossbedding	Horizontal permeability parallels direction of inclination; the steeper the dip of the foreset, the weaker the horizonal vector of permeability.
Ripple marks	Little data, but fine grain and more laminations combine to cause low permeability and hence ripple zones are commonly barriers to flow.
Grooves and flutes	As judged by fabric, permeability parallels long dimension.
Slump structures	No data, but probably always greatly reduce horizontal permeability.
Biogenic structures	Destroy depositional fabric and bedding and thus drastically reduce permeability and reduce horizontal anisotropy of permeability. Effect on porosity is unknown, but may be negligible.
LITHOLOGY	
Sandstone	Thicker beds tend to be coarser-grained and thus more permeable than finer ones. If weakly cemented or uncemented, ratio of maximum to minimum permeability is perhaps less than 5 to 1; if cement controlled, ratio may reach 100 to 1 or more.
Shale	The prime barrier to flow that outshadows all others by far. Thus it is the arrangement of sand and shale much more than permeability variation within the sand that controls flow in most reservoirs.

can be of the order of 100,000 times. Moreover, small-scale spatial variation is also great, especially in cemented sandstones, so that the permeability of a sandstone is one of its most variable properties, its variability far exceeding that of porosity. This variability has made permeability difficult to predict.

Transmission of fluid in a sandstone depends on a combination of features formed by both primary and secondary processes. Fabric and orientation of the framework grains, sedimentary structures, and the bedding facies of a sand, and to some extent its cementation, are all linked to primary deposition. The controls over permeability form a hierarchy (Table 10-12): texture and fabric are the fundamental building blocks of sedimentary structures and individual beds, which together define a bedding facies. One can, in turn, consider a sandbody as nothing more than a particular arrangement of different bedding facies. Superimpose upon these are secondary patterns of cementation, decementation (secondary porosity), fracture systems, and faulting, all of which may locally transcend the importance of the primary controls on fluid flow.

A sandstone reservoir can be viewed as a simple tank, as a layered tank, or as a tank with both complex geometry and internal barriers to flow (Fig. 10-64). As we have seen, depositional environment determines geometry, vertical sequence, sedimentary structures, texture, and kinds and proportions of interbedded nonsandy strata. Impermeable interbedded strata—chiefly shales—are also major barriers to flow within a sandstone and thus need careful consideration. Variations in porosity, permeability, water saturation, and pore throat diameter of a sandstone are all linked to depositional environment as well (Fig. 10-65), and on a smaller scale, individual sedimentary structures (Fig. 10-66) and individual laminae (Fig. 10-67) affect permeability. See Roach and Thompson (1959, Figs. 61 and 62) for an example of localization

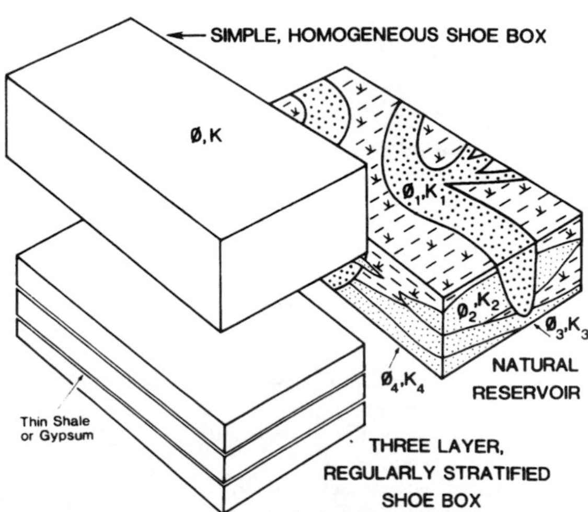

FIGURE 10-64. Simple tank, stacked tank, and natural reservoir.

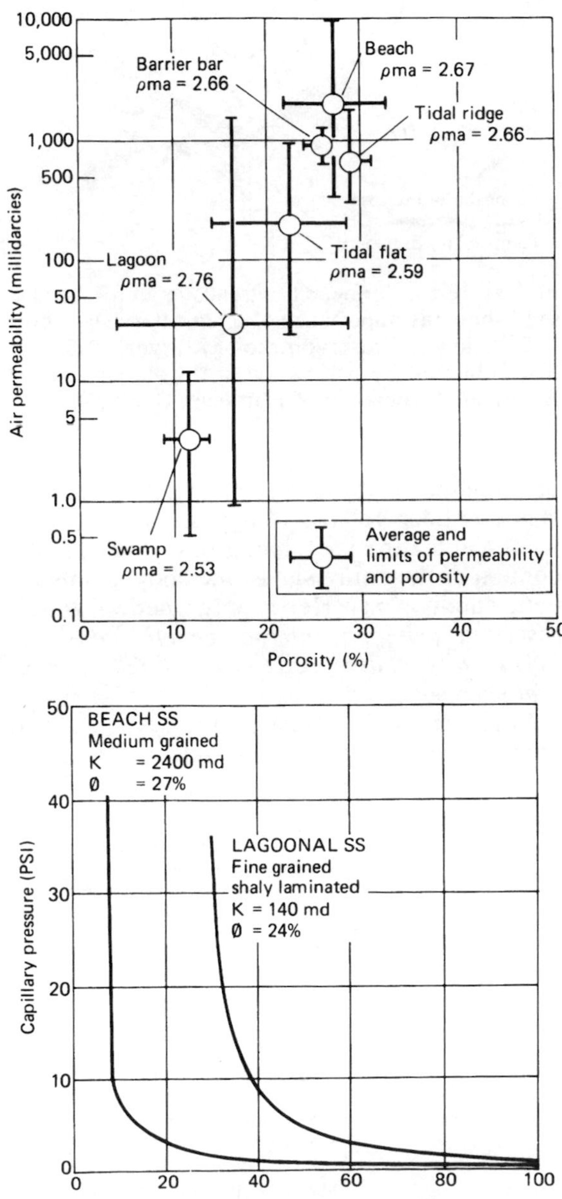

FIGURE 10-65. Porosity, permeability, and pore throat radius and their relation to depositional environment (Keelan, 1982, Figs. 4, 8, and 9). Published by permission of the author and the Society of Petroleum Engineers.

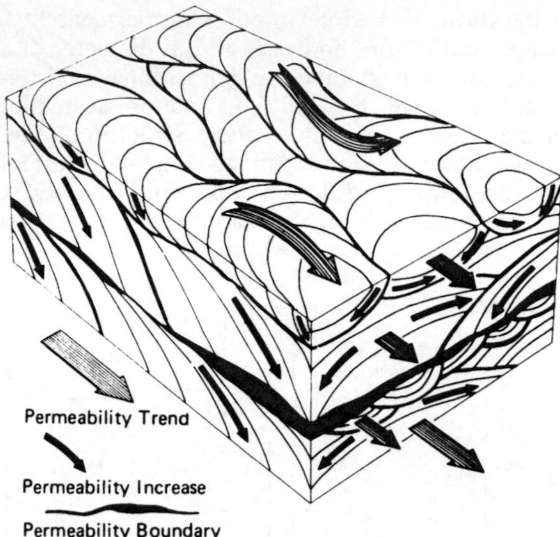

FIGURE 10-66. Permeability trends in a crossbedded sand show the importance of sedimentary structures to fluid flow and reservoir geology (Pryor, 1973, Fig. 15). Published by permission of the author and the American Association of Petroleum Geologists.

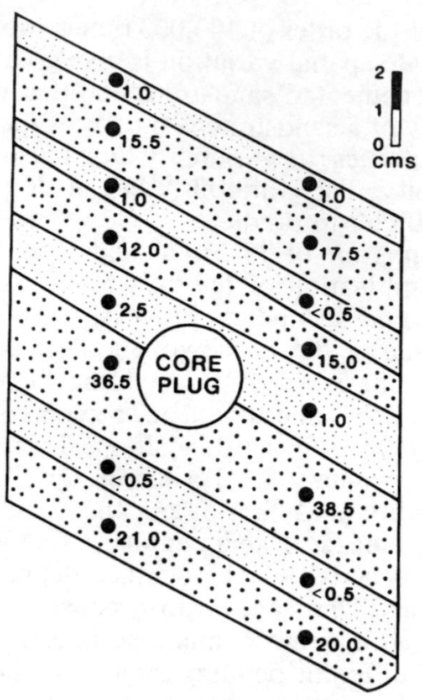

FIGURE 10-67. Small scale—laminae by laminae—variation of permeability and plug permeability in core from the eolian Permian Rotliegendes Sandstone of the North Sea. (Redrawn from van Veen, 1975, Fig. 9). Spot permeabilities measured by micropermeameter.

of uranium ore by crossbedding. Cementation also has been related to depositional environment and texture so it, too, is best studied in the context of the entire sandstone body and its genetic subdivisions. Hence only when we zone a reservoir *using its internal genetic units can best results be achieved for both fluid and mineral recovery.*

In all reservoir problems, the first step is to identify those rock properties relevant to porosity, permeability, production history and, for petroleum, possible formation damage. Below we examine some of the major problems.

Distinction between primary and secondary controls is essential (Table 10-13).

A fundamental, primary property is geome-

TABLE 10-13. Primary and secondary factors affecting quality of sandstone reservoirs.

Primary	Secondary
Provenance: Mineral composition of framework affects cements and clay mineralogy.	Framework mineralogy: Affects both mechanical strength and resistance to solution to framework grains and thus influences cement and primary and secondary porosity.
Primary depositional environment: Transport controls sizing, sorting, grain packing, and sedimentary structures plus bed thickness, geometry, and continuity as well as amount and type of bioturbation, all of which vary between depositional environments	Cementation: Depends on depositional environment and texture, kinds of framework grains, and burial history.
	Compaction: Depends on cement types and their timing.
	Fractures: Brittleness of host rock and tectonics.
	Burial history: Factors such as depth, duration of burial, temperature, and deformation, any one of which can be overriding.

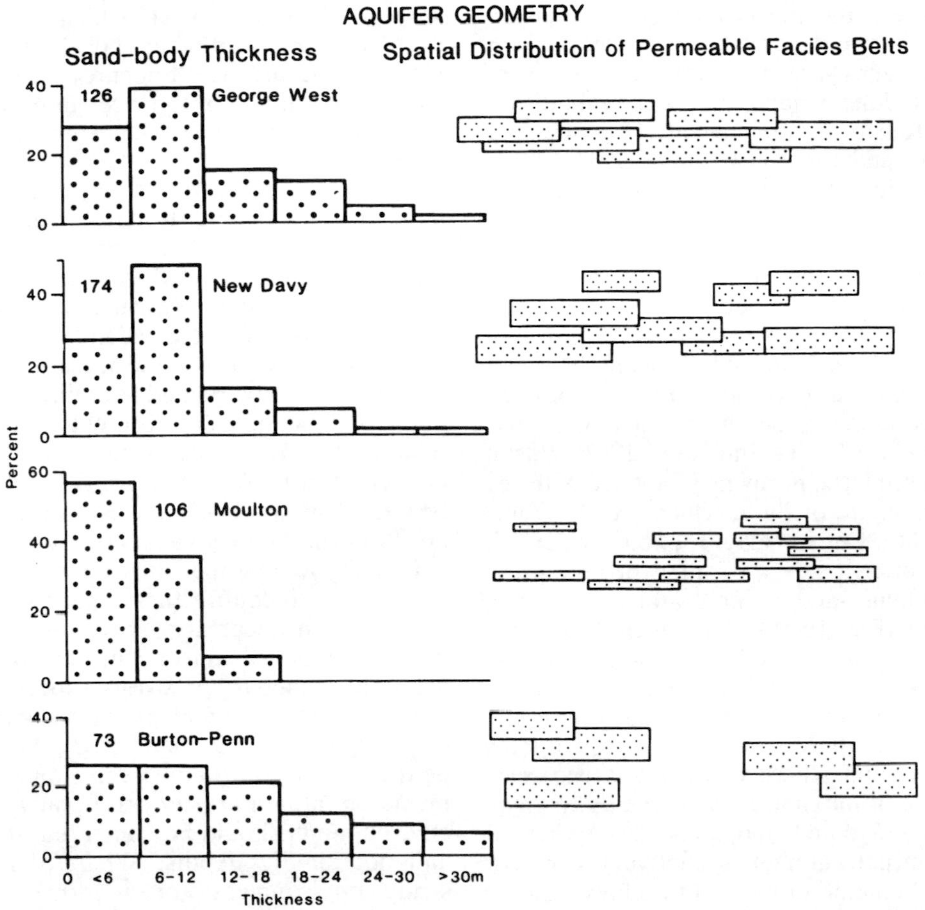

FIGURE 10-68. Statistical summary of sandbody thickness and diagrammatic spatial distribution of permeable facies belts of Miocene Oakville Sandstone of Texas Coastal Plain. (Redrawn from Galloway *et al.*, 1982, Fig. 21). Numbers indicate number of sandstone bodies.

try. Careful attention to the external and internal structure of individual sandstone bodies is always needed—their types, size, shape, and orientation. The reviews by LeBlanc (1977) and Sneider and others (1977) and the study by Bloomer (1977) all emphasize sandbody geometry as the first-order control of reservoir behavior, and statistical summaries based on careful subsurface mapping convey much information (Fig. 10-68). On the other hand, a common problem is the widespread, blanket sandstone or a thick channel that in reality consists of discrete reservoirs, each with independent production histories. A number of geologic factors may cause this.

First, thin, widespread, impermeable shale, evaporite, or dolomite beds may be responsible for the isolation of individual reservoirs in an otherwise continuous sandstone body. There has been some study of the continuity and shape of thin, shale beds in sandstones and how they are related to depositional environment (Richardson *et al.*, 1978; Weber, 1982, Fig. 2). Thin shales in marine sandstones are the most continuous—the "mud blanket" of a subsea fan is a good example as are the thin, widespread shales found in many shelf sequences. Both of these marine shales are better barriers to vertical flow and are more mappable than are the thin, discontinuous shales of fluvial deposits. More data on the characteristics of interbedded shales would help engineers better formulate reservoir models for the different sandy environments. Or a blanket sandstone may consist of a widespread, multistory, meandering or braided deposit with poor communication between each genetic sandy unit (Figs. 10-16 and 10-18) or be a mixture of eolian and fluvial environments (Fig. 3-19). The sandy turbidites of a mid- or upper subsea fan may consist of similar multistory reservoirs. The eolian Nugget Formation in Wyoming provides an example where the permeabilities of sandstone formed in its two major subenvironments—dune and interdune—are in great contrast; sandstones of dune origin have permeabilities of three to four orders of magnitude greater than interdune sandstones (Lindquist, 1983).

The distribution and kinds of clay minerals in a reservoir can have a dramatic effect on its performance and, today, thanks largely to near-routine study by the SEM, thin-section petrography, and mercury injection, we know much more about their role (Gaida *et al.*, 1973; Stalder, 1973; Wilson and Pittman, 1977); for example, clays are considered the major cause of low permeability in many tight gas sands (Walls, 1982; Gautier and Rice, 1982; Wilson, 1982), where large pores may be filled with secondary kaolinite or illite, which greatly reduce permeability. Detrital clays that reduce permeability occur as coats on framework grains, as burrow fillings and linings, and as dispersed clay flakes (Fig. 10-69). Once identified as primary, these clays should be related to depositional environment and mapped. In a muddy sea, best permeability should occur where sorting was best, where winnowing removed interstitial clay, and where bioturbation was minimal. A site of maximum wave energy, perhaps where the sea joined an open ocean or where an incipient structural high was active, favors winnowing. On the other hand, if the clays reducing the permeability of a sandstone are mostly authigenic, perhaps because they were derived from unstable framework grains such as plagioclase, volcanic rock fragments, or glass, better permeabilities should be sought where provenance was different or perhaps burial was shallower. Distinguishing detrital from authigenic clays is the needed first step toward improved recovery and exploration in sandstones where permeability is dependent on clay minerals.

Authigenic clays reduce the open space in pores, increase microporosity and ratio of surface area to volume, and increase the physical resistance to fluid flow as crystals grow outward from pore walls, especially the long, fiberous, illitic crystals. The abundance and kinds of authigenic clays, like the primary clays, should always be related to the subenvironments of a sandstone body.

Whether primary or secondary, clays can also react with introduced fluids. Iron-rich chlorites dissolve with acid, and when pH returns to normal, large ferric oxide crystals precipitate and can block pores. In addition, "illitic whiskers" can aggregate together in fresh water. Mixed-layer clays and others swell and, if pressure gradients become too high, actual transport and sedimentation of clays in the pore system can occur. See Laughrey (1984) for an excellent example of the roles of clay minerals in the Silurian Clinton Sandstone of Pennsylvania.

Differential cementation may occur on both a large and small scale. It may be identifiable on geophysical logs so that zones of variable cementation can be traced throughout the reservoir, as was done by Morgan *et al.* (1977) in the Pennsylvanian Tensleep Sandstone in Wyoming. Or it may vary on the scale of centimeters and meters and be related to grain size and sorting. Good examples of differential cementation related to different depositional environments are reported by Van Veen (1975) in the North Sea and Taylor (1978) and Kasino and Davies (1979) in the Pennsylvanian of Oklahoma.

Porosity and permeability decrease, as a general rule, with depth either exponentially or linearly; it can be preserved by the early introduction of petroleum or can develop later by solution (secondary porosity), especially of carbonate cements or feldspar and some types of rock fragments (see Chaps. 3 and 11). Because of the initial control of depositional environments on initial porosity and permeability, like environments should be compared, if possible, and not simply depths. Where the different sandy environments of a sequence extend through significantly different depths, basin porosity, permeability, and diagenesis will vary in the same environment, as has been shown by Selley (1978) in the North Sea.

Once controlling rock properties are recognized, they should be identified on logs and mapped throughout the reservoir (Table 10-14). To do so, one or two cores from the reservoir are commonly needed to "calibrate" fully a wireline log with its rocks. Keelan (1982) discusses and illustrates most of the techniques needed to characterize cores.

The foregoing emphasizes reservoir zonation

TABLE 10-14. Strategy for study of sandy reservoirs.

IDENTIFY rock properties relevant to variations of porosity, permeability, formation damage, and production history of contained fluids or mineralization of reservoir.

SELECT those related to logs or seismic.

MAP and ZONE the reservoir (wireline logs and cores) adding needed structure (closure, fractures, and faults).

COMPARE initial maps with production history of each well to confirm or modify barriers to flow.

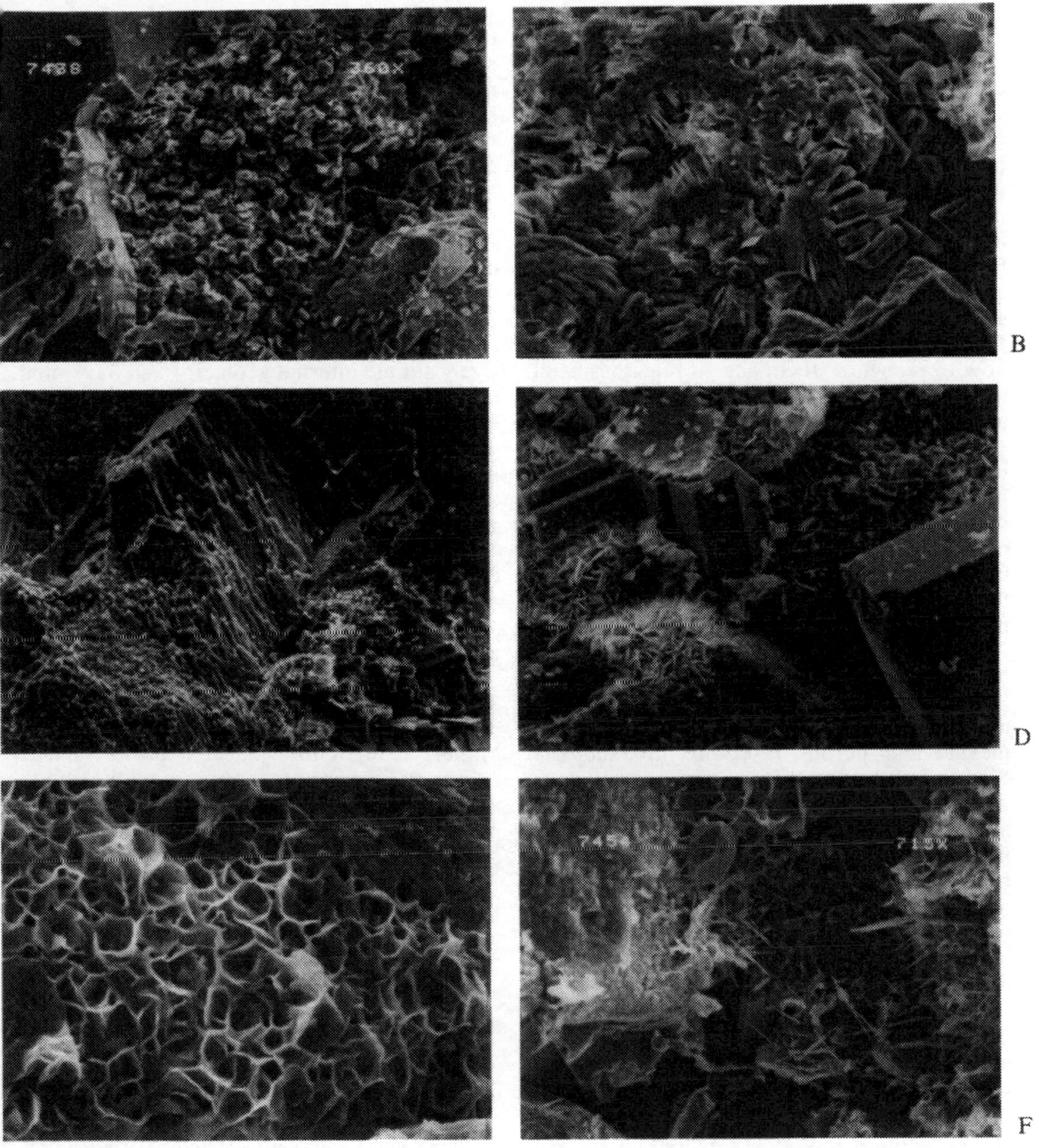

FIGURE 10-69. SEM photographs of clays, quartz, feldspar, and anhydrite, which form some of the authigenic minerals of sandstones: A) Many kaolinite books and minor fibers of illite almost completely occlude a large pore in the Travis Peak Fm. (Cretaceous) of Texas; B) Enlarged view of kaolinite and minor quartz. Isolated books of kaolinite are also common in many pores; C) Dissolution of feldspar creates secondary porosity in many sandstones as in the Norphlet Fm. (Cretaceous of Alabama; D) Chlorite and anhydrite in the Itararé Fm. (Permian of the Paraná Basin of Brazil; E) Smectite from the Frio Fm. (Tertary) of the Texas Gulf Coast (2668 m); and F) Fiberous illite and euhedral quartz cement in the Travis Peak Fm. of Texas. Photographs A, B, C, and F courtesy of R. D. Bray and Jane Collamer NL ERCO Industries, Houston Texas; photograph D courtesy of A. Saad of Instituto Pesquisas Tecnológias, Sao Paulo, Brazil; and Photograph E courtesy of William L. Harmon of Core Laboratories, Dallas Texas.

by recognition of its genetic units, a procedure we strongly recommend. Statistical zonation, however, is also useful and should always be considered as a secondary alternative (Alpay, 1972; Haralick and Shanmugam, 1973). Complementary, and usually following statistical zonation, is reservoir modeling wherein different genetic units are each assigned an average permeability and porosity and estimates of crossflow at boundaries are made. In addition, factors such as grain size and cementation, geometry and spacing of internal barriers, and sedimentary structures may also be included. Fractured reservoirs pose special problems—how to recognize them and to separate the effects of fracturing from those related to original deposition.

In all such studies close collaboration between engineers and geologists is needed for both successful field studies and for reservoir modeling (Harris and Hewitt, 1977). Examples of modeling include the computation of directional permeability from concave crossbedding (Weber, 1982) and the simulation of water flooding in a coastal barrier (Weber *et al.*, 1975). The general paper by Harris (1975) gives other examples.

Some sandstones contain secondary mineral deposits where impermeable seals divide a permeable and porous reservoir into subparts. Shales and anhydrites are the most common low-permeability barriers that inhibit flow by mineralizing fluids. The frequency and continuity of such barriers vary with depositional environments (Weber, 1982) and *thus exploration for stratiform ore deposits in sandstones begins with the study of depositional systems.* Facies-controlled, secondary uranium mineralization is described by Rackley (1976), DeVoto (1978, pp. 148–219), and Galloway and Hobday (1983, Chap. 13). The detailed tabulation of rock type and its inferred depositional environment and content of U_3O_8 in the Triassic Lower Dockum of the Texas Panhandle by McGowen *et al.* (1979, Table 2) is a good example of this approach. Copper mineralization in sandstones is fairly common, and occurrences are described by Garlick (1972), Smith (1974), and Mann (1979). Bjöerlykke and Sangster (1981, pp. 180–86) report on lead mineralization in a widespread Cambrian quartz arenite above an unconformity cut on the Precambrian rocks of the Baltic Shield and summarize other occurrences of galena-rich sandstones. Willdén (1980) describes the paleoenvironment of this Cambrian sandstone at a mine and how it relates to ore grade. Lead and zinc mineralization associated with volcanic basins are facies related and also depend on the distance from the volcanic vent, so that proximal and distal ore facies are recognized; in the distal facies, sediments become more abundant than volcanics, and turbidites with detrital lead and zinc occur (Plimer 1978, pp. 349–50).

Reservoir characteristics also influence our exploration for groundwater as is well shown by Jobin (1962), who contrasted the transmissive characteristics of eolian and marine sandstones with sandstones of fluvial origin in the Colorado Plateau. He noted that eolian and marine sandstones occur as blankets and wedges with relatively simple internal geometry and have moderate to high permeability. In contrast, fluvial sandstones are multistory and internally more complex and have low to moderate permeability. See Hall and Frank (1975) for another example of use of depositional systems in groundwater exploration.

Holocombe (1980) shows examples of paleoflow modeling of sedimentery ore bodies. In such models geometry, internal barriers to flow and the permeabilities of different subfacies are all important.

Other mineral deposits are detrital. These include gold and uranium placers, especially in South Africa (Pretorius, 1966, 1976); Smith and Minter, 1980). Other placer deposits such as cassiterite, diamonds, garnet, magnetite, monazite, illmenite, platinum, and zircon are summarized by Hails (1976).

References

Alpay, O.A.: A practical approach to defining reservoir heterogeneity. Jour. Petrol. Tech. 24, 841–848 (1972).

Baker, V.R.: Paleohydrology and sedimentology of Lake Missoula flooding in eastern Washington. Geol. Soc. America Spec. Pub. 144, 79 pp. (1973).

Bjöerlykke, A., and Sangster, D.F.: An overview of sandstone lead deposits and their relationship to red-bed copper and carbonate-hosted lead-zinc deposits. In: Skinner, B.J. (Ed.): Economic geology: Seventy-fifth anniversary, pp. 179–213. Econ. Geol. Pub. Co. 1981.

Bloomer, R.R.: Depositional environments of a reservoir sandstone in west Texas. Am. Assoc. Petroleum Geologists Bull. 61, 344–359 (1977).

DeVoto, R. H.: Uranium geology and exploration: Lectures, notes and references, 396, pp. Golden, Colorado: R. H. DeVoto 1978.

Gaida, K.H.; Ruhl, W.; and Zimmerle, W.: Rasterelektronenmikroskopische Untersuchungen des Porenraumes von Sandsteinen. Erdöl-Erdgas Zeitschr. 89, 331–343 (1973).

Galloway, W.E., and Hobday, D.K.: Terrigeneous clastic depositional systems, 423 pp. New York–Berlin–Heidelberg–Tokyo: Springer-Verlag 1983.

Galloway, W.E.; Kreiter, C.W.; and McGowen, J.H.: Depositional and groundwater flow systems in the exploration for uranium: A research colloquium, Sept. 8–9, 1978, 267 pp. Austin, Texas: Texas Bur. Econ. Geol. 1979.

Galloway, W.E.; Henry, C.D.; and Smith G.E.: Depositional framework, hydrostratigraphy and uranium mineralization of the Oakville Sandstone (Miocene), Texas Coastal Plain. Texas Bur. Econ. Geol., Rept. Invs. 113, 51 pp. (1982).

Garlick, W.G.: Sedimentary environment of Zambian copper deposition. Geol. Mijnbouw 51, 277–298 (1972).

Gautier, D.L., and Rice, D.D.: Conventional and low-permeability reservoirs shallow gas in the northern Great Plains. Jour. Petrol. Tech. 34, 2600–2608 (1982).

Hails, J.R.: Placer deposits. In: Wolf, K.H. (Ed.): Handbook of stratabound and stratiform ore deposits, Vol. 3, Principles and general studies, pp. 213–244. Amsterdam–Oxford–New York: Elsevier Scientific Pub. Co. 1976.

Hall, W.D., and Frank, L.J.: Aquifer evaluation using depositional systems: An example in north central Texas. Groundwater 13, 472–483 (1975).

Haralick, R.M., and Shanmugam K.: Computer classification of reservoir sandstones. Trans. Geoscience Electronic 11, 171–177 (1973).

Harris, D.G.: The role of geology in reservoir simulation studies. Jour. Petrol. Tech. 27, 625–632 (1975).

Harris, D.G., and Hewitt, C.H.: Synergism in reservoir management—the geological perspective. Jour. Petrol. Tech. 29, 761–770 (1977).

Hewitt, C., and Morgan, J.T.: The Fry *in situ* combustion test—reservoir characteristics. Jour. Petrol. Techol. 17, 337–353 (1965).

Holocombe, Colin J.: Paleoflow modeling for sedimentary ore bodies. Econ. Geol. 80, 172–179 (1980).

International Atomic Energy Agency: Formation of uranium ore deposits, proceedings of a symposium on the formation of uranium ore deposits, Athens. Vienna: Intern. Atomic Energy Agency, 748 pp. 1974.

Jobin, D.A.: Relation of the transmissive character of the sedimentary rocks of the Colorado Plateau to the distribution of uranium deposits. U.S. Geol. Survey Bull. 1124, 151 pp. (1962).

Kasino, R.E., and Davies, D.K.: Environments and diagenesis, Morrow Sands, Cimarron County (Oklahoma), and significance to regional exploration, production and well completion practices. In: Hyne, N.J. (Ed.): Pennsylvanian sandstones of the Mid-Continent, Sp. Pub 1, pp. 169–194. Tulsa, Oklahoma: Tulsa Geol. Soc. 1979.

Keelan, D.K.: Core analysis for aid in reservoir description. Jour. Petrol. Tech. 34, 2483–2491 (1982).

Laughrey, C.D.: Petrology and reservoir characteristics of the Lower Silurian Medina Group sandstones, Athens and Geneva Fields, Crawford County, Pennsylvania: Penn. Geol. Survey, 4th Ser. Min. Res. Rept. 85, 126 p. (1984).

LeBlanc, R.J., Sr.: Distribution and continuity of sandstone reservoir—Parts 1 and 2. Jour. Petrol. Tech. 29, 776–804 (1977).

Lindquist, S.J.: Nugget Formation reservoir characteristics affecting production in the Overthrust Belt of southwestern Wyoming. Jour. Petrol. Tech. 35, 1355–1365 (1983).

Mann, A.G.: Sedimentological control of copper deposition in the Sijarira Group, western Umniati River area, Rhodesia. In: Anhaeusser, C.R.; Foster, R.P.; and Stratten, T. (Eds.): A symposium on mineral deposits and the transportation and deposition of metals. Geol. Soc. South Africa Spec. Pub. 5, 219–227 (1979).

Mast, R.F., and Potter P.E.: Sedimentary structures, sand shape fabrics, and permeability. II. Jour. Geology 71, 548–565 (1963).

Maynard, J. Barry: Geochemistry of sedimentary ores, 305 pp. Heidelberg–Berlin–New York: Springer-Verlag 1983.

McGowen, J.H.; Granata, G.E.; and Seni, S.J.: Depositional framework of the Lower Dockum Group (Triassic) Texas Panhandle: Texas Bur. Econ. Geol., Rept. Invs. 97 60 pp. (1979).

Minter, W.E.L.: A sedimentological synthesis of placer gold, uranium and pyrite concentrations in Proterozoic Witwatersrand sediments. In: Miall, A.D. (Ed.): Fluvial sedimentology. Canadian Soc. Petroleum Geologists Mem. 5, 801–829 (1978).

Morgan, J.T.; Cordiner, F.S.; and Livingston, A.R.: Tensleep reservoir study, Oregon Basin Field, Wyoming: Reservoir characteristics. Jour. Petrol. Tech. 29, 886–902 (1977).

Plimer, I.R.: Proximal and distal stratabound ore deposits. Mineral Deposita 13, 345–353 (1978).

Pretorius, D.A.: Conceptual geological models in the exploration for gold mineralization in the Witwatersrand Basin. Univ. Witwatersrand, Johannesburg, Econ. Geol. Res. Unit, Inf. Circ. 36, 39 pp. (1966).

Pretorius, D.A.: The nature of the Witwatersrand gold–uranium deposits. In: K.H. Wolf (Ed.): Handbook of strata-bound and stratiform ore deposits, pp. 29–88. Amsterdam: Elsevier Scientific Pub. Co. 1976.

Pryor, W.A.: Permeability–porosity patterns and variation in some Holocene sand bodies. Am. Assoc. Petroleum Geologists Bull. 57, 162–189 (1973).

Rackley, R.I.: Origin of western-states type uranium mineralization. In: Wolf, K.H. (Ed.): Handbook of strata-bound and stratiform ore deposits 7. Principles and general studies, pp. 89–156. Amsterdam–Oxford–New York: Elsevier Sci. Pub. Co. 1976.

Richardson, J.G., Harris, D.G., Rossen R.H., and van Hee, G.: The effect of small discontinuous shales on oil recovery. Jour. Petrol. Tech. (Nov.), pp. 1531–1537 (1978).

Roach, C.H., and Thompson, M.E.: Sedimentary structures and localization and oxidation of ore at the Peanut Mine, Montrose County, Colorado. In: Garrells, R.M., and Larsen, E.S. (Eds.): Geochemistry and mineralogy of Colorado Plateau uranium ores. U.S. Geological Survey Prof. Paper 320, 197–202 (1959).

Selley, R.C.: Porosity gradients in North Sea oil-bearing sandstones. Jour. Geol. Soc. (London)135, 119–132 (1978).

Smith, G.E.: Depositional systems, San Angelo Formation (Permian), North Texas—facies control of red-bed copper mineralization. Texas Bur. Econ. Geol., Rept. Invs. 80, 73 pp. (1974).

Smith, N.D., and Minter, W.E.L.: Sedimentological controls of gold and uranium in two Witwatersrand paleoplacers. Econ. Geology 75, 1–14 (1980).

Sneider, R.M., and others.: Prediction of reservoir rock geometry and continuity in Pennsylvanian reservoirs, Elk city Field, Oklahoma. Jour. Petrol. Tech. 29, 851–866 (1977).

Stalder, P.U.: Influence of crystallographic habit and aggregate structure of authigenic clay minerals on sandstone permeability. Geol. Mijnbouw 52, 217–220. (1973).

Taylor, J.C.M.: Sandstones as reservoir rocks. In: Hobson, G.E. (Ed.): Developments in petroleum geology—1, pp. 147–196. London: Applied Science Publishers Ltd. 1977.

Taylor, J.C.M.: Control of diagenesis by depositional environment within a fluvial sandstone sequence in the northern North Sea Basin. Jour. Geol. Soc. (London) 135, 83–91 (1978).

Todd, D.K.: Groundwater hydrology, 336 pp. New York: John Wiley and Sons 1960.

Van Veen, F.R.: Geology of Leman Gas-Field. In: Woodland, A.W. (Ed.): Petroleum and the continental shelf of northwest Europe, Vol. 1, Geology, pp. 223–231. New York: John Wiley and Sons 1975.

Walls, Joel D.: Tight gas sands—permeability, pore structure, and clay. Jour. Petrol. Tech. 34, 2708–2714 (1982).

Weber, K.J.: Influence of common sedimentary structures on fluid flow in reservoir models. Jour. Petrol. Tech. 34, 665–672 (1982).

Weber, K.J.; Klootwijk, P.H.; Konieczek, J.; and Van der Vlught, W.R.: Simulation of water injection in a barrier-bar-type, oil reservoir in Nigeria. Jour. Petrol. Tech. 27, 1555–1565 (1975).

Willden, M.Y.: Paleoenvironment of the autochthonous sedimentary rock sequence at Laiswell, Swedish Caledonides. Acta Stockholmiensis, Stockholm Contr. Geol. 33, 100 pp. (1977).

Wilson, M.D.: Origins of clays controlling permeability in tight gas sands. Jour. Petrol. Tech. 34, 2871–2876 (1982).

Wilson, M.D., and Pittman, E.D.: Authigenic clays in sandstones: Recognition and influence on reservoir properties and paleoenvironmental analysis. Jour. Sed. Petrology 47, 3–31 (1977).

Epilogue; Unresolved Problems and Future Research

Every major sandy environment presents problems that deserve more attention, but most fundamental is simply a lack of systematic data for each environment. Simple as it may seem, what is needed, perhaps more than anything else, is a *systematic inventory* for each sandy environment of properties such as the mean and standard deviation of geometry, bedding thickness, grain size, kinds and proportions of sedimentary structures, thickness and continuity of shale or carbonate interbeds, and framework petrology, to mention but a few. For each such environment, these data should be integrated with its vertical profile. How much progress can be expected toward this goal by 1990? By 2000? In addition, there are many unresolved problems today as well as research that is just beginning, such as stratigraphic modeling. Below we comment on some of these problems.

Distinguishing between the types of fluvial cycles—those related to a depositional process and those related to an exterior cause such as tectonic uplift or climatic change in the source area—deserves consideration, especially in thick fluvial sections of tectonic basins (Miall, 1980). A specific example occurs in the Old Red Sandstone of Norway, where Steel *et al.* (1978) and Bryhni (1978) distinguished sedimentary fining-upward cycles a few meters thick from tectonic ones that are tens of meters thick and coarsen upward. Miall (1981) has explored the relation between alluvial deposits and tectonics, a problem that deserves more attention.

Another problem that involves alluvial deposits, especially in the middle Precambrian, is the nature of the depositional environments of its thick, widespread, mature to submature sandstones—are these braided stream deposits or the deposits of a current-swept or tidal shelf? Certainly such homogeneous and thick sandstones that lack fossils put to the test the physical criteria used by today's sedimentologists to identify ancient sedimentary environments. In all such deposits, measurement of paleocurrents and construction of a basin model are needed to support a depositional environment.

Coastal barriers that define a strand plain system offer at least three challenges to sedimentologists—how to recognize ancient coastal barriers when their upper part has been eroded, how to distinguish them from a wave-dominated delta, and how to distinguish them from coalesced, elongate shelf sandstones. Still another problem is why coastal barriers seem to be absent in some sandy basins—for example, where fluvial sandstones seemingly pass directly into marine shelf sandstones (Dott and Byers, 1982, p. 333). One possibility for their absence is a narrow ocean developing from an embryonic rift—was inshore wave power too small to generate significant coastal barriers? Or has the basin margin simply been everywhere eroded? Certainly a recurring problem in ancient basins is how to separate coastal barriers and inshore tidal sandstones from those of an open shelf. More studies of modern nearshore sands, such as those of Aigner and Reineck (1982), and, if possible, of modern shelf sands are certainly needed.

How many types of shelf sandstones are there and how well can we distinguish ancient current-dominated from wave- and tide-dominated sandy shelves? Perhaps later Tertiary or Pleistocene deposits along present shorelines provide good models, especially where they are well exposed, because it may be possible to relate them to the dynamics of their adjacent contemporary shelf. In all such studies careful attention to vertical sequence, stratification types, sedimentary structures, and body and trace fossils is necessary plus careful mapping of sandbody geometry and orientation. And here as always, relating individual environments to the depositional system is all important.

The recognition of tempestites—storm-surge turbidity deposits derived from local erosion of the bottom— on a shelf is another example of increasing discrimination that can be expected to generate new interpretations of what have been considered deep-water turbidites (Valecka, 1984). Such studies contribute to the interpretation of sparingly fossiliferous sandstone deposits, as do trace fossils, which are still understudied (Crimes, 1981). Position of a sandy facies in the basin—was it up dip and on a shelf, or was it down dip, distal, and probably deep marine?—always needs to be considered rather than just physical sedimentology alone. In addition, bed by bed study of paleocurrents can be informative to distinguish between a tempestite and turbidite—the former possibly having more randomly oriented paleocurrents than turbidites (Gray and Benton, 1982). Bed by bed study of paleocurrents is also needed to distinguish between fair-weather and storm deposits on an ancient shelf. Finally, there is the role of worldwide inventories of particular environments and their relation to either paleolatitude or tectonic setting. Marsaglia and Klein (1983) provide an example of this approach. They marshaled the literature of Phanerozoic storm deposits and related them to paleolatitude and preferred paleohurricane paths.

How are depositional systems, especially those formed in tectonically stable basins, related to worldwide sea level changes? The exciting studies of Shanmugam and Moiola (1982) which related subsea fans to low stands of sea level and Klein's (1982) analysis of depositional sequences on cratons suggest a fruitful field of research. Both provide a good example of how the study of sandy depositional systems will increasingly be related to "big picture" geology. See, for example, how depositional sequences are related to basin development and type by Kingston *et al.* (1983, p. 2175–77).

Modeling by computer of the stratigraphic variability of either a sandy depositional system, such as an alluvial valley fill, or the interval variation of structures and grain size of a single sandstone body is only beginning but already there is modest literature. Foremost are modeling studies of alluvial deposits. The objective of this research is to better understand those factors that control the interconnectedness of alluvial sandstone bodies—whether a cross section will show isolated, randomly spaced channel sandstones or fairly continuous sheets. The dependent variable is usually avulsion, the sudden switching of the main channel during flood, while causative variables include downstream slope, aggradation rate, vertical tectonic movements, and compaction of mud. Papers by Allen (1978), Price (1976), and Bridge and Leeder (1979) illustrate this effort, which is also applicable to the turbidite channels of subsea fans. Delta modeling requires correlating riverine input, inshore wave power, tidal range, and sea level. Examples are provided by Komar (1973) and Horowitz (1976). There are also a few computer studies that simulate the migration of coastal barriers (George and Hand, 1977).

Seismic mapping, however, seems more profitable and much more likely to expand. This effort is based on seismic wave form analysis (Fig. 10-13) and cross sectional slab models and will be increasingly used in subsurface geology as geophysicists and geologists combine their exploration efforts. The great advantage of seismic mapping of sandy depositional environments is its ability to extrapolate between wells. See Raffalovich and Daw (1984), and Focht and Baker (1985) for examples of seismic exploration for stratigraphic traps.

In final appraisal, what can be said of the overall ability of sedimentologists to correctly identify the environments of sandy deposits? Using the best of today's depositional models and a perceptive group of sedimentologists, what reasonable success can be expected for the correct environmental identification of sandy deposits? Would a success ratio of 60 percent be too optimistic? Or can we do much better? We don't know the answer, of course, but we do know that there are many sandy sections over which experienced sedimentologists disagree and interpretation is far from unanimous. This suggests that there are many ancient sandy environments for which today's depositional models are still incomplete or lacking. Are too many of today's modern sandy facies models based mostly in North America and Europe—too narrow an environmental spectrum and possibly different from much of the ancient past? Thus, the study of sandy depositional systems is anticipated to be a lively one for many years to come and the best results will always be obtained when they are fully integrated with the sedimentary basin itself and worldwide sea level changes.

References

Aigner, T., and Reineck, H.-E.: Proximity trends in northern North Sea storm sands from Helgoland Bight (North Sea) and their implications for basin analysis. Senckenbergiana Marit. 14, 183–215 (1982).

Allen, J.R.L.: Studies in fluviatile sedimentation: An exploratory model for the architecture of avulsion-controlled alluvial suites. Sedimentary Geology 21, 129–147 (1978).

Bridge, J.S., and Leeder, M.R.: A simulation model for alluvial stratigraphy. Sedimentology 26, 617–644 (1979).

Bryhni, Inge: Flood deposits in the Hornelen Basin, west Norway (Old Red Sandstone). Norsk. geol. Tidsskr. 58, 273–300 (1978).

Crimes, T.P.: Lower Paleozoic trace fossils of Africa. In: Holland, C.H. (Ed.): Lower Paleozoic of the Middle East, Eastern and southern Africa, pp. 189–198. (Lower Paleozoic rocks of the world, Vol. 3). Chichester, England: John Wiley and Sons 1981.

Dott, R.H., Jr., and Byers, C.W.: SEPM research conference on modern shelf and ancient cratonic sedimentation—the orthoquartzite–carbonate suite revisited, August 11–16, 1980. Jour. Sed. Petrology 51, 330–346 (1982).

Ethridge, Frank G., and Flores, Romero M. (Eds.): Recent and ancient non-marine depositional environments: Models for exploration. Soc. Econ. Paleon. Mineral Spec. Pub. 31, 349 pp. (1981).

Focht, G.W., and Baker, F.E.: Geophysical case history of the Two Hills Colony gas field of Alberta. Geophysics 50, 1061–1076 (1985).

George, D.J., and Hand, D.M.: Computer simulation of barrier island migration. Computers and Geoscience 3, 469–473 (1977).

Gray, D.I., and Benton, J.J.: Multidirectional paleocurrents as indicators of shelf storm beds. In: Einsele, G., and Seilacher, A. (Eds.): Cyclic and event stratification, pp. 350–356. Berlin–Heidelberg–New York: Springer-Verlag 1982.

Horowitz, D.H.: Mathematical modeling of sediment accumulations in prograding deltic systems, pp. 105–120. In: Merrian, D.F., Ed.: Quantitative techniques for the analysis of the IX International Sedimentological Congress, Nice, France (on 8 July 1975). Oxford Pergamon Press, 1876.

Horne, J.C. Ferm, J.C.; Caruccio, F.T.; and Baganz, B.P.: Depositional models in coal exploration and mine planning in Appalachian region. Am. Assoc. Petroleum Geologists Bull. 62, 2379–2411 (1978).

Kingston, D.R.; Dishroon, C.P.; and Williams, P.A.: Global basin classification system. Am. Assoc. Petroleum Geologists Bull. 67, 2175–2193 (1983).

Klein, George deVries: Probable sequential arrangement of depositional systems on cratons. Geology 10, 17–22 (1982).

Komar, Paul D.: Computer models of delta growth due to sediment input from rivers and longshore transport. Geol. Soc. America Bull. 84, p. 2217–2226 (1973).

Marsaglia, K.M., and Klein, G. deVries: The paleogeography of Paleozoic and Mesozoic storm depositional systems. Jour. Geology 91, 117–142 (1983).

Miall, A.D.: Cyclicity and the facies model concept. Bull. Canadian Petroleum, Geology 28, 59–80 (1980).

Miall, A.D.: Alluvial sedimentary basins: Tectonic setting and basin architecture. In: Miall, A.D. (Ed.): Sedimentation and tectonics in alluvial basins. Geol. Assoc. Canada Spec. Paper 23, 1–33 (1981).

Price, W.E., Jr.: A random-walk simulation model of alluvial-fan deposition. In: Merriam, D.F. (Ed.): Random processes in geology, pp. 52–62. Springer-Verlag: New York–Heidelberg–Berlin 1976.

Raffalovich, F.D., and Daw, T.B.: Use of seismic stratigraphy for Minnelusa exploration, northeastern Wyoming. Geophysics 49, 715–721 (1984).

Shanmugam, G., and Moiola, R.J.: Eustatic control of turbidites and winnowed turbidites. Geology 10, 231–235 (1982).

Steel, R., and Aasheim, S.M.: Alluvial sand deposition in a rapidly subsiding basin (Devonian, Norway). In: Miall, A.D. (Ed.): Fluvial sedimentology. Canadian Soc. Petroleum Geologists Mem. 5, 385–412 (1978).

Valecka, Jaroslav: Storm surge versus turbidite origin on the Coniacian to Santonian sediments in the eastern part of the Bohemian Cretaceous Basin. Geol. Rundschau 73, 651–682 (1984).

CHAPTER 11
Diagenesis

Introduction

A freshly deposited sand is obviously different from an ancient sandstone in strength, coherence, porosity, composition, and texture. *Diagenesis* includes the many post-depositional processes that alter a sediment to a rock, a sand to a sandstone. The term diagenesis was coined in 1888 by von Gümbel (p. 334) and has been used in many senses since then. A complete review of the subject, its history, and its terminology has been given by Dunoyer de Segonzac (1968). We use the term diagenesis to include all those processes, chemical and physical, which affect the sediment after deposition and up to the lowest grade of metamorphism, the greenschist facies. *Authigenic* or *authigenesis* refers to specific minerals and their formation. Authigenic is to some extent synonymous with diagenetic.

The possible combinations of ways in which diagenetic processes may affect the composition and texture of a sandstone are extraordinarily many but from all the possibilities it is possible to pick general tendencies. The overall trend is towards an equilibrium chemical composition and a physical adjustment to the surrounding stress field. A freshly deposited sand is a porous, nonequilibrium mixture of detrital minerals, frequently with unstable grain packing. After diagenesis the sandstone has become reduced in porosity through compaction and cementation, has lost many unstable detritals, and has gained stable authigenic precipitates. The end-products of sandstone diagenesis are responses in composition and texture to the impress of pressure and temperature and changing composition of porewaters over varying lengths of time. The character of the final product may be dependent on provenance and environment as well as post-depositional history.

By understanding diagenesis we can "look through" obscuring diagenetic features to better judge the provenance and environment of deposition of the original sand. At the same time we can use diagenesis to construct a post-depositional geological history. Practical questions abound. How are porosity and permeability related to cementation? Is cementation related to depth in a basin? Can the diagenetic history of a sandstone lead to a reconstruction of its thermal history in relation to the maturation of petroleum hydrocarbons? Do the heavy minerals represent the original detrital assemblage and so reflect provenance or are they the residue of intrastratal solution? Can the mineralogy of a clay matrix of a graywacke be used to infer the nature of the precursor rock fragments and their provenance? Does authigenic kaolinite in a sandstone indicate that it was exposed to meteoric water?

To answer these questions we start from the composition and texture, with emphasis on the authigenic minerals, cements, and the nature of pore spaces. We then add inferences of temperature, pressure, and pore solution composition. Finally, we relate diagenetic effects to the geological history, stratigraphic and structural, of the sedimentary basin. The stages in the geologic history of a sandstone relate diagenesis to burial, metamorphism, and uplift (Fig. 11-1). In the section on burial diagenesis (p. 465) we use a plot of burial depth versus post-depositional time to show the timing of diagenetic changes.

In discussing the timing of diagenesis we differentiate in a simple way between "early" and "late." "Early" refers to those changes which take place within a few thousands to hundreds of thousands of years, presumably while the sediment is still buried at a rather shallow depth, less than 50 m (for a full discussion of this stage of marine sediments, see Berner, 1980). "Late" refers to all later events. Differ-

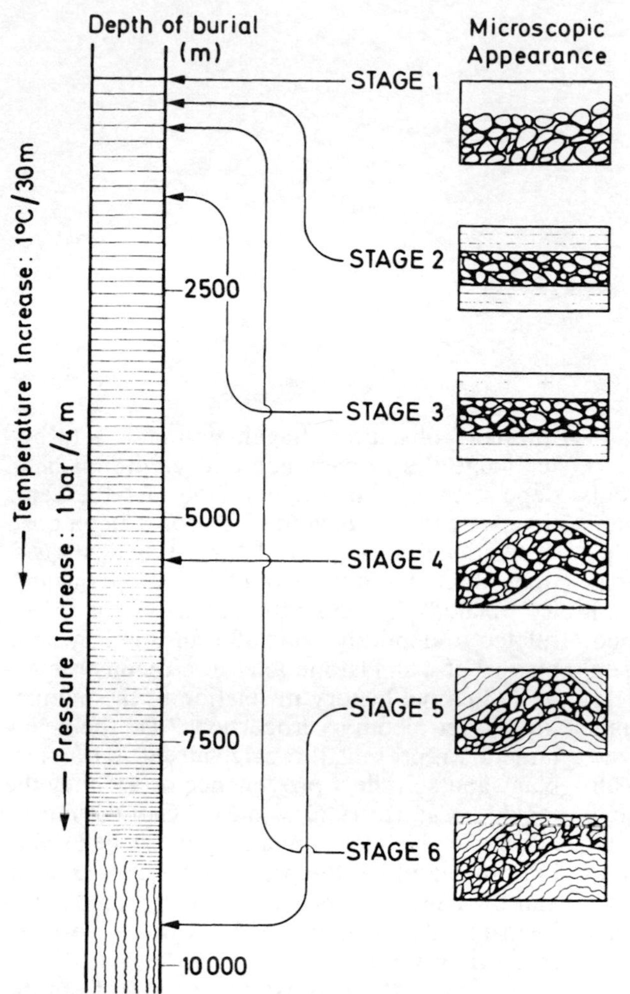

Immediately after deposition. Exposed to air or water of depositional environment. Original detritus, high porosity.

Buried a few meters to tens or meters. Exposed to interstitial waters. Some compaction, some early chemical precipitates possible. Commonly termed "early" diagenesis.

Buried to moderate depths of about 1000 m. Pore water may be a brine. Chemical cements and compaction may reduce porosity, clays may be altered.

Deep burial to thousands of meters, perhaps with folding. Porosity may be very low from chemical cement and pressure solution. Secondary porosity common.

Incipient metamorphism. Growth of chlorite and other metamorphic minerals with extensive pressure solution and quartzitic texture. Termed "high-grade" diagenesis or "anchimetamorphism."

After uplift and erosion, within tens of meters of land surface. Invasion by meteoric water, decementation and "weathering" of clays may increase porosity.

FIGURE 11-1. The stages of diagenesis in relation to depth of burial and increase of pressure and temperature. Temperature increase is variable, depending primarily on the geothermal gradient. Pressure increase is less variable but dependent on porosity of the overlying section.

entiating between early and late in this way, we emphasize differences between the rapid initial processes of lithification and compaction and the slower changes which come later and over a much longer time scale.

Aspects of Diagenesis

We use the term "aspect" to express the different points of view from which we may consider the ideas and evidences of diagenesis. One volume, *Aspects of Diagenesis* (Scholle and Schluger, 1979), presents a series of papers from two different viewpoints, one concerning clastic reservoirs, the other paleotemperature control. A prime aspect of diagenesis is the evidence of post-depositional change, for example, a euhedral authigenic tourmaline crystal projecting from a rounded detrital grain of slightly different composition (Fig. 11-2). From the evidence we may infer a process—the precipitation of authigenic tourmaline from a pore solution.

Another aspect is the time scale of the inferred process. Most authigenic tourmaline overgrowths are found in older sandstones, none in very young ones. Because of the extreme slowness of the precipitation of tourmaline at low temperatures, this process may have taken 10^7–10^8 years.

A final aspect of diagenesis is the ordering of a diagenetic event with respect to geologic events in the sandstone's history. The tourma-

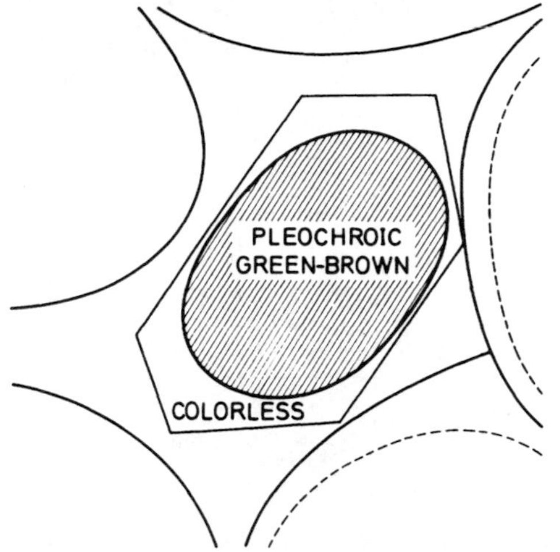

FIGURE 11-2. A colorless euhedral overgrowth of tourmaline precipitated in optical continuity with a rounded detrital core of pleochroic green-brown tourmaline. See Fig. 7-7 for a drawing of this kind of grain.

line overgrowth may have been precipitated during a long period of deep burial.

Though answers to some specific questions of diagenesis may be given by specialized studies, all of the different aspects of diagenesis can be discussed only with reference to the general geologic history of the sedimentary basin. Diagenesis is a historical geological subject. If we are able to infer geological events from diagenetic evidence, we increase our knowledge of the historical geology of a sandstone. Conversely, we may combine knowledge of the historical geology of sandstones and their diagenetic histories to learn much about geochemical processes in the sedimentary part of the earth's crust.

Diagenetic Textures and Compositions

Textures

Textural relations among mineral grains and crystals are the major direct evidence of diagenesis in sandstones (Table 11-1). The most obvious are the pseudomorphic replacements, for example, the replacement of a calcareous shell by quartz. The original, recognizable texture is preserved in a material that is known not to be the original. The more abundant replacements are those which preserve phantom structures, faint outlines of the original, in spite of the loss of grain boundaries. The recognition of phantom structures depends on the existence of many small inclusions. Crosscutting relationships are conclusive evidence of replacement. Most frequently encountered in sandstones is the embayment of detrital grains by a crystalline mosaic. Extensive replacement can result in "floating" inclusions of the detrital grain. Microstylolites that cut across detrital grains (Sloss and Feray, 1948; Heald, 1955) manifest postdepositional solution processes.

In most ancient sandstones the pores between detrital grains are filled with an aggregate or mosaic of minerals, most frequently carbonate, silica, or clay. The clay matrix of graywackes, only a small part of which may be primary, and pore fillings by carbonate, silica, zeolites, and other minerals point up to the contrast between modern and ancient sandstones. Pore filling may be partial or complete, may consist of only one mineral or several, and may or may not be associated with replacement of detrital grains. One mineral of the pore filling may replace detrital grains and a second mineral of the same pore filling not.

Some authigenic minerals display euhedral faces that are sharp and show no signs of abrasion during transportation. The distinction between the detrital and authigenic part of the crystal cannot always be made in ordinary microscopy and cathodo-luminescence must be used. Studies of sands of modern sediments have shown that all detrital grains over 0.1 mm in diameter are affected at least slightly by the abrasion process and sharp euhedral faces are rarely found.

In some sandstones, authigenic overgrowths cannot be distinguished; yet the pore space is almost completely filled. The observed interlocking crystalline mosaic texture can only result from chemical solution of detrital grains.

Diagenetic concretions and nodules in sandstones may be a combination of pore-filling with replacement or other textural elements. Such structures are formed by the greater cementation of grains in the nodule than in the surrounding matrix or may be due to cementation by a different kind of mineral. If the concretions were formed early in diagenesis, the original mineralogy and texture may be preserved better than in the surrounding sandstone because the concretion was less permeable and so was not transmissible to fluids responsible for solution or precipitation later in diagenesis (Bramlette, 1941).

TABLE 11-1. Evidence of sandstone diagenesis.

Type	Criterion	Examples	Appearance	Comments
TEXTURAL				
Pseudomorphic replacement	Original texture exactly replaced by foreign material	Calcareous shell replaced by quartz Volcanic glass shard replaced by zeolite	Quartz, Calcite, Zeolite	Cathodo-luminescence and SEM helpful
Phantom structure	Recognizable phantom shape or boundary	Calcareous skeletal grain replaced by sparry calcite	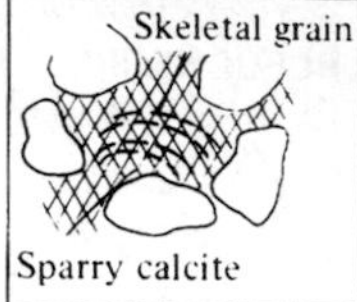 Skeletal grain, Sparry calcite	
Pore-filling	Former pore space between detrital grains filled	Sparry calcite cement	Cement	Allows calculation of minus-cement porosity
		Clay matrix of some lithic sandstones	Matrix	May be squashed rock fragments
Cross-cutting	Interruption of normal boundary	Rounded detrital quartz grain embayed by calcite Microstylolites	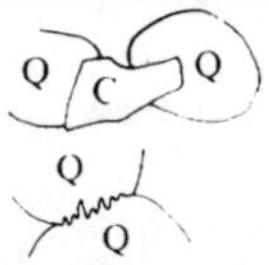 Q C Q Q Q	
Euhedral faces	Rational crystal faces show no rounding	Quartz overgrowths Dolomite rhombohedra	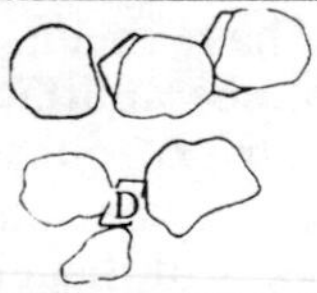 D	SEM helpful
Interlocking grain mosaic	Lack of detrital grain-pore space geometry	Sedimentary quartzite	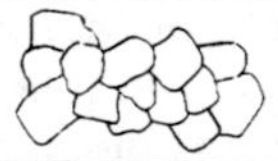	Original detrital boundaries may be absent or seen only by cathodo-luminescence
Concretions and nodules	Extent or kind of cement different from host sandstone	Hematite-cemented nodule in calcareous sandstone	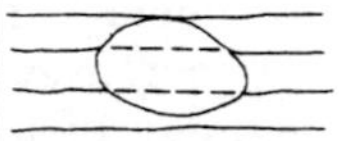	Infrequently show differential compaction in sandstone
MINERALOGICAL				
Compositional purity	Authigenic phases lack inclusions and trace elements of detritals	Colorless tourmaline overgrowth on colored detrital grain Authigenic albite	see Fig. 11-2 Ab_{85} Ab_{98}	Electron probe or cathodoluminescence best methods
Solubility	More soluble minerals uncommon as detrital	Anhydrite cement	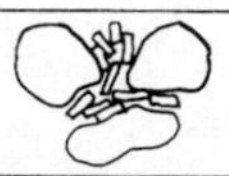	Avoid water in grinding thin-sections
PHYSICAL				
Porosity decrease	Porosity less than expected of ordinary packing of detritus	Permeability decrease Bulk density increase Seismic velocity increase	Porosity → Depth ↓	Also dependent on grain size and shape distribution secondary porosity can be significant at depth.
CHEMICAL				
Pore fluid composition	Composition different from that of sedimentary enviroment	Saline formation waters Oil and gas		Only appropriate for subsurface sampling
Isotopic composition	Isotope ratios different from sedimentary environment	$^{18}O/^{16}O$ $^{87}Sr/^{86}Sr$		Requires sampling of pure authigenic phases

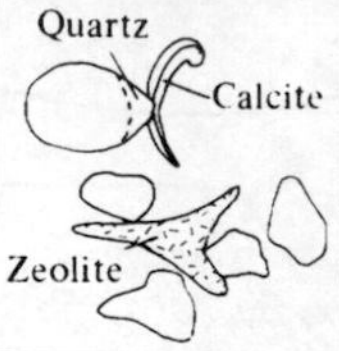

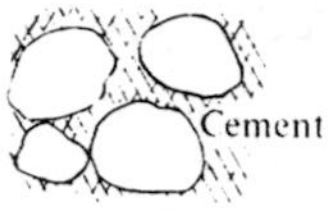

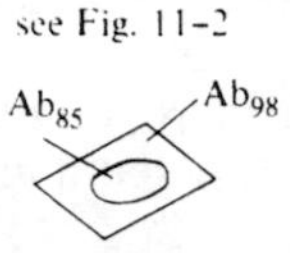

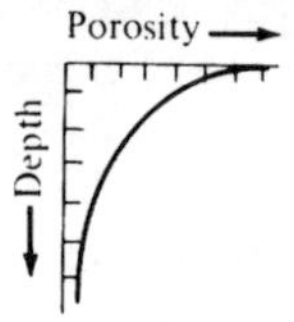

Mineralogy

The evidence for diagenesis that resides in the nature of the mineral substance rather than its shape, habit, or size is less conclusive than the textural evidence. A useful guide to diagenetic origin is purity of composition. Many authigenic overgrowths of quartz or other minerals are recognizable by the clarity and lack of crystalline, liquid, or gas inclusions so characteristic of the detrital contributions from igneous and metamorphic rocks. Cathodo-luminescence and electron probe studies show that authigenic minerals are much lower in trace element content and thus show lower luminescence than the detrital grains. Authigenic feldspars tend to be very pure Na- or K-forms and thus distinct from the wide range of plagioclases and alkali feldspars that are represented in the detrital fraction (Kastner and Siever, 1979). Figure 2-5 shows the phase relations of feldspars at low temperature.

The more soluble minerals are unlikely to be found as detritals in most sandstones that were deposited by water. Outside of carbonate depositional environments, clastic grains of carbonate rocks are unusual and represent somewhat special conditions of rapid mechanical erosion rather than chemical dissolution of a limestone terrain. Thus, most carbonate in sandstones is in the form of cement. Though there are gypsum sand dunes, most gypsum in sandstones is diagenetic. Halite, which can only be found if care is taken to preserve it by avoiding water during thin-section grinding, is even more probably of diagenetic origin. But caution is necessary and ultimate reliance should be placed on textural features.

Mineralogy in combination with texture is an important criterion of diagenesis. We are able to recognize the diagenetic nature of a dolomitized fossil precisely because it is dolomite and we know of no organism that secretes dolomite for its shell, though many deposit magnesian calcite. A zeolite mineral in the shape of a glass shard is diagnosed as a replacement because the shard shape of broken volcanic tephra is always associated with glass, and zeolites do not congeal from a melt.

Physical Properties

The major change in bulk physical properties that results from diagenesis is porosity reduction. Bulk density of sandstones increases as the porosity decreases. Permeability is commonly correlated with porosity (Fig. 3-15). Permeability is also related to the fabric or packing of the grains and is reduced in more compact, more tightly packed ancient sandstones. As the porosity decreases, other things being equal, the velocity of elastic waves propagated through the rock increases. This is the basis for using sonic logging of drill holes and seismic profiles to predict porosity, though mineralogy and fluid content affect the results. The physical changes in sandstones are primarily attributable to the precipitation of mineral cements in pore space during diagenesis. Only secondarily are such changes related to purely physical compaction.

Chemical Properties

Though there are some chemical analyses of modern sands available (see, for example, Potter, 1978; Franzinelli and Potter, 1983) there have been no statistically based reviews and comparisons of the composition of modern and ancient sands that would reveal the effects of diagenesis. We can infer some changes on the basis of mineralogical differences (Chap. 2). The general tendencies exhibited by ancient sandstones are to have more alkali metals, alkaline earths, and silica and less water than modern sands; these are the expected consequences of carbonate and silica cementation and clay mineral alteration.

Pore fluids of sandstones should be included in chemical analyses, for fluids are part of the rock when it is buried in the subsurface, as the great bulk of sandstones are. The differences between porewaters of modern and ancient sandstones reflect the extensive changes wrought by diagenesis. Porewaters of most modern sands are either fresh and similar in composition to river waters (alluvial sands) or are essentially seawater (beaches and offshore shelf sands, turbidites, deep-sea sands). Hypersaline and alkaline porewaters are associated with marine and terrestrial evaporite environments. In contrast, the porewaters of ancient sandstones may vary from potable water aquifers near the surface, not too dissimilar from the river water compositional range, to the concentrated brines of the deep subsurface. The brines are drastically different from seawater in composition not only because they may be much more concentrated in dissolved solids but because the ratios of various ions to each other may differ greatly from those in seawater. Many of the brines contain much more Ca than Mg; the reverse is true for seawater. Like waters of many evaporite lakes, formation waters may be enriched in sulfate, bicarbonate, or chloride ions. Silica is usually increased. All of these changes are evidence of the chemical reactions that have taken place between pore fluids and rocks after deposition. Composition and origin of many subsurface fluids are covered in Young and Galley (1965) and Collins (1975).

Oil, gas, or other organic concentrations such as asphalts or pitchlike substances, do not appear in modern sands, except for those places where such fluids enter the sand from external sources. Organic matter content in any form is very low in modern sands, though there always is some. The contrast of this picture with the frequent occurrence of all kinds of oil and gas accumulations in ancient sandstones is evidence of the diagenetic processes that have been responsible for the production and migration of the oil into the sandstone reservoir.

Summary: Two Typical Examples

The extent and nature of the differences between modern and ancient sands and the presence of textural elements, particularly the pseudomorphic replacements, indicate the post-depositional processes that have modified the rock. We can illustrate the rock properties that evidence diagenetic change by two composite examples, a quartz arenite and a graywacke.

A quartz arenite might show some interpenetration of detrital quartz and euhedral authigenic enlargement of detrital quartz grains, partly intergrown to give a strongly cemented rock (Fig. 11-3). The rock might have as much as 10–15 percent carbonate as pore filling, part of which has replaced both detrital and authigenic quartz. The carbonate may be calcite with a few rhombohedra of dolomite replacing the calcite. Some hematite cement may obscure both detrital grains and other diagenetic components. Heavy minerals may include little besides tourmaline, zircon, and rutile, and the tourmaline may show secondary enlargement by a colorless, non-pleochroic variety. The rock may have a porosity of 10 percent but its permeability may be as low as a few millidarcys. It will be hard and resistant to breakage. The pore fluids may include oil and gas and a concentrated brine.

A different set of properties may be shown by a graywacke in which there is a clay matrix of about 30 percent, at least part of which has the appearance of altered rock fragments (Fig. 11-4). Some of the matrix is of chloritic composition and appears to replace parts of detrital silicate grains. Scattered patchy areas of calcite fill pore space or replace both detrital grains and clay matrix. The carbonate areas may be associated with small rhombohedra of siderite or iron-rich dolomite. Shardlike fragments composed of zeolite, such as heulandite, are distributed through the rock, a few showing relics of an isotropic glass. The heavy minerals are abundant and varied, including many mafic minerals, and show little or no evidence of corrosion or dissolution. The rock's porosity may be as low as 2 percent and permeability may be vanishingly small. The rock is hard and, if thinly bedded, may tend to fracture along bedding planes. There is no oil and gas, but the

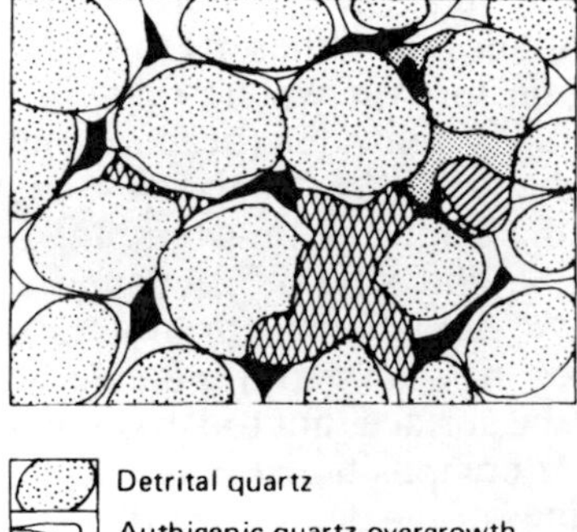

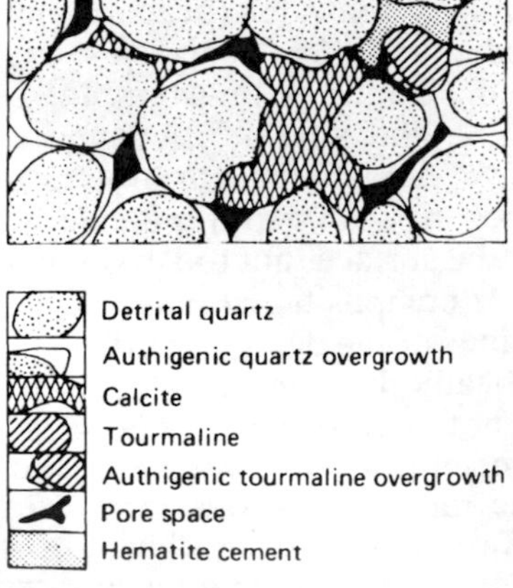

FIGURE 11-3. Composite view of diagenetic textures and minerals in a quartz arenite.

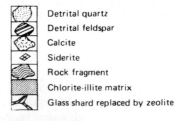

FIGURE 11-4. Composite view of diagenetic textures and minerals in a graywacke.

chemical analysis may show as much as 1 percent organic matter.

The genetic interpretation of such catalogues of evidence is unlikely to be made in terms of a single, simple process. Instead, these properties originate in a group of physical and chemical processes that interact and so affect the rock in different ways and to different extents depending on the subsurface geologic environment.

Physical Diagenetic Processes

Compaction

Physical processes that affect the nature of a sand after deposition lead to a loss of porosity and an increase in bulk density, primarily by an adjustment to the force of gravitation. The major way in which this happens is compaction through a change in packing by rotation, translation, fracturing, and plastic deformation of grains, which results in a loss of porosity. Physical rearrangement is greater for well-rounded than for angular grains (Füchtbauer, 1967). Rittenhouse (1971) has calculated the additional effect of varying proportions of ductile grains on compaction. Grain slippage is thought by Füchtbauer (1974) to be the chief mechanism of compaction of quartz arenite with small amounts of clay or rock fragments down to depths of 1 to 1.5 kilometers. Sands as deposited may have a wide range of packing densities, illustrated in the extreme by the difference between the windward and lee slope of a sand dune but easily observable also in various subenvironments of beaches and bars. So little has been done on packing that we cannot say whether such differences are erased as a result of compaction, but studies of imbrication indicate that there is inheritance and that the effects of compaction are small (Rusnak, 1957). Rittenhouse (1972) compared mean crossbedding dip angles in modern dunes and the Navajo (Triassic) Sandstone (a presumed eolian deposit) of the southwestern United States and estimated a thickness reduction and compaction of 24 to 27 percent. Brett (1955) found that the inclination of crossbeds was affected by folding, an effect attributed to grain rearrangement during shearing within the bed.

The porosities of sandstones that are relatively uncemented but have been buried to sufficient depths to have been compacted, such as the Tertiary sandstones of the Gulf Coast of North America, are high, fairly close to the values expected for modern sands. Dickinson (1953) quotes a general average porosity for 30 percent for 10,000 ft (3226 m) below the surface and believes this confirms the very slight compressibility of loose sand reported by Athy (1939). Some others, notably von Engelhardt (1960, pp. 16–21), believe that physical rearrangement alone can account for significant reductions in sandstone porosity but at the same time point to the importance of accompanying chemical dissolution. The weight of evidence, however, makes us conclude that the amount of physical compaction of uncemented sand grains achieved by rearrangement under load is small and would account for a drop in porosity of only a few percent. A full review of compaction of coarse-grained sediments is given by Chilingarian and Wolf (1975).

Bioturbation

Bioturbation of sands may extend many centimeters below the surface (Reineck and Singh, 1980). This mechanical action tends to decrease the sorting, which, together with mixing of clay from adjacent beds, will reduce porosity and increase compaction. Infiltration of clay on alluvial and deltaic plains may reduce porosity in sands near the surface.

Graywacke Matrix Formation

To the extent that some graywackes with high amounts of matrix clay are rocks intermediate between sandstones and shales, they may show compaction effects much like those of shales. The decrease in porosity under load of clayey materials is large and in the natural case has been summarized by Meade (1964) and Bryant *et al.* (1981). It is likely that some muddy turbidites may have been physically compacted from an initial porosity of 50 to 60 percent to 30 percent or less by rearrangement of clay particles and squeezing out of water. Because there are few examples of modern graywacke sands with high amounts of matrix (Heezen and Hollister, 1964) it is not possible to report more definite conclusions. Because many ancient graywackes present the appearance of clay matrix by virtue of the diagenetic modification and physical squashing of labile rock fragments, the role of compaction is doubly important, for it is largely responsible for both dewatering and the matrix. The difference between the load required to squeeze water from clay and that required to mechanically squash different kinds of rock fragments is unknown. For the latter it is likely that chemical alteration plays a role. Experiments that support the idea that the matrix originates by alteration of rock fragments were done by Hawkins and Whetten (1969). Samples of moderately well-sorted sands with abundant volcanic rock fragments from the Columbia River, sands that are similar in mineral and chemical composition to typical graywackes, were held at temperatures of 140–300°C and 1 kilobar water-pressure for 21 to 60 days, simulating burial to 3 or 4 km. Clay minerals and zeolites were produced from the volcanic fragments. Brenchley (1969) studied Ordovician graywackes from Wales and concluded that calcite cementation preserved detrital grains and inhibited matrix formation in some beds, but in noncalcareous beds, rock fragments were altered to matrix. The ratio of identifiable rock fragments to matrix of an Eocene graywacke, the Tyee Formation of Oregon, seems to decrease in the area of heaviest overburden, but stratigraphic control is insufficient to calculate the loading factor (Lovell, 1969).

Clay Mineral Dehydration and Compaction

Physical compaction may accompany the alteration of smectites to illite under conditions of deep burial as water is expelled from interlayer positions (see p. 457). In this process, as in the deformation of rock fragments, chemical change is an integral part of the process that results in physical deformation. The mechanics of migration of liquid water from clayey sandstones under deep burial pressures, when the rocks have already been compacted and lithified, are not clearly understood. Much of the water loss may take place through microscopic fractures.

Brittle and Ductile Deformation

The distribution and origin of fractured grains is a problem that has received little attention. Fractured or shattered grains in sandstones have been noted but the nature and number of such grains are reported infrequently. Fractured grains are found in relatively undeformed basins. They are certainly not restricted to strongly folded belts. Those that are seen in undeformed sandstones may result from burial stresses along lines of weakness inherited from the crystal in the source rock or induced during transport impact. Cathodo-luminescence studies have shown numbers of fractured and rewelded detrital grains that are not seen in ordinary microscopy (Sippel, 1968). In orogenic belts, strongly folded sandstones may show cataclastic textures, sheared grains, and other deformational structures that are clearly related to tectonic movements and so are not properly in the province of diagenesis. Hawkins (1978) has noted fracturing and crushing in coarse-grained Carboniferous sandstones of England that he believes is responsible for reducing the effective porosity by lowering the degree of sorting and by bridging pores. Nagtegaal (1978) reports breakage of grains at maximum stress points and partial healing in Miocene quartz arenites in the Niger delta but believes this has not led to significant loss of porosity and permeability. Microscopic fracturing may be associated with pressure solution at point-to-point contacts of detrital grains.

Ductile deformation of sand under moderate confining pressures, a few kilobars, at room temperature, followed by faulting at higher confining pressures has been reported by Byerlee and Brace (1969) and is an active field of investigation in rock mechanics related to earthquake mechanisms. Post-depositional physical changes such as slide brecciation, slump structures, and crumpling are discussed in Chap. 4.

Chemical Diagenetic Processes

Chemical changes are more important than physical changes in altering the character of many sandstones after deposition. Cementation and lithification are mainly the result of chemical precipitation of a binding agent or the chemical welding of adjacent detrital grains. Chemical processes, especially those involving complex alterations of labile rock fragments, involve the interaction between solid mineral grains and pore fluids. The bulk and mineral composition of the solids change slowly during diagenesis in response to changes in porewater chemistry.

The chemistry of diagenetic chemical changes is best analyzed by explicitly writing the balanced chemical equation and evaluating the change in free energy at constant pressure and temperature accompanying the reaction, as Curtis (1976) has done for weathering reactions, for example. This free energy is specifically the *Gibbs free energy* or Gibbs function, and is here abbreviated as G; the finite change in G is designated by ΔG. From the value of ΔG we can quantitatively evaluate an *equilibrium constant*, which can then be used to predict the entire range of concentrations of reactants necessary for dissolution, precipitation, or other chemical change.

We model this approach by writing the reversible equation, at a given temperature and pressure, that describes the precipitation of a pure ionic crystalline solid compound, AY, by reaction of two completely ionized electrolytes, AX and BY, in a pure aqueous solution

$$A^+ + X^- + B^+ + Y^- \xleftrightarrow[25°C]{1\ atm} AY_{(crystalline)} + B^+ + X^-, \quad (11\text{-}1)$$

where A and B are cations and X and Y anions.

We determined the standard free energy, ΔG^0, for this reaction by subtracting the sum of Gibbs free energies of formation, ΔG_f^0, of the reactants from the sum of the ΔG_f^0's of the products to obtain

$$\Delta G^0_{Reaction} = [\Delta G_f^0\,(AY) + \Delta G_f^0\,(B^+) + \Delta G_f^0\,(X^-)] - [\Delta G_f^0\,(A^+) + \Delta G_f^0\,(X^-) + \Delta G_f^0\,(B^+) + \Delta G_f^0\,(Y^-)]. \quad (11\text{-}2)$$

Then, we use another relation to get the equilibrium constant, K_{eq},

$$\Delta G_R^0 = -RT \ln K_{eq}, \quad (11\text{-}3)$$

where R is the gas constant, T is absolute temperature (degrees Kelvin), and ln is a natural base logarithm.

The equilibrium constant is formed in terms of the division of the product of the activities of the reaction products by the product of the activities of the reactants,

$$K_{eq} = \frac{(a_{AY})(a_{B^+})(a_{X^-})}{(a_{A^+})(a_{B^+})(a_{X^-})(a_{Y^-})} \quad (11\text{-}4)$$

which reduces in this case to

$$K_{eq} = \frac{a_{AY}}{(a_{A^+})(a_{Y^-})} \quad (11\text{-}5)$$

The *activity* of a substance, a_i, is frequently called *thermodynamic concentration* and is related to the true concentration of a substance, also referred to as *molarity* or *molality*. That relation is such that activity approaches concentration in very dilute solutions and deviates from it as total concentration of all substances in solution increases. By convention, the activity of any pure solid that is thermodynamically stable at the given temperature is taken as unity, and so Eq. (11-5) reduces to

$$K_{eq} = \frac{1}{(a_{A^+})(a_{Y^-})}\,. \quad (11\text{-}6)$$

In the case of very dilute solutions, the activities may be replaced by concentrations.

Equation (11-1) may be written from right to left to describe the dissolution of AY, in which case, B and X may be eliminated for dissolution in pure water and we get the simple result

$$K_{eq} = a_{A^+}a_{Y^-}\,. \quad (11\text{-}7)$$

If AY is a relatively slightly soluble substance, the solution will be dilute and K_{eq} will approach the *solubility product constant,* $K_{S.P.}$, which is given in terms of concentration, m, and is

$$K_{S.P.} = m_{A^+}m_{Y^-}\,. \quad (11\text{-}8)$$

Detailed expositions of this kind of calculation are given by Garrels and Christ (1965, Chap. 1), Krauskopf (1979, Chap. 1), and Drever (1982, Chaps. 2 and 5), as well as in textbooks of equilibrium chemistry and chemical thermodynamics. Through such calculations we deduce the chemical conditions needed for the various kinds of precipitation, dissolution, and other reactions discussed below.

In addition to these constraints of equilibrium chemistry, we must consider the effects of the kinetics of diagenetic reactions that may be so slow that equilibrium is not reached even over geologically significant times. A full discussion of the kinetics of diagenetic reasons is given for early diagenesis by Berner (1980). We discuss kinetics in relation to burial diagenesis in a later section of this chapter.

Precipitation

Petrographic evidence is strong for direct precipitation of various kinds of cements from pore solutions onto surfaces of the same or different mineral grains, a process frequently called "neoformation" (Füchtbauer, 1974, p. 130). This precipitation plays a major role in reducing porosity.

Many minerals have been identified as authigenic precipitates (Table 11-2). The most common diagenetic silicates and oxides are precipitates of minerals present in the primary detrital assemblage, such as quartz, feldspar, clays, zircon, and tourmaline. Diagenetically precipitated minerals are almost always purer than detrital phases. Diagenetic minerals may be of the same chemical composition but different crystal structure, as illustrated by the precipitation of opal on quartz surfaces and anatase overgrowths on brookite.

Other diagenetically precipitated minerals may be of phases not originally present in the

TABLE 11-2. Minerals found as pore-space precipitates in sandstones.

Silica minerals	Iron oxides
Opal-A	Hematite
Opal-CT	Goethite
Chalcedony	
Quartz	Titanium minerals
	Anatase
Zeolites	Brookite
Phillipsite	
Clinoptilolite	Tourmaline
Heulandite	
Laumontite	Zircon
Analcime	
	Carbonates
Feldspars	Calcite
K-feldspar	Dolomite
Albite	Ferroandolomite
	Siderite
Clay minerals	
Kaolinite	Sulfates
Illite	Gypsum
Smectite	Anhydrite
Chlorite	Barite

detrital mix. Secondary carbonate typically is of this kind. Others are anhydrite, iron oxides, halite, pyrite, and other normally nondetrital minerals. These minerals show crystalline forms impinging on or growing from detrital grain surfaces in encrusting, coating, or spiky outgrowths that would be difficult or impossible to explain as detrital. With some exceptions (tourmaline, zircon, brookite, anatase) most of the minerals listed in Table 11-2 have been synthesized experimentally at low or moderate temperatures. These experiments give us information on the chemical conditions necessary for precipitation.

Partial fillings of pores by chemical precipitates takes the form of overgrowths on detrital grains, precipitated clay linings of detrital grains, and isolated volumes of cement distributed in irregular patches. Many more fillings of the same mineralogy as the detrital grains, *syntaxial* growths, are precipitated in optical continuity with the original grain. The mineral that shows this behavior most commonly is quartz; feldspar overgrowths are also common but less obvious; calcite may also show such overgrowths. Among the minor constituents, zircon and tourmaline show optically continuous overgrowths. (See Figs. 5-23, 5-27, and 11-2 for examples of overgrowths.)

Minerals precipitated on substrates of a different mineralogy form *epitaxial* overgrowths, such as carbonates on quartz surfaces or clays on quartz. The growth of chalcedony on quartz surfaces is probably this kind of growth too. When originally precipitated, the chalcedony was probably opal-A or opal-CT, which was later recrystallized to microcrystalline quartz.

The size of secondary precipitate crystals is, from experience gained from artificial crystal growth technology, related to rapidity of crystal nucleation and crystal growth over those nuclei. It is also dependent on the mineralogy and form of the substrate. In general, the formation of many small crystals indicates relatively rapid nucleation compared to growth. Large single crystals are the result of slow nucleation compared with growth over the few nuclei. Large, optically continuous, quartz overgrowths represent growth over already present nuclei, the detrital grains. Large areas of calcite cement precipitated as single crystals extending over many pores show the effects of growth rate dominating over nucleation. There is an inevitable slowing down of crystal growth as pore space becomes progressively filled, permeability is reduced, and flow rate of solution is reduced (Fig. 11-5).

Related to the nature of crystal growth is the distribution of solid inclusions as well as crystal size. (Fluid inclusions, indicative of tempera-

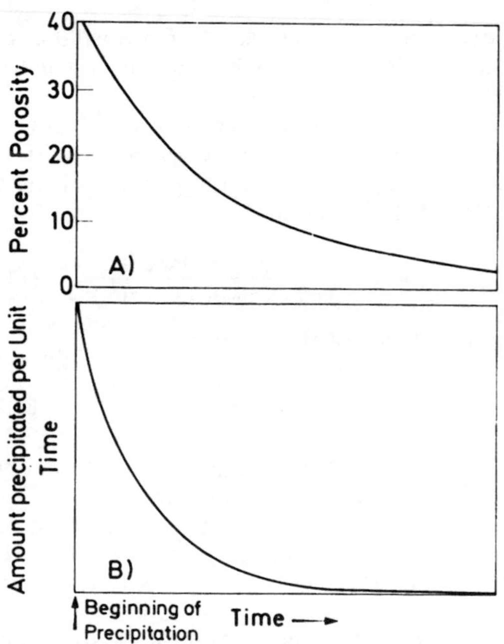

FIGURE 11-5. Exponential decrease in (A) porosity and (B) amount of cement precipitated as cementation of a sandstone proceeds. The precise shapes of such curves depend on the rate of fluid flow through the sand, the initial porosity and permeability, and the rate of precipitation per unit of fluid. The shapes of these curves are also affected by growth of secondary porosity by dissolution.

tures and compositions of pore solutions, are discussed under burial diagenesis.) Inclusions, randomly or regularly distributed, are remnants of the replaced rock. Secondary precipitates that have grown into voids are free of solid inclusions. The inclusion compositions and distribution can be a most useful guide to materials once present in the rock and now replaced. They also provide a guide to textures of those precursors preserved by the pattern of inclusions, such as those of oolites or fossils.

Some distinctive shapes of inclusions, such as the hourglass patterns seen on some authigenic feldspars (Kastner and Waldbaum, 1968), may be linked to the incorporation of inclusions by rapidly growing crystal faces and their being pushed aside by slower-growing faces. The ultimate in the incorporation of inclusions is the formation of large sand "crystals," in which growing calcite crystals have incorporated quartz grains making up to two-thirds of the weight of the whole crystal. The terms "poikiloblastic" and "poikilotopic" have been used for such textures.

Dissolution

We use dissolution here for *congruent dissolution*, a process by which the solid phase dissolves homogeneously, always leaving behind fresh surfaces of as yet undissolved solid unaltered in composition. Such dissolution is characteristic of pure NaCl, SiO_2, and $CaCO_3$. This process is distinguished from *incongruent dissolution*, in which there is a kind of selective dissolution whereby the solid that is left undissolved is changed in composition, because only some components of the crystal are leached into solution or because the ratio of the components in the dissolved fraction may be different from that in the original solid. For example, as in Eq. (11-9), a highly magnesian calcite will dissolve in such a way that the fraction dissolved has a much lower Ca/Mg ratio than the original solid, leaving undissolved a solid with a much higher Ca/Mg ratio (Chave *et al.*, 1962; Drever, 1982, pp. 54–55).

$Ca_{0.85}Mg_{0.15}(CO_3) \rightarrow$
Magnesian calcite
Ca/Mg = 5.66

$3.35\ Ca^{++} + 0.65\ Mg^{++} + 4CO_3^{--}$
Solution Ca/Mg = 5.15

$+\ Ca_{0.9}Mg_{0.10}(CO_3)$
Magnesian calcite (11-9)
Ca/Mg = 9

In this way, a magnesian calcite will gradually be converted to a purer calcite. This description of dissolution is not intended to imply a precise mechanism by which ions detach themselves from the crystal surface and solids become changed in composition or become reprecipitated from solution; it only describes the overall result.

Dissolution is an integral part of the recrystallization process, for at the lower end of the diagenetic temperature range, roughly from 0–200°C, "dry" recrystallization, that is, a restructuring of the mineral by solid diffusion without mediation by a surrounding fluid, would take so long that great lengths of geologic time would be needed for most transformations. Most recrystallizations proceed by dissolution of the precursor and precipitation of the final crystal. The whole process may take place in a thin film of solution immediately adjacent to the solid surface in such a way that the dissolved ions are immediately reprecipitated with restricted communication with the porewater at large.

The dissolution process is also a part of replacement of one mineral by another. Consideration of the geometric constraints on the nature of the replacement process indicates that in a large number of sandstones the substitution of one mineral for another takes place without volume change. The grains (and any cement) have to be in contact with each other to continue to support the rock without collapse during the process. This demands dissolution of the replaced mineral and precipitation of the replacing mineral in an exceedingly thin film between the bounding surfaces of the two phases (Fig. 11-6). Such a film is probably of the order of microns so that surface tension gives it enough strength to be mechanically supporting. The transport processes within this film must be adequate to move the dissolved material out and the precipitating material in from the open porewater reservoir. In the example shown in Fig. 11-6, a typical case of calcite replacing detrital quartz in an arenite, some idea of the complexities involved can be seen from the following sequence of reactions:

1. SiO_2 dissolves from the quartz surface into an undersaturated solution and hydrates to become monomeric silicic acid, H_4SiO_4:

$$SiO_{2(quartz)} + 2H_2O \rightarrow H_4SiO_4 \quad (11\text{-}10)$$

In dilute solutions at pH values below 9 and 25°C, a common geologic case:

$$K_{eq} = \frac{(a_{H_4SiO_4})}{(a_{SiO_2})(a^2_{H_2O})} \cong a_{H_4SiO_4} \cong 10^{-4} \quad (11\text{-}11)$$

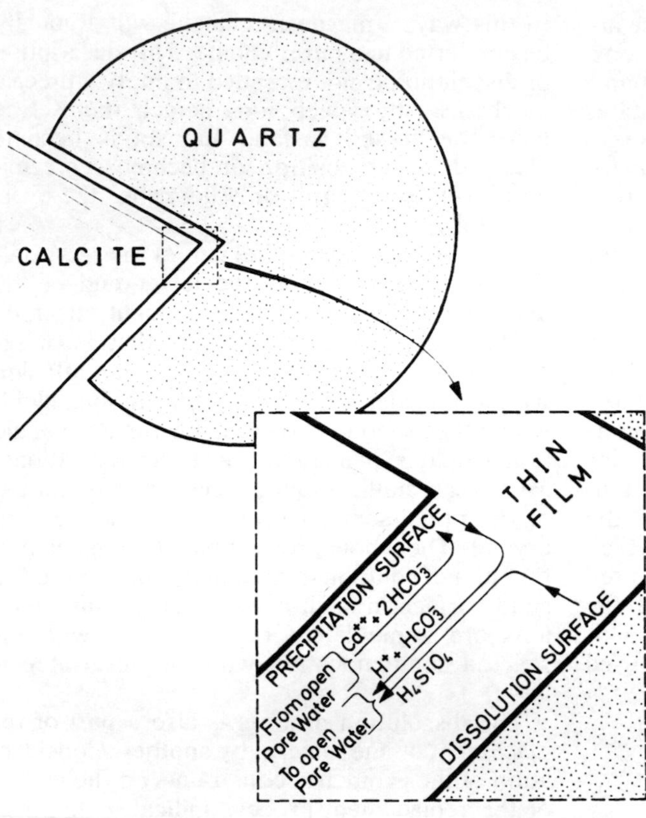

FIGURE 11-6. Chemical transport in thin film at a front of replacement of quartz by calcite. Inset shows details of process.

2. The concentration of H_4SiO_4 being higher in the film than in the pore fluid (necessary for dissolution to proceed), the H_4SiO_4 diffuses out into pores.
3. The concentration of Ca^{++} and HCO_3^- being higher in pore fluid than in thin film, the two kinds of ions diffuse into the film.
4. The product of the ion activities of Ca^{++} and HCO_3^- exceeding

$$K_{eq} = 10^{-8.35} = (a_{Ca^{++}})(a_{CO_3^{--}}), \quad (11\text{-}12)$$

$CaCO_3$ precipitates according to the equation:

$$Ca^{++} + 2HCO_3^- = CaCO_3 + H^+ + HCO_3^- \quad (11\text{-}13)$$

and, to maintain the concentration gradient that powers the diffusion, H^+ must diffuse out of the film into the pore space.

The critical matter in all of this transport is the *rate of diffusion in and out of the film,* which is a function of the thickness of the film, the temperature, and possibly other variables, for the behavior of such thin films is not well understood. In an analysis of a similar problem, the formation of stylolites, Weyl (1959) postulated a film including thin layers of clay such that the film included many subfilms or channels for diffusion, which seemed to him to be necessary to speed up the diffusion process sufficiently to accomplish much dissolution in reasonable geologic times. He proposed that such a postulate was justified by the common presence of clay films along stylolites. Though clay films are by no means uncommon in arenites and have been used to explain pressure solution (Heald, 1955; Thomson, 1959), there are too many replacement interfaces or interpenetration interfaces without any visible clay to accept this as a general relation. It is likely that the replacement mechanism may have several explanations: (1) the diffusion rate may indeed be slow and the geometries observed formed only after very long geologic times; (2) the diffusion rate is speeded up by clay films in those arenites where there is interstitial clay; and (3) the diffusion rate is speeded up by increases in temperature.

Accessory heavy minerals may disappear by dissolution, a process called *intrastratal solution* by Pettijohn (1941), who inferred such a process by a detailed analysis of the heavy minerals of sandstones as a function of geologic age. He noted the strong statistical tendency for some minerals to be absent in older rocks; for younger rocks consequently to have much more varied heavy mineral suites; and for certain authigenic minerals, in particular the titanium oxides, tourmalines, and zircons, to be much more common in ancient than in young sandstones. In general, the minerals absent in older rocks are those, such as pyroxenes and amphiboles, which are chemically least stable in earth surface environments. The chemically less stable minerals would gradually dissolve as a result of the passage of long geologic times; little or no trace of the dissolved minerals would be left, and some predominantly authigenic minerals, such as anatase, would form. This would obscure original differences in source areas, confuse attempts at stratigraphic correlation by heavy minerals, and make explicable the general tendency for Precambrian and early Paleozoic sandstones to have heavy mineral suites almost completely composed of zircon, tourmaline, and titanium and iron minerals. Support for the idea of diagenetic solution of heavy minerals comes from Bramlette's (1941) study that showed a much larger suite of heavy minerals in impermeable concretions in a sandstone than in their matrix of permeable sand.

An experimental dissolution study of light and heavy minerals reproduced natural dissolution effects corresponding to weathering and intrastratal solution, the former at pH values of 5.6, the latter at pH values from 5.6 to 10.6 (Nickel, 1973). Experimental studies of dissolution kinetics of mafic minerals showed more rapid reactions in anoxic solutions and slower rates in oxidizing solutions, in which armoring by ferric oxides hindered dissolution (Siever and Woodford, 1979). This result was used to explain the extent of intrastratal solution in deep porewaters, which are largely free of oxygen.

It has been argued by van Andel (1959) that a more likely explanation for heavy mineral sequences that go from an impoverished assemblage of zircon, tourmaline, and anatase at the base to a varied suite including garnets, amphiboles, pyroxenes, and others at the top is the gradual unroofing of tectonically active source areas that expose to erosion deep-seated igneous and metamorphic rocks containing unstable minerals in abundance. Such an unroofing, first proposed in the classic paper on the erosion of the Dartmoor granite (Groves, 1931), would be expected in the typical tectonic cyclical pattern of development of sedimentary basins linked to eroding source areas. Van Andel pointed to cyclical sequences of heavy minerals, each cycle going upwards from impoverished to varied, in some sedimentary sections. This repetition is strong evidence for the impossibility of intrastratal solution operating over long times as an explanation of the impoverished heavy mineral suites in terms of the relative amounts of zircon, tourmaline, and rutile—the ZTR index (Hubert, 1960, pp. 208–21). Hubert believes this index correlates with general mineralogical maturity of a sandstone and hence is the result of the action of weathering at the source and abrasion during transport rather than diagenesis.

The dichotomy between these two points of view that emphasize source area control, on the one hand, and diagenetic processes, on the other, is more apparent than real. The general statistical tendency noted by Pettijohn continues to be supported in a general way as more and more petrographic data become available, though there are many exceptions. The time scale over which the statistical data used by Pettijohn show definite trends is one of 10^8 years and the mechanism is presumably a slow chemical equilibration on a time scale with that order of magnitude. The time scale of geosynclinal or basinal tectonic cycles that would correlate with cyclical changes in heavy mineral suites controlled by source area evolution is smaller by one order of magnitude, that is, 10^7 years. It is therefore entirely probable that both processes operate simultaneously. In any given case the resolution of this argument lies in regional petrographic mapping that involves not only provenance but diagenesis, the latter with particular emphasis on burial history and the paleohydrologic regimes in which intrastratal solution might have taken place.

A variety of intrastratal solution that is directly observable in thin section is *decementation*. Just as the name implies, this process is one by which a former precipitate, such as a pore-filling calcite cement that had partially replaced detrital grains, dissolves, leaving behind its diagnostic relict texture of corroded grains and remnants of the pore space filling. Decementation implies a reversal from precipitation to dissolution that required some geological event which changed the composition of the pore fluids from supersaturated to undersaturated. Decementation under deep burial conditions has been given as the cause of secondary porosity in many sandstones (Schmidt and McDonald, 1979a,b). Secondary porosity, especially that produced by dissolution of calcite cement, may be the result of increases in calcite solubility controlled by decreases in temperature or increases in the salinity of porewaters. Also CO_2 is generated by oxidation of hydrocarbons so that calcite dissolves. The uplift of a sandstone from a deeply buried position to shallower depths may change the porewater from a slightly supersaturated or saturated brine to a cooler, less concentrated brine. If the temperature effect is greater than the brine salinity effect, calcite will dissolve. Invasion by undersaturated meteoric waters would have the same result. Such a change in position can be inferred by reconstruction of episodes of burial, uplift, and unconformity. If a formation is subjected to a number of such episodes, as those thin sandstones normally found in cratonic interior platforms have been, it may have had repeated reversals of precipitation and dissolution. This has been noted by Walker (1962) and, in relation to geomorphic history for a Mississippian sandstone of Indiana and Kentucky, by Hrabar and Potter (1969) and the Sedimentation Seminar (1979).

Recrystallization

The energetics behind any recrystallization of a substance from one form to another more stable form is the tendency towards a minimum in the Gibbs free energy of the chemical system. If we focus here on the equilibrium relations of a small system, that is, an assemblage of crystals only a few millimeters in size, and assume for the purposes of analysis that it is a more or less independent chemical system isolated from the rest of the rock, we can readily make explicit how such forces operate. Perhaps the simplest example is the recrystallization of a small group of intimately intergrown tiny crystals into a single large crystal. In this system:

$$G_{total} = G_{mass} + \sum_i G_{i_{interface}} \quad (11\text{-}14)$$

where G_{mass} is the contribution to the total made by the mass of all material in the system and $G_{i_{interface}}$ is the contribution of the interfacial ("surface") energy of each crystal. In growth by recrystallization the contribution to the total G of the assemblage that is made by the G_i term is reduced to a minimum as the total area of interface of crystals decreases to the point where it becomes the surface area of a single crystal. This describes the energetics of recrystallization of tiny crystals of quartz in the microcrystalline quartz mosaic of cherts to much larger crystals, or the formation of single large carbonate crystals from an aggregate of smaller ones. Another way of describing this process has been given by Smalley (1967), who calculated theoretical curves for the changes in topological properties and density as a result of cementation in a sandstone undergoing burial and diagenesis.

In the course of recrystallization, inclusions or impurities remain behind in their same relative positions so that relict structures may still be preserved. The recrystallization of small to big crystals might be thought to cause sandstones to become single large crystals, but this does not happen because the magnitude of $G_{interface}$ is negligible for most materials as grain size increases to magnitudes greater than 0.1 mm. This is a general lower limit for the size of preserved detrital grains in many metamorphic quartzites; finer particles have apparently recrystallized. Because the magnitude of $G_{interface}$ is a function of total surface area, it is clear why recrystallization will proceed in very fine-grained materials but not in sand-sized particles. An extended treatment of the thermodynamics of interfacial energy is given by Guggenheim (1967, pp. 45–58) and geological aspects are discussed by Verhoogen (1948) and, as applied to diagenesis, by Berner (1980, pp. 91–97).

In more complex recrystallizations one polymorph will invert to another without change in chemical composition. The most familiar is the recrystallization of aragonite to calcite. The push toward chemical equilibrium is provided by the difference between the G's of the two polymorphs at any given temperature and pressure where ΔG_R^0 = the standard ΔG of the reaction: aragonite → calcite and ΔG_f^0 = Gibbs free energy of formation in the standard state. Using thermodynamic data from Robie *et al.* (1978):

$$CaCO_{3(aragonite)} \rightarrow CaCO_{3(calcite)} \quad (11\text{-}15)$$

$$\Delta G^0_{f(calcite)} - \Delta G^0_{f(aragonite)} = \Delta G^0_R \quad (11\text{-}16)$$

$$(-269.80) - (-269.55) = -0.25 \text{ kilocalories (at 25°C, 1 atm pressure)} \quad (11\text{-}17)$$

Because it is only at one temperature at a given pressure that two such polymorphs can coexist at equilibrium, in general one is unstable with respect to the other and will be expected to recrystallize to it, if the system moves towards equilibrium. Aragonite is known from experimental studies to be the stable form of $CaCO_3$ only at high pressures (Simmons and Bell, 1963); under the temperature and pressure conditions of most sandstones, whether at the surface or deeply buried, calcite is the stable form. The observational confirmation of the tendency towards chemical equilibrium is given by the rarity of aragonite and the exclusivity of calcite in ancient rocks. In contrast, the lack of any strong tendency to equilibrium in the realm of modern sediments, where biological and nonbiological processes both seem to ignore chemical equilibrium over short time scales, is attested to by the abundance of aragonite. Aragonite crystals or cements in sandstones are found only in young rocks. The inference of the former presence of aragonite in older rocks is difficult if no relict structures are preserved, such as, for example, skeletal fragments of organisms known to produce aragonite.

Another example of polymorphic inversion is the recrystallization of opal-A, the amorphous form of silica that is precipitated by diatoms, radiolaria, and other organisms and precipitated abiogenically as the original silica precipitate in some sandstones. Opal-A changes to opal-CT, a form of disordered cristobalite, and then to quartz (for a discussion of the mineralogy and nomenclature of opal-A and -CT see Jones and Segnit, 1971). The free energy changes in this system are less precisely known than those in the carbonate system, mainly because opal includes a group of phases of varying crystallographic order, colloidal structure, and included water content. The overall change can be put in terms of the transformation of an amorphous silica to quartz. Using thermodynamic data for quartz from Robie *et al.* (1978) and for amorphous silica from Walther and Helgeson (1977):

$$SiO_{2(\text{amorphous silica})} \rightarrow SiO_{2(\text{quartz})} \quad (11\text{-}18)$$

$$\Delta G^0_{f(\text{quartz})} - \Delta G^0_{f(\text{amorphous silica})} = \Delta G^0_R \quad (11\text{-}19)$$

$$(-204.66) - (-202.89) = -1.77 \text{ kilocalories (at } 25°\text{C, 1 atm pressure)} \quad (11\text{-}20)$$

The individual steps include the dissolution of the original opal-A [essentially the same as Eq. (11-10)], reprecipitation of opal-CT, gradual ordering of opal-CT, dissolution of opal-CT, precipitation of chalcedony or microcrystalline quartz, and finally, grain growth and the formation of macrocystalline quartz. Each stage in this path involves a reduction in G for the silica as in the overall reaction given in Eqs. (11-18), (11-19), and (11-20), for at earth surface temperatures and pressures low quartz is stable with respect to all other forms and cristobalite has a lower G (is "more stable") than amorphous silica. The decrease in G due to polymorphic inversion is accompanied by a decrease in the contribution of $G_{\text{interface}}$ as the precipitate becomes more coarsely crystalline and, in the case of opal-CT, by a decrease in the contribution to G related to crystallographic disorder. The dissolution kinetics of opal-A have been studied by Hurd (1972), Stein (1977), and Lawson *et al.* (1978) and reviewed by Berner (1980, pp. 190–194). The dissolution rates of natural, untreated diatom surfaces are strongly dependent on temperature and on inhibitors, in particular, magnesium and organic coatings. The kinetics of opal-CT precipitation have been studied by Kastner *et al.* (1977) who found the process strongly favored by higher temperatures, magnesium, and alkalinity and inhibited by clay minerals. These factors involved in the polymorphic transformations of silica make understandable the restriction of opaline silica cement to young sandstones and the role of burial temperatures and porewater chemistry.

Experimental evidence (Osborn, 1953) and thermodynamic data are compatible with the inference from geological occurrences that the observed authigenic anatase in sandstones represents the stable form of TiO_2 at earth surface temperatures and pressures and that rutile and brookite—both of which are also known to have authigenic occurrences—will with time invert to anatase. A complicating factor in the stability of and inversion to anatase is the amount of iron present: anatase has the lowest amount in solid solution and is most stable in low-iron environments. This inversion is obviously such a slow one that it takes long geologic times, of the order of 10^8 years, to accomplish and may account in part for the increased abundance of anatase in older sandstones, a negative persistence related to diagenetic effects by Pettijohn (1941).

Partial Dissolution and Alteration

The reaction type affecting a great many rock-forming minerals, particularly the silicates, is partial or incongruent dissolution with alteration of the solid composition. The solid dissolves and reacts with water in such a way that both solute and residual solid are different in composition from the original solid phase. For one such reaction, the kaolinization of feldspar, the controls on the reaction are fairly well understood. Long known from weathering studies, the petrographic evidence for feldspar alteration during diagenesis is unequivocal; a grain with the geometry of a detrital cleavage fragment of feldspar is composed of small books and worms of kaolinite, a composition that would be highly unlikely to survive as an original detrital grain (Fig. 11-7). Equation (11-21a) describes the reaction for the hydrolysis—

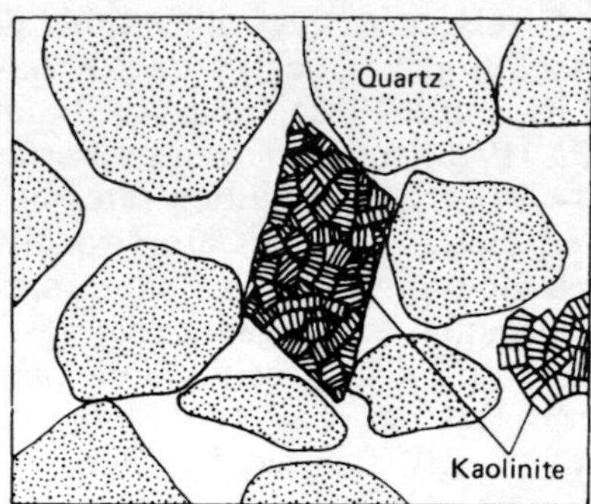

FIGURE 11-7. Alteration of a detrital feldspar grain to kaolinite after deposition. The origin as a feldspar is deduced from the typical cleavage-bounded shape and size and, if present, relics of the original feldspar. Also see altered feldspar in Fig. 5-5.

whether during weathering or diagenesis—of a K-feldspar and indicates the relevant chemical variables: the activities of the dissolved aqueous species K^+, H_4SiO_4, and H^+. Activity of H^+ may be converted to pH.

$$2KAlSi_3O_8 + 2H^+ + 2HCO_3^- + 9H_2O \longleftrightarrow Al_2Si_2O_5(OH)_4 + 4H_4SiO_4 + 2K^+ + 2HCO_3^- \quad (11\text{-}21a)$$

It is important to note that aluminum does not appear as a species in solution but is conserved as a solid in the aluminosilicate structure. The equilibrium constant is simplified if the kaolinite and feldspar are pure and so have unit activities; and because the solutions are sufficiently dilute, we can consider $a_{H_2O} \cong 1$ so that

$$K_{\text{feldspar} \rightarrow \text{kaolinite}} = \frac{(a_{K^+})^2 \, (a_{H_4SiO_4})^4}{(a_{H^+})^2}. \quad (11\text{-}21b)$$

Because the equilibrium constants for the equations that relate transformations among K-feldspar, K-mica (muscovite, illite), kaolinite, and gibbsite all involve one or more of the same activities, and the ratio a_{K^+}/a_{H^+} recurs, equilibrium diagrams can be drawn in terms of the logarithms of the activities that show the composition of aqueous solutions in equilibrium with the solid minerals in the system (Garrels and Christ, 1965, pp. 359–62; Hess, 1966; Helgeson *et al.*, 1969). From this kind of diagram (Fig. 11-8) one can predict that if log a_{K^+}/a_{H^+} values fall much below 5 or 6 and H_4SiO_4 concentrations drop much below 10^{-3} moles/liter, K-feldspar will kaolinize. A similar diagram represents the equilibria in the corresponding Na-system (Fig. 11-9).

It is the geologist's task to infer the geological environments, particularly the chemical environment of the subsurface, in which the vari-

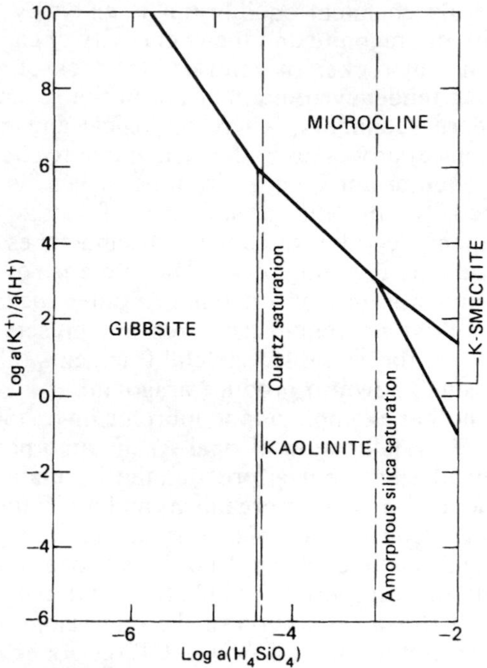

FIGURE 11-8. Diagram showing compositions of aqueous solutions in equilibrium with K-feldspar and clay minerals. The compositions are given in terms of the logarithm of the activity of H_4SiO_4 and the logarithm of the ratio of the activities of K^+ and H^+. (Modified from Helgeson *et al.*, 1969, Fig. 2).

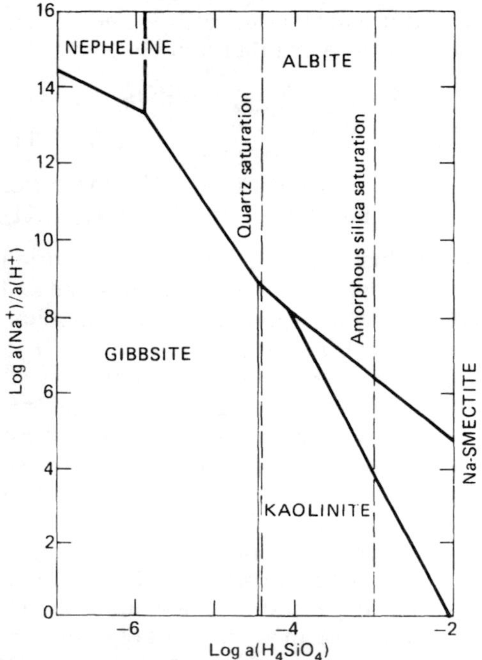

FIGURE 11-9. Diagram showing compositions of aqueous solutions in equilibrium with Na-feldspar and clay minerals. (Modified from Helgeson *et al.*, 1969, Fig. 3).

ables assume the values specified. In considering the in-place kaolinization of feldspar, one can infer a groundwater that is low in dissolved solids, including being low in K^+ and H_4SiO_4. A typical analysis might show the silica activity to be about $10^{-3.8}$, the K^+ activity about 10^{-3}, and the pH about 7. The most likely candidate is a groundwater of meteoric origin within a few hundred feet of the surface at most. Some late kaolinite cements or complete dissolution of feldspar may thus be related to meteoric water infiltration. We might also consider a more concentrated groundwater at greater depths and higher temperatures in which the K^+ activity was about 1 and the silica activity about 10^{-3}, figures more characteristic of some oil-field brines. In that case, the pH would need to be well below 5 to hydrolyze the feldspar, bearing in mind that the equilibrium diagram changes with temperature. Such acid porewaters are uncommon in deeply buried sandstones of most tectonically stable sedimentary basins; they would be expected mainly as a reflection of hydrothermal regimes and high CO_2 activities related to tectonism and very high heat flows, possibly related to plutonism and volcanism. In the more common case, where the pH is 7 or higher, the high K^+/H^+ ratio coupled with a high silica activity places the water into the illite field. At still higher SiO_2 and K^+/H^+ ratios authigenic K-feldspar might be expected to precipitate. Slightly different conditions are responsible for the dissolution, congruent or incongruent, of plagioclase, whose dissolution reactions are strongly dependent on the Na/Ca ratio of the feldspar as well as the porewater composition (see p. 442). In all of these alterations, shifts in acid balance, such as those to increased CO_2 activity, link the carbonate and silicate systems. Secondary porosity formed from dissolution of carbonate may become filled with later silicate phases such as kaolinite and illite.

Many of the clay minerals of diagenetic origin that are found in sandstones are derived from an alteration process that is more complex than simple precipitation or dissolution. The alteration of volcanic glass particles to smectite takes place by reactions that involve incongruent dissolution according to schemes such as the following:

$$\underset{\text{``volcanic glass''}}{Na_2KCaAl_5Si_{11}O_{32}} + MgSiO_3 + H_2O + \underset{\text{carbonic acid}}{4H^- + 4HCO_3^-} \rightarrow \underset{\text{smectite}}{Na(Al_5Mg)Si_{12}O_{30}(OH)_6} + Na^+ + K^+ + Ca^{++} + 4HCO_3^- \quad (11\text{-}22)$$

in which the composition of the glass is represented by a mixture of one part K-feldspar, one part anorthite, two parts albite, and one part enstatite; and the composition of the smectite is somewhat simplified.

These reactions forming diagenetic clays involve the preservation in solid form of much of the aluminosilicate, though not necessarily in the same crystallographic structure. One can simply visualize the process as one in which cations are exchanged, lost, or gained, and silica is gained or lost relative to alumina, by reaction of a detrital solid with the surrounding solution. The petrographic character of clay mineral alterations and replacements fits well with this idea, for they occur typically in sandstones with either clay matrix or argillaceous rock fragments; in fact, the idea that much clay matrix originates from alteration of rock fragments depends intrinsically on this mechanism.

The origins of pore-filling clay minerals authigenically precipitated in sandstones with little or no detrital clay or aluminosilicate minerals are less simply explained; such pore fillings appear to the petrographer to be precipitates from a homogeneous solution, for no relic of a different mineral as precursor is found. Typical of these pore fillings is the relatively coarsely crystalline kaolinite or dickite that is found in many quartz arenites. To form such a precipitate, the entering solution must have carried both dissolved silica and alumina. The form in which dissolved silica is carried is monomeric silicic acid, H_4SiO_4. Analyses of many groundwaters confirm the prediction based on the chemistry of H_4SiO_4 that silica can be carried in sufficient concentrations, from a few to a few tens of parts per million (ppm), to allow the precipitation of clay minerals (Siever, 1962, 1968a). The amount of dissolved alumina that can be carried in solution, on the other hand, is severely limited by the way in which its solubility is dependent on pH. In the pH region from about 4.5 to 8, the solubility of alumina is negligible for practical purposes (Fig. 11-10). Below pH 4.5 it becomes very soluble, with the dominant species in solution being Al^{+++} ion; above pH 8 it becomes very soluble with the dominant species being AlO_2^- or some polymerized variant of it. But the difficulty is that the pH range in which alumina is insoluble is just that pH range in which most natural waters, and apparently the great bulk of subsurface waters, fall.

The seeming contradiction between the chemical behavior of aluminum and petrographic evidence suggesting aluminum transport may be resolved if further study shows the geologic likelihood of waters of the composition

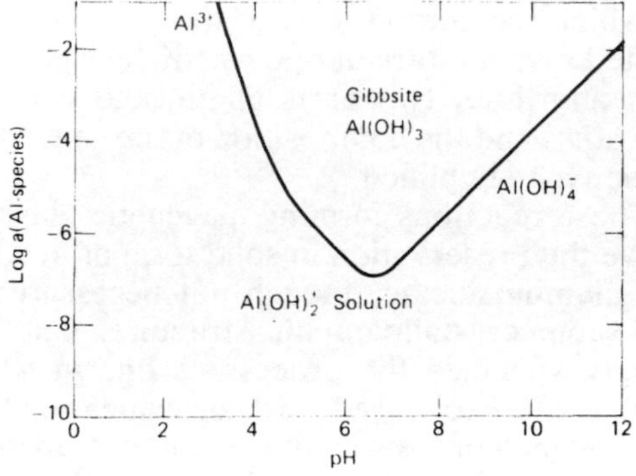

FIGURE 11-10. Solubility of gibbsite given in terms of the activity of Al species vs. pH. Different dissolved Al species dominate the solution in different regions. The curve shows the very low solubility of gibbsite, less than 10^{-5}, between pH 4 and 9. (Modified from Drever, 1982, Fig. 5-3).

needed. Such waters might be meteoric with low cation and moderate silica concentrations with pH values well below 5 (see kaolinite–solution boundary in Garrels and Christ, 1965, pp. 352–58). It is also possible that aluminum may form a soluble ion, such as an aluminum-hydroxy complex, that would permit transport. Without those possibilities, most groundwaters would need extraordinarily long times at reasonable flow rates, of the order of 10^8 years, to precipitate the quantities of kaolinite seen in some sandstones.

The incongruent dissolution of carbonates may play some role in diagenetic transformations in sandstones, though it is unlikely to be a major effect. The incongruent dissolution of dolomite or a magnesium-rich calcite will lead to a solution more enriched in Mg^{++} ion and a solid residue that is Ca-rich; from such a process we can get the formation of diagenetic calcite as a "replacement" of dolomite. Our knowledge of such diagenetic processes in carbonates in sandstones is small because we have few good chemical analyses of the precise compositions of carbonates. Optical determinations of the amount of Mg in calcite or excess Ca (or the amount of Fe) in dolomite are inadequate; X-ray diffraction or electron microprobe techniques must be used for this kind of information.

Porewater Reactions and Formation Waters

To study the chemical reactions that characterize diagenetic processes, we should consider the ways in which pore fluids—water, oil, and gas—change their composition and participate in the reactions. Porewater is an important phase making up a porous rock in its natural condition in the ground. Only at or near surface outcrops, in the vadose zone above the groundwater table, will sandstones be "dry." At the moderate to low temperatures and pressures under which all diagenetic reactions take place a fluid phase is always present, ranging from porewaters of modern sediments to concentrated brines, oil, and gas of deeply buried ancient sandstones. Because of the negligibility of purely solid–solid reactions under low-temperature conditions, the fluid is the important mediator in all reactions. A static pore fluid reacts with the rock just as does a fluid with a solid in a sealed laboratory vessel. The more common moving fluid reacts with the rock in such a way that one must consider the rates at which any dissolved reactants, the raw materials of reactions, are supplied and dissolved products removed from solid minerals, as well as the rate of the reaction itself. This leads to the necessity of evaluating possible reservoirs that supply the reactants and sinks that absorb the products, that is, the entire geological and geochemical system.

Naturally, geologists and petrologists working only with the solid phases of the rock will tend to ignore or insufficiently consider the chemical composition of the fluid phases. This tendency is increased by the ease of recovery of solid samples compared with the great difficulty of getting a representative sample of the fluid. Almost all of our information comes from formation water samples taken from oil wells. Information from springs and seepages is useful but is usually not traceable to any specific formation lying at any given depth. Even the information from oil well samples is subject to so much mixture from multiple sources, contamination from fluids in and out of the drill hole, and changes during sampling and storage that only a fraction of the analyses are reliable. The great bulk of the information on formation waters is not published although some is available (Young and Galley, 1965; Collins, 1975).

The composition of pore fluids is given by a chemical analysis, usually in terms of the major chemical species assumed to be in solution, measured in parts per million (ppm), more rarely as molarity. The species usually included are the abundant alkali metal ions: Na^+ and K^+ (in older analyses these may be lumped); the abundant alkaline earth elements: Ca^{++} and Mg^{++}; the balancing anions: HCO_3^-, Cl^-, and

$SO_4^=$; and sometimes dissolved SiO_2; pH may also be given. Other elements, such as I^- and Br^-, may be included in specific studies. The analysis of organic fluids, almost always hydrocarbons, leads to compositions given in terms of the elements C, H, S, N, and O, or in terms of the relative proportions of the major types of organic compounds present (Tissot and Welte, 1978; Hunt, 1979). In addition, specific compounds may be identified, such as any of the normal paraffins, ranging from methane to those with carbon numbers greater than 20, or specific members of the aromatic (benzene) group. A further refinement in the determination of the chemical composition of pore fluids is an analysis giving stable isotope ratios of specific elements. The most common are $^{18}O/^{16}O$, $^{12}C/^{13}C$, $^{32}S/^{34}S$, and D/H (deuterium/hydrogen). Such isotope ratios, in determining the fractionation between fluids and minerals of the sandstones, may help in deducing the marine, meteoric, or magmatic processes involved in the formation of the waters. Studies of some near-surface waters include C^{14} or tritium analyses that can be used to give the "age" and, therefore, recharge history of the water.

Much of the information and ideas about the origins of formation waters comes from the study of fluids in sandstone formations. There is now a consensus that, though many porewaters may have started as interstitial seawater in marine sediments, their present composition comes from reaction of water with the mineral components of the rock (see, for example, Nesbitt, 1980; Stoessel and Moore, 1983). The more soluble the minerals, evaporites being an extreme example (Carpenter, 1978), the more the effect on the composition. Meteoric waters migrate downward through soil zones and then down dip in permeable sandstones and mix with deeply buried waters that are partly inherited from trapped seawater. Brines moving upward from overpressured formations may enter shallower sandstones and react with their constituents. The sandstone petrologist needs to know the types and ranges of compositions of porewaters and how they relate to hydrodynamic movements of water through porous rocks. Those movements themselves are functions of the structure, surface geomorphology, climate, and geothermal regime of the sedimentary basin. Correlation of petrology and formation water composition leads to informed hypotheses about the origin and historical geology of the diagenesis of the sandstones studied (Merino, 1975b; Nesbitt, 1980; Land and Prezbindowski, 1981). For example, the albitization of feldspar shown in Eq. (11-27) may be promoted by invasion of brines containing Na^+ ions that had previously traversed and dissolved evaporites.

A special type of process which may alter the composition of porewaters is salt filtering by a semipermeable membrane, including both chemical and reverse chemical osmosis (DeSitter, 1947; Bredehoeft *et al.*, 1963; Graf, 1982). The osmotic process is dependent on the behavior of shales as semipermeable membranes and is powered by the pressure differential with depth. In any ordinary osmotic process there is a difference in concentration of salt between the two sides of a membrane, which gives rise to a pressure differential, the osmotic pressure, which induces movement of the solvent from the low- to the high-concentration side (Fig. 11-11A). In reverse osmosis there is the same concentration on both sides of the membrane but a pressure differential is impressed between the two sides of the membrane so the solvent flows from the high- to the low-pressure side (Fig. 11-11B). In a sedimentary column, the pressure differential is linked to varying depth of overburden. Thus a sandstone underlying a shale that acts as a semipermeable membrane will have a higher concentration of salts than a sandstone above the shale. Osmotic effects may also arise from differential pressures associated with overpressured formations (Jones, 1967; Graf, 1982). The osmotic mechanism has been demonstrated in the laboratory (McKelvey and Milne, 1962; von Engelhardt and Gaida, 1963) but, though an attractive hypothesis, is difficult to prove as the major process causing concentration differences in porewaters in sedimentary basins, for it is necessary to demonstrate an association of pressure differences with differences in salt content that cannot be ascribed to other origins. The geochemistry of oil-field brines and other formation waters has been reviewed by Collins (1975) and Hunt (1979, pp. 190–195).

Oil, gas, and other organic deposits are economically the most important subsurface diagenetic fluids found in sandstones. It is now a commonplace to note that oil and gas are normal, not unusual, components of a great many sandstones. The composition, distribution, and origin of oil, gas, and asphalts of various kinds in sandstone deposits is a subject which has an immense literature, much of which pertains to deposits found in sandstones (Levorsen, 1967; Tissot and Welte, 1978; Hunt, 1979). It is important to consider oil and gas in sandstones not only for their economic significance but because the presence of oil may have an effect on

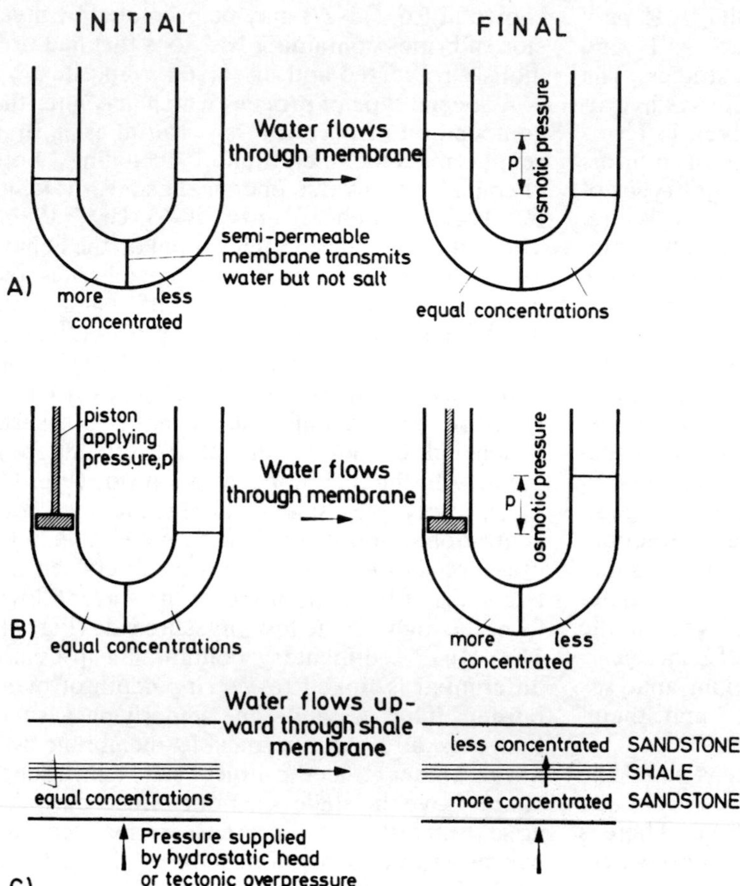

FIGURE 11-11. Initial and final (equilibrium) states of (A) normal and (B) reverse osmosis in a U-tube and (C) reverse osmosis in two sandstones separated by a thin shale acting as a semi-permeable membrane.

other diagenetic processes, such as the inhibition of quartz diagenesis noted by Philipp *et al.* (1963).

Sequence and Timing of Mineral Transformations

The ordering of many diagenetic events with respect to time is one of the most important keys to the ultimate interpretation of the ways in which the geological and geochemical history of a sandstone affected its composition and texture. Following that ordering we need to establish the real times—or absolute ages—of the various minerals and textures. Knowing the age of each diagenetic event could allow us to correlate diagenesis with geologic history. Equally important, it would allow correlation with time–temperature history, which is the key to oil maturation or degree and kind of cementation, to give two examples.

Cement Paragenesis

Replacement parageneses are the main guides to a relative time ordering of cements. A good example would be the implied history of a pseudomorph of chert (microcrystalline quartz) after dolomite (Fig. 11-12): the original diagenetic precipitate was a rhombohedron of dolomite—which itself may have replaced an earlier precipitate of calcite—followed by a later replacement, either directly by chalcedonic (microcrystalline) quartz or first by an amorphous or opaline silica that later recrystallized to microcrystalline quartz.

In the absence of pseudomorphs a replacement series is usually subject to alternative interpretations, for a rational crystal face bounding the interface between one mineral and another may mean either that one mineral precipitated first with its rational face and that the other grew around it anhedrally, or that the anhedral mineral precipitated as a euhedron first, later to be replaced by the euhedral mineral. A

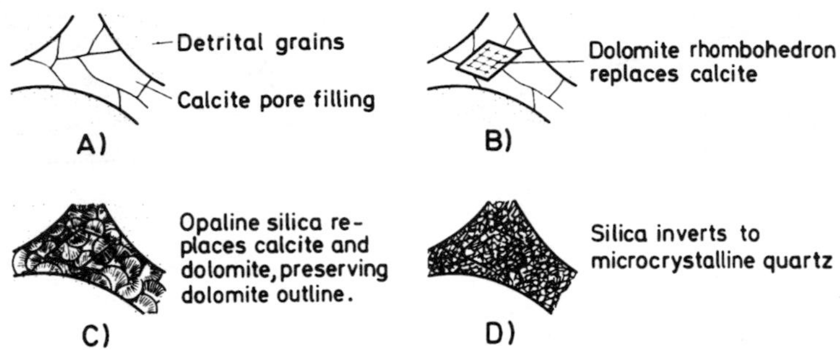

FIGURE 11-12. Possible evolution of a chert pseudomorph after dolomite.

first step in the study of such a boundary is to establish which mineral the euhedral face belongs to. An excellent example of how this is done is given by Gilbert (1949, p. 10) and Gilbert and Turner (1948), who point out the necessity of using the universal stage for proving the correct assignment of the face to the appropriate crystal. When such work is done, it usually confirms the preliminary guess based on petrographic common sense. Increasingly these textural relations in sandstones are being explored by scanning electron microscopy (e.g. Pittman, 1972; Wilson and Pittman, 1977; Glennie *et al.*, 1978; Davies *et al.*, 1979; Bjørlykke, 1979, 1984).

It is not always necessarily true that the replacing mineral will have its own rational faces, for studies of etch figures of minerals show that dissolution, a part of the replacement process, may result, in unusual cases, in the formation of crystal faces of the dissolved substance; in such a case the replacing mineral precipitates anhedrally around the rational face of the replaced mineral. This caution does not seem to apply to either quartz or carbonate, the two most abundant participants in replacement in sandstones; both of these minerals assume euhedral faces only when precipitated and not when dissolved under the chemical conditions found in sediments. This seems to be the case for many other replacing minerals, such as zeolites and feldspars, as well as clay minerals that display clear euhedral boundaries.

A wealth of petrographic studies illustrate the ways in which diagenetic parageneses in many different kinds of sandstones are evaluated; see, for example, *Sandstone Diagenesis* (Taylor, 1978) *Aspects of Diagenesis* (Scholle and Schluger, 1979), and *Clastic Diagenesis* (McDonald and Surdam, 1984). Loucks *et al.* (1977, Table 1) give an instructive table showing the basis for judgment of the paragenesis of the Frio Sandstone, shown in Fig. 11-13. In most contemporary studies, newer techniques of investigation, in addition to careful petrographic microscopy, are used in deciding the order of events. Cathodo-luminescence will reveal the presence of growth (zone) lines in carbonates growing into pore space that are absent when the carbonate has replaced another mineral. The use of the electron probe for chemical analysis may reveal the composition of inclusions too small to be resolved by ordinary microscopy. Electron-probe chemical analysis may also show that a portion of a mineral is so pure an end-member of a solid-solution series that it almost certainly is of low-temperature origin and therefore almost certainly of diagenetic origin.

Some early workers hypothesized a general "law" governing replacements (Cayeux, 1929, pp. 234–51; Waldschmidt, 1941). Waldschmidt, for example, stated that in the Rocky Mountain sandstones of various ages that he studied there was a definite order of precipitation, so that if there was one cementing agent, it was always quartz; if there were two, quartz was first and calcite second; if there were three, quartz was always first and dolomite second, followed by either anhydrite or calcite as the third; and so on. More recently, Schmidt and McDonald (1979b, Fig. 21) have proposed a general scheme for sequences of cements and porosity changes in quartz arenites. In contrast, other describers of cements have noted a great variety of sequences; most have found it difficult to see any overwhelming statistics for one kind of sequence or another. This is not to say that extensive future work may not reveal some general tendencies. For example, there does seem to be evidence that in the great majority of arenites in which quartz and calcite are diagenetic precipitates, the quartz was precipitated first and calcite second. In volcanogenic sandstones

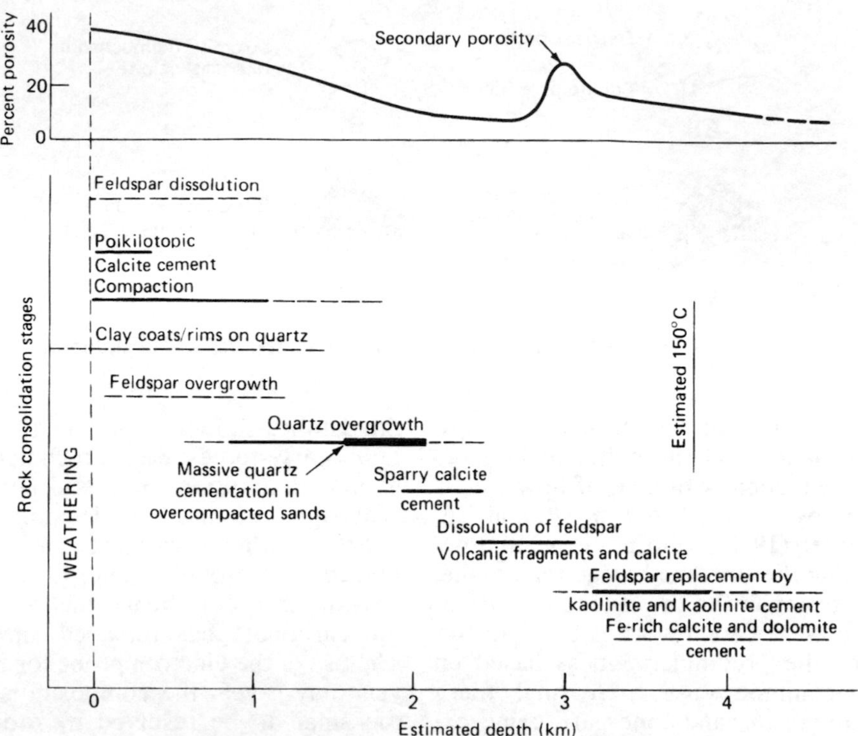

FIGURE 11-13. Paragenesis of cements in Frio Sandstone of Texas. Accompanying porosity changes in one region shown above. (Modified from Loucks *et al.*, 1977, Table 1 and Fig. 7).

the sequence glass to zeolite to feldspar–mica seems to be a general rule. The existence of such general tendencies should be powerful evidence for a generally applicable series of geologic events that determine the course of diagenesis of sandstones of various types and geologic origins.

A series of precipitates with a definite time order can be used to infer such mechanisms as fractional crystallization of salts from seawater. Thus precipitation of diagenetic gypsum, anhydrite, and carbonate in sandstones may be linked to evaporation and groundwater movement on a sabkha flat (Shearman, 1963; Illing *et al.*, 1965; Dean and Schreiber, 1978).

Though one series of diagenetic events may obliterate evidence of previous episodes, the experience of petrographers shows that some relics are frequent. The probability of some preservation dictates a careful petrographic search for the unusual that may prove to be a relic.

Diagenetic Ages

The "when" of diagenesis is a historical geological question. Mechanisms that account for the observed time-ordered sequence of diagenetic effects need to be tied to the period in the sandstone's history when such events took place. The post-depositional history of a sandstone may in fact be vital information needed before a mechanism can be proposed or validated. Radiometric dating of diagenetic precipitates is the best direct evidence but such age analyses are difficult because of the problem of extracting enough pure diagenetic precipitate to perform the analysis and because the mineral may have been precipitated slowly over a long geologic time; in that case the age measured would be an average. If separable, authigenic feldspars should be datable by $^{40}K/^{40}Ar$ decay. Authigenic clays were dated by implication by Hurley *et al.* (1963), who showed that the ages of some size fractions of shales were too young for their stratigraphic age and therefore these fractions had to be authigenic. Thus, $^{87}Rb/^{87}Sr$ methods can give some average authigenic ages for clay-rich sandstones but with no real resolution of the ages of different minerals. The development of high-precision mass spectrometry and the ability to analyze exceedingly small samples, especially as applied to newer dating schemes, such as $^{147}Sm/^{143}Nd$, promises to be of use in more precisely dating diagenetic phases (see discussion in Chap. 2).

The development of the magnetic reversal time scale in the 1960s allowed the dating of iron minerals by remanent magnetic stratigraphy (Van Houten, 1968). The magnetic reversal time scale remains most accurate for the Cenozoic but has been extended to the Mesozoic (McElhinny, 1978) and applied to ancient sandstones (Picard, 1964).

The dating of sandstones rests on the hypothesis that the magnetic iron oxides of red sandstones are diagenetic but formed so early that for practical purposes they are contemporaneous with the deposition of the detritus. Because the iron minerals are precipitated in place, they will have the magnetic orientation imposed by the polarity of the Earth at that time. A major difficulty in establishing a worldwide time scale for magnetic reversals for older sandstones is that the time scale for reversals is of the order of 10^4 to 10^5 years, much too short for matching with the stratigraphic time scale, where the resolution is almost never much better than 10^6 years. But the gradual strengthening of this hypothesis will reinforce the evidence for the very early diagenetic formation of secondary iron minerals in sandstones (see further discussion on p. 462).

Still another way of dating diagenetic events comes from study of pebbles in intraformational conglomerates. For example, the presence of pebbles of well-cemented, very late Mississippian sandstones in early Pennsylvanian conglomerates certainly indicates how early the cement was formed. But such occurrences of cemented sandstone pebbles are uncommon as compared with the variety of shale, limestone, and other lithologies that are found as pebbles in intraformational conglomerates. It may be that sandstones are not as a general rule cemented well enough early enough in their history to form pebbles resistant to stream abrasion. Observation of modern streams draining even well-cemented sandstone terrane rarely reveals significant quantities of small pebbles of sandstone; one finds either large blocks, boulders, and cobbles, or sand grains.

To the extent that a diagenetic event in a sandstone, such as the formation of a cement, can be related to an external geological event that can be dated, the age of the diagenetic event can be determined. If a thoroughly quartz-cemented facies of a sandstone were to be genetically related to tectonic deformation, then the cementation would be dated by reference to the structural event. Dating with respect to unconformities should also be possible. The correlation of diagenetic events and geologic history is further elaborated in the section below on burial diagenesis.

Major Diagenetic Effects

The most obvious diagenetic modification of sandstones is cementation. Though many minerals have been observed, carbonate and silica account for the bulk of cement of many quartz and other arenites. In contrast, in many lithic arenites and most graywackes and volcaniclastics the dominant cementing material is a combination of an altered matrix and precipitated cements composed of clay minerals, zeolites, and other silicates. In many arenites the iron oxides may be the dominant cements, in others gypsum and other evaporitic minerals. Tallman (1949) noted from his study—mostly of arenites—that there is a tendency for younger sandstones to be more cemented by carbonate and older ones to be dominated by silica. One might interpret this in terms of a secular change through geologic time but there seems to be no independent evidence for such a trend. It rather seems likely that the lesser importance of carbonate cement in older sandstones may be explained by a differential preservation similar in general nature to that based on solubility and ease of weathering which Garrels and Mackenzie (1969) proposed to account for the paucity of limestone in Precambrian sections. Because carbonate cement is much more soluble than silica, it is much more likely to disappear as a result of the statistical tendency to be exposed to undersaturated groundwaters at various times during the geologic history of a sandstone. Silica, on the other hand, once precipitated, dissolves less rapidly and is much less likely to become dissolved, and so, persisting, becomes relatively more abundant.

Some sandstones, typically quartz arenites with high original permeabilities, show pore spaces almost completely filled by cements, with no evidence of physical compaction. The source of that cement presents a problem. The most concentrated porewater brines known contain only a small amount of precipitable material so that a closed system cannot be responsible. Required is a flow system with continual replenishment of supersaturated solution, a deduction arrived at long ago by Van Hise (1904, p. 865–68) and discussed in detail by Berner (1980, pp. 118–26). As precipitation proceeds, the pore space is decreased, the permeability reduced, and the rate of precipitation slowed

(Fig. 11-5). The initial permeability and its rate of decrease is a function of grain-size distribution, so that a fine-grained sandstone would become more strongly cemented than a coarse-grained one (Adams, 1964, pp. 1575–77). All plausible flow models, including those complicated by pressure solution, have in common an exponential decrease in rate of precipitation as the process continues. Thus the time required for complete cementation is long, though it is not clear how long. We do know that few Recent or late Cenozoic sandstones are so cemented, even those that have been deeply buried. Unfortunately, too few quantitative data on the degree of cementation in individual sandstones are available (see, for example, Gilbert, 1949; Heald, 1956; Siever, 1959; Sibley and Blatt, 1976; Morris *et al.*, 1979). On the other hand, there is an increasing amount of data on porosity and permeability of sandstone that can be functionally linked to cementation, as reviewed in the section on burial diagenesis.

Carbonate Cementation

Though calcite is the most common carbonate mineral cementing sandstone, many examples are known of dolomite (Swineford, 1947; Sabins, 1962), ferroandolomite (ankerite), and siderite cementation (Siever, 1959). Rhodochrosite concretions are also known from sandstones. Aragonite cement is known only in Recent sands, apparently having inverted to calcite in older sandstones. The carbonate may be found as uniformly or patchily distributed pore fillings and replacements or segregated as concretions or in thin laminae. Well known are sand crystals, with large regions of calcite cement surrounding sand grains (Fuhrmann, 1968). Though every petrographic type of sandstone may be cemented in part by carbonate, the quartz arenites are the most typically carbonate-cemented. Older graywackes, Paleozoic and Precambrian, most typically contain little or no carbonate but many Mesozoic and Cenozoic graywackes have large amounts of carbonate cement. Recrystallized pelagic foraminiferal remains act as carbonate cement in many post-Cretaceous turbidites. Most arkoses contain some carbonate cement (see Fig. 5-6). Carbonate-cemented sandstones may grade laterally into sandy limestones, an occurrence which clearly points to original environmental conditions as responsible for the cement. The absence of carbonate cement in surface or near-surface exposures of some sandstones compared to its presence in the same horizon in the subsurface is clearly explainable in terms of weathering. The opposite pattern, the concentration of calcite cement in near-surface horizons of sandstones, is attributed to caliche precipitation (Nagtegaal, 1969). Because caliche is found in arid climates, such cements are paleoclimate indicators.

Many mechanisms may be responsible for carbonate cementation in various stages of diagenesis (Fig. 11-14). The idea that widespread carbonate cement is an early diagenetic precipitate related to a favorable (presumably marine) environment of sedimentation has long been popular. According to this view, sand was transported into a carbonate-precipitating environment and the cement was essentially a primary precipitate while the sand was exposed to the seawater. Support for this view has come from some modern sediment studies that have shown submarine lithification of carbonate and sandy sediments, especially the report of an aragonite-cemented sandstone from the Atlantic continental shelf (Allen *et al.*, 1969). The precipitation of aragonite or calcite, where it can be shown to be penecontemporaneous on geologic or petrographic grounds, comes as a result of prolonged exposure to supersaturated seawater, for of the nonmarine environments only some evaporitic lakes are carbonate supersaturated. But carbonate supersaturation, which in today's ocean is a function of temperature (and so latitude), may have been widespread and uniform in the more equable temperatures of the oceans of some past times. In any case, it is not easy from geologic or petrographic evidence to prove a primary or penecontemporaneous origin, and if modern sands are to be the basis for judgment, most carbonate cement is added during diagenesis, after at least slight burial.

The early diagenetic cementation of sand in the delta of the Fraser River by low magnesian calcite has been discussed by Garrison *et al.* (1969). They show that the composition of neither the Fraser River water nor the seawater of the delta is compatible with precipitation of carbonate and conclude that the responsible mechanism is the dissolution of carbonate shell material by porewater and reprecipitation higher in the sediment column by expressed water of compaction. Redistribution of shell material in other sandstones has been assumed on the basis of preservation of relict fossil structures in the carbonate cement.

Carbonate cement was the first diagenetic precipitate in many sandstones. Typical is the

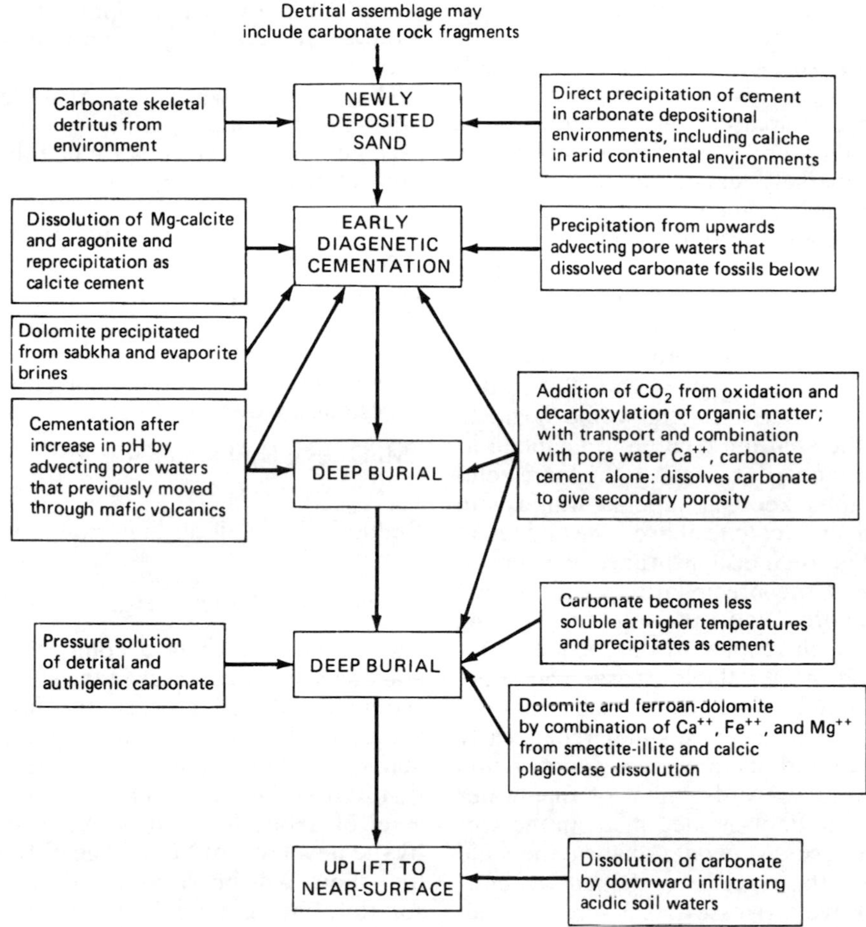

FIGURE 11-14. Carbonate cement precipitation and dissolution routes during diagenesis.

evidence given in an excellent summary by Glover (1963, p. 40): authigenic dolomite is molded directly against the rounded edge of a detrital quartz grain and is surrounded by a secondary quartz overgrowth. In other sandstones carbonate cement was precipitated second, after authigenic quartz overgrowths formed (Siever, 1959). Textures that show replacement of both authigenic and detrital quartz by carbonate that contains remnant inclusions of the replaced quartz are conclusive on this point.

There is no uniformity in the sequence of precipitation among the carbonates: calcite may be first, followed by dolomite, but dolomite as the first precipitate is also common. Sequences of calcite followed by dolomite followed by a second generation of calcite are also known. Though calcite is most common as fairly large anhedral crystals molded around or replacing detrital grains or other elements, dolomite and siderite are much more common as rhombohedra and can be quickly recognized as such by that habit. The fabric of carbonate cements in sandstones may be complex and has many of the characters shown by the cements of carbonate rocks.

Seawater that is trapped as porewater in marine sands may be carbonate supersaturated in carbonate environments such as the Bahama Banks, the Persian Gulf, and other well-known areas. But if the porewater were static and carbonate precipitated as cement, the amount would be negligible. For example, a volume of sand in a layer 1 cm thick and 50 × 50 cm in extent that has 40 percent porosity and contains seawater that is threefold supersaturated and precipitates all of the excess could only precipitate a single spherical grain of authigenic calcite of approximately 1.25 mm diameter.

Chemical mechanisms may affect carbonate solubility and so account for cementation. Aragonite and magnesian calcite, two common carbonate species that may be present as primary detritals in many sandstones, are more soluble than pure calcite. Many calcitic shells are composed of a mesh of extremely fine crystals and organic matter; this material has a higher solubility and dissolves more rapidly than more coarsely crystalline pure calcite. Given the many continental and marine sands that contain enough detrital carbonate to saturate the porewaters with respect to a carbonate phase (only a small amount is needed, due to the low solubilities of about 10^{-3} *M*), we would expect a recrystallization process to proceed. The more soluble materials would start to dissolve and pure sparry calcite would begin to precipitate, the solution tending to maintain itself at saturation with respect to the more soluble phases, thus keeping supersaturated with respect to the pure calcite phase. The end result would be the gradual disappearance of all unstable phases and the precipitation of all excess carbonate as pure calcite, the solution dropping to saturation with respect to pure calcite after the last of the more soluble phases had gone. Though this could happen in a closed system, it is more likely that most sands undergoing early diagenesis are conduits for water of compaction being expressed upwards, most of that water coming from the finer-grained muds in the section. In such cases we would analyze the state of saturation with respect to carbonate by reference to advective processes, as described in detail by Berner (1980).

Alterations in porewater chemistry may affect carbonate mineral solubility and thus dissolution and precipitation. Variations in pH, for example, may change carbonate solubility. Accounting for cementation in this way would seem reasonable at first sight but there is some question of the widespread geological applicability of such mechanisms. The relevant carbonate system equilibria are given by Garrels and Christ (1965, pp. 74–92) and can be used to show dependence of carbonate solubility at a given temperature and pressure on the partial pressure of CO_2 and the ionic strength. Solubility in this system is dependent on pH *only* if the acid–base buffering is controlled by a chemical system other than the carbonate one. If not externally controlled, the pH is dependent on the CO_2 pressure. In the two-phase system (solid–liquid), where no external gas pressure is present, pH depends on ΣCO_2, the total CO_2 present as all carbonate species present in solution (Berner, 1971, pp. 54–57). The solubility of the carbonate may be changed by alteration of the activities of the cations and anions of the pore fluid in a way that would change the ionic strength and alkalinity. Alternatively the ΣCO_2 may be changed.

One way in which the ΣCO_2 can change, other things being equal, is by the oxidation of organic matter according to the following equations (Berner, 1980, p. 82):

Reduction of ferric iron and oxidation of organic matter,

$$4Fe(OH)_3 + CH_2O + 7CO_2 \rightarrow 4Fe^{++} + 8HCO_3^- + 3H_2O \quad (11\text{-}23)$$

Reduction of manganese oxide and oxidation of organic matter,

$$2MnO_2 + CH_2O + 3CO_2 + H_2O \rightarrow 2Mn^{++} + 4HCO_3^- \quad (11\text{-}24)$$

Reduction of sulfate and oxidation of organic matter,

$$SO_4^{--} + 2CH_2O \rightarrow H_2S + 2\ HCO_3^- \quad (11\text{-}25)$$

As organic matter is decomposed by aerobic and anaerobic processes in the shallowly buried sediment, ΣCO_2 increases, tending to dissolve some of the primary carbonate present. As above, the most soluble carbonates will be the first to dissolve, and, in a closed system, a new level of saturation will be reached, controlled by the new level of ΣCO_2. The difference will be that there will be some net dissolution in the solution with greater ΣCO_2. Only if there is a mechanism for transport—advection—from one part of the section to another in which the state of saturation changes in the direction of precipitation can there be significant carbonate cementation. Carbonate cement that contains CO_2 derived from organic matter will have a distinctive $^{12}C/^{13}C$ signature, as the isotopic ratio is lighter in organic carbons than in carbonates precipitated from the ocean.

Advection, coupled with oxidation of organic matter, may be coupled to the accumulation of sedimentary sequences in which sands are buried interbedded with muds that contain some organic matter. Sand–mud associations that are common in fluvial fining-upward cycles or mid-fan turbidites are two examples. The expulsion of water of compaction of the muds drives the advection. The bacterial oxidation of organic matter may keep the steady state concentration of CO_2 poised at somewhat higher levels than those in equilibrium with the overlying fresh or seawaters for much of the section below the top 10–20 cm. In the top layers, however, ΣCO_2 may be significantly reduced by both diffusion as the concentration gradient is increased by virtue of the shorter path and by mixing by any bioturbation going on in the surface layers. This mechanism would predict the precipitation of calcite overgrowths and pore fillings in the surface layers.

Variations in pH of porewaters can occur in response to factors external to the immediate controls on pH exerted by the minerals of a sandstone and its contained porewater. The general control of pH of porewaters probably proceeds in much the same way as in the oceans, which we now recognize as a giant system in which mineral equilibria and input-output relations among the solution and silicates, oxides, carbonates, and other phases combine to buffer seawater at about pH 8 (Sillen, 1961; Garrels and Mackenzie, 1971; Holland, 1978). In the water–rock reactions of a sedimentary basin there are many variations and local inhomogeneities resulting from the diversity of lithologic sequences.

One way in which a particular lithology can cause an increase in pH would be downward circulation of groundwaters through an actively altering mafic volcanic section. A decrease might be caused by waters moving downward through acidic soils. These two are typical in that they are related to downward vadose water percolation through an active zone of weathering. Upward movement also has an effect. Carbonate cementation of channel sands in arid regions was noted by Glennie (1970) and is presumably related to a calichification process whereby evaporation of waters in the vadose zone results in carbonate precipitation.

Before high- or low-pH waters from surface or near-surface reactions will have traveled very far into the subsurface, they will have reacted with the mixture of minerals that generally tends to keep the pH stabilized around or a little above neutrality. An example of how such waters evolve has been given by Garrels and Mackenzie (1967). Dapples (1967, p. 118) suggests pH changes related to CO_2 changes near an outcrop may be important in the replacement of quartz by calcite in certain sandstones. In the evolution of the sedimentary fill of a basin, such groundwater movements may be related to unconformities. Sandstones that are deeply buried and remain so are not subject to such changing of downward infiltering waters from the surface. They are, however, subject to volcanic or hydrothermal invasions in some tectonic settings. Changes in CO_2 content of porewaters follow much the same pattern. High CO_2 pressures may come from waters percolating through soils but sources of CO_2 variation at great depth, other than from volcanic emanations, are difficult to envision. Isotopic evidence based on $^{12}C/^{13}C$ ratios has been used to infer an origin for some early carbonate cement whereby the CO_2 of the carbonate has been supplied by the decarboxylation of organic matter (Silverman, 1963; Spotts and Silverman, 1966).

Calcite may also precipitate because of the increase in Ca^{++} activity resulting from the dissolution of calcic plagioclase and the formation of a clay mineral, as shown here for the simple case of kaolinization of an anorthite.

$$CaAl_2Si_2O_8 + 2H^+ + H_2O \rightarrow Ca^{++} + Al_2Si_2O_5(OH)_4 \quad (11\text{-}26)$$

This reaction also decreases calcite solubility by shifting the pH to higher values. Alternatively, the conversion of the calcic plagioclase to authigenic albite in a porewater with NaCl is an ion-exchange reaction that increases Ca^{++} activity.

$$CaAl_2Si_2O_8 + 2Na^+ + 2Cl^- + 4SiO_2 \rightarrow 2NaAlSi_3O_8 + Ca^{++} + 2Cl^- \quad (11\text{-}27)$$

This reaction affects silica activity and thus might cause dissolution of silica, but would not affect pH.

Dolomite cementation, where it is established as the first early cement, can be explained with reference to the modern environments where dolomitization is now proceeding, such as the supratidal flats of the Persian Gulf (Illing *et al.*, 1965). The process involves the depletion of Ca relative to Mg by $CaCO_3$ precipitation from evaporating waters, followed by dolomite precipitation from the Mg-enriched solution. Much of the precipitation may take place in the subsurface as the result of sinking of the denser evaporite brine. The association of dolomite with anhydrite or gypsum cement, another product of this process, would strongly confirm the environmental inference made from cements. Sabins (1962) showed how one could distinguish among dolomite grains and crystals of detrital, primary, and diagenetic origin. Taylor (1978), in discussing dolomite cement in a fluviatile sandstone of the North Sea Basin, suggested diagenetic reorganization of calcium carbonate detritus in basal zones of the sandstone by Mg-rich porewaters. For such a Ca–Mg exchange to proceed, one would need steady contribution from a dissolving Mg-rich phase, such as a magnesian calcite, or advection of Mg-rich porewaters from elsewhere in the section. Because most sections tend to be depleted with respect to Mg and enriched in Ca,

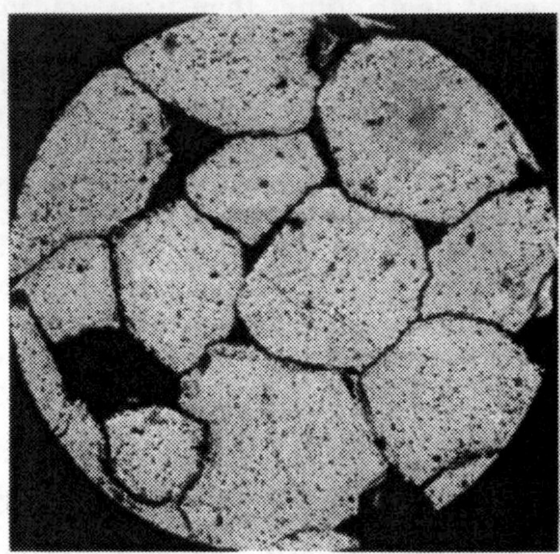
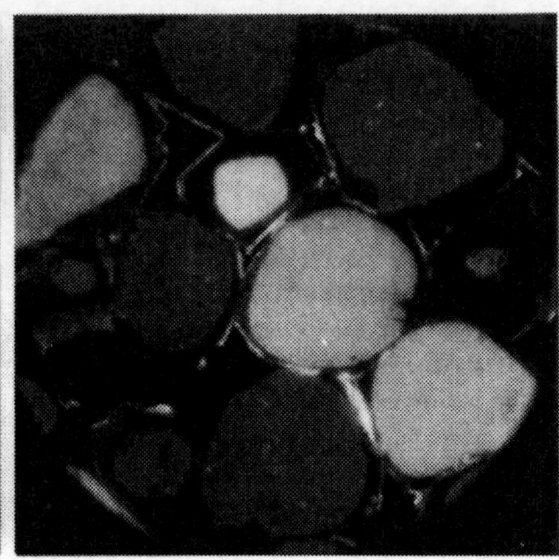

FIGURE 11-15. Comparison of photomicrographs of the Silurian Hoing Sandstone of Illinois taken in plane-polarized light (left) and by cathodo-luminescence (right). Detrital quartz grains appear to have sutured and cancavo-convex borders, indicating pressure solution in plane-polarized light, but cathodo-luminescence reveals that there is extensive secondary enlargement with little or no pressure solution. (Photos by R.F. Sippel).

this would have to be a special situation. Ferroandolomite—ankerite—cement in subsurface sandstones of the Eocene in Texas has been reported by Boles (1978), who proposed that the required Mg^{++} and Fe^{++} were derived from the conversion of detrital smectite to authigenic illite.

The calcite cement that is found replacing secondary quartz in so many sandstones is a later diagenetic product related to a redistribution of carbonate within the sedimentary rock pile. Schmidt and McDonald (1979b) suggest that carbonate dissolved in zones where secondary porosity is developing is transported upward, in part cementing higher sandstones. The solubility of carbonates decreases with increasing temperature, and with a much smaller effect, increases with increasing pressure. In most basins the net effect of burial is to decrease the solubility of carbonate and this may account for a small part of the cement but, unless large quantities of porewater are pumped through the sediment, relatively small amounts of carbonate would be produced. What does happen as a result of burial is the increase of pressure solution, which is usually hypothesized to explain silica cement (discussed in detail on p. 454). Pressure solution is also a factor in limestones and sandstones containing framework particles of carbonate which abut with small contact areas. Though the pressure-solution effect is not known quantitatively, its magnitude can be estimated, from the hydrostatic pressure effect on solubility, to be greater for carbonate than for quartz. If this is so, an explanation for some of the late carbonate cement lies in pressure solution of carbonate grains in the sandstone or in nearby formations, either limestones or sandstones, probably with porewater transport over limited distances.

Silica Cementation

The most common form of silica cementation is the secondary enlargement of quartz grains by optically continuous overgrowths that results in euhedral crystal faces or a mosaic of interlocking overgrowths (Fig. 11-15). There is no evidence of such cementation in modern sands, though Harder and Flehmig (1970) and Mackenzie and Gees (1971) have reported the experimental precipitation of quartz under simulated sedimentary conditions. Silcrete, formed by silica cementation of soil in arid lands, is normally opaline silica rather than quartz. Other than this, all of our clues must come from older sandstones. In a great many sandstones quartz is the first cement, though it is difficult to establish how early it started forming. In some it is the only cement. Quartz cement is most common in the arenites and much less conspicuous, if not entirely absent, in many wackes.

If quartz is the first cement and it formed under early diagenetic conditions, the precipitate must have come from supersaturated porewaters similar to those investigated in recent years for marine and nonmarine sediments buried to depths less than a few hundred meters. Such porewaters exceeding the equilibrium solubility of quartz, about 6 ppm (Fournier and Rowe, 1962; Fournier, 1983), have been shown to be widespread; extreme concentrations of dissolved SiO_2 may reach 80 ppm (Siever *et al.*, 1965; Sayles, 1979; Gieskes *et al.*, 1983). Natural waters are able to persist for long times at levels supersaturated with respect to quartz because of the extreme slowness with which quartz precipitates. Though amorphous silicas precipitate quickly, none of these waters is supersaturated with respect to opal-A, the material of diatoms and radiolarians, which has a solubility of 120–140 ppm—about $10^{-2.7}$ *M* (Siever, 1957). In intermediate states of concentration, the solution may exceed the solubility of opal-CT, which precipitates more rapidly than quartz. This disordered cristobalite variety of opaline silica has a solubility of about 70 ppm ($10^{-2.9}$ *M*) at 25°C (Fournier and Rowe, 1962). This material may later invert to chalcedony or microcrystalline quartz. Such authigenic chert rims on detrital grains may thus be a record of former high silica porewater values.

In the case of quartz precipitation, just as in that of carbonate, there is not enough supersaturation, only a few tens of parts per million, to account for more than a tiny fraction of the authigenic quartz present if porewater does not circulate extensively through the sand bed. To produce a porosity reduction of even 10 percent, waters must circulate many times while constantly precipitating the supersaturation excess as overgrowths. Such extensive circulation is characteristic of some alluvial sands, where shallow groundwaters of meteoric origin may be constantly moving in the course of supplying water to perennial streams. It may also happen in shoreline or coastal plain sands in areas where groundwater recharge is sufficient for infiltration to occur even in sands far from shore on the continental shelf. Blatt (1979) has discussed the kinds of flow rates required to reduce porosity by quartz cementation and concludes that flow rates in deeply buried formations are too slow and that the bulk of quartz cement in quartz arenites must be precipitated at shallow depths.

The dissolved silica in supersaturated porewaters comes from various sources (Fig. 11-16). The silica in porewaters of marine sediments is mainly derived from the dissolution of the amorphous silica skeletons of diatoms, radiolarians, siliceous sponges, and other silica-secreting organisms (Siever *et al.*, 1965; Hurd, 1972). The siliceous parts of these organisms start dissolving immediately after sedimentation and continue to dissolve as long as the waters are undersaturated with respect to amorphous silica. Thus a marine sand may become emergent on a coastal plain some time after sedimentation and still contain siliceous skeletal material which will continue to dissolve in circulating artesian groundwaters. In this way biogenic silica ultimately gets converted to authigenic quartz overgrowths after reprecipitation.

Mineral transformations of silicates, including clays, can also account for much of the silica (Siever, 1957, 1962; Towe, 1962). Weathering of feldspars and other silicates by meteoric water infiltration of the subsurface is a major source of dissolved silica that may get reprecipitated down the flow direction in the same sandstone or in other formations. In this way many alluvial sandstones may become depleted in feldspars by a form of weathering after deposition. The same process may result in authigenic quartz production. A large number of sandstones show feldspar dissolution; the process is now seen to be an important diagenetic effect (Land and Dutton, 1978; Pittman, 1979). Longer-term diagenetic transformations of clay minerals, as for example in the conversion of a silica-rich smectite to a mixed-layer or illite, can supply silica to porewaters. Many of the alteration reactions of volcanic glass, plagioclase, and mafic minerals to authigenic clays and zeolites result in the release of dissolved silica (Surdam and Boles, 1979) so that silicification in underlying sands is not uncommon and especially a chert cement.

All of these sources of silica can be found in sandstones; they are even more abundant in finer-grained sediments. This has prompted a common hypothesis that waters expressed from muds into sands may transport dissolved silica. The association of augmented silica cement adjacent to clay beds, layers, or galls has been frequently noted. For example, Füchtbauer (1967) showed how quartz cementation in a Dogger sandstone increases towards the shaly margin of the bed and ascribed it to an infiltration of silica-rich water from the shale into the sandstone. The largest part of the water of compaction commonly leaves muddy sediments relatively early in their diagenetic history. Thus it is likely that this kind of water movement correlates with the early diagenetic authigenic quartz that has been noted in so many places. Later transport of silica from diagenetic clay mineral and other silicate reactions would depend on the availability of water from dehydration of smectite to form mixed-layer or illitic clays. That water would be released only after pressures and temperatures associated with deep burial (Hower *et al.*, 1976). On the other hand, Blatt (1979) has argued on the basis of quantita-

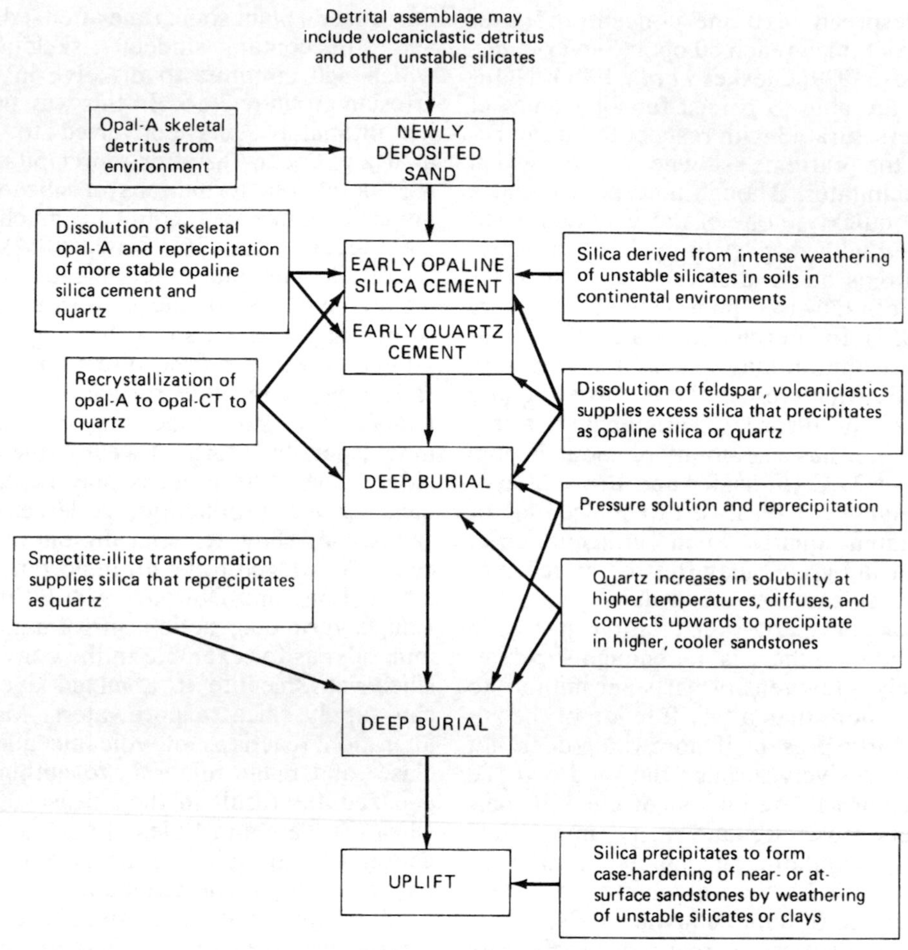

FIGURE 11-16. Silica cement precipitation and silicate dissolution routes during diagenesis.

tive estimates that this explanation cannot account for more than a few percent cement. Shales may act as open or closed systems with respect to water and other components depending on their permeability, but the evidence of dehydration of smectites is strong. The water produced has to go somewhere and the only appropriate hydration reactions would be those in which feldspar is hydrolyzed to form clay minerals. Yet the smectite–illite transformation is probably most closely associated with the opposite: authigenesis of feldspars (Kastner and Siever, 1979). It may be that there is a mass balance within the shale and no net gain or loss from an isochemical system.

The chemical conditions that affect silica solubility are now well known and have been summarized by Siever (1972). Increase of pH above 9 and increase of temperature are the two major factors that increase silica solubility. As previously noted, porewaters tend to have pH values around 8. It is most unusual to find groundwaters with pH values greater than 9; these are usually associated with terrains of altering volcanics or of alkaline lakes in arid regions. The zeolites associated with such sediments should be diagnostic of such an origin.

The temperature effect is of more general importance. In any section in which there are slowly rising groundwaters, or which becomes uplifted, porewaters will cool. As a result the waters become supersaturated and may be expected to precipitate the excess as quartz. The longer the circulation continues, the more quartz is precipitated. This may be one explanation for later diagenetic quartz that follows an earlier paragenesis of quartz and carbonate or other cements. Even under static conditions, diffusion will operate to transport dissolved silica from deeper, warmer regions to shallower, cooler formations. The rate of this transport depends on the temperature gradient and the permeability of the intervening rocks. Given the probability of convective movements of porewaters (Wood and Hewett, 1982) we may expect transport to be much more effective than diffusion alone. Wood and Surdam (1979) have applied convective-diffusion models to diagenetic processes to show how this may happen.

Opaline silica or chert authigenic rims and cement in sandstones are widespread and occur in a variety of sandstone types, most abundantly in volcanic-rich sands, where the opal is clearly derived from the alteration of volcanic glass. Most of the chert is now microcrystalline quartz, but reasoning by analogy with the formation of bedded and nodular chert, it appears that the original precipitate of chert in sandstones was probably opal-A. This original form recrystallized and inverted to opal-CT, followed by transformation to chalcedony or quartz with time and increase of temperature following burial. All of these occurrences imply the presence of porewaters that were supersaturated with respect to the successive phases of silica. Because of the relative rapidity of precipitation of opaline silica phases, it is doubtful if such porewaters would be stable for long times or could be transported over long distances without precipitating. This makes it likely that the source was nearby. The source in volcaniclastics is indigenous but in chert-cemented quartz arenites the source of the silica is exterior to the sandstone. The probable exterior source is soil waters in lateritic terrains, which are known to precipitate opal in the lower parts of the soil zone and in underlying permeable formations. Thus it is likely that chert-cemented sandstones were, at the time of cementation, placed not far beneath a lateritic soil, and thus below a future unconformity. This diagenetic process may thus prove to be a paleoclimate indicator. Alternatively, solutions from overlying volcanics, if present, could supply the silica.

Pressure Solution. The widespread distribution of quartz cement in moderate to deeply buried arenites suggests a generally operating process that is common to most of those sandstones whose framework detrital grains completely support the rock in the gravitational field. That process, which we now call pressure solution, was inferred at least as long ago as the late 18th century by James Hutton and has a long history in the geological literature. Heald's (1956) study of quartz cementation in relation to pressure solution was influential in calling renewed attention to the importance of the process as a major cause of cementation. Weyl (1959) provided a detailed analysis of how the process might work and proposed a functional relationship between time, depth of burial, and degree of pressure solution. Since then there have been contributions and reviews of petrographic evidence by Trurnit (1968), Sibley and Blatt (1976), and Blatt (1979); experimental work by Renton *et al.* (1969), Rutter (1976), Sprunt and Nur (1977a), and DeBoer *et al.* (1977); and thermodynamic analyses by Atkinson and Rutter (1975), DeBoer (1977), McClay (1977), Sprunt and Nur (1977b), and Robin (1978). Pressure solution at high pressures and temperatures has become recognized as an important factor in rock deformation (Rutter, 1976; McClay, 1977; Sprunt and Nur, 1977b). The basic idea of pressure solution is that the high effective pressures developed at point contacts of quartz grains increase the solubility at those points such that they preferentially dissolve. The solution process seems most advanced where clay films are present between grains. The dissolution liberates dissolved SiO_2 to the porewater, which then becomes supersaturated and reprecipitates the SiO_2 as quartz overgrowths, as shown in Fig. 11-17. Pressure solution simultaneously provides a source for all of the dissolved silica that had to be precipitated to make the cement and gives an explanation for the interpenetration of one quartz grain by another. Such interpenetration is a prime cause of porosity reduction. In lithic arenites some interpenetration results from ductile deformation of rock fragments but in quartz arenites almost all is caused by pressure solution. Grain contacts were classified by Taylor (1950) from tangential through concavo-convex to sutured to provide some index of interpenetration. She related these contacts to porosity reduction and depth of burial. A further evidence of pore space reduction that is related to pressure solution is the measurement of minus-cement porosity (Rosenfeld, 1949; Heald, 1956). This is the porosity which would be present if a specimen contained neither chemical cement nor secondary porosity, and does not include replacement cements. Minus-cement porosity measurements on many quartz arenites indicate significant compaction that could not be accomplished by simple mechanical rearrangement and so must be due to solution.

The petrographic evidence for pressure solution is excellent where one can clearly see the boundaries between detrital grains and pore-filling cement. In a great many sandstones, per-

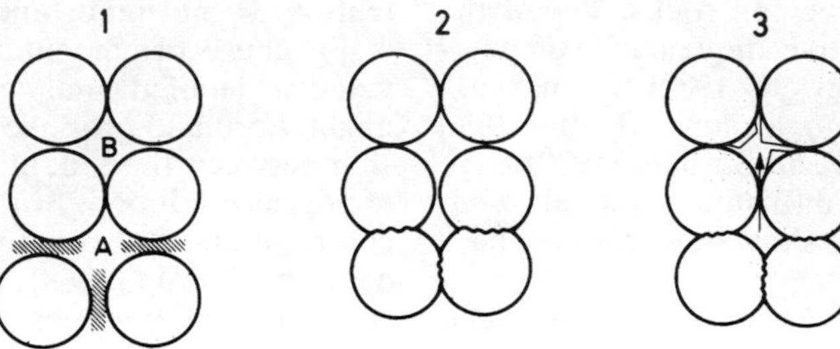

FIGURE 11-17. Sequence of stages in the development of sand grains cemented by pressure solution. In stage 1 the two beds differ only in the presence of interstitial clay. In stage 2 pressure solution has begun, decreasing the volume of pore space A. In stage 3 some fluid from A escapes to B, reduces pressure, and precipitates quartz. (Modified from Siever, 1962, Figure 6).

haps the majority, the boundaries cannot be seen, or such small portions can be seen that no quantitative measurement can be made. The best instrument for this study is a petrographic microscope equipped with cathodo-luminescence (alternatively an electron probe can be used for this purpose). Cathodo-luminescence allows the unambiguous distinction between detrital and authigenic (Sippel, 1968; Sibley and Blatt, 1976) (Fig. 11-15). Sandstones range from those that are relatively uncompacted, composed of detrital grains cemented by authigenic quartz with little or no pressure solution, to those in which there is extensive compaction of detrital grains by pressure solution (Fig. 11-18). Both processes operating separately or together will result in pore space reduction.

Though the evidence for pressure solution is firm, much work remains before we can relate the degree of pressure solution in a quantitative way to the fundamental parameter, the time duration of a given weight of overburden. That relation is now based on (1) theoretical models (Weyl, 1959; Füchtbauer, 1967; DeBoer, 1977; Sprunt and Nur, 1977b; Robin, 1978); (2) a limited amount of experimental data (Rutter, 1976; DeBoer *et al.*, 1977; Sprunt and Nur, 1977a); (3) a few quantitative measurements of minus-cement porosity by cathodo-luminescence (Sibley and Blatt, 1976); and (4) some attempts to link the above to a quantitative overburden history (Siever, 1959, 1983). There have been no detailed studies reported of pressure solution in unmetamorphosed sandstones in relation to large or small tectonic features. The relation of pressure solution to chemical composition of porewaters is still a subject for speculation. Petrographic evidence has suggested the importance of clay films in enhancing pressure solution (Heald, 1956). Theoretical considerations, the requirement of multiple paths for diffusion to speed up solute transfer from points of contact to open pore space, would support that conclusion. The rate at which pressure solution results in both detrital grain compaction and secondary quartz precipitation depends on the relative importance of three processes: the rate of dissolution at grain contacts, the rate of diffusion from contacts to pore space, and the rate of precipitation in pore space. Which of these is rate-limiting at different temperatures and lithostatic and pore pressures is not yet known. According to experiments on halite by Tada and Siever (1986), fracturing and deformation on a micro scale may speed up the process. There is no obvious relation between the degree of pressure solution and quartz cement in the same area. Some sandstones have more cement than can be accounted for by pressure solution. Others show less cement than would be produced by pressure solution; that is, they are net exporters of silica.

In summary, the results of many different types of investigation of pressure solution in sandstones indicate that the process is of moderate importance in arenites and is dependent on time, temperature, and depth of burial. A mass balance of pressure solution and quartz cement cannot be made for most sandstones. Pressure solution does not seem to account for the quartz cement of many sandstones. In others, under optimum conditions of burial temperature, pressure, and clay or other films that allow appreciable diffusion, pressure solution and its derived cement can be matched and result in porosity reduction.

Clay Minerals

The diagnostic alteration of clay minerals is a huge subject; much of the earlier literature was

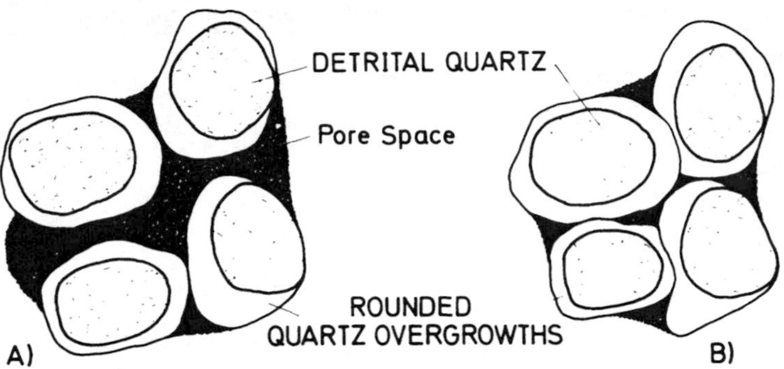

FIGURE 11-18. Alternative possibility for origin of interpenetrating authigenic overgrowths shown in Fig. 11-15. Original detrital grains may have had rounded quartz overgrowths from a previous sedimentary cycle. A) Original deposit; B) after pressure solution.

summarized by Millot (1963, 1970), Grim (1968, pp. 563–66), Dunoyer de Segonzac (1970), and, more recently, Velde (1979). Though much of this literature concerned shales there is an ever growing literature on clay mineral diagenesis in sandstones, reviewed by Wilson and Pittman (1977) and Pittman (1979). The major emphasis of studies on clays in sandstones has been on the formation of interstitial kaolinite or dickite, characteristically associated with quartz arenites; on the neoformation of clays from porewaters; and on the transformations of mixtures of detrital smectites and micaceous clays. The transformation and neoformation of clays associated with volcaniclastics are discussed below. Some of the authigenic kaolinite that is so common in sandstones is derived from the diagenetic dissolution of detrital feldspar. Another kind of kaolinite has the appearance of a well-crystallized, pure pore filling that was precipitated from a solution and shows no evidence of alteration from a precursor feldspar or other clay mineral (see p. 417). The presence of either kind of kaolinite may be evidence for an invasion of relatively fresh groundwaters entering the sandstone from a recharge outcrop belt in which there was a supply of distilled silica from chemical weathering. Mapping clays in a sandstone that changes from kaolinitic to micaceous may be a good way of establishing the limits of freshwater infiltration either in this or a paleohydrologic cycle.

The formation of a stable clay mineral assemblage of illite and chlorite in sandstones, as in shales, is a result of longtime and/or deep burial. This transformation is attested to by the illite–chlorite composition of both old—early Paleozoic and Precambrian—and deeply buried rocks. The process involves the long-term substitution of K and Mg for Na and Ca in clay minerals. The substitution takes place in moderately concentrated porewaters and is part of a general geochemical "maturation" process that sediments go through (Siever, 1968b; Garrels *et al.* 1971). The relation of this transformation to depth of burial was first explored by Kossovskaya and Shutov (1958), Phillip *et al.* (1963), and Füchtbauer (1967). Hower and coworkers (Hower *et al.* 1976; Aronson and Hower, 1976; Yeh and Savin, 1977) have accumulated mineralogical, chemical, and oxygen isotope evidence as well as experimental confirmation (Eberl and Hower, 1976) for the transformation of smectite to mixed-layer and illitic clays. They link clay mineralogical changes to temperatures and pressures related to burial of Cenozoic formations on the Gulf Coast of the United States. The relation of these transformations to other diagenetic phases during burial is treated in a separate section, below.

A generally observed inverse relationship between the amount of clay and the amount of cement in sandstones was shown explicitly for some Pennsylvanian sandstones by Siever (1959, Fig. 4). The explanation for the lack of cement lies partly in the decrease in permeability caused by clay matrix, both detrital and produced by alteration of rock fragments. For the same reason, lithic sandstones and graywackes tend to lack pressure solution, for there will be few grain to grain contacts between rigid quartz and/or carbonate grains and many between quartz and carbonate and the easily deformable clay or lithic fragments.

The chemical conditions for clay mineral authigenesis can be analyzed in the same way as those for authigenic feldspars (Table 11-3). Phyllosilicates are related to feldspars via cation and silica activities of the porewaters with which they are in contact. At silica levels some-

TABLE 11-3. Some clay mineral reactions during sandstone diagenesis.

Clay mineral formed	Precursor	Components added to (+) or subtracted from (−)	Stages of diagenesis (see Fig. 11-1)
Kaolinite	Feldspar	$-(K^+, SiO_2)$ $+H_2O$(Eqs. 11-21a and 11-26)	1, 2, 6
Kaolinite	Pore space	$+(Al_2O_3, SiO_2, H_2O)$	2, 6
Illite	Kaolinite	$+(K^+, SiO_2)$ $-(Al_2O_3, H_2O)$	3, 4, 5
Muscovite	Kaolinite	$+K^+$ $-H_2O$	5
Illite	Smectite	$+K^+$ $-(SiO_2, H_2O, Na^+, Ca^{++}, Mg^{++}, Fe^{++}$, etc.)	3, 4
Chlorite	Smectite	$+(Fe^{++}, Mg^{++})$ $-(SiO_2, H_2O, Na^+, Ca^{++})$	3, 4, 5
Smectite	Volcanic glass	$+H_2O$ $-(Na^+, K^+, Ca^{++})$	1, 2, 3, 4
Glauconite	Illite	$+(Fe^{++}, Fe^{+++})$ $-(K^+, Al_2O_3)$	1, 2

what lower than those required for feldspar authigenesis, clay minerals may form, the type of mineral depending on the cations involved: K^+ for illite, Mg^{++}, Na^+, and other cations for smectites, and Fe^{++} and Mg^{++} for chlorites. Though equilibrium conditions remain the best point of departure for this kind of analysis, the transformations take varying lengths of time, depending on the temperature, and both reactants and products may be present in the same thin section, a reflection of incomplete reaction. A precise reconstruction of the porewater chemistry in relation to clay mineral precipitation would depend on an accurate knowledge of the chemical compositions of the clay phases, especially of the smectites, which have highly variable compositions. These compositions are not easy to determine, even with the electron probe. Once known, the compositions can be put into the kind of diagenetic model for interactions of dynamic porewaters and sandstone minerals suggested by Wood and Surdam (1979).

Alteration of Volcaniclastics and Zeolites

Plate tectonics and its ramifications have stimulated much interest in the sandstones associated with volcanic island arcs. Attention has been focused on the diagenesis of these sandstones in relation to their provenance and environments of deposition. For these reasons, some aspects of the diagenetic modification of volcaniclastic sandstones have been covered in part in Chap. 6. Island arcs are not the only locus of volcaniclastic sandstones: they are also found on the continents in regions of crustal extension, such as the Rio Grande rift, in pull-apart basins of transform fault regions, and around continental hot spots, such as the Yellowstone region. In all of these tectonic settings, the ways in which the different kinds of volcanic materials were extruded into a range of sedimentary environments has much to do with their subsequent diagenetic histories.

Authigenic minerals of volcaniclastic sandstones include feldspars, zeolites, clay minerals, opaline silica, and quartz as major phases. The major source of all of these minerals is the devitrification of volcanic glass and alteration of the plagioclase feldspars and mafic minerals of the extrusive. The dissolution of these phases provides porewaters with all of the components needed for the authigenic growth of a multitude of silicate and other phases. The general nature of this kind of diagenesis was ably reviewed by Hay (1966). Since then specific examples have been discussed by Surdam (1973), Galloway (1974), Merino (1975a,b), and Boles and Coombs (1975); Surdam and Boles (1979) have reviewed the subject with emphasis on the chemistry of the reactions. Many of these petrologists have emphasized the role of zeolites in the diagenesis and metamorphism of sediments in volcanic-rich, thick sediments of eugeosynclinal—those associated with subduction zones—tracts, whose nature and importance was early pointed out by Coombs *et al.* (1959). These authors distinguished the zeolite facies as one of the lowest grades of metamorphism characteristic of deep burial of volcanic-rich sediments. Packham and Crook

(1960) designated these rocks as a diagenetic facies on the basis that their essentially sedimentary texture and fabric are preserved. Regardless of their name, sedimentary or metamorphic, the sandstones containing authigenic zeolites, feldspars, and other silicates such as prehnite and pumpellyite clearly show progressions of mineral facies that are, in part, functions of pressure and temperature and so of burial and thermal history. This facies owes its existence to the deposition of thermodynamically unstable, chemically reactive volcanics. Thick, deeply buried sedimentary sections lacking volcanics do not contain these minerals. Furthermore, there is a tendency for zeolite facies rocks to be transformed with increasing age; most rocks of this type are of Mesozoic and Cenozoic age and the facies is conspicuous by its infrequency in Paleozoic and Precambrian rocks. Rocks of the zeolite facies, typically located along accreting continental margins, tend to be involved in successive metamorphic episodes that blot out the previous stages. If they are not metamorphosed, they most frequently are subject to long-term diagenesis under deep burial conditions and normal thermal gradients, which alters them to a more stable assemblage of authigenic feldspars, quartz, and chlorite–illite.

Galloway (1974) has described the diagenetic minerals of a group of Tertiary sandstones of fore-arc and back-arc basins of the northeast Pacific. The sandstones range from plagioclase-rich arkoses to feldspathic lithic arenites; most are dominated by either plagioclase or plagioclase-rich volcanic rock fragments. Much of the detritus was derived from andesitic volcanics and associated plutonic rocks of intermediate compositions. He distinguished three stages of diagenesis: (1) early calcite pore-filling cement, (2) authigenic clay coats and rims on detrital grains, and (3) authigenic phyllosilicate and/or laumontite pore-filling cement. Also observed were quartz and K-feldspar overgrowths. A detrital mud matrix present in some sandstones may be altered by recrystallization of clay minerals to authigenic micas and disseminated chert; softer rock fragments have been crushed and altered. Galloway ascribes most of this paragenetic sequence to the increase in lithostatic and hydrostatic pressure as well as temperature and plotted the appearance of the several phases as a function of present burial depth. The early calcite cement may be related to the original sedimentary environment rather than depth. The second stage, burial from 300 to 1200 m (1000–4000 ft), induces chemical alteration of aphanitic volcanic debris and other unstable grains to clay coatings and rims; at greater depths and higher pressures a "disseminated pasty matrix" is formed in the areas of grain contacts. In stage 3, at depths from 900 to 3000 m (3000–10,000 ft), pore-filling zeolites, typically laumontite, or a well-crystallized chlorite or smectite are precipitated. In more advanced stages of burial, other kinds of authigenic minerals and cements appear, including chert and micas, a second generation of calcite, and, replacing feldspar and rock fragments, epidote, chlorite, and albite. This recital of phases and linkage to burial history is typical of many other sandstones of similar lineage; there are many variations and no two sequences are precisely the same (Fig. 11-19). In most of the sandstones described, there is imprecise information on former, as distinct from present, depths of burial and on the geothermal gradients characteristic of the basins, as is discussed below in the section on burial diagenesis.

Not all diagenesis of volcaniclastics is so simply related to depth of burial and in particular to the low geothermal gradients that are frequently assumed. Surdam and Boles (1979) have discussed in detail how many of these diagenetic reactions are more distinctively the result of particular chemical reactions among unstable phases and of fluid movements within the basin, especially early in burial history, when fluid/solid ratios are high. They limit the discussion to the pressure and temperature fields defined at the upper end by the laumontite–lawsonite and laumontite–wairakite boundaries, approximately 3 kb pressure and 300°C. They classify diagenetic reactions in volcaniclastics as (1) hydration: glass or plagioclase to zeolite [see a similar reaction for glass to smectite in Eq. (11-22)]; (2) carbonatization: Ca^{++} liberated by other reactions joins with HCO_3^- produced by oxidation of organic matter to form early calcite cement; and (3) dehydration reactions in later diagenesis. Dehydrations include many zeolite transformations:

heulandite to laumontite,

$$\begin{aligned} &Ca_3K_2Al_8Si_{28}O_{72} \cdot 23H_2O \rightarrow \\ &\quad 3CaAl_2Si_4O_{12} \cdot 4H_2O + 10SiO_2 \\ &\quad\quad + 2KAlSi_3O_8 + 11H_2O \end{aligned} \quad (11\text{-}28)$$

heulandite to prehnite,

$$\begin{aligned} &3Ca^{++} + Ca_3K_2Al_8Si_{28}O_{72} \cdot 23H_2O \rightarrow \\ &\quad 3Ca_2Al_2Si_3O_{10}(OH)_2 + 2KAlSi_3O_8 \\ &\quad\quad + 13SiO_2 + 17H_2O + 6H^+ \end{aligned} \quad (11\text{-}29)$$

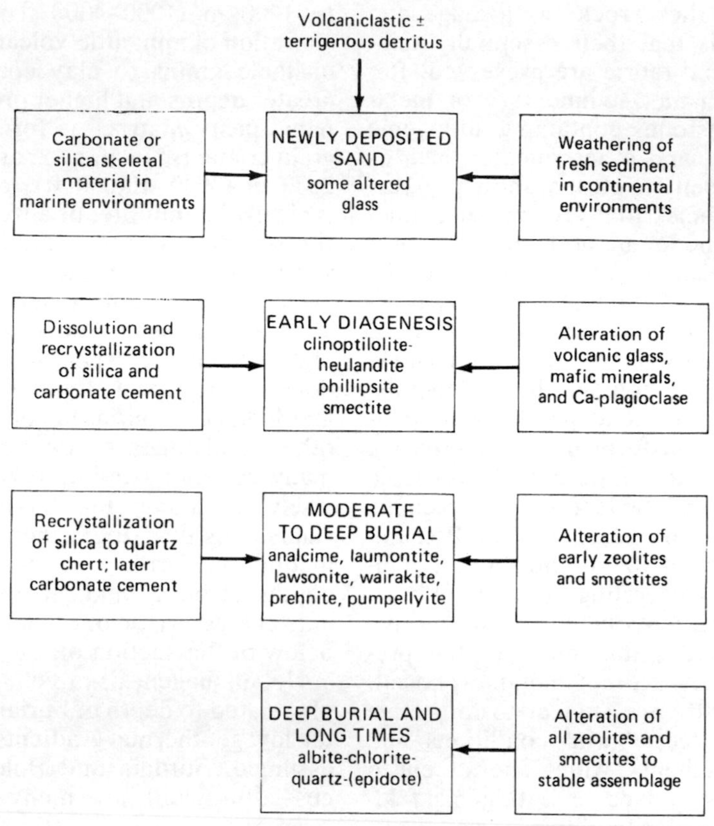

FIGURE 11-19. Routes of alteration of zeolites and other authigenic phases in the diagenesis of volcaniclastics.

heulandite to analcime,

$$8Na^+ + Ca_3K_2Al_8Si_{28}O_{72} \cdot 23H_2O \rightarrow 8NaAlSi_2O_6 \cdot H_2O + 12SiO_2 + 2K^+ + 3Ca^{++} + 15H_2O \quad (11\text{-}30)$$

heulandite to albite,

$$8Na^+ + Ca_3K_2Al_8Si_{28}O_{72} \cdot 23H_2O \rightarrow 8NaAlSi_3O_8 + 4SiO_2 + 23H_2O + 2K^+ + 3Ca^{++} \quad (11\text{-}31)$$

analcime to albite,

$$NaAlSi_2O_6 \cdot H_2O + SiO_2 \rightarrow NaAlSi_3O_8 + H_2O \quad (11\text{-}32)$$

laumontite to prehnite,

$$CaAl_2Si_4O_{12} \cdot 4H_2O + Ca^{++} \rightarrow Ca_2Al_2Si_3O_{10}(OH)_2 + SiO_2 + 2H_2O + 2H^+ \quad (11\text{-}33)$$

laumontite to pumpellyite,

$$2CaAl_2Si_4O_{12} \cdot 4H_2O + 2Ca^{++} + (Fe, Mg)^{++} + Fe^{+++} \rightarrow Ca_4(Fe,Mg)^{++}Fe^{+++}Al_4Si_6O_{21}(OH)_7 + 2SiO_2 + 9H^+ \quad (11\text{-}34)$$

Hydration reactions will decrease porosity and may liberate large amounts of heat from exothermic reactions, perhaps enough to significantly affect local thermal gradients. Surdam and Boles suggest that porewater compositions, especially pH and the activities of Ca^{++} and $Si(OH)_4$, are more important than simple temperature relations in controlling these reactions. Thus glass may alter to heulandite at the high silica activities characteristic of mafic glass hydrolysis but as the silica activity drops in response to mineral changes, the heulandite may convert to prehnite if the activity ratio of Ca^{++} to H^+ is high. They also point to the importance of the original detrital assemblage in determining the course of diagenesis, contrasting laumontite and heulandite formed from andesitic or basaltic detritus with the authigenic albite or K-feldspar formed from more silicic volcanic debris, rhyolitic or dacitic. Finally, they ascribe importance to the effects of fluid and lithostatic pressure on dehydration reactions, particularly as fluid pressure drops from equality with lithostatic pressure to lower values in newly fractured sandstones. Under these conditions, analcime may alter to albite and heulandite to laumontite.

Wood and Surdam (1979) proposed a model for combined convective mass transfer and diffusion processes in the diagenesis of volcaniclastic sandstones with alteration products distributed in well-defined zones. Their model explains these zones as a consequence of countercurrent mass flows in which different aqueous species diffuse in opposite directions. These species may react to precipitate a new phase whose position is a function of the rates of flow, diffusion constants, and kinetics of precipitation and dissolution, all cast in a simplified scheme. This is an example of our growing understanding of the various routes of diagenesis in these heterogeneous sandstones and the ways in which local and regional effects may arise. As more petrologists concentrate on the sediments of subduction zones, we will learn better how to infer provenance through the curtain of diagenesis.

Common Accessory Diagenetic Minerals

Many diagenetically formed minerals are quantitatively unimportant, but may have a bearing on the post-depositional history of a sandstone. Textural and mineralogical patterns are used in the same ways as those of the more abundant materials to infer parageneses and origins.

Feldspars. Authigenic feldspars in sandstones have been known for many years. Though most abundant in arkoses and volcaniclastic sandstones, they are also present in many quartz arenites and graywackes. The general conditions for low-temperature feldspar precipitation have been discussed by Kastner and Siever (1979). Feldspar overgrowths may be unnoticed in routine petrography, especially extremely small grains or overgrowths in clay-rich sandstones, and so be underestimated in abundance and occurrence. The feldspars, whether albite or K-feldspar, are extremely pure end-members of the alkali and plagioclase feldspar solid-solution series as established by electron probe and X-ray diffraction studies (Kastner, 1971). They are frequent as an early diagenetic mineral and predate tectonic deformation. They are clearly not related to hydrothermal or igneous activity in most, though not all, sandstones, They are formed almost exclusively as overgrowths on detrital feldspar grains.

The chemical conditions for authigenesis of feldspar, as noted earlier in this chapter, are a high enough concentration of dissolved silica and a high enough ratio of Na^+/H^+ or K^+/H^+ activities to make equations like Eq. (11-21a) go to the left (see Fig. 11-9). The geologic conditions that seem to be necessary are moderately elevated temperatures, a source of silica, either from skeletal parts of organisms or from hydrolyzing silicates, and abundant Na^+ and/or K^+ ions that come from porewaters. Some detrital feldspars, intermediates in the plagioclase or alkali feldspar series and so less stable than the pure end-members at low temperatures, first partially dissolve and then reprecipitate pure authigenic feldspars or exchange Ca^{++} for Na^+ as in Eq. (11-27). This "distillation" of an impure to a pure feldspar would explain the general dependence on the presence of detrital feldspars. The series of phase equilibria and chemical reactions that form authigenic feldspar are shown in Fig. 11-20 and Table 11-4. There are now many descriptions of authigenic feldspars in sandstones of all types (for example, Galloway, 1974; Waugh, 1978; Mankiewicz and Steidtmann, 1979; Odum *et al.*, 1979).

The comparison of chemical conditions for

TABLE 11-4. Geochemical reactions forming authigenic feldspars.

Kaolinite to Na- or K-feldspar:

$Al_2Si_2O_5(OH)_4 + 2K^+ + 4(H_4SiO_4) \rightarrow 2KAlSi_3O_8 + 9H_2O + 2H^+$

$Al_2Si_2O_5(OH)_4 + 2Na^+ + 4(H_4SiO_4) \rightarrow 2NaAlSi_3O_8 + 9H_2O + 2H^+$

Muscovite to K-feldspar:

$KAl_2(AlSi_3)O_{10}(OH)_2 + 2K^+ + 6(H_4SiO_4) \rightarrow 3KAlSi_3O_8 + 12H_2O + 2H^+$

Na–Al smectite to Na- or K-feldspar:

$Na(Al_5Mg)\ Si_{12}O_{30}(OH)_6 + 4Na^+ + 3(H_4SiO_4) \rightarrow 5NaAlSi_3O_8 + 8H_2O + 2H^+ + Mg^{++}$

$Na(Al_5Mg)\ Si_{12}O_{30}(OH)_6 + 5K^+ + 3(H_4SiO_4) \rightarrow 5KAlSi_3O_8 + 8H_2O + 2H^+ + Na^+ + Mg^{++}$

Analcime to albite:

$Na(AlSi_2O_6) \cdot H_2O + H_4SiO_4 \rightarrow NaAlSi_3O_8 + 3H_2O$

Analcime to K-feldspar:

$Na(AlSi_2O_6) \cdot H_2O + H_4SiO_4 + K^+ \rightarrow KAlSi_3O_8 + Na^+ + 3H_2O$

Phillipsite to K-feldspar:

$(\frac{1}{2}Ca,Na,K)_5(Al_5Si_{11}O_{32}) \cdot 10H_2O + 2K^+ + 4(H_4SiO_4) \rightarrow 5KAlSi_3O_8 + \frac{5}{2}Ca^{++} + 5Na^+ + 18H_2O$

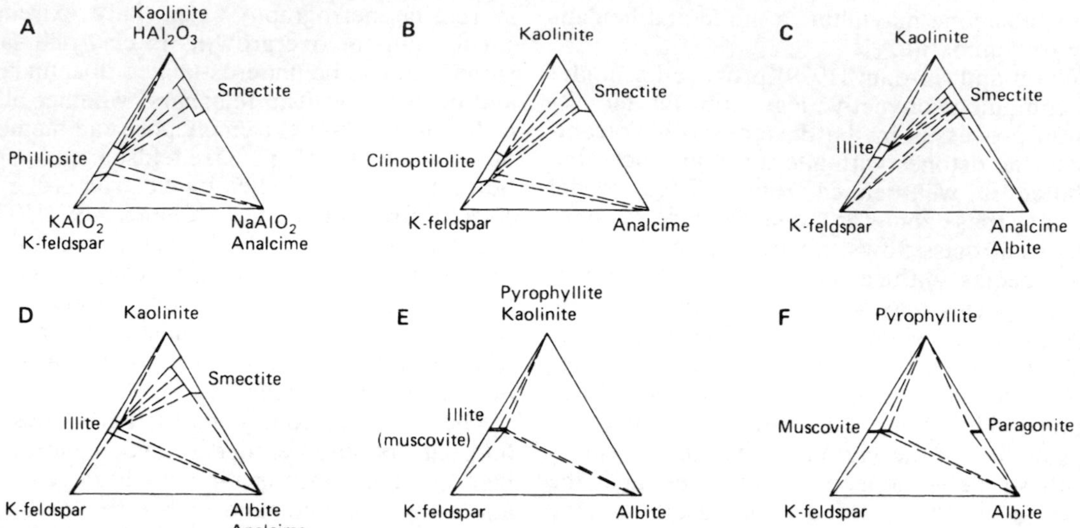

FIGURE 11-20. Phase equilibria including authigenic feldspars in the system Al_2O_3–$KAlO_2$–$NaAlO_2$ in the presence of quartz and other free silica phases. A) and B) Relations among zeolites, clay minerals, and K-feldspar; C) and D) clay minerals and two feldspars; E) two feldspars, muscovite (illite), and pyrophyllite (kaolinite); F) two feldspars, muscovite, paragonite, and pyrophyllite. (From Kastner and Siever, 1979, Fig. 2).

diagenetic dissolution of feldspar with those for authigenic feldspar growth is instructive. It appears that both processes are common (Heald and Larese, 1973; Pittman, 1979). Dissolution is related to porewater movements from undersaturated environments dominated by waters of meteoric origin that move downward and laterally in a basin. In these waters the Na^+/H^+ and K^+/H^+ activity ratios are low compared to silica activities. Authigenesis is related either to local isochemical changes or to porewater movements that transport waters high in Na^+/H^+ or K^+/H^+ ratio from lower formations. Though this is only a tentative generality based on the general nature of deeply buried porewaters in many sedimentary basins, it shows promise for the study of paleohydrology and its relation to diagenesis.

Iron Oxides. Authigenic iron oxides have been studied primarily in redbed sandstones though they have been known as a common outcrop weathering product in many other kinds of sandstones. Van Houten (1968) summarized much of the data and conclusions on origin and relation to magnetism. The mineralogy of the common black oxide grains is complex and may include not only hematite but magnetite, ilmenite, and titaniferous hematites, maghemites, and magnetites. The red pigment consists solely of hematite. The diagenetic process includes the aging and dehydration of the brown amorphous ferric oxide found in modern sediments to goethite ("limonite") and the dehydration of goethite to hematite. Warm climates promote the alteration of brown (limonite) to red (hematite) pigments covering sand grains. Walker (1967, 1974, 1976) has argued for the in-place weathering of iron-rich minerals as a major source of ferric oxide pigment in both moist climates and desert red sands. The dissolution of iron-rich silicates and precipitation of hematite is enhanced by circulation of oxidized meteoric waters. The hematite that forms as an early diagenetic product acquires its chemical remanent magnetism as it grows through critical crystal sizes, and continues to grow as long as the source, the preceding limonite, is available (Larson and Walker, 1982). This paleomagnetic dating of hematite may reflect the stratigraphic age only if diagenesis is early and rapid. Discrepancy between stratigraphic and paleomagnetic age may reveal the age of a later diagenetic episode of hematite formation.

Gypsum–Anhydrite. The occurrence of these minerals as cements in many sandstones is clearly related to evaporite conditions at the time of sedimentation or to movement of hypersaline porewaters from an overlying evaporite formation (Murray, 1964). An example of this is the Tensleep Sandstone of Wyoming, in which early diagenetic anhydrite is related to sabkha deposition (Mankiewicz and Steidtmann, 1979).

The origin is much the same as for dolomite cement that is related to sabkha flat sedimentation. This kind of occurrence may show anhydrite before or after dolomite, and both these before quartz. In any case, the original mineral is an early diagenetic product related to environmental conditions, made complicated by the fact that the conversion of gypsum to anhydrite is reversible and dependent on the salinity of the water, the temperature, and pressure (see p. 49). Higher salinities, temperatures, and pressures favor anhydrite, which is by far the most common cement. Later diagenetic anhydrite, such as found by Hoholick *et al.* (1984) for the St. Peter Sandstone in the Illinois basin, are probably related to changes in Ca^{++} and SO_4^{--} activities in warm, deeply buried brines, perhaps related to liberation of Ca^{++} by dolomitization or albitization of plagioclase.

Miscellaneous Diagenetic Minerals. Few other minerals have been investigated in such detail as the preceding. Though there have been many studies of diagenesis of various minerals, few have been applied specifically to sandstones. Pyrite and its less common polymorph, marcasite, are known as diagenetic minerals in sandstones and imply reducing conditions, presumably reflecting a deoxygenated environment such as the sandy tidal flats in which black FeS forms just below the surface (Berner, 1964).

Carnotite and uraninite, economically important uranium minerals, occur as diagenetic cement in the Mesozoic sandstones of the Colorado Plateau. Their origin is linked, on the one hand, to groundwater infiltration that transported uranium from a weathering primary granitic rock source into the permeable sandstones, and on the other, to warm, upward-moving saline solutions transporting dissolved uranium in deeper rocks. A discussion of the relation of the geochemistry of uranium and the geology of these deposits is given by Garrels and Christ (1965, pp. 388–94), Fischer (1970), and Adler (1974). The generation of ore deposits in sandstones as well as other sedimentary rocks has received increasing attention from economic geologists, who have been analyzing mineral parageneses in the context of basin analysis and hypothesizing formation water movements (see summaries in Maynard, 1983). Barite is another cementing mineral in sandstones but little is known of its geologic distribution and the kinds of geochemical regimes in which it forms.

Organic materials undergo many diagenetic changes during early middle and late diagenesis, many of them important in relation to the formation and migration of petroleum hydrocarbons. These changes are intimately related to burial diagenesis, which is described below.

Porosity Reduction and Production

Though we have discussed individual diagenetic effects on porosity, it is useful to consider together the many processes that affect porosity and its related property, permeability. Understanding and so being able to predict porosity trends is vital for oil and gas exploration, for reservoir engineering, and for underground waste disposal. In a general way, we have long known that porosity tends to decrease with depth because of compaction and cementation. We know that there are many sandstones in which porosity may be produced at some depths through decementation. Thus diagenesis may reduce porosity through compaction and cementation and produce it by dissolution of detrital grains and authigenic phases. The many studies that have revealed these porosity trends rely on the correlation of microscopic petrography of diagenetic textures and minerals with porosity and permeability changes with depth and position within a basin. Reviews of these problems have been given by, among others, Selley (1978), Pittman (1979), and Schmidt and McDonald (1979a,b). Porosity in relation to texture is covered in Chap. 3.

Porosity types have been divided into several categories: intergranular, dissolution, micro-, and fracture (Pittman, 1979). Intergranular porosity includes all pores between detrital grains. In newly deposited sands the primary intergranular porosity is high, over 40 percent in some sands. The pores are connected, pore apertures are large, and permeability is high. In buried, indurated sandstones intergranular porosity is reduced by physical rearrangement of grains during compaction and chemical precipitation of cements. If that cement, typically carbonate, is later dissolved, secondary intergranular porosity results.

Dissolution porosity arises by removal of soluble detrital or authigenic grains by porewaters. The grains may be of more soluble minerals such as carbonates or sulfates, or of less soluble phases such as feldspars. In addition to carbonate dissolution, feldspar dissolution is now known to be widespread as a source of secondary porosity. Pores may or may not be strongly connected and pore apertures may be small; thus permeability varies. Dissolution porosity depends on the primary intergranular porosity, the abundance of dissoluble grains, and the extent of secondary dissolution of earlier cements. Identifying dissolved grains may be difficult, if the grain is completely dissolved; then the petrographer relies on shapes and sizes of molds.

Microporosity derives from pores with very small apertures, less than 0.5 microns. Micropores may be associated not only with small pores but with some larger ones as well. They are common in sandstones in which abundant clay minerals, either primary or authigenic, occlude pore space and openings. Permeability is low as a result.

Natural fracture porosity, formed by breakage across detrital and/or authigenic phases, rarely amounts to more than a few percent but may strongly increase permeability, especially in clay-rich sandstones. The evolution of natural fractures follows the embrittlement of the sandstone by diagenetic cementation. The fracture system is an adjustment of the brittle rock to stresses on the rock originating through burial or tectonic deformation.

The general sequence of sandstone porosity changes with depth may start with primary intergranular porosity being slightly decreased by physical compaction and rearrangement of grains. Early cementation, typically by carbonate, may further reduce prosity. With increased burial, additional cements and authigenic minerals such as clays or zeolites may continue to reduce porosity, even while some dissolution, such as that of feldspar, is proceeding. Thus the rate of porosity decrease—or increase—may be determined by the balance between cementation and dissolution. At deeper levels of burial—several kilometers, but highly variable—secondary porosity formed by dissolution of carbonate cement may produce enough new pores to increase the porosity (see, for example, Loucks *et al.* 1977). This reversal of the normal porosity–depth trend may result in a sandstone with porosity and permeability high enough to make a productive oil reservoir. In the deepest zones of burial, porosity may again decrease with depth by cementation and chemical compaction.

The depth sequences of porosity changes are not uniform from basin to basin nor even within a basin. The depth trends depend not only on the complex course of diagenesis in different places and depths but a great deal on the original texture and mineralogy. We know much more about porosities of oil-producing quartz, arkosic, and lithic arenites than about those of less productive graywackes and other low-permeability sandstones. The primary porosities of the wackes may be high but compaction, deformation, and alteration of rock fragments and matrix reduce the porosity rapidly after burial. Perhaps permeability, and thus fluid movement becomes so low in these rocks that slight secondary dissolution produces little secondary porosity. Fracturing of wackes after lithification may produce some porosity and permeability, especially in tectonically deformed regions. Within the arenites there is much variation. Nagtegaal (1978) showed differences in framework grain stability among quartz, arkosic, and lithic arenites. The quartz frameworks were stable, subject only to pressure solution and slight mechanical compaction. Arkosic arenites are also subject to extensive dissolution of feldspars, increasing the general likelihood of diagenetic change. Lithic arenites are most unstable of all, being subject not only to all of the preceding but also to plastic deformation of the softer rock fragments. Porosity–depth profiles may also vary for sands deposited in different environments (Selley, 1978). The latter sedimentologic variable encompasses differing grain size and shape distributions, proportions of fine-grained sediments, and the permeability variations of different bedding sequences.

A general correlation of porosity types with diagenetic stages has been proposed by Schmidt and McDonald (1979b). They classify diagenetic stages as immature, semi-mature, mature, and supermature, based on the porosity-affecting process. Immature sands are affected only by mechanical compaction, semimature by chemical reduction of primary porosity, mature by secondary porosity, and supermature by disappearance of effective primary and secondary porosity. They correlate these stages with lithology, depth of burial, and temperature for a given subsidence rate, grain size, and geothermal gradient (Fig. 11-21). Though there is no doubt that these general relationships apply to a great many arenites, there is a large range and great variation of the exact time–temperature–depth interactions. Schmidt and McDonald ascribe most secondary porosity to dissolution of primary and secondary authigenic carbonate, largely in the mature stage. They believe the dissolution results from increased CO_2 produced by decarboxylation reactions of maturing organic matter in nearby sediments. The dissolved carbonate is thought to move upwards and, in part, precipitate in over-

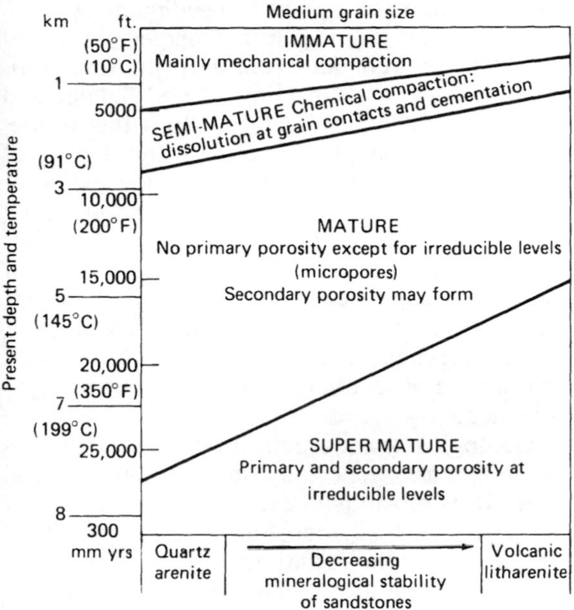

FIGURE 11-21. Correlation of lithology, depth of burial, and temperature for a given subsidence rate. (Modified from Schmidt and McDonald, 1979b, Fig. 10).

lying immature and semi-mature stage sandstones.

Deducing the course of diagenesis that produced a specific porosity–depth gradient in a sandstone section is a challenging test of our ability to use petrographic and geologic information for prediction. In the ideal case we want to know the values of the geologic variables: depth of burial as a function of post-depositional time, geothermal gradients as a function of time, and depositional environments of each formation. To this are added the petrographic variables: original mineralogies and textures, the nature and distribution of diagenetic textures and minerals, and especially the types and amounts of pore space. We are well started on the correlation of porosity, present depths of burial, and petrographic characters. We are beginning to get basinwide porosity maps (Hoholick *et al.*, 1984). We need more comprehensive reconstructions of basin geophysical and geochemical history to make the story complete.

Burial Diagenesis

The relationship between diagenesis and depth of burial is a reflection of the general relationship between diagenesis and metamorphism in the pressure–temperature field of the earth (Fig. 11-22). As clearly stated by Correns (1950), there is no simple and nonarbitrary dividing line between diagenesis and metamorphism. In practice there is no great difficulty in distinguishing between most metamorphic and sedimentary rocks. On the other hand, there is no clear-cut separation between low-grade metamorphic rocks or high-grade diagenetic rocks such as the zeolite facies (Coombs *et al.*, 1959). *Anchimetamorphic* has been used for those sediments extensively altered by diagenesis, transitional to low-grade metamorphic rocks (Kubler, 1964).

Though diagenesis and metamorphism represent a continuum in pressure and temperature,

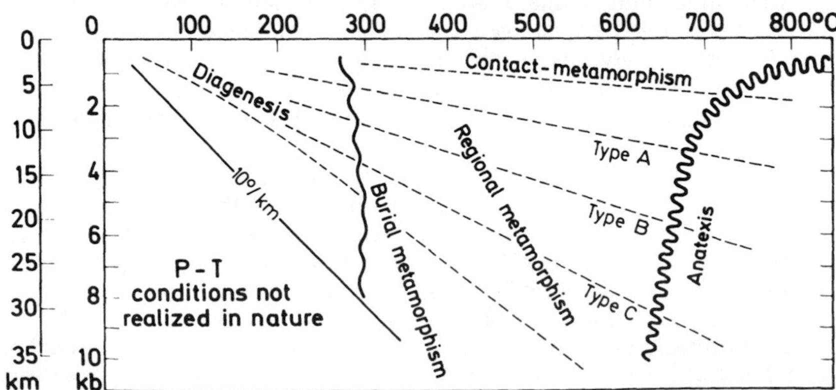

FIGURE 11-22. Schematic pressure–temperature diagram for diagenesis and different types of metamorphism. The *P–T* region below the lowest possible geothermal gradient of about 10°C/km is not realized in nature. (Modified from Winkler, 1979, Fig. 1).

there are important differences. We interpret the mineral assemblages of metamorphic rocks as indicators of pressure and temperature, assuming equilibrium (see, for example, Winkler, 1979). There is an order to the mineralogy of metamorphic rocks that is expressed either by isograds, metamorphic facies, or petrogenetic grids. But the mineral assemblages of many sandstones, even after extensive burial diagenesis, may still reflect to some extent the non-equilibrium detrital mixtures. The kinetics of mineral transformations are so slow at low temperatures that complete equilibrium is rarely reached in most sandstones.

As a sandstone is buried, its temperature and pressure increase along various paths related to the rapidity of burial, which depends both on rates of sedimentation and rates of subsidence. Both of these rates affect the geothermal gradient, which is primarily a function of the tectonic position of the basin and its proximity to mantle or lower-crust heat sources; thus temperatures are related to plate tectonic movements or to emplacement of magmas, which are also related to plate position (Siever, 1979; Bjørlykke, 1983). Pressure may be normal lithostatic on the grains and hydrostatic on the pore fluids, or pore fluids may be overpressured, that is, approach lithostatic pressures. Pressure, hence burial depth, and diagenetic changes in porosity are related. Pressure-induced decrease in porosity and increase in density of the sandstone result from chemical compaction through pressure solution and mechanical rearrangement, such as ductile deformation of rock fragments or fracturing. Filling of pore spaces by cement and/or clay matrix may also be affected by pressure. Abnormal pressures in sandstones are related to changes in permeability either in the sandstone or in adjacent formations that restrict free access of the pore fluids to the general porewater column that is at hydrostatic pressures. Both temperature and pressure strongly affect the mineral equilibria and the rates of mineral reactions and so profoundly affect both the course and the rates of diagenetic reactions.

Detailed studies of burial diagenesis of sandstones, started by works such as those of Philipp *et al.* (1963) and Füchtbauer (1967), are now proliferating in the literature, stimulated in no small part by the interaction of burial, diagenesis, and the maturation of petroleum hydrocarbons. These studies involve the correlation of burial history with diagenetic effects. The burial history is analyzed by plotting the depth of burial of a given horizon in relation to geologic time after formation. These diagrams (Fig. 11-23), variously termed sedimentation diagrams, geological model curves, geo-history diagrams, and many other names, are constructed for a given point in a sedimentary basin by inferring original uncompacted sediment thicknesses from an accurately known stratigraphic section (Van Hinte, 1978; Sclater and Christie, 1980; Siever, 1983). Into this framework fit the changes in porosity and density of shales and sandstones vs. depth, the extent of clay mineral diagenesis and cementation by quartz, carbonate, feldspar, zeolite, or other minerals, and the degree of thermal alteration of the rock independently measured by oil and gas maturation or other indices. The correlation of these diagenetic features with depth of burial and timing of tectonic events has been a major advance.

Geological thermometry of diagenetic phases in sediments, especially in relation to burial diagenesis of sandstones, is covered in a series of papers in the symposium volume *Aspects of Diagenesis* (Scholle and Schluger, 1979). Prominent as temperature indicators are vitrinite reflectance (Bostick, 1971, 1979), conodont color alteration (Epstein *et al.* 1977), oil and gas maturation indices (Tissot and Welte, 1978; Hunt, 1979), clay mineral alteration (Hoffman and Hower, 1979), and oxygen isotopes. Each of these methods depends on the proper calibration with other indicators and burial–time relations. All are more or less useful depending on the kinds of sandstones in which they are found; thermometry depends on the alteration of detrital constituents or, for fission track dating (Naeser, 1979), on the changes induced by radioactive matter in detritus. Oxygen isotopes of cements and diagenetic minerals have been used as temperature indicators (Land and Dutton, 1978; Eslinger *et al.*, 1979). $^{87}Sr/^{86}Sr$ ratios, while not temperature indicators, may be useful in determining the time and sequence of cementation, as suggested by Stanley and Faure (1979) in their study of cementation of Tertiary sandstones of the western plains of the United States.

Fluid inclusions have been the object of much study for geothermometry in metamorphic and hydrothermal rocks but relatively little has been done on sandstone cements or other diagenetic minerals, though there is much promise for such studies (Roedder, 1979). Burruss (1981) has reported on fluid hydrocarbon inclusions in quartz grains and overgrowths and compared them with inorganic fluid inclusions. Haszeldine *et al.* (1984) have combined authigenic quartz, fluid inclusion thermometry with thin-section petrography, organic geochemistry, and basin analysis to infer quartz precipitation from

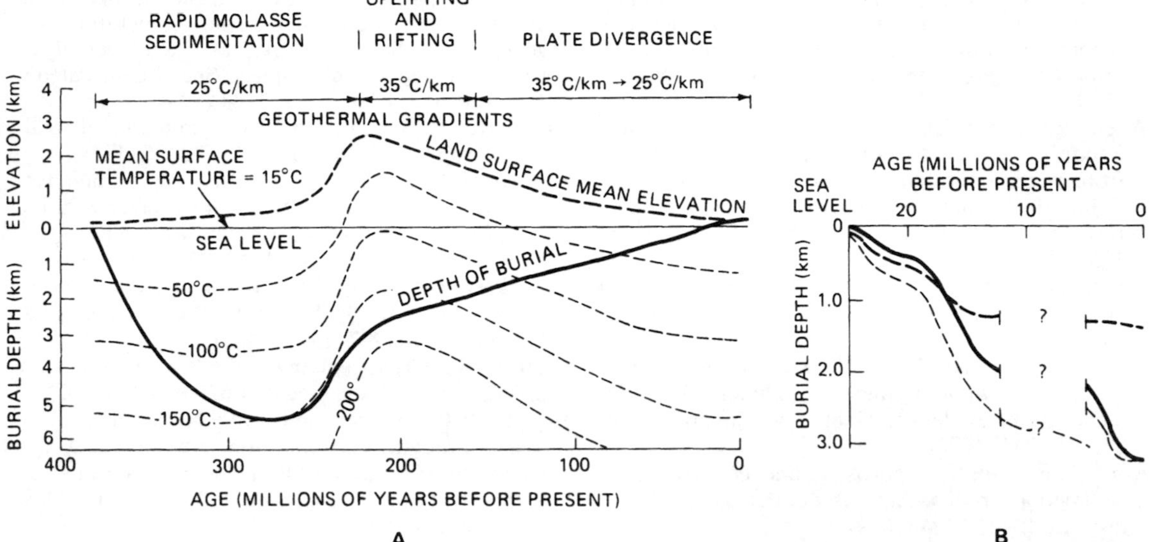

FIGURE 11-23. Sedimentation diagrams; the position of a bed with respect to sea level is shown as a function of time before the present. A) The base of the Catskill facies (Devonian) of New York State, uncorrected for compaction. Paleogeotherms (light dashed lines) are estimated assuming geothermal gradients modeled from tectonic history of the region. Modified from Gale (1985). B) Subsidence of the Lower Miocene of the Gulf of Lion. The heavy solid line gives the observed sediment thickness; the dotted line gives the reconstructed sediment accumulation after correction for compaction ("decompacting"); the dashed line gives the calculated tectonic subsidence after correcting for isostatic adjustment ("backstripping"). The interval marked with question marks represents a time during which some sediment was removed by erosion. Modified and generalized from Steckler and Watts (1980).

moving pore fluids at temperatures between 68°C and 94°C following a period of rapid subsidence. To date, there have been few too such analyses of sediments to be able to make generalizations other than the suggestion of Roedder that hot saline brines are frequently involved. His data, though they arise from rocks other than sandstones, are probably suggestive of the same conditions for many sandstones and are a tie to the porewater conditions inferred by a number of workers in sandstone diagenesis.

Conclusion

Diagenesis is a complex web of physical and chemical processes related to the geologic history of a sandstone. Our understanding of it depends both on a knowledge of the chemistry and physics involved and on careful observation of the geological distribution of diagenetic textures and minerals. Many excellent descriptions of diagenetic features of particular sandstones are available and petrologists are now doing what geologists profitably learned to do long ago—to map. Maps and cross sections of diagenetic characters give us the relationship to provenance and environment, on the one hand, and to burial history, on the other. Relevant questions include: How much stratigraphic control is there on cementation? How do different environmental facies control later diagenesis? How do thermal gradients relate to organic and inorganic diagenesis of sandstone? What effect does structural deformation have on diagenesis? Can we delineate petrographic provinces based on diagenesis as well as traditional ones based on provenance and can the two be related? How do diagenetic evolution patterns fit with general models of sedimentary basin development? The answers to these questions have to be based on quantitative measurements and so we need more work on how to do quantitative petrography of diagenetic textures and minerals. We need to couple petrography to the appropriate chemical and isotopic analyses of organic and inorganic constituents. The combination of quantitative petrography and geological mapping of diagenesis will teach us much more than we know now.

References

Adams, W.L.: Diagenetic aspects of Lower Morrowan, Pennsylvanian, sandstones, northwestern Oklahoma. Am. Assoc. Petroleum Geologists Bull. 48, 1568–1580 (1964).

Adler, H.: Concepts of uranium-ore formation in reducing environments in sandstones and other sediments. In: Formation of uranium deposits. pp. 141–168. Vienna: Intern. Atomic Energy Agency 1974.

Allen R.C.; Gavish, E.; Friedman, G.M.; and Sanders, J.E.: Aragonite-cemented sandstone from outer continental shelf off Delaware Bay: Submarine lithification mechanism yields product resembling beachrock. Jour. Sed. Petrology 39, 136–149 (1969).

Andel, Tj.H. van: Relections on the interpretation of heavy mineral analyses. Jour. Sed. Petrology 29, 153–163 (1959).

Aronson, J.L.; and Hower, J.: Mechanism of burial metamorphism of argillaceous sediment, 2: Radiogenic argon evidence. Geol. Soc. America Bull. 87, 738–744 (1976).

Athy, L.F.: Density, porosity, and compaction of sedimentary rocks. Am. Assoc. Petroleum Geologists Bull. 14, 1–24 (1939).

Atkinson, B.K.; and Rutter, E.H.: "Pressure Solution" or indentation? Geology 3, 477–478 (1975).

Bathurst, R.G.C.: Carbonate sediments and their diagenesis, 2nd Ed., 658 pp. New York: Elsevier 1975.

Berner, R.A.: Iron sulfides formed from aqueous solution at low temperatures and atmospheric pressure. Jour. Geology 72, 293–306 (1964).

Berner, R.A.: Principles of chemical sedimentology, 240 pp. New York: McGraw-Hill 1971.

Berner, R.A.: Early diagenesis, 241 pp. Princeton, New Jersey: Princeton University Press 1980.

Bjørlykke, K.: Diagenesis in Mesozoic sandstones from Spitsbergen and the North Sea—A comparison. Geol. Rundschau 68, 1152–1171 (1979).

Bjørlykke, K.: Diagenetic reactions in sandstones. In: Parker, A., and Sellwood, B.W. (Eds.): Sediment diagenesis, pp. 169–214. (NATO Advanced Study Inst. Series, C, 115). Dordrecht–Boston–Lancaster: D. Reidel Pub. Co. 1983.

Bjørlykke, K.: Formation of secondary porosity: how important is it? In: McDonald, D.A., and Surdam, R.C. (Eds.): Clastic diagenesis. Am. Assoc. Petroleum Geologists Mem. 37, 277–286 (1984).

Blatt, H.: Diagenetic processes in sandstones. In: Scholle, P.A., and Schluger, P.R. (Eds.): Aspects of diagenesis. Soc. Econ. Paleon. Mineral. Spec. Pub. 26, 141–157 (1979).

Boles, J.R.: Active ankerite cementation in the subsurface Eocene of southwest Texas. Contrib. Mineral. Petrol. 68, 13–22 (1978).

Boles, J.R., and Coombs, D.S.: Mineral reactions in zeolitic Triassic tuff, Hokonui Hills, New Zealand. Geol. Soc. America Bull. 86, 163–173 (1975).

Bostick, N.H.: Thermal alteration of clastic organic particles as an indicator of contact and burial metamorphism in sedimentary rocks. Proc., 2nd Am. Assoc. Stratigraphic Palynologists, Geoscience Man. 3, 83–92 (1971).

Bostick, N.H.: Microscopic measurement of the level of catagenesis of solid organic matter in sedimentary rocks to aid exploration for petroleum and to determine former burial temperatures—a review. In: Scholle, P.A., and Schluger, P.R. (Eds.): Aspects of diagenesis. Soc. Econ. Paleon. Mineral. Spec. Pub. 26, 17–44 (1979).

Bramlette, M.N.: The stability of minerals in sandstone. Jour. Sed. Petrology 11, 32–36 (1941).

Bredehoeft, J.D.; Blyth, C.; White, W.A.; and Maxey, G.B.: Possible mechanism for concentration of brines in subsurface formations. Am. Assoc. Petroleum Geologists Bull. 47, 257–269 (1963).

Brenchley, P.J.: Origin of matrix in Ordovician greywackes, Berwyn Hills, North Wales. Jour. Sed. Petrology 39, 1297–1301 (1969).

Brett, G.W.: Cross-bedding in the Baraboo quartzite of Wisconsin. Jour. Geology 63, 143–148 (1955).

Bryant, W.R.; Bennett, R.H.; and Katherman, C.E.: Shear strength, consolidation, porosity, and permeability of oceanic sediments. In: The sea, Vol. 7, The oceanic lithosphere, pp. 1555–1616. New York: Wiley 1981.

Burruss, R.C.: Hydrocarbon fluid inclusions in studies of sedimentary diagenesis. In: Hollister, L.S., and Crawford, M.L. (Eds.): Fluid inclusions: Applications to petrology. Mineral Assoc. Canada Short Course Handbook 6, 138–155 (1981).

Byerlee, J.D., and Brace, W.F.: High pressure mechanical instability in rocks. Science 164, 713–715 (1969).

Carpenter, A.B.: Origin and chemical evolution of brines in sedimentary basins. Oklahoma Geol. Survey Circ. 79, 60–77 (1978).

Cayeux, L.: Les roches sédimentaires de France, Roches siliceuses, 250 pp. Paris: Masson 1929.

Chave, K.E.; Deffeyes, K.S.; Weyl, P.K.; Garrels, R.M.; and Thompson, M.E.: Observations on the solubility of skeletal carbonates in aqueous solutions. Science 137, 33 (1962).

Chilingarian, G.V., and Wolf, K.H. (Eds.): Compaction of coarse-grained sediments, I, 552 pp. New York: Elsevier 1975.

Chilingarian, G.V., and Wolf, K.H. (Eds.): Compaction of coarse-grained sediments, II, 808 pp. New York: Elsevier 1976.

Collins, A.G.: Geochemistry of oilfield waters, 496 pp. New York: Elsevier 1975.

Coombs, D.S.; Ellis, A.J.; Fyfe, W.S.; and Taylor, A.M.: The zeolite facies, with comments on the interpretation of hydrothermal syntheses. Geochim. et Cosmochim. Acta 17, 53–107 (1959).

Correns, C.W.: Zur geochemie der diagenese. Geochim. et Cosmochim. Acta 1, 49–54 (1950).

Curtis, C.D.: Stability of minerals in surface weathering reactions: A general thermochemical approach. Earth Surface Processes 1, 63–70 (1976).

Dapples, E.C.: Diagenesis of sandstones. In: Larsen, G., and Chilingar, G.V. (Eds.): Diagenesis in sediments, pp. 91–125. (Developments in Sedimentology, Vol. 8). Amsterdam: Elsevier 1967.

Davies, D.K.; Almon, W.R.; Bonis, S.B.; and Hunter, B.E.: Deposition and diagenesis of Tertiary–Holocene volcaniclastics, Guatemala. In:

Scholle, P.A., and Schluger, P.R. (Eds.): Aspects of diagenesis. Soc. Econ. Paleon. Mineral. Spec. Pub. 26, 281–306 (1979).

Dean, W.E., and Schreiber, B.C. (Eds.): Marine evaporites. Soc. Econ. Paleon. Mineral. Short Course 4, 188 pp. (1978).

DeBoer, R.B.: On the thermodynamics of pressure solution—interaction between chemical and mechanical forces. Geochim. et Cosmochim. Acta 41, 249–256 (1977).

DeBoer, R.B.; Nagtegaal, P.J.C.; and Duyvis, E.M.: Pressure solution experiments on sand. Geochim. et Cosmochim. Acta 41, 257–264 (1977).

Degens, E.T.: Geochemistry of sediments, 342 pp. Englewood Cliffs, New Jersey: Prentice-Hall 1965.

DeSitter, L.U.: Diagenesis of oil-field brines. Am. Assoc. Petroleum Geologists Bull. 31, 2030–2040 (1947).

Dickinson, G.: Geological aspects of abnormal reservoir pressures in Gulf Coast, Louisiana. Am. Assoc. Petroleum Geologists Bull. 37, 410–432 (1953).

Drever, J.I.: The geochemistry of natural waters, 368 pp. Englewood Cliffs, New Jersey: Prentice-Hall 1982.

Dunoyer de Segonzac, G.: The birth and development of the concept of diagenesis (1866–1966). Earth Sci. Rev. 4, 153–201 (1968).

Dunoyer de Segonzac, G.: The transformation of clay minerals during diagenesis and low-grade metamorphism: A review. Sedimentology 15, 281–346 (1970).

Eberl, D.D., and Hower, J.: The kinetics of illite formation. Geol. Soc. America Bull. 87, 1326–1330 (1976).

Engelhardt, W. von: Der Porenraum der Sedimente, 207 pp. Berlin–Heidelberg–New York: Springer 1960.

Engelhardt, W. von: The origin of sediments and sedimentary rocks, 359 pp. Stuttgart: E. Schweizerbart'sche Verlagsbuchhandlung 1973. (Trans. Johns, W.D. New York: Wiley-Halsted Press 1977.)

Engelhardt, W. von, and Gaida, K.H.: Concentration of pore solutions during the compaction of clay sediments. Jour. Sed. Petrology 33, 919–930 (1963).

Epstein, A.G.; Epstein, J.B.; and Harris, L.D.: Conodont color alteration—an index to organic metamorphism. U.S. Geol. Survey Prof. Paper 995, 27 pp. (1977).

Ernst, W.G., and Blatt, H.: Experimental study of quartz overgrowths and synthetic quartzites. Jour. Geology 72, 461–470 (1964).

Eslinger, E.V.; Savin, S.M.; and Yeh, H.W.: Oxygen isotope geothermometry of diagenetically altered shales. In: Scholle, P.A., and Schluger, P.R. (Eds.): Aspects of diagenesis. Soc. Econ. Paleon. Mineral. Spec. Pub. 26, 113–124 (1979).

Fischer, R.P.: Similarities, differences, and some genetic problems of the Wyoming and Colorado plateau types of uranium deposits in sandstone. Econ. Geology 65, 778–784 (1970).

Fournier, R.O.: A method of calculating quartz solubilities in aqueous sodium chloride solutions. Geochim. et Cosmochim. Acta 47, 579–586 (1983).

Fournier, R.O., Rowe, J.J.: The solubility of cristobalite along the three-phase curve, gas plus liquid plus cristobalite. Am. Mineralogist 47, 897–902 (1962).

Franzinelli, E., and Potter, P.E.: Petrology, chemistry, and texture of modern river sands, Amazon River system. Jour. Geology 91, 23–39 (1983).

Füchtbauer, H.: Influence of different types of diagenesis on sandstone porosity. Proc., 7th World Petroleum Cong., 353–369 (1967).

Füchtbauer, H.: Sediments and sedimentary rocks I: Part II, 2nd. Ed., 464 pp. Stuttgart: E. Schweizerbart'sche Verlagsbuchhandlung. New York: Wiley-Halsted Press 1974.

Fuhrmann, W.: "Sandkristalle" und Kugelsandstein. Ihre Rolle bei der Diagenese von Sanden. Der Aufschluss 5, 105–111 (1968).

Gale, P.E.: Diagenesis of the Middle to Upper Devonian Catskill facies sandstones in southeastern New York State. Master's Thesis, Harvard University (1985).

Galloway, W.E.: Deposition and diagenetic alteration of sandstone in Northeast Pacific arc-related basins: Implications for graywacke genesis. Geol. Soc. America Bull. 85, 379–390 (1974).

Garrels, R.M., and Christ, C.L.: Solutions, minerals, and equilibria, 450 pp. New York: Harper and Row 1965.

Garrels, R.M., and Mackenzie, F.T.: Origin of the chemical composition of some springs and lakes. In: Stumm, W. (Ed.): Equilibrium concepts in natural water systems, pp. 222–242. (Advances in Chemistry Series, Vol. 67). New York: Reinhold 1967.

Garrels, R.M., and Mackenzie, F.T.: Sedimentary rock types: relative proportions as a function of geological time. Science 163, 560–571 (1969).

Garrels, R.M., and Mackenzie, F.T.: Evolution of sedimentary rocks, 397 pp. New York: Norton 1971.

Garrels, R.M.; Mackenzie, F.T.; and Siever, R.: Sedimentary cycling in relation to the history of the continents and oceans. In: Robertson, E.C. (Ed.): The nature of the solid earth, pp. 93–121. New York: McGraw-Hill 1971.

Garrison, R.E.; Luternauer, J.L.; Grill, E.V.; MacDonald, R.D.; and Murray, J.W.: Early diagenetic cementation of Recent sands, Fraser River delta, British Columbia. Sedimentology 12, 27–46 (1969).

Gieskes, J.M.; Elderfield, H.; and Nevsky, B.: Interstitial water studies, Leg 65, Deep Sea Drilling Project. In: Lewis, B.T.R., Robinson, P. et al, Init. Repts. DSDP 65, 441–449. Washington: U.S. Govt. Printing Office (1983).

Gilbert, C.M.: Cementation of some California Tertiary reservoir sands. Jour. Geology 57, 1–17 (1949).

Gilbert, C.M., and Turner, F.J.: Use of the universal stage in sedimentary petrography. Am. Jour. Sci. 247, 1–26 (1948).

Glennie, K.W.: Desert sedimentary environments, Developments in Sedimentology, No. 14, 222 p. Amsterdam: Elsevier 1970.

Glennie, K.W.; Mudd, G.C.; and Nagtegaal, P.J.C.: Depositional environment and diagenesis of Permian Rotliegendes sandstones in Leman Bank and Sole Pit areas of the U.K. southern North Sea. Jour. Geol. Soc. (London) 135, 25–34 (1978).

Glover, J.E.: Studies in the diagenesis of some Western Australian sedimentary rocks. Royal Soc. Western Australia Jour. 46, 33–56 (1963).

Graf, D.L.: Chemical osmosis, reverse chemical osmosis, and the origin of subsurface brines. Geochim. et Cosmochim. Acta 46, 1431–1448 (1982).

Grim, R.E.: Clay mineralogy, 2nd Ed., 596 pp. New York: McGraw-Hill 1968.

Groves, A.W.: The unroofing of the Dartmoor granite and distribution of its detritus in southern England. Geol. Soc. London Quart. Jour. 87, 62–96 (1931).

Guggenheim, E.A.: Thermodynamics, 5th, Revised Ed., 390 pp. Amsterdam: North-Holland 1967.

Gümbel, C.W. von: Grundzüge der Geologie, 1144 pp. Kassel: Fischer 1888.

Ham, W.E. (Ed.): Classification of carbonate rocks. Am. Assoc. Petroleum Geologists Mem. 1, 279 pp. (1962).

Harder, H., and Flehmig, W.: Quarzsynthese bei tiefen Temperaturen. Geochim. et Cosmochim. Acta 34, 295 (1970).

Haszeldine, R.S.; Samson, I.M.; and Comford, C.: Dating diagenesis in a petroleum basin, a new fluid inclusion method. Nature 307, 354–357 (1984).

Hawkins, J.W., Jr., and Whetten, J.T.: Graywacke matrix minerals: Hydrothermal reactions with Columbia River sediments. Science 166, 868–870 (1969).

Hawkins, P.J.: Relationship between diagenesis, porosity reduction, and oil emplacement in late Carboniferous sandstone reservoirs, Bothamsall Oilfield, F. Midlands. Jour. Geol. Soc. (London) 135, 7–24 (1978).

Hay, R.L.: Zeolites and zeolite reactions in sedimentary rocks. Geol. Soc. America Spec. Paper 85, 130 pp. (1966).

Heald, M.T.: Stylolites in sandstones. Jour. Geology 63, 101–114 (1955).

Heald, M.T.: Cementation of Simpson and St. Peter sandstones in parts of Oklahoma, Arkansas, and Missouri. Jour. Geology 64, 16–30 (1956).

Heald, M.T., and Larese, R.E.: The significance of solution of feldspar in porosity development. Jour. Sed. Petrology 43, 458–460 (1973).

Heald, M.T., and Renton, J.J.: Experimental study of sandstone cementation. Jour. Sed. Petrology 36, 977–991 (1966).

Heezen, B.C., and Hollister, C.: Deep-sea current evidence from abyssal sediments. Marine Research 1, 141–174 (1964).

Helgeson, H.C.; Garrels, R.M.; and Mackenzie, F.T.: Evaluation of irreversible reactions in geochemical processes involving minerals and aqueous solutions—II: Applications. Geochim. et Cosmochim. Acta 33, 455–481 (1969).

Hem, J.D.: Study and interpretation of the chemical characteristics of natural water, 2nd Ed. U.S. Geol. Survey Water-Supply Paper 1473, 363 pp. (1970).

Hess, P.C.: Phase equilibria of some minerals in the K_2O–Na_2O–Al_2O_3–SiO_2–H_2O system at 25°C and 1 atmosphere. Am. Jour. Sci. 264, 289–309 (1966).

Hoffman, J., and Hower, J.: Clay mineral assemblages as low grade metamorphic geothermometers: Application to the thrust faulted disturbed belt of Montana, U.S.A. In: Scholle, P.A., and Schluger, P.R. (Eds.): Aspects of diagenesis. Soc. Econ. Paleon. Mineral. Spec. Pub. 26, 55–80 (1979).

Hoholick, J.D.; Metarko, T.; and Potter, P.E.: Basinwide maps of porosity and cement: St. Peter and Mount Simon sandstones of the Illinois basin. Am. Assoc. Petroleum Geologists Bull. 68, 753–764 (1984).

Holland, H.D.: The chemistry of the atmosphere and oceans, 351 pp. New York: Wiley 1978.

Hower, J.; Eslinger, E.; Hower, M.E.; and Perry, E.A.: Mechanism of burial metamorphism of argillaceous sediment, mineralogical and chemical evidence. Geol. Soc. America Bull. 87, 725–737 (1976).

Hrabar, S.V., and Potter, P.E.: Lower West Baden (Late Mississippian) sandstone body of Owen and Greene Counties, Indiana. Am. Assoc. Petroleum Geologists Bull. 53, 2150–2160 (1969).

Hubert, J.F.: Petrology of the Fountain and Lyons Formations. Front Range, Colorado. Colorado School Mines Quart. 55, No. 1, 1–242 (1960).

Hunt, J.M.: Petroleum geochemistry and geology, 617 pp. San Francisco: W.H. Freeman 1979.

Hurd, D.C.: Factors affecting dissolution rate of biogenic opal in seawater. Earth Planet. Sci. Letters 15, 411–417 (1972).

Hurley, P.M.; Hunt, J.M.; Pinson, W.H., Jr.; and Fairbairn, H.W.: K–Ar age values on the clay fractions in dated shales. Geochim et Cosmochim. Acta 27, 279–284 (1963).

Illing, L.V.; Wells, A.J.; and Taylor, J.C.M.: Penecontemporaneous dolomite in the Persian Gulf. In: Pray, L.C., and Murray, R.C. (Eds.): Dolomitization and limestone diagenesis. Soc. Econ. Paleon. Mineral. Spec. Pub. 13, 89–111 (1965).

Jones, J.B., and Segnit, E.R.: The nature of opal I, nomenclature and constituent phases. Geol. Soc. Australia Jour. 18, 57–68 (1971).

Jones, P.H.: Hydrology of Neogene deposits in the northern Gulf of Mexico basin. Louisiana Water Res. Inst. Bull. GT-2, 105 pp. (1967).

Kastner, M.: Authigenic feldspars in carbonate rocks. Am. Mineralogist 56, 1403–1442 (1971).

Kastner, M.; Keene, J.B.; and Gieskes, J.M.: Diagenesis of siliceous oozes-I. Chemical controls on the rate of opal-A to opal-CT transformation-an experimental study. Geochim. et Cosmochim. Acta 41, 1041–1059 (1977).

Kastner, M., and Siever, R.: Origin of authigenic feldspars in carbonate rocks. Geol. Soc. America Spec. Paper 121, 155–156 (1968).

Kastner, M., and Siever, R.: Low temperature feldspars in sedimentary rocks. Am. Jour. Sci. 279, 435–479 (1979).

Kastner, M., and Waldbaum, D.R.: Authigenic albite from Rhodes. Am. Mineralogist 53, 1579–1602 (1968).

Kossovskaya, A.G., and Shutov, V.D.: Zonality in the structure of terrigenous deposits in platform and geosynclinal regions. Eclogae géol. Helvetiae 51, 656–666 (1958).

Krauskopf, K.: Introduction to geochemistry, 2nd. Ed., 617 pp. New York: McGraw-Hill 1979.

Kubler, B.: Les argiles, indicateurs de métamorphisme. Rev. Inst. Français Pétrol. 19, 1093–1112 (1964).

Land, L.S., and Dutton, S.P.: Cementation of a Pennsylvanian deltaic sandstone: Isotopic data. Jour. Sed. Petrology 48, 1167–1176 (1978).

Land, L.S., and Prezbindowski, D.R.: The origin and evolution of saline formation water, Lower Cretaceous carbonates, south central Texas, U.S.A. Jour. Hydrology 54, 51–74 (1981).

Larson, E.E., and Walker, T.R.: A rock magnetic study of the lower massive sandstone, Moenkopi Formation (Triassic), Gray Mountain area, Arizona. Jour. Geophys. Research 87, 4819–4836 (1982).

Lawson, D.S.; Hurd, D.C.; and Pankratz, H.S.: Silica dissolution rates of decomposing phytoplankton assemblages at various temperatures. Am. Jour. Sci. 278, 1373–1393 (1978).

Levorsen, A.I.: Geology of petroleum, 2nd Ed., 724 pp. San Francisco: W.H. Freeman 1967.

Loucks, R.G.; Bebout, D.G.; and Galloway, W.E.: Relationship of porosity formation and preservation to sandstone consolidation history—Gulf Coast Lower Tertiary Frio Formation. Gulf Coast Assoc. Geol. Soc. Trans. 27, 109–116 (1977).

Lovell, J.P.B.: Tyee Formation: A study of proximality in turbidites. Jour. Sed. Petrology 39, 935–953 (1969).

Mackenzie, F.T., and Gees, R.: Quartz synthesis at earth-surface conditions. Science 173, 533–535 (1971).

Mankiewicz, D., and Steidtmann, J.R.: Depositional environments and diagenesis of the Tensleep Sandstone, Eastern Big Horn basin, Wyoming. In: Scholle, P.A., and Schluger, P.R. (Eds.): Aspects of diagenesis. Soc. Econ. Paleon. Mineral. Spec. Pub. 26, 319–336 (1979).

Maynard, J.B.: Geochemistry of sedimentary ore deposits, 305 pp. Berlin–Heidelberg–New York: Springer-Verlag 1983.

McClay, K.R.: Pressure solution and Coble creep in rocks and minerals: A review. Jour. Geol. Soc. (London) 134, 57–75 (1977).

McDonald, D.A., and Surdam, R.C. (Eds.): Clastic diagenesis. Am. Assoc. Petroleum Geologists Mem. 3, 434 pp. (1984).

McElhinny, M.W.: The magnetic polarity time scale: Prospects and possibilities in magnetostratigraphy. In: Cohee, G.V.; Glaessner, M.F.; and Hedberg, H.D. (Eds.): The geologic time scale. Am. Assoc. Petroleum Geologists Studies in Geology 6, 57–66 (1978).

McKelvey, J.G., and Milne, I.H.: The flow of salt solutions through compacted clay. In: Clays and clay minerals, pp. 248–259. (Proc. 9th Natl. Clay Conf.). London: Pergamon 1962.

Meade, R.H.: Removal of water and rearrangement of particles during the compaction of clayey sediments—a review. U.S. Geol. Survey Prof. Paper 497B, 23 pp. (1964).

Merino, E.: Diagenesis in Tertiary sandstones from Kettleman North Dome, California—I. Diagenetic mineralogy. Jour. Sed. Petrology 45, 320–336 (1975a).

Merino, E.: Diagenesis in Tertiary sandstones from Kettleman North Dome, California—II. Interstitial solutions: Distribution of aqueous species at 100°C and chemical relation to the diagenetic mineralogy. Geochim. et Cosmochim. Acta 39, 1629–1645 (1975b).

Millot, G.: Géologie des argiles, 499 pp. Paris: Masson 1963.

Millot, G.: Geology of clays, 429 pp. (Trans. by Farrand, W.R., and Paquet, H.). New York: Springer-Verlag 1970.

Morris, R.C.; Proctor, K.E.; and Koch, M.R.: Petrology and diagenesis of deep-water sandstones, Ouachita Mountains, Arkansas and Oklahoma. In: Scholle, P.A., and Schluger, P.R. (Eds.): Aspects of diagenesis. Soc. Econ. Paleon. Mineral. Spec. Pub. 26, 263–279 (1979).

Murray, R.C.: The origin and diagenesis of gypsum and anhydrite. Jour. Sed. Petrology 34, 512–523 (1964).

Naeser, C.W.: Thermal history of sedimentary basins: Fission-track dating of subsurface rocks. In: Scholle, P.A., and Schluger, P.R. (Eds.): Aspects of diagenesis. Soc. Econ. Paleon. Mineral. Spec. Pub. 26, 55–80 (1979).

Nagtegaal, P.J.: Microtextures in Recent and fossil caliche. Leidse Geol. Mededel. 42, 131–142 (1969).

Nagtegaal, P.J.: Sandstone-framework instability as a function of burial diagenesis. Jour. Geol. Soc. (London) 135, 101–105 (1978).

Nesbitt, H.W.: Characterization of mineral-formation water interactions in Carboniferous sandstones and shales of the Illinois sedimentary basin. Am. Jour. Sci. 280, 607–630 (1980).

Nickel, E.: Experimental dissolution of light and heavy minerals in comparison with weathering and intrastratal solution. Contrib. Sedimentology (Stuttgart) 1, 1–68 (1973).

Odum, I.E.; Willand, T.N.; and Lassin, R.J.: Paragenesis of diagenetic minerals in the St. Peter sandstone (Ordovician), Wisconsin and Illinois. In: Scholle, P.A., and Schluger, P.R. (Eds.): Aspects of diagenesis. Soc. Econ. Paleon. Mineral. Spec. Pub. 26, 425–443 (1979).

Opdyke, N.D.: The paleomagnetism of oceanic cores. In: Phinney, R.A. (Ed.): The history of the earth's crust, pp. 61–72. Princeton, New Jersey: Princeton Univ. Press 1968.

Osborn, E.F.: Subsolidus reactions in oxide systems in the presence of water at high pressures. Jour. Am. Cer. Soc. 36, 147–151 (1953).

Packham, G.H., and Crook, K.A.W.: The principle of diagenetic facies and some of its implications. Jour. Geology 68, 392–407 (1960).

Pettijohn, F.J.: Persistence of minerals and geologic age. Jour. Geology 49, 610–625 (1941).

Philipp, W.; Drong, H.J.; Füchtbauer, H.; Haddenhorst, H.G.; and Jankowsky, W.: The history of migration in the Gifhorn trough (NW Germany). 6th World Petroleum Cong., Frankfort, Sec. I, Paper 19, PD 2, 457–481 (1963).

Picard, M.D.: Paleomagnetic correlation of units within Chugwater (Triassic) Formation, west-central Wyoming. Am. Assoc. Petroleum Geologists Bull. 48, 269–291 (1964).

Pittman, E.D.: Diagenesis of quartz in sandstones as revealed by scanning electron microscopy. Jour. Sed. Petrology 42, 507–519 (1972).

Pittman, E.D.: Recent advances in sandstone diagenesis. Ann. Rev. Earth Planet. Sci. 7, 39–62 (1979).

Potter, P.E.: Petrology and chemistry of modern big river sands. Jour. Geology 86, 423–449 (1978).

Reineck H.E., and Singh, I.B.: Depositional sedimentary environments, 2nd Ed., 549 pp. New York: Springer-Verlag 1980.

Renton, J.J.; Heald, M.T.; and Cecil, C.B.: Experimental investigation of pressure solution of quartz. Jour. Sed. Petrology 39, 1107–1117 (1969).

Rittenhouse, G.: Pore space reduction by solution and cementation. Am. Assoc. Petroleum Geologists Bull. 55, 80–91 (1971).

Rittenhouse, G.: Mechanical compaction of sands containing different percentages of ductle grains: A theoretical approach. Am. Assoc. Petroleum Geologists Bull. 55, 92–96 (1971).

Rittenhouse, G.: Cross-bedding dip as a measure of sandstone compaction. Jour. Sed. Petrology 42, 682–683 (1972).

Robie, R.A.; Hemingway, B.S.; and Fisher, J.R.: Thermodynamic properties of minerals and related substances at 298.15K and 1 bar (10^5 pascals) pressure and at higher temperatures. U.S. Geol. Survey Bull. 1452, 456 pp. (1978).

Robin, P.Y.: Pressure solution at grain-to-grain contacts. Geochim. et Cosmochim. Acta 42, 1383–1389 (1978).

Roedder, E.: Fluid inclusion evidence on the environments of sedimentary diagenesis, a review. In: Scholle, P.A., and Schluger, P.R. (Eds.): Aspects of diagenesis. Soc. Econ. Paleon. Mineral. Spec. Pub. 26, 55–80 (1979).

Rosenfeld, M.A.: Some aspects of porosity and cementation. Producers Monthly, Pennsylvania Oil Prod. Assoc. 13, 39–42 (1949).

Rusnak, G.A.: A fabric and petrologic study of the Pleasantview sandstone. Jour. Sed. Petrology 27, 41–55 (1957).

Rutter, E.H.: The kinetics of rock deformation by pressure solution. Phil. Trans. Royal Soc. London A. 283, 203–219 (1976).

Sabins, F.F., Jr.: Grains of detrital, secondary, and primary dolomite from Cretaceous strata of the western interior. Geol. Soc. America Bull. 73, 1183–1196 (1962).

Sayles, F.L.: The composition of interstitial solutions I. Fluxes across the seawater–sediment interface in the Atlantic Ocean. Geochim. et Cosmochim. Acta 43, 526–546 (1979).

Schmidt, V., and McDonald, D.A.: Texture and recognition of secondary porosity in sandstones. In: Scholle, P.A., and Schluger, P.R. (Eds.): Aspects of diagenesis. Soc. Econ. Paleon. Mineral. Spec. Pub. 26, 209–225 (1979a).

Schmidt, V., and McDonald, D.A.: The role of secondary porosity in the course of sandstone diagenesis. In: Scholle, P.A., and Schluger, P.R. (Eds.): Aspects of diagenesis. Soc. Econ. Paleon. Mineral. Spec. Pub. 26, 175–207 (1979b).

Scholle, P.A.: A color illustrated guide to constituents, textures, cements, and porosities of sandstones and associated rocks. Am. Assoc. Petroleum Geologists Mem. 28, 201 pp. (1979).

Scholle, P.A., and Schluger, P.R. (Eds.): Aspects of diagenesis. Soc. Econ. Paleon. Mineral. Spec. Pub. 26, 443 pp. (1979).

Sclater, J.G., and Christie, P.A.F.: Continental stretching—an explanation of the post-Early Cretaceous subsidence of the central graben of the North Sea. Jour. Geophys. Research 85, 3711–3739 (1980).

Sedimentation Seminar: Bethel Sandstone (Mississippian) of western Kentucky and south-central Indiana, a submarine-channel fill. Kentucky Geol. Survey, Rept. Inv. 11, Ser. X, 24 pp. (1969).

Selley, R.C.: Porosity gradients in North Sea oil-bearing sandstones. Jour. Geol. Soc. (London) 135, 119–132 (1978).

Shearman, D.J.: Recent anhydrite, gypsum, dolomite and halite from the coastal states of the Arabian shore of the Persian Gulf. Geol. Soc. London Proc. 607, 63–65 (1963).

Sibley, D.F., and Blatt, H.: Intergranular pressure solution and cementation of the Tuscarora orthoquartzite. Jour. Sed. Petrology 46, 881–896 (1976).

Siever, Raymond: The silica budget in the sedimentary cycle. Am. Mineralogist 42, 821–841 (1957).

Siever, Raymond: Petrology and geochemistry of silica cementation in some Pennsylvanian sandstones. In: Ireland, H.A. (Ed.): Silica in sediments. Soc. Econ. Paleon. Mineral. Spec. Pub. 7, 55–79 (1959).

Siever, Raymond: Silica solubility, 0°C–200°C, and

the diagenesis of siliceous sediments. Jour. Geology 70, 127–150 (1962).

Siever, Raymond: Establishment of equilibrium between clays and sea water. Earth Planet. Sci. Letters 5, 106–110 (1968a).

Siever, Raymond: Sedimentological consequences of a steady-state ocean-atmosphere. Sedimentology 11, 5–29 (1968b).

Siever, Raymond: The low temperature geochemistry of silicon. In: Wedepohl, K.H., and Turekian, K. (Eds.): Handbook of geochemistry, Vol. II-14. Berlin–Heidelberg–New York: Springer 1972.

Siever, Raymond: Plate tectonic controls on diagenesis. Jour. Geology 87, 127–155 (1979).

Siever, Raymond: Burial history and diagenetic reaction kinetics. Am. Assoc. Petroleum Geol. Bull. 67, 684–691 (1983).

Siever, Raymond, and Hager, J.L.: Paleogeography, tectonics and thermal history of some Atlantic margin sediments. In: Kerr, J.W. (Ed.): Geology of the North Atlantic borderlands. Canadian Soc. Petroleum Geologists Mem. 7, 95–117 (1981).

Siever, Raymond, and Woodford, N.: Dissolution kinetics and the weathering of mafic minerals. Geochim. et Cosmochim. Acta 43, 717–724 (1979).

Siever, Raymond; Beck, K.C.; and Berner, R.A.: Composition of interstitial waters of modern sediments. Jour. Geology 73, 39–73 (1965).

Sillen, L.G.: The physical chemistry of sea water. In: Sears, M. (Ed.): Oceanography. Am. Assoc. Adv. Sci. Pub. 67, 549–581 (1961).

Silverman, S.R.: Carbon isotope geochemistry of petroleum and other natural organic materials. 3rd Intern. Wissenschaftlichen Konferenz für Geochemie, Mikrobiologie und Erdölchemie Vorträge, Budapest 2, 328–341 (1963).

Simmons, G., and Bell, P.: Calcite–aragonite equilibrium. Science 139, 1197–1198 (1963).

Sippel, R.F.: Sandstone petrology, evidence from luminescence petrography. Jour. Sed. Petrology 38, 530–554 (1968).

Sloss, L.L., and Feray, D.E.: Microstylolites in sandstone. Jour. Sed. Petrology 18, 3–14 (1948).

Smalley, I.J.: A simple model of a diagenetic system. Sedimentology 8, 7–26 (1967).

Spotts, J.H., and Silverman, S.R.: Organic dolomite from Point Fermin, California. Am. Mineralogist 51, 1144–1155 (1966).

Sprunt, E.S., and Nur, A.: Destruction of porosity through pressure solution. Geophysics 42, 726–741 (1977a).

Sprunt, E.S., and Nur, A.: Experimental study of the effects of stress on solution rate. Jour. Geophys. Research 82, 3013–3022 (1977b).

Stanley, K.O., and Benson, L.V.: Early diagenesis of high plains Tertiary vitric and arkosic sandstone, Wyoming and Nebraska. In: Scholle, P.A., and Schluger, P.R. (Eds.): Aspects of diagenesis. Soc. Econ. Paleon. Mineral. Spec. Pub. 26, 401–423 (1979).

Stanley, K.O., and Faure, G.: Isotopic composition and sources of strontium in sandstone cements: The high plains sequence of Wyoming and Nebraska. Jour. Sed. Petrology 49, 45–54 (1979).

Steckler, M.S., and Watts, A.B.: The Gulf of Lion: subsidence of a young continental margin. Nature 287, 425–429 (1980).

Stein, C.L.: Dissolution of diatoms and diagenesis in siliceous sediments, 150 p., Cambridge (Mass.): Harvard Univ. Ph.D. thesis (1977).

Stoessel, R.K., and Moore, C.H.: Chemical constraints and origins of four groups of Gulf Coast reservoir fluids. Am. Assoc. Petroleum Geologists Bull. 67, 896–906 (1983).

Surdam, R.C.: Low-grade metamorphism of tuffaceous rocks in the Karmutzen Group, Vancouver Island, British Columbia. Geol. Soc. America Bull. 84, 1911–1922 (1973).

Surdam, R.C., and Boles, J.R.: Diagenesis of volcanic sandstones. In: Scholle, P.A., and Schluger, P.R. (Eds.): Aspects of diagenesis. Soc. Econ. Paleon. Mineral. Spec. Pub. 26, 227–242 (1979).

Swineford, A: Cemented sandstones of the Dakota and Kiowa formations in Kansas. Kansas Geol. Survey Bull. 70, Pt. 4, 53–104 (1947).

Tada, R., and Siever, R.: Experimental knife-edge pressure solution of halite. Geochim. et Cosmochim. Acta 50, 29–36 (1986).

Tallman, S.L.: Sandstone types: Their abundance and cementing agents. Jour. Geology 57, 582–591 (1949).

Taylor, J.C.M.: Sandstone diagenesis, introduction. Jour. Geol. Soc (London) 135, 3–5, and following papers, pp. 6–135 (1978).

Taylor, J.M.: Pore-space reduction in sandstones. Am. Assoc. Petroleum Geologists Bull. 34, 701–716 (1950).

Thomson, A.: Pressure solution and porosity. In: Ireland, H.A. (Ed.): Soc. Econ. Paleon. Mineral. Spec. Pub. 7, 92–111 (1959).

Tissot, B.P., and Welte, D.H.: Petroleum formation and occurrence, 538 pp. New York: Springer-Verlag 1978.

Towe, K.M.: Clay mineral diagenesis as a possible source of silica cement in sedimentary rocks. Jour. Sed. Petrology 32, 26–28 (1962).

Trurnit, P.: Pressure solution phenomena in detrital rocks. Sedimentary Geology 2, 89–114 (1968).

Van Hinte, J.E.: Geohistory analysis—application of micropaleontology in exploration geology. Am. Assoc. Petroleum Geologists Bull. 62, 201–222 (1978).

Van Hise, C.R.: A treatise on metamorphism. U.S. Geol. Survey Mono. 47, 1286 pp. (1904).

Van Houten, F.B.: Iron oxides in red beds. Geol. Soc. America Bull. 79, 399–416 (1968).

Velde, B.: Clays and clay minerals in nature and synthetic systems, 218 pp. New York: Elsevier 1979.

Verhoogen, J.: Geological significance of surface tension. Jour. Geology 56, 210–217 (1948).

Waldschmidt, W.A.: Cementing materials in sandstones and their probable influence on migration and accumulation of oil and gas. Am. Assoc. Petroleum Geologists Bull. 25, 1839–1879 (1941).

Walker, T.R.: Reversible nature of chert-carbonate

replacement in sedimentary rocks. Geol. Soc. America Bull 73, 237–242 (1962).

Walker, T.R.: Formation of red beds in modern and ancient deserts. Geol. Soc. America Bull. 78, 353–368 (1967).

Walker, T.R.: Formation of red beds in moist tropical climates: A hypothesis. Geol. Soc. America Bull. 85, 633–638 (1974).

Walker, T.R.: Diagenetic origin of continental red beds. In: Falke, H. (Ed.): The continental remain in Central, West and South Europe, pp. 240–282. (NATO Advanced Study Inst. Series C. Math. and Phys. Sci.) Dordrecht, Holland: D. Reidel Pub. Co. 1976.

Walther, J.V., and Helgeson, H.: Calculation of the thermodynamic properties of aqueous silica and the solubility of quartz and its polymorphs at high pressures and temperatures. Am. Jour. Sci. 277, 1315–1351 (1977).

Waugh, B.: Authigenic K-feldspar in British Permo-Triassic sandstones. Jour. Geol. Soc. (London) 135, 51–56 (1978).

Weyl, P.K.: Pressure solution and the force of crystallization—a phenomenological theory. Jour. Geophys. Research 64, 2001–2025 (1959).

Wilson, M.D., and Pittman, E.D.: Authigenic clays in sandstones. Recognition and influence on reservoir properties and paleoenvironmental analysis. Jour. Sed. Petrology 47, 3–31 (1977).

Winkler, H.G.F.: Petrogenesis of metamorphic rocks, 5th Ed., 348 pp. Berlin–Heidelberg–New York: Springer 1979.

Wollast, R.: Kinetics of the alteration of K-feldspar in buffered solutions at low temperature. Geochim. et Cosmochim. Acta 31, 635–648 (1967).

Wood, J.R., and Hewett, T.A.: Fluid convection and mass transfer in porous sandstones—a theoretical model. Geochim. et Cosmochim. Acta 46, 1707–1713 (1982).

Wood, J.R., and Surdam, R.C.: Application of convective-diffusion model to diagenetic processes. In: Scholle, P.A., and Schluger, P.R. (Eds.): Aspects of diagenesis. Soc. Econ. Paleon. Mineral. Spec. Pub. 26, 243–250 (1979).

Yeh, H.W., and Savin, S.M.: Mechanism of burial metamorphism of argillaceous sediment, 3.0-isotope evidence. Geol. Soc. America Bull. 88, 1321–1330 (1977).

Young, A., and Galley, J.E. (Eds.): Fluids in subsurface environments. Am. Assoc. Petroleum Geologists Mem. 4, 414 pp. (1965).

CHAPTER 12
Sandstones, Tectonics, and Continental Evolution

Introduction

At some stage in the geological investigation of a sandstone, we need to place the formation in a larger perspective—to see it in relation to the basin fill as a whole and even to think about it in the longer view of geologic history and continental evolution. We need to go from a provincial to a cosmopolitan outlook. Only by placing the local study in a larger framework can we really understand the sandstone in question. Hence in this chapter we consider the larger questions related to sands and sandstones.

Sands have proved to be the most useful sediment in unraveling the geologic history of the large majority of sedimentary basins on the continents and, increasingly, of oceanic basins off continental margins. As we have seen in Chap. 7, the kind and proportions of the detrital minerals provide clues to the character of the source rocks and the location of the source area and tell us something about the relief and climate of that region. The paleocurrent record is largely reconstructed from the primary sedimentary structures of sandstones. Neither shales nor limestones shed as much light on the interrelated problems of provenance, paleocurrents, and tectonics. Sandstones, especially when considered with associated shales and limestones, are most useful for environmental interpretations and thus the patterns of filling of basins.

There are also such broad questions as the relations between the sands of the stable cratons and those of mobile belts. Are there basic differences? What is the relation between sands, tectonics, and the architecture of the depositional basin? And what of the sands of the deep ocean basins? Can they be related to tectonics, provenance, and paleocurrents in the same ways as sands on the continents?

Geologists have long held that the evolution—structural, stratigraphic, and sedimentologic—of sedimentary basins proceeds in an orderly fashion. Beginning with the definition and characterization of geosynclines by James Hall, E. S. Dana, and others in the mid-19th century, continuing with Bertrand's ideas in 1897, up to a modern restatement by Aubouin in 1965, the ideas proposed were based on reasoning from ancient rocks. Great series of deformed sediments, metasediments, and igneous rocks in major continental mountain belts of the world were deduced to have originated by mechanisms that were as yet unknown in the modern world.

Starting in 1965 a major shift occurred as plate tectonic theory was enunciated. The new theory was a synthesis of contemporary, geophysically sensed tectonic movements, and emphasized the geology of the ocean basins. The late 1960s and the 1970s were times of intense activity in fitting the history of the oceans and continental margins with the newer ideas of plate tectonics. Patterns of continental drift were subsumed and justified by the kinematics of plate motions and correlated with observations and ideas of the previous century. The 1980s have seen the wholesale reexamination of the continental record and newer ideas on the evolution of orogenic belts. We review some of these ideas in this chapter as they relate to the deposition of sandstones. We believe that the reevaluation of earlier ideas, integrated with current geophysical theory, is leading to a new synthesis. Current views not only tie the stratigraphic and structural ideas of Stille (1936) and Kay (1951) to the petrologic theories relating tectonics and sandstone compositions of Krynine (1941) and Pettijohn (1943), but accord well with the knowledge of the stratigraphy and sedimentology of the sea floor and continental margins acquired since the Deep Sea Drilling Project began in 1968.

Because sand and sandstone are our main concern, we begin with a consideration of the factors needed to produce sandy basins, those that are characterized by abundant sandstone deposits. We follow with a discussion of plate tectonics, and conclude with a discussion of sandstone in the history of the earth.

Sandy Basins

General controls on the fill of a basin are four in number—availability of terrigenous detritus, sea level, climate, and tectonics (Table 12-1). These four general controls can be recast in terms of other variables: the relief and climate of the basin source areas, possible sinks intervening between sources and basin that could trap sand, and sediment transport regimes within the basin. Below we explore these factors in relation to the evolution of sandy basins, those containing more than 75 percent sandstone.

The combination of high relief and high rainfall always generates the most sand and mud; the combination of high relief and low rainfall produces much less. The combination of low relief and rainfall yields the least detritus (Chap. 5; Potter, 1978, Fig. 14). Both temperature and rainfall, interacting with relief, influence vegetation, which modifies the effects of rainfall alone. The high relief may be proximal or distal to the basin. Certainly, distance between source and sink is not a key variable in the origin of sandy basins—sand transported by short rivers may accumulate near its source or sand may be transported hundreds and even thousands of kilometers from its source by a large river.

The rock types of the source area influence sediment yield. For example, recently uplifted, poorly consolidated sediments yield more debris than well-consolidated sediments, and recent volcanics probably yield more debris than granite and metamorphic terranes, given comparable relief and rainfall. Relief and rainfall also control sand composition (Chap. 5). High-relief sources may yield abundant immature lithic or sublithic sands or lesser volumes of arkosic and subarkosic sands near basement uplifts. Low-relief sources, regardless of rainfall, yield small volumes of sediment. If the low-relief source has high rainfall, sands derived from it will be mature.

An intervening, subsiding basin near the source should also be considered because it may trap so much sand that mud is the principal sediment transported across it. Chemical weathering of immature detritus may be significant in intervening areas where warm, rainy climates lead to extensive dissolution of feldspar and other unstables, producing a mature sand from an immature one. Recognition of possible intervening basins requires a broad grasp of stratigraphy and paleogeography, and is vital in basin analysis and reconstruction of evolving continental margins.

Within a sandy basin, currents separate sand from mud and silt in various environments: braided streams, eolian dune fields, high-energy coastlines and shelves, and proximal parts of subsea fans. Where high relief is closely coupled to a depositional basin, alluvial fans, fan deltas with turbidites, and narrow sandy coastal plains and shelves will be the dominant sand-rich environments. The sandy fill of the Ridge Basin is an example of the dominance of sand in the alluvial fill of a transform fault basin (Fig. 12-1). Where the high relief source is far removed from the basin, there will be a major river with a large, sandy delta. If this delta is bordered by deep water, associated subsea fans will be present. If not, a sandy shelf develops.

Sea level fluctuation is also a factor in basin fill (Vail *et al.* 1977; Hardenbol *et al.* 1981; Klein, 1982; Shanmugam and Moiola, 1982). Although none of these studies directly focus on sandy basins, they do emphasize interactions of global and local sea level, sediment supply, basin subsidence, and basin fill. A high stand of sea level may flood much of a continent's interior and its passive margins, whereas a low stand favors emergent continental interiors and

TABLE 12-1. Factors that control the fill of a basin.

Availability of terrigenous mud and sand—depends on uplift of source area, rainfall, and possible sediment trapping in intervening basins.

Sea level—determines if the surface of sedimentation is at, above, or below sea level and is thus the major control on the depositional systems that fill a basin.

Climate—two major controls: (1) high rainfall needed for rivers to carry appreciable mud and sand and (2) carbonates require a low-latitude, warm-water, marine basin.

Tectonics—controls relief of source area, basin geometry, and rate of basin subsidence.

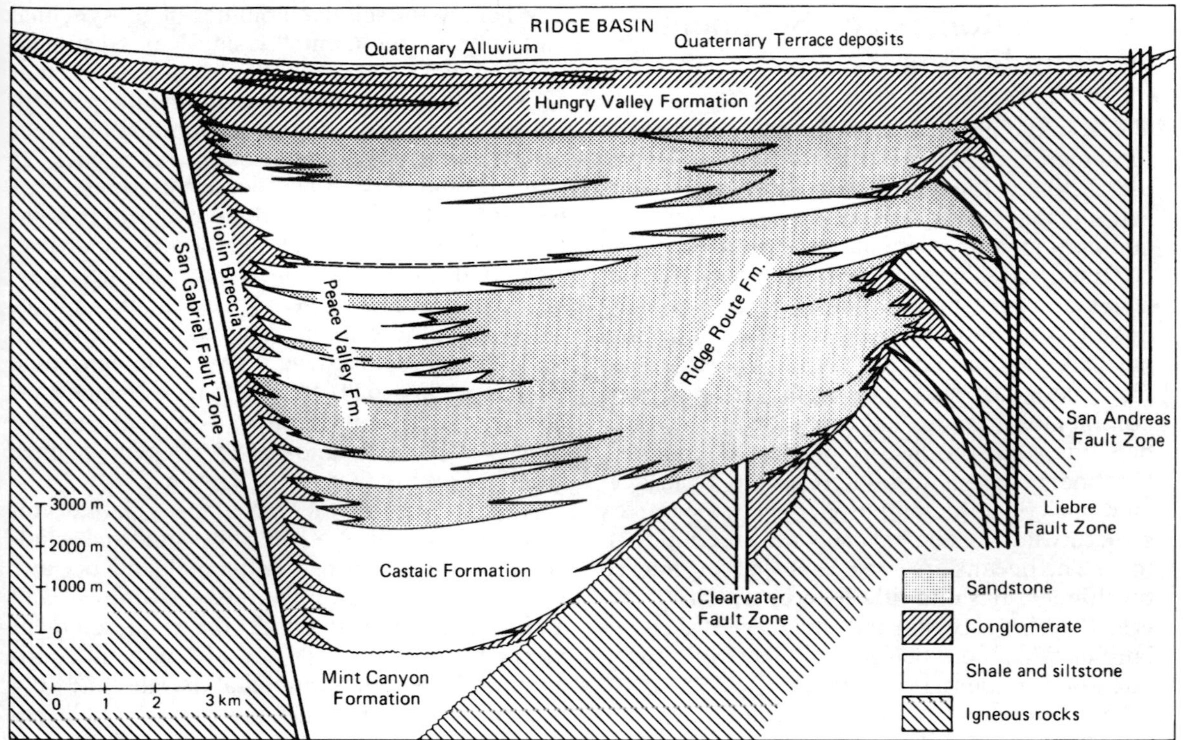

FIGURE 12-1. An example of a sandy basin. Diagrammatic cross section of Ridge Basin showing major lithologic units and stratigraphic and structural relations. This transform basin, located along the San Andreas fault, is dominated by sand as a consequence of nearby uplifted eroding borders, short transport distance, and a climate that allowed fast erosion with relatively little chemical weathering. (Modified and simplified from Crowell, 1982).

narrow submerged continental shelves. Shanmugam and Moiola (1982) have suggested that high stands of sea level cause sediment trapping on continental shelves and in interior seas whereas low stands favor eolian and fluvial deposits on the continent and large turbidite fans at the base of the continental slope.

Climate also plays a role in the filling of a basin. Where a basin is sand- or mud-starved, a cold climate will inhibit carbonate accumulation whereas the same shelf with a warm climate can sustain high carbonate production. Thus climate interacts with terrigenous supply to determine if the basins of passive continental margins contain few or appreciable carbonates. It is, of course, climate that causes some continental interior basins to have been filled chiefly by eolianites rather than fluvial deposits (Glennie, 1970, Fig. 3). See Fontaine (1983, p. 9) for the possible role of climatic contrasts along South America's west coast in the Tertiary, when sandy basins were produced where the paleoclimate was dry and shale-rich basins where it was wet. Underlying this interpretation is the idea that a wet climate favors mud production over sand, all other factors being equal, because of greater weathering (Potter *et al.*, 1980, p. 75).

As yet, we have no overall theory that would fully relate plate tectonics to the interacting combinations of sea level, sediment supply, climate, and local tectonics that produce a sandy basin. These basins occur in a variety of plate tectonic settings—on passive and active margins, in transform basins, on cratonic interiors, and in Andean-type convergences. This variety reflects the as yet poorly spelled out interactions of plate tectonics with paleoclimate, sea level, and sediment transport regimes to and within the basin. Thus accidents of climate and paleolatitude of a basin may be of great local importance. Ultimately, we believe, the apparently nonsystematic variations of paleoclimates, sea level, and local tectonics will be meaningfully linked to plate tectonics and its geophysical causes. At this time, however, plate tectonics is the major systematic control on provenance, environment, and diagenesis of sand in sedimentary basins. It is that system we outline in the following sections.

Plate Tectonics and Sedimentary Accumulations

The relative movement of the first-order structural elements of the Earth's crust, lithospheric plates 100 to 150 km thick, creates the tectonic settings that control the nature of sediments accumulating in depositional environments on the continents and in the oceans. Major plate geometries and margins are not necessarily closely linked to second-order features, the continental margins, which may or may not coincide with plate boundaries. As we shall see, however, continental margins and evolution are strongly dependent on the history of plate boundaries. Continental areas are differentiated from oceanic ones by a thick lithosphere with a complex structure and an ancient and varied tectonic history. The oceans are somewhat simpler, floored by thinner, basaltic lithosphere, which is covered by relatively thin and undeformed accumulations of oceanic pelagic sediments and turbidites of continental provenance.

The continents' major structural elements have long been defined as cratons and mobile belts or geosynclines. Cratons are large continental areas on which relatively thin, essentially flat-lying sediments completely or partially cover a tectonically stable basement. The basement is commonly Precambrian, although it may be younger. The craton thus includes the shields, where basement rock is exposed, and platforms, where basement is covered by flat-lying rocks of variable thickness. Mobile belts or geosynclines are linear, thick accumulations of sediments, metasediments, and igneous rocks, now deformed and uplifted, that show evidence of a complex history of uneven subsidence, varied bathymetries, and heterogeneous uplifts. Current ideas favor the origin of mobile belts as remnants of former continental margins that have become involved in various types of plate convergences. Evolution of mobile belts is linked to the opening and closing of oceanic basins by plate convergences and divergences and so to continental drift. Plate motions may involve accretions of variously sized continental blocks and oceanic plateaus which may be moved along continental margins for long distances by transform fault motion (Jones, 1978; Nur and Ben-Avraham, 1982). The search for order in the complex sequence of events in the evolution of a deformed mobile belt continues, but now with definite geophysical mechanisms tied to tectonic and sedimentologic histories.

What are the relative volumes of geosynclinal and cratonic sediments? Using Kay's estimate (1951, p. 92) for area and thickness, we find that 82 percent by volume of the sediments in North America are geosynclinal accumulations (eugeosynclines 59 percent; miogeosynclines 23 percent) and only 18 percent are cratonic. If 25 percent of the cratonic sediment and 15 percent of the geosynclinal accumulation is sand, it is clear that most sandstone of the geologic past has been trapped in geosynclines, the ratio of geosynclinal to cratonic sand being at least 2.5 to 1 and probably near 4 to 1. Though some (Gilluly *et al.* 1970) have estimated somewhat different sediment volumes, it is clear that the total volume of sandstones of geosynclines greatly exceeds that of the cratons.

Another way of looking at the relative volumes of sediments, especially sand and sandstone, is to compare continental and oceanic sedimentary accumulations. Ronov and Yaroshevsky (1969) divide the total sediment volume, 654×10^6 km^3, into 61.6 percent continental, 28 percent continental margin, and 10.4 percent oceanic basin. Gilluly *et al.* (1970) divide a total of 715×10^9 km^3 into 44.8 percent continent and 34.3 percent offshore, the remaining 20.9 percent being oceanic. However one divides a somewhat uncertain total, it is clear that most sediment is deposited on or near continents and their margins and relatively little in the ocean basins. Considering that only a small part of oceanic sediment is sand, we conclude that a very small fraction of the world's sand is dispersed to the oceans far from the continents.

Once deposited, sand is recycled in different ways, depending on its depositional site. Estimates of recycling rates of sediments based on erosion rates and sediment volumes range from a little over 100 million years to almost 300 million years for mean sediment residence times (Gregor, 1970; Garrels and Mackenzie, 1971; Li, 1972; Blatt and Jones, 1975; Blatt *et al.*, 1980, pp. 32–38). These estimates represent mostly continental and continental margin processes—that is where most sediment is—but a figure based on oceanic sediment alone would give a figure of about the same magnitude. Cycling through the lower crust and mantle via subducting slabs may involve some continental margin sediment that is sandy, but the volumes of sand are small compared to other sediments found on or at the edge of oceanic lithosphere. We conclude that the operations of plate tectonics together with other sedimentary processes keep sand at or near the surface of the Earth in continental crust once it arrives by sedimentary and igneous differentiation.

Geosynclines—The Classical View

It is convenient, as has been pointed out by a number of geologists, such as Cobbing (1978), to continue some tradition of geosynclinal terminology, that is, to refer to linear belts of thick sedimentary and metasedimentary accumulations, usually deformed, as geosynclines. And, as Bally and Snelson (1980, p. 24) point out, "familiarity with the old concepts and terms is necessary to appreciate the voluminous literature based on these concepts."

As summarized by Aubouin (1965, p. 104), a geosyncline was conceived as normally consisting of a mio-/eugeosynclinal couple (Fig. 12-2). From the foreland or craton to the interior, the ideal geosyncline consisted of a miogeosynclinal basin, a medial ridge which separated the miogeosyncline from the eugeosyncline, and an interior tectonic land or ridge. This geosynclinal domain was thought to have been localized at or near the margin of the continent or, as in the case of the Urals, within the continent.

Sedimentation and volcanism within the geosyncline and its ultimate destruction by orogenesis was considered to proceed according to a pattern. A eugeosynclinal furrow and an associated ridge, the first to form, were characterized by ophiolitic emissions (pillow basalts and autointrusions now altered to greenstones) and flysch deposits. The miogeosynclinal domain was theorized to become isolated from the eugeosyncline by development of a medial geanticline and was filled with sediment derived from the geanticline and the deformed eugeosynclinal assemblage. The miogeosynclinal assemblage, upon uplift, shed coarse debris into a molasse trough in front of the elevated mountain chain (foredeep).

The above view of geosyncline origin and development was built by reasoning from rocks of the past and assuming geophysical mechanisms such as crustal subsidence, horizontal compressive stresses, tensional stresses, and vertical uplifts, each process more or less independent with no dynamical mechanisms justifying them. Sedimentary basins were defined by the kinds and thicknesses of sandstones and other sediments contained in them, their presumed position with respect to ancient continental margins, and their geometry and size. Eugeosynclines were differentiated from miogeosynclines first of all on the basis of the presence of volcanic materials and secondarily on the basis of their greater thickness of section, the abundance of turbidites, and their greater structural complexity.

Sandstone compositions were linked in a general way to various basinal elements, particularly through the tectonic cycle that seemed to characterize them. Thus quiescent stages and cratonic sources of detritus were inferred for the early sandstones of cratonic basins and miogeosynclines, leading to quartz arenites. Graywackes were found almost exclusively in eugeosynclines and were seen to be pre-orogenic deposits in foredeeps. Arkoses were described as the products of intense orogenic phases of basin deformation, primarily as alluvial continental deposits. Lithic sandstones were the aftermath of orogenic uplifts. The terms flysch and molasse were used to describe pre-orogenic and post-orogenic sequences.

Lithologic associations and sandstone types could rarely be tied to basin type by a simple correlation so that definition of those types by unique stratigraphic and petrographic characters and sequences was not possible (Potter and Pettijohn, 1977). At the same time, the classification of basins—geosynclines—by Kay (1951) depended largely on their form and presumed relation to uplifts and continental position. They were specifically seen not to have any uniform evolution.

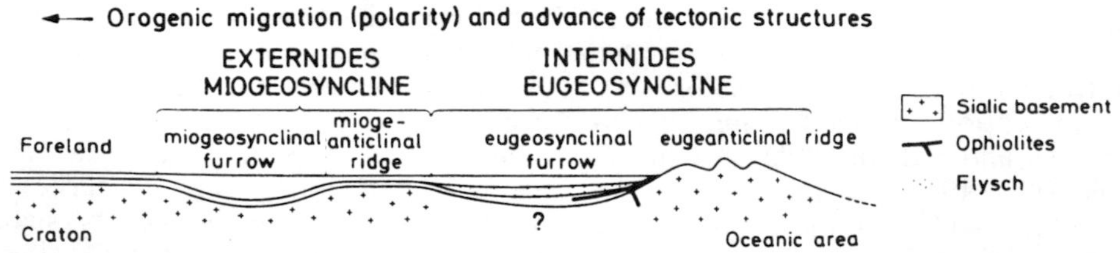

FIGURE 12-2. Organizational pattern of ideal geosyncline and related tectonic elements. (Modified from Aubouin, 1965, Fig. 16).

The many attempts that have been made over the last century to rationalize all of these sometimes conflicting ideas have led to recognition that there are subsiding basins (1) whose structural architecture and history are the result of deep-seated geophysical forces; (2) whose stratigraphic architecture is related to the patterns of subsidence and deformation plus shifts in sea level that may be external, such as eustatic sea level changes; (3) whose evolving petrologic constitution results from the structural and geographic relationship of the basin to nearby or distant source areas; and (4) whose evolution is also determined by geophysical forces operating on a previously deposited sedimentary sequence. Below we attempt to tie all these, sometimes apparently disparate, elements to the framework of the working hypothesis of today, plate tectonic theory.

Plate Tectonic Settings

We use "plate tectonic setting" to describe the geophysical and bathymetric or topographic setting of sedimentary basins associated with plate boundaries or intraplate areas. The reader is referred to any book on plate tectonics (for example, LePichon *et al.*, 1973) for a discussion of plate tectonic theory, characterizations of plate movements and geometries of the present, and how continental drift motions can be deduced from paleomagnetic and geologic data. The relation of sedimentary basins and sedimentation to plate tectonics was explored in groundbreaking papers by Bird and Dewey (1970) and Dickinson and his coworkers (Dickinson, 1970, 1971, 1974; followed by Dickinson and Seely, 1979; Dickinson, Ingersoll, and Graham, 1979; Dickinson and Valloni, 1980; Maynard, 1984; Valloni and Mezzadri, 1984) and many others since the late 1960s. Their views were summarized by Mitchell and Reading (1978) and, with reference to a more specialized application to diagenetic processes, by Siever (1979). A steady flow of papers has continued to link sedimentation patterns to varied paths of subduction zone complexes, and has attempted to characterize the petrology of sandstones associated with various plate tectonic basins (e.g., Dickinson and Valloni, 1980; Dickinson, 1982; Crook, 1980, 1982; Schwab, 1981; Ingersoll, 1982). At the same time there has been an extraordinary addition to our knowledge of deep-sea sands and their relation to tectonic and sedimentary environments by the Deep Sea Drilling Program (DSDP) and its successor International Program of Ocean Drilling (IPOD) (for example, Bartolini *et al.* 1975; Fujioka *et al.* 1980; Iijima *et al.* 1980; Murdama and Kazakova, 1980; Buchman and Leggett, 1982).

The kinematics of plate movements give rise to a number of simple plate interactions: convergence, divergence, and transform. Convergence, where the surface of one plate is continental and the other oceanic, is maintained by subduction of the oceanic plate beneath the continental plate; the oceanic lithosphere is consumed deep in the mantle below the continent margin. This convergence type is frequently referred to as *Andean*. Where convergence takes place between two oceanic plates, an intra-oceanic island-arc subduction zone develops, sometimes called *Marianas*-type (Uyeda, 1981). Convergence between two continental plates leads to orogenic mountain belts, sometimes called *Himalayan*, or continental collision. More simply, some use the explicit terms: continent-ocean, ocean-ocean, or continent-continent. Divergence takes place as two oceanic plates grow by accretion at mid-ocean ridges or as two continental plates move apart along continental rift valleys. Transform plate boundaries give rise to strike-slip, horizontal relative motions characterized by transform faults; lithosphere is neither consumed nor formed. The type example is the San Andreas fault of California between the North American Plate and the Pacific Plate.

It should be stressed that modern plate boundaries are defined by geophysical parameters: seismic slip patterns, depth of earthquakes, relative motions, and the like. Sea floor plate boundaries are mapped by magnetic striping of the sea floor. Continental plate boundaries are associated with major fault or suture zones. Former plate boundaries and plate motions are inferred from the magnetic anomalies of the sea floor, patterns of oceanic fracture zones and aseismic ridges, and matching geologic patterns on the continents, as well as stratigraphic patterns in deep-sea sediments. However, these methodologies can be used only for the history of the past 150 million years and work well only for the oceans. Because no ocean floors older than Jurassic exist now, reconstruction of plate boundaries and movements of early Mesozoic and earlier times depends wholly on the linkage of sedimentary, metamorphic, and igneous rock types, structural patterns, and plate kinematics as deduced from the record of rocks now found on the continents. Thus the accuracy of our deductions of pre-Mesozoic paleo-plate tectonics is a function of our knowledge of the sedimentary patterns associated with modern, Meso-, and Cenozoic plate tectonic settings. Even for these more recent times the deduction of geologic histories of continents by simple plate tectonic schemes is not appropriate, as continental deformation is affected by some nonrigid plate behavior and basin subsidence patterns (England and McKenzie, 1982).

An attempt at a simple classification of sedimentary environments correlated with plate tectonic settings is fraught with difficulty but bears dividends: the organization of our data and thinking that allows us to formalize relationships in a meaningful way. Two thoughtful essays on this kind of classification and its relation to geosynclinal terminology (Curray, 1975; Bally and Snelson, 1980) illustrate the varying approaches. Curray correlates plate tectonic environments and collision events with some emphasis on the modern world and ocean floor processes; Bally and Snelson give more attention to the development of basins on the continents and continental margins. These are only two of the many basin classifications now in use, some of them complex schemes (see, for example, Halbouty *et al.*, 1970; Perrodon, 1971; Klemme, 1975; Kingston *et al.*, 1983; Miall, 1984). Some schemes emphasize structural evolution of the basin, others the sedimentary fill; all relate in a general way to the geophysical framework of the basin in plate tectonic terms.

The scheme that we adopt, as shown in Table 12-2, seeks to group sedimentary environments simply by their location with respect to plate boundaries and intraplate positions. Naturally these criteria are nonexclusive and overlapping.

Obviously, it is not possible to make one-to-one correlations of plate tectonic settings with classificatory systems based on the older geosynclinal formulations. Though the assignment of cratonic basins is obvious, the assessment of mio- and eugeosynclinal with reference to the more complex division of tectonic basins associated with plate tectonics would be forcing a comparison. To the extent that origins are implied by the eu- and miogeosyncline terminology, the terms continue to be imprecise and even misleading in some ways. The same applies to the derivative terms "eugeocline" and "miogeocline." In a general way, one can correlate eugeosyncline with any of several tectonic environments associated with subduction zones, the ones that contain varying amounts of volcanics and volcaniclastics. But, as in the history of the Appalachian–Caledonian–Hercynian mountain belts, an early eugeosyncline associated with subduction may later be the site of a nonvolcanic accumulation traceable to a continent–continent collision.

The miogeosyncline may likewise be seen as the site of the continent-inboard section of a subsiding passive continental margin associated with the rifting of a continent and sea floor spreading. This designation, however, would ignore the fact that many sections of miogeosynclines are continent-inboard accumulations of sediment that may be associated with outboard subduction zones or other tectonic provinces, the miogeosynclines simply representing areas too far distant to be affected. The other terms that have been applied to various sedimentary basins, such as taphro-, auto-, paralia-, and zeugo-, have not received much use and it would seem better to dispense with them in favor of more common basin terminology, whether or not they are linked with plate tectonic environments.

In some sense, all of the sedimentary environments that have been discussed in Chap. 10 can be partitioned among the tectonic environments discussed here. The characters that environments impart to sand are thus in large part determined by the geomorphology of tectonic environments. On the other hand, the provenance of sands, not necessarily predicted by the sedimentary environment, though having some relationship to it, is linked to the tectonic setting (Chap. 7). This view of provenance both carries on the modern view of sedimentary tectonics, now 50 years old, and recognizes the relation of sedimentary, igneous, and metamorphic rocks of the continents to plate tectonic settings. The spelling out of the tectonics is now more distinct than ever before and tied to large-scale crustal movements that in the modern world are detected by geophysical methods.

In the following sections we discuss in sequence: the range of sedimentary environments, the types of provenance, the dispersal patterns, and the diagenetic minerals and textures that may be characteristic of each of the major plate tectonic settings. We stress that this attempt is tentative, for much remains to be done in this area, which, though its tectonic ideas are new, depends on a reinterpretation of classic sedimentologic and stratigraphic data, admittedly seen with new eyes.

TABLE 12-2. Sedimentary environments associated with plate tectonic settings.

Plate tectonic setting	Tectonic and geomorphic subdivisions	Sediments and other rock types
Subduction: intra-oceanic (Marianas-type)	Trench	Slumps, turbidites, and pelagics
Fore-arc	Accretionary prism, fore-arc island	Deformed sediment pile, melange metamorphics, ophiolite
	Fore-arc basin	Shelf, slope, and basinal
Arc	Arc islands, arc massif	volcaniclastics, some differentiated massif plutonics
Back-arc *	Shallow to deep marginal or interarc seas	Shelf, slope, and basinal to deep-sea pelagic
Subduction: oceanic-continental (Andean-type)	Trench	Slumps, turbidites, and pelagics
Fore-arc	Accretionary prism, fore-arc island	Deformed sedimentary pile, melange metamorphics, ophiolite
	Fore-arc basin: coastal plain, continental shelf, and slope	Alluvial, deltaic, shelf, slope, and basinal
Arc	Continental volcanic mountain chain	Volcaniclastic, alluvial, plutonic massif
Back-arc	High plateaus, intermontane basins, foreland basins	Alluvial, lacustrine, and marine epicontinental
Continent-continent collision (Himalayan-type)	Suture and thrust belt mountain chains and intermontane basins	Alluvial and lacustrine
	Remnant ocean basins	Deltaic, submarine fans, hemi-pelagics
Intra-continental rift active	Fault valley basins	Alluvial, lacustrine, volcaniclastic
Failed	Intracontinental basins	Alluvial, marine epicontinental
Inter-continental rift, narrow, early stage	Narrow gulfs, oceans (Red Sea-type)	Shelf, slope, basinal, hemi-pelagic
Mid-ocean ridges	Central rift valley, ridge flank, ponded basins	Pelagic over nascent oceanic crust (basalt flows, sheeted dikes, layered gabbros)
Passive continental margins	Coastal plain, continental shelf, slope, and rise	Alluvial, shelf, slope, and rise
Transform faults continental (San Andreas-type)	Extensional (pull-apart) and compressional basins	Alluvial, lacustrine, volcaniclastic
Oceanic (Vema-type)	Fracture zone basins	Pelagic, turbidite, volcaniclastic
Intraplate continental	Cratonic basins, foreland basins	Alluvial, marine epi-continental
Oceanic	Abyssal hills, abyssal plains, oceanic plateaus and rises	Pelagic, turbidite

* Arrows indicate general evolution of one plate boundary to another.

Plate Convergence Settings

Intra-oceanic Convergences

The subduction zones associated with oceanic island arcs traditionally were the most ignored of sedimentologic provinces, though Stille (1936) and Kay (1951) both pointed to the importance of island arcs in their view of geosynclines. But since the advent of plate tectonics, renewed attention has been drawn to such areas, primarily those of the western Pacific Ocean. The tectonic setting is the subduction of oceanic lithosphere beneath an adjacent oceanic plate, the consequence being the complex of structures typified by Tonga-Kermadec, the Marianas, and others (Fig. 6-27B). These zones can be divided into the subzones of Table 12-2.

The trench is the site of the deep-sea pelagic sedimentation, mixed in some places with turbidites and slumped material derived from the arc side of the trench and perhaps mixed with detritus derived from the seaward side. The turbidites contain abundant volcaniclastic material derived ultimately from the volcanoes on the arc itself. Turbidite fan sands may be interbedded with some wind-transported ash falls, depending on prevailing winds. Some shallow-shelf carbonate material of the inner (arc side) slope may be included in the resedimented shallow-water detritus transported in canyons, gullies, and across the slope down to the trench. Near the foot of the slope, slumped sediments may be included.

The relatively thin layer of pelagic sediment dominating the ocean side of the trench, the outer part, is a record of pelagic sedimentation over all of the ocean floor between the subduction zone and the mid-ocean ridge (MOR) where the lithosphere was formed. In general, the sedimentary layer overlying oceanic basalt will consist in upward order of (1) sediment initially deposited on newly created lithosphere at the MOR; (2) pelagic sediment, usually carbonate-rich from the ridge flanks where the sea floor is above the carbonate compensation depth (CCD); (3) carbonate-free pelagic sediment—red clay and siliceous ooze from abyssal hill regions below the CCD; and (4) the mixed pelagic and turbidite deposits of an abyssal plain. This sequence is illustrated by a section across the northeast Pacific, where lithosphere formed at the East Pacific Rise (EPR) travels northwest through successive provinces to the Aleutian trench. Sand is a significant fraction of the sedimentary layer only where sandy abyssal plain turbidites are included. As the sedimentary layer travels with the underlying lithosphere down and across the trench to the inner slope, it becomes juxtaposed with and overlain by inner-slope trench turbidites. Abyssal plain turbidites, especially sandy ones, if present, are relatively thin in most regions of the oceanic lithosphere far from continental terrigenous sources and/or separated from them by sediment transport barriers, mainly the trenches of subduction zones. In some places, such as the northeastern Pacific, turbidite sands are extensive and thick.

At the intersection of the inner slope and trench floor the oceanic lithosphere is underthrust beneath the inner slope. The general picture of the inner slope is that of an imbricated accretionary prism, an accumulation of sediment scraped off the downward-moving plate surface. Though there is much discussion about the proportion of sediment subducted (transported downward into the mantle) as opposed to the material scraped off (accreted to the surface part of the arc), it is clear that both processes operate to some extent, the proportions varying with the type of subduction zones (Uyeda, 1981; Lundberg, 1983; Moore *et al.*, 1982b). Underthrusting at the accretionary prism produces some of the deformation patterns associated with melange deposits. Sediments that are already lithified and embrittled by diagenesis (before deformation) are fractured and faulted whereas softer sediments, primarily the muds, deform plastically.

The sediments of Nias, an Indonesian island that is the surface expression of the accretionary prism on the inner side of the Sunda trench, are some of the better investigated of this type (Moore, 1979; Moore and Karig, 1980; Hamilton, 1973). Much of the sediment on the sea floor northwest of the trench is fine-grained turbidite detritus of the Bengal and Nicobar fans (Fig. 12-3); this material is derived from the major rivers draining the Himalayan mountains (Curray and Moore, 1974; Graham *et al.*, 1975). Graham *et al.* used this information to infer this orogenic-pelagic sediment as the major component of the Nias accretionary prism but Moore (1979) has argued that the provenance is the arc terrane of Sumatra and the fore-arc sediments of the Sumatra shelf. Moore's major evidence is the contrast in framework mineralogy of the Bengal and Nicobar fans (Ingersoll and Suczek, 1979) and the sandstones of Nias. The latter show their arc derivation by their greater content of volcanic and sedimentary lithic fragments, polycrystalline quartz, and smaller amounts of feldspar. The fan sandstones have higher quartz and lower volcanic and sedimentary lithic fragments.

Lundberg (1983) has contrasted the sands of the Nicoya peninsula in Costa Rica, landward of the Middle America trench, with the sediments of the Marianas trench. He suggested that if there is no significant longitudinal transport of terrigenous material, the intra-oceanic fore-arc regions are dominated by the pelagic component. They lack the melange characteristics ordinarily applied to this environment. Both these fore-arcs thus are characterized by pelagics, mudstones, carbonates, and cherts instead of thick trench and inner-slope terrigenous turbidites.

The accretionary prism merges with fore-arc basins in the direction of the arc, commonly separated from it by uplifted ridges of melange that mark the crest of the prism between the trench and fore-arc basin (Dickinson and Seely, 1979). The fore-arc basin region may be one of subsidence between two positive areas, the arc region being elevated by magmatic activity and

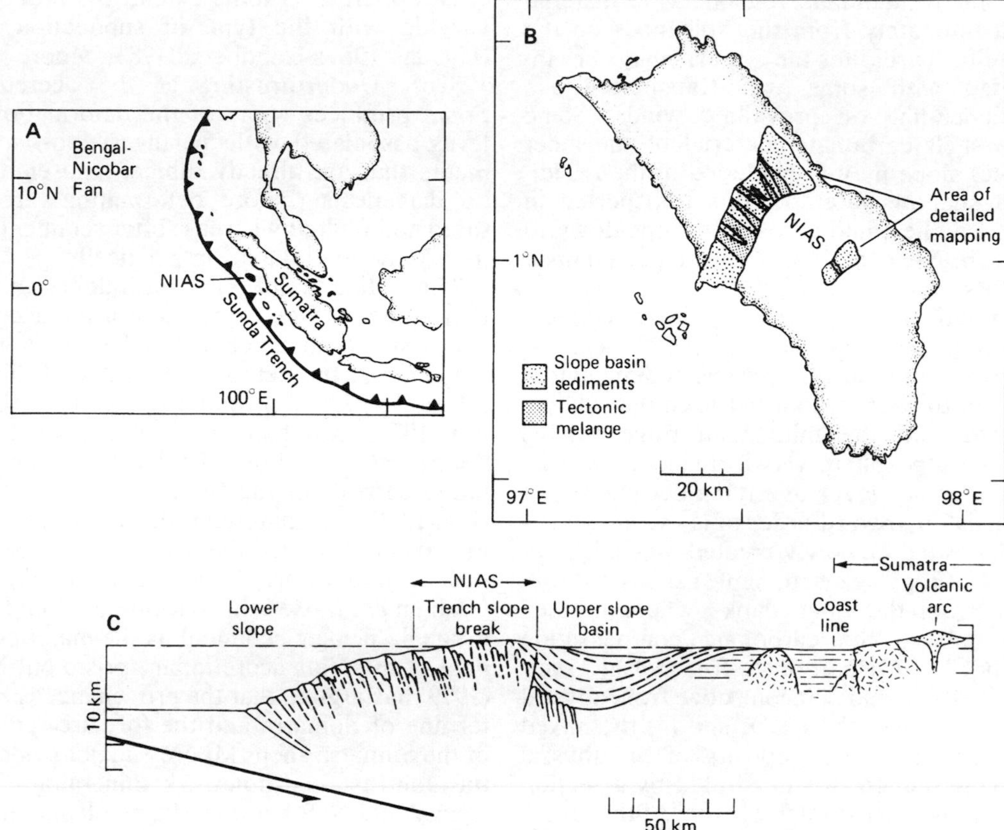

FIGURE 12-3. Simplified geologic map and cross section of the island of Nias showing the imbricate structure of an accretionary prism. Some of the detritus of the Bengal–Nicobar submarine fan is being accreted to the Sumatran fore-arc along the Sunda arc subduction zone. (Modified from Moore and Karig, 1980). (A) General location map of Nias, the Sunda Trench and the Bengal-Nicobar fan. B) Map of Nias showing areas of slope-basin sediments and tectonic mélange. C) Cross-section of Nias across mapped area showing relations to trench and arc.

the accretionary prism being thrust up as successive packets of sediment are underthrust. The fore-arc basin region may contain volcaniclastic sediments, subaqueous ash flow and welded tuffs such as those described by Fiske and Matsuda (1964), Bond (1973), and Howells *et al.* (1979). If water depths are relatively shallow, as in the western Pacific Sunda shelf, sands with some of the characteristics of shelf sands of passive margins interfinger with reef and other shelf carbonates.

The arc region itself is the scene of extensive volcanism and deeper-seated magmatic intrusions. The sands found in these regions are volcaniclastics of all types, typically of andesitic or other intermediate compositions found on the land part of the arc. Ash deposits on land occur with beach sands interfingering with the marine sands of the fore-arc region. Sediments of the arc include eroded older extrusives and underplated intrusives, and may be metamorphosed. Continued arc volcanism leads to the construction of a larger arc massif that resembles continental crust. In this stage the provenance of detritus from the arc becomes more difficult to distinguish from that of major continental sources. This is particularly so if, as in the Japanese Islands and Indonesia, fragments of older continental terranes are incorporated in the arc (Uyeda and Miyashiro, 1974). This kind of evolution, coupled with the idea of back-arc spreading proposed by Karig (1974) and Karig and Sharma (1975), has been suggested for the incorporation of arc terranes into stable cratonic continental crust and continental growth by Crook (1980) and Crook and Feary (1982).

The back-arc region, frequently a marginal sea between the arc and a continent, is a subsidiary spreading region of oceanic lithosphere whose motions may force migration of the arc complex and subduction zone in the direction of the advancing lithosphere on the far side of the subduction zone. The sediments of this region include the normal range of pelagics with a strong volcaniclastic component derived from the arc. Depths of water range from the shallow shoreline of the back of the arc to oceanic depths of the middle of the back-arc region. The depositional environments vary from those of near-shore sands to distal turbidites. The typical examples of spreading back-arc basins are the seas in back of the Mariana arc (the Mariana Trough) and the Tonga arc (the Lau Basin) (Uyeda, 1981). The Sea of Japan, which lies between the Japan arc system and the Asian mainland, is a back-arc basin, the scene of mixing between the volcanic arc source and the terrigenous mainland source. The arc source detritus is diverse, for the Japanese Islands are complex, including fragments of older continental terranes as well as the modern arc (Uyeda and Miyashiro, 1974).

The provenance of intra-oceanic arc sands reflects the mixture of volcanism of basaltic and intermediate compositions and pelagic—ultimately terrigenous—sources. On QFL diagrams they occupy a region roughly midway between F and L: they are more toward F with a higher plutonic component and more toward L with a higher volcaniclastic component (Fig. 7-8). On the $Q_pL_{vm}L_{sm}$ diagrams of Ingersoll and Suczek (1979) the arc component lies near the base (away from the Q_p component). With mixing of arc and continental sources in back-arc basins, the sands move more towards the Q pole.

One may compare two different modern settings, the fore-arc region of the Nias Island west of Sumatra (Moore, 1979), already discussed above (p. 484), and the insular shelf of Puerto Rico (Breyer and Ehlmann, 1981) (Table 12-3). In both regions the arcs themselves are complex, including fragments of continental blocks

TABLE 12-3. Comparison of the provenance of two fore-arc region sands, those from the island of Nias Indonesia (Moore, 1979) and the insular shelf of Puerto Rico (Bryer and Ehlmann, 1981).

NIAS: slope basin (Miocene rock outcrops)	*PUERTO RICO:* shelf (Modern sediments)

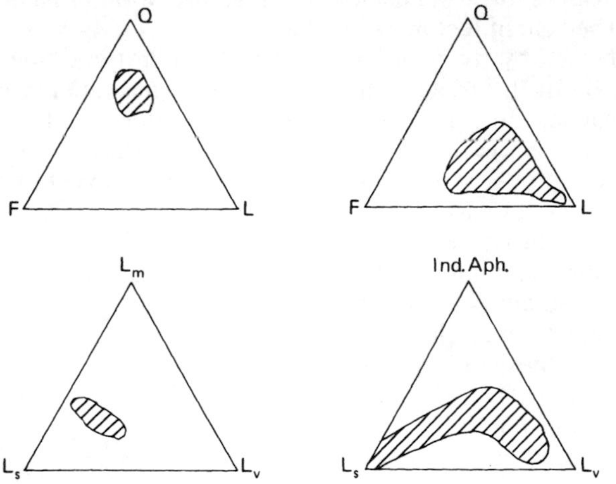

Q = quartz; F = feldspar; L = lithic grains; L_m = metamorphic lithics; L_s = sedimentary lithics; L_v = volcanic lithics; Ind. Aph = indeterminate aphanitic lithics.

NIAS	PUERTO RICO
Major component: quartz (polycrystalline av. 6.5%; most monocrystalline are undulose).	Major component: igneous rock grains, including microphaneritic, felsitic, microlitic, microphenocrystic, and microcryptocrystalline. Many grains of reworked marine muds.
Volcanics mainly microlitic and lathwork; some granitic, slate, phyllite, mica schist lithics.	Feldspar > quartz. monocrystalline quartz > polycrystalline quartz; non-undulose quartz > undulose quartz. plagioclase >> K-feldspar.
K-feldspar dominant.	Serpentine abundant; + amphibole, pyroxene, and opaques = 5–10%.
Glauconite.	Provenance: arc derivation includes coarse-grained plutonics, serpentinized peridotite, some Tertiary limestone.
Amphibole + pyroxene + zircon + opaque < 1%.	
Inferred provenance: arc terrane of Sumatra includes Tertiary sedimentary, metamorphic, and igneous plutonic rocks as well as Sumatran shelf (glauconite). Some may be from Himalayan orogen via Bengal–Nicobar fans and subduction.	

that include plutonic and metamorphic rocks. On Nias the quartz-rich nature of the sands reflects their derivation from the Sumatran coast, quartz-rich sediments and older igneous and metamorphic terranes. The relative lack of volcanic materials is ascribed to the continental-type rock derivation or to volcanic inactivity during the late Tertiary. The sands off Puerto Rico reflect the mixed volcanic, plutonic, and metamorphic terranes of the island as well as the nature of the intense weathering that selectively destroys unstable components on the rainy north side of the island in comparison with the dryer south side. These two studies point to the importance of recognizing the effects of climate and weathering on inference of provenance and how loss of unstable mineral and rock fragments during weathering and transport may lead to more mature sands than might otherwise be inferred from a volcanic arc source.

The diagenesis of arc-derived sandstones is also affected by their tectonic environment as well as their provenance (see Chap. 11). As Galloway (1974) pointed out, diagenetic reactions of silicates and carbonates are strongly influenced by the unstable nature of the detritus, on the one hand, and deep burial in fore- and back-arc basins, on the other. The role of tectonics in the burial process is expressed by the geothermal gradient (Siever, 1979). In the trench, the gradients may be high because, though the heat flow is low, the lithospheric crust is thin. Here gradients typical of oceanic regions, between 40 and 60°C/km, are found. But as soon as the sediment is transported by underthrusting into the accretionary prism, it encounters a much lower thermal gradient. The low gradient is a product of the lower heat flow over the descending slab and the greatly increased thickness of sediment overlying it. The thermal gradient may be even lower than that predicted from purely conductive heat flow if there is appreciable groundwater convection movement and therefore convective loss of heat.

The fore-arc region is thus a region where, though burial environments are deep, they may not be as hot as in other tectonic environments. The arc region is of course a hot one, the heat from the rising magma being the source. The back-arc regions also have high geothermal gradients because of proximity to a spreading axis and the slow cooling of newly crystallized basaltic lithosphere. Though no comparative studies have been made, it would seem that the diagenetic state of fore- and back-arc sandstones may differ for an equivalent depth of burial in a sense analogous to the paired metamorphic belts that contrast high-pressure–lower-temperature with high-pressure-higher-temperature assemblages in inner and outer metamorphic belts such as those of Japan.

The dispersal of volcaniclastic material is particularly sensitive to the effects of the wind as transporting agent. Where there are strong prevailing winds, most of the airborne debris will be dropped downwind. If the winds are orthogonal to the strike of the arc, the bulk of the airborne ash will be either in fore- or back-arc regions. This is a potential clue to paleowinds and explains why the two regions may show different provenance. The submarine dispersal of fore- and back-arc detritus is dependent on near-shore and deeper-water transport agents characteristic of continental margins. The trench and inner-slope deposits may show a combination of downslope and longitudinal directions; shallow shelf deposits of both fore- and back-arcs are controlled by waves and tides. The back-arc region will show analogous dispersal patterns as diagrammed on Fig. 12-4. Windblown and submarine dispersal patterns when combined with provenance offer the best chance of choosing between the fore- and back-arc interpretation of an ancient subduction zone basin.

The temporal and spatial sequences of sediment lithologies and patterns of transport of arc basins are determined by their tectonic settings. The oceanic lithosphere is overlain by pelagic sediments; the lowest parts of these oceanic se-

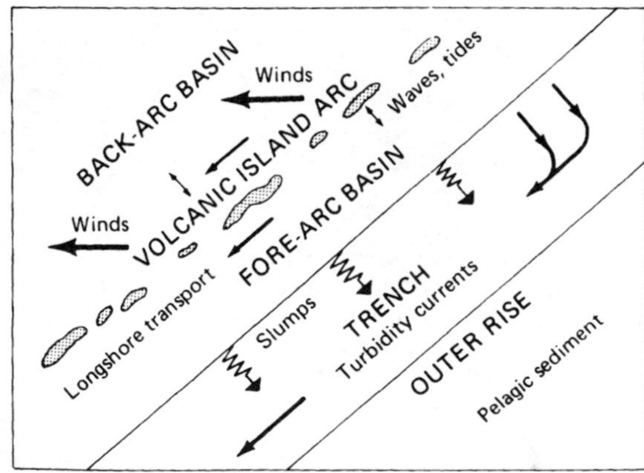

FIGURE 12-4. Dispersal patterns of fore- and back-arc regions. The dispersal of airborne volcaniclastics is governed by prevailing wind patterns, of fore- and back-arc regions by shelf and slope currents, and of trenches by slumps and longitudinal and transverse turbidity currents.

quences may include fragments of the basaltic lithosphere, usually lumped under the category of ophiolite (Coleman, 1981). Overlying this is the accretionary prism sequence, dominantly a deformed mélange of pelagics, trench deposits, and overlying slope deposits (Fig. 12-5). If the subduction zone migrates toward the subducting plate, the mélange may be overlain by much less deformed fore-arc basin deposits, which may range from turbidite fans to shallow shelf sands and carbonates. The fore-arc region will be overridden by the arc itself, leading to extensive igneous intrusion and consequent metamorphism of the sedimentary section. This would be superimposed on the high-pressure–low-temperature metamorphism characteristic of the mélange belt. With continued migration, a back-arc section might overlie the entire section, the back-arc sediments relatively undeformed and unmetamorphosed.

True intra-oceanic arc basin sandstones are not commonly found on the continents and continental margins. First, most contain little sand, only that derived from terrigenous abyssal plains that they may have traversed. Second, by the time they are accreted to a continental margin they start to assume the characteristics of an oceanic-continental convergence. The arc has become a massif with some of the characteristics of continental block plutonic provinces. The convergence with a continental mass also changes the nature of the sedimentary sequence, its paleocurrents, and its provenance. The many examples of fore-arc and other arc-related sequences, such as the modern central American region east of the Middle America trench (Moore *et al.*, 1982a), the late Mesozoic and Paleogene Great Valley sequence of central and northern California (Ingersoll, 1978, 1982), and others (Leggett, 1982), have in common derivation from an oceanic-continental convergence. The examples of intra-oceanic convergence are probably limited to ophiolite sequences, such as those of Newfoundland, the Alps, and Oman (Coleman, 1977; Clague and Stanley, 1977; Coleman, 1981). Ultimately, of course, all intra-oceanic arcs will become transformed to an oceanic-continental convergence if and as plate motions transport a continental mass to the subduction zone.

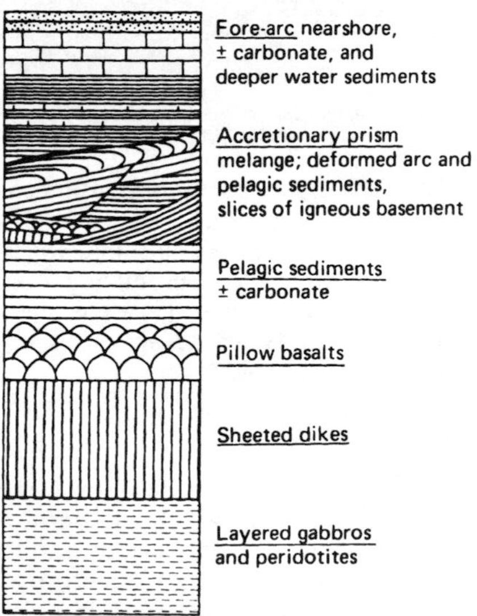

FIGURE 12-5. Columnar section of oceanic crust interbedded with and overlain by accretionary prism sediments and fore-arc deposits. This section would be generated by the seaward progradation of a fore-arc.

Oceanic-Continental (Andean-Type) Convergences

The sedimentary environments and tectonics of an oceanic-continental convergence differ in many ways from an intra-oceanic arc convergence. The interaction with a continental block may involve two different models: (1) an arc offshore with a marginal sea between the arc and the continent and (2) the true Andean-type, an arc located on the continent itself (Fig. 6-27C). The latter has both fore-arc and back-arc regions on the continent and only a strip of marine fore-arc between the marine trench and the onshore fore-arc basin. These different geometries impose distinctive styles of deformation and metamorphism, sedimentary provenance, paleocurrent regimes, subsidence patterns of basins, and diagenetic pathways.

In Andean-type convergences the trench may lie close to the shore as does the Peru-Chile trench off the coast of South America or somewhat farther, as the Aleutian trench off Alaska (Fig. 12-6). The accretionary prism and fore-arc basins become part of the continental block converging with the oceanic plate. The arc is inland, the magmas rising through thick continental crust to form a continental volcanic mountain chain. Though there is no spreading back-arc, at least in the sense of intra-oceanic arcs, there may be high plateaus such as the Altiplano of the Andes. As is obvious, many of our ideas of this kind of orogenic and sedimentary pattern come from this particular modern example and there is little agreement on the essential features of a general model. The portion

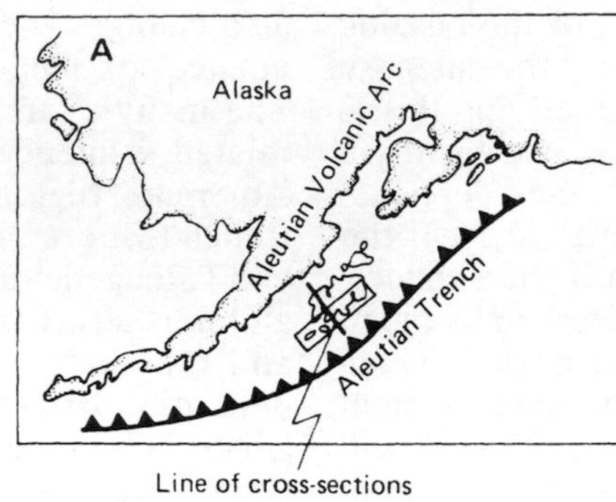

FIGURE 12-6. Generalized structure of Eocene to Miocene sediments of the Kodiak shelf of the Aleutian fore-arc region. An extensive turbidite abyssal plain was accreted to the fore-arc, producing a broad belt of faulted deformed sediment between the Aleutian trench and the islands of the arc. (Modified from Moore and Allwardt, 1980).

of Central America opposite the Middle America trench in its northern region in Mexico can be seen as this type of convergence whereas that part of the Middle America trench opposite Guatemala, although not intra-oceanic, is more of the Marianas-type (Von Huene and Aubouin, 1980; Moore *et al.*, 1982a). The difference between the two is determined by the subduction beneath an essentially oceanic plate, the Caribbean, in Guatemala and under a continental plate, the North American, under Mexico. The Hellenic trench to the south and southwest of Greece is another example (LePichon and Angelier, 1979).

The patterns of sedimentation in Andean-type convergences remain the same as those in intra-oceanic arcs. The trench is the scene of pelagics overlain by and interfingering with trench turbidites derived from the fore-arc region, in this case from the narrow continental borderland (Fig. 12-7). Much of the accretionary prism and fore-arc may be on land and so the complex of sediment that constitutes a mélange may be bordered and overlain by a deltaic plain or alluvial sequence. Though there is no back-arc basin as such, high plateaus may be the site of alluvial and lacustrine sedimentation. Elsewhere, foreland basins on the continent, lying behind the volcanic and/or faulted moun-

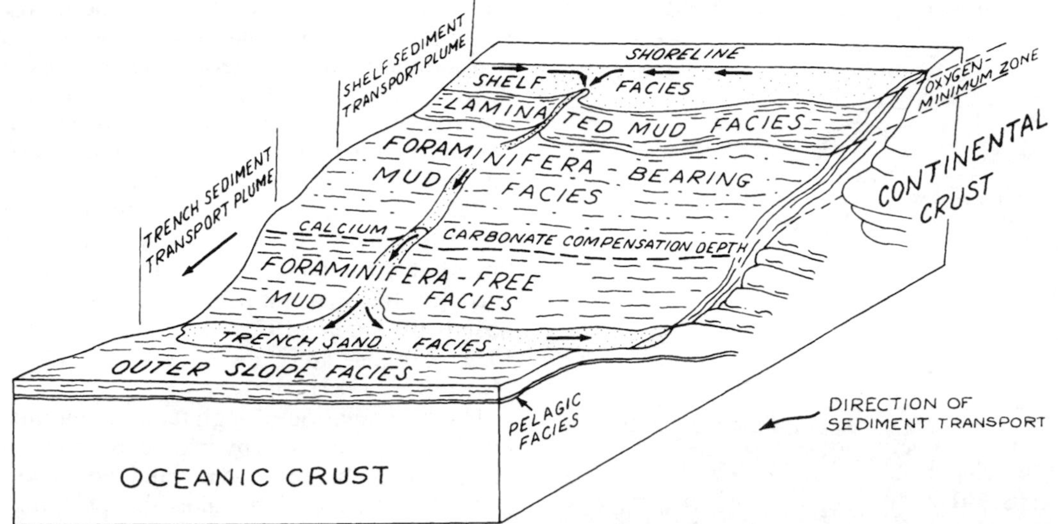

FIGURE 12-7. Diagram of the Middle America fore-arc and trench off Costa Rica, showing sediment dispersal and depositional patterns in relation to submarine topography. (Modified from Moore *et al.*, 1982a, Fig. 9).

tains of the convergent belt, receive much detritus and subside in response to such loading. These basins, typified by basins such as the Cretaceous of the western U.S., thus provide a sedimentary record of the orogenic activity of the convergent belt.

Volcaniclastics of Andean-type convergences are variable (Miyashiro, 1975) (see Chap. 6). Dickinson (1970) related the evolution of magmatic compositions to the provenance of volcaniclastic and plutonic detritus in sandstones derived from erosion of arc terranes. Subaerial dispersal of volcaniclastics dominates, from ash falls to water-laid tuffs.

The studies of the Mount St. Helens eruption in 1980 have been instructive (Decker and Decker, 1981). The windblown and water-dispersed debris of the eruption was widely distributed downwind and mixed with locally derived alluvial, lacustrine, and glacial materials. Many of these sediments were one sedimentary cycle removed from their ultimate source, the older parts of the volcanic terrane of the Cascade Mountains. The strong easterly flow of the prevailing winds in this region produced an ash fallout far to the east over the continental interior, hundreds of kilometers away. Meanwhile, the region to the west, in the direction of the subduction zone, received negligible quantities. There was little contribution to the sands of the Willamette River in Oregon, a broad valley lying between the active arc of the Cascades and the Coast range that contains subduction zone sediments of Mesozoic and Cenozoic age. Willamette sands contain a varied suite of detrital minerals coming from plutonic and metasedimentary sources in the Klamath Mountains to the south, young volcaniclastics from the Cascades, and older volcaniclastics and turbidites from the Coast ranges.

Dispersal patterns of the continental parts of Andean-type convergences are typical of continental orogenic belts and show little similarity to the marine and eolian patterns of intra-oceanic arcs (Fig. 12-8). In addition to the vagaries of wind directions, the river drainage systems may be largely interior, as in some intermontane basins. Paleoslope directions may be determined by the strike of the volcanic arc and associated deformation patterns. Thus a form of longitudinal transport along tectonic axes may play as important a role in dispersal patterns as in some turbidites. The general slope of much of a continental margin may be towards the interior; then much of the volcaniclastic and alluvial debris of the convergence is transported to the interior of the continent rather than toward the nearby ocean. Dispersal in marine parts of Andean-type subduction zones is similar to that of intra-oceanic convergences, producing shallow shelf deposits and slope and basinal (trench) turbidites.

Diagenesis of Andean-type convergence sediments depends largely on the composition and dispersal of unstable, chemically reactive volcaniclastic detritus and the burial history of the basin in which the sequence is deposited (Galloway, 1974; see Chap. 11). It is difficult to generalize about such varied continental basin

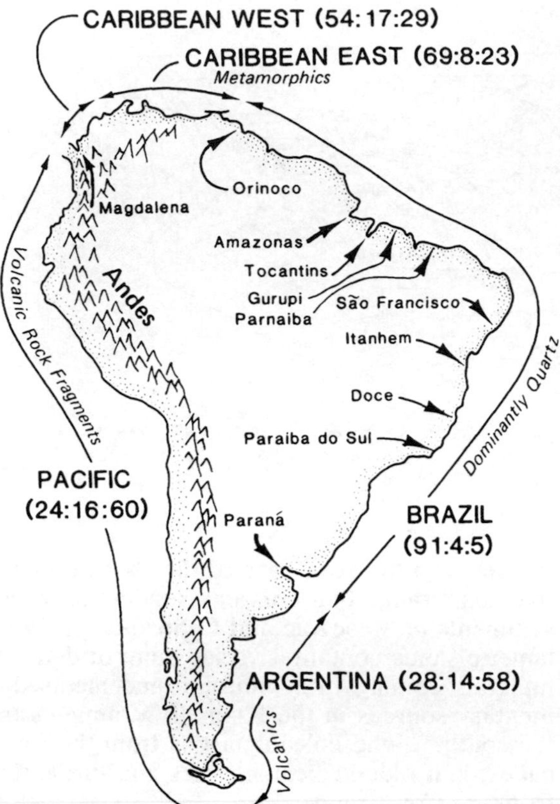

FIGURE 12-8. Dispersal of continental detritus on the beaches of South America. The four principal mineral associations show how tectonic control of sand mineralogy is overwhelming despite differences in climate. (Modified from Potter, 1986, Fig. 8).

sequences but we infer somewhat lower geothermal gradients over thickened continental crust and a large component of plutonic and sedimentary detritus. This detritus, more stable than the volcaniclastics, would make these sandstones somewhat less susceptible to pervasive diagenetic changes. Where the volcaniclastic component is high and basins are in close proximity to the high geothermal gradient of the volcanic arc, we may expect patterns of diagenesis similar to those of marine intra-oceanic arc basins, except for the absence of seawater. Patterns of zeolite and other silicate and carbonate diagenesis may resemble the nonmarine sequences of intermontane basins of continent-continent convergences.

Continent-Continent Convergences (Himalayan-type)

A continent-ocean convergence changes to a continent-continent convergence as a continental mass on the plate that is subducting approaches and then collides with the overriding continental plate. The geophysics of these convergences is not well worked out: the point at issue is the extent of subduction of less dense continental crust into a denser mantle. Continued subduction of large volumes of continent violates ideas of buoyancy of continental crust, and alternative views of continental deformation that do not involve much subduction but may involve strike-slip motion have been proposed (Bird *et al.*, 1975; Molnar and Tapponier, 1975; Molnar and Gray, 1979; England and McKenzie, 1982).

The combination of high tectonic mountains, the heterogeneous provenance of continental blocks, and the pre-, syn-, and post-orogenic history of the collision makes this province one of the least easily abstracted into a simple model. Yet the history of continents is written largely as a succession of orogenies, their forerunners and their epilogues; each is best described in terms of the successive steps by which it is constructed.

We hypothesize an initial convergence between two plates, one with a subduction zone at the boundary of a continental mass on the overriding plate (plate A) and a subducting plate (plate B) carrying a continental mass (Fig. 12-9). As the continent on B approaches the convergence, the sediments of the continental rise begin to accumulate as the accretionary prism. Because these sediments, a mixture of turbidite fans and hemi-pelagic deposits, are much thicker than pelagic sediments, the accretionary prism quickly grows to a larger imbricate underthrust fault belt. Continued subduction of plate B, even for only a small distance, piles up this prism belt and also results in uplift in back of it, the adjoining former arc and fore-arc of the overriding plate A. Continued convergence results in continued uplift of the collision zone accompanied by deformation, frictional heating along shear zones, metamorphism, and partial melting. As the entire belt is highly elevated, erosion and sedimentation are active in a mountainous continental setting. Once the convergence grinds to a halt, erosion dominates and relaxation to lower elevations and quiescence continues a long post-orogenic history.

In general outline this scenario does not differ greatly from earlier 20th century ideas of orogenic belt evolution. What is different is recognition of both the kinematics of plate motions and some notions of the dynamics of the process. The kinematics is revealing of the range of pre-orogenic histories that are possible and how to use provenance and dispersal of sandstones

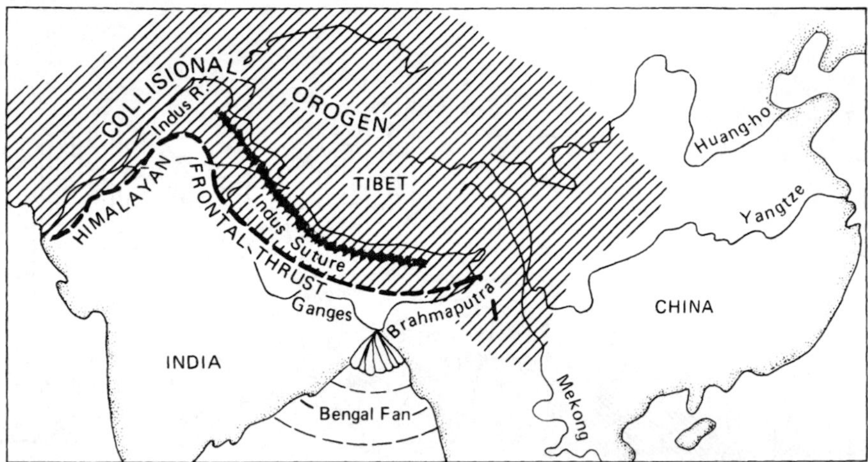

FIGURE 12-9. Sketch map of the Himalayan collisional orogen. Eroding from the high mountains of the orogen are a diverse group of igneous, metamorphic, and sedimentary rocks. The major river systems draining the orogen disperse the sediment to the western Indian Ocean, the Bay of Bengal, and far distant, to the China Seas. To the north, drainage may be interior or integrated with major rivers flowing north to the Arctic Ocean. Compare with Fig. 12-10.

to deduce that history. We continue to use words like flysch and molasse in the general sense both of distinct facies and of pre- and post-orogenic but now place them in the context of their tectonic settings. Thus the turbidites or flysch of a continental convergence are deposited on the continental rise of an advancing subducting plate. The coarse alluvial sands of the molasse of a continental orogen are seen as the erosion products of the mountainous collision of two continental blocks at a plate boundary.

The provenance of sediment eroded from a continent-continent convergence derives from the nature of the margins of the two continents involved (Fig. 12-9). These continental borderlands are the site of earlier sedimentary and metasedimentary sequences that become involved in a thrust belt. Lesser in volume but of diagnostic value are the fragments of ophiolite and the mélange belt inherited from an intraoceanic or Andean-type convergence. Plutonic terranes associated with older parts of the continental blocks or from the evolution of arc massifs accreted ahead of the continental convergence may supply diverse igneous material. The combination of these sources leads to the provenance classifications termed "recycled orogens" and "suture belts" by Dickinson and Suczek (1979) and Ingersoll and Suczek (1979). The detritus is richer in quartz than feldspar and rock fragments and richer in rock fragments, especially of the sedimentary and metasedimentary varieties, than in feldspar. The effects of maturity must also be taken into account; climate, as well as topography and source materials, affects composition of the detritus that finally reaches the depositional environment (Potter, 1978; Franzinelli and Potter, 1983).

Material eroded from orogenic suture belts is distinguished not only on the basis of the QFL plot but by the rarity of primary volcaniclastic debris, although recycled volcaniclastics of earlier arc episodes may be present. The major hallmark of these orogenic belts is their association with deep-seated igneous activity and regional metamorphism rather than abundant venting of extrusives. Associated with such igneous activity may be abundant hot springs and high heat flow but these may have little effect on the detritus eroded from highlands.

The interrelationship between dispersal patterns and the sedimentary environments associated with continental collision orogens is complex. In the immediate region of the suture and thrust belt, the dominant sedimentary mode is intermontane basin fills, alluvial valleys filled with the debris of the mountain erosion, tectonically produced lakes, and in arid regions, high deserts with eolian deposits. Given the provenance, all of these may be moderately to very sandy deposits. If the drainage is poorly integrated one may distinguish provenance from different sub-sources of sedimentary, metasedimentary, and plutonic material. Once an integrated drainage system incorporating a major river system is formed, transport exterior to the mountainous region is established. The broad alluvial plains below the mountain belt and the major deltas associated with these river sys-

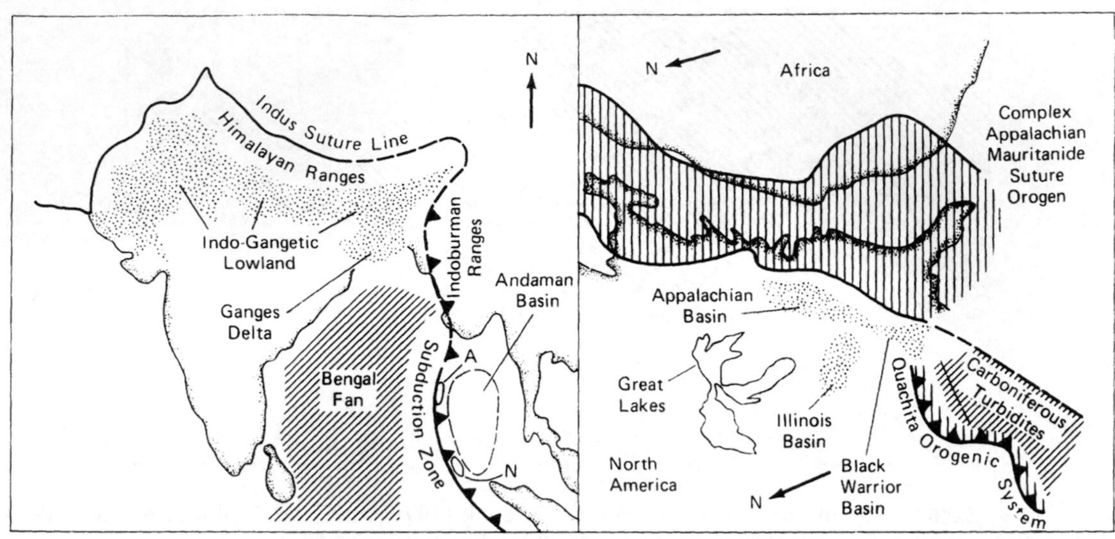

FIGURE 12-10. Comparison of dispersal in the Himalayan orogen with that in the Appalachian–Ouchita orogen after Graham *et al.* (1975, Fig. 1). Stippled areas show nonmarine and shallow marine deposits and diagonal rulings show turbidites.

tems become the major sites of sedimentation either in foreland basins or at nearby continental margins. In the latter, because of mixing of different provenances one can rarely hope to precisely establish both the direction and nature of the source areas. The deltas typical of this origin are the Ganges–Brahmaputra and Indus of the Himalayan convergence, the Tigris–Euphrates of the Alpine system, and the Mackenzie of Alaska (Audley-Charles *et al.*, 1977). The large size and strong association of deltas with continental convergence are in strong contrast to those found in continent-ocean convergences, which are smaller and less common, the sizable Snake–Columbia and Fraser river deltas along the Juan de Fuca subduction convergence being significant exceptions.

It is also common that the deltas associated with major orogens of the continent-continent convergence type spill over onto the deep-sea floor as the delta builds out and beyond the continental shelf and slope. This situation with respect to the Ganges delta has been explored by Curray and Moore (1974), Graham *et al.* (1975), and Ingersoll and Suczek (1979). The Ganges subsea fan and Bengal abyssal plain sediments are ultimately caught up in the convergence at the Java trench and reappear in the accretionary prism on the island of Nias. This giant pattern of orogeny and transport at two plate boundaries has been compared with the Appalachian–Ouachita system (Graham *et al.*, 1975, 1976) (Fig. 12-10). Deltas are deposited in foreland basins by river systems draining collision orogens that debouch into a shallow epicontinental sea away from the mountain belt, such as the Rhine River delta carrying the detritus of the Alpine orogeny. The ancient examples of these deltas are abundant in the Appalachian–Caledonian–Hercynian belt of the Paleozoic, typified by the Catskill–Portage facies deltas of New York and Pennsylvania, the Old Red of England and the continent, and somewhat similar deltaic deposits of the Carboniferous of both North America and Europe. These are categorized as foredeep or foreland basins by Dickinson (1974) and Bally and Snelson (1980).

Thus a continental orogen produces both molasse—the alluvium and deltaic deposits on the continents—and flysch—the turbidite deposits fed by deltas extended over the continental slope—at the same time in different places. The terms flysch and molasse, if used at all, should reflect only general categories of associations of sediment types and environments with no implications of pre- and post-orogenic origin.

The post-depositional evolution of sediments derived from continent-continent convergences may be as varied as the environments in which they are deposited. Those of intermontane continental alluvial basins may be associated with high or moderate heat flow depending on the association with plutonic activity; those associated with continental margin or cratonic deltas may be deeply buried in lower geothermal gradient marine environments. Sandstones deposited on oceanic crust may have relatively high thermal gradients related to the thinner crust overlying mantle heat sources. The paucity of highly reactive mafic material of volcaniclastics puts such sandstones on a course of diagenesis similar to those of cratonic and passive margin sequences with similar materials. Again, the role of provenance in supplying materials of differing stabilities is important in determining the future course of diagenesis.

The preservation of sediments derived from continental orogens depends on age and the geographic position of the environments. Sediments of intermontane basins will survive only for a few tens to a few hundred millions of years, as high rates of erosion tend to strip the sedimentary cover from metamorphic and igneous basements. The sediments of continental margin deltas and continental rise turbidites are likely to become reinvolved in plate convergences in a few hundred million years and after that will be seen only as deformed and metamorphosed fragments in a new thrust belt. Distinguishing these recycled turbidites from the first-cycle turbidites of an intraoceanic or Andean convergence presents a formidable challenge to the sedimentary petrologist.

Plate Divergence Environments

Oceanic plate divergences occur where basaltic magmas rise from the mantle along a mid-ocean ridge (MOR) (Fig. 6-27A). If the divergence takes place under continental crust, the result is a rift valley. Rifts on continents may further evolve to a true oceanic spreading center, from which the two parts of the former continent drift away, widening the ocean between. The formerly adjoining continental margins at a plate boundary then assume the nature of passive margins as they move from plate boundary to an intraplate position. The divergent boundaries vary greatly as sites for sand deposition and preservation, ranging from sand-dominated rift valleys to sand-deficient environments of the MOR.

Mid-Ocean Ridges

The MOR is perhaps the least interesting of tectonic settings from the point of view of the sandstone petrologist, for there is virtually no noncarbonate sand to be found. Pelagic biogenic carbonate or red clays may contain minor amounts of silt and fine sand blown from upwind continents. Locally derived sand-sized materials from the primary igneous and metamorphic rocks of the ridge itself can be found and traced to MOR-rock origin but they are minor in volume (Kastner and Siever, 1967). The insoluble residues of foraminiferal oozes of MORs consist largely of feldspar, mafic minerals, basaltic glass, and volcanic rock fragments, as would be expected from a terrane that is almost exclusively tholeiitic basalt.

Most ridge provinces are too far from terrigenous sources to receive sand. Even MORs near a continent, such as the East Pacific Rise (EPR), are denied sand from the continent by intervening sediment traps. This is not always so: ridges proximal to terrigenous sand sources or turbidity current dispersal mechanisms will allow the overlay of some sand on the ridge province. But in general, spreading rates, normally from 2 to 10 cm/year, are too fast in relation to normal rates of deep-sea sedimentation to allow any significant accumulation. As the oceanic plate moves away from the MOR, it cools and contracts along the curve modeled by Parsons and Sclater (1977) and so, as ridge flanks are reduced to lower levels, there is a greater possibility of receiving abyssal plain turbidites in an intraplate environment.

Intra-Continental Rifts

Intra-continental rifts are associated with incipient plate divergences and other extensional crustal movements. Long noted for their significance for the origin of sedimentary basins and the formation of distinctive kinds of sandstones are the grabens, half-grabens, normal-faulted valleys, and taphro-geosynclines of various authors (e.g., Krynine, 1950; Kay, 1951). The designation rift now is applied to all such extensions of the crust.

The basic ideas of geophysical mechanisms that form rifts on continents were enunciated by, among others, McKenzie (1967), Sclater and Francheteau (1970), and Sleep (1971) and are reviewed by Kinsman (1975). Mantle convective upwelling and the intrusion of basaltic magma localized along an axis leads to heating, bulging, and then stretching and extension by normal faulting of the continental lithosphere. As the extension proceeds, the rift begins to widen, oceanic lithosphere continues to be produced at a mid-ocean ridge, and the rifted edges of the continent's margin become intraplate ("passive" or "trailing" continental margins). Uplift of the rifted continent edges, frequently the site of an earlier orogenic belt, supplies a large amount of detritus. The rifted edges, now removed from the mid-ocean ridge, begin to cool and contract, initiating subsidence. Ero-

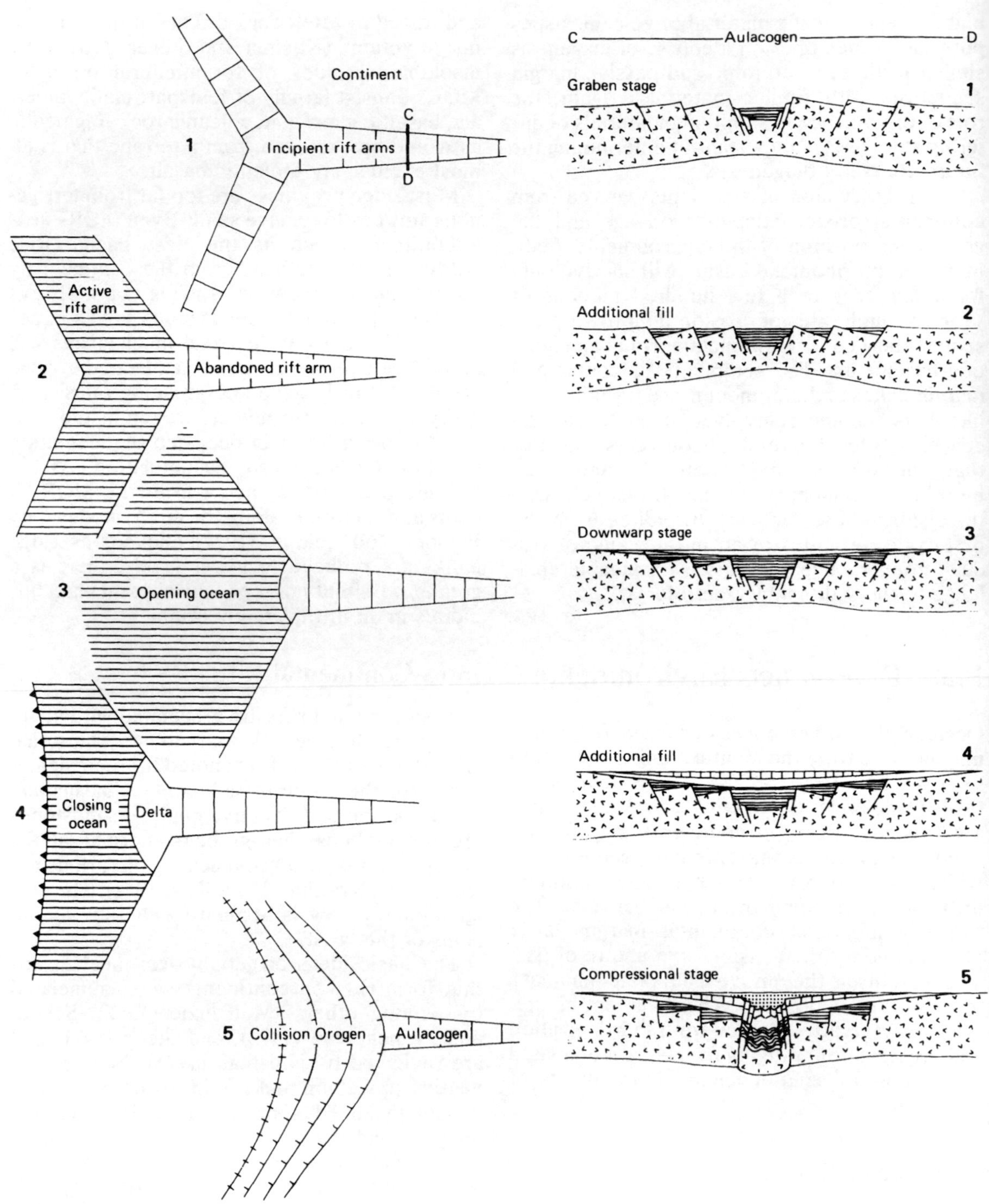

FIGURE 12-11. Stages in the evolution of an aulacogen after Hoffman *et al.* (1974, Fig. 10). At stage 1 a three-armed rift is formed. At stages 2 and 3, two of the arms spread to form an opening ocean while the third rift, failing to open, is abandoned, remaining as a subsiding transverse trough at a reentrant on the continental margin. In stages 4 and 5, the ocean closes by subduction, ultimately forming a collision orogen, while the abandoned rift is preserved as an aulacogen.

sion of older rocks—normally sediments, metasediments, and some plutonics—at the rifted edges of the continent leads to deposition of that debris on the aprons of the continent. This sediment load induces an isostatic subsidence that adds to the downward movement. The geophysical models roughly coincide with the subsidence history of the continental shelves and provide a reasonable explanation for the origin of continental shelf sedimentary accumulations (Royden *et al.*, 1980).

Intra-continental rifts such as the East African, the Rio Grande in North America, and the Rhine graben in Europe are—at least thus far in their history—"failed rifts." That is, they continue as active sites of extension but have not progressed to a stage of construction of oceanic lithosphere and a widening ocean as have the Red Sea and the Gulf of Aden. Some failed rifts have been modeled as one arm of a triple junction that develops into an aulacogen (Burke and Dewey, 1973; Hoffman *et al.*, 1974; Burke, 1977) (Fig. 12–11). Rifts may be transitional to transform plate boundaries as in the Gulf of California and the Gulf of Aqaba, where rifts alternate with long transform lengths and rhombic grabens may form (Blake *et al.*, 1978; Ben-Avraham *et al.*, 1979; Crowell, 1979). These are discussed below under transform boundary environments.

Rifts are associated with volcanism and plutonism, the composition of the igneous rocks ranging from basalt to rhyolite. There is some question whether the igneous activity follows the extension or initiates it. In the first case, the crustal stretching and extension thins the crust and heats it to the point where melting begins and igneous activity follows. In the second, the igneous intrusions at depth cause heating, bulging, and consequent stretching and extension. For the sandstone petrologist it may not be possible to distinguish between the two idealized models. In actual rifts, both plutonic and volcanic igneous activity, however initiated, are concurrent with normal faulting, uplift of the rift borders, and sedimentation in the valley.

Modern rifts are illustrated by the East African rift system, about 3,000 km long, that is connected to the Afar region and the Red Sea spreading axis at its northern end (Baker *et al.*, 1972). The rift is about 2 km deep and 40 to 50 km wide. The main sedimentary slope, the ramp, is away from the borders of the rift. The valleys themselves are filled with up to 2 km of sediment deposited as alluvial fans, alluvial plains, and freshwater lakes. Volcanics of a wide range of compositions are intercalated. Other examples of rifts are the Rhine Valley Graben (Illies, 1977); Lake Baikal, which contains up to 5 km of sediment (Logatchev and Florentsov, 1978); and the Rio Grande rift in the southwest of the United States (Cordell, 1978).

Ancient rifts include the Triassic–Jurassic fault basins of the borders of the North Atlantic Ocean and the Niger–Benue system formed as South America separated from Africa (Burke and Dewey, 1973). In North America, Hoffman (1973) describes a long-lived aulacogen in the Precambrian of Canada (Fig. 12-11) and Braile *et al.* (1982) document four late Precambrian–early Paleozoic rifts in the United States. In the Lake Superior region there is another Precambrian rift system of Keweenawan age (Wallace, 1981). Marginal Mesozoic rift basins of Brazil are other good examples (Ojeda, 1981).

The sedimentation patterns of rift valleys are dominated by the continental mode of sedimentation, alluvial, eolian, and lacustrine. Uplifted borders usually give an adequate supply of debris to continually fill the valley together with volcaniclastics and lava flows. When supply is low, starved basins can result. The ratio of detrital supply to subsidence and sedimentation is reflected in the relative proportions of alluvial valley fill and lacustrine deposits; in some cases the lakes—or dry playas—lie below sea level, Lake Tanganyika, for example. The facies patterns of alluvial fans and fanglomerates may grade laterally into alluvial valleys, then to lake delta and shoreline sands, and finally to fine-grained lacustrine deposits. The lacustrine deposits and interior sabkhas in arid regions contain evaporites, and organic fractions of sediments may be high. Turbidites may form in the deeper lakes (Degens *et al.*, 1971).

The provenance of the sands reflects the uplifted continental basement at the rift borders (Dickinson and Suczek, 1979; Ingersoll and Suczek, 1979) (Fig. 12-12). The detritus tends to occupy the region of a QFL plot that is quartz–feldspar-rich and contains relatively small proportions of rock fragments, as would be expected from recycled sediments, deeper crustal metasediments, and plutonic rocks. The typical result is arkosic sands, with quartz-rich sands more characteristic of deeply weathered terranes. As rift valley sediments accumulate, they may show an upward shift from the basal sequences that contain much of the sedimentary cover of the rift border regions to upper sequences that contain more feldspar and minerals from unroofed plutons and high-grade metamorphic rocks, a classical story in sedimentary petrology. Ingersoll and Suczek sug-

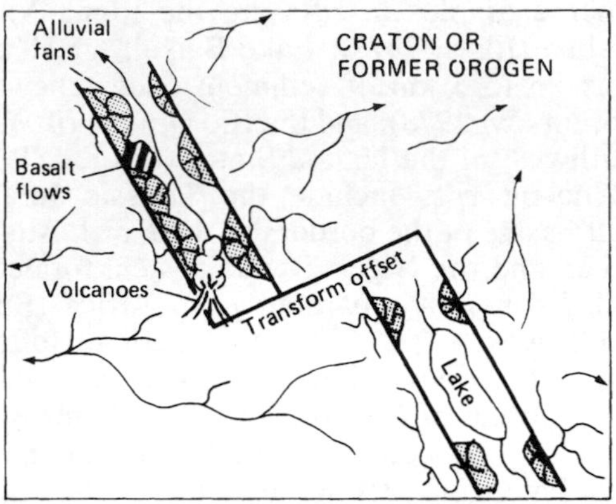

FIGURE 12-12. Provenance and dispersal of rift valley sediments. The craton or orogen may include different kinds of igneous, metamorphic, and sedimentary rocks, which are eroded to form rift-valley filling detritus in alluvial fans and valley floor plains and lakes. Drainage is variable depending on the magnitude of rift border uplift.

gest that the subdued tectonism of rifted continental margins will give rise to more quartz-rich detritus as a result of heterogeneous sedimentary-plutonic source rocks and intense weathering. However, such quartz-rich materials may also be produced by the effects of climate.

The dispersal patterns of rift-derived sediments are determined by the internal drainage and strike of the valley itself and the regional slope, if any, away from the rift. The valley drainage follows the classical picture of fanglomerates locally interfingering with the strike-determined alluvial plain at the floor of the valley. The upland regional slopes away from the rift borders would dip in opposite directions normal to the strike and then follow patterns determined by local geologic controls, the influence of the rift dying away after several hundred kilometers from the rift. If a narrow sea has invaded part of the rift, such as the Red Sea and Gulf of Aden that opened in the later Tertiary for the northern part of the African rift system, deltas may form at an end of the rift.

The diagenesis of rift sandstones is determined, first, by the combination of quartz–feldspar detritus with volcaniclastic glass and lithic fragments and, second, by the high heat flow that is associated with the rift basin. Thick sedimentary piles such as the 10 km of Cenozoic sediment in the Benue trough of West Africa, formed as a failed rift of the triple junction by which Africa drifted from South America (Burke and Dewey, 1973), are subject to extensive cementation, clay mineral alteration, and authigenesis of carbonates and silicates in the lower sections of the sequence. Unfortunately we have few detailed and extensive studies of these kinds of sediments beyond those of the Newark Series of North America. As sandstones associated with rifting episodes of the North Sea encountered in the active oil drilling program of that province become better known, we learn more about the diagenesis of these environments (Taylor, 1978).

The North Sea is a sedimentary basin (Fig. 12-13) whose subsidence origin is related to a

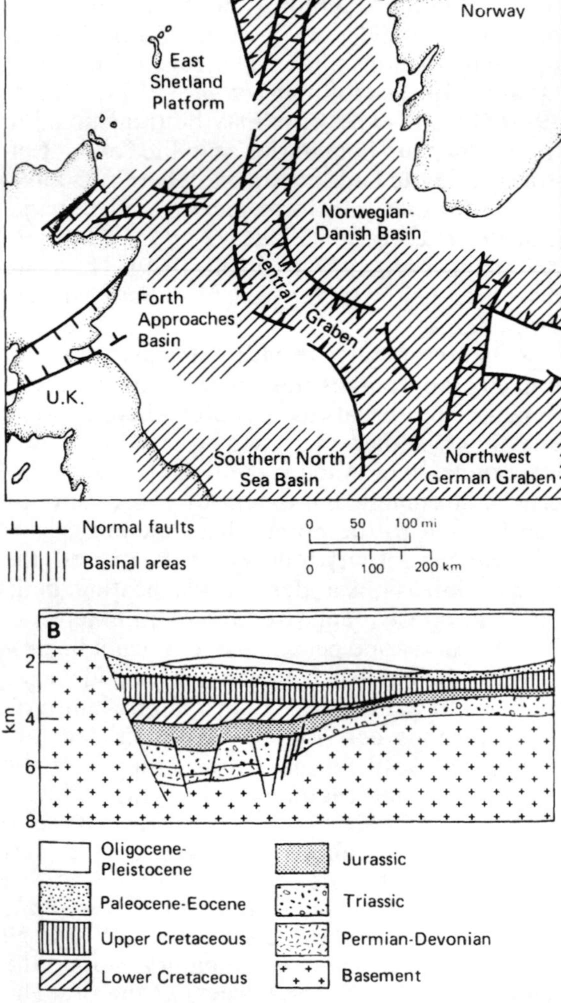

FIGURE 12-13. Map (A) and cross section (B) of the North Sea Basin, a failed rift that formed during the opening of the North Atlantic in Mesozoic time. (After Ziegler, 1975).

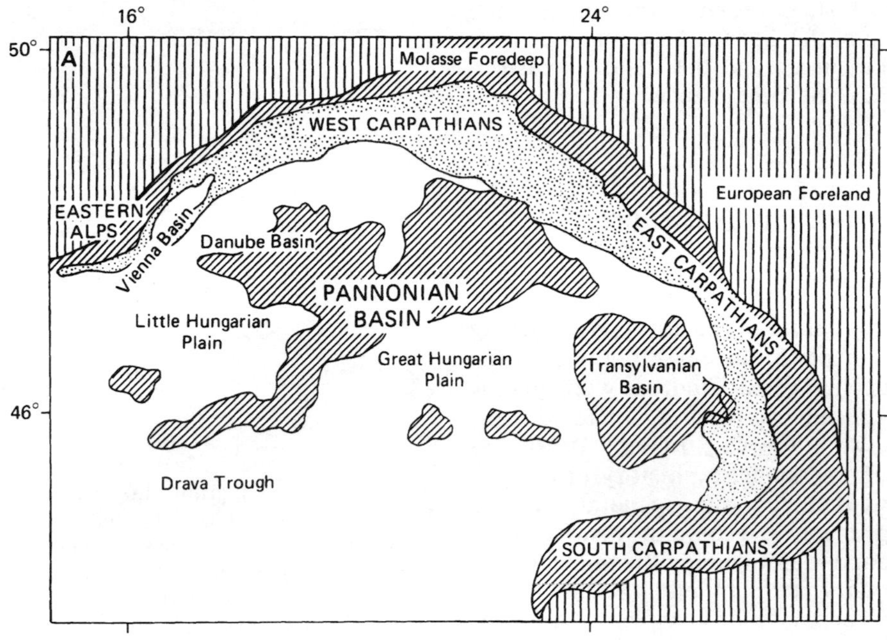

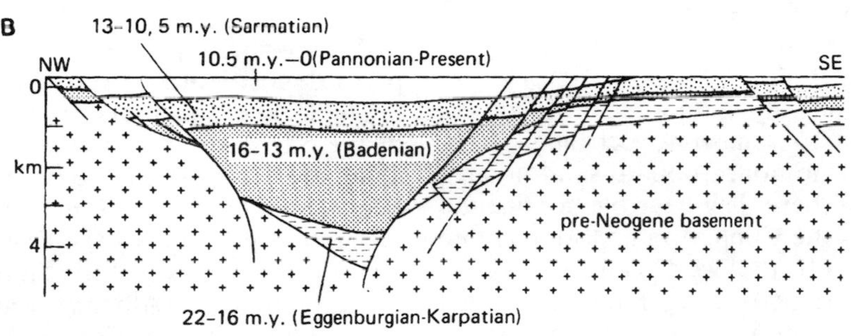

FIGURE 12-14. A) Map of the Pannonian Basin of Eastern Europe. This basin and the smaller, deeper depressions of the Transylvanian and Vienna Basins were formed by normal faulting collapse after the major thrust belts of the Carpathian Alps were formed during the Alpine plate convergence. B) Cross section of the Vienna Basin. Age of stratigraphic units given in millions of years before the present. (After Burchfiel and Royden, 1982, Figs. 1, 2 and 10).

rifting episode associated with crustal stretching and extensional faulting which occurred with the opening of the Atlantic Ocean (Ziegler, 1975). The Pannonian Basin in Eastern Europe is associated with the Alpine collision (Burchfiel and Royden, 1982; Royden *et al.*, 1983) (Fig. 12-14). Many other extensional basins at present or former rifted margins of the continents have subsidence histories related to heating associated with extension, followed by cooling and contraction and loading by sediment. Important aspects of sandstones in these basins are fault-controlled sedimentation and continental provenance of the detritus. Many intracratonic basins such as those of the North American Interior are now thought to be controlled by deep-seated rifts such as that of the New Madrid fault zone and the Michigan Basin (see p. 495). The lowest part of the post-extension sequence includes rift valley sandstones and conglomerates, locally directly overlying a much older metamorphic and igneous basement, in other places overlying an earlier and undeformed sedimentary sequence.

Similarly, the later sandstones of a narrow rift that has evolved to a narrow sea such as the Red Sea–Gulf of Aden will be deposited in accord with mixing and dispersal patterns of the marine invasion. Thus the fanglomerate-alluvial-lacustrine sequences of early rifting are succeeded by deltaic and shoreline sands. As spreading continues and areas of deep ocean floor open, turbidite sands transported from the narrow continental margin are deposited as submarine fans and, to a limited extent, plain deposits. With further opening, the sea assumes the condition of the Atlantic Ocean in Mesozoic and Cenozoic times and the major sandstone accumulations become those of a passive margin.

Plate Transform Boundary Settings

The transform nature of the extensive zones of lateral relative plate movements along strike-slip faults such as the San Andreas were quickly recognized at the beginning of plate tectonic theory. Crowell (1974a, 1974b, 1979) has spelled out the sedimentary history of the San Andreas and deduced some general relationships among structure, geological history of the two plate boundaries adjoining the fault system, and sedimentary and provenance patterns of sedimentary accumulations associated with the fault. As noted above (see p. 496) there are transitions between rifts and transforms and between sutures and transforms as, for example, east and west of India.

Transform faults along mid-ocean ridges are of little interest to sandstone petrologists for there is little or no noncarbonate sand in such environments. Where they near the continental margin, such as the Clipperton and Mendocino fracture zones off the west coast of North America, fine-grained terrigenous sediments become an important part of their fill. Sands are a relatively minor part of these fracture-zone basin fills except where they impinge on turbidite fans at the edge of the continent.

The sedimentary sequences associated with the San Andreas are those expected from the geometry of straight and sinuous strike-slip faults (Crowell, 1974a). The nearly flat topography and verticality of the fault of some sections of the San Andreas lead to no differential elevation as a result of fault movement and little or no sedimentary consequences. Curves, bends, or varying elevations along other sections of the fault, however, accompanied by branching and braiding of the fault system itself, lead to fault-bounded sedimentary basins. The most striking of these are the pull-apart basins or "holes" (see Fig. 12-15). The sedimentary fills of these basins are derived from active erosion of the surrounding high-standing blocks and the dispersal pattern is toward the interior. At the edges, fault-bounded conglomerates, fanglomerates, and breccias related to fault movement are deposited. Along the axis of the basin, muds and sands are deposited as alluvium with some playa deposits in arid regions.

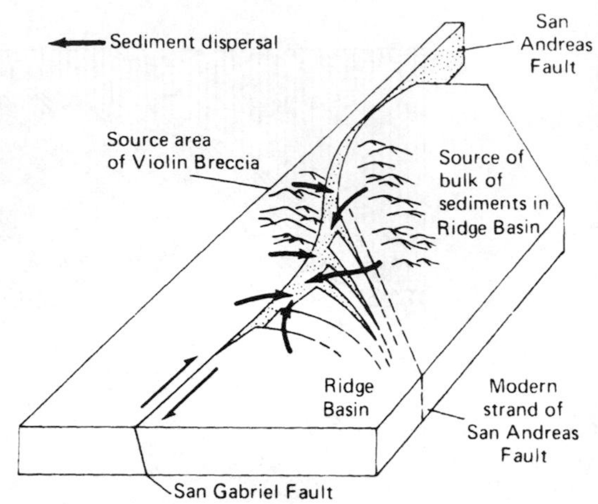

FIGURE 12-15. Diagram showing the origin of the Ridge Basin at a curve in the San Andreas Fault during the Pliocene, when the San Gabriel Fault was probably the main San Andreas Strand. The basin is filled with sediment derived from bordering highlands. (After Crowell, 1974b, Fig. 5).

The largest accumulation of sandy sediments of the San Andreas zone is at the delta of the Colorado River at the south end of the Salton Sea trough. Here one of the major river systems of the western part of the continent transports far-traveled debris into the Salton–Baja California depressions (Allison, 1964; Muffler and Doe, 1968). Nonmarine and marine deltaic muds and sands grade to coarser comglomerates and breccias associated with the steep fault-bounded basin margins. The total accumulation in the Salton trough, 6000 m, includes a basal 1000 m of claystone and sandstone thought to have been deposited at the northern end of the Gulf of California. Overlying the basal deposits are nonmarine fan sandstones and conglomerates derived from the basin margins, probably in excess of 3000 m. Opening during this period of sedimentation, the Salton trough produced complex relations of overlaps and unconformities. Metamorphism of the Salton sediments is going on now with very high thermal gradients. The lower sediments in this trough are now greenschist facies rocks (Muffler and White, 1969; Elders *et al.*, 1972).

Another type of accumulation of sediments associated with the San Andreas system is the Ridge Basin, a narrow fault-bounded depression at the intersection of the San Gabriel and San Andreas faults (Fig. 12-1). About 12,000 m of late Miocene to early Pleistocene sediment, both marine and nonmarine, were deposited as large amounts of detritus were spread over alluvial fans, playas, lakes, and marine bays. Deposition and deformation of the basin were contemporaneous, older beds being more deformed than younger ones. Sediments are coarse-grained sandstone, conglomerate, and breccia near the two fault zones, grading to shale along the axis of the basin. The rocks beneath and surrounding the basin are granites, gneisses, metasediments, and earlier Tertiary sandstone, shale, and conglomerate. The Violin Breccia, accumulated at the foot of a steep scarp of the San Gabriel fault, is an earthy, poorly sorted material that contains angular gneissic blocks up to 2 m.

As would be expected, the provenance of continental transform basins is a mixture of local and regional sources. The more abundant local detritus is derived from the immediately surrounding upfaulted areas that may include many compositions from earlier sediments to basement crystalline rocks. Regional contributions are significant, where the fault intersects a major river drainage system, such as the Colorado River intersecting the Salton trough. The general cast of sandstone compositions, classed as part of an uplifted basement provenance by Dickinson and Suczek (1979), is quartz- and feldspar-rich. Sands are arkosic given the classical picture of high uplift and rapid erosion that exposes basement rocks. Lithic sandstones, in which the rock fragments are of earlier sediments, are typical of the areas in which uplift is less extreme or in younger stages of development when the sedimentary cover is largely intact. The composition of the detritus is more difficult to interpret if a large river system is involved. Though this tectonic province is not defined as having a close relationship to volcanism, "leaky" transforms with large extensions of the crust may be the site of invading magmas. The results are dikes, sills, and irregular plutons as well as lava flows and volcaniclastics at the surface. Thus volcanic rock fragments in the sandstones, primarily basaltic, may reflect locally derived material rather than transport from an island arc or continental hot spot.

The stratigraphic sequences and dispersal patterns of transform basins may be complex. Determined largely by their relation to local geology, the sedimentary fills of different basins along the length of a transform may have little direct relationship to each other. Most dispersal is to the interior, and paleocurrents are strongly linked to size gradients away from fault scarps. Where the basins are in part marine, dispersal is related to shoreline and deeper continental margin transportation. Coarse shoreline sands may grade over short distances into the turbidites of deeper offshore basins. Sedimentation in offshore basins of California typifies this kind of regime (Emery, 1960; Crowell, 1974a).

The diagenesis of these kinds of basins has some affinity to those of early stages of rifted continental margins, that is, thick alluvial and lacustrine sections that may be associated with high geothermal gradients, especially if the transform is leaky and contains igneous bodies. Marine transform basins may contain diagenetically altered chemical sediments such as carbonates, cherts, and siliceous shales deposited at times when sandstones were trapped in updip basins. The Monterey Formation of California (Miocene), well known because of extensive offshore oil drilling programs, is interpreted to be this kind of deposit (Garrison and Douglas R. G., 1981).

Narrow marine gulfs containing sand may be associated with transform boundaries and incipient continental breakup. Two examples are the Gulf of California separating Baja California from Mexico and the Gulf of Aqaba, a part of the Syrian-African rift system. In both of these places a mid-ocean ridge spreading center with a large transform component runs into a continent and becomes dominantly transform (Larson *et al.* 1968; Crowell, 1974b, 1979; Ben-Avraham *et al.*, 1979). In both gulfs, ponded marine basins contain turbidites. Where the basins are near the borders of the adjacent continents, alluvial fans may build out into the basin as submarine cones. Some Cenozoic basins of California are of this nature (Nilsen and Clarke, 1975; Van de Kamp *et al.*, 1976).

Ballance and Reading (1980) and Biddle and Christie-Blick (1985) provide sedimentologic overviews of strike-slip basins.

Intra-Plate Settings

Though plate boundary and former plate boundary settings are important, intra-plate regions are also the sites of a great many sedimentary basins, some of which contain much sand and sandstone.

Oceanic Intra-Plate

This setting is the site of abyssal plain turbidites. Most are silty and only contain fine sand near their submarine fan and channel feeders.

Later Mesozoic and Cenozoic sediments of the sea floor are currently explored by surface sampling, DSDP drilling, and seismic profiling. Our only information on sandy deposits of the intra-oceanic regions of earlier times is from accreted pelagic sediments now found in the deformed belts of oceanic plate convergences.

The provenance of present-day ocean intraplate environments has been discussed by Dickinson and Valloni (1980) and Valloni and Maynard (1981). Dickinson and Valloni discuss only one intra-plate oceanic setting, that of the volcanic archipelagoes such as the central Pacific island chain (e.g., Hawaiian Islands, Emperor sea mounts). These settings, as would be expected from their derivation from basaltic sources, contain little or no quartz but appreciable plagioclase feldspar and volcanic lithic grain fragments. Valloni and Maynard include the intra-plate oceanic setting primarily in their trailing-edge category, which includes intra-plate volcanic island types. Other than these volcanic-rich sands, they found no obvious subdivisions of the intra-plate trailing-edge category. The general average composition may be linked to higher amounts of quartz and feldspar and lower amounts of lithic fragments of sand eroded from continents and continental margins and transported as turbidites to abyssal plains. Differences from one abyssal plain to another can be related to provenance from specific continental margins and interiors.

Dispersal patterns are related to turbidity current patterns of abyssal plains that are made up of material derived from nearby continental slope-fan regions and transported to distant mid-ocean channel paths. Diagenesis of these turbidites, while they are still in the intra-oceanic environment is slight, judging from cores of the DSDP. Sands are uncemented except where there is a significant component of carbonate or silica which may, through diagenetic alterations in mineralogy, provide some weak cementation. Changes in clay mineral composition are relatively slight, mainly partial alteration from smectite to mixed-layer. Where volcaniclastic materials are present, particularly glass, zeolites may form and alter along paths similar to those of volcaniclastic sands of arc-related basins (see Chaps. 6 and 11).

Passive Continental Margins

The most important intra-plate setting is that of the passive margin of a spreading ocean such as the Atlantic, where oceanic and continental crusts meet. This passive margin is the product of initial continental rifting followed by formation of a mid-ocean ridge and continental drift away from it (Fig. 12-16). The trailing edge of the continent is the site of extensive sedimentation on continental shorelines, shelves, slopes, and rises.

The general sedimentary sequence is tied, on the one hand, to the subsidence mechanism, cooling and contraction as the rifted margin moves away from the mid-ocean ridge. On the other, it is related to erosion of the initially uplifted continental margin at the time of rifting. Delivery of continental block detritus is accomplished by major river systems and dispersal systems associated with the shorelines, shelves, slopes, and rises of the continental margins. Volcanism is absent after the continental margin has moved away from the mid-ocean ridge.

The basement, known from seismic profiling, is a complex of the continental material that was rifted, typically an earlier continental orogen. The initial rift belts may be several hundred kilometers wide: most rift blocks represent failed areas of crustal extension as one major rift grew and became the mid-ocean ridge. The rift valleys are filled with continental material that grades upward into the basal marine sequences that spread over the lower parts of the subsiding continental margin. Thus starts a long continued period of gradually decreasing rates of subsidence and sedimentation on the shelves and slope-rise environments. Differential subsidence and dispersal patterns produce a continental margin sequence of slightly deformed sediments with pronounced facies transitions controlled by depth of water, type and amount of detrital supply, eustatic sea level changes, and current patterns.

The well-investigated continental shelf of eastern North America is the best known example of a passive margin evolved since the early Mesozoic rifting that initiated the Atlantic Ocean (Schlee *et al.*, 1976; Klitgord and Behrendt, 1979) (Fig. 12-17). A rifted basement of Paleozoic sedimentary, metasedimentary, and igneous terrane underlies Triassic and early Jurassic arkoses, redbeds, and volcanics that fill the rifts and comprise the lowest stratigraphic units. Succeeding this sequence is the record of a narrow ocean, marked by marine evaporites in tropical latitudes. As the ocean spread and terrigenous sands and muds built up shoreline and shelf accumulations, carbonate banks and reefs formed at the shelf edge. These shoreline and shelf facies prograded toward the ocean

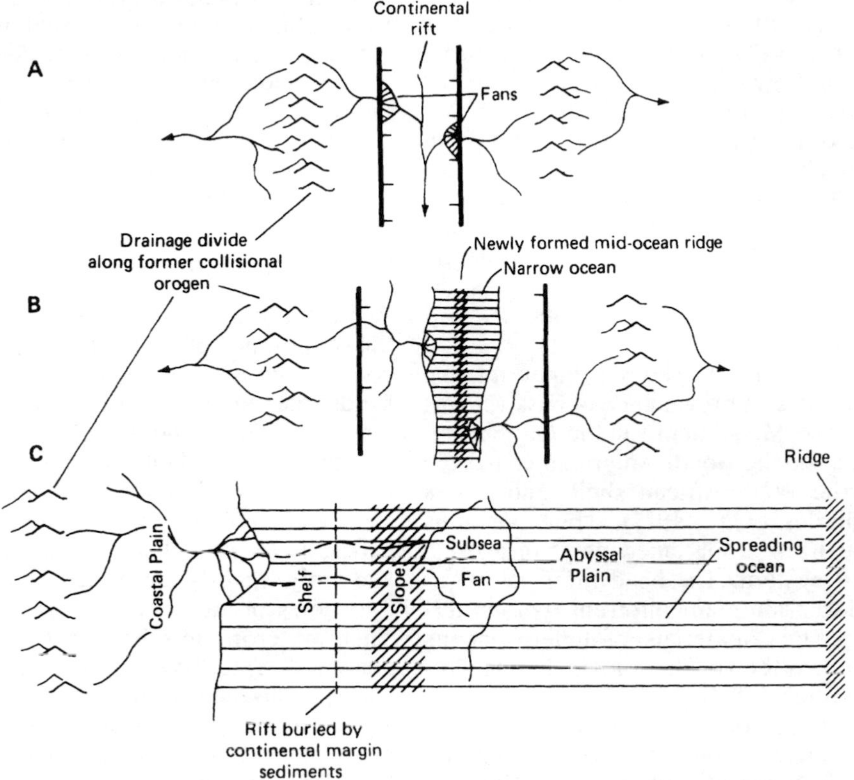

FIGURE 12-16. Schematic evolution of a passive continental margin. A) A continental rift valley forms along the line of an earlier collisional orogen. B) The rift begins to spread along a newly formed mid-ocean ridge and a narrow arm of the sea occupies the former rift valley. C) Continued spreading has formed an ocean and the rift is buried beneath continental margin sediments. Such rifts are typically broad zones rather than a single graben.

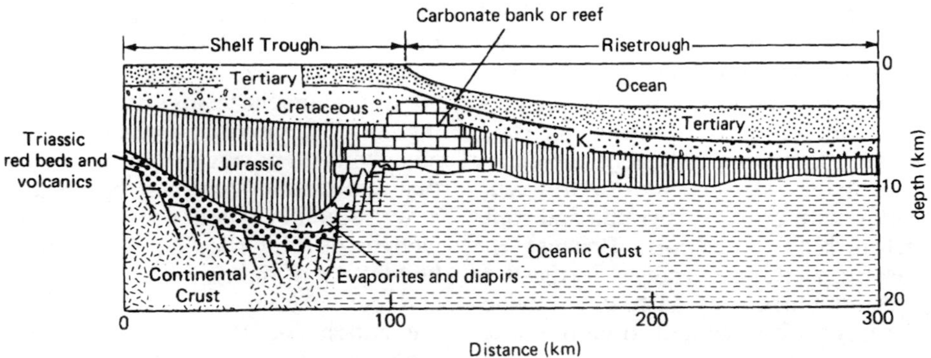

FIGURE 12-17. Schematic cross section of the Atlantic continental margin of the United States across the Baltimore Canyon trough off southern New Jersey. Mesozoic and Tertiary sediments in the shelf trough overlie rifted, thinned, and intruded continental crust and in the rise trough overlie oceanic crust. (After Grow *et al.*, 1979, Figs. 10, 11 and 13).

during the Jurassic and Early Cretaceous. Contemporaneous with these facies was a slope-rise turbidite fan and plain accumulation. The turbidites were fed by siliciclastic outer-shelf and slope deposits related by shelf dispersal systems to major river influxes. These systems operated through breaches in the outer-shelf carbonate belt. Above this section are shelf limestones interbedded with sandstones and shales. Enough is known of the passive margin of South Africa (von Rad *et al.*, 1982) to show that it has some of the same elements. See also Reyre (1984).

Deltas of large rivers play an important role in the evolution and provenance of passive margin sands—the Mississippi for the Gulf Coast, the Amazon for the South American coast, the Niger for the West African shelf, and others (Audley-Charles *et al.*, 1977). Their positions may be localized along failed parts of a triple junction rift system. The headwaters of major rivers tend to drain many different types of terrane and so mix materials of heterogeneous provenance (Potter, 1978; Franzinelli and Potter, 1983; Potter, 1984).

The provenance of passive margin sediments has a wide range of variability (Dickinson and Suczek, 1979; Ingersoll and Suczek, 1979; Valloni and Maynard, 1981). Quartz and feldspar dominate but lithic fragments may be an important component. Volcaniclastic lithic fragments and glass are absent, as would be expected from the nature of the provenance, except for the basal rift valley sections.

Dispersal patterns of passive margin sediments follow the two directions associated with shorelines, shelves, and rises: the downslope, normal-to-shore general transport direction and the longshore or longitudinal direction. The partition between these depends on wave regimes, tidal currents, and other transport mechanisms. Off the edge of the continental shelf, submarine canyons, downslope slumps, and turbidity currents interplay with contour currents and some longitudinal turbidite transport. Because the sediment is primarily waterborne terrigenous clastics and autochthonous carbonates, windblown materials are too diluted to be of consequence.

Diagenesis is extensive in deeper sections of a thick terrigenous and carbonate sequence on older continental shelf margins, where thicknesses of 10 km or more are common. Even with thermal gradients less than 30°C/km and as low as the 16.2°C/km for the COST GE-1 Well off the coast of Florida (Robbins, 1979), temperatures reach levels of 200°C or greater. Consequently there is a great loss in porosity and permeability, extensive cementation, and maturation of organic matter in the deepest sections. These sedimentary sequences may contain large quantities of petroleum hydrocarbons, stimulating interest in possible source beds and the degree of diagenesis that would allow the maturation to oil.

Continental Intra-Plate

A literature almost as extensive as that for all the other kinds of sedimentary basins combined covers almost every aspect of the origin of sandstones in basins located on the continents, far from plate boundaries. Some are at the edge or near orogenic belts related to contemporary or former plate boundaries. These foreland basins can be contrasted with those on the craton, generally at some distance from orogenic belts, and only slightly deformed (Fig. 12-18). The two types of basin may grade into each other. Their outcrop rims have been mapped in the field, and extensive oil field development has led to an abundance of subsurface information. Stratigraphic correlations are excellent, and many detailed and extensive paleocurrent maps have been prepared.

Foreland basins evolve at the continental borders of orogenic belts that result from plate boundary deformation. These basins are related to the flood of detritus transported inboard to the continent from mountain belts. The outboard highlands may result from plate movements ranging from Andean-type convergences to continental collisions. The main mechanisms of foreland subsidence appear to be isostatic: the great load of detritus shed from the nearby mountains depresses the crust. Alternative or additional mechanisms, such as boundary faulting, may be involved. Some foreland basins have long histories as the result of orogenies that rejuvenated quiescent basins and generated new sediment sequences. The thick clastic wedges of the Ordovician, Devonian, and Carboniferous of the Appalachians are of this type.

Recent geophysical investigations have led to the conclusion that many cratonic basins lie over rifted basements. This has given rise to a model of formation that follows the general picture of heating and distension, followed by rifting and normal fault crustal extension, which creates the initial rift basin. That basin then continues to subside in response to cooling and crustal contraction plus the isostatic adjustment to the weight of sediment deposited. These models have been built from aeromagnetic and

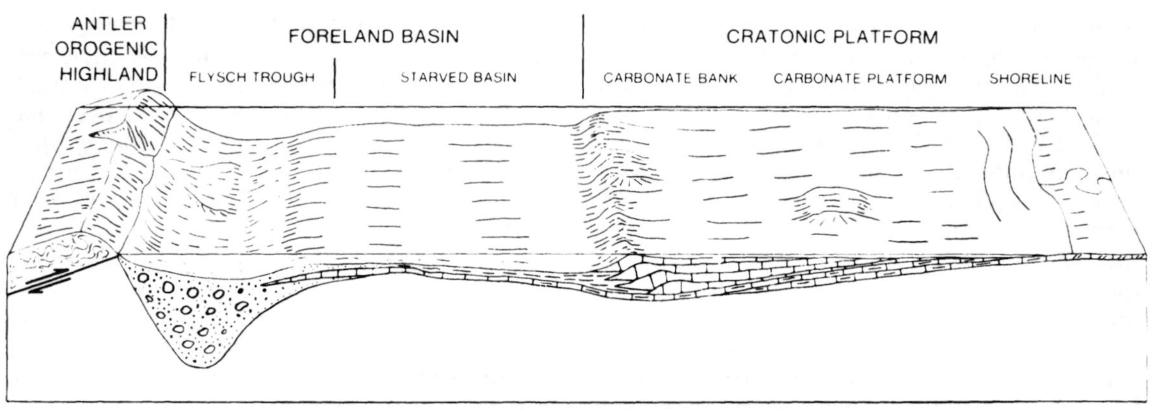

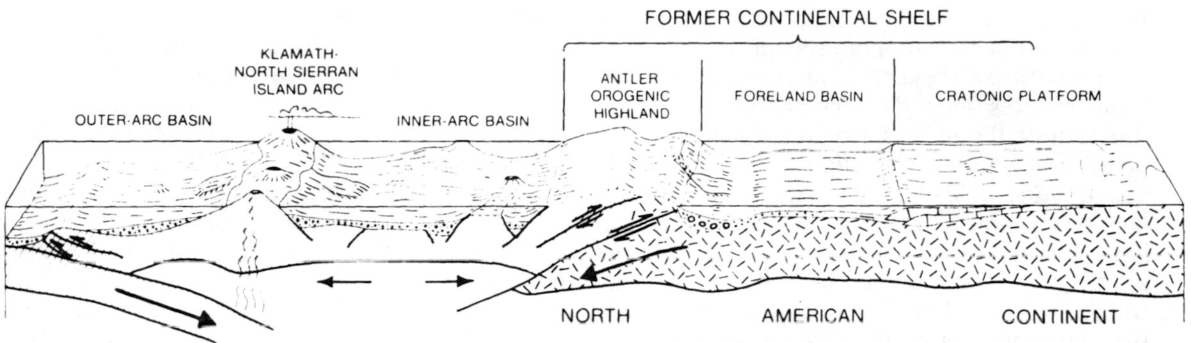

FIGURE 12-18. Paleozoic foreland and cratonic basins in the western United States in relation to the Antler orogenic belt and the Klamath-North Sierran Island arc. (From Poole and Sandberg, 1977).

gravity surveys, seismic profiles, especially those of the U.S. COCORP (Consortium for Continental Reflection Profiling), the Canadian LITHOPROBE, and deep seismic sounding programs of other countries (e.g., Zoback *et al.*, 1980; Kane *et al.*, 1981; Brown *et al.*, 1982). Support for this origin has come from the fragmentary knowledge about basal arkosic sandstones whose composition would presumably fit this origin in parts of the Illinois, Michigan, and other basins.

Regardless of the mechanism of subsidence of many foreland and cratonic basins, the sedimentary sequences tend to show varying courses of evolution that include early clastic accumulations followed by carbonate sequences and finishing with an upper clastic sequence. The lower sandstones overlying initial arkosic phases tend to be quartz arenites and the upper ones lithic arenites. This model was explored by Potter and Pettijohn (1977, Table 9-1) for the Appalachian basin, where several such sequences are recorded. Foreland basins inboard of Andean-type convergences may be rich in volcaniclastics and/or plutonics as well as the sediments of arc-related basins.

The provenance of intracratonic basins tends to reflect their continental block origin and, with respect to the quartz arenites, sources of recycled and well-weathered quartzose detritus. The infrequent turbidites that are found in foreland basins close to continental margins reflect transient involvement in deeper seas that may be the result of marginal sea or back-arc development in an Andean-type convergence episode. In these there may be some volcaniclastic contribution if the paleowinds permit. The lithic sandstones reflect continental orogenic belt detritus that is shed into the interior of the continent rather than to the margin. An example is the way that the Alpine molasse has spread north onto shallow cratonic basinal areas of north Europe. In North America much of the Carboniferous lithic sandstone of the interior reflects transport from the distant mountains thrown up by the Acadian–Caledonian continental convergence orogeny. Though Sloss (1966) had earlier argued against any clear relation of cratonic to continental margin events, later reconsideration has led to ideas of geophysical mechanism that link tectonic behavior of the two different kinds of provinces with consequent linkage of their sedimentary histories (Sloss and Speed, 1974).

The dispersal patterns of foreland and cratonic basins has been covered in depth by Potter and Pettijohn (1977). Patterns of deposition on a continental interior paleoslope tend to be persistent and in the direction from the cratonic high—a Precambrian shield in many cases—towards passive or active continental margins, the passive being more common. Quartz arenites may be dispersed by a coastal plain, longshore, shelf regime whereas later lithic arenite dispersal may be dominated by the patterns of large river systems. Dispersal patterns may vary sympathetically with provenance as source regions change from continental interiors to continental margin orogenies.

The paths of diagenesis of these quartz and lithic arenites depend on the depth of burial—a function of the subsidence and detrital supply history—and the alteration of fairly stable quartz- and feldspar-rich debris and general lack of volcaniclastic material. The paleohydrologic circulation of these basins may play an important role in transport of dissolved material from one part of the section to another. The thermal gradients of such basins are generally relatively low, though in their initial stages, if rifting is the origin, the thermal gradients may be elevated. Because depths of burial are variable, the degree of cementation, and thus porosity and permeability, ranges widely.

Evolutionary Sequences and Successor Basins

A clear signal comes from the analysis of the evolution of a portion of the sedimentary sequence on the Earth's continental crust over a significant length of geologic time: the record is one of deposition in a series of successive basins whose tectonic environments change over time (Fig. 12-19). The changing tectonics can be linked to changing patterns of plate motions. Intra-oceanic island arc convergences may collide with continents or become Andean-type convergences. Then the convergence may become continent-continent as a continent on the subducting plate approaches the convergence. The orogen produced by the continent-continent convergence may rift to produce first a continental rift valley and then an opening, spreading ocean with passive margins. The above sequence in a general way describes the evolution of the Paleozoic–Mesozoic history of the Appalachian–Caledonian–Hercynian orogenic belt at the margins of the North American, South American, European, and African continents. Other evolutionary sequences may be different: the complex events leading up to the Alpine orogeny or to the evolution of the Tasman geosyncline in Australia illustrate different paths.

Some theorists advocate a general evolutionary path followed more or less by all mobile belts. Others argue forcefully that there may be almost any sequence of plate motions giving rise to different sorts of histories. An example of the first is the Wilson cycle by which a mantle plume produces a triple junction which then leads to an opening ocean along two of the arms and an aulacogen along the failed third arm. These first steps, however, include only the beginning of ocean spreading formation and do not necessarily imply how convergences follow and in what order.

Accretionary tectonics may complicate interpretations of continental margin basin evolution. Whole terranes may be rifted along transform fault continental margins and welded to the continent. These exotic terranes are bounded by faults and characterized by significant differences in sedimentary and structural histories between them and their neighboring terranes (Beck *et al.*, 1980). The Cache Creek is representative of the terranes thought by some to constitute over 70 percent of the North American Cordillera (Coney *et al.*, 1980). Most such terranes, termed "suspect" for their uncertain paleogeographic position with respect to both the North American continent and other terranes, appear to be of oceanic origin, allochthonous to the continent. The exotic nature of such terranes is evidenced by anomalous paleomagnetic pole positions that are discordant with similarly aged rocks from the North American craton. These terranes are also marked by different sedimentary sequences and faunal assemblages now in juxtaposition (Churkin and Eberlein, 1977; Irving, 1979; Beck, 1980; Jones *et al.*, 1982). Allochthonous accretionary processes have been suggested as the common process responsible for mountain building and continental growth along active margins (Ben-Avraham *et al.*, 1981; Nur and Ben-Avraham, 1982). Sedimentary analysis provides valuable information on the nature and evolution of suspect terranes (e.g., Dickinson, 1979; Dickinson, Helmold, and Stein 1979; Ward and Stanley, 1982). Particularly important in this kind of analysis is the study of provenance, which can reveal important differences between

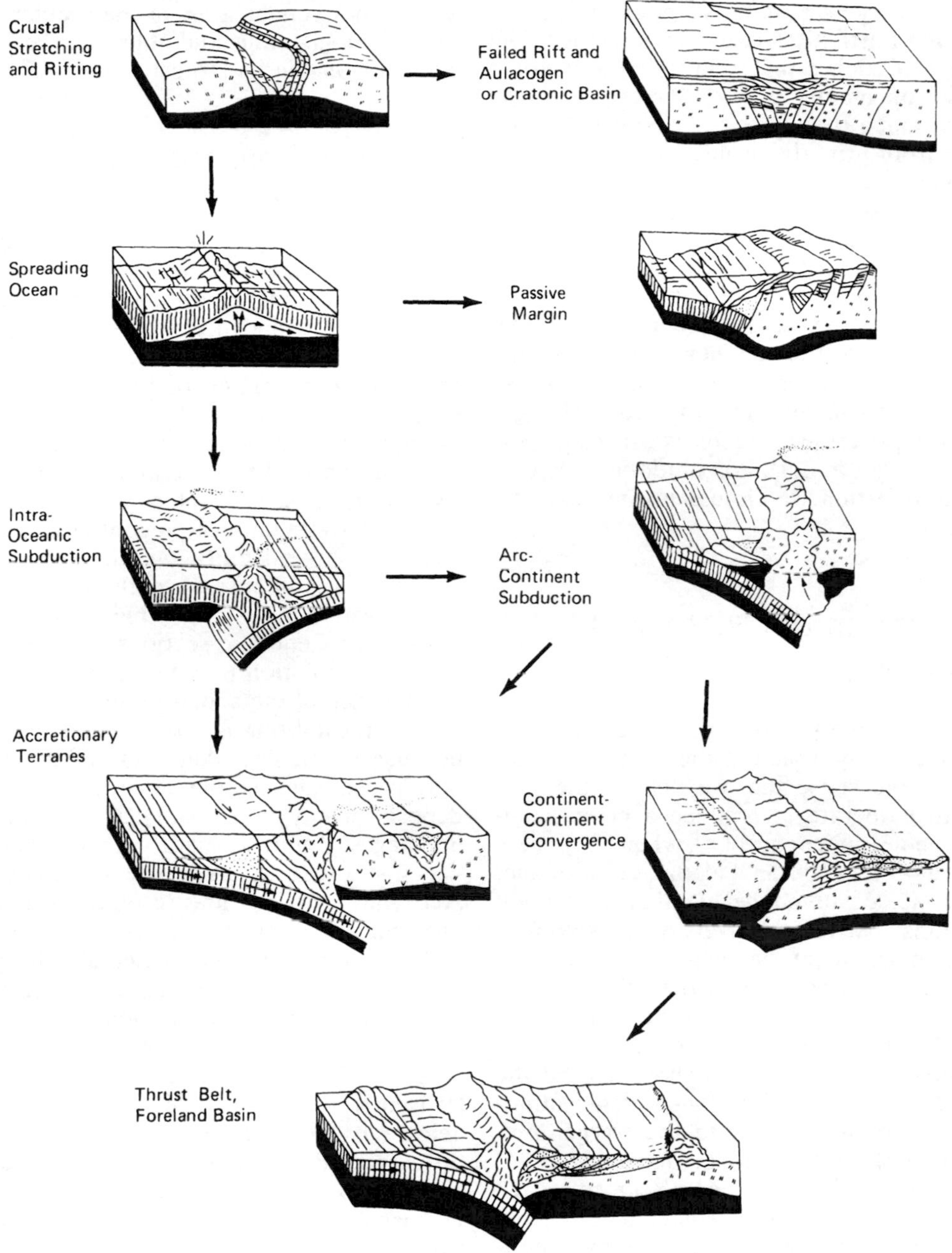

FIGURE 12-19. Some paths of evolution of successor basins. (Modified from A. Green, Exxon Production Research).

adjoining terranes with respect to areal and temporal variation that reflects plate tectonic setting.

In the absence of geophysical theory predicting long-term evolutionary patterns of plate motions, we can only try to extract such patterns from the geologic history of individual belts. That geological history is read from the rock record, and sandstones can be used as one of its most important components. We cannot tell at this time what pattern(s), if any, will serve as an important constraint on geophysical theories for long-term plate motions. How such patterns related to the evolution of continents is also the subject of debate. Continental evolution is now an arena of discussion in which igneous petrologists, isotope geochemists, sedimentary petrologists, and Earth historians come together. The elements of such a theory are (1) the isotopic evidence on mantle sources for crustal rocks and how much they have been recycled or are derived from primitive mantle, (2) the igneous rock evidence for evolution of plutonic additions to the continental crust by magmatic processes at subduction zones, and (3) the sedimentary evidence for sedimentary and crystalline rock sources of continental margin sediments and how such sediments are permanently welded to the continents as part of the plate convergence or other processes. These matters lead to old questions of secular changes in geologic processes, and for us especially, the secular changes in Earth's evolutionary history. It is that question we discuss in the next section.

Sandstone in the History of the Earth

In almost all of this book we have implied a uniformitarian approach to the origin of sandstones, in the sense that we believe the processes that operate to produce a given sandstone type are independent of when they were deposited—Precambrian, Paleozoic, Mesozoic, or Cenozoic. We do not believe that the general interactions between provenance, tectonics, climate, environment, and diagenesis were any different in the Precambrian than they are today. Yet we also know that the Earth has been going through an evolutionary process since its formation about 4.7 billion years ago. Did the change from the primitive anoxygenic atmosphere of the early planet to the oxygen-rich atmosphere of the Phanerozoic have any effect on the nature of the sandstone produced? Have there been later periods of higher partial pressure of CO_2? As the size and geography of continents and continental crust evolved through geologic time, did the average composition of sandstones change? And did the course of evolution of life and subsequent proliferation of the animal and plant kingdoms affect the character of sandstones?

Is there any evidence from mineral or chemical composition that sandstones of different ages are different? Or do the older sandstones exhibit particular textures or sedimentary structures that distinguish them from the younger sandstones? Here we briefly review the data available and then suggest ways in which sandstones may reflect some of the secular changes over geologic time and thus be used to map Earth history in its largest sense.

Sandstone Composition in Relation to Age

One way to check an age effect is to compare averages of mineralogical and/or chemical analyses of sandstones of different ages. Averaging the diverse compositions of sandstones of different types all formed at one time might show changes in the ratios of the major elements. Over how large or small a time interval can we or must we calculate averages? The smallest time interval will be determined by the reliability of time–stratigraphic correlations. The practical lower limit of paleontologically based stratigraphic reliability within most sedimentary basins is about one million years, though better resolution can be achieved in some Mesozoic and Cenozoic sections. In the early Paleozoic, resolution may be poorer. Because of the absence of metazoan fossils, the resolution in the Precambrian is poor. This is unfortunate because we suspect that many of the most important changes in the environments of sand deposition may have taken place during the earlier stages of Earth's history. Paleomagnetic reversal stratigraphy, primarily in the later Cenozoic, may provide time resolutions as little as one hundred thousand years.

Sandstones from like facies should be compared to properly assess age differences. Comparing sandstones of an Archean flysch with Paleozoic cratonic sandstones would reveal mostly facies differences. To discern any secular trends, we should select an easily recognizable facies and compare the sands of this facies throughout the history of the Earth. Such a facies, for example, is the molasse facies, which we associate with continental detritus shed from plate convergence orogenies. The time interval represented by such a facies may be of the order of 20 to 45 million years. This facies, as we have pointed out in the first part of this chapter, is only one component of basin evolution sequences, such as those of the Caledonian, Hercynian, and Alpine basins of Europe. Molasse facies of ages from Precambrian to Tertiary are recognizable in the geologic record. It is likely that significant secular changes in sandstone composition, if they exist, could be brought out by appropriate sampling and averaging of these facies alone. Care would be needed to avoid the many trends in sandstone composition within an evolutionary sequence and along the depositional strike of a linear orogenic belt.

Significant secular changes in composition might better be demonstrated by averaging over orogenic sequences than by taking only one of their constitutent facies. Here we must define the boundaries of the sequence uniformly and allow for differential preservation of sandstones of different tectonic environments. For example, it is only in the youngest orogenic episodes that a significant volume of the uppermost intermontane sandstones are preserved. In older sequences there is a tendency for sediments deposited along the outer continental margins, the flysch, to be involved in later deformation and metamorphism. Sandstones belonging to this facies may thus be lost for averaging purposes. We might average very large segments of geologic time that correspond to major episodes in Earth history. Thus we might compare the Archean (pre–2500 million years), the earliest definable times, with the Proterozoic (2500 to 570 million years), during which the Earth's dynamic system began to assume its present general form. We might compare these with the Paleozoic (570 to 245 million years), which started with a global rifting of continents and ended with the assembly of Pangaea, and the post-Paleozoic (245 to 0 million years), a time of continued continental divergence bordering the Atlantic and convergence along the Tethyan belt. This might be more revealing of the influence of the largest scale of tectonics on sandstone composition.

The immense amount of data required to reveal such trends and the way the analysis proceeds is illustrated by the work on the Russian Platform of Ronov (1968) and Ronov *et al.* (1963, 1965, 1969). So far no other comparable compilations are available, either in terms of chemical or petrographic analyses, for large areas of other continents. Until that kind of information becomes available, the best that we can do is to use the incomplete data we have to see if any obvious trends can be discerned. One way to look at chemical data is in terms of the element ratios, particularly the SiO_2/Al_2O_3 and Na_2O/K_2O ratios (see Chap. 2).

If sandstones are distinguished only on the basis of eras (Precambrian, Paleozoic, Mesozoic, and Cenozoic) without respect to petrography, there appears to be no detectable relation between age and the SiO_2/Al_2O_3 or Na_2O/K_2O ratios (Fig. 12-20). The Na_2O/K_2O ratio is affected by several different mineral parameters, as discussed in Chap. 2. Because source factors must be distinguished from environment and diagenesis factors, sandstones have to be considered by petrographic type. If we consider only arkoses, for example, there does seem to be a dividing line, at Na_2O/K_2O = 0.6, between Mesozoic and Cenozoic feldspathic sands which are more sodic and the Paleozoic and Precambrian which are less sodic. But the differences are not great. Furthermore,

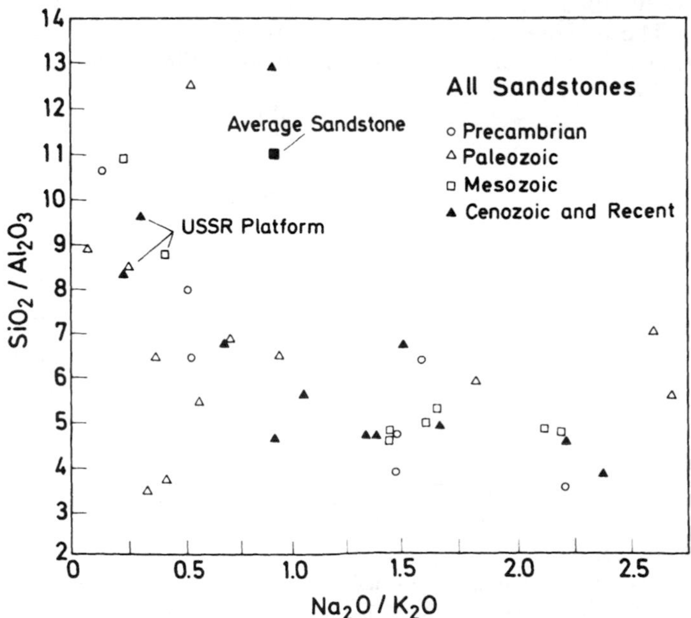

FIGURE 12-20. SiO_2/Al_2O_3 versus Na_2O/K_2O ratios in sandstones. Data from various sources, mainly from Pettijohn, 1963. Average sandstone not based on these data.

the differences may not be an original attribute. Arkoses tend to be highly permeable formations, and there may be some gradual dissolution of the more calcic plagioclase relative to K-feldspar that over a long time would decrease the Na_2O/K_2O ratio.

If one considers only graywackes, the sandstones that typically have the highest Na_2O/K_2O ratios, there appears to be no age effect. Some of the highest Na_2O/K_2O ratios are those of two classic Paleozoic graywackes, the Kulm and the Tanner, and the Precambrian Knife Lake (Table 5-7). We can compare these with the modern sandy, volcanic-rich sediments of the Columbia River. Lacking primary matrix, these were shown by Whetten (1966) to be similar to graywackes in almost every other respect. Matrix-rich graywackes when artificially metamorphosed (Hawkins and Whetten, 1969), these sands have lower Na_2O/K_2O ratios than most other graywackes, even though they include many basaltic fragments. Labile mineral compositions of graywackes change during long-term diagenesis—and low-grade metamorphism—from smectite, zeolite, mafic glass, and plagioclase to chlorite and more albitic plagioclase. It appears that much of this change is isochemical. The differences among some graywackes may be more due to original source contributions than to diagenesis. It may be that the proportion and kind of originally detrital volcanic debris, determined by the relation of the depositional environment to island arcs and oceanic or continental hot spots, is more important than any other factor in determining graywacke alkali metal ratios.

There is one other barrier to a proper evaluation of the analyses—the size effect. Fine-grained graywackes approach shales in composition. The coarser-grained ones have the higher Na content, which is undoubtedly located in the framework-fraction feldspars (albites) rather than the matrix. Obviously, meaningful comparisons can be made only if the rocks compared have similar grain size.

It is hard to see any significant age effect in SiO_2/Al_2O_3 ratios in sands, though no Paleozoic and Precambrian samples have ratios lower than 5.

Other ratios show some age differences. The CaO/MgO ratio appears to increase with decreasing age. The data of Vinogradov and Ronov (1956) show ratios of 0.78 in pre-Devonian sandstones, 1.39 in Devonian, 3.56 in Carboniferous, and 3.90 in Mesozoic sandstone. This change is similar to that found in carbonate rocks by R. A. Daly as long ago as 1909 and discussed in relation to other elemental ratios by Garrels and Mackenzie (1971) (Fig. 12-21). The change in CaO/MgO is related to progressive dolomitization of carbonate cement with time. Other chemical components that show secular trends in clays and shales, such as the Fe_2O_3/FeO ratio, which increases with decreasing age (Nanz, 1953), do not appear to show such clear trends in sandstones.

Analyses of petrographic data on sandstones show few obvious tendencies. There is some evidence that the feldspar content of sandstones increases with decreasing age, the Paleozoic sandstones on both the North American and Russian platforms containing less feldspar than those of Mesozoic or Cenozoic age (Tables 2-2 and 2-3). Likewise, as discussed in Chap. 11, the number of heavy mineral species increases with decreasing age. The clay mineral-

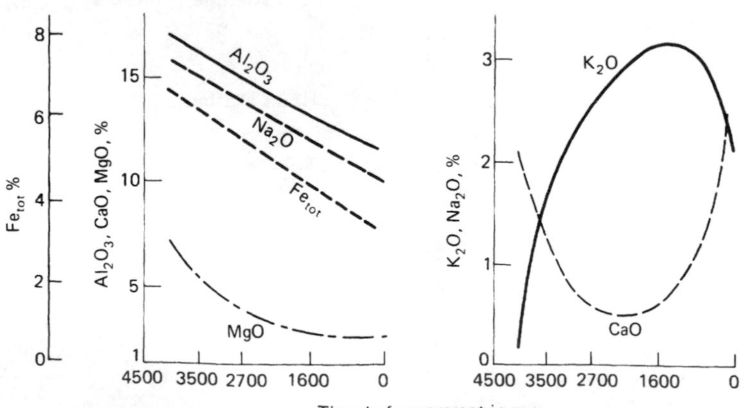

FIGURE 12-21. Changes through time in the chemical composition of sediment derived from erosion of shields, platforms, and geosynclines after Ronov (1972).

ogy of sandstones seems to show the same general tendencies as in shales, Mesozoic and Cenozoic samples being richer in smectites and kaolinite and the older rocks richer in chlorite and illite (Weaver, 1967).

These apparent secular changes, if not the result of biased sampling, are probably more the consequence of long-term diagenetic changes than they are of original differences in sand compositions. As noted in Chap. 11, the secular variations in carbonate and silica cement, the diversity of heavy mineral suites, the changes from illite–chlorite to smectite–kaolinite clays, and the proportion of graywacke matrix can all be directly related to diagenesis.

The change in feldspar content can also be related to gradual alteration during diagenesis, especially the common dissolution of detrital feldspar that is volumetrically far more important than the appearance of new feldspar by authigenesis. If loss of feldspar through sedimentary recycling were dominant, feldspar would decrease with increased recycling as age decreased—that is the heart of the maturity concept. But detrital feldspar abundance is related to tectonism and climate too (see Chaps. 2 and 8). If there were little recycling loss, implying less chemical weathering, we might suspect a tendency for increased tectonism and more arid or cold climates to the present time. The countervailing tendencies of feldspar loss by diagenetic dissolution and weathering-transport during recycling, on the one hand, and feldspar gain by tectonism and climate, on the other hand, cannot easily be assessed. We need a much sounder statistical analysis of a larger petrographic data base to test hypotheses. Because feldspars have differing chemical stabilities, it would be important to determine the proportion of different feldspars as a function of age to check any diagenetic effect which may be the chief cause of feldspar variation. The variations observed could, of course, be due to sample bias—the older sandstones being mainly cratogenic and the younger ones being derived from orogenic belts.

There is some evidence that the very oldest sandstones, those over 2.5 billion years old, may be different. Engel (1963) and Ronov (1964) believe that there is a higher proportion of graywackes and arkoses relative to quartz arenites or quartzose lithic arenites in such old sections. Another perplexing occurrence is the detrital gold, uraninite, and pyrite in sandstones of the Witwatersrand in South Africa, the Blind River district in Canada, and the Jacobina district in Brazil. All seem to be in the time interval of 1.7 to 2.0 billion years old and no others like them are known of any other age. Another seeming "non-uniformitarian" kind of sandstone is the extremely thick quartz arenite that seems to be widespread in the upper Precambrian. Quartzites such as the Lorrain of Ontario, the Baraboo of Wisconsin, the Athabaska of Saskatchewan, and the Uinta of Utah are all very pure and well over 1000 m thick whereas Phanerozoic quartz arenites tend to be very thin, rarely over a few tens of meters thick. Are all these seemingly unique sandstones the result of special events or conjunctions of evolutionary patterns, never again to be repeated?

To summarize, most apparent secular changes that we now deduce from only a crude statistical evaluation may be traced to long-term diagenetic changes but some, particularly special characters of very old sandstones, may reflect evolutionary changes in the overall history of the crust.

Sandstones in Relation to Evolutionary Changes in Earth History

Have certain events in Earth history been reflected in sandstone compositions (Fig. 12-22)? One such event already mentioned is the evolution of oxygen in the Earth's atmosphere, which took place sometime in the Precambrian (Holland and Schidlowski, 1982; Holland, 1984). Sandstones formed in a reducing environment would be expected to contain much more detrital pyrite, magnetite, and other oxides of reduced metals. They might also be expected to have a greater abundance of detrital ferrous silicates, for part of the present rapid weathering of iron-rich olivines and pyroxenes is related to oxidation of the iron. The difficulty is that very low levels of oxygen, pressures of about 10^{-4} or 10^{-5} atmospheres, are probably sufficient to oxidize ferrous iron (Holland, 1984), and this low level might have been reached as early as 3 billion years ago. The minerals that might be found in sandstones which might be diagnostic of oxygen pressures close to or a little less than that of the present atmosphere are the opaque heavy minerals of oxidizable heavy metals, such as uraninite (Grandstaff, 1976). The information that we need is the precise mineralogical identification of these heavy opaques; for that, one needs polished ore sections of heavy mineral concentrates—an unusual technique for the sedimentary petrologist.

Another development on the surface of the

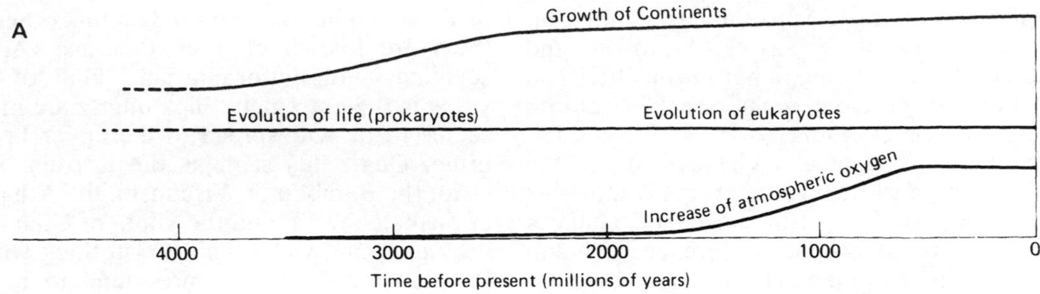

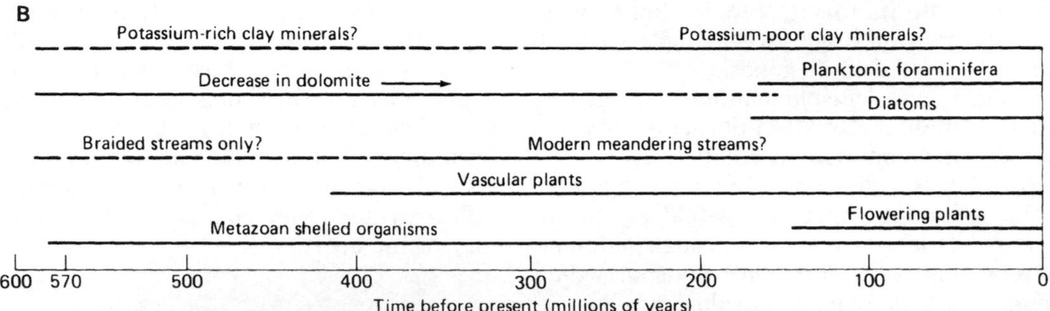

FIGURE 12-22. Some secular changes in the Earth's history that may have affected the processes by which sand was produced, dispersed, and deposited. A) Geologic time, B) Phenerozoic time.

Earth was the evolution of the vascular land plants in the Upper Silurian, earlier Devonian. It is clear that there is no profound chemical or petrologic difference between pre-Devonian and later sandstones but may there not be some differences in erosion and denudation rates that would be reflected in sedimentation rates of sandstones? Or was there an extensive cover of primitive nonvascular plants? Or is the rate of weathering and erosion not as greatly influenced by vegetation as has been thought? Cawley *et al.* (1969) studied chemical weathering of vegetated and nonvegetated areas in Iceland and concluded that the vegetation effect was not overwhelming. Variable rates of oxidative weathering of Precambrian soils of various ages have been inferred from studies of paleosols and linked to oxygen levels in the atmosphere (Holland, 1984). We would expect that this effect might have operated in combination with a growing nonvascular land bacterial and fungal flora in the later Precambrian to bring pre-Devonian denudation rates close to those of later times. In trying to estimate rates of denudation in the distant past, we may have to rely mainly on an analysis of sedimentary volumes produced by denudation of land surfaces. A good statistical comparison of this kind of data, pre- and post-Devonian, difficult as it may be, will be needed to demonstrate that the evolution of the higher land plants made a significant difference. The alternative is to use indirect tracers in sediments such as Sr and Nd isotopes that would give clues to changes in the kind and/or amount of denudation.

It has been suggested that the organization of strata involving sandstones, particularly fining-upward alluvial cycles, might not have been developed and that the alluvial deposits would be noncyclic sheets of coarse immature materials in the absence of a land plant cover (Schumm, 1968, p. 1583). Others have suggested a change from braided to meandering streams with the advent of a vascular plant cover (Cotter, 1978).

Has the development of a rich benthonic fauna in later times affected the character of sands and sandstones? Bioturbation, the disturbance of deposited sediment by burrowing organisms, leads to destruction of primary sedimentary features—especially laminations. Are the older sandstones generally better laminated?

How has the evolution of sialic continental crust affected sandstone compositions? Do we have any sandstones dating from the time when there was no such crust? If so, we should recognize them from their chemical and petrographic composition as having been derived solely from mafic terranes. They might have been largely volcaniclastics. And, when the sialic crust was thin, might we still have had much more of a contribution from mafic rocks and less quartz-rich detritus? Sandstones of that stage could be recognized by an average composition intermediate between that expected from the earliest stage and that of all later sandstones derived mainly from sialic crust. Or the earliest sialic material might have been volcanic rather than plutonic and so be found in volcaniclastics.

Given this kind of crustal evolution, were tectonic styles different in early stages when continents were small and/or thin? We do not know when the current style of plate tectonics evolved, but there are suggestions that the Archean was a pre–plate tectonic time, with small continental land areas and limited continental shelf areas (Windley, 1976). Tectonic activity might have been much more generally distributed over the surface. This might have led to a more general distribution of sandstones reflecting that tectonism, rather than the later, more familar localization of tectonic activity along plate boundaries and continental margins. At the inception of plate tectonics, perhaps in the Proterozoic, plates may have been very small, so that the lines along which they converged, subduction zones and continental convergences, were more numerous and formed a much larger proportion of Earth than today. As continental areas were smaller, was much more of the geologic record lost by oceanic subduction than today? The answer to most of these questions lies in the petrologic study of the oldest sandstones to determine their provenance, including careful determination of the amounts and precise compositions of K-feldspar and plagioclase as well as of mafic minerals, parameters that might be used to distinguish among growing continental blocks and volcanic piles.

Opinion differs on these matters, some workers interpreting the Archean sands as silicic volcaniclastic materials (Ayers, 1969) whereas others consider them to be derived from sialic crust by normal weathering and erosion (Donaldson and Jackson, 1965; Walker and Pettijohn, 1971).

Other secular changes in Earth history are known, but have not been shown to have a great effect on the character and distribution of sandstones. One such effect is the gradual slowing down of the Earth's rotation rate in response to the tidal friction induced by the moon, the quantitative measurement of the variation in length of the day and month being based on growth patterns in corals, molluscs, and stromatolites (Pannella *et al.*, 1968). The more rapid rotation rate and higher tides from a closer moon might have produced a greater effectiveness of tidal currents. Would the marine crossbedding related to such currents be on a significantly larger scale than that of later times? An analysis of paleotidal sedimentation by Klein (1971; Klein and Ryer, 1978) led him to conclude that the tidal range and frequency of tidal deposits have not changed much through geological time, though much remains to be done in the quantitative analysis of such sediments.

Mention of the moon raises the possibility that sandstones exist on the moon and can be used to interpret its history. The results of the Apollo 11 and Apollo 12 missions to the moon do not indicate any deposits that would be sensibly called sand, though there are abundant fragmental materials of all sizes (e.g., King et al, 1970; Murray et al, 1981, p. 202–218). These particulates have been interpreted in terms of "lunar stratigraphy and sedimentology" (Lindsay, 1976). Our conventional definition of sand requires that materials of sand size (1/16 to 2 mm) constitute a major proportion of the deposit. Neither the regolith nor breccias found on the moon meet these requirements. We may yet discover ash falls and flows. We would consider them in the same way as volcaniclastics, differing only in the value of gravity and the dependence of any current on the entrapment of enough gas to provide a fluid transport medium.

Mars, with an active atmosphere, does have sands. The Viking Mars Lander took photographs that clearly show sand accumulations formed by the Martian winds. Wind speeds have been estimated from their bedforms, ripples and dunes, and scours and buildups adjacent to rocky protuberances (Murray *et al.*, 1981, pp. 101–12). Channel patterns are taken as evidence of former river flows that transported all sizes of detritus (Murray *et al.*, 1981, pp. 113–21). The extensive photographic coverage of the Martian surface and some partial chemical analyses performed by the lander are our only data base for the study of Martian sands.

Venera 13 and 14 landed on Venus in March, 1982, and transmitted panoramic television pictures to the Earth. These pictures have been interpreted to indicate layered, lithified loose material, and perhaps ripple marks and crossbedding (Florensky *et al.*, 1983). Transport is inferred to be eolian or ballistic-eolian. But the results of planetary exploration thus far lead us to believe that those interested in sandstones will probably be most happy studying the Earth.

Conclusions and a Look at the Future

What can we say about the usefulness of sandstones for the study of the larger aspects of the Earth? To the degree that the minerals of sandstones reflect the chemistry of the atmosphere and oceans, we may get some better information on their evolution. More important, though, is the use of sandstones as indicators of provenance and paleocurrents, for these are the keys to paleogeography and ultimately to paleotectonics. But in the study of the larger elements of paleotectonics we must map large regions, even continents. We hope that sandstone petrologists in the future will put together paleogeographic and provenance maps of many different basins over several continents or the whole world. Just as the 1950s saw sandstone petrologists study whole basins, their descendants in the 1980s and 1990s will need to put together continents and simultaneously as they refine ways of analyzing the details of individual basin development.

At the same time much more work on statistical averaging of petrographic data, chemical analyses, and relative abundances of different sandstone types will have to be done if we are to get a clearer picture of long-term trends, both within and between basinal sequences and over the whole history of the Earth.

The interpretation of the compositional data will have to be based on better quantitative data on chemical stabilities and reaction kinetics. The analysis of minerals that might reflect the oxidation potential of primitive earth environments will help us only if we can answer the following questions. How fast does pyrite oxidize as a function of oxygen pressure? What is the reliable value for the equilibrium oxygen pressure needed to convert magnetite to hematite? What chemical species influence the kinetics of the magnetite–hematite conversion?

Interpretation of provenance, environment, and diagenesis cast in terms of plate tectonics has been an intense activity in the 1970s and 1980s. We have no doubt that specific deductions will change as that theory changes and as newer ideas of the geophysics of continental margins and interiors become part of it. What plate tectonics has done is to bring the interpretation of sandstones into much closer conjunction with geophysics, geochemistry, and structural geology. That closeness is, in a way, a return to the work of pioneers of the field in the late 19th and early 20th centuries, studies that put sandstone origin squarely in the middle of the general geology of the time.

We again state our conviction that the answers to most of the questions we have posed in this book will come not from unimaginable new techniques, but from the intelligent application of methods and tools that we now have at our disposal. We rely on the coupling of quantitative field and laboratory analysis—primarily but not exclusively microscopy—with the ideas and data of chemistry, on the one hand, and fluid dynamics, on the other, to give us the solution. But because we are concerned with the geology of sand—and so with the actual distribution of sands on the Earth—we use one of the most quantitative techniques ever developed by geologists—*the map*. The decision to map implies an intelligent choice of parameters to map, and the ability to measure them quantitatively. We believe that there is an order to the Earth and that through sagacious mapping we can discern much of it.

References

Allison, E.C.: Geology of areas bordering Gulf of California. In: van Andel, Tj.H., and Shor, G.G. Jr. (Eds.): Marine geology of the Gulf of California—a symposium. Am. Assoc. Petroleum Geologists Mem. 3, 3–29 (1964).

Aubouin, J.: Geosynclines, 335 pp. (Developments in Geotectonics, Vol. 1). Amsterdam: Elsevier 1965.

Audley-Charles, M.G.; Curray, J.R.; and Evans, G.: Location of major deltas. Geology 5, 341–344 (1977).

Ayers, D.L.: Early Precambrian metasandstone from Lake Superior Park, Ontario, and its implications for the origin of the Superior Province (Abstract). Geol. Soc. America Abs. with Programs, part 7, p. 5 (1969).

Bachman, S.B., and Leggett, J.K.: Petrology of Middle America Trench and trench slope sands, Guerrero margin, Mexico. In: Init. Rept. Deep Sea Drilling Proj. 66, 429–436 (1982).

Baker, B.H.; Moler, P.A.; and Williams, L.A.J.: Geology of the eastern rift system of Africa. Geol. Soc. America Spec. Paper 136, 67 pp. (1972).

Ballance, P.F., and Reading, H.G., Eds.: Sedimentation in oblique-slip mobile zones. Int. Assoc. Sedimentologists Sp.Pub. 4, 266 p. (1980).

Bally, A.W., and Snelson, S.: Facts and principles of world petroleum occurrence: Realms of subsidence. In: Miall, A.D. (Ed.): Facts and principles of world petroleum occurrence. Canadian Soc. Petroleum Geologists Mem. 6, 9–94 (1980).

Bartolini, C.; Malesani, P.G.; Manetti, P.; and Wezel, F.C.: Sedimentology and petrology of Quaternary sediments from the Hellenic Trench, Mediterranean Ridge and the Nile Cone from D.S.D.P. Leg 13 cores. Sedimentology 22, 205–236 (1975).

Basu, A.: Weathering before the advent of land plants: Evidence from unaltered detrital K-feldspars in Cambrian–Ordovician arenites. Geology 9, 132–133 (1981).

Beck, M.E.: Paleomagnetic record of plate margin tectonic processes along the western edge of North America. Jour. Geophys. Research 85, 7115–7131 (1980).

Beck, M.E.; Cox, A.; and Jones, D.: Mesozoic and Cenozoic microplate tectonics of western North America. Geology 8, 454–456 (1980).

Ben-Avraham, Z.; Garfunkle, Z.; Almagor, G.; and Hall, J.K.: Continental breakup by a leaky transform: The Gulf of Elat (Aqaba). Science 206, 214–216 (1979).

Ben-Avraham, Z.; Nur, A.; Jones, D.; and Cox, A.: Continental accretion: From oceanic plateaus to allochthonous terranes. Science 213, 45–54 (1981).

Bertrand, M.: Structure des alpes français et recurrence de certains facies sédimentaires. Compt. Rend. Congr. Intern. Géol., 6th Sess., p. 1-3-177 (1897).

Biddle, K.T., and Christie-Blick, Nicholas, Eds.: Strike-slip deformation, basin formation, and sedimentation. Soc. Econ. Paleon. Mineral. Sp. Pub. 37, 386 p. (1985).

Bird, J.M., and Dewey, J.F.: Lithosphere plate-continental margin tectonics and the evolution of the Appalachian orogen. Geol. Soc. America Bull. 81, 1–3 (1970).

Bird, P.; Toksoz, M.N.; and Sleep, N.H.: Thermal and mechanical models of continent-continent convergent zones. Jour. Geophys. Research 80, 4405–4416 (1975).

Blake, M.C. Jr.; Campbell, R.H.; Dibblee, T.W. Jr.; Howell, D.G.; Nilsen, T.H.; Normark, W.R.; Vedder, J.C.; and Silver, E.A.: Neogene basin formation in relation to plate tectonic evolution of San Andreas fault system, California. Am. Assoc. Petroleum Geologists Bull. 62, 344–372 (1978).

Blatt, H., and Jones, R.L.: Proportions of exposed igneous, metamorphic, and sedimentary rocks. Geol. Soc. America Bull. 86, 1085–1088 (1975).

Blatt, H.; Middleton, G.; and Murray, R.: Origin of sedimentary rocks, 2nd Ed., 782 pp. Englewood Cliffs, New Jersey: Prentice-Hall 1980.

Bond, G.C.: A late Paleozoic volcanic arc in the eastern Alaska Range, Alaska. Jour. Geology 81, 557–575 (1973).

Braile, L.W.; Keller, G.R.; Hinze, W.J.; and Lidiak, E.G.: An ancient rift complex and its relation to contemporary seismicity in the New Madrid seismic zone. Tectonics 1, 225–237 (1982).

Breyer, J.A., and Ehlmann, A.J.: Mineralogy of arc-derived sediment: Siliciclastic sediment on the insular shelf of Puerto Rico. Sedimentology 28, 61–74 (1981).

Brown, L.D.; Jensen, L.; Oliver, J.; Kaufman, S.; and Steiner, D.: Rift structure beneath the Michigan Basin from COCORP profiling. Geology 10, 645–649 (1982).

Burchfiel, B.S., and Royden, L.: Carpathian foreland fold and thrust belt and its relation to Pannonian and other basins. Am. Assoc. Petroleum Geologists Bull. 66, 1179–1195 (1982).

Burke, K.: Aulacogens and continental breakup. In: Donath, F.A. (Ed.): Ann. Rev. Earth Planet. Sci. 5, 371–396 (1977).

Burke, K., and Dewey, J.F.: Plume-generated triple junctions: Key indicators in applying plate tectonics to old rocks. Jour. Geology 81, 406–433 (1973).

Cowley, J.L.; Burruss, R.C.; and Holland, H.D.: Chemical weathering in central Iceland: an analog of pre-Silurian weathering. Science 165, 391–392 (1969).

Churkin, M., and Eberlein, G.D.: Ancient borderland terranes of the North American Cordillera: Correlation and microplate tectonics. Geol. Soc. America Bull. 88, 769–786 (1977).

Clague, D.A., and Stanley, P.F.: Petrologic nature of the oceanic Moho. Geology 5, 133–136 (1977).

Cobbing, E.J.: The Andean geosyncline in Peru and its distinction from Alpine geosynclines. Jour. Geol. Soc. (London) 135, 207–218 (1978).

Coleman, R.G.: Ophiolites, ancient oceanic lithosphere, 229 pp. Berlin: Springer-Verlag 1977.

Coleman, R.G.: Tectonic setting for ophiolite obduction in Oman. Jour. Geophys. Research 86, 2497–2508 (1981).

Coney, P.J., Jones, D.L., and Monger, J.W.H.: Cordilleran suspect terranes. Nature 288, 329–333 (1980).

Cordell, L.: Regional geophysical setting of the Rio Grande rift. Geol. Soc. America Bull. 89, 1073–1090 (1978).

Cotter, E.: The evolution of fluvial style with special reference to the Appalachian Paleozoic. In: Miall, A.D. (Ed.): Fluvial sedimentology. Canadian Soc. Petroleum Geologists Mem. 5, 361–384 (1978).

Crook, K.A.W.: Fore-arc evolution and continental growth: A general model. Jour. Struct. Geol. 2, 289–303 (1980).

Crook, K.A.W., and Feary, D.A.: Development of New Zealand according to the fore-arc model of crustal development. Tectonophys. 87, 65–107 (1982).

Crowell, J.C.: Origin of late Cenozoic basins in southern California. In: Dickinson, W.R. (Ed.): Tectonics and sedimentation. Soc. Econ. Paleon. Mineral. Spec. Pub. 22, 190–204 (1974a).

Crowell, J.C.: Sedimentation along the San Andreas fault, California. In: Dott, R.H. Jr., and Shaver, R.H. (Eds.): Modern and ancient geosynclinal sedimentation. Soc. Econ. Paleon. Mineral. Spec. Pub. 19, 292–303 (1974b).

Crowell, J.C.: The San Andreas fault system through time. Jour. Geol. Soc. (London) 136, 293–302 (1979).

Crowell, J.C.: Explanation to accompany geologic map of Ridge Basin, southern California. Soc. Econ. Paleon. Mineral., Pacific Section, 5 pp. (1982).

Curray, J.: Marine sediments, geosynclines and orogeny. In: Fischer, A.G., and Judson, S. (Eds.):

Petroleum and global tectonics, pp. 157–222. Princeton: Princeton Univ. Press 1975.

Curray, J., and Moore, D.G.: Sedimentary and tectonic processes in the Bengal deep-sea fan and geosyncline. In: Burke, C.A., and Drake, C.L. (Eds.): Geology of continental margins, pp. 617–627. New York: Springer-Verlag 1974.

Daly, R.A.: First calcareous fossils and the evolution of the limestones. Geol. Soc. America Bull. 20, 153–164 (1909).

Decker, R., and Decker, B.: Volcanoes, 244 pp. New York: W.H. Freeman 1981.

Degens, E.; von Herzen, R.P.; and Wong, H.: Lake Tanganiyika: Water chemistry, sediments, geological structure. Naturwissenschaften 58, 229–241 (1971).

Dickinson, W.R.: Relations of andesites, granites, and derivative sandstones to arc-trench tectonics. Rev. Geophys. Space Phys. 8, 813–860 (1970).

Dickinson, W.R.: Sedimentation within and beside ancient and modern magmatic arcs. In: Dott, R.H., Jr., and Shaver, R.H. (Eds.): Modern and ancient geosynclinal sedimentation. Soc. Econ. Paleon. Mineral. Spec. Pub. 19, 230–239 (1971).

Dickinson, W.R.: Plate tectonics and sedimentation. In: Dickinson, W.R. (Ed.): Tectonics and sedimentation. Soc. Econ. Paleon. Mineral. Spec. Pub. 22, 1–27 (1974).

Dickinson, W.R.: Mesozoic fore-arc basin in central Oregon. Geology 7, 166–170 (1979).

Dickinson, W.R.: Compositions of sandstones in circum-Pacific subduction complexes and fore-arc basins. Am. Assoc. Petroleum Geologists Bull. 66, 121–137 (1982).

Dickinson, W.R., and Seely, D.R.: Structure and stratigraphy of fore-arc regions. Am. Assoc. Petroleum Geologists Bull. 63, 2–31 (1979).

Dickinson, W.R., and Suczek, C.A.: Plate tectonics and sandstone compositions. Am. Assoc. Petroleum Geologists Bull. 63, 2164–2182 (1979).

Dickinson, W.R., and Valloni, R.: Plate settings and provenance of sands in modern ocean basins. Geology 8, 82–86 (1980).

Dickinson, W.R.; Helmold, K.P.; and Stein, J.A.: Mesozoic lithic sandstones in central Oregon. Jour. Sed. Petrology 49, 501–516 (1979).

Dickinson, W.R.; Ingersoll, R.V.; and Graham, S.A.: Paleogene sediment dispersal and paleotectonics in northern California. Geol. Soc. America Bull. 90(1), 897–898; 90(2), 1458–1528 (1979).

Donaldson, J.A., and Jackson, G.D.: Archean sedimentary rocks of North Spirit Lake area, northwestern Ontario. Canadian Jour. Earth Sci. 2, 622–647 (1965).

Elders, W.A.; Rex, R.W.; Meidav, T.; Robinson, P.T.; and Biehler, S.: Crustal spreading in southern California. Science 178, 15–24 (1972).

Emery, K.O.: The sea off southern California, 366 pp. New York: Wiley 1960.

Engel, A.E.J.: Geologic evolution of North America. Science 140, 143–152 (1963).

England, P.C., and McKenzie, D.P.: A thin viscous shell model for continental deformation. Roy. Astron. Soc. (England), Jour. 70, 295–321 (1982).

Eriksson, K.A.: Transitional sedimentation styles in the Moodies and Fig Tree groups, Barberton Mountain Land, South Africa: Evidence favoring an Archean continental margin. Precambrian Research 12, 141–160 (1980).

Fiske, R.S., and Matsuda, T.: Submarine equivalent of ash flows in the Tokiwa Formation, Japan, Am. Jour. Sci. 262, 76–106 (1964).

Florensky, C.P.; Basilevsky, A.T.; Krychkov, V.P.; Kusmin, R.O.; Nikolaeva, D.V.; Pronin, A.A.; Chernaya, I.M.; Tyuflin, Y.S.; Selivanov, H.S.; Naraeva, M.K.; and Roncat, L.B.: Venera 13 and Venera 14: Sedimentary rocks on Venus? Science 221, 57–59 (1983).

Fontaine, D.A.: Source and reservoir rock factors—Tertiary hydrocarbon accumulations and basins of Pacific coastal South America—part II. Houston Geol. Soc. Bull. 26, 8–14 (1983).

Franzinelli, E., and Potter, P.E.: Petrology, chemistry, and texture of modern river sands, Amazon River system. Jour. Geology 91, 23–39 (1983).

Fujioka, K.; Honze, E.; and Nasu, N.: Sandstone and siltstone beds overlying conglomerate at Deep Sea Drilling Project site 439, Japan Trench. In: Init. Rept. Deep Sea Drilling Proj. 56, 1067–1073 (1980).

Galloway, W.E.: Deposition and diagenetic alteration of sandstone in northeast Pacific arc-related basins: Implications for graywacke genesis. Geol. Soc. America Bull. 85, 379–390 (1974).

Garrels, R.M., and Mackenzie, F.T.: Evolution of sedimentary rocks, 394 pp. New York: W.W. Norton 1971.

Gilluly, J.; Reed, J.C., Jr.; and Cady, W.: Sedimentary volumes and their significance. Geol. Soc. America Bull. 81, 353–376 (1970).

Glennie, K.W.: Desert sedimentary environments. In: Developments in Sedimentology, Vol. 14, 222 pp. Amsterdam: Elsevier 1970.

Graham, S.A.; Dickinson, W.R.; and Ingersoll, R.V.: Himalayan–Bengal model for flysch dispersal in the Appalachian–Ouachita system. Geol. Soc. America Bull. 86, 273–286 (1975).

Graham, S.A.; Ingersoll, R.V.; and Dickinson, W.R.: Common provenance for lithic grains in Carboniferous sandstones from Ouachita Mountains and Black Warrior Basin. Jour. Sed. Petrology 46, 620–632 (1976).

Garrison, R.E., and Douglas, R.G., (Eds.): The Monterey Formation and related Siliceous rocks of California. Pacific Section, Soc. Econ. Paleon. and Mineral, 327 p. (1981).

Grandstaff, D.E.: A kinetic study of the dissolution of uraninite. Econ. Geology 71, 349–373 (1976).

Gregor, C.B.: Denudation of the continents. Nature 228, 273–275 (1970).

Grow, J.A.; Mattick, R.F.; and Schlee, J.S.: Multichannel seismic depth sections and interval velocities over outer continental shelf and upper continental slope between Cape Hatteras and Cape

Cod. In: Watkins, J.S.; Montadert, L.; and Dickerson, P.W. (Eds.): Geological and geophysical investigations of continental margins. Am. Assoc. Petroleum Geologists Mem. 29, 65–83 (1979).

Halbouty, M.T.; Meyerhof, A.A.; King, R.E.; Dott, R.H., Sr; Klemme, H.D.; and Shabad, T.: World's giant oil and gas fields, geologic factors affecting their formation, and basin classification, part II. Factors affecting formation of giant oil and gas fields. In: Halbouty, M.T. (Ed.), Geology of giant petroleum fields. Amer. Assoc. Petroleum Geologists Mem. 14, 528–555 (1970).

Hamilton, W.: Tectonics of the Indonesian region. Geol. Soc. Malaysia Bull. 6, 3–10 (1973).

Hamilton, W.: Subduction in the Indonesian region. In: Talwani, M., and Pittman, W.C. III (Eds.): Island arcs, deep sea trenches and back-arc basins. Am. Geophys. Union, M. Ewing Ser. 1, 15–31 (1977).

Hardenbol, J.; Vail, P.R.; and Ferrer, J.: Interpreting paleoenvironments, subsidence history and sea-level changes of passive margins from seismic and biostratigraphy. Oceanolog. Acta 5, 33–44 (1981).

Hawkins, J.W., Jr., and Whetten, J.T.: Graywacke matrix minerals: Hydrothermal reactions with Columbia River sediments. Science 166, 868–870 (1969).

Hempton, M.R.; Dunne, L.A.; and Dewey, J.F.: Sedimentation in an active strike-slip basin: Southeastern Turkey. Jour. Geology 91, 401–412 (1983).

Henderson-Sellers, A., and Cogley, J.G.: The Earth's early hydrosphere. Nature 298, 832–835 (1982).

Hoffman, P.F.: Evolution of an early Proterozoic continental margin: the Coronation geosyncline and associated aulacogens of the northwestern Canadian Shield. Royal Soc. London, Phil. Trans., Ser. A, 273, 547–581 (1973).

Hoffman, P.F.: Wopmay orogen: A Wilson cycle of early Proterozoic age in the northwest of the Canadian shield. In: Strangway, P.W. (Ed.): The continental crust and its mineral deposits. Geol. Assoc. Canada Spec. Paper 20, 523–549 (1980).

Hoffman, P.; Dewey, J.F.; and Burke, K.: Aulacogens and their genetic relationship to geosynclines, with a Proterozoic example from Great Slave Lake, Canada. In: Dott, R.H., Jr., and Shaver, R.M. (Eds.): Modern and ancient geosynclinal sedimentation. Soc. Econ. Paleon. Mineral. Spec. Pub. 19, 38–55 (1974).

Holland, H.D.: The chemical evolution of the atmosphere and oceans, 522 p. Princeton: Princeton Univ. Press (1984).

Holland, H.D., and Schidlowski, M. (Eds.): Mineral deposits and the evolution of the biosphere, 334 pp. New York: Springer-Verlag 1982.

Howells, M.F.; Leveridge, B.E.; Addison, R.; Evans, C.D.S.; and Nutt, M.J.C.: The Capel Curig volcanic formation, Snowdonia, North Wales: Variations in ash flow tuffs related to emplacement environment. In: Harris, A.L.; Holland, C.H.; and Leake, B.E. (Eds.): The Caledonides of the British Isles reviewed. Geol. Soc. London Spec. Pub. 8, 611–618 (1979).

Iijima, A.; Matsumoto, R.; and Tada, R.: Zeolite and silica diagenesis and sandstone petrography at sites 438 and 439 off Sanribu, northwest Pacific, Leg 57, Deep Sea Drilling Project. In: Init. Rept. Deep Sea Drilling Proj. 56, 57, 1005–1009 (1980).

Illies, J.H.: Ancient and recent rifting in the Rhinegraben. Geol. Mijnbouw 56, 329–350 (1977).

Ingersoll, R.V.: Petrofacies and petrologic evolution of the Late Cretaceous fore-arc basin, northern and central California. Jour. Geology 86, 335–353 (1978).

Ingersoll, R.V.: Initiation and evolution of the Great Valley forearc basin of northern and central California, U.S.A. Geol. Soc. London Spec. Pub. 10, 459–467 (1982).

Ingersoll, R.V., and Suczek, C.A.: Petrology and provenance of Neogene sand from Nicobar and Bengal fans, DSDP sites 211 and 217. Jour. Sed. Patrology 49, 1217–1228 (1979).

Irving, E.: Paleopoles and paleolatitudes of North America and speculations about displaced terranes. Canadian Jour. Earth Sci. 16, 669–694 (1979).

Jones, D.L.: Microplate tectonics of Alaska—significance for the Mesozoic history of the Pacific Coast of North America. In: Howell, D.G., and McDougall, K.A. (Eds.): Mesozoic paleogeography of the western United States. Soc. Econ. Paleon. Mineral., Pacific Section Symp. 2, 71–74 (1978).

Jones, D.L.; Cox, A.; Coney, P.J.; and Beck, M.E.: The growth of western North America. Scientific American 247 (5), 70–84 (1982).

Kane, M.F.; Hildenbrand, T.G.; and Hendricks, J.D.: Model for the tectonic evolution of the Mississippi embayment and its contemporary seismicity. Geology 9, 563–569 (1981).

Karig, D.E.: Evolution of arc systems in the Western Pacific. Ann. Rev. Earth Planet. Sci. 2, 51–55 (1974).

Karig, D.E., and Sharma, G.F. III: Subduction and accretion in trenches. Geol. Soc. America Bull. 86, 377–389 (1975).

Kay, M.: North American geosynclines. Geol. Soc. America Mem. 48, 143 pp. (1951).

Keto, L.S., and Jacobsen, S.B.: The causes of ^{87}Sr/^{86}Sr variations in seawater of the past 750 million years. Geol. Soc. America, Abstracts with Program 17, 628 (1985).

King, E.A., Jr.; Carman, M.F.; and Butter, J.C.: Mineralogy and petrology of coarse particulate material from lunar surface at Tranquillity Base. Science 167, 650–652 (1970).

Kingston, D.R.; Dishroon, C.P.; and Williams, P.A.: Global basin classification system. Am. Assoc. Petroleum Geologists Bull. 67, 2175–2193 (1983).

Kinsman, D.J.J.: Rift valley basins and sedimentary history of trailing continental margins. In: Fischer, A.G., and Judson, S. (Eds.): Petroleum and global tectonics, pp. 83–126. Princeton: Princeton Univ. Press 1975.

Klein, G. deVries: A sedimentary model for determining paleotidal range. Geol. Soc. America Bull. 82, 2585–2592 (1971).

Klein, G. deVries: Probable sequential arrangement of depositional systems on cratons. Geology 10, 17–22 (1982).

Klein, G. deVries, and Ryer, T.A.: Tidal circulation patterns in Precambrian, Paleozoic, and Cretaceous epeiric and mioclinal seas. Geol. Soc. America Bull. 89, 1050–1058 (1978).

Klemme, H.D.: Giant oil fields related to their geologic setting: a possible guide to exploration. Can. Petrol. Geol. Bull. 23, 30–66 (1975).

Klitgord, K.D., and Behrendt, J.C.: Basin structure of the U.S. Atlantic margin. In: Watkins, J.S.; Montadert, L.; and Dickerson, P.W. (Eds.): Geological and geophysical investigations of continental margins. Am. Assoc. Petroleum Geologists Mem. 29, 85–112 (1979).

Krynine, P.D.: Differentiation of sediments during the life history of a landmass (Abstract). Geol. Soc. America Bull. 52, 1915 (1941).

Krynine, P.D.: Petrology, stratigraphy and origin of the Triassic sedimentary rocks of Connecticut. Connecticut Geol. Survey Bull. 73, 239 pp. (1950).

Larson, W.A.; Menard, H.W.; and Smith, S.M.: Gulf of California: A result of ocean floor spreading and transform faulting. Science 161, 781–784 (1968).

Leggett, J.K. (Ed.): Sedimentation and tectonics on modern and ancient active plate margins. Geol. Soc. London Spec. Pub. 10, 576 pp. (1982).

LePichon, X., and Angelier, J.: The Hellenic arc and trench system: A key to the neotectonic evolution of the Eastern Mediterranean area. Tectonophys. 60, 1–42 (1979).

LePichon, X.; Francheteau, J.; and Bonnin, J.: Plate tectonics, 300 pp. New York: Elsevier 1973.

Li, Yuan-Hui: Geochemical mass balance among lithosphere, hydrosphere, and atmosphere. Am. Jour. Sci. 272, 119–137 (1972).

Lindsay, J.F.: Lunar stratigraphy and sedimentology, 302 pp. New York: Elsevier 1976.

Logatchev, N.A., and Florentsov, N.A.: The Baikal system of rift valleys. Tectonophys. 45, 1–13 (1978).

Lowe, D.R.: Archean sedimentation. Ann. Rev. Earth Planet. Sci. 8, 145–167 (1980).

Lundberg, N.: Development of forearcs of intraoceanic subduction zones. Tectonics 2(1), 51–61 (1983).

Maynard, J.B.: Composition of plagioclase feldspar in modern deep-sea sands: Relationship to tectonic setting. Sedimentology 31, 493–501 (1984).

McKenzie, D.P.: Some remarks on heat flow and gravity anomalies. Jour. Geophys. Research 72, 6261–6273 (1967).

Miall, A.D.: Principles of sedimentary basin analysis, 490 pp. New York–Heidelberg: Springer-Verlag 1984.

Mitchell, A.H.G., and Reading, H.: Sedimentation and tectonics. In: Reading, H. (Ed.): Sedimentary environments and facies, 557 pp. New York: Elsevier 1978.

Miyashiro, A.: Volcanic rock series and tectonic setting. Ann. Rev. Earth Planet. Sci. 3, 251–269 (1975).

Molnar, P., and Gray, D.: Subduction of continental lithosphere: Some constraints and uncertainties. Geology 7, 58–62 (1979).

Molnar, P. and Tapponier, P.: Cenozoic tectonics of Asia: Effects of a continental collision. Science 189, 419–426 (1975).

Moore, G.F.: Petrography of subduction zone sandstones from Nias Island, Indonesia, Jour. Sed. Petrology 49, 71–84 (1979).

Moore, J.C., and Allwardt, A.: Progressive deformation of a Tertiary trench slope, Kodiak Islands, Alaska. J. Geophys. Res. 85, 4741–4756 (1980).

Moore, G.F., and Karig, D.E.: Structural geology of Nias Island, Indonesia: Implications for subduction zone tectonics. Am. Jour. Sci. 280, 193–223 (1980).

Moore, J.C.; Watkins, J.S.; McMillew, K.J.; Bachman, S.B.; Leggett, J.K.; Lundberg, N.; Shipley, T.H.; Stephan, J.F.; Beghtel, F.W.; Butt, A.; Didyk, B.M.; Miitsuma, N.; Shepard, L.E.; and Stradner, H.: Facies belts of the Middle America trench and fore arc region, southern Mexico: Results from Leg 66 DSDP. In: Leggett, J.K. (Ed.): Sedimentation and tectonics on modern and ancient active plate margins. Geol. Soc. London Spec. Pub. 10, 77–96 (1982a).

Moore, J.C.; Watkins, J.S.; and Shipley, T.H.: Summary of accretionary processes, Deep Sea Drilling Project, Leg 66: Offscraping, underplating, and deformation of the slope apron. In: Init. Rept. Deep Sea Drilling Proj. 66, 825–836 (1982b).

Muffler, L.J.P., and Doe, B.R.: Composition and mean age of detritus of the Colorado River delta. Jour. Sed. Petrology 35, 384–399 (1968).

Muffler, L.J.P., and White, D.E.: Active metamorphism of upper Cenozoic sediments in the Salton Sea geothermal field and the Salton Trough, southeastern California. Geol. Soc. America Bull. 80, 157–182 (1969).

Murdmaa, I., and Kazakova, V.: Coarse-silt fraction mineralogy of Japan Trench sediments, Deep Sea Drilling Project Legs 56 and 57. In: Init. Rept. Deep Sea Drilling Project 56, 57, 1005–1009 (1980).

Murray, B.; Malin, M.C.; and Greeley, R.: Earthlike planets, 387 pp. New York: W.H. Freeman 1981.

Mutch, T.A.; Arvidson, R.E.; Head, J.W. III; Jones, K.L.; and Saunders, R.S.: The geology of Mars, 400 pp. Princeton: Princeton Univ. Press 1976.

Nanz, R.H., Jr.: Chemical composition of Precambrian slates with notes on the geochemical evolution of lutites. Jour. Geology 61, 51–64 (1953).

Nilsen, T.H., and Clarke, S.H., Jr.: Sedimentation and tectonics in the early Tertiary continental borderland of central California. U.S. Geol. Survey Prof. Paper 925, 64 pp. (1975).

Nur, A., and Ben-Avraham, Z.: Oceanic plateaus,

the fragmentation of continents, and mountain-building. Jour. Geophys. Research 87, 3644–3661 (1982).

Ojeda, H.A.D.: Estrutura, estratigrafia e evolucão das bacias marginais brasileiras. Rev. Brasileira Geociências 11(4), 257–273 (1981).

Palmason, G. (Ed.): Continental and oceanic rifts. Am. Geophys. Union, Geodyn. Ser. 8, 309 pp. (1982).

Pannella, G.; MacClintock, C.; and Thompson, M.N.: Paleontologic evidence of variations in length of synodic month since Late Cambrian. Science 162, 792–796 (1968).

Parsons, B., and Sclater, J.G.: An analysis of the variation of ocean floor bathymetry and heat flow with age. Jour. Geophys. Research 82, 803–827 (1977).

Perrodon, A.: Essai de classification des bassins sedimentaires. Sciences de la Terre, 16 197–227 (1971).

Pettijohn, F.J.: Archean sedimentation. Geol. Soc. America Bull. 54, 925–972 (1943).

Pettijohn, F.J.: Sedimentary rocks, 2nd Ed., 718 pp. New York: Harper 1957.

Pettijohn, F.J.: Chemical composition of sandstones. U.S. Geol. Survey Prof. Paper N-40S, 19 pp. (1963).

Pettijohn, F.J.: The Archean of the Canadian shield: A resume. In: Doe, B.R., and Smith, D.K. (Eds.): Studies in mineralogy and Precambrian geology. Geol. Soc. America Mem. 135, 131–149 (1972).

Poole, F.G., and Sandberg, C.A.: Mississippian paleogeography and tectonics of the western United States. In: Stewart, J.H.; Stevens, C.H.; and Fritsche, A.E. (Eds.): Paleozoic paleogeography of the western United States. Soc. Econ. Paleon. Mineral., Pacific Section, Los Angeles, pp. 67–85 (1977).

Potter, P.E.: Petrology and chemistry of modern big river sands. Jour. Geology 86, 423–449 (1978).

Potter, P.E.: South America and a few grains of sand. Part 1 Beach Sand. Jour. Geol. 94 (1986).

Potter, P.E., and Pettijohn, F.J.: Paleocurrents and basin analysis, 2nd Ed., 425 pp. New York: Springer-Verlag 1977.

Potter, P.E.; Maynard, J.B.; and Pryor, W.A.: Sedimentology of shale, 303 pp. New York–Heidelberg–Berlin: Springer-Verlag 1980.

Reyre, Dominique: Caractères pétroliers et évolution géologique d'une marge passive. Le cas du bassin Bas Congo-Gabon Bull. Centres Rech. Explor.-Prod. Elf Aquitaine 8, 303–332 (1984).

Robbins, E.I.: Geothermal gradients. In: Scholle, P.A. (Ed.): Geological studies of the COST GE-1 well, United States South Atlantic Outer Continental Shelf Area. U.S. Geol. Survey Circ. 800, 72–73 (1979).

Ronov, A.B.: Common tendencies in the chemical evolution of the earth's crust, ocean, and atmosphere. Geokhimiya 8, 715–743 (1964).

Ronov, A.B.: Probable changes in the composition of sea water during the course of geological time. Sedimentology 10, 25–43 (1968).

Ronov, A.B.: Evolution of the composition of the rocks and the geochemical processes in the Earth's sedimentary shell. Geokhimiya 2, 137–147 (1972).

Ronov, A.B., and Yaroshevsky, A.A.: Chemical composition of the Earth's crust. Am. Geophys. Union Mono. 13, 37–57 (1969).

Ronov, A.B.; Girin, Y.P.; Kazakov, G.A.; and Ilyukhim, M.N.: Comparative geochemistry of the platform and geosynclinal sedimentary rocks. Geokhimiya 8, 961–979 (1965).

Ronov, A.B.; Migdiscov, A.A.; and Barskaya, N.B.: Tectonic cycles and regularities in the development of sedimentary rocks and paleographic environments of sedimentation of the Russian Platform (an approach to a quantitative study). Sedimentology 13, 179–212 (1969).

Ronov, A.B.; Mikhailovskaya, M.S.; and Solodkova, I.I.: Evolution of the chemical and mineralogical composition of sandy deposits. In: Vinogradov, A.P. (Ed.): Chemistry of the Earth's crust 1, 201–252. Moscow: Akad. Nauk S.S.S.R. (1963).

Royden, L., Horvath, F., and Rumpler, J.: Evolution of the Pannonian Basin system: Part I, Tectonics; Part II, Subsidence and thermal history tectonics. Tectonics 2, 63–90, 91–137 (1983).

Royden, L.; Sclater, J.G.; and Von Herzen, R.P.: Continental margin subsidence and heat flow: Important parameters in formation of petroleum hydrocarbons. Am. Assoc. Petroleum Geologists Bull. 64, 173–187 (1980).

Schlee, J.S.; Behrendt, J.C.; Grow, J.A.; Robb, J.M.; Mattick, R.E.; Taylor, P.T.; and Lawson, B.J.: Regional geologic framework of northeastern United States. Am. Assoc. Petroleum Geologists Bull. 60, 926–951 (1976).

Schumm, S.A.: Speculations concerning paleohydrologic controls of terrestrial sedimentation. Geol. Soc. America Bull. 79, 1573–1588 (1968).

Schwab, F.L.: Evolution of the western continental margin, French-Italian Alps: Sandstone mineralogy as an index of plate tectonic setting. Jour. Geology 89, 349–368 (1981).

Sclater, J.G., and Francheteau, J.: The implications of terrestrial heat flow observations on current tectonic and geochemical models of the crust and upper mantle of the earth. Royal Astron. Soc., Geophys. Jour. 20, 509–537 (1970).

Shanmugam, G., and Moiola, R.J.: Eustatic control of turbidites and winnowed turbidites. Geology 10, 231–235 (1982).

Sheridan, R.E.: Sedimentary basins of the Atlantic margin of North America. Tectonophys. 36, 113–132 (1976).

Siever, R.: Plate tectonic controls on diagenesis. Jour. Geology 87, 127–155 (1979).

Siever, R., and Kastner, M.: Mineralogy and petrology of some Mid-Atlantic Ridge sediments. Jour. Marine Res. 25, 263–278 (1967).

Sleep, N.: Thermal effects of the formation of Atlantic continental margins by continental break-up.

Royal Astron. Soc., Geophys. Jour. 24, 325–350 (1971).

Sloss, L.L.: Orogeny and epeirogeny: The view from the craton. N.Y. Acad. Sci. Trans., Ser. II, 28, 579–587 (1966).

Sloss, L.L., and Speed, R.C.: Relationships of Cratonic and continental margin tectonic episodes. In: Dickinson, W.R. (Ed.): Tectonics and sedimentation. Soc. Econ. Paleon. Mineral. Spec. Pub. 22, 98–119 (1974).

Stille, H.: Present tectonic state of the earth. Am. Assoc. Petroleum Geologists Bull. 20, 849–880 (1936).

Tankard, A.J.; Jackson, M.P.A.; Erikson, K.A.; Hobday, D.K.; Hunter, D.R.; and Minter, W.E.L.: Crustal evolution of southern Africa, 523 pp. New York: Springer-Verlag 1982.

Taylor, J.C.M.: Sandstone diagenesis, introduction. Jour. Geol. Soc. (London) 135, 3–5, and following papers, 6–135 (1978).

Turner, C.C., and Walker, R.G.: Sedimentology, stratigraphy, and crustal evolution of the Archean greenstone belt near Sioux Lookout, Ontario. Canadian Jour. Earth Sci. 10, 817–845 (1973).

Uyeda, S.: Subduction zone and back arc basins—a review. Geol. Rundschau 70, 552–569 (1981).

Uyeda, S., and Miyashiro, A.: Plate tectonics and the Japanese Islands: A synthesis. Geol. Soc. America Bull. 85, 1159–1170 (1974).

Vail, P.R.; Mitchum, R.M., Jr.; Todd, R.G.; Widmier, J.M.; Thompson, S., III; Sangree, J.B.; Bubb, J.N.; and Hatlelid, W.G.: Seismic stratigraphy and global sea-level changes. In: Payton, C.A. (Ed.): Seismic stratigraphy—applications to hydrocarbon exploration. Am. Assoc. Petroleum Geologists Mem. 26, 49–212 (1977).

Valloni, R., and Maynard, J.B.: Detrital modes of recent deep-sea sands and their relation to tectonic setting: A first approximation. Sedimentology 28, 75–83 (1981).

Valloni, R., and Mezzadri, G.: Compositional suites of terrigenous deep-sea sands of the present continental margins. Sedimentology 31, 353–364 (1984).

Van de Kamp, P.C.; Keake, B.E.; and Senior, A.: The petrography and geochemistry of some Californian arkoses with application to identifying gneisses of sedimentary origin. Jour. Geology 84, 195–212 (1976).

Veizer, J.: Sedimentation in geologic history: Recycling vs. evolution or recycling with evolution. Contrib. Mineral. Petrol. 38, 261–278 (1973).

Vinogradov, A.P., and Ronov, A.B.: Composition of the sedimentary rocks of the Russian platform in relation to the history of its tectonic movements. Geokhimiya 6, 3–24 (1956).

Von Huene, R., and Aubouin, J.: Leg 67: The Deep Sea Drilling Project Mid-America Trench transect off Guatemala. Geol. Soc. America Bull 91, 421–432 (1980).

von Rad, U.; Hinx, K.; Sarnthein, M.; and Seibold, E. (Eds.): Geology of the northwest African continental margin, 704 pp. New York: Springer-Verlag 1982.

Walker, J.C.G.: Possible limits on the composition of the Archaean ocean. Nature 302, 518–520 (1983).

Walker, R.G., and Pettijohn, F.J.: Archean sedimentation: Analysis of the Minnitaki basin, northwestern Ontario. Geol. Soc. America Bull. 82, 2099–2130 (1971).

Wallace, H.: Keeweenawan geology of the Lake Superior basin. In: Campbell, F.H. (Ed.): Protozoic basins of Canada. Geol. Survey of Canada, p. 399–417 (1981).

Ward, P., and Stanley, K.O.: The Haslam formation: A Late Santonian–Early Campanian fore arc basin deposit in the Insular Belt of southwestern British Columbia and adjacent Washington. Jour. Sed. Petrology 52, 975–990 (1982).

Warner, J.L.: Sedimentary processes and crustal cycling on Venus. In: Boynton, W.V. and others (Eds.): Proc., 13th Lunar Planet. Sci. Conf., March 15–19, 1982, Houston, Texas, part 2, JGR. Jour. Geophys. Res. B, 88, Suppl. (2) pp. A495–A500 (1983).

Watchorn, M.B.: Fluvial and tidal sedimentation in the 3,000 Ma Mozaan Basin, South Africa. Precambrian Research 13, 27–42 (1980).

Weaver, C.E.: Potassium, illite, and the ocean. Geochim. et Cosmochim. Acta 31, 2182–2196 (1967).

Whetten, J.T.: Sediments from the lower Columbia River and origin of graywacke. Science 152, 1057–1058 (1966).

Windley, B.F. (Ed.): The early history of the earth: Based on the proceedings of a NATO Advanced Study Institute held at the University of Leicester, 5–11 April, 1975, 619 pp. London–New York: John Wiley and Sons 1976.

Worzel, J.L., and Burk, C.A.: Margin of Gulf of Mexico. Am. Assoc. Petroleum Geologists Bull. 62, 2290–2307 (1978).

Ziegler, P.A.: North Sea basin history in the tectonic framework of northwestern Europe. In: Woodland, A.C. (Ed.): Petroleum and the continental shelf on northwest Europe pp. 131–162. New York: John Wiley and Sons 1975.

Zoback, M.D.; Hamilton, R.M.; Crone, A.J.; Russ, D.P.; McKeown, F.A.; and Brockman, S.R.: Recurrent intraplate tectonism in the New Madrid seismic zone. Science 209, 971–976 (1980).

APPENDIX

Petrographic Analysis of Sandstones

Introduction

The petrographic analysis of sediments, sandstones in particular, became an organized discipline with the development of thin-section techniques and the polarizing microscope—a development attributed to Henry Clifton Sorby. Sorby was making thin sections in 1849, published a paper on the microscopical structure of "calcareous grit" in 1851, and his paper in 1880 on the non-calcareous stratified rocks was a major milestone in the thin-section analysis of sandstones. Folk (1965) has ably summarized Sorby's petrographic contributions as has Summerson (1978). Until shortly before World War II, most students of sedimentary rocks, unlike those of igneous and metamorphic rocks, failed to "follow through" on Sorby's auspicious beginnings. One brilliant exception was Lucien Cayeux (1929). In more recent years, however, the thin-section analysis of sedimentary rocks has become commonplace and is now a fully exploited tool for research.

The prime object of the study of a thin section is, or should be, the reading of *rock history*. The microscope is the most useful, general method for close study of the mineral composition, fabric, and general makeup of a rock. Such close study is a necessary complement to field studies in interpreting the origin of sands and sandstones.

The study of sand and sandstone in the laboratory has proceeded in other directions also, and there has been a proliferation of methods for study of grain size, grain shape and roundness, porosity and permeability, and the like. Many of these methods, however, are applicable only to unconsolidated deposits and, useful as they may be, the thin-section approach remains the single most effective means of investigation of sandstones in the laboratory. Its effectiveness, however, depends on the imagination and skill of the operator.

Rock Description and Analysis

Rock description and analysis are based on study of outcrops, cores, hand specimens, and thin sections. Thin sections can be easily prepared for unconsolidated sands also (Middleton and Kraus, 1980). Emphasis here is placed on hand-specimen and thin-section study of either core or outcrop samples.

The art of rock description and analysis is learned by doing and by the study of published examples. Good descriptions of rocks of all types have been published in Bulletin 150 of the U.S. Geological Survey. A number of abbreviated descriptions are given in Grout (1932, pp. 22–28). See also the approach to rock descriptions of Ferm and Weisenfluh (1981), an approach that uses colored pictures of rock types with a code name and number.

The use of a well-designed petrographic form develops a regular pattern of description and maximizes the efficiency and effectiveness of any microscopic study. The level of description, however, may vary widely—from a concise paragraph based largely on qualitative observation and semiquantitative visual estimates to a three- or four-page typed report based on counts of 200 to 500 or more grains; or in place of a written report, the data may be directly entered on a computer disk for later processing.

Choice between semiquantitative and quantitative estimates depends on the investigator's objectives, his judgment, and the available time. If very large numbers of samples are to be studied and little time is available for the task, semiquantitative estimates must suffice. Mineral percentages can be estimated by compari-

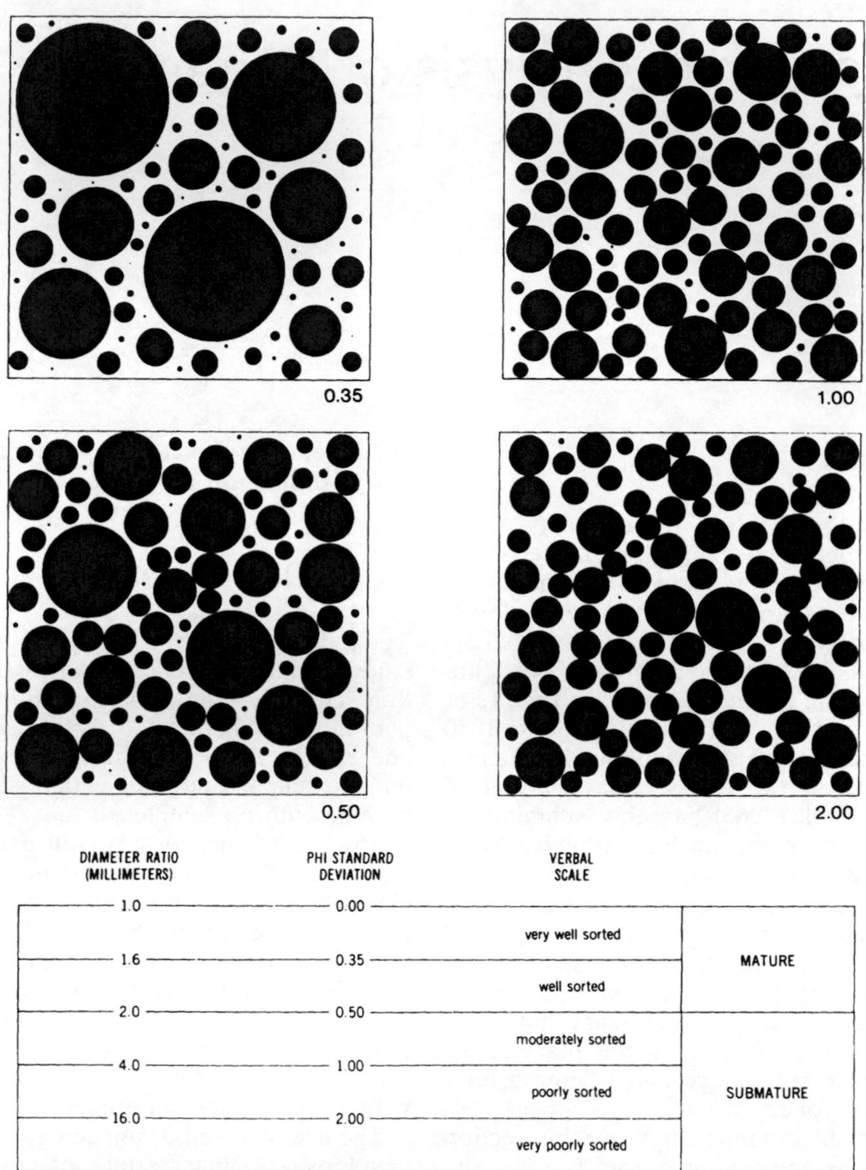

FIGURE A-1. Sorting images from Harrell (1984, Figs. 3, 4, 5, and 6) and sorting classes from Folk (1968, p. 102). Published by permission of the authors and the Society of Economic Paleontologists and Mineralogists.

son charts with a reticle (Terry and Chilingar, 1955). Sorting can also be estimated by comparison charts (Fig. A-1). Roundness of individual grains is always so estimated (Fig. A-2). Counts of 50 to 100 grains have usually been made to estimate either average grain roundness or percentage of angular grains.

But the earmark of the modern petrographer is the point count of 200 to 500 grains per slide for estimates of composition. Quantitative estimates are needed for many petrographic classifications and for most subsequent statistical analysis. By using binomial and Poisson confidence charts (Pearson and Hartley, 1954, Tables 40 and 41), one can determine the reliability of an estimate based on a given number of counts. Van der Plas and Tobi (1965) provide a comparable chart (Fig. A-3). Dennison (1962) also gives useful charts, all of which are based on the normal and binomial distributions. In practice, it is not uncommon for a petrographer to utilize semiquantitative estimates for some of the petrographic variables and reserve system-

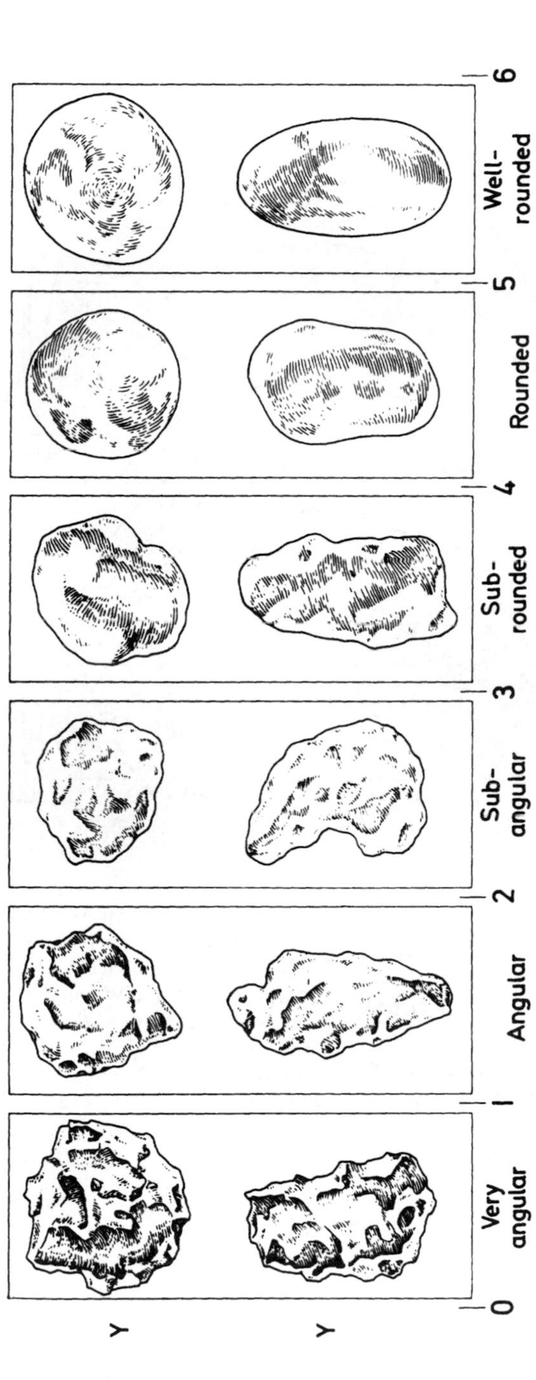

FIGURE A-2. Roundness images and classes. Columns show grains of similar roundness but different sphericity. (Redrawn from Powers, 1953, Fig. 1).

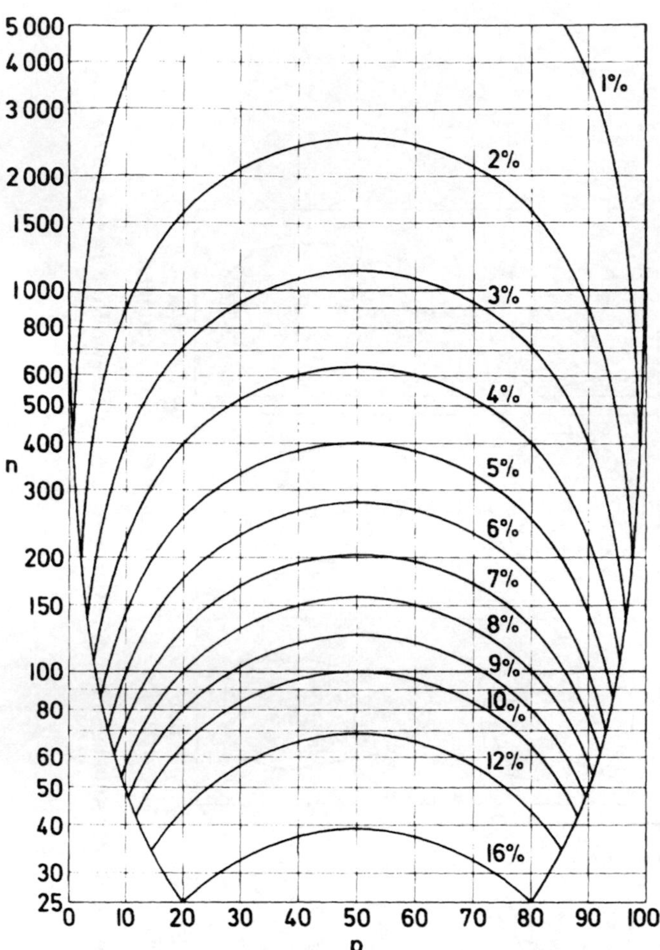

FIGURE A-3. Ninety-five percent confidence limits for mineral proportions, where n is total number of grains counted and p is the estimated proportion of a particular mineral. Curved contours in percent give confidence limits. Worked example: n is 500, p is 28 percent, and the confidence limit is 4 percent so that in repeated sampling the true proportion will lie within 24 and 32 percent. (Modified from van der Plas and Tobi, 1965, Fig. 1).

atic point counting for the one or two most significant ones. An automatic electronic stage and counter is both a convenience and an efficient time saver and back scatter election image analysis offers promise of automated point counts (Dilko and Graham, 1985).

Sorting, angularity, and clay content define textural maturity, which should be specified. Table A-1 gives a flow chart for this procedure. Sorting can be estimated by comparison with Fig. A-1 or one can estimate it by determining the ratio of two representative grain diameters: the diameter in millimeters of a grain of the 84th percentile and the diameter of a grain of the 16th percentile. Conversion to phi units, subtraction, and division by two yields the sorting. Clay of authigenic origin should be ignored, when clay content is determined.

To facilitate petrographic analysis, we have included two petrographic forms. Table A-2 is a skeletal form and Table A-3 a very detailed one modified from Folk (1968, pp. 133–38). We recognize, of course, that in many studies all the items in these report forms will not be appropriate and that there may be special comments that we have not included. The detailed form emphasizes, however, the comprehensive nature of a full description. The analysis of the Trivoli Sandstone presented below follows this form.

We have included a few amplifying remarks for the most effective use of these forms. After looking at the hand specimen with the naked eye, and hand lens, or binocular, it is best to scan the thin section with low power to appraise its general characteristics. This should be followed by a grain size analysis of its terrigenous

TABLE A-1. *Textural maturity flow chart. (Modified from Folk, 1968, p. 102).*

STEP 1	Clay content (micaceous material less than 30 μ, excluding authigenic material) a) If greater than 5 percent, sand is *immature*. b) If less than 5 percent, determine sorting.
STEP 2	Sorting (See Fig. A-1) a) If sorting is greater than 0.5ø (diameter ratio over 2.0), sand is *submature*. b) If sorting is less than 0.5ø determine roundness.
STEP 3	Roundness (See Fig. A-2) a) If sand-size grains are subangular to angular (3.0 or less on the Powers scale, Table 3-3), sand is *mature*. b) If roundness exceeds 3.0, sand is *supermature*.

TABLE A-2. *Short petrographic report for sandstones.*

HAND SPECIMEN
Color, grain size, sorting, induration, bedding, etc. and field name.

THIN-SECTION DESCRIPTION

Abstract:	Digest and condense all the petrography and summarize in 25 words.
Texture:	Modal size, sorting, and nature of grain-to-grain contacts. Bedding. Percent sand, silt, and clay. Roundness.
Mineralogy:	Give percent of terrigenous, orthochemical (precipitated cement) and allochemical (transported grains formed within the basin of deposition) material plus description and amounts of different types of terrigenous debris. A brief paragraph for each constituent.
Interpretation:	Character of source area plus type of transportation and character of depositional basin, if possible. Diagenesis.

GENERAL COMMENT
Always keep description and interpretation separate. At times you may estimate abundances with 100 point counts or by using comparison charts.

components under medium or high power. Size analysis is commonly the best way to become acquainted with both the texture and mineralogy of the section. Counts of as low as 50 grains to 100 grains are satisfactory for the mean and sorting for many problems, or images can be used (Beard and Weyl, 1973).

Modal mineralogical analysis follows. Counts of 200 to 500 grains are generally sufficient for all but minor constituents (less than 1 percent) using high or medium power. It is best to estimate proportions of varietal types by separate counts. The modal analysis should be always supplemented by a qualitative observation of each component: its median and size range, angularity, inclusions, alteration products, etc. Such descriptions should avoid useless detail such as, for example, enumeration of ordinary optical properties, and should instead be directed to useful special features. It is, for example, pointless to note that the quartz is uniaxial but it may be significant to observe that quartz is composite and well rounded. The relation between grain size and composition and the distribution of cementing agents within the section should always be observed. The textural relations of one mineral to another are most important and should be recorded.

Any comprehensive study of thin sections should be supplemented by X-ray analysis of interstitial matrix, (Wilson and Clark, 1978) and microscopic heavy mineral analysis. Thin sections should always be stained for feldspar (Houghton, 1980), but electron probe analysis is desirable for feldspar and other mineralogical compositions. If a chemical bulk analysis is available, it should certainly be cited and related to petrographic features. Any many more methods can be utilized (Table A-4).

Methods vary depending upon the question to be asked. For provenance studies, outcrop samples are satisfactory, and both the microprobe and SEM-EDX can help significantly with more precise mineralogical determinations, as is well shown by Maynard's (1984) study of the detrital plagioclase feldspars. Diagenetic and reservoir studies of sandstones are best made, however, only with subsurface samples and, if possible, only from the plugs from which porosity and permeability were measured. By so doing, diagenetic mineralogical and textural features such as cement history

TABLE A-3. Detailed petrographic report form. (Modified from Folk, 1968, pp. 133–38).

I. SAMPLE IDENTIFICATION

Formation name, age, and precise geographic location of outcrop and full well name, location, and depth of a core sample.

II. FIELD RELATIONS OR SUBSURFACE DATA

Outcrop or core thickness and position of outcrop or core with respect to formational boundaries: associated lithologies, bedding characteristics, sedimentary structures, fossils, deformation, and mineralization.

III. HAND SPECIMEN DESCRIPTION

Concise, simple description including a field name consistent with petrographic analysis. Include color, grain size, sorting, roundness, mineral composition, fossils, induration, sedimentary structures, bedding, tectonic deformation, and weathering.

IV. THIN-SECTION DESCRIPTION

A. Abstract

Brief comments, perhaps 50 words. Prepare only after all other aspects of the report are complete. Include rock name, summarized modal analysis, interpretation, possible economic significance, and relevance to scientific problems.

B. Texture

1. Fabric

Grain-to-grain relations, grain orientation, cementation, and porosity.

2. Grain size

Specify mean, median, and range sorting and percent of gravel, sand, silt, and clay. Plot cumulative cure on log-probability paper if 100 or more grains are counted.

3. Angularity and sphericity

Describe and comment on how they vary with size.

4. Textural maturity

C. Porosity, permeability, and water saturation from subsurface plugs

D. Mineralogy

Separate into terrigenous, allochemical, and orthochemical constituents and give percent of each. Reproduce modal analysis in a compact table.

V. INTERPRETATION

Here one integrates the data from the thin section with all other evidence: Field or subsurface observations, chemical or microprobe analysis, and the literature. Remember that the best interpretation is one that uses all the relevant facts and exploits their significance to the fullest—*stopping just short of the point of unjustified conclusions.*

A. Source area

Estimate lithologic composition and maturity of sandstones in source plus tectonic state and weathering, location, and distance from depositional site. Plot on QFR or other diagram. How many sedimentary cycles involved? Did more than one source region contribute? Plate tectonic significance?

B. Depositional basin

Estimate nature and strength of currents and water depth. Identify environment of deposition as fully as evidence permits. If marine, how far from shore? Utilize biotic constituents and trace fossils as much as possible. Tectonic setting of basin.

C. Diagenesis

State and interpret diagenetic history and emphasize the pore system. Are diagenetic effects major or minor? How are they related to porosity and permeability? Ratio of primary to secondary porosity?

VI. ECONOMIC IMPORTANCE

Discuss economic importance and give industrial name. Comment on possible market value and problems of development.

VII. BEARING ON SCIENTIFIC PROBLEMS

How does the interpretation of this sample relate to the historical development of the sandstone body or the sedimentary basin? To paleotectonics? To thermal history of basin?

---

and pore distribution can be *directly related* to porosity, permeability, and water saturation using straightforward statistical methods. See Keelan (1982) for a summary of the petrophysical study of sandstone cores and Sneider *et al.* (1983) on how to estimate some laboratory measurements with careful binocular observation. Impregnation under pressure with blue dye is always essential for thin-section study of pore systems. Blue fluorescent dye is also of value (Yangus and Dravis, 1985).

Good report forms clearly separate description from interpretation and one should always strive to maintain this separation. But how does one make the interpretation? Table A-5, modified from van Andel (1958, Table 1), relates petrographic and other properties to the major objectives of the study of sand and sandstone. This table summarizes the essential information of much of the previous material of the book. Table A-5 emphasizes that the fullest interpretation requires some knowledge of the size, shape, and orientation of the sandstone body, its associated sediments, and position of the sandstone body in the basin. This underscores the fact that petrographic interpretation is greatly enhanced by other information about the sandstone body from which the sample was obtained.

Table A-5 also underscores another important point—most petrographic properties are the response to the joint effect of both inheritance and depositional environment. For example, angularity of quartz may be related to the maturity of the sandstones in the source area as well as the effectiveness of rounding in the last depositional environment. Nor should later

TABLE A-4. Instrumental methods for the study of sandstones.

Method	Information obtained
Standard petrographic microscope	Thin sections and heavy separates; QFR, roundness, % matrix, size, fabric, and cement stratigraphy. For subsurface studies use samples impregnated with dye to define average size and sorting of pores and, if possible, study thin sections from plugs to correlate petrologic parameters with porosity, permeability, and water saturation.
Microprobe	Chiefly feldspar types and composition of inclusions and of minerals in rock fragments.
SEM	Mineralogy and texture of detrital and authigenic phases. Surface textures of grains. On subsurface sample, examine interiors of pores and their fill for cement stratigraphy and compare and contrast samples of high and low permeability.
SEM-EDX	Exact automated mineralogy and porosity.
Cathodo-luminescence	Precise determination of overgrowths and enhanced cement stratigraphy.
Isotopes	
Oxygen ($^{18}O/^{16}O$)	Salinity and temperature of water from which cements and clays were precipitated and exchange with formation waters.
Carbon ($^{13}C/^{12}C$)	Carbonate cements from oxidation of organic carbon or primary carbonate.
Strontium	Detrital provenance, ultimate age, and recycling.
Radioactive ($^{87}Rb/^{87}Sr$; $^{40}K/^{40}A$; $^{147}Sm/^{143}Nd$)	Age of detrital minerals.
Porosity, permeability and water saturation determined from core plugs	Fundamental to all reservoir studies. Study petrography on core plugs, if possible; use micropermeameter to investigate small-scale variations of permeability.
Geophysical logs	Essential for subsurface studies to define porosity and permeability as well as sandbody geometry and both local and regional distribution. Many types now routinely available; gamma-ray, neutron, density, sonic, self-potential, resistivity, dipmeter, etc.

diagenesis be ignored. In short, a *history* is involved and the petrographer's problem is to decipher where and when the particular effect took place: in the source area, in transport, at the site of deposition, or afterward?

Although quantitative petrography is essential, there is not substitute for penetrating qualitative observation. It *alone* is the key to what may be worthwhile to count.

A systematic, comprehensive question set can be helpful (Sedimentation Seminar, 1978).

A Comprehensive Petrographic Analysis: The Trivoli Sandstone of Southern Illinois

I. Sample Identification

Trivoli Sandstone Member of the Modesto Formation, McLeansboro Group, Pennsylvanian System. Basal member of the Trivoli Cyclothem and named as such by Wanless (1931). Outcrops in parts of western Illinois, especially Peoria and adjacent counties, and southern Illinois, especially Williamson and adjacent counties. Known in the subsurface over wide areas of the Illinois Basin. Description is a composite from samples collected from outcrops and the subsurface in the Illinois Basin.

II. Field Relations

Maximum sandstone thickness is 157 ft (48 m) in Franklin County. Thick or channel phase generally varies from 40–100 ft (12–31 m) and mainly in erosional valleys cut in underlying beds. Thin or sheet phase, ranging from 0–40 ft. (0–12 m), is distributed more uniformly. Sandstone has a sharp unconformity separating it from the gray marine shale of the West Franklin Limestone of the underlying cyclothem. Grades upward into silty shale and then to the underclay below the Chapel (No. 8) coal bed. Basal part of sandstone may be conglomeratic with locally derived pebbles.

On the outcrop the sandstone weathers into large slabs and flags, and is uniform in appearance. Ripple mark and crossbedding are characteristic throughout. Crossbedding varies from planar to large troughs. Mean direction of crossbedding is 158° based on 81 observations (Andresen, 1961, p. 25). Plant fossils are characteristic, large fragments of abraded stems and twigs occur in the conglomerates, and smaller particles, not easily identifiable, in the finer-grained beds. Beds are gently deformed, conformable with the structure of other Pennsylva-

TABLE A-5. Objectives of study, relevant petrographic properties, and their interpretation. (Modified from van Andel, 1958, Table 1).

Objective	Property	Remarks
Source	Roundness	Generally modified but little in a single cycle and hence useful in assessing character of source rocks. Rounded quartz generally, but not always, implies recycling.
	Directional structures	Regional mapping outlines current system in basin and thus helps locate source region. May also have some environmental significance.
	Mineralogical maturity	Mature mineralogy commonly reflects cratonic source, recycling, and appreciable weathering; immature mineralogy indicates uplift and rapid erosion of crystalline rocks. Absence of polycrystalline quartz indicates preexisting sediments.
Transportation	Grain size	Generally not diagnostic of environment except for presence or absence of gravel; thick conglomeratic sections indicate strong gradients and proximity to source, but pebbly sands may be transported hundreds of kilometers by large streams. Vertical size profile contains environmental information.
	Mineralogy	Abrasion minimal for sand in all but steepest mountain streams.
Depositional environment	Associated sediments	Knowledge of lateral equivalents plus preceding and following units essential for maximum interpretation.
	Fossils	Establish environment of deposition.
	Mineralogy	Argillaceous rock fragments may be eliminated on beaches with high wave energy; or, if buried, may result in pseudomatrix, enhance physical compaction, and reduce permeability. Detrital feldspar may be dissolved later to enhance secondary porosity.
	Vertical sequence	Vertical variation of grain size, stratification, and lithology is essential key to identification of depositional environment.
	Sandbody shape and orientation	Map pattern of sand accumulation tells much about process and environment of sand dispersal in basin.
	Textural maturity	Clay content and sorting of framework reflect final depositional environment of sand dispersal in basin.
Post-depositional history	Authigenic minerals	Early and late cements reduce porosity and permeability.
	Permeability and porosity	Rough overall measure of the extent of diagenesis. Secondary porosity favored by early carbonate cement and detrital feldspar.

nian beds in the basin. Dips are so low that they are rarely accurately measurable in the field.

III. Hand Specimen Description

The sandstone, in outcrop, is brown to reddish brown, but in the subsurface it is light gray. It is uniformly fine grained; and well sorted, with shaly, micaceous (muscovite), and carbonaceous partings. It can readily be identified as a lithic arenite by the abundance of non-quartz grains and the lack of much matrix. Cementation is normally moderate (can easily be disaggregated without breaking grains), but some specimens show extensive cementation (disaggregation difficult and breaks grains). Grains appear angular; sparkling facets of euhedral quartz overgrowths and calcite cement crystals hide detrital outlines. Small-scale crossbedding and ripples can be seen in some specimens.

IV. Thin-Section Description

A. *Abstract*

A typical moderately well-sorted, sublithic to lithic arenite characteristic of many continental Coal Measures (Fig. A-4). Mono- and polycrystalline quartz are subangular to subrounded and commonly form about 50 percent of the sand, feldspar averages 4 percent, and rock fragments about 9 percent. Clay and badly squashed rock fragments (7 percent) plus calcite and ferroan dolomite (11 percent) bind

FIGURE A-4. Typical appearance of Trivoli Sandstone as determined by thin section, partially crossed nicols, × 100.

the framework together. Outcrop and extensive subsurface mapping indicate that the Trivoli Sandstone was mostly derived from the Appalachian mobile belt and was deposited in the Illinois Basin as a coastal plain–deltaic complex.

B. Texture

1. Fabric is typical of lithic arenites. Most of the rock is supported structurally by the quartz–feldspar–chert framework, but deformation of argillaceous rock fragments has squashed the latter to "clay matrix" which intrudes into pore space and surrounds competent framework grains. Grain contacts are more numerous than expected for a well-sorted sand, indicating some compaction and rearrangement of grains accompanied by the squashing of rock fragments.

Because most grains tend to be equant, a preferred shape orientation is not obvious, but orientation of large detrital mica flakes is excellent and parallel to bedding. Quantitative studies of a similar sandstone lower in the Pennsylvanian, the Pleasantview of western Illinois (Rusnak, 1957), show that there is both a general preferred direction of elongate grains in a section parallel to bedding and imbrication normal to the bedding.

The sandstone is cemented by a combination of the clay matrix with precipitated mineral cements of calcite, ferroan dolomite, siderite, quartz, carbonate, and, on the outcrop, hematite and limonite. Cement, most of which is carbonate, averages about 20 percent and this, coupled with clay matrix, reduces porosity to about 20 percent from an original porosity that must have been about 40 to 45 percent.

2. Grain size distribution is that of a moderately well-sorted fine to medium-grained sand (Fig. A-5). Size distributions are skewed to fine sizes, because of the abundance of clay. The break in the size curve suggests a mixture of two size distributions. If most of the clay were assumed to come from the degradation of argillaceous rock fragments and removed, the sediment would be very well-sorted, fine- to medium-grained sand, corresponding to the fraction coarser than about 3 in Figure A-5.

3. Large numbers of framework grains are subrounded to rounded, but roundness is difficult to estimate in many grains because of lack of preserved border between detrital core and either overgrowth or replacing cement, but visual roundness of quartz is estimated at 0.25;

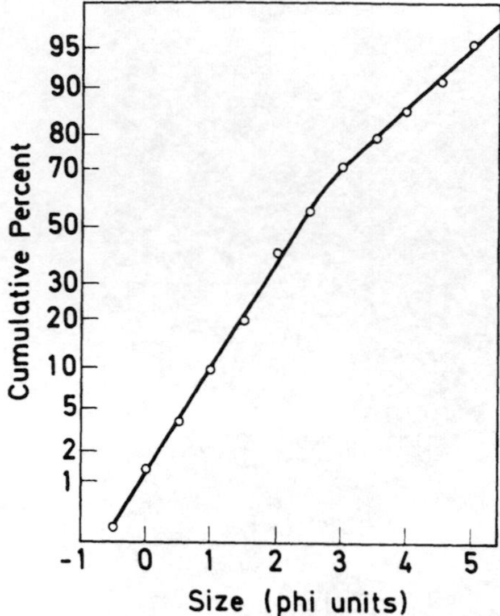

FIGURE A-5. Grain size analysis of Trivoli Sandstone as determined by thin section (300 counts). Note two populations on arithmetic probability paper.

TABLE A-6. *Some typical modal analyses of the Trivoli sandstone (Andresen, 1961, Table 2).*

Minerals	Samples		
	1E	C15B	C21A
Quartz			
Monocrystalline	23	38	24
Polycrystalline	28	23	27
Chert	—	T	T
Feldspar			
Microline	T	T	T
Plagioclase	—	1	—
Untwinned	1	5	2
Mica	3	3	1
Rock fragments			
Metamorphic	4	7	8
Sedimentary	4	2	5
Clay matrix	7	5	5
Cement	23	16	21
Average quartz	0.08	0.18	0.10

the feldspar average is 0.5. Sphericity is in the range 0.60–0.85 and is highly variable because of the heterogeneous composition. Coarser samples tend to be better rounded and more spherical.

4. The Trivoli is texturally immature as indicated by the sorting and roundness. If most of the clay matrix were not counted as detrital, then many samples would be texturally mature.

C. *Mineralogy (Table A-6)*

1. Terrigenous detritus

a) Quartz is separable into mono- and polycrystalline, chert, and secondary overgrowths. Cathodo-luminescence shows that some of the monocrystalline quartz has low luminescence and may be of relatively low-temperature origin—sedimentary to low-grade metamorphic, or low-temperature hydrothermal. Authigenic overgrowth quartz exceeds 5 percent in only a few samples; in most it is about 1 percent. Second-cycle grains are present in some samples.

b) Chert is present in small or trace amounts in most samples. A few grains show faint suggestions of fossil outlines.

c) Feldspars present are sodic varieties of plagioclase, orthoclase, microcline, and a few grains of anorthoclase. Some grains are kaolinized, but it is not certain that they have been altered in place. More microcline grains are altered than plagioclase. Most orthoclase is untwinned. The ratio of plagioclase to K-feldspar is 1.7 based on X-ray diffraction patterns of the $>2\mu$ fraction.

d) Rock fragments are dominantly argillaceous varieties, both low-grade metamorphic and sedimentary, which grade into clay matrix. Many appear deformed and have corroded edges. Many siltstone and shale fragments look like Pennsylvanian rocks lower in the section and may be of local origin. A few limestone fragments were found.

e) Mica is almost all muscovite, with small amounts of biotite and chlorite, and occurs in shreds and plates, many bent and broken, many deformed around quartz grains. Mica flakes are all oriented parallel to bedding, and are very abundant in thin shaly partings.

f) Clay minerals were analyzed by X-ray diffraction after separation of the $<2\mu$ fraction. This fraction probably includes not only matrix but some of the rock fragment clays as well. Clays present are kaolinite, illite, chlorite, and a mixed-layer clay. Kaolinite is well crystallized and some can be seen to be an authigen pore filling with under the SEM. Under cathodo-luminescence kaolinite appears a bright blue, indicating a low-temperature origin. Chlorite is poorly crystallized and appears to be an iron-rich variety on the basis of relative intensities of different order of basal spacing (first order low, second order high). Illite shows typical grading into mixed-layered varieties. The mixed-layered varieties are abundant and seem to be randomly interlayered. Based on peak intensities, the ratio of kaolinite to chlorite is about 10, chlorite to illite about 0.2, and quartz to illite about 2.2. Analyses of ground samples of the $>2\mu$ fraction show a higher ratio of kaolinite to chlorite, 12.3; a higher ratio of quartz to illite, 4.9; and the same ratio of chlorite to illite, 0.2.

g) Heavy minerals are dominantly zircon, tourmaline, and rutile with lesser amounts of apatite and garnet. The ZTR index is very high, indicating high maturity. Opaque heavy minerals are mainly leucoxene with lesser amounts of pyrite, hematite, and limonite. Roundness of the heavy minerals varies: garnet is angular, rutile is rounded, tourmaline and zircon have both rounded and angular varieties. Some tourmalines show small spikes of authigenic overgrowths.

h) Conglomerate pebbles at base of sandstone in channel deposits are almost entirely sideritic clay concretion fragments. These are the same type as found in shales overlying coal beds in the Illinois Basin. Other pebbles include hematite- and limonite-cemented sandstone similar to other Pennsylvanian sandstones in that part of the section, a few limestone pebbles, and rarely a coal pebble.

2. Chemical Constituents

a) Calcite is present as a clear untwinned mosaic of interlocking crystals that fills pore space and replaces detrital grains of quartz and feldspar and, much more rarely, rock fragments or clay matrix. The only inclusions are of the quartz or other framework grain replaced. Calcite is present either in small amounts, about 5 percent, as irregular patchy areas or, in a few samples, as an abundant constituent making up more than 20 percent of the rock. Calcite not only replaces detrital quartz, but is itself replaced by some sharply euhedral authigenic quartz.

b) Iron carbonates include two varieties, siderite and ferroan dolomite, both present as small individual or clusters of rhombohedra which replace calcite in some areas and appear to be intergrown in other areas. X-ray diffraction of ground samples indicates both siderite and dolomite; the slight stain caused by oxidation of the iron that is ubiquitous in all of the rhomobohedra in outcrop samples implies the presence of iron in the dolomite.

c) Authigenic quartz is present as secondary overgrowths deposited in optical continuity with detrital grains. Some replace calcite. This quartz variety is shown by cathodo-luminescence to be more abundant than indicated by estimates based on ordinary microscopy and in one section is about 8 percent. The difference in estimates is traceable to the fact that the detrital outlines of original grains are not always distinguishable and so anhedral overgrowths are not recognized.

d) Authigenic kaolinite is present as well as crystallized, small, vermicular aggregates in pore spaces. A few grains of glauconite, probably authigenic but possibly detrital, are present in a few samples. Colorless authigenic tourmaline overgrowths are rare. Anatase is idiomorphic and can probably be authigenic.

V. Interpretation

The Trivoli is one of the best known mappable sand bodies of the upper part of the Pennsylvanian section of the Illinois Basin, primarily from the subsurface but also from its outcrop belt (Fig. A-6). It was the first Pennsylvanian sandstone of the basin to be given systematic petrographic study (Siever, 1954), and one of the few individual sandbodies for which there is a large and useful body of stratigraphic and petrologic data. The only other Pennsylvanian sandbody to be studied in such detail, and a useful comparison, is the Anvil Rock Sandstone lower in the section in Carbondale Formation of the Kewanee Group (Hopkins, 1958).

a) The composition and texture of the Trivoli are consistent with the general interpretation of Pennsylvanian sandstones of the Illinois Basin as part of an alluvial–deltaic complex derived by a major river system primarily from sedimentary and metamorphic highlands of the Appalachian mobile belt far to the east and northeast. The lack of much feldspar and the abundance of metamorphic and sedimentary rock fragments imply the absence of any large area of eroding igneous rock, either intrusive or extrusive. The many sedimentary rock fragments that can be identified easily with rocks of the Illinois Basin section, such as the clay ironstones and coal, coupled with the abundance of relatively soft pelitic rock fragments, demonstrate that some material was being supplied by local source areas within the basin. Some of the material is multicycled as indicated by the roundness of quartz and heavy minerals and the presence of a few second-cycle grains; but how much of this is from nearby and how much from distant sources is not possible to say. It is possible that a careful comparison of samples from the minor tributaries mapped by Andresen (1961, p. 26) with those of major channels would show differences that would be interpretable in terms of local versus distant sources.

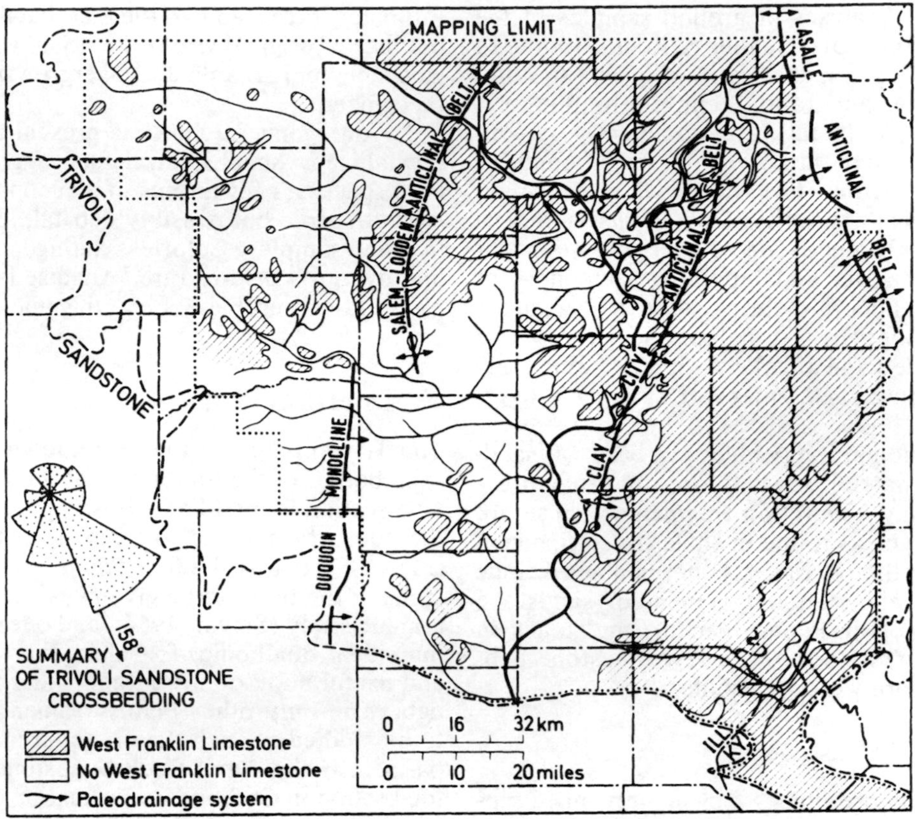

FIGURE A-6. Paleodrainage at base of Trivoli Sandstone and crossbedding. (Modified from Andresen, 1961, Fig. 9).

The Trivoli has much the same composition everywhere; local variance introduced by sampling sheet- or channel-phases or coarse- and fine-grained beds is much greater than regional variance. This homogeneity indicates a well-mixed contribution from distant sources, such as would be characteristic of a large river system, with a contribution from homogenous nearby rocks lower in the section—rocks that we know show little regional variance.

The direction and paleogeography of the source lands can be deduced from paleocurrent and subsurface channel studies that indicate a major source to the east and northeast in the Applachian mobile belt and a lesser source to the north and northwest, from the Transcontinental Arch and Canadian Shield. Tectonically, the framework composition of the Trivoli (Table 6A) indicates the tectonic setting of a recycled orogen (Dickinson, 1985, Fig. 1).

b) All of the characteristics of the Trivoli indicate a dominantly alluvial origin—the plant remains, the crossbedding, the valley system of the channel sandstones, the sorting of the detritus, and the lack of any marine fossils. The shoreline must have been to the south but the record of constant marine transgressions and the increasingly marine nature of the section to the southwest in Missouri and Kansas indicate that shoreline to have been no more than a few hundred miles away at any time and probably much closer at many times. Thus the sedimentary framework of the Trivoli is that of a deltaic–coastal plain complex.

The difference between coarser channel and finer sheet phases is explainable in terms of higher current shear stress in the high discharge channels and lower shear stress in the lower discharge streams, perhaps braided, of the sheet phase that followed the aggradation of the channels.

c) Two major changes in the sand took place as a result of diagenesis, the transformation of many soft rock fragments into matrix and the precipitation of secondary quartz and carbonate cement. The alteration of the rock fragments was mainly a mechanical process rather than chemical, for the composition of the matrix is much the same as the rock fragments. Implied is a more or less plastic flow of soft argillaceous material around and between the rigid, competent framework quartz and feldspar grains.

The precipitation of the cement follows the order: calcite, perhaps contemporaneous with iron carbonates but perhaps earlier, followed by secondary quartz. After being brought close to or at the surface by uplift and erosion, hematite and limonite cement were added. The calcite may have been precipitated following sand deposition when it was buried only by the overlying silt and clay and a coal swamp [Chapel (No. 8) coal]. The probable groundwater movement at this time would have been a circulation through the Trivoli upward to the swamp—typical pattern for such swamps—perhaps carrying with it some of the waters still being squeezed from the underlying thick marine shale sequence. Those waters, originally meteoric, may have become acid and depleted in oxygen as a result of passage through strata containing abundant pyrite, leading to the dissolution of carbonate fossils and part of limestone beds and to the disappearance of pyrite. The resulting water, passing through the sand and mixing with other less carbonate-saturated waters, might have thus become less acid and so precipitated carbonate, including iron-rich varieties. Alternatively, some of the calcite may simply have been detrital or freshwater fossil calcite redistributed by dissolution and reprecipitation in the immediate local area where the cement now is. This alternative is suggested by the irregular patchy distribution of some cement.

The authigenic quartz overgrowths may owe their origin to a more general pressure-solution process operating later in the sand's history, when it was deeply buried. Because of later erosion, it is difficult to estimate that overburden. If some of the clay minerals were altered in such a way as to liberate siliica, as, for example, from a more siliceous montmorillonite to an illite, then that silica may have been contributed to the groundwater too, later to be precipitated as the groundwater slowly equilibrated with quartz.

VI. Economic Importance

The Trivoli itself is not an economic resource. Its composition makes it unsuitable for molding or foundry sands. It shows no evidence of any oil or gas, now or at any former time. Yet its study has economic significance in relation to exploration for and exploitation of coal beds in the Illinois basin, for sands such as the Trivoli act as cutouts of coal beds and interfere seriously with mining operations.

VII. Bearing on Scientific Problems

The Trivoli is very typical of many sandstones in the Illinois Basin and of many alluvial sandstones in other places, particularly Coal Measure sandstones. The many kinds of information available—outcrop and subsurface core samples, extensive subsurface stratigraphic data, and surface paleocurrent mapping—make its origin one of the best-supported interpretations we have. Problems, such as the relation of composition to near and distant sources and any correlation between that and minor tributaries and major channels, remain to be studied. A detailed analysis of the history of diagenetic events has yet to be made. Such detailed studies will help give a broader base for future general models.

References

Andel, Tj.H. van: Origin and classification of Cretaceous, Paleocene and Eocene sandstones of western Venezuela. Am. Assoc. Petroleum Geologists Bull. 42, 734–763 (1958).

Andersen, M.J.: Geology and petrology of the Trivoli Sandstone in the Illinois Basin. Illinois Geol. Survey Circ. 316, 31 p. (1961).

Beard, D.C., and Weyl, P.K.: Influence of texture on porosity and permeability of unconsolidated sand. Am. Assoc. Petroleum Geologists Bull. 57, 349–369 (1973).

Cayeux, L.: Les roches sédimentaires de France—roches siliceuses, 774 pp. Paris: Impr. Nationale 1929.

Dennison, J.M.: Graphical aids for determining reliability of sample means and an adequate sample size. Jour. Sed. Petrology 32, 743–750 (1962).

Dickinson, W.R.: Interpretating provenance relations from detrital modes of sandstones. In: Zuffa, G.G. (Ed.), Provenance of arenites. Dordrecht: D. Reidel Pub. Co. (NATO ASI Ser. C 148, 333–362, 1985).

Dilko, A., and Graham, S.C.: Quantitative mineralogic characterization of sandstones by back-scattered election image analysis. Jour. Sed. Petrology 55, 347–355 (1985).

Dott, R.L., Jr.: Wacke. graywacke and matrix—what approach to immature sandstone classification? Jour. Sed. Petrology 34, 625–632 (1964).

Ferm, J.C., and Weisenfluh, G.A.: Cored rocks of the southern Appalachian coal field, 93 pp. Lexington, Kentucky: Univ. Kentucky, Dept. Geology 1981.

Folk, R.L.: Henry Clifton Sorby (1826–1908), the founder of petrography. Jour. Geol. Education 8, 43–47 (1965).

Folk, R.L.: Petrology of sedimentary rocks, 170 pp. Austin, Texas: Hemphill's 1968.

Grout, F.F.: Petrography and petrology, 552 pp. New York: McGraw-Hill 1932.

Harrell, James: A visual comparator for degree of sorting in thin and plane sections. Jour. Sed. Petrology 54, 648–650 (1984).

Hopkins, M.E.: Petrography and petrology of the Anvil Rock Sandstone of southern Illinois. Illinois Geol. Survey Cir. 256, 49 pp. (1958).

Houghton, H.F.: Refined techniques for staining plegioclase and alkali feldspar in thin section: Jour. Sed. Petrology 50, 629–631 (1980).

Keelen, D.K.: Core analysis for aid in reservoir description. Jour. Petrol. Tech. 34, 2483–2491 (1982).

Maynard, J. Barry: Composition of plagioclase feldspars in modern deep-sea sands. Sedimentology 13, 493–501 (1984).

Middleton, L.T., and Kraus, M.J.: Simple techniques for thin section preparation of unconsolidated materials. Jour. Sed. Petrology 50, 622–623 (1980).

Pearson, E.S., and Hartley, H.D.: Biometrika tables for statisticians, Vol. 1, 238 p. York: Cambridge Univ. Press, 1954.

van der Plass, and Tobi, A.C.: A chart for judging one reliability of point counting results. Amer. Jour. Sci. 263, 87–90 (1965).

Powers, M.C.: A new roundness scale for sedimentary particles. Jour. Sed. Petrology 23, 117–119 (1953).

Rusnak, G.A.: A fabric and petrologic study of the Pleasantview Sandstone. Jour. Sed. Petrology 27, 41–55 (1957).

Sedimentation Seminar: Studies for students: A question set for sands and sandstones. Brigham Young University Geology Studies 24(2), 1–8 (1978).

Siever, Raymond: Trivoli Sandstone of Williamson County, Illinois. Jour. Geology 57, 614–618 (1949).

Sneider, R.M.; King, H.R.; Hawkes, H.E.; and Davis, T.B.: Methods for detection and characterization of reservoir rock, Deep Basin Gas Area, Western Canada. Jour. Petrol. Tech. 35, 1725–1734 (1983).

Sorby, Henry Clifton: On the microscopical structure of the calcareous grit of the Yorkshire coast. Geol. Soc. London Quart. Jour. 1–6 (1851).

Sorby, Henry Clifton: On the structure and origin of the non-calcareous stratified rocks. Geol. Soc. London Proc. 36, 46–92 (1880).

Summerson, Charles H., (Ed.): Sorby on geology. University of Miami, Comparative Sedimentology Laboratory, Geological Milestones III, 241 pp. (1978).

Terry, R.D., and Chilingar, G.V.: Summary of "Concerning some additional aids in studying sedimentary formations" by M. S. Shvetsov. Jour. Sed. Petrology 25, 229–234 (1955).

Wanless, H.R.: Pennsylvanian cycles in western Illinois. In: Papers presented at the quarter centennial celebration of the Illinois State Geological Survey. Illinois Geol. Survey Bull. 60, 182–193 (1931).

Wilson, M.J., and Clark, D.R.: X-ray identification of clay minerals in thin sections. Jour. Sed. Petrology 48, 656–660 (1978).

Yavgus, J.E., and Dravis, J.J.: Blue fluorescent dye technique for recognition of microporosity in sedimentary rocks. Jour. Sed. Petrology 55, 600–602 (1985).

Author Index

Numbers in *italics* denote complete citations in the References and Annotated Bibliographies.

Grayson, D. K., 216, *247*
Green, J., 216, *245*

I

J

K

T

U

V

W

Y

Z

Subject Index

W

X

Z

Lightning Source UK Ltd.
Milton Keynes UK
UKOW02f1705061113

220553UK00003B/6/P